VOLUME FIVE HUNDRED AND SEVENTY

METHODS IN ENZYMOLOGY

Chemokines

METHODS IN ENZYMOLOGY

VOLUME FIVE HUNDRED AND SEVENTY

METHODS IN ENZYMOLOGY

Chemokines

Edited by

TRACY M. HANDEL

Department of Pharmacology,
Skaggs School of Pharmacy and
Pharmaceutical Sciences,
University of California,
San Diego, La Jolla, California, USA

AMSTERDAM • BOSTON • HEIDELBERG • LONDON
NEW YORK • OXFORD • PARIS • SAN DIEGO
SAN FRANCISCO • SINGAPORE • SYDNEY • TOKYO
Academic Press is an imprint of Elsevier

Academic Press is an imprint of Elsevier
50 Hampshire Street, 5th Floor, Cambridge, MA 02139, USA
525 B Street, Suite 1800, San Diego, CA 92101–4495, USA
The Boulevard, Langford Lane, Kidlington, Oxford OX5 1GB, UK
125 London Wall, London, EC2Y 5AS, UK

First edition 2016

Notices
Knowledge and best practice in this field are constantly changing. As new research and experience broaden our understanding, changes in research methods, professional practices, or medical treatment may become necessary.

Practitioners and researchers must always rely on their own experience and knowledge in evaluating and using any information, methods, compounds, or experiments described herein. In using such information or methods they should be mindful of their own safety and the safety of others, including parties for whom they have a professional responsibility.

To the fullest extent of the law, neither the Publisher nor the authors, contributors, or editors, assume any liability for any injury and/or damage to persons or property as a matter of products liability, negligence or otherwise, or from any use or operation of any methods, products, instructions, or ideas contained in the material herein.

ISBN: 978-0-12-802171-2
ISSN: 0076-6879

For information on all Academic Press publications
visit our website at http://store.elsevier.com/

CONTENTS

Contributors xiii
Preface xxi

1. Chemokine Detection Using Receptors Immobilized on an SPR Sensor Surface **1**

José Miguel Rodríguez-Frade, Laura Martínez-Muñoz, Ricardo Villares, Graciela Cascio, Pilar Lucas, Rosa P. Gomariz, and Mario Mellado

1. Surface Plasmon Resonance 2
2. Chemokine Receptors: Members of the GPCR Family 3
3. Chemokine Receptor Immobilization on the Sensor Chip 4
4. Viral Particles as Chemokine Receptors Carriers in SPR 6
5. SPR-Based Applications for Chemokine Receptors 12
6. Conclusions 14

Acknowledgments 15
References 15

2. Study of Chemotaxis and Cell–Cell Interactions in Cancer with Microfluidic Devices **19**

Jiqing Sai, Matthew Rogers, Kathryn Hockemeyer, John P. Wikswo, and Ann Richmond

1. Introduction 20
2. Methods 24
3. Limitations 37
4. Perspectives 40

Acknowledgments 42
References 43

3. Generating Chemokine Analogs with Enhanced Pharmacological Properties Using Phage Display **47**

Karim Dorgham, Fabrice Cerini, Hubert Gaertner, Astrid Melotti, Irène Rossitto-Borlat, Guy Gorochov, and Oliver Hartley

1. Introduction 48
2. Methods 50
3. Limitations 65
4. Perspectives 68

References 70

4. Methods for the Recognition of GAG-Bound Chemokines 73

Pauline Bonvin, Franck Gueneau, Nicolas Fischer, and Amanda Proudfoot

1. Introduction 73
2. Methods 74
3. Summary 85

References 85

5. Monitoring Scavenging Activity of Chemokine Receptors 87

Barbara Moepps and Marcus Thelen

1. Introduction 88
2. Fluorescent Chemokines 89
3. Monitoring Scavenging with Radiolabeled Chemokines 92
4. Monitoring Scavenging with Fluorescent Chemokines 93
5. Monitor Chemokine Uptake by Microscopy and Flow Cytometry (FACS) 108

References 114

6. Dual-Color Luciferase Complementation for Chemokine Receptor Signaling 119

Kathryn E. Luker and Gary D. Luker

1. Introduction 120
2. Methods 121

Acknowledgment 128

References 128

7. Analysis of Arrestin Recruitment to Chemokine Receptors by Bioluminescence Resonance Energy Transfer 131

J. Bonneterre, N. Montpas, C. Boularan, C. Galés, and N. Heveker

1. Introduction 132
2. Methods 133
3. Interpretation and Limitations of BRET Data 145
4. Perspectives 149

Acknowledgments 150

References 150

8. Probing Biased Signaling in Chemokine Receptors 155

Roxana-Maria Amarandi, Gertrud Malene Hjortø, Mette Marie Rosenkilde, and Stefanie Karlshøj

1. Introduction 156
2. Types of Bias in the Chemokine System 157

3. Chemokine System-Mediated Intracellular Signaling 162
4. Methods 165
Acknowledgments 180
References 181

9. Mutagenesis by Phage Display 187

Pauline Bonvin, Christine Power, Amanda Proudfoot, and Steven Dunn

1. Introduction 187
2. Investigating the Pharmacophore of Chemokine Binders by Phage Display 188
3. Phage Display Selection to Modulate the Selectivity of Chemokine Binders 197
4. Summary 205
References 205

10. Studying Chemokine Control of Neutrophil Migration *In Vivo* in a Murine Model of Inflammatory Arthritis 207

Yoshishige Miyabe, Nancy D. Kim, Chie Miyabe, and Andrew D. Luster

1. Introduction 208
2. Methods 211
3. Conclusions 229
References 230

11. Production of Chemokine/Chemokine Receptor Complexes for Structural Biophysical Studies 233

Martin Gustavsson, Yi Zheng, and Tracy M. Handel

1. Introduction 234
2. Methods 236
3. Summary and Conclusions 257
Acknowledgments 257
References 258

12. *In Vivo* Models to Study Chemokine Biology 261

F.A. Amaral, D. Boff, and M.M. Teixeira

1. Introduction 262
2. Methods 264
3. Limitations 275
4. Other Perspectives 277
References 277

13. Monitoring Chemokine Receptor Trafficking by Confocal Immunofluorescence Microscopy **281**

Adriano Marchese

1. Introduction 282
2. Methods 284
3. Concluding Remarks 290

References 291

14. Active Shaping of Chemokine Gradients by Atypical Chemokine Receptors: A 4D Live-Cell Imaging Migration Assay **293**

Kathrin Werth and Reinhold Förster

1. Background 294
2. Introduction 294
3. Materials and Equipment 296
4. Preparation of Chambers 297
5. Preparation of Cells 299
6. Filling of Chambers 300
7. Time-Lapse Imaging 302
8. Analysis 302

Acknowledgments 307

References 307

15. The Role of Chemokine and Glycosaminoglycan Interaction in Chemokine-Mediated Migration *In Vitro* and *In Vivo* **309**

Irene del Molino del Barrio, John Kirby, and Simi Ali

1. Introduction 310
2. *In Vitro* Chemotaxis 311
3. *In Vivo* Chemotaxis 318
4. Generation of Mammalian Transfectants Expressing Chemokine Receptors 322
5. Discussion 329

Acknowledgments 330

References 331

16. Examining Roles of Glycans in Chemokine-Mediated Dendritic–Endothelial Cell Interactions **335**

Xin Yin, Scott C. Johns, Roland El Ghazal, Catherina L. Salanga, Tracy M. Handel, and Mark M. Fuster

1. Introduction 336
2. Methods 339

3. Perspectives 350
Acknowledgments 353
References 353

17. Preparation and Analysis of N-Terminal Chemokine Receptor Sulfopeptides Using Tyrosylprotein Sulfotransferase Enzymes 357
Christoph Seibert, Anthony Sanfiz, Thomas P. Sakmar, and Christopher T. Veldkamp

1. Introduction 358
2. Methods 361
3. Caveats and Limitations 380
4. Perspectives 381
Acknowledgments 385
References 385

18. Disulfide Trapping for Modeling and Structure Determination of Receptor:Chemokine Complexes 389
Irina Kufareva, Martin Gustavsson, Lauren G. Holden, Ling Qin, Yi Zheng, and Tracy M. Handel

1. Introduction 390
2. Architecture of Receptor:Chemokine Interfaces 392
3. Cysteine as a Natural Crosslinking Agent 395
4. Disulfide Trapping 396
5. Conclusion and Perspectives 416
Acknowledgments 417
References 417

19. Analysis of G Protein and β-Arrestin Activation in Chemokine Receptors Signaling 421
Alessandro Vacchini, Marta Busnelli, Bice Chini, Massimo Locati, and Elena Monica Borroni

1. Introduction 422
2. G Proteins Signaling 423
3. β-Arrestins Signaling: Detection of β-Arrestins Recruitment by Protein Conformational Changes 431
4. Calculation of Biased Signaling 434
5. Summary 436
References 437

20. Flow Cytometry Detection of Chemokine Receptors for the Identification of Murine Monocyte and Neutrophil Subsets **441**

Ornella Bonavita, Matteo Massara, Achille Anselmo, Paolo Somma, Hilke Brühl, Matthias Mack, Massimo Locati, and Raffaella Bonecchi

1. Introduction 442
2. Blood Collection and Preparation of Blood Cells from Mice 444
3. Staining of Cell Surface Markers and Chemokine Receptors 445
4. Flow Cytometry Analysis 449
5. Discussion 453
6. Concluding Remarks 454

Acknowledgments 455
References 455

21. Molecular Pharmacology of Chemokine Receptors **457**

Raymond H. de Wit, Sabrina M. de Munnik, Rob Leurs, Henry F. Vischer, and Martine J. Smit

1. Introduction 458
2. Pharmacological Quantification of Chemokine Receptor Binding 462
3. Pharmacological Quantification of Chemokine Receptor Signaling 482
4. Conclusions and Future Perspectives 505

Acknowledgments 507
References 507

22. Preparation and Characterization of Glycosaminoglycan Chemokine Coreceptors **517**

Nikola Kitic, Martha Gschwandtner, Rupert Derler, Tanja Gerlza, and Andreas J. Kungl

1. Introduction 518
2. Preparation and Characterization of GAGs 520
3. Methods for Studying Chemokine–GAG Interactions 529
4. Concluding Remarks 536

Acknowledgment 537
References 537

23. Production of Recombinant Chemokines and Validation of Refolding **539**

Christopher T. Veldkamp, Chad A. Koplinski, Davin R. Jensen, Francis C. Peterson, Kaitlin M. Smits, Brittney L. Smith, Scott K. Johnson, Christina Lettieri, Wallace G. Buchholz, Joyce C. Solheim, and Brian F. Volkman

1. Introduction 540
2. Methods 544
3. Caveats and Limitations 558
4. Perspectives 558

Acknowledgments 562
References 562

24. Quantitative Analysis of Dendritic Cell Haptotaxis **567**

Jan Schwarz and Michael Sixt

1. Introduction 567
2. Methods 570
3. Perspectives 580

Acknowledgments 581
References 581

Author Index *583*
Subject Index *615*

CONTRIBUTORS

Simi Ali
Institute of Cellular Medicine, Medical Faculty, Newcastle University, Newcastle upon Tyne, United Kingdom

F.A. Amaral
Immunopharmacology, Department of Biochemistry and Immunology, Universidade Federal de Minas Gerais, Belo Horizonte, Brazil

Roxana-Maria Amarandi
Laboratory for Molecular Pharmacology, Department of Neuroscience and Pharmacology, Faculty of Health and Medical Sciences, The Panum Institute, University of Copenhagen, Copenhagen, Denmark; Faculty of Chemistry, Alexandru Ioan Cuza University of Iaşi, Iaşi, Romania

Achille Anselmo
Humanitas Clinical and Research Center, Rozzano, Italy

D. Boff
Immunopharmacology, Department of Biochemistry and Immunology, Universidade Federal de Minas Gerais, Belo Horizonte, Brazil

Ornella Bonavita
Humanitas Clinical and Research Center, Rozzano; Department of Medical Biotechnologies and Translational Medicine, Università degli Studi di Milano, Milano, Italy

Raffaella Bonecchi
Humanitas Clinical and Research Center; Department of Biomedical Sciences, Humanitas University, Rozzano, Italy

J. Bonneterre
Department of Biochemistry and Research Centre, Sainte-Justine Hospital, Université de Montréal, Montreal, Quebec, Canada

Pauline Bonvin
NovImmune S.A.,; Geneva Research Centre, Merck Serono S.A., Geneva, Switzerland

Elena Monica Borroni
Department of Medical Biotechnologies and Translational Medicine, Università degli Studi di Milano, Milano; Humanitas Clinical and Research Center, Rozzano, Italy

C. Boularan
Institut des Maladies Métaboliques et Cardiovasculaires, Institut National de la Santé et de la Recherche Médicale, U1048, Université Toulouse III Paul Sabatier, Toulouse, France

Hilke Brühl
Universitätsklinikum Regensburg, Regensburg, Germany

Wallace G. Buchholz
Biological Process Development Facility, College of Engineering, University of Nebraska—Lincoln, Lincoln, Nebraska, USA

Marta Busnelli
CNR Institute of Neuroscience, Milan, Italy

Graciela Cascio
Department of Immunology and Oncology, Centro Nacional de Biotecnología (CNB/CSIC), Madrid, Spain

Fabrice Cerini
Department of Pathology and Immunology, Faculty of Medicine, University of Geneva, Geneva, Switzerland

Bice Chini
CNR Institute of Neuroscience, Milan, Italy

Sabrina M. de Munnik
Amsterdam Institute for Molecules Medicines and Systems (AIMMS), Division of Medicinal Chemistry, Vrije Universiteit, Amsterdam, The Netherlands

Raymond H. de Wit
Amsterdam Institute for Molecules Medicines and Systems (AIMMS), Division of Medicinal Chemistry, Vrije Universiteit, Amsterdam, The Netherlands

Irene del Molino del Barrio
Institute of Cellular Medicine, Medical Faculty, Newcastle University, Newcastle upon Tyne, United Kingdom

Rupert Derler
Antagonis Biotherapeutics G.m.b.H., Graz, Austria

Karim Dorgham
Sorbonne Universités, UPMC Univ Paris 06, Inserm UMRS1135, Centre d'Immunologie et des Maladies Infectieuses (CIMI-Paris), Paris, France

Steven Dunn
NovImmune S.A.,; Geneva Research Centre, Merck Serono S.A., Geneva, Switzerland

Roland El Ghazal
VA San Diego Healthcare System, Medical and Research Sections; Department of Medicine, Division of Pulmonary and Critical Care, University of California San Diego, La Jolla, California, USA

Nicolas Fischer
NovImmune S.A., Geneva, Switzerland

Reinhold Förster
Institute of Immunology, Hannover Medical School, Hannover, Germany

Mark M. Fuster
VA San Diego Healthcare System, Medical and Research Sections; Department of Medicine, Division of Pulmonary and Critical Care, University of California San Diego, La Jolla, California, USA

Hubert Gaertner
Department of Pathology and Immunology, Faculty of Medicine, University of Geneva, Geneva, Switzerland

C. Galés
Institut des Maladies Métaboliques et Cardiovasculaires, Institut National de la Santé et de la Recherche Médicale, U1048, Université Toulouse III Paul Sabatier, Toulouse, France

Tanja Gerlza
Antagonis Biotherapeutics G.m.b.H., Graz, Austria

Rosa P. Gomariz
Department of Cell Biology, Universidad Complutense de Madrid, Madrid, Spain

Guy Gorochov
Sorbonne Universités, UPMC Univ Paris 06, Inserm UMRS1135, Centre d'Immunologie et des Maladies Infectieuses (CIMI-Paris), Paris, France; Department of Pathology and Immunology, Faculty of Medicine, University of Geneva, Geneva, Switzerland

Martha Gschwandtner
Department of Pharmaceutical Chemistry, Institute of Pharmaceutical Sciences, Karl-Franzens-University Graz, Graz, Austria

Franck Gueneau
NovImmune S.A., Geneva, Switzerland

Martin Gustavsson
Skaggs School of Pharmacy and Pharmaceutical Sciences, University of California San Diego, La Jolla, California, USA

Tracy M. Handel
Department of Pharmacology, Skaggs School of Pharmacy and Pharmaceutical Sciences, University of California San Diego, La Jolla, California, USA

Oliver Hartley
AP-HP, Groupement Hospitalier Pitié-Salpêtrière, Département d'Immunologie, Paris, France

N. Heveker
Department of Biochemistry and Research Centre, Sainte-Justine Hospital, Université de Montréal, Montreal, Quebec, Canada; Institut des Maladies Métaboliques et Cardiovasculaires, Institut National de la Santé et de la Recherche Médicale, U1048, Université Toulouse III Paul Sabatier, Toulouse, France

Gertrud Malene Hjortø
Laboratory for Molecular Pharmacology, Department of Neuroscience and Pharmacology, Faculty of Health and Medical Sciences, The Panum Institute, University of Copenhagen, Copenhagen, Denmark

Kathryn Hockemeyer
Vanderbilt Institute for Integrative Biosystems Research and Education, Vanderbilt University, Nashville, Tennessee, USA

Lauren G. Holden
Skaggs School of Pharmacy and Pharmaceutical Sciences, University of California San Diego, La Jolla, California, USA

Davin R. Jensen
Department of Biochemistry, Medical College of Wisconsin, Milwaukee, Wisconsin, USA

Scott C. Johns
VA San Diego Healthcare System, Medical and Research Sections; Department of Medicine, Division of Pulmonary and Critical Care, University of California San Diego, La Jolla, California, USA

Scott K. Johnson
Biological Process Development Facility, College of Engineering, University of Nebraska—Lincoln, Lincoln, Nebraska, USA

Stefanie Karlshøj
Laboratory for Molecular Pharmacology, Department of Neuroscience and Pharmacology, Faculty of Health and Medical Sciences, The Panum Institute, University of Copenhagen, Copenhagen, Denmark

Nancy D. Kim
Center for Immunology and Inflammatory Diseases, Division of Rheumatology, Allergy and Immunology, Massachusetts General Hospital, Harvard Medical School, Boston, Massachusetts, USA

John Kirby
Institute of Cellular Medicine, Medical Faculty, Newcastle University, Newcastle upon Tyne, United Kingdom

Nikola Kitic
Department of Pharmaceutical Chemistry, Institute of Pharmaceutical Sciences, Karl-Franzens-University Graz, Graz, Austria

Chad A. Koplinski
Department of Biochemistry, Medical College of Wisconsin, Milwaukee, Wisconsin, USA

Irina Kufareva
Skaggs School of Pharmacy and Pharmaceutical Sciences, University of California San Diego, La Jolla, California, USA

Andreas J. Kungl
Department of Pharmaceutical Chemistry, Institute of Pharmaceutical Sciences, Karl-Franzens-University Graz; Antagonis Biotherapeutics G.m.b.H., Graz, Austria

Christina Lettieri
Department of Pediatrics, Children's Hospital and Medical Center, University of Nebraska Medical Center, Omaha, Nebraska, USA

Rob Leurs
Amsterdam Institute for Molecules Medicines and Systems (AIMMS), Division of Medicinal Chemistry, Vrije Universiteit, Amsterdam, The Netherlands

Massimo Locati
Department of Medical Biotechnologies and Translational Medicine, Università degli Studi di Milano, Milano; Humanitas Clinical and Research Center, Rozzano, Italy

Pilar Lucas
Department of Immunology and Oncology, Centro Nacional de Biotecnología (CNB/CSIC), Madrid, Spain

Gary D. Luker
Department of Radiology, Center for Molecular Imaging; Department of Biomedical Engineering; Department of Microbiology and Immunology, University of Michigan, Ann Arbor, Michigan, USA

Kathryn E. Luker
Department of Radiology, Center for Molecular Imaging, University of Michigan, Ann Arbor, Michigan, USA

Andrew D. Luster
Center for Immunology and Inflammatory Diseases, Division of Rheumatology, Allergy and Immunology, Massachusetts General Hospital, Harvard Medical School, Boston, Massachusetts, USA

Matthias Mack
Universitätsklinikum Regensburg, Regensburg, Germany

Adriano Marchese
Department of Pharmacology, Loyola University Chicago, Health Sciences Division, Maywood, Illinois, USA

Laura Martínez-Muñoz
Department of Immunology and Oncology, Centro Nacional de Biotecnología (CNB/CSIC), Madrid, Spain

Matteo Massara
Humanitas Clinical and Research Center, Rozzano; Department of Medical Biotechnologies and Translational Medicine, Università degli Studi di Milano, Milano, Italy

Mario Mellado
Department of Immunology and Oncology, Centro Nacional de Biotecnología (CNB/CSIC), Madrid, Spain

Astrid Melotti
Department of Pathology and Immunology, Faculty of Medicine, University of Geneva, Geneva, Switzerland

Chie Miyabe
Center for Immunology and Inflammatory Diseases, Division of Rheumatology, Allergy and Immunology, Massachusetts General Hospital, Harvard Medical School, Boston, Massachusetts, USA

Yoshishige Miyabe
Center for Immunology and Inflammatory Diseases, Division of Rheumatology, Allergy and Immunology, Massachusetts General Hospital, Harvard Medical School, Boston, Massachusetts, USA

Barbara Moepps
Institute of Pharmacology and Toxicology, University of Ulm Medical Center, Ulm, Germany

N. Montpas
Department of Biochemistry and Research Centre, Sainte-Justine Hospital, Université de Montréal, Montreal, Quebec, Canada

Francis C. Peterson
Department of Biochemistry, Medical College of Wisconsin, Milwaukee, Wisconsin, USA

Christine Power
NovImmune S.A.,; Geneva Research Centre, Merck Serono S.A., Geneva, Switzerland

Amanda Proudfoot
NovImmune S.A.,; Geneva Research Centre, Merck Serono S.A., Geneva, Switzerland

Ling Qin
Skaggs School of Pharmacy and Pharmaceutical Sciences, University of California San Diego, La Jolla, California, USA

Ann Richmond
Department of Veterans Affairs, Tennessee Valley Healthcare System; Department of Cancer Biology, School of Medicine; Vanderbilt Institute for Integrative Biosystems Research and Education, Vanderbilt University, Nashville, Tennessee, USA

José Miguel Rodríguez-Frade
Department of Immunology and Oncology, Centro Nacional de Biotecnología (CNB/CSIC), Madrid, Spain

Matthew Rogers
Vanderbilt Institute for Integrative Biosystems Research and Education, Vanderbilt University, Nashville, Tennessee, USA

Mette Marie Rosenkilde
Laboratory for Molecular Pharmacology, Department of Neuroscience and Pharmacology, Faculty of Health and Medical Sciences, The Panum Institute, University of Copenhagen, Copenhagen, Denmark

Irène Rossitto-Borlat
Department of Pathology and Immunology, Faculty of Medicine, University of Geneva, Geneva, Switzerland

Jiqing Sai
Department of Veterans Affairs, Tennessee Valley Healthcare System; Department of Cancer Biology, School of Medicine, Vanderbilt University, Nashville, Tennessee, USA

Thomas P. Sakmar
Laboratory of Chemical Biology and Signal Transduction, The Rockefeller University, New York, USA

Catherina L. Salanga
Department of Pharmacology, Skaggs School of Pharmacy and Pharmaceutical Sciences, University of California San Diego, La Jolla, California, USA

Anthony Sanfiz*
Laboratory of Chemical Biology and Signal Transduction, The Rockefeller University, New York, USA

* Current address: Department of Microbiology, New York University Medical Center, New York, NY, USA.

Jan Schwarz
Institute of Science and Technology Austria (IST Austria), Klosterneuburg, Austria

Christoph Seibert
Laboratory of Chemical Biology and Signal Transduction, The Rockefeller University, New York, USA

Michael Sixt
Institute of Science and Technology Austria (IST Austria), Klosterneuburg, Austria

Martine J. Smit
Amsterdam Institute for Molecules Medicines and Systems (AIMMS), Division of Medicinal Chemistry, Vrije Universiteit, Amsterdam, The Netherlands

Brittney L. Smith
Department of Biochemistry and Molecular Biology, University of Nebraska Medical Center; Department of Pathology and Microbiology; The Eppley Institute and the Fred and Pamela Buffett Cancer Center, University of Nebraska Medical Center, Omaha, Nebraska, USA

Kaitlin M. Smits
Department of Biochemistry and Molecular Biology, University of Nebraska Medical Center; Department of Pathology and Microbiology; The Eppley Institute and the Fred and Pamela Buffett Cancer Center, University of Nebraska Medical Center, Omaha, Nebraska, USA

Joyce C. Solheim
Department of Biochemistry and Molecular Biology, University of Nebraska Medical Center; Department of Pathology and Microbiology; The Eppley Institute and the Fred and Pamela Buffett Cancer Center, University of Nebraska Medical Center, Omaha, Nebraska, USA

Paolo Somma
Humanitas Clinical and Research Center, Rozzano, Italy

M.M. Teixeira
Immunopharmacology, Department of Biochemistry and Immunology, Universidade Federal de Minas Gerais, Belo Horizonte, Brazil

Marcus Thelen
Institute for Research in Biomedicine, Università della Svizzera italiana, Bellinzona, Switzerland

Alessandro Vacchini
Department of Medical Biotechnologies and Translational Medicine, Università degli Studi di Milano, Milano; Humanitas Clinical and Research Center, Rozzano, Italy

Christopher T. Veldkamp
Department of Biochemistry, Medical College of Wisconsin, Milwaukee; Department of Chemistry, University of Wisconsin–Whitewater, Whitewater, Wisconsin, USA

Ricardo Villares
Department of Immunology and Oncology, Centro Nacional de Biotecnología (CNB/CSIC), Madrid, Spain

Henry F. Vischer
Amsterdam Institute for Molecules Medicines and Systems (AIMMS), Division of Medicinal Chemistry, Vrije Universiteit, Amsterdam, The Netherlands

Brian F. Volkman
Department of Biochemistry, Medical College of Wisconsin, Milwaukee, Wisconsin, USA

Kathrin Werth
Institute of Immunology, Hannover Medical School, Hannover, Germany

John P. Wikswo
Vanderbilt Institute for Integrative Biosystems Research and Education; Department of Biomedical Engineering; Department of Molecular Physiology and Biophysics; Department of Physics and Astronomy, Vanderbilt University, Nashville, Tennessee, USA

Xin Yin
Jiangsu Key Laboratory of Marine Pharmaceutical Compound Screening, School of Pharmacy, Huaihai Institute of Technology, Lianyungang, China; VA San Diego Healthcare System, Medical and Research Sections; Department of Medicine, Division of Pulmonary and Critical Care, University of California San Diego, La Jolla, California, USA

Yi Zheng
Skaggs School of Pharmacy and Pharmaceutical Sciences, University of California San Diego, La Jolla, California, USA

PREFACE

The last 25 years have featured many significant discoveries related to the role of chemokines and their receptors as the maestros of cell migration. They are involved in numerous biological processes starting with immune responses that allow successful conception and fetal development. They subsequently provide cues that regulate development of many organ systems such as the cardiovascular system and the central nervous system. And they are involved in the development and function of both the innate and adaptive arms of our immune system that guard us against physiological insults as we navigate through life. Unfortunately, these otherwise developmentally important and protective molecules can contribute to many diseases as we age. Unbalanced or unresolved chemokine/receptor activities are hallmarks of many inflammatory and autoimmune diseases such as rheumatoid arthritis, inflammatory bowel disease, and multiple sclerosis. In 1996, two chemokine receptors were discovered to be the coreceptors that facilitate the entry of HIV into cells and thereby contribute to AIDS. In 2001, the connection between chemokines and cancer metastasis was established, and we now know that chemokines contribute to cancer via numerous additional mechanisms: by promoting angiogenesis, by providing survival and proliferation signals, and by establishing an immunosuppressive microenvironment that facilitates tumor growth and thwarts chemotherapy and immunotherapy. They are involved in cardiovascular disease, diabetes, pain, and the list goes on.

Because of their unusually broad clinical significance, they have been aggressively pursued as therapeutic targets, and the scientific community has realized the importance of understanding the complexities of how these proteins function in order to bring more chemokine-targeted drugs to the clinic. This understanding has been accompanied by increasingly sophisticated methods for studying the molecular and cellular mechanisms of chemokines in normal physiology and disease. It is instructive to look back at the protocols documented in the first *Methods of Enzymology* volumes from 1997 compared to this volume. While many of these early methods remain as standard laboratory practice, others have been replaced by more efficient methods, or methods that were never feasible before. As documented herein, there are improved ways to carry out assays of binding and signaling that in many cases utilize biosensor or imaging technologies. For more sophisticated assays of cell migration than classic Boyden chambers, microfluidic devices have been developed that allow real-time monitoring of cell migration with control of gradient steepness and flow. Moreover, imaging methods for monitoring migration of specific cell types in *in vivo* models of disease have been developed. It has become possible not only to make chemokines routinely but also to rapidly generate novel modified chemokines with defined pharmacological properties that hold promise as biologics. Whereas the concept of determining structures of chemokine receptors in complex with their natural ligands and with small molecule antagonists was a dream in 1997, it is now possible, and in fact several such structures have now been solved because of the ability to express and purify receptors. This volume covers these and a wide variety of other methods that I believe will be useful to the chemokine community at large. It represents the efforts of many researchers whom I am deeply grateful to for taking the time to share their expertise. I would also like to thank the many other researchers whose work laid the foundation for the evolution of the methods described herein.

Tracy M. Handel

CHAPTER ONE

Chemokine Detection Using Receptors Immobilized on an SPR Sensor Surface

José Miguel Rodríguez-Frade*, Laura Martínez-Muñoz*, Ricardo Villares*, Graciela Cascio*, Pilar Lucas*, Rosa P. Gomariz†, Mario Mellado*,[1]

*Department of Immunology and Oncology, Centro Nacional de Biotecnología (CNB/CSIC), Madrid, Spain
†Department of Cell Biology, Universidad Complutense de Madrid, Madrid, Spain
[1]Corresponding author: e-mail address: mmellado@cnb.csic.es

Contents

1. Surface Plasmon Resonance 2
2. Chemokine Receptors: Members of the GPCR Family 3
3. Chemokine Receptor Immobilization on the Sensor Chip 4
4. Viral Particles as Chemokine Receptors Carriers in SPR 6
    4.1 Materials 7
    4.2 Method I: Generation of Retroviral Particles 8
    4.3 Method II: Generation of LVPs 9
    4.4 Method III: Titration of Viral Proteins 10
    4.5 Method IV: Determination of Chemokine Receptor Levels on the VP 11
    4.6 Method V: Quantitation of Receptor Number on VPs 11
    4.7 Method VI: Attachment of VPs to Biosensor Surfaces 12
5. SPR-Based Applications for Chemokine Receptors 12
6. Conclusions 14
Acknowledgments 15
References 15

Abstract

Chemokines and their receptors take part in many physiological and pathological processes, and their dysregulated expression is linked to chronic inflammatory and autoimmune diseases, immunodeficiencies, and cancer. The chemokine receptors, members of the G protein-coupled receptor family, are integral membrane proteins, with seven-transmembrane domains that bind the chemokines and transmit signals through GTP-binding proteins. Many assays used to study the structure, conformation, or activation mechanism of these receptors are based on ligand-binding measurement, as are techniques to detect new agonists and antagonists that modulate chemokine function. Such methods require labeling of the chemokine and/or its receptor, which

Methods in Enzymology, Volume 570
ISSN 0076-6879
http://dx.doi.org/10.1016/bs.mie.2015.09.013

can alter their binding characteristics. Surface plasmon resonance (SPR) is a powerful technique for analysis of the interaction between immobilized receptors and ligands in solution, in real time, and without labeling. SPR measurements nonetheless require expression and purification steps that can alter the conformation, stability, and function of the chemokine and/or the chemokine receptor. In this review, we focus on distinct methods to immobilize chemokine receptors on the surface of an optical biosensor. We expose the advantages and disadvantages of different protocols used and describe in detail the method to retain viral particles as receptor carriers that can be used for SPR determinations.

1. SURFACE PLASMON RESONANCE

Their ability to monitor the interaction between a molecule immobilized on the surface of a sensor and its interacting molecular partner in solution have made surface plasmon resonance (SPR) techniques powerful tools for analysis of the interplay between biological molecules (Rich & Myszka, 2006). SPR is based on the generation of an electromagnetic evanescent wave at a metal surface when polarized light strikes it at a critical angle of incidence (Homola, 2008). SPR biosensors measure the change in refractive index of the solvent during formation and dissociation of complexes near the assay surface (Rich & Myszka, 2000). Signals depend on mass alteration of the surface analyzed and are expressed as resonance units (RUs). The RU signal is thus directly proportional to mass variations on the sensor surface.

SPR-based techniques are commonly used for detailed analysis of the interaction between immobilized receptors and ligands in solution, in real time, and without analyte labeling. They allow determination of binding affinity and of the binding kinetics of a specific interaction, and are thus widely used in high-throughput screening procedures (Cooper, 2002). Alternative strategies for this type of analysis require fluorescent or radiolabeling of ligand to report its binding to the receptor; these approaches have extra time requirements and in some cases, due to altered binding properties of the partners, cause interference with molecular interactions.

The use of purified proteins facilitates SPR determinations. Experiments with membrane proteins offer additional complexity, as their expression and purification can alter conformation, stability and/or function outside the cell environment. Solubilization of membrane proteins solves the problem in some cases, although detergents must be selected carefully (le Maire,

Champeil, & Moller, 2000). Some lipid/detergent mixtures have been used in nuclear magnetic resonance (NMR) and crystallization experiments (Seddon, Curnow, & Booth, 2004); in structural and functional studies of the extremely light-sensitive G protein-coupled receptor (GPCR), rhodopsin, Reeves et al. used a mixture of the lipid 1,2-dimyristoyl-sn-glycero-3-phosphocholine and the detergent 3-[(3-cholamidopropyl)-dimethylammonio]-1-propane-sulfonate (CHAPS) (Reeves, Hwa, & Khorana, 1999).

Proteins are often expressed at low levels in the cell membrane. Overexpression can be attempted for a protein of interest if alterations in cell localization, protein conformation, and/or ligand binding can be ruled out with confidence. Proteins can also be overexpressed in heterologous systems such as in bacteria, although aggregation and correct posttransductional modifications must be monitored to avoid alteration in binding properties. In the case of the chemokine receptors and other GPCR, an additional problem is their complex structure. These receptors are formed by highly hydrophobic, seven-transmembrane domains that require a lipid bilayer to adopt a stable, functional conformation. Inclusion of these proteins in the lipid environment is no simple matter, and has been the object of several studies (Seddon et al., 2004; Soubias & Grawrisch, 2011).

A new technique combines plasmonic and fluorescence imaging and allows determination of membrane protein distribution in single living cells as well as evaluation of local binding kinetics in the context of the native environment (Wang et al., 2012). Other possibilities include use of a gold film bearing a periodic metallic nanopore array that supports free-standing lipid bilayers with membrane proteins that mimic a biological membrane (Maynard et al., 2009).

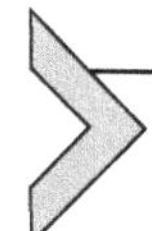

2. CHEMOKINE RECEPTORS: MEMBERS OF THE GPCR FAMILY

GPCR, the largest family in the human genome, include receptors for hormones, neurotransmitters, chemokines, calcium ions, and light (Fredriksson, Lagerstrom, Lundin, & Schioth, 2003; Vassilatis et al., 2003). It is probably the most clinically relevant receptor class (Overington, Al-Lazikani, & Hopkins, 2006) and is thus the main focus of drug discovery research (Takakura, Hattori, Tanaka, & Ozawa, 2015). The structure of these receptors is complex, with a seven-transmembrane spanning hydrophobic α-helix, and requires a lipid environment to maintain native conformation (Drake, Shenoy, & Lefkowitz, 2006; Tan, Brady, Nickols, Wang, & Limbird, 2004).

The members of the GPCR subfamily of chemokine receptors bind a group of low-molecular-weight pro-inflammatory cytokines (Rossi & Zlotnik, 2000). The chemokines were first described as specific mediators of leukocyte directional movement (Griffith, Sokol, & Luster, 2014); since then, they have been associated with a much wider variety of cell types and functions (Anders, Romagnani, & Mantovani, 2014; Guo et al., 2015; Maeda, 2015; Sozzani, Del Prete, Bonecchi, & Locati, 2015). Chemokines are essential elements in a variety of diseases characterized by inflammation and immune cell infiltration (Murdoch & Finn, 2000) such as asthma, atherosclerosis, rheumatoid arthritis, multiple sclerosis, colitis, Crohn's disease, and psoriasis. Two of these receptors, CXCR4 and CCR5, are the two main coreceptors for HIV-1 infection (Bleul, Wu, Hoxie, Springer, & Mackay, 1997), and some chemokine receptors also participate in tumor metastasis (Muller et al., 2001; Strieter, 2001) and transplant rejection (el-Sawy, Fahmy, & Fairchild, 2002).

The nearly 50 chemokines described to date are classified according to functional criteria as constitutive or inducible (Proudfoot, 2002). Constitutive chemokines are generally implicated in immune system homeostasis, whereas inducible chemokine expression is regulated mainly during inflammatory processes. In addition, several viruses encode highly selective chemokine receptor ligands that act as agonists or antagonists, and might thus have a role in viral dissemination or evasion of a host immune response (Alcami & Lira, 2010; Murphy, 2015). Given the physiological relevance of the chemokine receptors, several SPR assays have been developed to study their purification, solubilization, and reconstitution, to analyze their function, to evaluate allosterism, and to improve pharmacological studies.

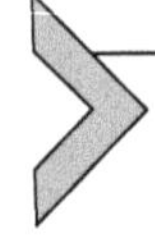

3. CHEMOKINE RECEPTOR IMMOBILIZATION ON THE SENSOR CHIP

The first SPR experiments with chemokine receptors analyzed binding to β-arrestin 1 of C-terminal-derived CCR5 peptides immobilized on a CM5 sensor chip via an N-terminal cysteine thiol group (Huttenrauch, Nitzki, Lin, Honing, & Oppermann, 2002). For SPR, peptides and proteins, including the receptors, must be attached to the solid support of the sensor chip without altering native conformation or binding activity. This attachment must be stable over the course of a binding assay, and a sufficient number of receptors must be presented to the solvent phase to interact with the ligand. It is also essential to avoid nonspecific binding of the sample, which influences RU detection and can mask the specific binding signal.

The most common strategy is covalent attachment to an *N*-hydroxysuccinimide/(1-ethyl-3-(3-dimethylamino-propyl)-carbodiimide (NHS/EDC)-derivatized dextran matrix, which allows coupling to amino groups. Low pH is sometimes necessary, although it increases possible receptor denaturation. Multiple amino groups on the receptor can lead to its random coupling and orientation on the sensor surface (O'Shannessy, Brigham-Burke, & Peck, 1992; Stein & Gerisch, 1996). Other modified matrices can be used to allow more specific, oriented receptor coupling (Alves, Park, & Hruby, 2005).

Another strategy is based on the noncovalent attachment via a histidine tag, avidin/biotin, or antibody-based designs that allow unidirectional coupling of membrane proteins (Fig. 1). These approaches offer several advantages, as receptor purification before capture is unnecessary, the orientation of the immobilized receptor is homogeneous due to directed capture through a probe, and high densities are achieved that enable study of low-molecular mass ligands; nonetheless, cell transfection, receptor solubilization and, when using avidin/biotin systems, specific receptor labeling might be necessary. The use of an anti-CXCR4 mAb immobilized to the gold surface of an immunochip facilitates CXCR4 concentration and correct orientation prior to reconstitution of a lipid bilayer environment on the chip (Stenlund, Babcock, Sodroski, & Myszka, 2003) (Fig. 1A). The

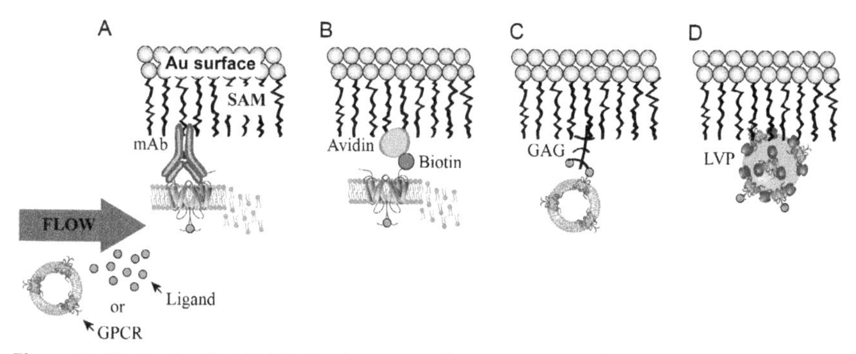

Figure 1 Strategies for GPCR attachment to dextran matrix. The interactive surface of a sensor chip consists of a self-assembled dextran monolayer (SAM) that bears functional groups to allow interaction of different compounds. For GPCR capture on the sensor chip surface, several strategies are used including direct binding to the dextran surface of monoclonal antibodies (A), avidin (B), GAG (C), or LVP (D). (See the color plate.)

structural and functional integrity of the captured receptor can be tested using conformation-sensitive monoclonal antibodies as well as its natural ligand. In the case of CXCR4, this strategy also illustrated the importance of the detergent in maintaining its structural integrity (Stenlund et al., 2003). The detergents CHS/DDM/CHAPS, combined with the lipids DOPC/DOPS at a 7:3 ratio, appear to be the best mixture for SPR studies of CCR5 and CXCR4 activity (Navratilova, Sodroski, & Myszka, 2005).

This antibody capture method was used to study natural ligand/small molecule binding to the chemokine receptors CXCR4 and CCR5. In these conditions, both receptors retained their ability to bind their ligands CXCL12 and CCL5, as well as the small molecule inhibitors, JM-2987 and TAK-779, respectively (Navratilova, Dioszegi, & Myszka, 2006), which demonstrates the value of this technique for biophysical and structural analyses. A similar approach was used to screen cocrystallization conditions for CCR5 and to characterize its binding to nine HIV gp120 variants (Navratilova, Pancera, Wyatt, & Myszka, 2006).

Some approaches use biotinylated ligands immobilized on streptavidin-functionalized sensor chips (Harding, Hadingham, McDonnell, & Watts, 2006) (Fig. 1B). Although biotin-labeled chemokines detect their receptors as effectively as commercially available antibodies and can be used, e.g., cell sorting (Le Brocq et al., 2014), biotinylation could alter the binding properties of these ligands, given their small size. Chemokines act *in vivo* by binding to cell surface glycosaminoglycans (GAG), which present them to their receptors for binding and subsequent biological activity (Hamel, Sielaff, Proudfoot, & Handel, 2009) (Fig. 1C). SPR methods are also used to determine GAG/chemokine-binding kinetics (Tanino et al., 2010), and sensor chip-immobilized GAG were used to retain chemokines at the SPR surface and to determine their ability to associate with viral chemokine-binding proteins (Alcami & Viejo-Borbolla, 2009).

Biosensors permit automated screening of receptor activity in distinct solubilization conditions, allow quantitation of the total amount of receptor captured on the surface, and assess receptor activity by assaying its response to conformation-sensitive ligands (Navratilova et al., 2005).

4. VIRAL PARTICLES AS CHEMOKINE RECEPTORS CARRIERS IN SPR

All methods described above are based on the capture of detergent-solubilized GPCR prior to reconstitution in a lipid environment

(Navratilova et al., 2005). Use of detergents and the need for purification are nonetheless variable elements that can affect functional characteristics of a receptor on the sensor surface. Given their intrinsic structural complexity, this is especially important in the case of GPCR. The chemokine receptors are found in many conformations at the cell membrane, including preformed homodimers, heterodimers, and even oligomers (Munoz, Holgado, Martinez, Rodriguez-Frade, & Mellado, 2012; Thelen, Munoz, Rodriguez-Frade, & Mellado, 2010). These distinct conformations and the way a given ligand modulates these structures dictates its functional effect. Receptor conformation is also influenced by the lipid composition of specific membrane regions and/or other cell surface molecules (Fallahi-Sichani & Linderman, 2009; Inagaki et al., 2012; Martinez-Munoz et al., 2014), which makes it difficult to establish reliable methods that reproduce native GPCR conformation(s) on the biosensor surface.

The use of viral particles (VPs) as receptor carriers could solve these problems (Fig. 1D). Particle immobilization eliminates the need to purify and reconstitute membrane proteins for ligand-binding studies; the technique allows functional characterization of interactions with membrane proteins that cannot be studied using standard equilibrium binding assays. Viruses incorporate plasma membrane proteins into their lipid envelope when they bud from the cell (Fig. 2). Both retro- and lentiviral particles (LVPs) have been immobilized for SPR determinations (Hoffman, Canziani, Jia, Rucker, & Doms, 2000; Suomalainen & Garoff, 1994; Vega et al., 2011); by modifying expression of the proteins of interest on the surface of the packaging cells, their use facilitates generation of adequate positive and negative controls for each assay.

4.1 Materials

- Plasmids pLVTHM, psPAX2, pMD2G, pCMV-dR8.91, pCMV-dR8.74 (Addgene; Cambridge, MA, https://www.addgene.org/Didier_Trono/)
- HEK-293 T cells (CRL-11268 ATCC; Manassas, VA)
- JetPEI (Polyplus-transfection SA; Illkirch, France)
- Aldehyde/sulfate latex beads (Invitrogen, Thermo Fisher Scientific; Waltham, MA)
- ^{35}S-methionine (Perkin Elmer; Waltham, MA)
- Biacore C1, F1, and CM5 sensor chips (GE Healthcare Life Sciences; Pittsburgh, PA)

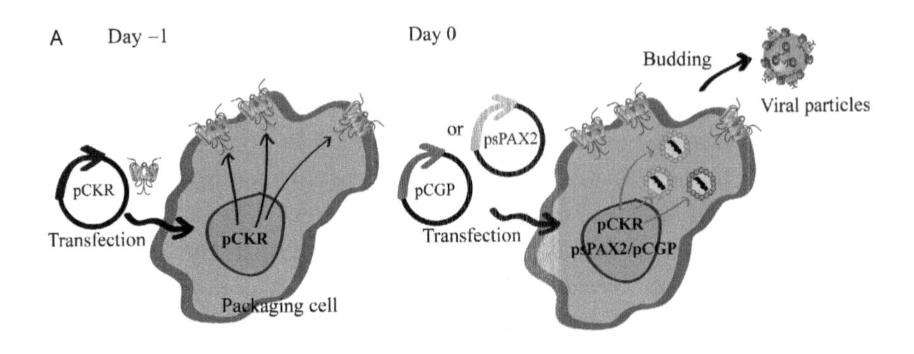

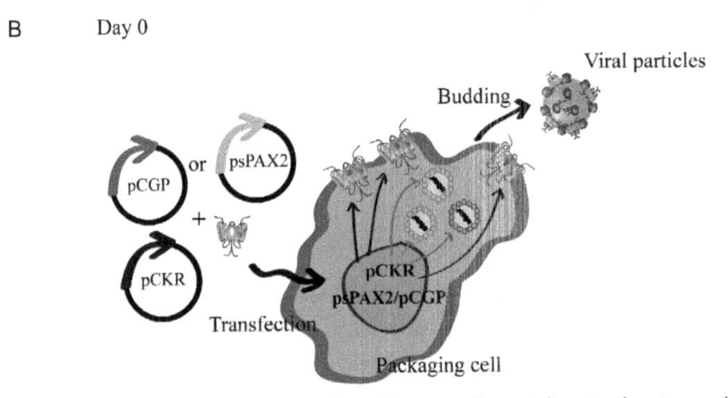

Figure 2 Generation of retroviral and lentiviral particles. Packaging cells are transfected with a chemokine receptor-containing plasmid (pCKR) in advance (A) or simultaneously (B) with the vectors needed to generate retroviral (pCGP) or lentiviral particles (psPAX2). During budding, viral particles incorporate the cell membrane as a coating that retains the structure and functional properties of the receptor. (See the color plate.)

- (Met)-free DMEM and DMEM media (Thermo Fisher Scientific; Waltham, MA)
- Anti-p24 antibody (Abcam; Cambridge, UK)

4.2 Method I: Generation of Retroviral Particles

To generate retroviral particles, murine leukemia virus (MLV) pseudotypes are produced by calcium phosphate-mediated transfection of HEK-293 T cells with a 3:1 ratio of receptor plasmid to pCGP, which encodes the MLV *gag* and *pol* genes. When a specific chemokine receptor must be evaluated, HEK-293 T cells transfected transiently or stably with the receptor should be used. At 4 h posttransfection, fresh medium supplemented with 10 m*M* *n*-butyric acid is added to increase protein expression; supernatant is harvested at 48 h posttransfection, and cell debris is removed by low speed centrifugation and 0.45 μm filtration. The supernatant is pelleted (140,000 × *g*, 90 min) through 20% sucrose/PBS, washed in PBS, ultracentrifuged a second time (280,000 × *g*, 45 min) through 20% sucrose/PBS, and the pellet resuspended in 100 μl 10 m*M* Hepes, pH 7.4. The pseudotypes obtained can be stored at 4 °C or aliquoted and frozen at −20 °C. MLV pseudotypes must be analyzed by SDS-PAGE and Western blot for MLV gag and receptor expression, and purified by equilibrium density ultracentrifugation on a 15–45% sucrose gradient (217,000 × *g*, 16 h) (Fig. 2).

4.3 Method II: Generation of LVPs

Lentiviruses are also Retroviridae, but infect both dividing and nondividing cells (Frimpong & Spector, 2000). Lentiviral vectors are based on modified HIV-1 virus and are usually generated by transient cell transfection of several plasmids. The packaging unit (e.g., pCMV-dR8.91, pCMV-dR8.74, psPAX2) bears the genes that encode structural proteins and enzymes (gag, pol, tat, rev) needed to generate the LVP, the transfer vector (pLVTHM) bears the genetic material to be transduced into the target cell, and an envelope plasmid (pMD2G) derived from a heterologous virus, usually vesicular stomatitis virus (VSV), determines vector tropism. In the packaging unit, genes that encode crucial virulence proteins (env, vif, vpr, vpu, nef) are eliminated previously. In fact, only the packaging unit is needed to generate LVPs, which will not be infective. In this case, pMD2G plasmid can be excluded from the process, although viral protein and receptor levels must be titrated using alternative methods, as particles that lack the pMD2G plasmid will neither infect cells nor replicate, both necessary processes for LVP titration (see Section 4.4).

Although the packaging cell might express certain chemokine receptors (i.e., hCXCR4 in the case of HEK-293 T cells), a gene of interest can be included if needed in the transfer plasmid (pLVTHM) to increase receptor levels at the target cell surface and thus, at the LVP surface. It is also possible to express the receptor of interest using alternative plasmids under the control of more efficient promoters. In all cases, optimal conditions must be established for effective transfection of the packaging cells and expression of the desired protein in the LVP. Several procedures should be tested, including transfection of the plasmid containing the protein of interest prior to psPAX2 transfection, as well as the simultaneous transfection of both plasmids (Fig. 2). LVP-containing supernatants should be collected initially at different time points and target protein expression on the LVP should be analyzed (see Sections 4.5 and 4.6) to establish the most effective, reproducible conditions.

As for retroviral particles, receptor expression is evaluated before LVP generation, and control LVPs are made using mock-transfected packaging cells. GPCR-expressing LVP can be attached to SPR biosensor surfaces by standard immobilization strategies. LVP can be produced by JetPEI-mediated transfection of HEK-293 T cells with pLVTHM or alternative transfer vectors, psPAX2 and pMD2G plasmids. At 4 h posttransfection, fresh medium is added (DMEM supplemented with 10% fetal calf serum, 2 m*M* L-glutamine and 1 m*M* sodium pyruvate), supernatant is harvested at 48 h posttransfection, and LVP processed as for retroviral particles. LVPs are aliquoted and stored at −80 °C. LVP must be analyzed by SDS-PAGE and Western blot for VSV envelope glycoprotein, VSV-G, and receptor expression prior to use in SPR determinations.

4.4 Method III: Titration of Viral Proteins

Prior to use in SPR experiments and to standardize the method, VPs should be titrated by PCR using specific primers for viral genes in the plasmids (Scherr, Battmer, Blomer, Ganser, & Grez, 2001). Virus can also be titrated by transducing HEK-293 T cells with serial dilutions of purified particles. In this case, HEK-293 T cells (3×10^4 cells/well) are plated in 1 ml complete DMEM in 24-well plates (day 1) and cultured (overnight, 37 °C, 5% CO_2), to adhere the plate. Serial dilutions of purified VP in 250 μl complete DMEM are prepared in parallel. On day 2, cells are counted ($6–8 \times 10^4$ cells/well) as a reference, and medium is aspirated from the plates before adding the VP dilutions; incubate (24 h) and add complete DMEM (1 ml/well). At 96 h after VP addition, detach transduced cells using 0.05% trypsin/EDTA. Evaluate the percentage of transduced cells by flow cytometry, measuring GFP expression (included in the transfer plasmid for LVP). VPs that lack VSV-G or alternative envelope plasmids cannot be titrated by transducing HEK-293 T cells, as the resulting particles are neither infective nor replicative. Titer is expressed as transforming units/ml, calculated as the number of cells transduced by a given volume of virions. This method cannot be used in packaging cells not transfected with pMD2G plasmid.

4.5 Method IV: Determination of Chemokine Receptor Levels on the VP

Chemokine receptor expression on the VP can be determined by Western blot using specific antibodies, by ELISA on VP-coated plates, or by flow cytometry. In this last case, small particle size limits measurement of the surface proteins by conventional techniques, and particles are coupled to latex beads to allow detection.

Aldehyde/sulfate latex beads have numerous aldehyde groups grafted to the surface of the polymer particle; this high density allows easy coupling of proteins and other materials to the latex particles in a one-step process. The sulfate groups on the microsphere surface maintain stability during covalent coupling. Use of fluorescently labeled ligands or conformation-dependent antibodies allows detection of the target GPCR in its native conformation on the VP surface. Latex beads are sonicated (5 min, room temperature—RT) and then incubated (15 min, RT) with the VPs. The reaction mixture is then diluted (1 ml PBS) and incubated (60 min, 4 °C, with continuous shaking). Finally, 100 μl glycine buffer (100 m*M* final concentration) is added to block free reactive groups. Incubate (30 min, RT, with continuous shaking), centrifuge ($1950 \times g$, 3 min, RT), and wash once in PBS with 0.5% BSA. Resuspend beads in PBS with 0.5% BSA; 15 μl of beads are usually sufficient for eight different staining conditions. Chemokine receptors expressed on the surface of VPs bound to latex beads can then be stained with appropriate antibodies by standard techniques. As a negative control for flow cytometry, include uncoupled beads or beads coupled to VPs obtained from mock-transfected HEK-293 T cells. Protein expression in CKR-expressing and control VP can be monitored using GFP expressed by pLVTHM or by transfection with other GFP-bearing plasmids.

4.6 Method V: Quantitation of Receptor Number on VPs

To quantitate receptor number on the VPs surface, incubation media of the packaging HEK-293 T cells is removed 24 h posttransfection and cells are cultured in methionine (Met)-free DMEM supplemented with ^{35}S-Met (50 mCi/ml; 48 h, 37 °C, 5% CO_2). ^{35}S-Met-labeled VPs are collected, lysed, and proteins separated by SDS-PAGE. ^{35}S-Met incorporation can be visualized with a Phosphoimager, and relative receptor stoichiometry calculated using Quantity One software (BioRad, Hercules, CA) by normalizing the receptor of interest to a viral protein (such as p24 in lentivirus) with a known number of Met residues and a known number of molecules in the virus; we assume 2100 molecules of p24/LVP (Briggs, Wilk, Welker, Krausslich, & Fuller, 2003). Values are interpolated on a regression curve for p24 data as internal standard. For this procedure, the packaging cells must be transfected with an envelope plasmid.

4.7 Method VI: Attachment of VPs to Biosensor Surfaces

For binding studies using purified receptor-bearing VPs, the virions must be retained on a derivative gold surface suitable for use in the optical biosensor (Fig. 3). A number of sensor surfaces are available from Biacore, each with different properties. A gold surface with a carboxylated alkane thiol (Biacore C1 chip) or a short carboxydextran matrix (Biacore F1 chip or CM5 sensor chip) must first be derivatized by a 10-min activation of surface carboxyl groups using a 1:1 mixture of EDC (1 *M*) and NHS (0.25 *M*). The buffer containing the VPs must be determined individually; pH range and ionic strength must be carefully controlled for immobilization and regeneration processes. Whereas MLV pseudotypes and LVPs are both injected in acetate buffer, composition and pH vary (0.1 *M* sodium acetate pH 5.5 and 10 m*M* sodium acetate pH 4.0, respectively); free surface carboxyl groups remaining after attachment are quenched with 1 *M* ethanolamine (pH 8.5, 5 μl/min, 7 min, RT).

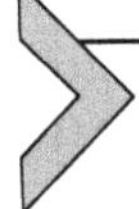

5. SPR-BASED APPLICATIONS FOR CHEMOKINE RECEPTORS

Allosteric modulators are compounds that associate with receptors through binding sites different from those used by the specific ligands. They can modulate the endogenous agonists positively or negatively, or have no influence on receptor activity. Allosteric processes are particularly relevant in the case of the chemokine receptors, which can be homo- or heterodimers, or oligomers. In lymphoblasts that express CCR2, CCR5 and CXCR4 simultaneously, their specific antagonists (AMD3100 for CXCR4 and TAK779 for CCR5) affect ligand binding as well as CCL2-triggered cell migration toward CCR2 and CXCL12-triggered migration toward CXCR4 (Sohy, Parmentier, & Springael, 2007).

A SPR-based method was developed for high-throughput, label-free screening to identify ortho- and allosteric CCR5 ligands (Navratilova, Besnard, & Hopkins, 2011). The method is based on immobilization to the sensor gold surface of an antibody to a tag peptide (TETSQVAPA) included at the CCR5 C-terminal end. The antibody retains purified

A

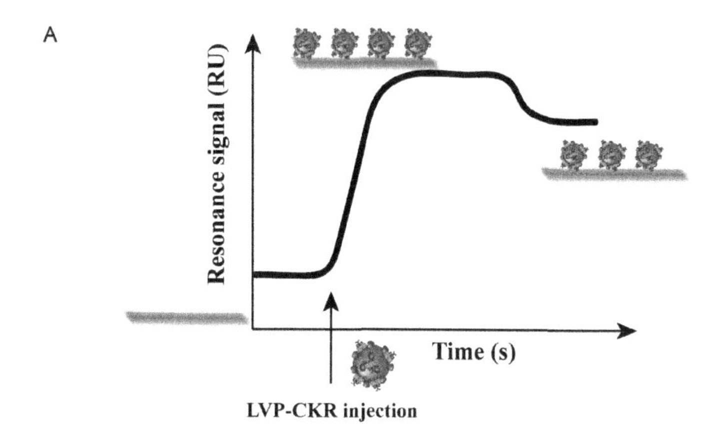

B

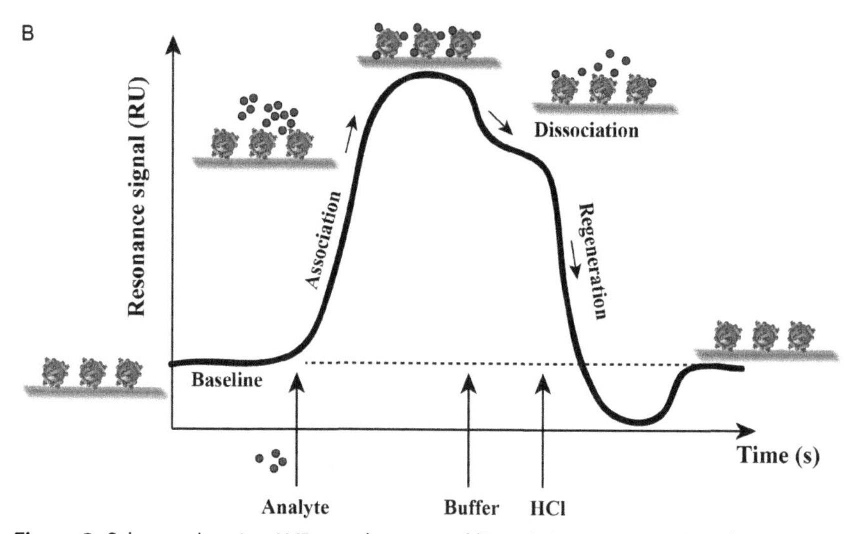

Figure 3 Scheme showing LVP attachment and ligand detection in a SPR device. As for purified proteins, LVPs are injected in the sensor in established salt and pH conditions to achieve maximum attachment to the dextran surface (A). The flow through the microfluidic chamber is maintained until the resonance signal stabilizes (baseline). The test analyte can now be injected (B) and changes in the resonance signal during the association and dissociation phases are recorded. Finally, the analyte is removed using appropriate buffers (regeneration) and the SPR-bound LVP is ready for further experiments.

CCR5. As a control, a reference chamber is used in which a CCR5 with a blocked binding site is retained in the same way. Although five CCR5-binding compounds were identified, with affinities in the 8.2–49 μ*M* range ligands (Navratilova et al., 2011), the method uses solubilized receptors and therefore excludes the receptor's natural context. At least in theory, screening based on immobilized, receptor-bearing VPs would identify additional compounds that modulate chemokine responses *in vivo*, such as GAG, other chemokine receptors, or proteins that can bind to the receptor being studied.

As key mediators in many diseases, the chemokines have attracted considerable attention from researchers, and their detection in biological fluids could offer new possibilities for diagnosis of inflammatory and autoimmune disorders. A LVP bearing CXCR4 in its native plasma membrane context was used to develop a SPR method for rapid detection and quantification of analytes such as CXCL12 in biological samples, with no need for pretreatment (Vega et al., 2013); proof of concept was shown using a homemade biosensor (Mauriz et al., 2006). Gold chips are cleaned with trichloroethylene, acetone, and ethanol, immersed in piranha solution [H_2SO_4: H_2O_2 (3:1)], rinsed with water, ultrasonicated (5 min), and dried with an N_2 stream. The chip is placed over the flow cells and the prism is optically coupled to the chip using refractive index matching oil. A constant flow speed of 20 μl/min is maintained throughout the immobilization process. A carboxyl-terminated alkanethiol self-assembled monolayer (SAM) is then formed on the gold chip by flowing mercaptoundecanoic acid (0.05 m*M* solution in ethanol; 20 min); alkanethiol excess is rinsed with ethanol followed by water before the next immobilization step. The carboxylic surface is activated with 0.2 *M* EDC and 0.05 *M* NHS to form an NHS ester intermediate, followed immediately by injection of CXCR4-bearing LVP in acetate buffer (10 m*M* NaOAc). LVPs are thus coupled covalently to the ester via amine groups. The surface is blocked with 1 *M* ethanolamine pH 8.5. HCl (5 m*M*) is used to regenerate the surface completely, which permits more that 150 regeneration cycles with no notable signal loss relative to initial binding values (Fig. 3). The method allows detection of 5–40 n*M* CXCL12 in urine, is optimal in a 6.8–7.5 pH range, and although statistical significance must be corroborated with a larger sample cohort, it is clearly effective for detecting small molecules in urine samples.

6. CONCLUSIONS

SPR-based technologies provide an excellent method for monitoring interactions between ligands and their receptors. SPR studies of GPCR nonetheless present special difficulties, as these receptors are expressed at the cell membrane at relatively low levels and require a hydrophobic environment to maintain their native structure. In recent years, several protocols have allowed successful solubilization, purification, reconstitution, and retention of correctly oriented GPCR at the surface of a sensor chip; these methods have demonstrated their value for evaluation of ligand-binding properties and have improved pharmacological studies. Many analyses have been developed to determine the role of solubilizing detergents, to select the best lipid composition for receptor reconstitution, and to orientate the receptor to increase the ligand-binding sites. Current evidence indicates that GPCR at the cell surface are not individual entities, but form dynamic oligomeric complexes that are modulated by receptor expression, ligand binding, the expression of other membrane proteins, and even the lipid composition. These factors obviously affect not only ligand binding, but also the pharmacological properties of the receptors. VPs offer new possibilities for these types of analyses. When a virus buds from the cell, it incorporates plasma membrane proteins into its lipid envelope, a unique means of reproducing native GPCR conformation(s) on the biosensor surface. Proof of concept using chemokine receptors demonstrates its validity and illustrates the potential for the use of VPs immobilized on SPR biosensors in ligand quantification, binding analysis, and drug screening.

ACKNOWLEDGMENTS

We specially thank the present and former members of the DIO chemokine group who contributed to some of the work described in this review. We also thank C. Bastos and C. Mark for secretarial support and editorial assistance, respectively. This work was partially supported by grants from the Spanish Ministry of Economy and Competitiveness (SAF2011-27370 and SAF2014-53416-R), the RETICS Program (RD 12/0009/009 RIER), and the Madrid regional government (S2010/BMD-2350; RAPHYME).

REFERENCES

Alcami, A., & Lira, S. A. (2010). Modulation of chemokine activity by viruses. *Current Opinion in Immunology*, *22*, 482–487.

Alcami, A., & Viejo-Borbolla, A. (2009). Identification and characterization of virus-encoded chemokine binding proteins. *Methods in Enzymology*, *460*, 173–191.

Alves, I. D., Park, C. K., & Hruby, V. J. (2005). Plasmon resonance methods in GPCR signaling and other membrane events. *Current Protein and Peptide Science*, *6*, 293–312.

Anders, H. J., Romagnani, P., & Mantovani, A. (2014). Pathomechanisms: Homeostatic chemokines in health, tissue regeneration, and progressive diseases. *Trends in Molecular Medicine*, *20*, 154–165.

Bleul, C. C., Wu, L., Hoxie, J. A., Springer, T. A., & Mackay, C. R. (1997). The HIV coreceptors CXCR4 and CCR5 are differentially expressed and regulated on human T lymphocytes. *Proceedings of the National Academy of Sciences of the United States of America*, *94*, 1925–1930.

Briggs, J. A., Wilk, T., Welker, R., Krausslich, H. G., & Fuller, S. D. (2003). Structural organization of authentic, mature HIV-1 virions and cores. *EMBO Journal*, *22*, 1707–1715.

Cooper, M. A. (2002). Optical biosensors in drug discovery. *Nature Reviews. Drug Discovery*, *1*, 515–528.

Drake, M. T., Shenoy, S. K., & Lefkowitz, R. J. (2006). Trafficking of G protein-coupled receptors. *Circulation Research*, *99*, 570–582.

el-Sawy, T., Fahmy, N. M., & Fairchild, R. L. (2002). Chemokines: Directing leukocyte infiltration into allografts. *Current Opinion in Immunology*, *14*, 562–568.

Fallahi-Sichani, M., & Linderman, J. J. (2009). Lipid raft-mediated regulation of G-protein coupled receptor signaling by ligands which influence receptor dimerization: A computational study. *PLoS ONE*, *4*, e6604.

Fredriksson, R., Lagerstrom, M. C., Lundin, L. G., & Schioth, H. B. (2003). The G-protein-coupled receptors in the human genome form five main families. Phylogenetic analysis, paralogon groups, and fingerprints. *Molecular Pharmacology*, *63*, 1256–1272.

Frimpong, K., & Spector, S. A. (2000). Cotransduction of nondividing cells using lentiviral vectors. *Gene Therapy*, *7*, 1562–1569.

Griffith, J. W., Sokol, C. L., & Luster, A. D. (2014). Chemokines and chemokine receptors: Positioning cells for host defense and immunity. *Annual Review of Immunology*, *32*, 659–702.

Guo, F., Wang, Y., Liu, J., Mok, S. C., Xue, F., & Zhang, W. (2015). CXCL12/CXCR4: A symbiotic bridge linking cancer cells and their stromal neighbors in oncogenic communication networks. *Oncogene*. http://dx.doi.org/10.1038/onc.2015.139.

Hamel, D. J., Sielaff, I., Proudfoot, A. E., & Handel, T. M. (2009). Chapter 4: Interactions of chemokines with glycosaminoglycans. *Methods in Enzymology*, *461*, 71–102.

Harding, P. J., Hadingham, T. C., McDonnell, J. M., & Watts, A. (2006). Direct analysis of a GPCR-agonist interaction by surface plasmon resonance. *European Biophysics Journal*, *35*, 709–712.

Hoffman, T. L., Canziani, G., Jia, L., Rucker, J., & Doms, R. W. (2000). A biosensor assay for studying ligand-membrane receptor interactions: Binding of antibodies and HIV-1 Env to chemokine receptors. *Proceedings of the National Academy of Sciences of the United States of America*, *97*, 11215–11220.

Homola, J. (2008). Surface plasmon resonance sensors for detection of chemical and biological species. *Chemical Reviews*, *108*, 462–493.

Huttenrauch, F., Nitzki, A., Lin, F. T., Honing, S., & Oppermann, M. (2002). Beta-arrestin binding to CC chemokine receptor 5 requires multiple C-terminal receptor phosphorylation sites and involves a conserved Asp-Arg-Tyr sequence motif. *The Journal of Biological Chemistry*, *277*, 30769–30777.

Inagaki, S., Ghirlando, R., White, J. F., Gvozdenovic-Jeremic, J., Northup, J. K., & Grisshammer, R. (2012). Modulation of the interaction between neurotensin receptor NTS1 and Gq protein by lipid. *Journal of Molecular Biology*, *417*, 95–111.

Le Brocq, M. L., Fraser, A. R., Cotton, G., Woznica, K., McCulloch, C. V., Hewit, K. D., et al. (2014). Chemokines as novel and versatile reagents for flow cytometry and cell sorting. *Journal of Immunology*, *192*, 6120–6130.

le Maire, M., Champeil, P., & Moller, J. V. (2000). Interaction of membrane proteins and lipids with solubilizing detergents. *Biochimica et Biophysica Acta*, *1508*, 86–111.

Maeda, N. (2015). Proteoglycans and neuronal migration in the cerebral cortex during development and disease. *Frontiers in Neuroscience*, *9*, 98.

Martinez-Munoz, L., Barroso, R., Dyrhaug, S. Y., Navarro, G., Lucas, P., Soriano, S. F., et al. (2014). CCR5/CD4/CXCR4 oligomerization prevents HIV-1 gp120IIIB binding to the cell surface. *Proceedings of the National Academy of Sciences of the United States of America*, *111*, E1960–E1969.

Mauriz, E., Calle, A., Abad, A., Montoya, A., Hildebrandt, A., Barcelo, D., et al. (2006). Determination of carbaryl in natural water samples by a surface plasmon resonance flow-through immunosensor. *Biosensors & Bioelectronics*, *21*, 2129–2136.

Maynard, J. A., Lindquist, N. C., Sutherland, J. N., Lesuffleur, A., Warrington, A. E., Rodriguez, M., et al. (2009). Surface plasmon resonance for high-throughput ligand screening of membrane-bound proteins. *Biotechnology Journal, 4,* 1542–1558.

Muller, A., Homey, B., Soto, H., Ge, N., Catron, D., Buchanan, M. E., et al. (2001). Involvement of chemokine receptors in breast cancer metastasis. *Nature, 410,* 50–56.

Munoz, L. M., Holgado, B. L., Martinez, A. C., Rodriguez-Frade, J. M., & Mellado, M. (2012). Chemokine receptor oligomerization: A further step toward chemokine function. *Immunology Letters, 145,* 23–29.

Murdoch, C., & Finn, A. (2000). Chemokine receptors and their role in inflammation and infectious diseases. *Blood, 95,* 3032–3043.

Murphy, P. M. (2015). Viral chemokine receptors. *Frontiers in Immunology, 6,* 281.

Navratilova, I., Besnard, J., & Hopkins, A. L. (2011). Screening for GPCR ligands using surface plasmon resonance. *ACS Medicinal Chemistry Letters, 2,* 549–554.

Navratilova, I., Dioszegi, M., & Myszka, D. G. (2006). Analyzing ligand and small molecule binding activity of solubilized GPCRs using biosensor technology. *Analytical Biochemistry, 355,* 132–139.

Navratilova, I., Pancera, M., Wyatt, R. T., & Myszka, D. G. (2006). A biosensor-based approach toward purification and crystallization of G protein-coupled receptors. *Analytical Biochemistry, 353,* 278–283.

Navratilova, I., Sodroski, J., & Myszka, D. G. (2005). Solubilization, stabilization, and purification of chemokine receptors using biosensor technology. *Analytical Biochemistry, 339,* 271–281.

O'Shannessy, D. J., Brigham-Burke, M., & Peck, K. (1992). Immobilization chemistries suitable for use in the BIAcore surface plasmon resonance detector. *Analytical Biochemistry, 205,* 132–136.

Overington, J. P., Al-Lazikani, B., & Hopkins, A. L. (2006). How many drug targets are there? *Nature Reviews. Drug Discovery, 5,* 993–996.

Proudfoot, A. E. (2002). Chemokine receptors: Multifaceted therapeutic targets. *Nature Reviews. Immunology, 2,* 106–115.

Reeves, P. J., Hwa, J., & Khorana, H. G. (1999). Structure and function in rhodopsin: Kinetic studies of retinal binding to purified opsin mutants in defined phospholipid-detergent mixtures serve as probes of the retinal binding pocket. *Proceedings of the National Academy of Sciences of the United States of America, 96,* 1927–1931.

Rich, R. L., & Myszka, D. G. (2000). Advances in surface plasmon resonance biosensor analysis. *Current Opinion in Biotechnology, 11,* 54–61.

Rich, R. L., & Myszka, D. G. (2006). Survey of the year 2005 commercial optical biosensor literature. *Journal of Molecular Recognition, 19,* 478–534.

Rossi, D., & Zlotnik, A. (2000). The biology of chemokines and their receptors. *Annual Review of Immunology, 18,* 217–242.

Scherr, M., Battmer, K., Blomer, U., Ganser, A., & Grez, M. (2001). Quantitative determination of lentiviral vector particle numbers by real-time PCR. *BioTechniques, 31,* 520–524.

Seddon, A. M., Curnow, P., & Booth, P. J. (2004). Membrane proteins, lipids and detergents: Not just a soap opera. *Biochimica et Biophysica Acta, 1666,* 105–117.

Sohy, D., Parmentier, M., & Springael, J. Y. (2007). Allosteric transinhibition by specific antagonists in CCR2/CXCR4 heterodimers. *The Journal of Biological Chemistry, 282,* 30062–30069.

Soubias, O., & Gawrisch, K. (2011). The role of the lipid matrix for structure and function of the GPCR rhodopsin. *Biochimica et Biophysica Acta, Biomembranes, 1818,* 234–240.

Sozzani, S., Del Prete, A., Bonecchi, R., & Locati, M. (2015). Chemokines as effector and target molecules in vascular biology. *Cardiovascular Research, 107,* 364–372.

Stein, T., & Gerisch, G. (1996). Oriented binding of a lipid-anchored cell adhesion protein onto a biosensor surface using hydrophobic immobilization and photoactive crosslinking. *Analytical Biochemistry, 237*, 252–259.

Stenlund, P., Babcock, G. J., Sodroski, J., & Myszka, D. G. (2003). Capture and reconstitution of G protein-coupled receptors on a biosensor surface. *Analytical Biochemistry, 316*, 243–250.

Strieter, R. M. (2001). Chemokines: Not just leukocyte chemoattractants in the promotion of cancer. *Nature Immunology, 2*, 285–286.

Suomalainen, M., & Garoff, H. (1994). Incorporation of homologous and heterologous proteins into the envelope of Moloney murine leukemia virus. *Journal of Virology, 68*, 4879–4889.

Takakura, H., Hattori, M., Tanaka, M., & Ozawa, T. (2015). Cell-based assays and animal models for GPCR drug screening. *Methods in Molecular Biology, 1272*, 257–270.

Tan, C. M., Brady, A. E., Nickols, H. H., Wang, Q., & Limbird, L. E. (2004). Membrane trafficking of G protein-coupled receptors. *Annual Review of Pharmacology and Toxicology, 44*, 559–609.

Tanino, Y., Coombe, D. R., Gill, S. E., Kett, W. C., Kajikawa, O., Proudfoot, A. E., et al. (2010). Kinetics of chemokine-glycosaminoglycan interactions control neutrophil migration into the airspaces of the lungs. *Journal of Immunology, 184*, 2677–2685.

Thelen, M., Munoz, L. M., Rodriguez-Frade, J. M., & Mellado, M. (2010). Chemokine receptor oligomerization: Functional considerations. *Current Opinion in Pharmacology, 10*, 38–43.

Vassilatis, D. K., Hohmann, J. G., Zeng, H., Li, F., Ranchalis, J. E., Mortrud, M. T., et al. (2003). The G protein-coupled receptor repertoires of human and mouse. *Proceedings of the National Academy of Sciences of the United States of America, 100*, 4903–4908.

Vega, B., Calle, A., Sanchez, A., Lechuga, L. M., Ortiz, A. M., Armelles, G., et al. (2013). Real-time detection of the chemokine CXCL12 in urine samples by surface plasmon resonance. *Talanta, 109*, 209–215.

Vega, B., Munoz, L. M., Holgado, B. L., Lucas, P., Rodriguez-Frade, J. M., Calle, A., et al. (2011). Technical advance: Surface plasmon resonance-based analysis of CXCL12 binding using immobilized lentiviral particles. *Journal of Leukocyte Biology, 90*, 399–408.

Wang, W., Yang, Y., Wang, S., Nagaraj, V. J., Liu, Q., Wu, J., et al. (2012). Label-free measuring and mapping of binding kinetics of membrane proteins in single living cells. *Nature Chemistry, 4*, 846–853.

CHAPTER TWO

Study of Chemotaxis and Cell–Cell Interactions in Cancer with Microfluidic Devices

Jiqing Sai*,†, Matthew Rogers‡, Kathryn Hockemeyer‡, John P. Wikswo‡,§,¶,||, Ann Richmond*,†,‡,1

*Department of Veterans Affairs, Tennessee Valley Healthcare System, Nashville, Tennessee, USA
†Department of Cancer Biology, School of Medicine, Vanderbilt University, Nashville, Tennessee, USA
‡Vanderbilt Institute for Integrative Biosystems Research and Education, Vanderbilt University, Nashville, Tennessee, USA
§Department of Biomedical Engineering, Vanderbilt University, Nashville, Tennessee, USA
¶Department of Molecular Physiology and Biophysics, Vanderbilt University, Nashville, Tennessee, USA
||Department of Physics and Astronomy, Vanderbilt University, Nashville, Tennessee, USA
[1]Corresponding author: e-mail address: ann.richmond@vanderbilt.edu

Contents

1. Introduction 20
2. Methods 24
   2.1 Chemotaxis in Microfluidic Chemotaxis Chamber 24
   2.2 Gradient Switching in Microfluidic Device 27
   2.3 Cell–Cell Interaction in Tumor Microenvironment 31
3. Limitations 37
4. Perspectives 40
   4.1 Shearing Force and Calculation of Chemotactic Force 40
   4.2 Cell Polarity Change Versus Cell Turning During a Gradient Change of Direction 40
   4.3 Potential Future Uses of Microfluidic Devices: Analysis of Circulating Tumor Cells 42
   4.4 Advantages of Using 3D Microbioreactors to Investigate Factors that Influence the Tumor Microenvironment 42
Acknowledgments 42
References 43

Abstract

Microfluidic devices have very broad applications in biological assays from simple chemotaxis assays to much more complicated 3D bioreactors. In this chapter, we describe the design and methods for performing chemotaxis assays using simple microfluidic chemotaxis chambers. With these devices, using real-time video microscopy we can examine the chemotactic responses of neutrophil-like cells under conditions of varying gradient steepness or flow rate and then utilize software programs to calculate the

Methods in Enzymology, Volume 570
ISSN 0076-6879
http://dx.doi.org/10.1016/bs.mie.2015.09.023

speed and angles of cell migration as gradient steepness and flow are varied. Considering the shearing force generated on the cells by the constant flow that is required to produce and maintain a stable gradient, the trajectories of the cell migration will reflect the net result of both shear force generated by flow and the chemotactic force resulting from the chemokine gradient. Moreover, the effects of mutations in chemokine receptors or the presence of inhibitors of intracellular signals required for gradient sensing can be evaluated in real time. We also describe a method to monitor intracellular signals required for cells to alter cell polarity in response to an abrupt switch in gradient direction. Lastly, we demonstrate an *in vitro* method for studying the interactions of human cancer cells with human endothelial cells, fibroblasts, and leukocytes, as well as environmental chemokines and cytokines, using 3D microbioreactors that mimic the *in vivo* microenvironment.

1. INTRODUCTION

Microfluidic devices can be designed to control the flow of liquid inside cell-sized channels and to thereby enable a variety of biological studies. The dimensions of the channels in microfluidic devices are typically 10s–100s of microns, and hence with appropriate fluid controls and sensors, can support the manipulation and analysis of very small volumes. Fabrication of these microdevices requires the use of techniques adapted from semiconductor microfabrication and plastic molding, such as photolithography or micromachining to create molds, and replica casting or embossing or glass etching to create the actual devices. Many of the devices are ideally suited to high-resolution microscopic imaging of chemotaxis.

Chemotaxis is a directional cell movement during which cells sense a chemical gradient in a chemokine or chemoattractant and move toward the chemical source. Many types of cells use chemotaxis to actively move to specific locations. The inflammatory process provides an excellent example of chemotaxis, wherein immune cells respond to a gradient of chemokines or chemoattractants, and move up the gradient to reach the site of infection. Once the immune cells "sense" the gradient, they extravasate from vascular vessels and move toward the infection site within the adjacent tissue to destroy bacteria, remove dead cells, and heal the wound area. To set up an *in vitro* chemotaxis assay requires generation of a reliable chemokine/chemoattractant gradient. Traditional *in vitro* chemotaxis assays use a passive diffusion mechanism to generate the gradients, such as a modified Boyden chamber (Boyden, 1962) or agarose- or collagen-based assays (Haddox, Knowles, Sommers, & Pfister, 1994; Haddox, Pfister, & Sommers, 1991;

John & Sieber, 1976; Nelson, Quie, & Simmons, 1975), and other techniques like Zigmond or Dunn chambers (Zicha, Dunn, & Brown, 1991; Zigmond, 1977). With the Boyden chamber or modified Boyden chamber, transwells covered with polycarbonate filters with tiny pores (from 3 to 10 μm in diameter) are used to separate two different concentrations of chemokine. The assay relies on diffusion between the two chambers to generate a gradient across the membrane. The Zigmond and Dunn chambers generate the gradient through a very small bridge area between two chemokine reservoirs. Assays based upon agarose or collagen rely on chemokine diffusion through the agarose or collagen gel and require cells to crawl through or under the agarose or collagen up the gradient of chemotactic factors.

All of these traditional chemotaxis assays have common disadvantages. (1) They can generate only linear gradients and cannot provide either a variety of gradient shapes or rapid alterations of gradient direction or gradient profiles, all of which occur in the tissues *in vivo*. (2) Changing the type or concentration of chemokines within the gradient region is not feasible during experiments. After the gradient has been established, any change to the chemokine requires a significant amount of time before the new gradient can be established and this change most likely will disturb cell migration. (3) The steepness and range of the gradient with these assays are determined by both the difference in the concentration of the chemokines in the two solutions on either side of the barrier and the thickness of the barrier that separates them, and this steepness cannot be altered without replacing one of the two solutions, with concomitant hydrodynamic forces. For all of these reasons, it is desirable to develop a new chemotaxis system that can create and maintain a stable concentration gradient and also allow manipulation of the slope and direction of the gradient to better understand the factors involved in "gradient sensing." Microfluidic devices are ideal tools for creating and controlling chemokine gradients. Microfabrication methods enable creation of arbitrary microchannel designs through which cells can chemotax. Microchannel dimensions can be created small enough that diffusion occurs in minutes to seconds, thus reducing the waiting time for a gradient to become established (Li Jeon et al., 2002; Lin, Nguyen et al., 2004; Lin, Saadi et al., 2004).

In this chapter, we first demonstrate a method using a simple microfluidic device that allows the recording of cell migration in response to a chemokine gradient (Fig. 1) (Walker et al., 2005). The design of this device includes a Christmas tree-like mixer and splitter, in which the concentrated chemokine is injected through one input (IN_B) and the buffer (without

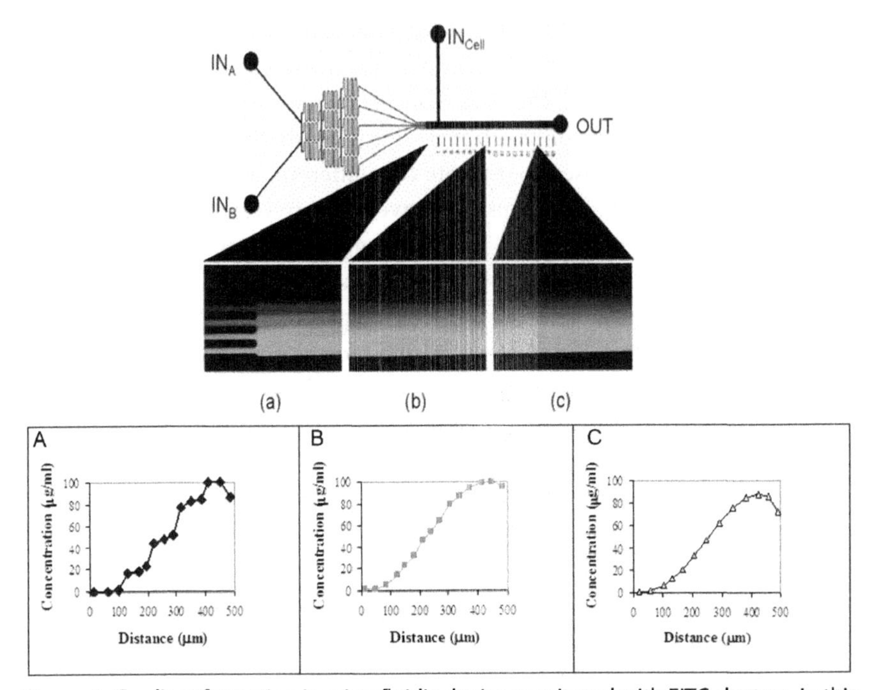

Figure 1 Gradient formation in microfluidic device monitored with FITC-dextran. In this device, a constant 1 µl/m flow of 100 µg/ml FITC-dextran (MW 10,000) in Hank's buffer from one syringe and a flow at the same rate of buffer alone was driven by a single-, dual-syringe pump. The fluorescent images were taken at different locations along the main channel. The intensity of the fluorescence was quantitated across the channel for each image and is shown in the plots. *Adapted from Walker et al. (2005).*

chemokine), is input from the other (IN_A), both driven by a single-syringe pump with two syringes. Within the mixer/splitter, whenever two input flows with different concentrations merge, the resulting laminar flows mix by lateral diffusion in the serpentine section and the concentration of the mixed solutions becomes uniform with a concentration equal to the average of the two inputs and twice the flow rate. When a single flow is split into two flows, the concentration in each branch is equal to the concentration before the split but the flow rates are halved. The splitting and merging of the two input flows can produce three different streams, which are split and then merged into four, and, for the device shown in Fig. 1, these flows in turn create five flows. Through three tiers (or stages) of this splitting and mixing, five streams of chemokine fluids with different concentrations are generated in the order of increasing concentrations that span those of the two original flow streams from the two syringes. Then these five streams of chemokine fluids are merged again into the main channel, where a chemokine gradient is formed from low to high across the main channel all the way to the output end (OUT). Depending on the location in the main channel, the chemokine gradient shows a step-wise gradient at the entrance of the main channel and becomes a relatively smoother gradient with the help of lateral chemokine diffusion within the laminar flow of the channel (Fig. 1). The cells are seeded in the main channel through the cell input (IN_{cell}), and time-lapse video microscopy is taken in the mid-location of main channel.

The second microfluidic device used in chemotaxis studies shown in this chapter is a microfluidic switching system (Fig. 2) (Liu et al., 2008). This device allows one to change the direction of the gradient in less than

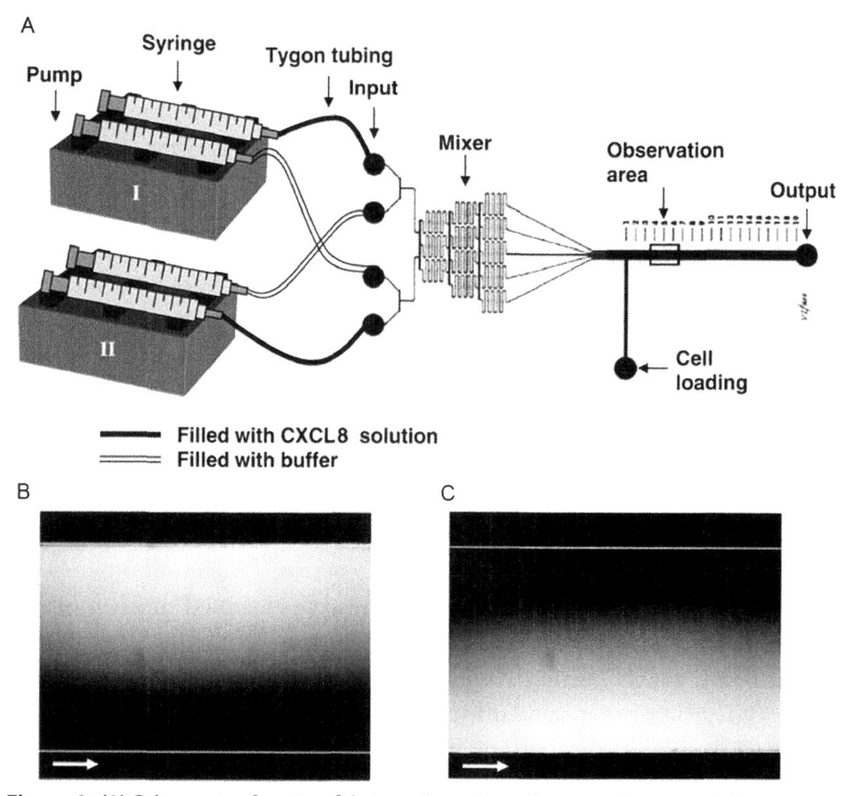

Figure 2 (A) Schematic of a microfabricated gradient device with a pair of dual-syringe pumps for gradient switching; (B) chemokine gradient generated by running pump I; (C) switching of the chemokine gradient is generated by running pump II. The white arrows in (B) and (C) indicate the flow direction in the main channel. *Originally published in Liu, Sai, Richmond, and Wikswo (2008).*

1 min and is a very useful tool to study how the cells respond to the directional change of the gradient. Similar to the previous microfluidic chemotaxis chamber, the two pairs of input fluid streams are driven separately by two pumps and three tiers of mixer are used to generate gradients. Only one pump runs at a time: when pump I runs, it generates a low-to-high gradient across the main channel from the bottom, and when pump II starts running and pump I stops, a reversed gradient is generated.

Our third microfluidic device described in this chapter is a miniature tumor microenvironment system in which a cluster of tumor cells (i.e., a spheroid), fibroblasts, and endothelial cells are cocultured in a system that mimics the tumor microenvironment (Fig. 6) (Hockemeyer et al., 2014). This device provides an ideal tool to study these cell–cell interactions in a simulated tumor microenvironment.

2. METHODS

2.1 Chemotaxis in Microfluidic Chemotaxis Chamber

2.1.1 Making Microfluidic Chemotaxis Chamber

Microfluidic devices that consist of three tiers of "divide and mix" microchannels that enter into the main microchannel (Fig. 1) were fabricated in a class-100 clean room (less than 100 particles per cubic foot of air). The splitting/mixing microchannels were designed to generate five streams of that span a range of concentrations of fluid as described above. A stable gradient of chemical thus forms across the streams in the main microchannel. In the models and experiments presented here, the main microchannel dimensions are 500 μm wide, 100 μm tall, and 1 cm long. Polydimethylsiloxane (PDMS) was chosen as the construction material because it allows devices to be fabricated rapidly and is optically transparent, facilitating observation of both the concentration gradient and the cells involved in chemotaxis. Microfluidic devices were made by molding microchannels in the PDMS (Sylgard 184, Dow Corning, Midland, MI) and then bonding the mold to glass coverslips using standard methods in soft lithography (Whitesides, Ostuni, Takayama, Jiang, & Ingber, 2001).

Materials:

Class 100 clean room
PDMS (Sylgard 184, Dow Corning, Midland, MI)
Glass coverslips, No. 1, 24 mm × 50 mm
SU-8-2100 (MicroChem Corp., Newton, MA)

Programmable hotplate (Barnstead/Thermolyne International, Dubuque, IA)
16-gauge blunt needle: outer diameter (OD) 1.65 mm; inner diameter (ID) 1.19 mm (Becton, Dickinson and Company, Franklin Lakes, NJ)
Plasma cleaner (Harrick Scientific Corporation, Ossining, NY)
Tygon tubing, OD: 1.52 mm; ID: 0.508 mm (Cole Parmer, Vernon Hills, IL)

Procedures:

1. Curing agent and prepolymer are mixed together (1:10 wt/wt) and placed in a desiccator for 45 min. The mixture is then poured over a SU-8-2100 (MicroChem Corp., Newton, MA) master in a tissue culture dish which contains the positive relief of the microchannel design. The PDMS mold is cured for 2.5 h at 80 °C on a programmable hotplate.
2. Once the mold has cooled, it is peeled off the master, cut to size, and access holes are punched for tubing. A blunt 16-gauge needle with an OD of 1.65 mm and an ID of 1.19 mm is used as the punching tool.
3. Each PDMS mold is then treated with a plasma cleaner for 20 s and placed on a number 1 24 mm × 50 mm glass coverslip.
4. The bonded device is treated again with the plasma cleaner for 15 s to facilitate filling with liquid.
5. Tygon tubing with an OD of 1.52 mm and an ID of 0.508 mm, cut into four 30.5 cm lengths, is inserted into each access hole.
6. Sterilized DI water is injected into the device via the waste line until the device and the remaining three tubes are filled with water. Bubbles are removed by pinching the three lines closed and applying pressure to the waste line syringe. The increased fluid pressure within the device causes the air in trapped bubbles to diffuse through the PDMS.

2.1.2 Chemotaxis Assay for Differentiated HL-60 Cells

The HL-60 cell line is a promyelocytic cell line derived from human leukemia (Collins, Gallo, & Gallagher, 1977). The cells are premyelocytes, but can be induced *in vitro* to differentiate into different lineages of mature myeloid cells depending on the reagents used for induction (Collins, Ruscetti, Gallagher, & Gallo, 1978). If dimethyl sulfoxide at 1–1.5% is provided for HL-60 cell culture, the cells will differentiate into granulocyte-like cells, or neutrophils. Chemotaxis is one of the most important characteristics of neutrophils in inflammatory response. Differentiated HL-60 cells are widely used to study neutrophil chemotaxis, since they are readily available and easy to genetically modify. Although it has been reported that the differentiation of HL-60 may boost the expression of chemokine receptor, CXCR2, a major receptor to the inflammatory chemokine CXCL8 (Collins, 1987), the expression level is too low to drive an efficient chemotactic response to CXCL8 (Elvin, Kerr, McArdle, & Birnie, 1988). Therefore, HL-60 cells were stably transfected with a CXCR2 expression vector and the transduced cells exhibited a robust chemotaxis in response to a CXCL8 gradient (Sai, Walker, Wikswo, & Richmond, 2006). The response of these cells to a chemotactic gradient of CXCL8 can be visually observed with a microfluidic device, as described in Liu et al. (2008), Sai et al. (2006), and Walker et al. (2005).

Materials:

Stable HL-60 cells expressing CXCR2
Modified Hank's balanced salt solution (HBSS): 150 m*M* sodium chloride, 4 m*M* potassium chloride, 1.2 m*M* magnesium chloride, 10 mg/ml glucose, and 20 m*M* HEPES, pH 7.2
Dimethyl sulfoxide (Endotoxin-low, Sigma-Aldrich)
RPMI-1640 medium supplemented with 10% fetal bovine serum (FBS) (Gibco)
CO_2 incubator
Fibronectin (human) (BD Biosciences, San Jose, CA)
Bovine serum albumin (BSA) (Sigma-Aldrich)
CXCL8 or MIP-2 (Recombinant protein from a commercial source)
Syringe, 1 ml (Becton, Dickinson)
Syringe pump, Harvard PHD 2000 (Harvard Apparatus, Holliston, MA)
Blunt end needles, 25 gauge, ID = 0.3 mm, and OD = 0.5 mm (Howard Electronic Instruments, Inc., El Dorado., KS)
Inverted microscope with CCD camera (Axiovert 200M, Zeiss, Germany)

Procedures:

1. To prepare the differentiated HL-60 cells 2×10^5, HL-60 cells are inoculated into 10 ml antibiotic-free RPMI medium supplemented with 10% FBS and 1.3% endotoxin-low DMSO is added. Cells are incubated at 37 °C, 5% CO_2 for 5–7 days. The morphology of cells will change during the process of differentiation from a perfectly round shape to a randomly irregular shape with multiple protrusions on the cell membrane.
2. To coat the microfluidic chemotaxis chamber, the microfluidic device is refilled with HBSS buffer and the device is viewed under an inverted microscope to verify that no air bubbles are trapped; if bubbles are present,

they are removed by the method described in step 6 above (see Section 2.1.1). 100 μg/ml fibronectin solution is injected into the device through the cell injection line (Fig. 1) with inputs A and B blocked so that the fibronectin can coat the slide at room temperature for 1 h. The excess fibronectin solution is then washed out from the device by injecting 1 ml HBSS buffer through the waste line just prior to loading the cells.

3. The differentiated HL-60 cells are washed with serum-free RMPI-1640 medium and resuspended in serum-free RPMI-1640 medium at 2×10^6 cells/ml. The cells are injected into the microfluidic device and incubated for 20 min at 37 °C and 5% CO_2 to allow the cells to settle and attach to the microchannel floor. The cells should be uniformly distributed on the microchannel floor prior to exposing them to the chemokine.
4. The microfluidic device containing the cells is placed on the inverted microscope (Zeiss Axiovert 200M) with a controlled environmental chamber set at 37 °C and 5% CO_2. A 1 ml syringe is loaded with chemokine (CXCL8) in serum-free RPMI-1640 medium at a concentration of 50 ng/ml and another syringe is loaded with medium only as control, and then the syringes are connected to the syringe pump. The input lines A and B are connected to the chemokine syringe or the medium syringe, respectively. Pumping is applied at a high flow rate (50 μl/min) for 1.5 min with the cell loading line open and waste line closed to allow quick replacement of HBSS buffer in the input lines with chemokine or medium.
5. The pumping is paused, the waste line is opened while the cell loading line is closed, and then a slower pump rate of 1 μl/min is utilized for the remainder of experiment.
6. Time-lapse video recording is initiated once the slower pump rate is initiated.

2.2 Gradient Switching in Microfluidic Device

One of the advantages of microfluidic devices is that they are capable of providing a diverse range of gradients. The controlled and reproducible generation of a spatiotemporally complicated chemoattractant gradient is very important in the study of the mechanisms of cellular chemotaxis. For example, microfluidic devices are excellent for studying how cells respond to two opposing chemokine gradients, how cells change their morphology when the direction of gradient switches, and how the movement of cells alters as the steepness of the gradient changes. In the methods listed below, we describe a modified microfluidic chemotaxis device that allows switching of the direction of gradient in seconds (Fig. 2).

2.2.1 Making a Gradient Switching Microfluidic Device

Materials:

Class 100 clean room

PDMS (Sylgard 184 Dow Corning, Midland, MI)

Glass coverslips, No. 1, 50 mm

SU-8-2050 (MicroChem Corp., Newton, MA)

Programmable hotplate (Barnstead/Thermolyne International, Dubuque, IA)

16-gauge blunt needle, OD: 1.6 mm; ID: 1.2 mm (Becton, Dickinson, Franklin Lakes, NJ)

Plasma cleaner (Harrick Scientific Corporation, Ossining, NY)

Tygon tubing, ID: 0.02 in.; OD: 0.05 in. (Saint-Gobain Performance Plastics Corporation)

Procedures:

1. A 100 μm thick layer of photoresistant SU8-2050 is exposed to UV light through a chrome mask to generate negative master patterns on a silicon substrate.
2. PDMS mixed with a curing agent at a ratio of 10:1 is cast onto the master to replicate the master patterns and cured for 2.5 h at 80 °C.
3. The cured PDMS is peeled from the master, and with a 16-gauge blunt needle, holes are punched for connecting the Tygon tubing, which serve as the inputs and outputs.
4. The surfaces of the channel side of the cured PDMS and a clean glass cover slide (No. 1) are treated with plasma for 20 s and bonded together to form an irreversible seal.
5. Approximately, 30 cm long Tygon tubing is inserted into the input holes to connect the device to a pair of two syringe pumps.

2.2.2 Chemotaxis Assay for Differentiated HL-60 Cells with Gradient Switch

Materials:

HL-60 cells stably expressing CXCR2

Modified HBSS: 150 m*M* sodium chloride, 4 m*M* potassium chloride, 1.2 m*M* magnesium chloride, 10 mg/ml glucose, and 20 m*M* HEPES, pH 7.2

Dimethyl sulfoxide (Endotoxin-low, Sigma-Aldrich)
RPMI-1640 medium supplemented with 10% FBS (Gibco)
CO_2 incubator
Fibronectin (human) (BD Biosciences, San Jose, CA)
BSA, Sigma-Aldrich
CXCL8 or MIP-2 (Commercially available)
Syringe (1 ml, Becton-Dickinson)
Syringe pump, Harvard PHD 2000 (Harvard Apparatus, Holliston, MA)
Blunt end needles (25 gauge, ID = 0.3 mm, and OD = 0.5 mm) (Howard Electronic Instrument, Inc., El Dorado., KS)
Inverted microscope (Axiovert 200M, Zeiss, Germany), with a charge-coupled device (CCD) camera (Hamamatsu, Japan)
MetaMorph software (Molecular Devices, Inc., Sunnyvale, CA)

Procedures:

1. The main channel of the device is coated with human fibronectin at a 100 μg/ml concentration for 1 h at room temperature and rinse with HBSS buffer to remove excess fibronectin in the channels before loading cells.
2. HL-60 cells are differentiated by culturing 2×10^5 CXCR2-expressing HL-60 cells/ml in antibiotic-free medium that contains 1.3% DMSO for a week.
3. The differentiated CXCR2-HL-60 cells are washed and resuspended with serum-free RPMI-1604 medium at a concentration of 4×10^6 cells/ml. The prepared cells are injected into the device through the loading channel and seed the cells for 5 min at 37 °C and 5% CO_2.
4. The device with cells is set on the stage of an inverted microscope with a temperature and CO_2 controlled chamber. The four tubing inputs are connected to syringes filled with either CXCL8 solution or buffer alone. The solutions are injected into the device driven by two syringe pumps, as indicated in Fig. 2A. The key feature is the use of two pumps and four syringes, such that one pump drives one gradient and the second pump drives the reverse gradient.
5. Initially, both pumps are run at a high flow rate (50 μl/min) to quickly fill the four segments of tubing (about 1 min). A single pump is then run at low flow rate (0.5 μl/min) to maintain a chemokine gradient that is high on the side of the channel of the top (Fig. 2B). At the same time, time-lapse video recording is initiated. After running the pump for 20 min, stop the first pump and run the second to generate a gradient with the high concentration on the opposite side of the channel

(Fig. 2C), and continue time-lapse video recording for an additional 20 min.

2.2.2.1 Cell Tracking and Data Analysis

1. Time-lapse video data are generated by taking pictures every 20 s using a CCD camera. MetaMorph software is used to track and analyze cell movement. The HL-60 cells that migrate more than 20 μm in the initial gradient and keep moving after the gradient switching are defined as exhibiting a directionally biased migration response to the chemokine gradient and are included in the quantification of the chemotaxis, while the cells that remain within a 20 μm radius of their original positions are excluded from the analysis, as are cells that stopped or detached from the substrate.
2. As shown in Fig. 3, the first 5 min following initiation of flow (the cell accommodation interval) was excluded from data extraction because during this time the cells respond to the chemokine gradient, become polarized, and begin to move toward the gradient. Thus, counting the events during this time frame would adversely affect the precise analysis of cell movement. The initial response interval was defined as the time

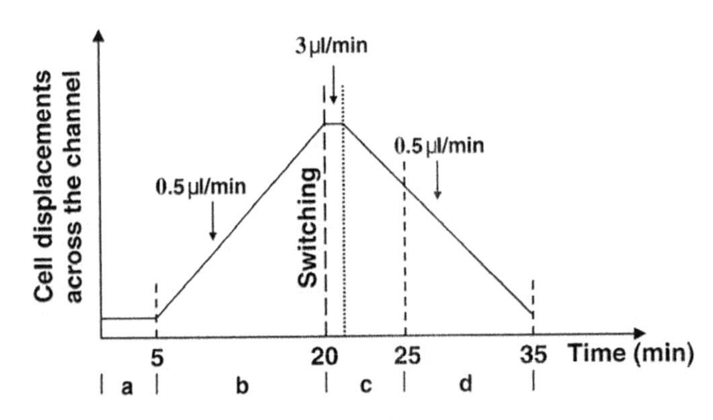

Figure 3 Summary of intervals used to quantify HL-60 response to changing chemoattractant gradients. (a) Cell accommodation interval for 5 min; (b) initial response interval toward a forward gradient for 15 min; (c) prompt response interval toward the switched/reversed gradient for 5 min, which includes 1 min of high flow during gradient reversal; (d) later response interval after the gradient reversed for 10 min. *Originally published in Liu et al. (2008).*

between the beginning of cell movement and the time the gradient direction was shifted (15 min). The gradient was then switched, a process which required approximately 1 min. The 5-min interval that began with the reversal of two pumps was termed the "prompt response interval." During this period of time, the response of cells to the directional change of the gradient was taken into account. The later response interval was the next 10 min, which was continued until the end of the experiment. Chemotactic indexes (CI) before and after switching were calculated as mean and standard error, and the cell response time and average moving angles were also determined for comparison (Fig. 4).

3. The CI is defined as the displacement along the direction of the gradient divided by the total migration distance and is used to quantify the cell motility toward the chemokine gradient. The results were evaluated by the Student's *t*-test and single factor analysis of variance (ANOVA). The number (*N*) of cells is indicated in Fig. 5.
4. Conclusions: In summary, we present a simple microfluidic system for studying cell migration in a time-dependent gradient environment and describe the response of PI3K-inhibited neutrophil-like HL-60 cells to such a gradient condition. The inhibited HL-60 cells showed normal chemotaxis but reduced cell motility, as indicated by slower response time to the gradient switch, less polarization, and a reduced number of cells that turned upon reversal of the CXCL8 gradient.

2.3 Cell–Cell Interaction in Tumor Microenvironment

Microfluidic devices can support cell–cell signaling analyses with more analytical depth than can other coculture methods. For example, cancer is not a cellular or genetic disease of one cell type, but a complex cascade of interactions among several different cell types, proteins, cytokines, and circumstances in a microenvironment. The process of cancer progression and metastasis is complicated, encompassing recruitment of fibroblasts as well as leukocytes to the developing tumor or the premetastatic niche. Numerous pathways are activated simultaneously across several different types of cells. In order to study these interactions, microfluidics offers what other current models cannot: a method to mimic and observe an *in vivo*-like microenvironment, while still maintaining control over the system as a whole. Herein, we describe a microfluidic device that is capable of a creating such a microenvironment, complete with spheroids (aggregates of cancer cells that

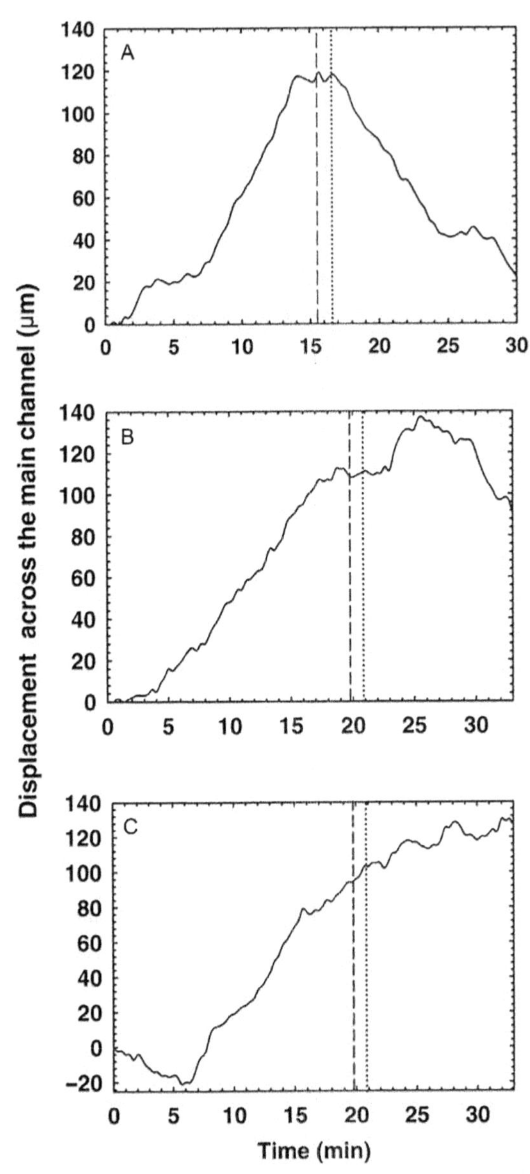

Figure 4 Examples of displacement of the HL-60 cells and the change in their morphology in response to a switch in the direction of the gradient. Displacement is measured relative to the initial position of the cells before exposure to the gradient. The time of the onset of switching is indicated by the dashed vertical line, and the return to low velocity by the dotted line. (A) An HL-60 cell quickly changed direction; (B) a Wortmannin-inhibited HL-60 cell turned slowly; (C) a Wortmannin-inhibited HL-60 cell kept moving in the initial direction without turning. *Adapted from Liu et al. (2008).*

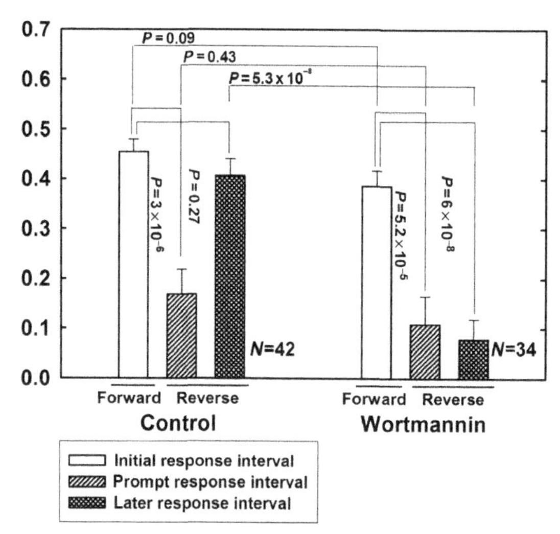

Figure 5 Chemotaxis of HL-60 cells as measured by the chemotaxis index (CI) in control experiments without PI3K inhibitor as compared to that in the presence of PI3K inhibitor. Data are for the three intervals of the gradient switching shown in Fig. 4. The gradient direction is indicated either as forward or as reverse. *Originally published in Liu et al. (2008).*

behave more like an *in vivo* tumor than cells on 2D surfaces), fibroblasts, and endothelial cells that form a blood vessel-type structure.

2.3.1 Creation of the Microfluidic Device

The microfluidic device (see Fig. 6) was designed in AutoCAD and fabricated on a planar silica chip using a computer-controlled, femtosecond laser machining station. The chip was then etched in 10 *M* KOH at 80 °C for 1 h to form channels within the silica. Fabrication and assembly were conducted in an ISO 1000 stage clean room (Hockemeyer et al., 2014). Once the device is built, it can be reused several times for repeat experiments. To sterilize before initiating an experiment, 70% ethanol is flown through the device, followed by flowing 1 × PBS supplemented with 400 μg/ml penicillin, 400 μg/ml streptomycin, and 1 μg/ml amphotericin B with a rate of 20 μl/min.

Materials for spheroid formation:

MDA MB 231-GFP^{+} cells (MDA MB 231 cells from ATCC were transduced with a lentiviral GFP vector and selected by antibiotic resistance and expression of GFP by flow cytometry)

Trypsin/EDTA (Sigma)

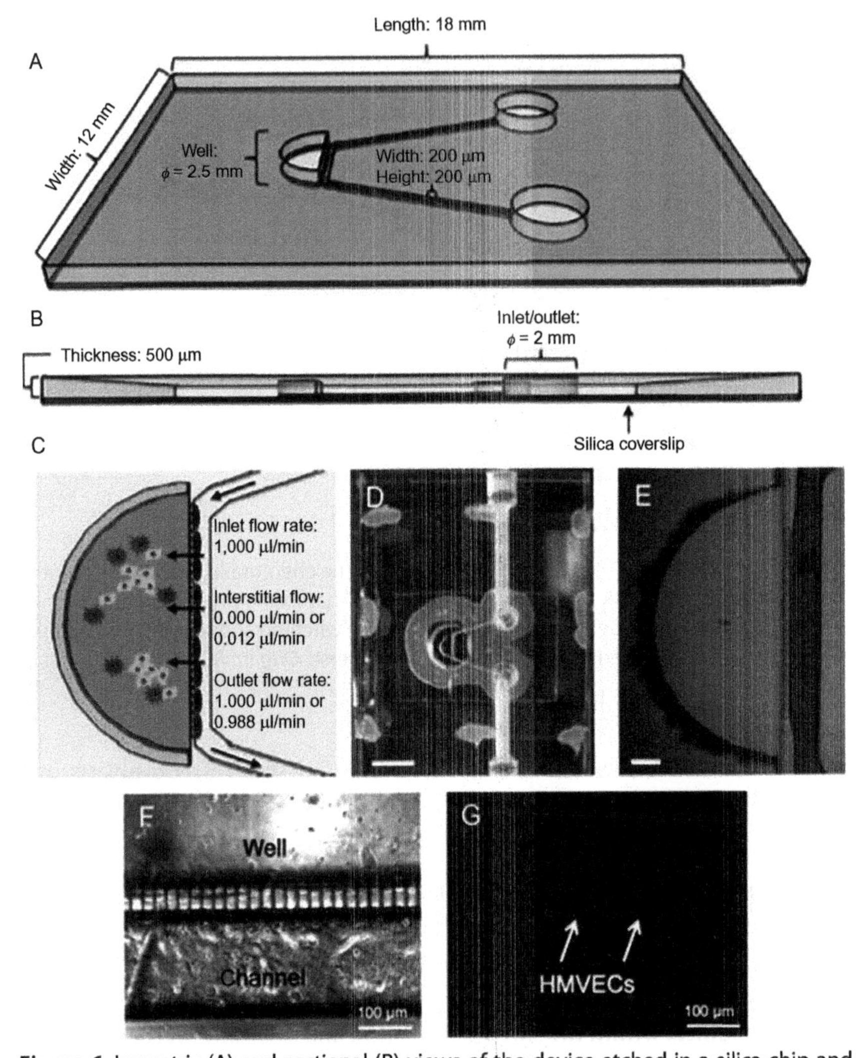

Figure 6 Isometric (A) and sectional (B) views of the device etched in a silica chip and sealed to a coverslip. A zoomed in orthogonal view of the semicircular well and channel (C) displays spatial orientation of cancer cell spheroids (green), fibroblasts (red), and microvascular endothelial cells (blue) during an experiment along with rates of inlet flow, outlet flow, and interstitial flow. Images of the fabricated device (D and E) give a size reference with scale bars 5 mm and 200 μm, respectively. (D) Assembled device glued to the acrylic manifold. Inlet and outlet port supply is sealed by red o-rings on the top and bottom of the diagram. (E) A view of the cell well, interstitial flow channel, and bridge between the two containing 6 μm flow channels every 30 μm. (F) A simulated vascular wall through endothelial cell monolayer is created by HMVECs labeled with Cell Tracker Blue cultured on channel side of porous membrane spatially mimic blood vessel wall. 20 × magnification of channel in bright field shows some visibility of cells in monolayer. (G) 20 × magnification of DAPI channel visually confirms HMVECs are beginning to form a monolayer. *Originally published in Hockemeyer et al. (2014).* (See the color plate.)

DMEM (10% FBS plus 2 m*M* L-glutamine) (Gibco) with 20% methylcellulose (Sigma)
96-well sterile non-tissue-culture-treated U-bottom plate

Procedures for spheroid formation:

1. MDA MB 231-GFP cells are detached from a culture flask using Trypsin/EDTA.
2. Live cells are collected by centrifugation and resuspended in DMEM (10% FBS, L-glutamine) and 20% methylcellulose solution to a density of 1000 cells/ml.
3. One milliliter of this cell suspension is pipetted into each well of the 96-well U-bottom plate.
4. The culture is placed in the water-jacketed incubator (37 °C) for a period of 2–3 days, until a small spheroid forms in each well.

Materials for endothelial cell loading:

HMVECad cells (Life Technologies)
Cell Tracker Blue (CTB) (Invitrogen)
MCD-131 medium supplemental with microvascular growth supplement (Life Technologies)
Trypsin (0.05%)/EDTA (0.53 m*M*) in HBSS, without calcium, magnesium (Corning)
25 m*M* HEPES-containing medium
100 μl gas-tight syringe
Blunt 23-gauge dispensing needle

Procedures for endothelial cell loading:

1. Endothelial cells are stained with 8 μ*M* CTB in serum-free MCDB-131 for 1 h at 37 °C.
2. CTB is removed and cells are incubated in normal media for 30 min.
3. Cells are detached using Trypsin (0.05%)/EDTA (0.53 m*M*) in HBSS solution.
4. Cells are centrifuged at 800 g for 5 min.
5. Cells are resuspended in 25 m*M* HEPES-containing MCDB-131 media at density of 5×10^6 cells/ml.
6. 500 μl HEPES containing MCDB-131 medium is allowed to flow through the channel.
7. 100 μl HEPES containing MCDB-131 medium is added to the semicircular chamber.
8. Endothelial cells are loaded into the device's channel using the gas-tight syringe and a 23-gauge needle.

9. The device is positioned on an angle for 3 h so cells will seed on interface between the channel and semicircular well.

Materials for extracellular matrix (ECM) preparation:

HEPES buffer solution (NaOH, HEPES, and 10× PBS in water)—pH 7.2
Matrigel (BD Biosciences)
Rat tail collagen type 1 (BD Biosciences)

Procedures for preparation of ECM:

1. The ECM is prepared on ice; once the Matrigel is added, subsequent steps/protocols are completed swiftly to ensure the ECM does not start to polymerize before the fibroblasts/spheroids are added.
2. A desired amount of ECM is prepared, using a 1:1 ratio of collagen and Matrigel, with a final concentration of 1.5 mg/ml collagen and 10% Matrigel. The collagen is added to an Eppendorf tube first, then buffer with the HEPES solution. The Matrigel is added last and then the cells are loaded into this ECM suspension.

Materials for fibroblast loading:

Fibroblasts (NAFs and/or CAFs,)—normal mammary-associated fibroblasts (NAFs) and breast cancer-associated fibroblasts (CAFs) were the generous gifts of Drs. Harold L. Moses and Simon Hayward (Vanderbilt)
MCDB-131 medium supplemented with 10% FBS, 1× insulin amp in, 1× nonessential amino acids, 1.4 m*M* L-glutamine (Gibco), 13% amniomax basal medium, and 2.1% amniomax C100 supplement (Gibco)
Cell Tracker Red (Life Technologies)
TrypLE (Life Technologies)
ECM solution

Procedure fibroblast loading:

1. Fibroblasts are incubated in 7 μ*M* Cell Tracker Red for 1 h at 37 °C.
2. Fibroblasts are detached using TrypLE.
3. Detached fibroblasts are collected by centrifugation at 800 × *g* for 2 min.
4. Cells are resuspended and counted.
5. Cells are diluted to a concentration of 2.5×10^5 cells/ml and added to the ECM solution.
6. The fibroblast/ECM solution is set aside until it is ready for spheroid integration.

Materials for tumor spheroid loading:

1.5 ml centrifuge tube
Cut pipette tips
Experimental medium (DMEM, 10% FBS, LG, and 25 m*M* HEPES)

Fibroblast loaded, ECM solution

Spheroids

Procedures for tumor spheroid loading:

1. Spheroids to the 1.5 ml centrifuge tube carefully, using the cut pipette tips so as not to disrupt their integrity.
2. Spheroids are washed twice with experimental medium and placed on ice.
3. Supernatant is aspirated from spheroids and aliquots are placed into the premade fibroblast/ECM suspension to achieve a final concentration of 50–60 spheroids in 200 μl of the ECM.
4. 7.5 μl of this fibroblast/spheroid/ECM solution is loaded into the semicircular chamber.
5. The device is incubated at 37 °C and 5% CO_2 for 1.5 h to allow polymerization of the ECM, inverting the orientation intermittently to ensure an even cell density spread within the chamber.
6. After 1.5 h, 100 μl experimental medium is added to the top of the semicircular chamber. Then, depending on the type of experiment, one may add 100 μl of CXCL12 ligand diluted to 10 ng/ml, or 100 μl vehicle solution to the top of the semicircular chamber as well. This part of the protocol can be altered to incorporate a variety of chemokines/stimuli, as it is not limited solely to CXCL12. Alternatively, inhibitors of chemokines/chemokine receptor interactions may be added.
7. Flow can be established by connecting tubing to the inlet and outlet of the channel and to two syringe pumps. Depending on the rates of the pumps, one can establish either interstitial or regular perfusion flow through the channel.
8. The device is placed on a microscope stage (Zeiss Axiovert 200M) equipped with an incubated stage set at 37 °C and CO_2 chamber set at 5% CO_2 and imaged at 15 min intervals over 20 h under a 10 × objective lens in bright field and for fluorescence at excitations of 488 nm and 546 nm to visualize GFP-expressing cancer cells and Cell Tracker Red-labeled fibroblasts, respectively.
9. The time-lapse imaging data are quantitated using MetaMorph software (Figs. 7 and 8).

3. LIMITATIONS

Despite the many advantages of microfluidic devices as applied in chemotaxis assays, some problems have appeared. First, a constant flow rate is required to maintain a stable chemokine gradient, and the shear force

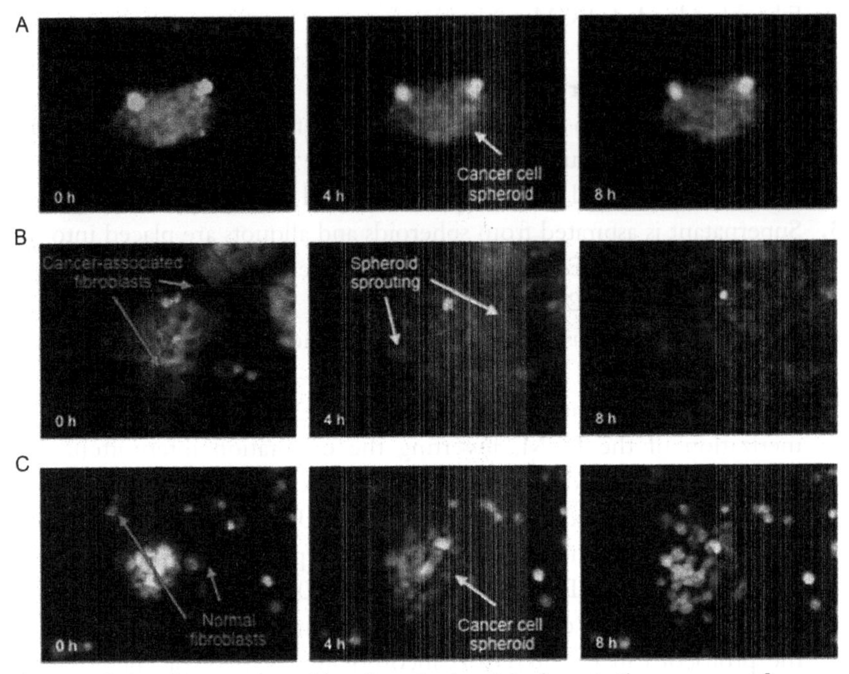

Figure 7 Spheroid sprouting with cultured spheroids alone, in the presence of cancer-associated fibroblasts, or in the presence of normal tissue-associated fibroblasts. Shortly after polymerization of the reconstituted basement membrane, the device was imaged over 12 h. Cancer cells are GFP-expressing and fibroblasts were labeled with Cell Tracker Red. (A) Cancer cell spheroids alone did not exhibit sprouting. (B) Cancer cells sprouted into the surrounding matrix when cultured with CAFs. (C) When cultured with NAFs, cancer cells moved around the spheroid or within small clusters, but very little migration occurred outside of local movement. *Originally published in Hockemeyer et al. (2014).* (See the color plate.)

consequently generated within a very small volume places some stress on the cells. Thus, it is very important to carefully regulate the flow rate. To resolve this problem, some devices have been designed to generate a gradient without using an external pump (Gao, Sun, Lin, Webb, & Li, 2012; Xu et al., 2012). These devices use either gravity-based passive pumping to generate and maintain the gradient (Gao et al., 2012), a low flow-rate osmotic pump (Xu et al., 2012), pure diffusion within very small grooves, or even a flow-free design (Abhyankar, Lokuta, Huttenlocher, & Beebe, 2006; Kim & Kim, 2010; Zhu et al., 2012).

Another disadvantage of microfluidic devices is the limited number of samples that may be assayed at the same time due to the complexity of the assay. Additional modifications are needed to allow the testing multiple

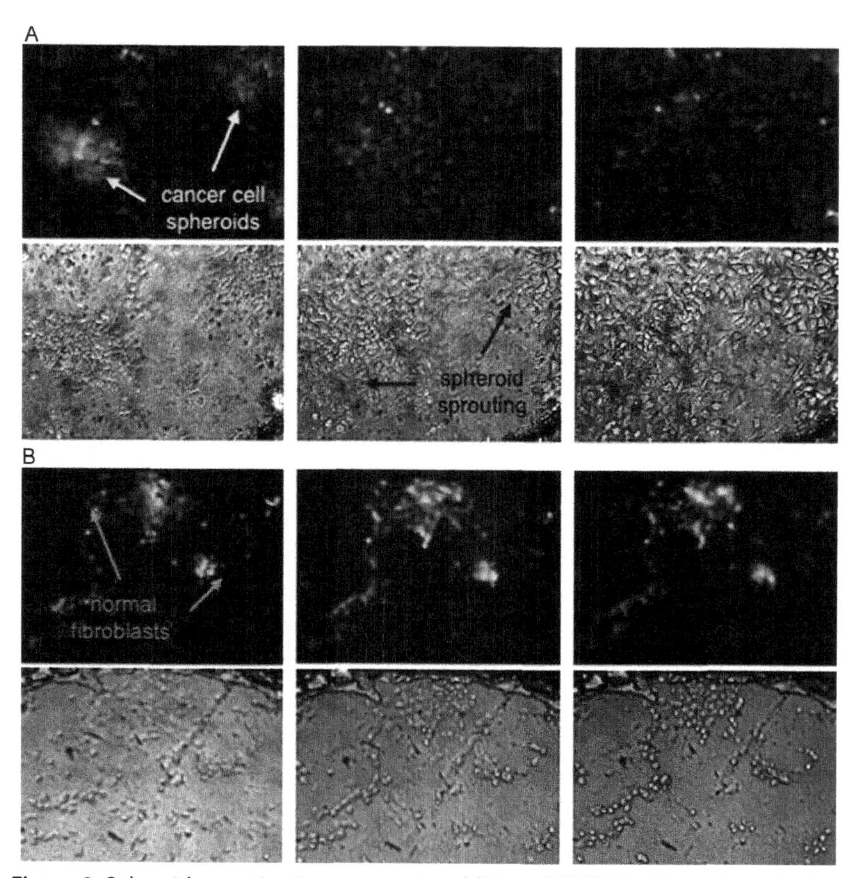

Figure 8 Spheroid sprouting in response to addition of CXCL12. After polymerization, medium supplemented with 10 ng/ml CXCL12 was added to the well and the device was imaged over 20 h. (A) When stimulated with CXCL12 without the presence of fibroblasts, cancer cells sprouted extensively from the spheroid and were highly mobile within the surrounding matrix. (B) When stimulated with CXCL12 in the presence of normal fibroblasts, spheroid sprouting was abrogated and cancer cells remained localized to the spheroid. *Originally published in Hockemeyer et al. (2014).*

samples under multiple conditions. New designs with multiple repeating units, such as arrayed microdevices (Berthier, Surfus, Verbsky, Huttenlocher, & Beebe, 2010), allow the testing of multiple samples or multiple conditions. These devices have the potential to be used to analyze a patient's neutrophil chemotaxis in response to multiple chemokine/chemoattractant gradients and will be very useful in the diagnosis of a primary immunodeficiency disorders. This type of device also demonstrates the capability for increased throughput in microenvironmental studies designed for screening targeted therapies for specific human diseases. Moreover, 3D microbioreactors are currently being designed for high-throughput screening (Wen, Zhang, & Yang, 2010; Yang, Zhang, & Wen, 2008), and we hope to utilize a simplified microbioreactor for this purpose in the future.

4. PERSPECTIVES

4.1 Shearing Force and Calculation of Chemotactic Force

As described above, the shearing force generated by constant flow places some stress on the cells being analyzed. The value of this is that it can mimic the flow that leukocytes experience in circulation, and with a microfluidic device the impact of the flow force on chemotaxis can be calculated. A typical cell in the microfluidic chemotaxis device as described in Section 2.1 encounters two different forces from orthogonal directions: the shearing force applies externally along the direction of flow and the internal chemotactic force generated from the chemokine gradient. Under these two different forces applied at a 90° angle, the trajectory of the cell migration will be the result from the balance between these two forces (Fig. 9). If the cell moves along the 45° angle line between these two forces, it is predicted that the chemotactic force generated under this specific chemokine gradient will be equivalent to the shearing force by the flow (Fig. 9). Of course, the final calculation will need to evaluate the surface area of the cell that encounters the shearing force and the friction between the cell and the chamber surface (Sai et al., 2006).

4.2 Cell Polarity Change Versus Cell Turning During a Gradient Change of Direction

With the gradient switching device, we tested the ability of the dHL-60 cells to respond to a switch in the direction of the gradient in the presence or absence of inhibition of PI3K. In the presence of PI3K inhibitor, the dHL-60 cells exhibited a slower change in the direction of cell migration. Moreover, when the cells that were not treated with PI3K inhibitor changed the direction of migration in response to the switch in the direction of the gradient, the majority of the cells changed their polarity (head became tail and tail became head) and migrated in the opposite direction, and only a minor proportion of cells turned around without changing polarity. This change of polarity is probably the most efficient way for a cell to change the direction of cell migration (Liu et al., 2008).

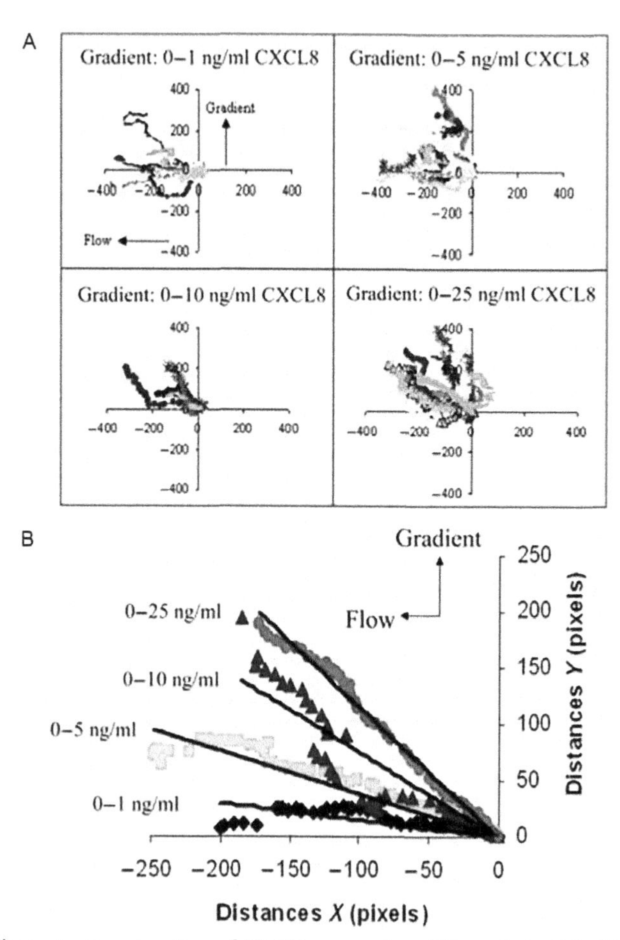

Figure 9 Chemotactic migration of CXCR2-expressing differentiated HL-60 cells under different levels of CXCL8 gradient steepness. Different levels of CXCL8 gradient steepness were generated by constant flow of a certain concentration of CXCL8 in Hank's buffer and buffer alone at a rate of 1 μl/min. The real-time video of cell movement was recorded and the data were analyzed using the MetaMorph program to trace movement of 8–12 cells for each experiment. The left panel shows the trajectories of the cells under each level of gradient steepness, and the right panel shows the mean trajectory of cell movement for each experiment. Since the cell movement was driven by two forces, flow and chemotactic force, the trajectory of cell movement represents the net result of these two forces. With the increase of the steepness of chemokine gradient, the angle of the trajectories was more in line with the direction of the gradient. *Originally published in Sai et al. (2006).*

4.3 Potential Future Uses of Microfluidic Devices: Analysis of Circulating Tumor Cells

A more promising application of microfluidic devices is for the purification of circulating tumor cells (CTCs) from the peripheral blood of cancer patients. The earliest steps of metastasis involve the shedding of malignant tumor cells from the primary tumor followed by intravasation of these tumor cells into the blood vasculature. Careful monitoring of the number of CTCs can provide very important index for the prognosis of cancer patients as well as an indicator for response to ongoing therapy. However, among millions of different types of other cells in the bloodstream, it is not easy to identify the sparse tumor cells. With the help of microfluidic devices, the tumor cells can be concentrated by binding to an antibody that specifically detects CTCs. The antibody is precoated on the surface of micropoles in the microfluidic channels (Lim et al., 2012; Nagrath et al., 2007; Stott et al., 2010; Zheng, Yang, & Li, 2010; also see review in Qian, Zhang, & Chen, 2015), and as CTCs bind to the antibodies they are concentrated for identification and quantitation by flow cytometry.

4.4 Advantages of Using 3D Microbioreactors to Investigate Factors that Influence the Tumor Microenvironment

A major problem that preclinical translational studies in oncology deal with is the fact that mouse models utilizing human tumor cells do not allow testing the interaction of human tumor cells with human immune cells, endothelial cells, and other stromal cells. Microbioreactors allow one to test the effects of therapeutic drugs on the interaction of immune cells with the tumor and its stroma in a manner that can readily be imaged in real time and analyzed. Immunotherapeutic agents can be evaluated using these microbioreactors, where the activity of cytotoxic T cells can be monitored visually and biochemically. Moreover, organoid cultures of human tumors interacting with the patient's own immune cells, fibroblasts, mesenchymal stem cells, and endothelial cells can readily be examined and tumor metabolomics can be analyzed in the presence and absence of combinations of therapeutic agents. Once these microbioreactors can be scaled for moderate throughput, we will have a unique method for preclinical studies that utilize all human cells in a microenvironment that mimics the patient's tumor.

ACKNOWLEDGMENTS

We thank Allison Price for her editorial assistance. For their early contributions to the design and production of the gradient microfluidic devices, we thank Glenn W. Walker (Vanderbilt

Institute for Integrative Biosystems Research and Education (VIIBRE) and Department of Molecular Physiology and Biophysics, Vanderbilt University, and currently at University of North Carolina, Chapel Hill and North Carolina State University, Raleigh, NC), Mark Stremler (Department of Mechanical Engineering, Vanderbilt University, and currently at Virginia Tech), and Chang Y. Chung (Department of Pharmacology, Vanderbilt University). We thank William Hofmeister, Alexander Terekhov, and Lino Costa (Center for Laser Applications, University of Tennessee Space Institute, Tullahoma, TN and the Department of Materials Science and Engineering, University of Tennessee, Knoxville, TN), as well as Chris Janetopoulos (Department of Biological Sciences, Vanderbilt University, and currently at the University of the Sciences, Philadelphia), for their help in the design and production of the microbioreactor. We are indebted to Hal Moses and Simon Hayward at Vanderbilt University for the human cancer-associated fibroblasts and normal tissue-associated fibroblasts isolated and cultured from breast tissue. We appreciate the efforts of Anna E. Vilgelm (Vanderbilt University) in the characterization of factors opposing CXCL12 in the microbioreactor studies. We are appreciative of Kevin Seale and the VIIBRE and SyBBURE (Systems Biology and Bioengineering Undergraduate Research Experience) teams for all their support, guidance, and encouragement. We thank Melody Swartz (Laboratory of Lymphatic and Cancer Bioengineering, EPFL Institute, Lausanne, Switzerland, and currently at the University of Chicago) for her advice and guidance in the initial development of the devices. Thanks to Tammy Sobolik, Yingchun Yu, and Linda Horton (Department of Cancer Biology, Vanderbilt University), who provided technical support and guidance, and Ricardo Richardson (Meharry Medical College, and currently at North Carolina State University) and Jingshong Xu (formerly at University of California, San Francisco, and currently at the University of Illinois at Chicago) for their advice and help with chemotaxis assays. The development and application of the devices reviewed in this chapter was supported by grants from the TVHS and the Department of Veterans Affairs through MERIT and Senior Research Career Scientist Awards to A.R.; NIH grants R01 CA34590 (A.R.) and K12-CA90625 (AEV); Tennessee Higher Education Commission grant to the Center for Laser Applications at UT Space Institute (W.H.); RO1GM080370 (C.J.), a Vanderbilt Discovery Award (C.J. and W.H.), the Vanderbilt Academic Venture Capital Fund (J.P.W.), and the Vanderbilt Ingram Cancer Center Support Grant (CA68485) for core facilities.

REFERENCES

Abhyankar, V. V., Lokuta, M. A., Huttenlocher, A., & Beebe, D. J. (2006). Characterization of a membrane-based gradient generator for use in cell-signaling studies. *Lab on a Chip*, *6*(3), 389–393. http://dx.doi.org/10.1039/b514133h.

Berthier, E., Surfus, J., Verbsky, J., Huttenlocher, A., & Beebe, D. (2010). An arrayed high-content chemotaxis assay for patient diagnosis. *Integrative Biology*, *2*(11-12), 630–638. http://dx.doi.org/10.1039/c0ib00030b.

Boyden, S. (1962). The chemotactic effect of mixtures of antibody and antigen on polymorphonuclear leucocytes. *The Journal of Experimental Medicine*, *115*, 453–466.

Collins, S. J. (1987). The HL-60 promyelocytic leukemia cell line: Proliferation, differentiation, and cellular oncogene expression. *Blood*, *70*(5), 1233–1244.

Collins, S. J., Gallo, R. C., & Gallagher, R. E. (1977). Continuous growth and differentiation of human myeloid leukaemic cells in suspension culture. *Nature*, *270*(5635), 347–349.

Collins, S. J., Ruscetti, F. W., Gallagher, R. E., & Gallo, R. C. (1978). Terminal differentiation of human promyelocytic leukemia cells induced by dimethyl sulfoxide and other polar compounds. *Proceedings of the National Academy of Sciences of the United States of America, 75*(5), 2458–2462.

Elvin, P., Kerr, I. B., McArdle, C. S., & Birnie, G. D. (1988). Isolation and preliminary characterisation of cDNA clones representing mRNAs associated with tumour progression and metastasis in colorectal cancer. *British Journal of Cancer, 57*(1), 36–42.

Gao, Y., Sun, J., Lin, W. H., Webb, D., & Li, D. (2012). A compact microfluidic gradient generator using passive pumping. *Microfluidics and Nanofluidics, 12*(6), 887–895. http://dx.doi.org/10.1007/s10404-011-0908-0.

Haddox, J. L., Knowles, I. W., Sommers, C. I., & Pfister, R. R. (1994). Characterization of chemical gradients in the collagen gel-visual chemotactic assay. *Journal of Immunological Methods, 171*(1), 1–14.

Haddox, J. L., Pfister, R. R., & Sommers, C. I. (1991). A visual assay for quantitating neutrophil chemotaxis in a collagen gel matrix. A novel chemotactic chamber. *Journal of Immunological Methods, 141*(1), 41–52.

Hockemeyer, K., Janetopoulos, C., Terekhov, A., Hofmeister, W., Vilgelm, A., Costa, L., et al. (2014). Engineered three-dimensional microfluidic device for interrogating cell-cell interactions in the tumor microenvironment. *Biomicrofluidics, 8*(4), 044105. http://dx.doi.org/10.1063/1.4890330.

John, T. J., & Sieber, O. F., Jr. (1976). Chemotactic migration of neutrophils under agarose. *Life Sciences, 18*(2), 177–181.

Kim, M., & Kim, T. (2010). Diffusion-based and long-range concentration gradients of multiple chemicals for bacterial chemotaxis assays. *Analytical Chemistry, 82*(22), 9401–9409. http://dx.doi.org/10.1021/ac102022q.

Li Jeon, N., Baskaran, H., Dertinger, S. K., Whitesides, G. M., Van de Water, L., & Toner, M. (2002). Neutrophil chemotaxis in linear and complex gradients of interleukin-8 formed in a microfabricated device. *Nature Biotechnology, 20*(8), 826–830. http://dx.doi.org/10.1038/nbt712.

Lim, L. S., Hu, M., Huang, M. C., Cheong, W. C., Gan, A. T., & Looi, X. L. (2012). Microsieve lab-chip device for rapid enumeration and fluorescence in situ hybridization of circulating tumor cells. *Lab on a Chip, 12*(21), 4388–4396. http://dx.doi.org/10.1039/c2lc20750h.

Lin, F., Nguyen, C. M., Wang, S. J., Saadi, W., Gross, S. P., & Jeon, N. L. (2004). Effective neutrophil chemotaxis is strongly influenced by mean IL-8 concentration. *Biochemical and Biophysical Research Communications, 319*(2), 576–581. http://dx.doi.org/10.1016/j.bbrc.2004.05.029.

Lin, F., Saadi, W., Rhee, S. W., Wang, S. J., Mittal, S., & Jeon, N. L. (2004). Generation of dynamic temporal and spatial concentration gradients using microfluidic devices. *Lab on a Chip, 4*(3), 164–167. http://dx.doi.org/10.1039/b313600k.

Liu, Y., Sai, J., Richmond, A., & Wikswo, J. P. (2008). Microfluidic switching system for analyzing chemotaxis responses of wortmannin-inhibited HL-60 cells. *Biomedical Microdevices, 10*(4), 499–507. http://dx.doi.org/10.1007/s10544-007-9158-z.

Nagrath, S., Sequist, L. V., Maheswaran, S., Bell, D. W., Irimia, D., Ulkus, L., et al. (2007). Isolation of rare circulating tumour cells in cancer patients by microchip technology. *Nature, 450*(7173), 1235–1239. http://dx.doi.org/10.1038/nature06385.

Nelson, R. D., Quie, P. G., & Simmons, R. L. (1975). Chemotaxis under agarose: A new and simple method for measuring chemotaxis and spontaneous migration of human polymorphonuclear leukocytes and monocytes. *The Journal of Immunology, 115*(6), 1650–1656.

Qian, W., Zhang, Y., & Chen, W. (2015). Capturing cancer: Emerging microfluidic technologies for the capture and characterization of circulating tumor cells. *Small*, *11*(32), 3850–3872. http://dx.doi.org/10.1002/smll.201403658.

Sai, J., Walker, G., Wikswo, J., & Richmond, A. (2006). The IL sequence in the LLKIL motif in CXCR2 is required for full ligand-induced activation of Erk, Akt, and chemotaxis in HL60 cells. *The Journal of Biological Chemistry*, *281*(47), 35931–35941. http://dx.doi.org/10.1074/jbc.M605883200.

Stott, S. L., Hsu, C. H., Tsukrov, D. I., Yu, M., Miyamoto, D. T., Waltman, B. A., et al. (2010). Isolation of circulating tumor cells using a microvortex-generating herringbone-chip. *Proceedings of the National Academy of Sciences of the United States of America*, *107*(43), 18392–18397. http://dx.doi.org/10.1073/pnas.1012539107.

Walker, G. M., Sai, J., Richmond, A., Stremler, M., Chung, C. Y., & Wikswo, J. P. (2005). Effects of flow and diffusion on chemotaxis studies in a microfabricated gradient generator. *Lab on a Chip*, *5*(6), 611–618. http://dx.doi.org/10.1039/b417245k.

Wen, Y., Zhang, X., & Yang, S. T. (2010). Microplate-reader compatible perfusion microbioreactor array for modular tissue culture and cytotoxicity assays. *Biotechnology Progress*, *26*(4), 1135–1144. http://dx.doi.org/10.1002/btpr.423.

Whitesides, G. M., Ostuni, E., Takayama, S., Jiang, X., & Ingber, D. E. (2001). Soft lithography in biology and biochemistry. *Annual Review of Biomedical Engineering*, *3*, 335–373. http://dx.doi.org/10.1146/annurev.bioeng.3.1.335. 3/1/335 [pii].

Xu, C., Poh, Y. K., Roes, I., O'Cearbhaill, E. D., Matthiesen, M. E., Mu, L., et al. (2012). A portable chemotaxis platform for short and long term analysis. *PLoS One*, 7(9), e44995. http://dx.doi.org/10.1371/journal.pone.0044995.

Yang, S. T., Zhang, X., & Wen, Y. (2008). Microbioreactors for high-throughput cytotoxicity assays. *Current Opinion in Drug Discovery & Development*, *11*(1), 111–127.

Zheng, X. T., Yang, H. B., & Li, C. M. (2010). Optical detection of single cell lactate release for cancer metabolic analysis. *Analytical Chemistry*, *82*(12), 5082–5087. http://dx.doi.org/10.1021/ac100074n.

Zhu, X., Si, G., Deng, N., Ouyang, Q., Wu, T., He, Z., et al. (2012). Frequency-dependent Escherichia coli chemotaxis behavior. *Physical Review Letters*, *108*(12), 128101.

Zicha, D., Dunn, G. A., & Brown, A. F. (1991). A new direct-viewing chemotaxis chamber. *Journal of Cell Science*, *99*(Pt. 4), 769–775.

Zigmond, S. H. (1977). Ability of polymorphonuclear leukocytes to orient in gradients of chemotactic factors. *The Journal of Cell Biology*, *75*(2 Pt. 1), 606–616.

CHAPTER THREE

Generating Chemokine Analogs with Enhanced Pharmacological Properties Using Phage Display

Karim Dorgham[*,1], Fabrice Cerini[†,1], Hubert Gaertner[†], Astrid Melotti[†], Irène Rossitto-Borlat[†], Guy Gorochov[*,†,2], Oliver Hartley[‡,2]

[*]Sorbonne Universités, UPMC Univ Paris 06, Inserm UMRS1135, Centre d'Immunologie et des Maladies Infectieuses (CIMI-Paris), Paris, France
[†]Department of Pathology and Immunology, Faculty of Medicine, University of Geneva, Geneva, Switzerland
[‡]AP-HP, Groupement Hospitalier Pitié-Salpêtrière, Département d'Immunologie, Paris, France
[2]Corresponding authors: e-mail address: guy.gorochov@upmc.fr; oliver.hartley@unige.ch

Contents

1. Introduction 48
   1.1 Chemokines in Health and Disease 48
   1.2 Chemokine Structure and Activity 49
   1.3 Applying Phage Display Technology to Chemokine Receptors 50
2. Methods 50
   2.1 Library Design and Construction 52
   2.2 Selection of Libraries on Cells 60
3. Limitations 65
   3.1 The Atypical Chemokine Receptor DARC (ACKR1) 65
4. Perspectives 68
   4.1 A Chemokine-Based HIV Prevention Strategy 68
   4.2 New Tools to Study CCR5 Pharmacology and Cell Biology 68
   4.3 An Intrakine to Protect Cells from HIV Infection and a Vaccine Adjuvant 69
   4.4 A Prototypic Inhibitor of CX3CR1, the Fractalkine/CX3CL1 Receptor 69
References 70

Abstract

Phage display technology, which allows extremely rare ligands to be selected from libraries of variants according to user-defined selection criteria, has made a huge impact on the life sciences. In this chapter, we describe phage display methods for the discovery of chemokine analogs with enhanced pharmacological properties. We discuss strategies for chemokine library design and provide a recommended technique for library construction. We also describe cell-based library selection approaches that we have

[1] These authors made an equal contribution to the manuscript.

Methods in Enzymology, Volume 570
ISSN 0076-6879
http://dx.doi.org/10.1016/bs.mie.2015.09.014

used to discover chemokine analogs, not only receptor antagonists but also variants with unusual effects on receptor signaling and trafficking. By providing a survey of the different phage chemokine projects that we have undertaken, we comment on the parameters most likely to affect success. Finally, we discuss how phage display-derived chemokine analogs with altered pharmacological activity represent valuable tools to better understand chemokine biology, and why certain among them have the potential to be developed as new medicines.

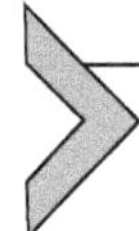

1. INTRODUCTION

1.1 Chemokines in Health and Disease

Chemokines (reviewed in Griffith, Sokol, & Luster, 2014; Raman, Sobolik-Delmaire, & Richmond, 2011) are a family of small (7–10 kDa) proteins that engage chemokine receptors, which are themselves members of the G protein-coupled receptor (GPCR) superfamily. There are approximately 50 human chemokines and 20 human chemokine receptors. The major role of the chemokine–chemokine receptor system is in immunity, where chemokine activity underlies processes ranging from coordinating the circulation of leukocytes around the lymphatic system to driving the egress of effector cells from the bloodstream into sites of inflammation (Griffith et al., 2014). Additional roles for chemokines include embryogenesis, wound healing, and angiogenesis (Raman et al., 2011).

Chemokines are key players in inflammatory diseases such as rheumatoid arthritis, atherosclerosis, and asthma (reviewed in Koelink et al., 2012) and in the evolution of and metastatic spread of tumors (reviewed in Balkwill, 2012). The strategy of adding chemokine receptor agonists to vaccine adjuvants in order to boost immune reactions is being explored (Bobanga, Petrosiute, & Huang, 2013), and the chemokine receptors that are used by HIV as entry coreceptors represent important targets for antiviral prevention and therapy (Kuhmann & Hartley, 2008).

Hence, the chemokine–chemokine receptor system presents a number of potential targets for combating pathology (Viola & Luster, 2008), but in spite of significant investment in small molecule drug development programs since the 1990s, only two drugs that target chemokine receptors have been licensed so far: maraviroc, a CCR5 inhibitor used for HIV therapy (Kuritzkes, Kar, & Kirkpatrick, 2008), and plerixafor (Uy, Rettig, & Cashen, 2008), a CXCR4 inhibitor used to mobilize hematopoietic stem cells. In addition to representing valuable research tools for the study of chemokine receptor structure, pharmacology, and cell biology, engineered chemokine analogs have the potential be developed into new medicines. In this chapter, we describe the use of phage display to isolate chemokine analogs with promising pharmacological activity.

1.2 Chemokine Structure and Activity

Chemokines share a characteristic tertiary structure featuring a folded core region, stabilized by disulfide bridges, from which appends a flexible N-terminal domain (Fernandez & Lolis, 2002) (Fig. 1). Numerous structure–activity studies have revealed a two-site mechanism for chemokine–chemokine receptor engagement (Fernandez & Lolis, 2002; Wells et al., 1996; Allen, Crown, & Handel, 2007). Initially, high-affinity and high-specificity engagement is achieved through an "address" interaction between the folded core region of the chemokine and the extracellular region of the chemokine receptor (also referred to as chemokine recognition site 1, CRS1 (Scholten et al., 2012)). Subsequently, the flexible N-terminal domain of chemokine engages the transmembrane domain of the chemokine receptor (now referred to as CRS2 (Scholten et al., 2012)) providing a "message" interaction that leads to signal transduction. Numerous

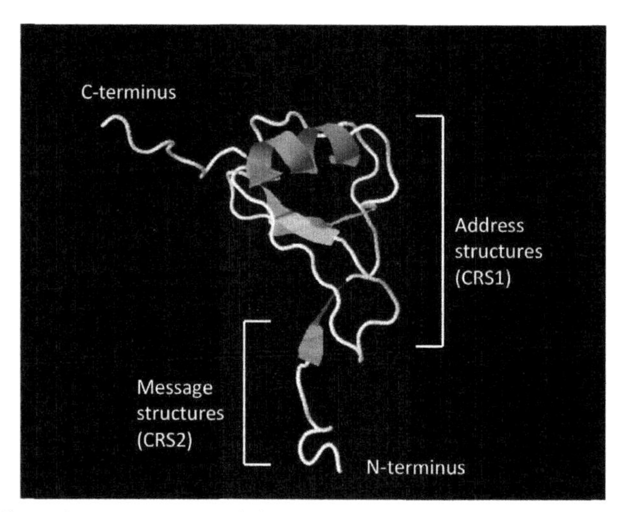

Figure 1 Chemokine structure and function. Chemokines share a common tertiary structure (here MCP-1/CCL2 is shown as an example—ribbon structure produced using jmol from PDB structure 1DOM). The rigid, folded core of the protein carries "address" structures (CRS1) that provide high-affinity and high-specificity interactions with chemokine receptors, and the flexible N-terminal region carries "message" structures (CRS2) that affect receptor function. The C-terminus does not participate in receptor interaction and can be modified without affecting pharmacology.

examples of chemokine analogs featuring modified N-terminal regions exhibiting modified functional activity have been described, including antagonists, partial agonists, and superagonists (Allen et al., 2007; Hartley & Offord, 2005).

1.3 Applying Phage Display Technology to Chemokine Receptors

Phage display technology (Sidhu, 2001) involves the insertion of libraries encoding ligands, most frequently antibodies or peptides, in frame with filamentous phage coat proteins so that the encoded gene is expressed on the phage surface. Coupling of genotype and phenotype in this way makes it possible to use *in vitro* evolution approaches to isolate extremely rare phage clones encoding ligands with desirable characteristics from vast libraries.

While phage display has been used quite extensively to isolate ligands to receptors for which the extracellular domains can be produced as isolated soluble proteins, its application to integral membrane proteins such as GPCRs is more challenging (Hartley, 2002). First, expressing, purifying, and presenting integral membrane proteins in their physiological conformations are an obstacle well known to the GPCR structural biology field, and an added challenge with conventional phage display is to efficiently attaching receptors to the solid phase without affecting their conformation. One solution is to bypass purification and presents the receptor in the context of the cell in which it is expressed. Screening unbiased peptide or antibody libraries on intact cells carries a major risk of isolating ligands that bind to other irrelevant targets on the cell surface (Hartley, 2002) but is more straightforward when the phage libraries are based on the receptor's own ligands, because the libraries are naturally biased toward the target. Chemokine receptors, with their two-site mechanism of ligand interaction, are particularly attractive targets, because by targeting mutations to the flexible N-terminal domain, it is possible in theory to selectively manipulate pharmacological activity (CRS2) without affecting receptor affinity and specificity (CRS1) (Fig. 2).

2. METHODS

Of the five components involved in a typical chemokine phage display selection program (Fig. 3), only the library design and construction and the panning on cells components will be detailed in this chapter. Methods for the production and purification of phage are common to the majority of

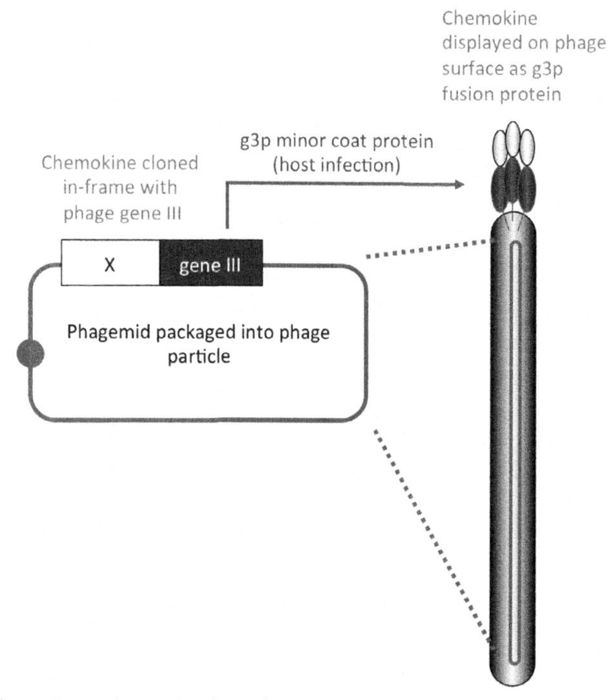

Figure 2 Chemokine phage display. Libraries encoding chemokine variants are cloned for expression as C-terminal fusions to the phage minor coat protein, g3p.

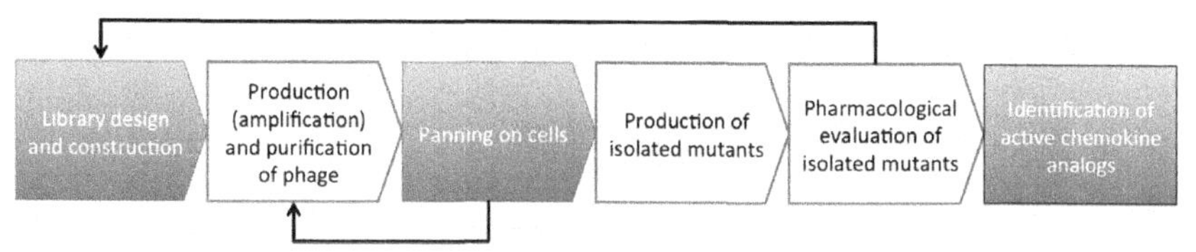

Figure 3 Components of a complete chemokine phage display program. Libraries are designed and constructed, and the encoded phage clones are produced and purified. Selection involves cycles of panning on cells followed by amplification of captured phage. After several such rounds, variants isolated from the library are produced and subjected to pharmacological evaluation. New structure–activity insights can be used to design next-generation libraries, ultimately leading to the identification of chemokine analogs with the desired pharmacological properties.

phage display strategies and are well described elsewhere (e.g., Barbas, 2000; Kay, Winter, & McCafferty, 1996; Clackson & Lowman, 2004). The production of chemokine analogs using both recombinant (Horuk, Reilly, & Yansura, 1997; Offord, Gaertner, Wells, & Proudfoot, 1997) and chemical synthesis (Dawson, 1997; Clark-Lewis, Vo, Owen, & Anderson, 1997) platforms has been similarly well documented. Finally, choice of the method or methods used for the pharmacological evaluation of chemokine variants depends on the target receptor and the pharmacological output of interest.

2.1 Library Design and Construction

2.1.1 Choice of Phage Display System

Since the C-terminus of chemokines can generally be modified without affecting pharmacological activity, we chose a phage display strategy in which chemokine variants are fused via their C-termini to the bacteriophage g3p minor coat protein. From the numerous phage display vectors have been designed for g3p fusion expression, we opted to use the well-characterized pHEN-1 phagemid (Hoogenboom et al., 1991), using *Nco*I and *Not*I as cloning sites (Fig. 4).

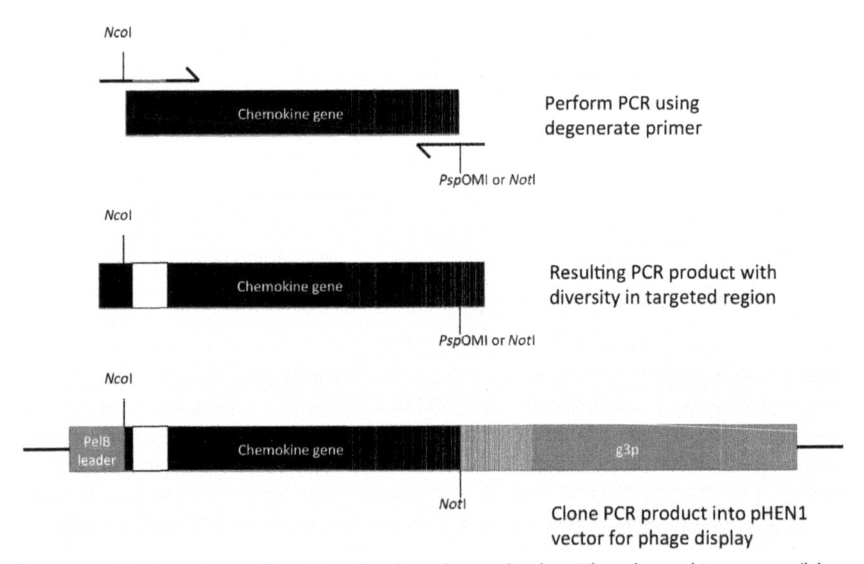

Figure 4 Cloning chemokine libraries for phage display. The chemokine gene (blue; black in the print version) is amplified by oligonucleotide primers that introduce diversity into the targeted region (red; white in the print version) and append appropriate restriction sites. The amplified fragment is digested and cloned upstream and in-frame with the phage minor coat protein g3p (green; gray in the print version). For libraries based on RANTES/CCL5, the PCR fragment was digested using *Psp*OMI as the 3′ restriction enzyme. *Psp*OMI and *Not*I generate compatible cohesive ends.

Table 1 Examples of Degenerate Oligonucleotide Primers Used to Generate Chemokine Phage Display Libraries

Library	5′ Oligonucleotide Primer	References
RANTES/ CCL5 Library 2	5′ CAG CCG GCC ATG GCC **NNK NNK** CCA **NNK NNK NNK** CAA **NCC** ACA CCC TGC TGC TTT GCC TAC ATT GCC CGC 3′	Gaertner et al. (2008)
RANTES/ CCL5 Library 5	5′ CAG CCG GCC ATG GCC CAG GGT CCA CCT TTG ATG **NNK NNK NNK NNK** TGC TGC TTT GCC TAC ATT GCC CGC 3′	Gaertner et al. (2008)
Fractalkine/ CX3L1 Library 1	5′ CCG GCC ATG GCC **NNKCANNNKNNKGNCNTGNCA** AAA TGC AAC ATC ACG TGC 3′	Dorgham et al. (2009)

The primers shown range from 51 to 69 bases in length and encode theoretical diversities from 1.6×10^5 (RANTES/CCL5 Library 5) to 1.3×10^7 (RANTES/CCL5 Library 2).

2.1.2 Introducing Library Diversity

Diversity can be readily introduced to a specific region of a gene by PCR amplification using degenerate oligonucleotide primers (some examples are shown in Table 1). Most commonly, complete randomization of a specific amino acid position is achieved by encoding the degenerate codon NNK (where N is A, C, G, or T and K is G or T), covering all 20 amino acids, plus 1 stop codon (the amber codon, TAG), which can be suppressed using an appropriate *Escherichia coli* host strain (generally the *SupE*-positive TG1 strain is used).

In our experience, the quality of the primers used for library amplification is the key to success. In many cases, the required upstream primers will be in excess of 60 bases in length, and we recommend that these should be ordered from providers offering premium quality synthesis and purification. Libraries generated using low quality primers will include a higher proportion of truncation and frame shift mutants, wasting valuable library space, and potentially having an adverse effect on the outcome of selection experiments (see Section 2.2 below).

2.1.3 How Much Diversity Is Feasible?

Work with antibody phage display has shown that the larger the size of initial library that can feasibly be sampled, the higher the probability of isolating a high-affinity ligand against a given target (Griffiths et al., 1994). While PCR fragments encoding very high diversity can be readily obtained and ligated into phage display vectors, the factor limiting final library size is the number of host *E. coli* cells that can be transformed with the ligated phage display vector, and if this figure is not in excess of the theoretical diversity (i.e., the total number of encoded amino acid combinations), only a fraction of the diversity will be sampled, and potentially active sequences will not be isolated. Large-scale transformation of *E. coli* to generate large libraries is an onerous task. For laboratories not specialized in phage display, we would recommend a maximum library size of 3.2×10^6 (i.e., five positions fully randomized, 20^5).

2.1.4 Partial Diversity

A greater number of residues can be mutagenized without generating unfeasibly high theoretical diversity when partial randomization approaches are used. In partial randomization, codon sets more restrictive than NNK, encoding fewer amino acids, are used (e.g., NCA, covering ACA, CCA, GCA, and TCA—Thr, Pro, Ala, and Ser). The choice of the restricted codon set can be driven by the size, charge, polarity, or hydrophobicity of the encoded amino acids, according to sequence alignments of related chemokines, and/or structure–activity clues emerging from selection of a previous library. Several examples of libraries encoding partial diversity at certain positions are shown in Table 2.

2.1.5 Chemokine Walking: Exploring Diversity Step by Step

"Chemokine walking" is a strategy that we have used to fully explore diversity over a relatively long stretch of amino acids without the need to generate very large libraries. A relatively small initial library is created, and selection of this library provides structure–activity insights that can be used to inform the design of a next-generation library, in which amino acids identified as important for activity are fixed, and diversity is introduced to neighboring positions. Moving from the modest anti-HIV chemokine analogs isolated from a first generation RANTES/CCL5 library (anti-HIV IC_{50} 1600 p*M*), we were able to design a second-generation library (Library 2 in Table 2) that provided analogs with further improved anti-HIV activity (650 p*M*) (Gaertner et al., 2008). Third-generation libraries developed according to these improved sequences (Libraries 5 and 6 in Table 2) ultimately yielded fully optimized anti-HIV chemokines (<30 p*M*) (Gaertner et al., 2008). It is noteworthy that these molecules were obtained by exploring a region of 10 amino acids in length; it would not have been technically feasible to generate and select a single library of diversity 20^{10}.

A complete summary of the phage chemokine libraries that we have produced and selected, including unpublished work, is shown in Table 2.

Table 2 Phage Chemokine Libraries Constructed and Selected by the Authors to Date

Chemokine Library	N-Terminal Sequence	Theoretical Diversity	Library Size Obtained	Examples of Selected Sequences	References
RANTES/ CCL5	SPYSSDTTPCC				
Library 1	XS#XSSX###CC	2.0×10^6	5×10^6	LSPVSSQSSACC (P1-RANTES) FSPLSSQSSACC (P2-RANTES) MSPLSSQASACC (P7-RANTES)	Hartley et al. (2003) and Dorgham et al. (2008)
Library 2	XXPXXXQ#TPCC	1.28×10^7	3.1×10^7	QGPPLMQTTPCC (2P3-RANTES) QGPLSGQSTPCC (2P8-RANTES)	Gaertner et al. (2008)
Library 3	LSPVSSQSSACCXXXXX	3.2×10^6	3.1×10^7	LSPVSSQSSACCFAYIA (P1-RANTES)	Oliver Hartley, Astrid Melotti, and Irène Rossitto-Borlat (unpublished)
Library 4	LSPVSSQSSACC... XXXSXXC	3.2×10^6	2.2×10^7	LSPVSSQSSACC... FYTSGKC (P1-RANTES)	Oliver Hartley, Astrid Melotti, and Irène Rossitto-Borlat (unpublished)
Library 5	QGPPLMXXXXCC	1.6×10^5	1×10^6	QGPPLMATQSCC (5P12-RANTES) QGPPLMSLQVCC (5P14-RANTES)	Gaertner et al. (2008)

Continued

Table 2 Phage Chemokine Libraries Constructed and Selected by the Authors to Date—cont'd

Chemokine Library	N-Terminal Sequence	Theoretical Diversity	Library Size Obtained	Examples of Selected Sequences	References
Library 6	QGP‡∘XXXXXCC	3.8×10^7	5.5×10^7	QGPPGDIVLACC (6P4-RANTES)	Gaertner et al. (2008)
Library 7	QGPPXXWXQXCC	1.6×10^5	1.2×10^7	QGPPLQWMQACC (7P14-RANTES)	Gaertner et al. (2008)
Library 8	QGPLXXXXQVCC	1.6×10^5	1.0×10^6	QGPLSGWAQVCC (8P6-RANTES)	Gaertner et al. (2008)
Library 9	XXXXXSDTTPCC	3.2×10^6	3.1×10^7	QWVMGSDTTPCC (9P10-RANTES) QGQRISDTT-CC (9P1-RANTES)	Oliver Hartley, Astrid Melotti, and Irène Rossitto-Borlat (unpublished)
IL-8/CXCL8	SAKELRCQC				
Library 1	XXXXXLRCQC	3.2×10^6	9.0×10^6	Selection failed	Oliver Hartley, Astrid Melotti, and Irène Rossitto-Borlat (unpublished)
Fractal-kine/ CX3CL1	.QHHGVTKCNITC				
Library 1	X&XX$§#KCNITC	1.5×10^6	5×10^6	ILDNGVSKCNITC (F1-CX3CL1)	Dorgham et al. (2009)
Library 2	.XXX$§#KCNITC	4×10^5	2×10^6	-QPGGVSKCNITC (F3-CX3CL1)	Karim Dorgham (unpublished)

Mip3a/CCL20	.ASNFDCC				
Library 1	XX!+>+CC	1.1×10^6	1×10^6	GRMQQECC (M1-CCL20) QGTYLQCC (M2-CCL20)	Karim Dorgham (unpublished)
Library 2	.X#+£XCCX	6.6×10^5	1×10^6	.VAQVQCCE (M3-CCL20)	Karim Dorgham (unpublished)
MCP-1/CCL2	.QPDAINAPVTCC				
Library 1	XXXXAINAPVTCC	1.6×10^5	1×10^6	Selection failed	Karim Dorgham (unpublished)
Library 2	.XXXXINAPVTCC	1.6×10^5	1×10^6	Confidential	Karim Dorgham (unpublished)
IP10/CXCL10	.VPLSRTVRCTC				
Library 1	XXXXSRTVRCTC	1.6×10^5	1×10^6	Selection failed	Karim Dorgham (unpublished)
Library 2	.XXXXRTVRCTC	1.6×10^5	1×10^6	Selection failed	Karim Dorgham (unpublished)
SDF-1/ CXCL12	.KPVSLSYRCPC				
Library 1	XXPXXXSYRCPC	3.2×10^6	1×10^6	Selection failed	Karim Dorgham (unpublished)

The following single character codes are used to describe possible amino acid combinations at given positions: X (any amino acid), # (S, P, T, or A), ‡ (G, L, or P), ∘ (G, L, or M), & (L, P, Q, or R), $ (V, A, D, or G), § (L, M, or V), + (N, Y, Q, H, K, D, or E), £ (F, M, L, I, or V), ! (S, I, M, T, N, K, or R), and > (F, S, Q, W, L, Y, or C).

2.1.6 Required Materials

- pHEN-1 phagemid vector for digestion
- Restriction enzymes *Nco*I, *Psp*OMI, *Not*I (NEB R0193S, R0653S, R0189S)
- 10 × digestion buffers (NEBuffer 3.1 and NEBuffer 4, NEB B7004S and B7203S)
- Calf-intestinal alkaline phosphatase (NEB M0290S)
- Forward and reverse primers for PCR
- Taq polymerase (Thermo Fisher 18038-042)
- DNA purification minicolumn kit (Promega A9282)
- T4 DNA ligase (NEB M0202S)
- Electrocompetent *E. coli* TG1 cells (Agilent 200123)
- Gene Pulser®/MicroPulser™ Electroporation Cuvettes, 0.1 cm gap (Bio-Rad 1652083)
- MicroPulser™ Electroporator (Bio-Rad 1652100)

2.1.7 Library Construction

1. Perform *Nco*I/*Not*I digestion of 15 μg of pHEN-1 plasmid DNA.

10 × digestion buffer NEBuffer 3.1	10 μL
pHEN-1	15 μg
*Nco*I (NEB R0193S)	3 μL
*Not*I (NEB R0189S)	3 μL
ddH_2O	To 100 μL

Incubate for 2 h at 37 °C.

2. Add 2.5 μL of calf intestine alkaline phosphatase to the digestion reaction and incubate for a further 30 min at 37 °C.
3. Gel purify the digested plasmid backbone using preparative scale 0.8% agarose gel. Excise the band corresponding to the digested plasmid backbone and cut the gel slice into eight pieces of approximately equal size. Extract DNA from the gel using the DNA purification minicolumn kit using one column per gel slice. Elute each column with 50 μL water and pool the eluates. Verify the quality of the purification on an analytical gel (0.8% agarose).
4. PCR amplification using degenerate oligonucleotides. We empirically optimize PCR conditions (amount of template DNA and annealing

temperature) prior to scale-up. As an example, the optimized reaction conditions used to generate the PCR fragment for Library 5 (Gaertner et al., 2008). Depending on target library size, between five and twenty 50 μL reactions are performed in parallel and then pooled.

10 × PCR buffer	5 μL
Template DNA	60 ng
$MgCl_2$ (50 m*M*)	1.5 μL
dNTP (10 m*M*)	1 μL
Primer forward (25 μ*M*)	3 μL
Primer reverse (25 μ*M*)	3 μL
Taq polymerase	0.25 μL
ddH_2O	To 50 μL

Thermocycling conditions: 94 °C for 5 min, then 30 cycles (94 °C for 1 min, 62 °C for 1 min, 72 °C for 1 min).

5. Purify each PCR product using a DNA purification minicolumn kit, loading five 50 μL reactions per column, eluting each column with 50 μL water. Pool eluted material.
6. Perform *Nco*I/*Psp*OMI digestion of the purified PCR fragment.

10 × digestion buffer (NEBuffer 4)	30 μL
Purified PCR product from step 5	100 μL
*Nco*I	20 μL
*Psp*OMI	20 μL
ddH_2O	130 μL

Incubate for 3 h at 37 °C.

7. Gel purify the digested PCR product using a 1.5% agarose gel. Excise the band corresponding to the digested PCR product and cut the gel slice into six pieces of approximately equal size. Extract DNA from the gel with the DNA purification minicolumn kit, using one column per gel slice. Verify the quality of the purification on an analytical gel (1.5% agarose).

8. Perform test ligations using 2 μL of digested backbone and varying volumes of digested PCR product. Below is an example of the test ligations performed for the preparation of Library 5.

10 × ligation buffer	2 μL
Digested vector backbone	2 μL
Digested PCR product	1, 2, or 5 μL
T4 DNA ligase	1 μL
ddH_2O	To 10 μL

Incubate overnight at 15 °C.

Evaluate the test ligations by electroporating them into electrocompetent *E. coli* TG1 and monitoring the number of transformants obtained. Define the optimal vector: insert ratio.

9. Perform a 10-fold scaled-up ligation reaction using the optimized vector: insert ratio. Use a DNA purification minicolumn kit to remove salts from the ligation reaction. Elute with 40 μL water.
10. Electroporate the purified ligation mixture into electrocompetent *E. coli* TG1 using 2 μL ligation mixture per electroporation (i.e., 20 electroporations in total).
11. Pick 40 colonies from the titration plates for insert sequencing.
12. Harvest the colonies growing on the bioassay dishes by adding 10 mL of prewarmed 2YT ampicillin glucose and scraping. Pool the bacterial cell suspensions.
13. Remove a 100 μL aliquot of the pooled bacterial cell suspension to produce an initial phage stock. Centrifuge the remainder at 3345 × *g*, resuspend in 50 mL of freeze medium, and store 1 mL aliquots at −80 °C.

For library quality control, 20 colonies are picked at random and PCR screened with flanking primers LMB3 (5′CAG GAA ACA GCT ATG AC3′) and pHENseq (5′CTA TGC GGC CCC ATT CA3′) to ensure inserts of the expected size are present. Additionally, 40 colonies are picked at random and sequenced (using pHENseq as a sequencing primer) to verify library diversity and the levels of frameshift mutations.

2.2 Selection of Libraries on Cells

2.2.1 Considerations

In phage display experiments, the outcome of selection is the net result of two distinct selection pressure components (Fig. 5). The first is the

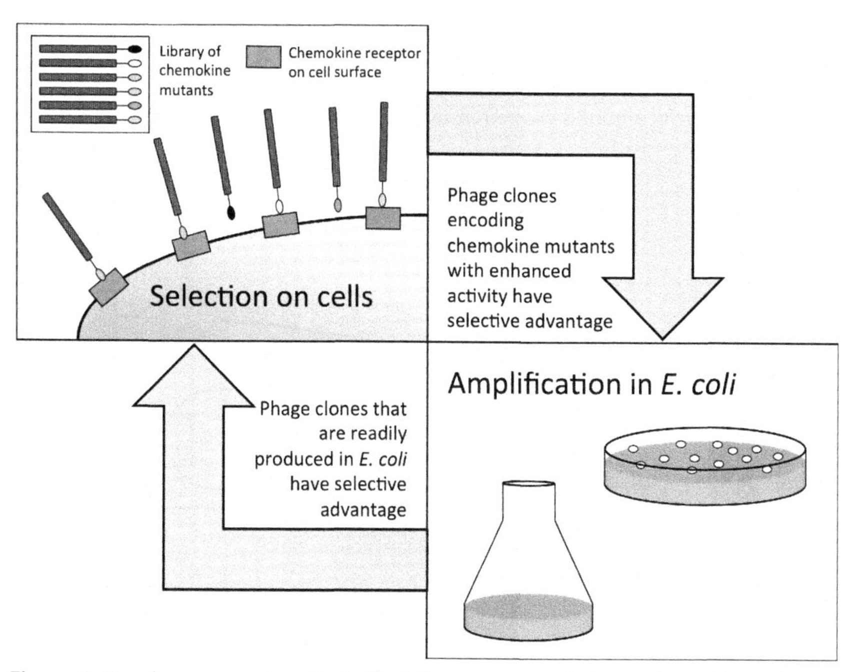

Figure 5 Two key parameters that affect the outcome of phage chemokine selection experiments. The outcome of selection depends on both the stringency of the user-defined panning step and the influence of growth advantage during the amplification step.

user-defined selection component, which in chemokine phage display is based on the capacity of phage clones to interact with chemokine receptors expressed on living cells. The second is a consequence of the amplification process that is required between rounds of selection. Any phage clones with a growth advantage will be enriched during this phase.

In order for useful phage clones to be selected, the user-defined selection component must be able to outrun the growth advantage component. Two criteria are essential for success. First, the initial library should be of sufficiently high quality, with as low a proportion as possible of deletion and frame shift mutants, both of which will have an inherent growth advantage compared to clones expressing full-length fusion proteins. This criterion is addressed during initial library production (see Section 2.1 above). Second, the user-defined selection step must be sufficiently stringent to eliminate a large proportion of insufficiently active phage clones at each round of selection. Below, we present as examples two chemokine phage display approaches that have worked well in our hands, one based on the selection of phage clones capable of inducing chemokine receptor internalization, the

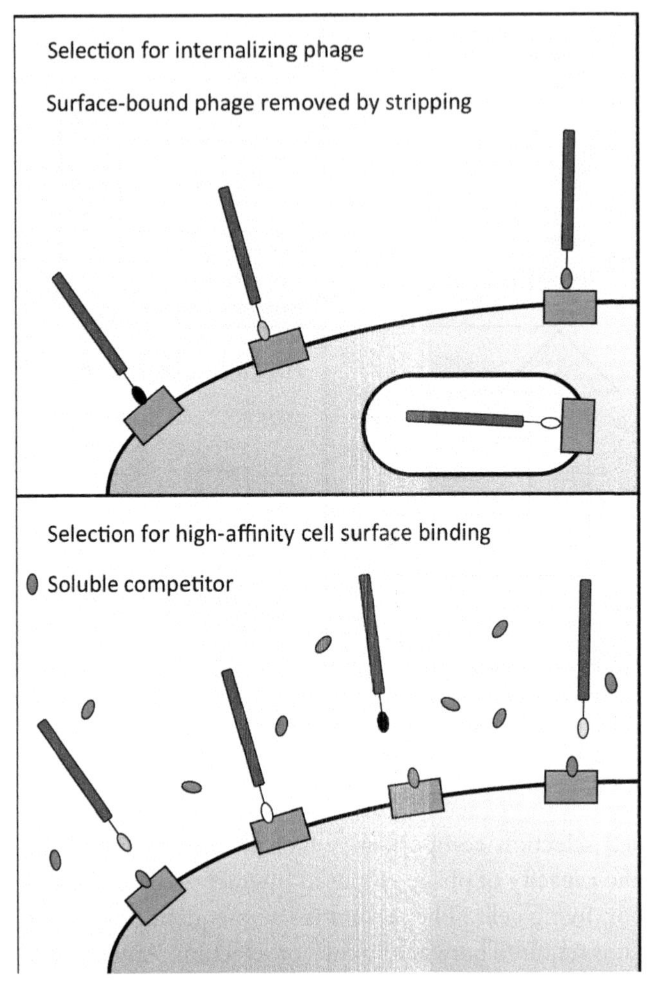

Figure 6 Two cell-based panning approaches for phage chemokine selection. In both cases, the phage chemokine library is first incubated with live adherent cells. Subsequently, either internalizing phage are selected by using harsh washing conditions to strip surface-associated phage from the cell surface and then recovering cell-associated phage (*Option 1*, upper panel), or high-affinity cell surface-binding phage are selected by extensively washing the cell surface and then eluting by adding an excess of native chemokine as a soluble competitor (*Option 2*, lower panel).

other based on selection of phage that are eluted from the cell surface in the presence of an excess of soluble chemokine competitor (Fig.6).

2.2.2 Cell Lines

Cell lines expressing target chemokine receptors are obtained by stable transfection (Hartley et al., 2003) or lentiviral vector transduction (Gaertner et al.,

2008) of commonly available adherent cell lines (e.g., CHO, HEK-293, NIH-3T3, etc.). The resulting cell populations are purified to clonality, with selected clones chosen according to evaluation of surface expression by flow cytometry. Additional characterization by immunofluorescence microscopy can be used to identify clones showing low levels of chemokine receptors in intracellular structures prior to ligand addition.

2.2.3 Required Materials

- Cell lines (see Section 2.2.2 above)
- Phage library stock (approximately 10^{14} cfu/mL) amplified, rescued, and purified from initial *E. coli* library (see Section 2.1)
- Culture medium for cell lines (RPMI or DMEM as appropriate, Gibco GlutaMAX™) supplemented with penicillin–streptomycin (Gibco 15140–122) and 10% fetal bovine serum (Sigma-Aldrich F0926)
- Phosphate-buffered saline (PBS) (ThermoFisher 10010–023)
- PBS supplemented with 1% bovine serum albumin (Sigma-Aldrich A3912; PBS-BSA)
- 30 m*M* Tris–HCl buffer, pH 8.0 supplemented with 1 m*M* EDTA (cell lysis buffer)
- 245 × 245 mm^2 bioassay dishes (Thermofisher 240835)
- 2xTY liquid medium obtained by dissolving 12.4 g 2xTY powder (Sigma-Aldrich Y2627) in 400 mL milliQ water and autoclaving
- 2xTY agar prepared by dissolving 12.4 g 2xTY powder and 6.9 g bacto agar powder (BD 214010) in 400 mL milliQ water and autoclaving
- Ampicillin (Sigma-Aldrich A9393)

  Option 1:
  - Corning six-well cell culture plate (Sigma-Aldrich CLS3516)
  - Sterile disposable cell scraper (Sigma-Aldrich CLS3010-100EA)

  Option 2:
  - Corning 25 cm^2 cell culture flasks (Sigma-Aldrich CLS3289)
  - 2 *M* NaCl (high salt wash solution)
  - 10 m*M* glycine–NaOH buffer pH 11.4 (basic wash solution)
  - PBS supplemented with 1 m*M* EDTA (PBS-EDTA)

2.2.4 Option 1: Selection for Internalizing Ligands

1. (Day 1) Plate 10^6 cells per well in a 25 cm^2 culture flask in 5 mL culture medium. Grow cells overnight at 37 °C, 5% CO_2.
2. (Day 2) Dilute phage stock in PBS to generate 100 μL suspension at either 10^{10} cfu/mL (low complexity libraries, e.g., Library 5) or

10^{11} cfu/mL phage (higher complexity libraries, e.g., Library 2). Preincubate with 100 μL culture medium for 1 h at 37 °C.

3. Add the phage suspension to the medium covering the cells. Incubate 1 h at 37 °C, 5% CO_2.
4. Wash cells 5 × at room temperature with 10 mL PBS-BSA.
5. Add 5 mL culture medium and incubate cells for a further 30 min at 37 °C, 5% CO_2.
6. Wash cells 3 × with high-salt wash solution, once with basic wash solution and once with PBS.
7. Detach cells with 5 mL PBS-EDTA and centrifuge for 5 min at 500 × *g*.
8. Resuspend the cell pellet in 500 μL of cell lysis buffer.
9. Perform three freeze–thaw cycles by rapidly freezing the cell suspension in liquid N_2 and then thawing at ambient temperature, vortex the suspension thoroughly between cycles.
10. Centrifuge the suspensions for 5 min at 500 × *g* and transfer the supernatant to a 15 mL Falcon tube.
11. Add 3 mL log-phase *E. coli* TG1 to the Falcon tube, incubate for 45 min at 37 °C with gentle agitation.
12. Remove a 100 μL aliquot for colony titration and spread the remaining bacterial culture on to a 245 × 245 mm^2 bioassay dish containing 2xTY agar supplemented with ampicillin (50 μg/mL) and 1% glucose. Incubate overnight at 37 °C.
13. (Day 3) Pick up to 40 colonies from the titration plates for insert sequencing. Recover the colonies growing on the bioassay dish by adding 10 mL of prewarmed 2xTY liquid medium supplemented with ampicillin (50 μg/mL) and 1% glucose to the surface and scraping.
14. Remove 100 μL of the scraped bacterial cell suspension to produce phage stocks for the following selection round. Centrifuge the remainder at 3345 × *g*, resuspend in 5 mL of freeze medium, and store at −80 °C.

2.2.5 Option 2: Selection for High-Affinity Cell Surface Binding

1. (Day 1) Plate 10^6 transfected cells per well of a six-well cell culture plate in 4 mL cell culture medium. Grow cells overnight at 37 °C, 5% CO_2.
2. (Day 2) Remove medium in the culture well and add 2 mL of fresh cell culture medium. Transfer an aliquot of phage library stock to provide a phage concentration of 10^{10} cfu/mL and incubate 1 h at 37 °C, 5% CO_2.

3. Wash cells 10 × at room temperature with 10 mL of PBS and then scrape cells from the plate into 10 mL PBS-BSA.
4. Centrifuge the scraped cell suspension (450 × *g*, 5 min at 4 °C) and resuspend the pellet 100 μL PBS-BSA supplemented with 10 μ*M* native chemokine. Incubate for 20 min on ice.
5. Centrifuge the cell suspension (450 × *g*, 5 min at 4 °C) and add 3 mL of a log-phase *E. coli* TG1 culture to the supernatant. Incubate for 45 min at 37 °C with gentle agitation. For subsequent steps, see steps 12, 13, and 14 from Section 2.2.4, above.

3. LIMITATIONS

Using the methods described in this chapter, we have isolated a range of valuable new chemokine analogs acting on CCR5 (Gaertner et al., 2008; Hartley et al., 2003; Dorgham et al., 2008) and CX3CR1 (Dorgham et al., 2009; Poupel et al., 2013) (summarized in Table 3). Recently, Hanes et al. used similar methods to isolate promising SDF-1/CXCL12 analogs that act as modulators of CXCR4 and ACKR3 (Hanes et al., 2015). However, we have encountered problems isolating analogs against several other target chemokine receptors including ACKR1, CCR6, CXCR3, and CXCR4. Below, we discuss the case of ACKR1, using it as an example to discuss potential explanations for selection failure in chemokine phage display.

3.1 The Atypical Chemokine Receptor DARC (ACKR1)

The atypical chemokine receptor DARC (Duffy antigen receptor for chemokines; known as ACKR1) (Horuk, 2015) is a highly promiscuous receptor with a number of natural ligands, including RANTES/CCL5 and IL-8/CXCL8 (Neote, Mak, Kolakowski, & Schall, 1994). We constructed N-terminally extended libraries based on these two chemokines of the form X-X-X-X-X-RANTES/CCL5(5–68) and X-X-X-X-X-IL-8/CXCL8(10–77) (see Table 2) and selected them on CHO cells stably expressing ACKR1 using the *Option 1* method described in Section 2.2. Several independent selection experiments were carried out using the IL-8/CXCL8 library but in each case frameshift and deletion mutants accumulated. The first selection we performed using the RANTES/CCL5 library yielded a clone capable of blocking binding of an anti-ACKR1 antibody on ACKR1-expressing cells, but its activity was only marginally superior to that of native CCL5. Interestingly, this clone featured a deletion mutation eliminating Pro^9. Subsequent selection experiments carried out

Table 3 A Summary of Chemokine Analogs That We Have Isolated Using the Methods Described in This Chapter

Receptor	Ligand	N-Terminal Sequence	Properties	References
CCR5	RANTES/ CCL5	SPYSSDTTPCC	Natural ligand	
	P1-RANTES	LSPVSSQSSACC	Anti-HIV, antagonist	Hartley et al. (2003)
	P2-RANTES	FSPLSSQSSACC	Anti-HIV, superagonist	Hartley et al. (2003)
	P7-RANTES	MSPLSSQASACC	Anti-HIV, superagonist, vaccine adjuvant	Dorgham et al. (2008)
	5P12-RANTES	QGPPLMATQSCC	Anti-HIV, antagonist	Gaertner et al. (2008)
	5P14-RANTES	QGPPLMSLQVCC	Anti-HIV, biased ligand? Modified intracellular trafficking	Gaertner et al. (2008), Zidar (2011), and Bonsch, Munteanu, Rossitto-Borlat, Furstenberg, and Hartley (2015)
	6P4-RANTES	QGPPGDIVLACC	Anti-HIV, superagonist	Gaertner et al. (2008)
	9P10-RANTES	QWVMGSDTTPCC	Anti-HIV, superagonist	Oliver Hartley, Astrid Melotti and Irène Rossitto-Borlat (unpublished)
CCR1	RANTES/ CCL5	SPYSSDTTPCC	Natural ligand	
	R1-1-RANTES	ASTSSSGASACC	Antagonist	Karim Dorgham (unpublished)
	R1-2-RANTES	LSSTSSQSPPCC	Antagonist	Karim Dorgham (unpublished)
CX3CR1	Fractalkine/ CX3CL1	QHHGVTKC	Natural ligand	
	F1-CX3CL1	ILDNGVSKC	Antagonist, anti-inflammatory	Dorgham et al. (2009) and Poupel et al. (2013)
	F3-CX3CL1	QPGGVSKC	Agonist	Karim Dorgham (unpublished)

using the same library led to accumulation of frameshift and deletion mutants, and the project was abandoned.

There are several potential explanations for our ability to isolate promising analogs acting on ACKR1. First, the libraries we constructed may have been of suboptimal quality. When we sequenced clones picked at random from the starting IL-8/CXCL8 library, we noted that almost 50% of the clones encoded frameshift mutations in the N-terminal region encoded by the degenerate oligonucleotide that was used to construct the library. The problems we encountered with this library were most likely due to the poor quality of the custom synthesized degenerate oligonucleotide primer. This was not the explanation for selection failure with the RANTES/CCL5 library, however. Sequence analysis of the starting library showed a much higher proportion of full-length, in-frame clones, and when the library was used in a selection experiment on CHO cells expressing CCR5, we readily isolated several clones with anti-HIV potencies comparable with those obtained in previous selection on CCR5 of a primary library (Hartley et al., 2003), including 9P10-RANTES (Table 3).

A second potential explanation is that our initial library design was not appropriate for selecting analogs active on ACKR1. In this regard, it is interesting to note that the one potential lead we obtained in this selection experiment was a clone encoding a rare codon deletion corresponding to Pro^{9} that must have been introduced during the PCR library amplification step. With this result in mind, it may have been better to have started with a nonextended library i.e., X-X-X-X-X-RANTES/CCL5 (6–68) rather than the one-residue N-terminally extended library that we used.

We suspect that the two issues outlined above, (i) poor library quality and/or (ii) inappropriate library design, are the most likely explanations for failures with chemokine phage display that we have experienced. While we cannot eliminate the possibility that some chemokine receptors may be intrinsically less susceptible to modulation by chemokine analogs than others, or that some chemokines are intrinsically less suitable for phage display, our recommendation is that researchers should (i) be ready to restart library construction if the quality of the library appears to be compromised by mutations and deletions and (ii) be aware that their initial library design may not be the best, and that several false starts might be required before a selection program can really gain momentum.

4. PERSPECTIVES

In this section, we will briefly highlight the successes that we have achieved with chemokine phage display and the perspectives for further use of the analogs that were isolated.

4.1 A Chemokine-Based HIV Prevention Strategy

Starting with promising ligands isolated from an initial selection (Hartley et al., 2003), we used a chemokine walking approach involving second- and third-generation libraries to isolate a range of highly potent anti-HIV chemokine analogs (Gaertner et al., 2008). Two of the most promising analogs, 5P12-RANTES and 6P4-RANTES shown to be fully effective in the macaque vaginal challenge model for HIV prevention (Veazey et al., 2009), and both showed excellent stability to temperature, to vaginal pH, in the presence of human vaginal lavage and in the presence of human semen (Cerini et al., 2008). 5P12-RANTES, which was also shown to present an exceptionally high barrier to the development of HIV escape mutants *in vitro* (Nedellec et al., 2011), was taken forward for further development. A scalable, low-cost cGMP-compliant production process for clinical grade 5P12-RANTES has been developed (Oliver Hartley, unpublished), and work required to bring 5P12-RANTES vaginal gel into a first clinical trial is almost complete.

4.2 New Tools to Study CCR5 Pharmacology and Cell Biology

The selection strategy used to isolate anti-HIV analogs such as 5P12-RANTES and 6P4-RANTES was designed to favor analogs pharmacologically similar to the chemically synthesized prototype PSC-RANTES i.e., CCR5 superagonists capable of strongly inducing receptor sequestration. While the selection program did indeed yield analogs pharmacologically equivalent to PSC-RANTES such as 6P4-RANTES, it also provided highly potent HIV entry inhibitors with strikingly different pharmacological profiles (Gaertner et al., 2008).

One example is 5P12-RANTES, which belongs to a group of highly potent analogs that neither activate receptor signaling nor elicit receptor internalization. Another example is 5P14-RANTES, which elicits receptor internalization in the absence of G protein-mediated signaling. It has been suggested that 5P14-RANTES represents a strongly biased agonist of CCR5

(Zidar, 2011), and that CCR5 internalized by 5P14-RANTES follows a distinct intracellular trafficking itinerary from that of CCR5 internalized by natural ligands and superagonists such as PSC-RANTES (Bonsch et al., 2015).

Hence aside from their potential for development as anti-HIV medicines, this set of analogs represents a valuable toolbox that can be used to better understand chemokine structure–activity relationships, particularly now that crystal structures of chemokine–chemokine receptor complexes are becoming available (Qin et al., 2015; Burg et al., 2015).

4.3 An Intrakine to Protect Cells from HIV Infection and a Vaccine Adjuvant

P2-RANTES (Hartley et al., 2003; Jin, Kagiampakis, Li, & Liwang, 2010) was incorporated into a gene therapy approach to inhibit HIV infection. It was engineered to carry a C-terminal KDEL motif that prevents export from the endoplasmic reticulum during synthesis. When introduced into a T cell line, the trapped intracellular chemokine, or "intrakine," was shown to reduce surface expression of CCR5, thereby protecting the target cells from HIV infection (Petit et al., 2014). The P2-CCL5 intrakine was subsequently tested as a component of a combination gene therapy approach. Coding sequences for the P2-CCL5 intrakine and the C46 peptide inhibitor of viral fusion (Egelhofer et al., 2004) were introduced into the same lentiviral vector which was used to genetically modify T cells or CD34+ progenitors. Gene-modified cells were then used in a model of humanized mice challenged with HIV. This treatment protected CD4+ T cells against HIV-induced depletion and reduced HIV load (Petit et al., 2015).

P7-RANTES, another analog directly isolated from the initial RANTES/CCL5 library (Hartley et al., 2003), was encoded as an adjuvant in a DNA vaccine strategy and demonstrated to be capable of enhancing anti-tumor immune responses by increasing levels of local leukocyte recruitment (Dorgham et al., 2008).

4.4 A Prototypic Inhibitor of CX3CR1, the Fractalkine/CX3CL1 Receptor

CX3CR1 and its only known chemokine ligand, CX3CL1, are implicated in a range of inflammatory diseases (D'Haese, Friess, & Ceyhan, 2012). We generated a Fractalkine/CX3CL1 library (Table 2) and subjected it to selection on CX3CR1-expressing cells using the *Option 2* method described in Section 2.2. We isolated a variant called F1-CX3CL1, which binds to

CX3CR1 with an affinity comparable to that of the native ligand without inducing chemotaxis, receptor internalization, or intracellular calcium response. F1-CX3CL1 functions as an antagonist, inhibiting CX3CL1-induced calcium flux, chemotaxis and cell adhesion *in vitro*, as well as macrophage recruitment in a noninfectious murine model of peritonitis (Dorgham et al., 2009). F1-CX3CL1 is a prototypic CX3CR1 blocker that is now commonly used to demonstrate the involvement of the CX3CR1/CX3CL1 axis in *in vitro* disease models (e.g., Gaudin et al., 2011; Ren et al., 2014). F1-CX3CL1 has also been shown to be active in a murine model of neovascularization (Kumar et al., 2013) and to modulate monocyte recovery after myeloablation (Jacquelin et al., 2013). Finally, F1-CX3CL1 was shown to be a potent inhibitor of the progression of atherosclerotic lesions through its capacity to selectively inhibit the activity of inflammatory monocytes (Poupel et al., 2013).

REFERENCES

Allen, S. J., Crown, S. E., & Handel, T. M. (2007). Chemokine: Receptor structure, interactions, and antagonism. *Annual Review of Immunology, 25*, 787–820.

Balkwill, F. R. (2012). The chemokine system and cancer. *The Journal of Pathology, 226*, 148–157.

Barbas, C. F. (2000). *Phage display: A laboratory manual.* Cold Spring Harbor, NY: Cold Spring Harbor Labortory Press.

Bobanga, I. D., Petrosiute, A., & Huang, A. Y. (2013). Chemokines as cancer vaccine adjuvants. *Vaccines (Basel), 1*, 444–462.

Bonsch, C., Munteanu, M., Rossitto-Borlat, I., Furstenberg, A., & Hartley, O. (2015). Potent anti-HIV chemokine analogs direct post-endocytic sorting of CCR5. *PLoS One, 10*, e0125396.

Burg, J. S., Ingram, J. R., Venkatakrishnan, A. J., Jude, K. M., Dukkipati, A., Feinberg, E. N., et al. (2015). Structural biology. Structural basis for chemokine recognition and activation of a viral G protein-coupled receptor. *Science, 347*, 1113–1117.

Cerini, F., Landay, A., Gichinga, C., Lederman, M. M., Flyckt, R., Starks, D., et al. (2008). Chemokine analogues show suitable stability for development as microbicides. *Journal of Acquired Immune Deficiency Syndromes (1999), 49*, 472–476.

Clackson, T., & Lowman, H. B. (2004). *Phage display: A practical approach.* Oxford: Oxford University Press.

Clark-Lewis, I., Vo, L., Owen, P., & Anderson, J. (1997). Chemical synthesis, purification, and folding of C-X-C and C-C chemokines. *Methods in Enzymology, 287*, 233–250.

Dawson, P. E. (1997). Synthesis of chemokines by native chemical ligation. *Methods in Enzymology, 287*, 34–45.

D'Haese, J. G., Friess, H., & Ceyhan, G. O. (2012). Therapeutic potential of the chemokine-receptor duo fractalkine/CX3CR1: An update. *Expert Opinion on Therapeutic Targets, 16*, 613–618.

Dorgham, K., Abadie, V., Iga, M., Hartley, O., Gorochov, G., & Combadiere, B. (2008). Engineered CCR5 superagonist chemokine as adjuvant in anti-tumor DNA vaccination. *Vaccine, 26*, 3252–3260.

Dorgham, K., Ghadiri, A., Hermand, P., Rodero, M., Poupel, L., Iga, M., et al. (2009). An engineered CX3CR1 antagonist endowed with anti-inflammatory activity. *Journal of Leukocyte Biology*, *86*, 903–911.

Egelhofer, M., Brandenburg, G., Martinius, H., Schult-Dietrich, P., Melikyan, G., Kunert, R., et al. (2004). Inhibition of human immunodeficiency virus type 1 entry in cells expressing gp41-derived peptides. *Journal of Virology*, *78*, 568–575.

Fernandez, E. J., & Lolis, E. (2002). Structure, function, and inhibition of chemokines. *Annual Review of Pharmacology and Toxicology*, *42*, 469–499.

Gaertner, H., Cerini, F., Escola, J. M., Kuenzi, G., Melotti, A., Offord, R., et al. (2008). Highly potent, fully recombinant anti-HIV chemokines: Reengineering a low-cost microbicide. *Proceedings of the National Academy of Sciences of the United States of America*, *105*, 17706–17711.

Gaudin, F., Nasreddine, S., Donnadieu, A. C., Emilie, D., Combadiere, C., Prevot, S., et al. (2011). Identification of the chemokine CX3CL1 as a new regulator of malignant cell proliferation in epithelial ovarian cancer. *PLoS One*, *6*, e21546.

Griffith, J. W., Sokol, C. L., & Luster, A. D. (2014). Chemokines and chemokine receptors: Positioning cells for host defense and immunity. *Annual Review of Immunology*, *32*, 659–702.

Griffiths, A. D., Williams, S. C., Hartley, O., Tomlinson, I. M., Waterhouse, P., Crosby, W. L., et al. (1994). Isolation of high affinity human antibodies directly from large synthetic repertoires. *The EMBO Journal*, *13*, 3245–3260.

Hanes, M. S., Salanga, C. L., Chowdry, A. B., Comerford, I., McColl, S. R., Kufareva, I., et al. (2015). Dual targeting of the chemokine receptors CXCR4 and ACKR3 with novel engineered chemokines. *The Journal of Biological Chemistry*. http://dx.doi.org/10.1074/jbc.M115.675108.

Hartley, O. (2002). The use of phage display in the study of receptors and their ligands. *Journal of Receptor and Signal Transduction Research*, *22*, 373–392.

Hartley, O., Dorgham, K., Perez-Bercoff, D., Cerini, F., Heimann, A., Gaertner, H., et al. (2003). Human immunodeficiency virus type 1 entry inhibitors selected on living cells from a library of phage chemokines. *Journal of Virology*, *77*, 6637–6644.

Hartley, O., & Offord, R. E. (2005). Engineering chemokines to develop optimized HIV inhibitors. *Current Protein & Peptide Science*, *6*, 207–219.

Hoogenboom, H. R., Griffiths, A. D., Johnson, K. S., Chiswell, D. J., Hudson, P., & Winter, G. (1991). Multi-subunit proteins on the surface of filamentous phage: Methodologies for displaying antibody (Fab) heavy and light chains. *Nucleic Acids Research*, *19*, 4133–4137.

Horuk, R. (2015). The Duffy antigen receptor for chemokines DARC/ACKR1. *Frontiers in Immunology*, *6*, 279.

Horuk, R., Reilly, D., & Yansura, D. (1997). Expression, purification, and characterization of *Escherichia coli*-derived recombinant human melanoma growth stimulating activity. *Methods in Enzymology*, *287*, 3–12.

Jacquelin, S., Licata, F., Dorgham, K., Hermand, P., Poupel, L., Guyon, E., et al. (2013). CX3CR1 reduces Ly6Chigh-monocyte motility within and release from the bone marrow after chemotherapy in mice. *Blood*, *122*, 674–683.

Jin, H., Kagiampakis, I., Li, P., & Liwang, P. J. (2010). Structural and functional studies of the potent anti-HIV chemokine variant P2-RANTES. *Proteins*, *78*, 295–308.

Kay, B. K., Winter, J., & McCafferty, J. D. (1996). *Phage display of peptides and proteins: A laboratory manual*. San Diego, CA: Academic Press.

Koelink, P. J., Overbeek, S. A., Braber, S., de Kruijf, P., Folkerts, G., Smit, M. J., et al. (2012). Targeting chemokine receptors in chronic inflammatory diseases: An extensive review. *Pharmacology & Therapeutics*, *133*, 1–18.

Kuhmann, S. E., & Hartley, O. (2008). Targeting chemokine receptors in HIV: A status report. *Annual Review of Pharmacology and Toxicology*, *48*, 425–461.
Kumar, A. H., Martin, K., Turner, E. C., Buneker, C. K., Dorgham, K., Deterre, P., et al. (2013). Role of CX3CR1 receptor in monocyte/macrophage driven neovascularization. *PLoS One*, *8*, e57230.
Kuritzkes, D., Kar, S., & Kirkpatrick, P. (2008). Fresh from the pipeline: Maraviroc. *Nature Reviews. Drug Discovery*, 7, 15–16.
Nedellec, R., Coetzer, M., Lederman, M. M., Offord, R. E., Hartley, O., & Mosier, D. E. (2011). Resistance to the CCR5 inhibitor 5P12-RANTES requires a difficult evolution from CCR5 to CXCR4 coreceptor use. *PLoS One*, *6*, e22020.
Neote, K., Mak, J. Y., Kolakowski, L. F., Jr., & Schall, T. J. (1994). Functional and biochemical analysis of the cloned Duffy antigen: Identity with the red blood cell chemokine receptor. *Blood*, *84*, 44–52.
Offord, R. E., Gaertner, H. F., Wells, T. N., & Proudfoot, A. E. (1997). Synthesis and evaluation of fluorescent chemokines labeled at the amino terminal. *Methods in Enzymology*, *287*, 348–369.
Petit, N., Baillou, C., Burlion, A., Dorgham, K., Levacher, B., Amiel, C., et al. (2015). Gene transfer of two entry inhibitors protects CD4+ T cell from HIV-1 infection in humanized mice. *Gene Therapy*, in press.
Petit, N., Dorgham, K., Levacher, B., Burlion, A., Gorochov, G., & Marodon, G. (2014). Targeting both viral and host determinants of human immunodeficiency virus entry, using a new lentiviral vector coexpressing the T20 fusion inhibitor and a selective CCL5 intrakine. *Human Gene Therapy Methods*, *25*, 232–240.
Poupel, L., Boissonnas, A., Hermand, P., Dorgham, K., Guyon, E., Auvynet, C., et al. (2013). Pharmacological inhibition of the chemokine receptor, CX3CR1, reduces atherosclerosis in mice. *Arteriosclerosis, Thrombosis, and Vascular Biology*, *33*, 2297–2305.
Qin, L., Kufareva, I., Holden, L. G., Wang, C., Zheng, Y., Zhao, C., et al. (2015). Structural biology. Crystal structure of the chemokine receptor CXCR4 in complex with a viral chemokine. *Science*, *347*, 1117–1122.
Raman, D., Sobolik-Delmaire, T., & Richmond, A. (2011). Chemokines in health and disease. *Experimental Cell Research*, *317*, 575–589.
Ren, J., Hou, X. Y., Ma, S. H., Zhang, F. K., Zhen, J. H., Sun, L., et al. (2014). Elevated expression of CX3C chemokine receptor 1 mediates recruitment of T cells into bone marrow of patients with acquired aplastic anaemia. *Journal of Internal Medicine*, *276*, 512–524.
Scholten, D. J., Canals, M., Maussang, D., Roumen, L., Smit, M. J., Wijtmans, M., et al. (2012). Pharmacological modulation of chemokine receptor function. *British Journal of Pharmacology*, *165*, 1617–1643.
Sidhu, S. S. (2001). Engineering M13 for phage display. *Biomolecular Engineering*, *18*, 57–63.
Uy, G. L., Rettig, M. P., & Cashen, A. F. (2008). Plerixafor, a CXCR4 antagonist for the mobilization of hematopoietic stem cells. *Expert Opinion on Biological Therapy*, *8*, 1797–1804.
Veazey, R. S., Ling, B., Green, L. C., Ribka, E. P., Lifson, J. D., Piatak, M., Jr., et al. (2009). Topically applied recombinant chemokine analogues fully protect macaques from vaginal simian-human immunodeficiency virus challenge. *The Journal of Infectious Diseases*, *199*, 1525–1527.
Viola, A., & Luster, A. D. (2008). Chemokines and their receptors: Drug targets in immunity and inflammation. *Annual Review of Pharmacology and Toxicology*, *48*, 171–197.
Wells, T., Proudfoot, A., Power, C. A., Lusti-Narasimhan, M., Alouani, S., Hoogewerf, A. J., et al. (1996). The molecular basis of the chemokine/chemokine receptor interaction—Scope for design of chemokine antagonists. *Methods*, *10*, 126–134.
Zidar, D. A. (2011). Endogenous ligand bias by chemokines: Implications at the front lines of infection and leukocyte trafficking. *Endocrine, Metabolic & Immune Disorders Drug Targets*, *11*, 120–131.

CHAPTER FOUR

Methods for the Recognition of GAG-Bound Chemokines

Pauline Bonvin, Franck Gueneau, Nicolas Fischer, Amanda Proudfoot[1]
NovImmune S.A., Geneva, Switzerland
[1]Corresponding author: e-mail address: amandapf@orange.fr

Contents

1. Introduction 73
2. Methods 74
 2.1 Material 74
 2.2 Recognition of Heparin-Bound Chemokine by ELISA 75
 2.3 Recognition of HS-Bound Chemokine by Biolayer Interferometry 77
 2.4 Binding to Chemokines Displayed on Endothelial Cells 80
 2.5 Inhibition of Chemokine Binding to GAGs by Biolayer Interferometry (BLI) 81
3. Summary 85
References 85

Abstract

Chemokines play a pivotal role in the multistep cascade of cellular recruitment, where they provide the directional signal. They activate cells through a high-affinity interaction with their receptors, members of the large family of heptahelical G protein-coupled receptors. In order to provide the directional signal, they bind to cell surface proteoglycans through a low-affinity interaction with the glycosaminoglycan (GAG) moiety. While several methods have been described to measure the chemokine–GAG interaction, this chapter describes methods to identify whether anti-chemokine antibodies or chemokine-binding proteins recognize the GAG-bound chemokine.

1. INTRODUCTION

The importance of chemokines binding to endothelial surfaces was suggested soon after the family was identified by the demonstration that neutrophils migrated toward immobilized IL-8/CXCL8 by a mechanism called haptotaxis (Rot, 1993). Interestingly, the chemokine field actually started with the identification of a protein by heparin-sepharose affinity chromatography, platelet factor 4 (PF4)/CXCL4 (Deuel, Keim, Farmer, & Heinrikson, 1977).

Methods in Enzymology, Volume 570
ISSN 0076-6879
http://dx.doi.org/10.1016/bs.mie.2015.10.001

While it was always well accepted that chemokines need to be immobilized on the endothelial surface through their interaction with glycosaminoglycans (GAGs) in order to provide a directional signal, direct evidence was provided by the loss of the ability to recruit cells *in vivo* by chemokine mutants with abrogated GAG-binding capacity (Proudfoot et al., 2003).

Chemokine binding to GAGs can be determined by several methods. These include affinity chromatography to heparin sepharose, binding assays, isothermal fluorescence titration, and surface plasmon resonance to name a few, which are described in detail in Hamel, Sielaff, Proudfoot, and Handel (2009). It should be noted that the majority of these assays use heparin since it is readily available commercially and is less expensive than the other classes of GAGs. There are six major classes of GAGs which include heparin, heparan sulfate (HS), chondroitin sulfate (CS), dermatan sulfate (DS), keratin sulfate (KS), and hyaluronic acid (HA). Heparin and HA are soluble GAGs, whereas HS, DS, KS, and CS are usually covalently attached to a protein core, referred to as proteoglycans. It should be noted that heparin is more highly sulfated than HS, the most abundant GAG which is found on almost every cell in the body.

Chemokines therefore exist both in the fluid phase in the circulation and as a bound form immobilized on proteoglycans expressing GAGs. While there are many methods for establishing whether a chemokine binds to GAGs, to our knowledge, there is to date only one report of recognition of GAG-bound chemokine by an antibody using a fluorescence-activated cell sorting (FACS)-based method (Burns, Gallo, DeVico, & Lewis, 1998). In order to establish the mechanism of action of anti-chemokine antibodies and chemokine-binding proteins, it would be necessary to determine whether they recognize either the free or GAG-bound chemokine or both. This is particularly important if such molecules are being developed for testing in animal models of disease or even more importantly, for therapeutic use. Whether antibodies that have reached clinical trials recognize the free or bound form of the chemokine, or both, has not been reported. We describe here three methods to determine whether such proteins recognize GAG-bound chemokines, and a method to inhibit the binding of chemokines to GAGs.

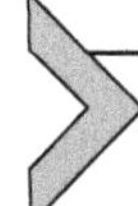

2. METHODS

2.1 Material

Chemokines were produced in the prokaryotic *Escherichia coli* as previously described either in inclusion bodies (Proudfoot & Borlat, 2000) or as NusA fusions (Magistrelli et al., 2005) or purchased from Peprotech (Rocky Hill, NJ).

2.2 Recognition of Heparin-Bound Chemokine by ELISA

Synthetic production of GAGs is difficult and therefore only a few GAGs are commercially available. Low-molecular-weight heparin is a soluble GAG which is easily purified from sources such as porcine intestinal mucosa. Heparin is therefore readily available and commonly used for *in vitro* experiments. In this section, we report a method allowing the determination of whether an anti-chemokine monoclonal antibody is able to bind to its target when the chemokine is complexed with heparin (Fig. 1). This assay has been developed for use with IgG molecules but could easily be adapted for other antibody formats or chemokine-binding proteins by using an appropriate coating reagent.

Required materials

Goat anti-human-IgG Fcγ (Jackson ImmunoResearch, cat. no. 109-005-098)

Sterile phosphate-buffered saline (PBS, prepared in house or obtained from Sigma-Aldrich, cat. no. D8537)

Transparent Maxisorp plate (Nunc, cat. no. 442404)

Tween-20 (Sigma-Aldrich, cat. no. P7949)

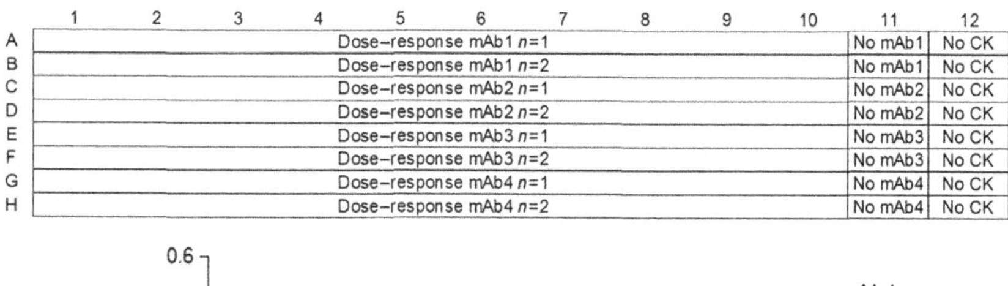

<table>
<tr><th></th><th>1</th><th>2</th><th>3</th><th>4</th><th>5</th><th>6</th><th>7</th><th>8</th><th>9</th><th>10</th><th>11</th><th>12</th></tr>
<tr><td>A</td><td colspan="10">Dose–response mAb1 n=1</td><td>No mAb1</td><td>No CK</td></tr>
<tr><td>B</td><td colspan="10">Dose–response mAb1 n=2</td><td>No mAb1</td><td>No CK</td></tr>
<tr><td>C</td><td colspan="10">Dose–response mAb2 n=1</td><td>No mAb2</td><td>No CK</td></tr>
<tr><td>D</td><td colspan="10">Dose–response mAb2 n=2</td><td>No mAb2</td><td>No CK</td></tr>
<tr><td>E</td><td colspan="10">Dose–response mAb3 n=1</td><td>No mAb3</td><td>No CK</td></tr>
<tr><td>F</td><td colspan="10">Dose–response mAb3 n=2</td><td>No mAb3</td><td>No CK</td></tr>
<tr><td>G</td><td colspan="10">Dose–response mAb4 n=1</td><td>No mAb4</td><td>No CK</td></tr>
<tr><td>H</td><td colspan="10">Dose–response mAb4 n=2</td><td>No mAb4</td><td>No CK</td></tr>
</table>

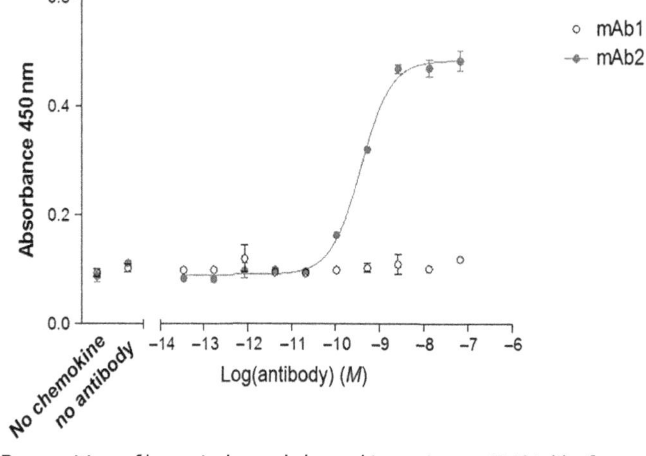

Figure 1 Recognition of heparin-bound chemokine using an ELISA-like format. Example of plate layout (upper panel) and of results (lower panel). In this example, mAb2 binds to the heparin-bound chemokine, whereas mAb1 does not interact with the heparin-bound chemokine.

Bovine serum albumin (BSA; Sigma-Aldrich, cat. no. A6003)
Clear 96-Well Polystyrene Conical Bottom Microwell Plates (Nunc, cat. no. 249570)
Specific human IgG monoclonal antibody targeting the chemokine of interest (mAb1)
Chemokine of interest (1 μ*M* solution)
Heparin, fluorescein conjugate (Life technologies, cat. no. H7482) resuspended at 55 μ*M* in H_2O
anti-Fluorescein-POD, Fab fragment (Roche, cat. no. 11 426 346 910), stock solution at 150 U/mL
3,3′,5,5′-Tetramethylbenzidine Liquid Substrate, Super Slow (Sigma, cat. no. T5569)
Sulfuric acid H_2SO_4 2 *M*

Methods

1. Dilute the anti-human-IgG to 4 μg/mL in PBS. Add 50 μL/well of this antibody solution to rows A and B of a 96-well transparent Maxisorb microplate and incubate stationary overnight at 4 °C.
2. On the next day, remove the coating solution by inverting the plate. Wash the plate three times with a solution of PBS containing 0.05% (v/v) of Tween-20. At the end of the washes, remove any remaining wash buffer by inverting plate and blotting it against paper towels.
3. To avoid nonspecific interactions, block the plate by adding 200 μL/well of PBS containing 3% (w/v) of BSA and incubate at room temperature for 2 h on a horizontal orbital microplate shaker (450 rpm).
4. Wash the Maxisorb plate as described in step 2.
5. Dilute mAb1 to 100 μg/mL in PBS + 3% BSA in wells A1 and A12 of a 96-well plastic plate (conical bottom) in a final volume of 200 μL/well. Add 120 μL PBS + 3% BSA to the remaining wells A2–A11. Mix the solution in A1 and serially transfer 60 μL (dilution 1:3) from A1 to A10. The well A11 will serve as negative ctrl (PBS + 3% BSA only). Transfer the prepared dilution of mAb1 to the Maxisorb plate, in duplicate (rows A and B, 50 μL/well). Incubate at room temperature for 1 h on a horizontal shaker (450 rpm).
6. In the meantime, prepare the chemokine:heparin complexes: mix 10 μL of 1 μ*M* chemokine solution with 2 μL of 55 μ*M* heparin conjugated to fluorescein (heparin-FITC) in 2 mL of PBS + 3% BSA to obtain a final concentration of 5 n*M* chemokine in presence of 55 n*M* heparin (11-fold excess of heparin, tube 1). As a negative control, dilute 1 μL of heparin-FITC in 1 mL of PBS + 3% BSA (tube 2). Protect both tubes with aluminium foil and incubate in the dark for 1 h at room temperature.

7. Wash the Maxisorb plate as described in step 2.
8. Add 50 μL/well of the chemokine:heparin solution (tube 1) to each well of the Maxisorb plate, except wells A12 and B12. Add 50 μL/well of the control heparin solution (tube 2) to wells A12 and B12. Protect the plate with an aluminium foil and incubate at room temperature for 1 h on a horizontal shaker (450 rpm).
9. Wash the Maxisorb plate as described in step 2.
10. Dilute the anti-fluorescein Fab 1:5000 in PBS + 3% BSA and add 50 μL/well to the Maxisorb plate. Protect the plate with aluminium foil and incubate at room temperature for 2 h on a horizontal shaker (450 rpm).
11. Wash the Maxisorb plate as described in step 2.
12. Add 50 μL/well of tetramethylbenzidine substrate (brought to room temperature before use) to the Maxisorb plate. Incubate in the dark until the desired coloration is reached (15–30 min).
13. Add 50 μL/well of H_2SO_4 2 *M* to stop the reaction. Read immediately the absorbance at 450 nm in an appropriate device. Data should be analyzed using GraphPad Prism nonlinear regression curve fitting or an equivalent software.

2.3 Recognition of HS-Bound Chemokine by Biolayer Interferometry

HS is a GAG naturally expressed by mammalian cells and is therefore more representative of physiological GAGs that immobilize chemokines than heparin. It is commercially available as a sodium salt isolated from kidney extract. We report here the biotinylation of HS and its use in a biolayer interferometry (BLI) assay to monitor the binding of monoclonal antibodies to their chemokine target displayed on HS (Fig. 2). This assay has also been used successfully with chemokine-binding proteins.

Required materials

BupH MES buffered saline packs (Thermo Scientific, cat. no. 28390)

EZ-Link hydrazide–PEG_4–biotin (Thermo Scientific, cat. no. 21360)

Dimethyl sulfoxide (DMSO; Sigma, cat. no. D8418)

Heparan sulfate sodium salt from bovine kidney (Sigma, cat. no. H7640-1MG)

1-Ethyl-3-(3-dimethylaminopropyl)carbodiimide hydrochloride (EDC; Life technologies, cat. no. 77149)

Amicon Ultra-0.5 mL 3K Ultracel-3K membrane (Merck Millipore, cat. no. UFC500308)

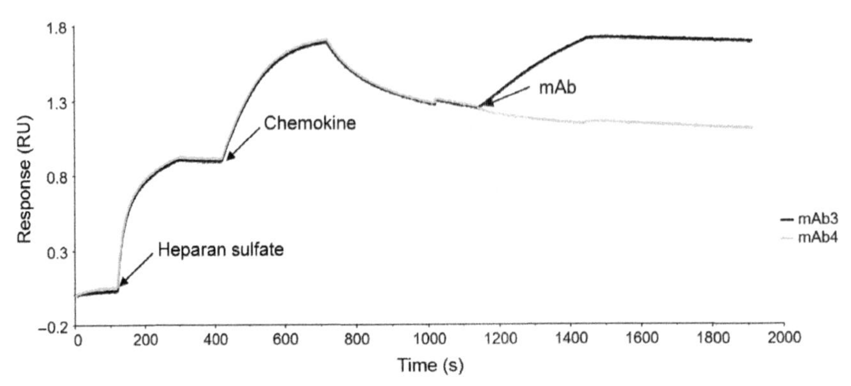

Figure 2 Recognition of heparan sulfate-bound chemokine by biolayer interferometry. Biotinylated heparan sulfate, the chemokine, and the antibody are successively added onto streptavidin-coated biosensors, demonstrating that mAb3 binds to the chemokine coated on the heparan sulfate surface, whereas mAb4 does not bind to the chemokine in this context.

Octet RED96 system (Pall ForteBio, cat. no. 30–5048) or an equivalent system

Dip and Read streptavidin (SA) biosensors (Pall ForteBio, cat. no. 18–5019)

Kinetics buffer 10 × (Pall ForteBio, cat. no. 18–1092)

Specific monoclonal antibody targeting the chemokine of interest (mAb1)

Chemokine of interest

NaCl 5 *M* solution

Biotinylation of HS

1. Dissolve one pack of MES buffer into 500 mL of deionized water to obtain 500 mL of 100 m*M* MES, 0.9% NaCl, pH 4.7.
2. Resuspend the hydrazide–PEG_4–biotin at 250 m*M* in DMSO: add 396 μL of DMSO to the vial of hydrazide. Dilute this stock solution 1:5 in DMSO to obtain 50 m*M* hydrazide–PEG_4–biotin. Store the 250 m*M* stock solution at −20 °C.
3. Resuspend the HS at 5 mg/mL in MES buffer: add 200 μL of MES to the vial containing 1 mg heparin sulfate, mix by inversing the vial 8–10 times and leave on the bench for 10–15 min. Transfer the entire volume to a clean 1.7-mL Eppendorf tube.
4. Resuspend EDC at 400 m*M* in H_2O: add 130 μL of H_2O to 10 mg of EDC, mix by inversing the vial 8–10 ×, and leave on the bench for 10–15 min.

5. Add 5.25 μL of 50 m*M* hydrazide–PEG_4–biotin to the HS (final concentration 1.25 m*M*). Add 3.4 μL of EDC to the HS (final concentration 6.5 m*M*). Incubate overnight at room temperature on a rotary shaker.
6. On the next day, use a 3K centrifugal filter device to remove the excess of biotin: transfer the HS solution to a clean filter and fill the filter with H_2O. Centrifuge at 14,000 × *g* for 15 min. Discard the filtrate, fill the filter with H_2O, and centrifuge at 14,000 × *g* for 30 min. Repeat this step.
7. Following the last centrifugation, recover the protein by transferring the solution into a clean 1.7 mL tube (spin for 2 min at 1000 × *g*). Complete to 200 μL with H_2O and store at 4 °C.

BLI assay

1. Prepare Kinetics buffer 1 × (KB) by diluting the 10 × stock solution 1:10 in PBS.
2. Dilute biotinylated HS (see paragraph above) 1:2000 in KB. Dilute the chemokine to be tested to 100 n*M* in KB. Dilute mAb to 1 μg/mL in KB. Transfer 200 μL/well of the adequate solution to a black flat bottom 96-well microplate to have the following solution in each well:
   - Columns 1, 3, 5, and 8: KB
   - Column 2: biotinylated HS 1:2000
   - Column 4: 100 n*M* chemokine
   - Column 6: 1 μg/mL mAb
   - Column 7: NaCl 5 *M*

   Use one row per antibody to be tested.
3. Configure the following experiment on the Octet RED96 (kinetics experiment), using SA biosensors:
   - Baseline 120 s in column 1
   - Loading 180 s in column 2
   - Baseline 120 s in column 3
   - Association 300 s in column 4
   - Dissociation 300 s in column 3
   - Association 300 s in column 6
   - Dissociation 300–600 s in column 5

   During the experiment, maintain a constant shaking (1000 rpm) and temperature (30 °C). To determine whether the chemokine or the antibody interact nonspecifically with the biosensors, a control experiment should be performed, by replacing the biotinylated HS with an irrelevant biotinylated protein.

4. Biosensors can be regenerated by 3 cycles of 120 s NaCl 5 *M* (column 7)/10 s KB (column 8). Do not use biosensors that have been regenerated more than four times.

2.4 Binding to Chemokines Displayed on Endothelial Cells

To investigate whether anti-chemokine antibodies interact with GAG-bound chemokine in a physiologically more relevant context, human umbilical vein endothelial cells (HUVEC) were used as an example of endothelial cells naturally expressing a variety of GAGs on their surface. Cells are cultured in a transparent bottom well plate and the signal observed following binding of the antibody to chemokine-coated cell surfaces is read using the fluorometric microvolume assay technology.

Required materials

HUVEC can be obtained from various sources. We used those from Lonza (cat. no. C2519A).

EGM-2 BulletKit (Lonza, cat. no. CC-3162)

Cell dissociation solution 1 × (Sigma-Aldrich, cat. no. C5914-100ML)

HEPES buffer (Sigma-Aldrich, cat. no. 83264)

Sterile black microplates with transparent flat bottom (Corning, ref: 3603)

AF647 $F(ab')_2$ fragment goat anti-human IgG, Fcγ fragment specific (Jackson ImmunoResearch, ref: 109-606-170)

Specific monoclonal antibody targeting the chemokine of interest (mAb1)

Chemokine of interest

Cell culture

1. HUVEC should be cultured in EGM-2 medium supplemented with growth factors according to manufacturer's instructions. Split the cells once they are at 80–90% confluence (usually every 5–8 days) and seed them at 10,000 cells/cm^2.
2. To prepare the cells for the experiment, rinse a T75 flask (80–90% confluence) with 5 mL HEPES buffer. Remove the HEPES buffer.
3. Add 5 mL of the cell dissociation solution to the flask and incubate for 4–7 min, until all the cells are rounded. Gently tap the flask to detach the cells from the surface.
4. Transfer the cells to a 15 mL sterile tube. Wash the flask with 5 mL HEPES buffer and add this wash to the 15 mL tube.
5. Centrifuge at 220 × *g* for 5 min. Discard the supernatant. Resuspend the pellet in 5 mL culture medium and count the cells.

6. Dilute the cells to 0.2×10^6 cells/mL in the EGM-2 medium. Add 100 μL/well (20,000 cells/well) of this suspension to a sterile transparent bottom 96-well plate. Incubate at 37 °C, 5% CO_2 for 48 h.

Binding assay

1. Prior to the assay, centrifuge each mAb that has to be tested (10 min at 16,000 × *g*) in order to pellet any aggregate/precipitate. Immediately after the centrifugation, dilute mAb1 to 7.5 n*M* in PBS. Dilute the chemokine to 100 n*M* in PBS. Dilute the AF647 anti-hIgG 1:2000 in PBS.
2. Remove the culture medium from the 96-well plate. Wash the cells twice with 200 μL/well PBS.
3. Add 50 μL/well of the chemokine solution and incubate without agitation for 1 h at 4 °C.
4. Remove the chemokine solution. Wash the cells twice with 200 μL/well PBS.
5. Add 50 μL/well of mAb1 solution and 25 μL/well of AF647 anti-hIgG. The final concentration of antibody in the well is therefore 5 n*M*. Incubate without agitation for 1 h at 4 °C.
6. Read the plate immediately or up to 2 h after the end of the incubation on a 8200 cellular detection system (Applied Biosystems) or on an equivalent analyzer. Positive signals (fluorescence) indicate that mAb successfully bound to GAG displayed chemokines.

2.5 Inhibition of Chemokine Binding to GAGs by Biolayer Interferometry (BLI)

It might also be crucial to characterize whether an inhibitor prevents the interaction between GAGs and the target chemokine. Based on the protocol described in Section 2.3, we developed a BLI assay allowing to determine whether chemokine:antibody complexes interact with HS or not (Fig. 3). This format has also been successfully used with chemokine-binding proteins.

Required materials

Octet RED96 system (Pall ForteBio, cat. no. 30–5048) or an equivalent system

Dip and Read SA biosensors (Pall ForteBio, cat. no. 18–5019)

Kinetics buffer 10× (Pall ForteBio, cat. no. 18–1092)

Biotinylated heparan sulfate (see Section 2.3)

Specific monoclonal antibody targeting the chemokine of interest (mAb1)

Chemokine of interest

NaCl 5 *M* solution

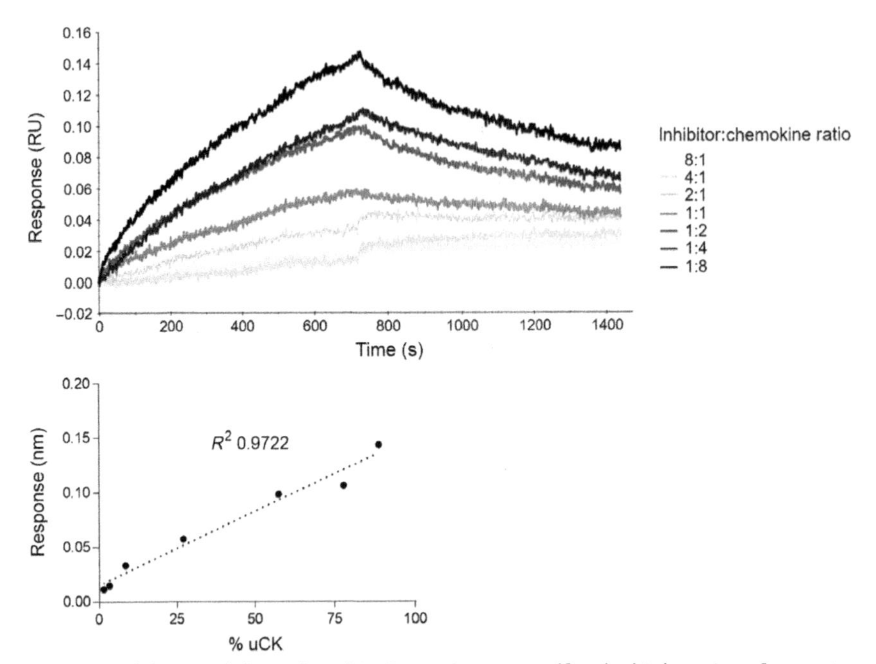

Figure 3 Inhibition of chemokine binding to heparan sulfate by biolayer interferometry. Upper panel: example of the signals obtained with various inhibitor:chemokine ratios. Lower panel: linear regression analysis of the response obtained by biolayer interferometry as a function of the percentage of unbound chemokine, indicating that the antibody prevents the interaction of the chemokine with GAGs.

BLI experiment

1. Prepare Kinetics buffer 1 × (KB) by diluting the 10 × stock solution 1:10 in PBS.
2. Dilute biotinylated HS 1:2000 in KB. Dilute the chemokine to be tested to 60 n*M* in KB. Dilute mAb1 to 50 μg/mL in the 60 n*M* chemokine solution (tube 1, final volume 700 μL). Add 300 μL of 60 n*M* chemokine solution in seven test tubes labeled 2–7. Mix the solution in tube 1 and serially transfer 300 μL (dilution 1:2) from tube 1 to tube 7 to obtain a dose–response of antibody in presence of 30 n*M* chemokine. Incubate these solutions for 30 min at 30 °C.
3. In the meantime, prepare a dose–response of mAb equivalent to that prepare in the step 2 but in KB. Dilute the chemokine to 30 n*M* in KB (final volume 1700 μL).
4. Transfer 200 μL/well of the adequate solution to a black flat bottom 96-well microplate to have the following solution in each well:

- Columns 1, 3, 5, 7, 10: KB
- Column 2: biotinylated HS 1:2000
- Column 4: dose–response of mAb in KB
- Column 6: dose–response of mAb in presence of 30 n*M* chemokine
- Column 8: chemokine 30 n*M*
- Column 9: NaCl 5 *M*

The row H is used as a negative control: load an irrelevant biotinylated protein instead of HS on the biosensor (column 2) and use the highest concentration of the dose–response in columns 4 and 6 (50 μg/mL).

5. Configure the following experiment on the Octet RED96 (kinetics experiment), using SA biosensors:
   - Baseline 120 s in column 1
   - Loading 180 s in column 2
   - Baseline 120 s in column 3
   - Association 720 s in column 4
   - Dissociation 720 s in column 3
   - Regeneration: 3 cycles of 10 s in NaCl 5 *M* (column 9) and KB (column 10)
   - Baseline 120 s in column 5
   - Association 720 s in column 6
   - Dissociation 720 s in column 5
   - Regeneration: 3 cycles of 10 s in NaCl 5 *M* (column 9) and KB (column 10)
   - Baseline 120 s in column 7
   - Association 720 s in column 8
   - Dissociation 720 s in column 7
   - Regeneration: 3 cycles of 10 s in NaCl 5 *M* (column 9) and KB (column 10)

   During the experiment, maintain a constant shaking (1000 rpm) and temperature (30 °C).
6. Columns 4 and 8 are, respectively, negative and positive controls. To validate the experiment, no signal should be observed in column 4 (i.e., the antibody should not bind to HS on its own) and equivalent signals should be obtained for all biosensors in column 8 (i.e., binding of the chemokine to HS).

Data analysis

If the dissociation constant K_D characterizing the chemokine:mAb complex had been determined, it is possible to calculate the percentage of free chemokine in solution in presence of increasing concentrations of mAb. If the antibody inhibits the binding of the chemokine to HS, the maximal signal observed by BLI will be proportional to the percentage of uncomplexed chemokine.

1. Substract the data from the row H (irrelevant loading) from data obtained in rows A–G.
2. From these data, determine the signal maximal (R_{max}) obtained for each antibody concentration (column 6).
3. At the equilibrium, we have the following situation:

$$\frac{[\mathrm{Ab}]_\mathrm{f}[\mathrm{CK}]_\mathrm{f}}{[\mathrm{AbCK}]}=K_\mathrm{D}\quad [\mathrm{Ab}]_\mathrm{f}+[\mathrm{AbCK}]=[\mathrm{Ab}]_\mathrm{i}\quad [\mathrm{CK}]_\mathrm{f}+[\mathrm{AbCK}]=[\mathrm{CK}]_\mathrm{i}$$

where [Ab] is the concentration of the antibody, [CK] the concentration of the chemokine, [AbCK] the concentration of antibody: chemokine complexes, $[xx]_i$ the initial concentration of the protein (i.e., when the solution is prepared), and $[xx]_f$ the final concentration of the protein (i.e., at the equilibrium).

Therefore:

$$\frac{([\mathrm{Ab}]_\mathrm{i}-[\mathrm{AbCK}])([\mathrm{CK}]_\mathrm{i}-[\mathrm{AbCK}])}{[\mathrm{AbCK}]}=K_\mathrm{D}$$

Knowing K_D, $[CK]_i$, and $[Ab]_i$, we can thus calculate the concentration of each species at the equilibrium. The percentage of unbound chemokine at the equilibrium is defined by:

$$\%\mathrm{uCK}=\frac{[\mathrm{CK}]_\mathrm{f}}{[\mathrm{CK}]_\mathrm{i}}$$

4. Using GraphPad Prism or an equivalent software, plot R_{max} in function of %uCK. Perform a linear regression of the data to determine whether the signal observed is proportional to the concentration of unbound chemokine. We usually consider that a $R^2 \geq 0.9$ indicates that the antibody inhibits the binding of the chemokine to GAGs.

Remark: depending on the K_D, the concentrations of antibody used in the assay may require adjustments. The above experiment is ideal for K_D values of 1–10 nM but not for higher affinities which would require lower excess of antibodies.

3. SUMMARY

Monoclonal antibodies recognize a defined epitope and can either bind without neutralizing or neutralize the biological activity of the target. Chemokines have two distinct binding interactions, with specific G protein-coupled receptor(s) or with GAGs present on cell surfaces. The epitopes for these binding partners, receptor, or cell surface proteoglycans may be distinct or overlapping and mAbs can prevent one or the other or both. The methods described here allow the establishment of whether a mAb or chemokine-binding protein recognizes only the free chemokine or also the bound chemokine displayed on endothelial surfaces which is presumed to be the biologically active form. We also report assays allowing to determine whether antibodies or chemokine-binding proteins can prevent the interaction between the chemokine and GAGs, therefore interfering with the capture of the chemokine on the endothelium.

REFERENCES

Burns, J. M., Gallo, R. C., DeVico, A. L., & Lewis, G. K. (1998). A new monoclonal antibody, mAb 4A12, identifies a role for the glycosaminoglycan (GAG) binding domain of RANTES in the antiviral effect against HIV-1 and intracellular Ca2+ signaling. *The Journal of Experimental Medicine*, *188*, 1917–1927.

Deuel, T. F., Keim, P. S., Farmer, M., & Heinrikson, R. L. (1977). Amino acid sequence of human platelet factor 4. *Proceedings of the National Academy of Sciences of the United States of America*, *74*, 2256–2258.

Hamel, D. J., Sielaff, I., Proudfoot, A. E., & Handel, T. M. (2009). Chapter 4. Interactions of chemokines with glycosaminoglycans. *Methods in Enzymology*, *461*, 71–102.

Magistrelli, G., Gueneau, F., Muslmani, M., Ravn, U., Kosco-Vilbois, M., & Fischer, N. (2005). Chemokines derived from soluble fusion proteins expressed in *Escherichia coli* are biologically active. *Biochemical and Biophysical Research Communications*, *334*, 370–375.

Proudfoot, A. E., & Borlat, F. (2000). Purification of recombinant chemokines from E. coli. *Methods in Molecular Biology*, *138*, 75–87.

Proudfoot, A. E., Handel, T. M., Johnson, Z., Lau, E. K., LiWang, P., Clark-Lewis, I., et al. (2003). Glycosaminoglycan binding and oligomerization are essential for the *in vivo* activity of certain chemokines. *Proceedings of the National Academy of Sciences of the United States of America*, *100*, 1885–1890.

Rot, A. (1993). Neutrophil attractant/activation protein-1 (interleukin-8) induces *in vitro* neutrophil migration by haptotactic mechanism. *European Journal of Immunology*, *23*, 303–306.

CHAPTER FIVE

Monitoring Scavenging Activity of Chemokine Receptors

Barbara Moepps*, Marcus Thelen[†,1]
*Institute of Pharmacology and Toxicology, University of Ulm Medical Center, Ulm, Germany
[†]Institute for Research in Biomedicine, Università della Svizzera italiana, Bellinzona, Switzerland
[1]Corresponding author: e-mail address: marcus.thelen@irb.usi.ch

Contents

1. Introduction 88
2. Fluorescent Chemokines 89
3. Monitoring Scavenging with Radiolabeled Chemokines 92
4. Monitoring Scavenging with Fluorescent Chemokines 93
   4.1 Part A: Expression and Purification of Fluorescent Protein-Tagged Chemokines from Baculovirus-Infected Insect Cells 93
   4.2 Part B: Expression and Purification of Tagged Chemokines from Bacteria for Enzymatic Labeling 100
5. Monitor Chemokine Uptake by Microscopy and Flow Cytometry (FACS) 108
   5.1 Materials 109
   5.2 Monitoring Scavenging by Flow Cytometry 109
   5.3 Monitoring Scavenging by Confocal Microcopy Using Live Cultures or Fixed Samples 109
   5.4 Monitoring Scavenging in Whole Tissue 113
   5.5 Monitoring Scavenging in Migrating Cells 113
References 114

Abstract

Migration and positioning of cells is fundamental for complex functioning of multicellular organisms. During an immune response, cells are recruited from remote distances to a distinct location. Cells that are passively transported leave the circulation stimulated by locally produced signals and follow chemotactic cues to reach specific destinations. Such gradients are short (<150 μm) and require a source of production where the concentration is the highest and a sink in apposition where the attractant dissipates and the concentration is the lowest. Several straight forward methods exist to identify *in vitro* and *in vivo* cells producing chemoattractants. This can be achieved at the transcriptional level and by measuring secreted proteins. However, to demonstrate the activity of sinks *in vitro* and *in vivo* is more challenging. Cell-mediated dissipation of an attractant must be revealed by measuring its uptake and subsequent destruction. Elimination of chemoattractants such as chemokines can be monitored *in vitro* using radiolabeled ligands or more elegantly with fluorescent-labeled chemoattractants.

Methods in Enzymology, Volume 570
ISSN 0076-6879
http://dx.doi.org/10.1016/bs.mie.2015.11.003

The latter method can also be used *in vivo* and enables to monitor the process in real time using time-lapse video microscopy.

In this chapter, we describe methods to produce fluorescently labeled chemokines either as fusion proteins secreted from insect cells or as recombinant bacterial proteins that can enzymatically be labeled. We discuss methods that were successfully used to demonstrate sink activities of scavenger receptors. Moreover, fluorescent chemokines can be used to noninvasively analyze receptor expression and activity in living cells.

1. INTRODUCTION

Chemokines are important regulators of leukocyte trafficking and play a critical role during development by orchestrating migration and positioning of cells. To mediate directional cell movement, chemokines form gradients along which cells migrate. The sources of chemoattractants and juxtaposed sinks for their dissipation are equally important for gradient formation (Crick, 1970). While local retention of chemokines is accomplished by binding to cell-surface proteoglycans, elimination of chemokines is achieved by proteases and/or by receptor-mediated uptake. After interaction, a receptor and its bound ligand cointernalize and both molecules are subsequently degraded, or, as in some instances, only the chemokine is degraded in lysosomes while the receptor becomes sorted to a recycling compartment and becomes eventually reexpressed at the cell surface. In the case that receptors efficiently perform many cycles of internalization and reexpression, thereby depleting the ligand from the environment, the process is also referred to as scavenging.

Binding of chemokines to chemokine receptors induces G protein-dependent signaling cascades leading to cell migration (Thelen, 2001). The process is usually accompanied by receptor phosphorylation and the subsequent recruitment of β-arrestins, a key step for receptor internalization. During migration receptors are endocytosed and take up chemokine, which they deliver for degradation in lysosomes (Volpe et al., 2012). However, in order to be able to continue to sense the chemokine gradient and to move over time receptors must recycle. Two-dimensional migration assays of human monocytes stimulated with fluorescent-labeled CCL2 showed CCR2-mediated ligand internalization and subsequent degradation of the chemokine, whereas the receptor recycles to the cell surface (Volpe et al., 2012). Similarly, it was reported that chemokine receptors expressed on apoptotic leukocytes can sequester chemokines during resolution of inflammation (Ariel et al., 2006). However, extensive exposure of chemokines to cognate receptors induces their internalization and down-regulation *in vivo* (Sanchez-Alcaniz et al., 2011) and *in vitro* (Thelen, 2001).

Recently, a family of atypical chemokine receptors (ACKRs) has been defined comprising ACKR1 (DARC), ACKR2 (D6), ACKR3 (CXCR7), and ACKR4 (CCRL1) (Bachelerie, Graham, et al., 2014). The receptors are phylogenetically related and share the overall heptahelical structure of chemokine receptors, but do not couple to G proteins and hence, profoundly differ in their signaling capacity. Atypical receptors can recruit β-arrestins and rapidly cycle between the cell surface and intracellular compartments, but do not stimulate cell migration (Bachelerie, Ben-Baruch, et al., 2014). Cycling of the receptors was shown to be ligand dependent as well as ligand independent (Bonecchi et al., 2008; Haraldsen & Rot, 2006; Leick et al., 2010; Naumann et al., 2010). The main function of the atypical receptors lays in removing chemokines, thereby establishing local gradients of homeostatic chemokines, as in the case of ACKR3 and ACKR4 (Boldajipour et al., 2008; Dambly-Chaudiere, Cubedo, & Ghysen, 2007; Dona et al., 2013; Haraldsen & Rot, 2006; Ulvmar et al., 2014; Venkiteswaran et al., 2013). ACKR2 actively scavenges most inflammatory CC chemokines, hence dampening immune responses and contributing to the resolution of inflammation (de la Torre et al., 2005; Jamieson et al., 2005). By contrast, ACKR1 actively transcytoses chemokines taking them up at the basal side and presenting them on the luminal surface of endothelial cells. In this way, ACKR1 can promote leukocyte recruitment (Pruenster et al., 2009).

The scavenging activity of typical and ACKRs can be measured quantitatively through the degradation of radiolabeled chemokines (Borroni et al., 2009; Pruenster et al., 2009) or in a more comparative, but real-time fashion through uptake of fluorescent chemokines (Brennecke et al., 2013; Luker, Steele, Mihalko, Ray, & Luker, 2010; Volpe et al., 2012). Measuring degradation or uptake of receptor-specific chemokines represents also a powerful alternative to reveal surface expression of chemokine receptors in the absence of specific antibodies. Moreover, the method provides information on the functional state of a surface-expressed receptor.

2. FLUORESCENT CHEMOKINES

In the past years, the use of fluorescent chemokines more and more evolved to characterize their interactions with chemokine receptors, and their internalization and intracellular trafficking (Boldajipour et al., 2008; Ford et al., 2014; Hatse et al., 2004; Martinez-Martin et al., 2015). Additional applications for fluorescent chemokines include FACS analysis of chemokine receptor expression and studies of chemokine presentation on the cell surface by glycosaminoglycans or extracellular matrix components (Kawamura et al., 2014; Martinez-Martin et al., 2015). Recombinant fluorescent chemokines were used to determine leukocyte subsets with distinctive receptor profiles and scavenging capacities and for evaluating receptor antagonist activity (Ford et al., 2014). Exogenous application of fluorescent chemokines led also to the observation that scavenging of chemokines is not restricted to atypical receptors, but also includes typical receptors, e.g., CCR2 (Ford et al., 2014; Volpe et al., 2012). Fluorescent chemokines are particularly useful to determine the scavenging activity of ACKRs *in vivo*, *ex vivo*, and *in vitro*. When compared to immunostaining of surface receptors, fluorescent chemokines enable to monitor both receptor expression and receptor activity.

Random labeling of chemokines, e.g., with NHS-esters, a method which is often used to conjugate antibodies, is not recommended to generate fluorescent chemokines because of potential loss-of-function and the uncontrolled mixtures that are obtained. Moreover, the N-terminus of chemokines is critical for binding and activation of cognate receptors. Therefore, modifications of chemokines are usually introduced at the C-terminus (Bachelerie, Ben-Baruch, et al., 2014; Clark-Lewis et al., 1995; Weber, Uguccioni, Baggiolini, Clark-Lewis, & Dahinden, 1996). An exception makes the fully synthetic CXCL4 variant CXCL4L1, a ligand for CXCR3a and CXCR3b, which was labeled with a fluorescent dye (TAMRA) at the N-terminus (Struyf et al., 2011). Various methods are available for site-directed labeling of chemokines. Chemokines can be prepared through full chemical synthesis (Clark-Lewis, 2000) or semi-synthesis (Beck-Sickinger & Panitz, 2014; Dawson, Muir, Clark-Lewis, & Kent, 1994), which allows site-specific introduction of substituted amino acids. Synthetic chemokines typically carry a modified lysine near the C-terminus that is either directly conjugated with a fluorescent dye or biotinylated to be subsequently labeled with streptavidin-coupled dyes (chemokines are commercially available labeled with Alexa647 or biotin from ALMCAC, Craigavon, UK). Synthetic chemokines maintain their biological properties such as receptor affinity and selectivity and were successfully used to reveal receptor binding in completion assays and to measure chemokine uptake (Ford et al., 2014; Ford, Hansell, & Nibbs, 2013; Hansell et al., 2011; Hatse et al., 2004; Postea et al., 2012; Strong, Thierry, Cousin, Moulon, & Demotz, 2006; Watts et al., 2013). Similarly, biotinylated chemokines can be used to measure receptor surface expression and affinity, by allowing the chemokine to bind and in a second step detecting bound chemokine with streptavidin (Balabanian et al., 2005). Commercially available quantum dots conjugated with streptavidin and coupled to biotinylated chemokines were employed to measure scavenging activity of ACKR3 expressed on B cells (Humpert et al., 2012) and the modulation of the functional state of CXCR4 by herpes simplex virus glycoproteins (Martinez-Martin et al., 2015). Next to fully synthetic also recombinant approaches were taken to generate fluorescent-labeled chemokines. Chemokines can be expressed in bacteria, insect, or mammalian cells (Imai, 2000; Proudfoot & Borlat, 2000). While eukaryotic cells mostly yield soluble secreted proteins with correctly processed N-termini, expression in bacteria usually leads to amorphous precipitates in inclusion bodies from which chemokines must be extracted, folded, and their N-termini processed. Both approaches have their advantages and will be described in detail below. Refolding of fluorescent protein fusion constructs from solubilized inclusion bodies is complicated, if not impossible. However, fluorescent proteins with broad spectral variability can be fused to the C-termini of chemokines and readily expressed in eukaryotic cells. Typically, secreted proteins are isolated from cell culture supernatants to acceptable yields, however, with considerably higher expenses as produced in bacterial systems. On the other hand, one has to keep in mind that fluorescent proteins are naturally expressed in the cytosol were reducing conditions are prevalent. Thus, in the extracellular milieu under oxidizing conditions the fusion proteins tend to aggregate and are susceptible to repeated freeze thawing cycles. Interestingly, the relative large molecular mass of fluorescent proteins (~25 kDa) compared with chemokines (8–12 kDa) usually does not create problems with respect to their functionality. Such fusion proteins were successfully used to show chemokine uptake during cell migration (Volpe et al., 2012), in cell cultures (Boldajipour et al., 2011, 2008; Brennecke et al., 2013; Mahabaleshwar, Tarbashevich, Nowak, Brand, & Raz, 2012), and in tissue sections (Naumann et al., 2010). Chemokines fused to the bioluminescent protein Gaussia luciferase was used in whole animal settings to measure the scavenging activity of ACKR3 (Luker et al., 2012).

Already from the beginning of their discovery, recombinant chemokines were expressed in bacteria and shown to possess equivalent activity as their naturally expressed counterparts (Lindley et al., 1988). Several general protocols for purification are available (Edgerton, Gerlach, Boesen, & Allet,

2000; Proudfoot & Borlat, 2000; Veldkamp et al., 2007; Yang et al., 1999). By contrast to chemokines expressed in eukaryotic cells recombinant bacterial chemokines must be refolded and their N-termini processed to be functional. Also initial yields of chemokines in bacterial inclusion bodies are high their refolding efficiency (see below) varies considerably. However, expression of chemokines in bacteria is usually more economic then expression in eukaryotic cells. Bacterial expressed chemokines can be used for site-directed labeling by fusing a short tag at the C-terminus comprising a recognition sequence for enzymatic labeling (Allen, Hamel, & Handel, 2011; Kawamura et al., 2014; Yin, Liu, Li, & Walsh, 2004; Yin et al., 2005; Zhou et al., 2007). Special care must be taken to correctly process the N-termini of nonsecreted proteins and to remove any lipopolysaccharide (LPS) from the preparations. Nevertheless, bacterial expressed chemokines have been successfully purified to high yields and enzymatically labeled with phosphopantetheinyl transferases which use fluorescent-labeled coenzyme A (CoA) as substrate (Kawamura et al., 2014) or were expressed with an avitag and enzymatically biotinylated with BirA (Allen et al., 2011). Most reagents are commercially available or can easily be produced in a generic laboratory (George, Pick, Vogel, Johnsson, & Johnsson, 2004; Yin, Lin, Golan, & Walsh, 2006). *In vitro*-labeled recombinant bacterial fluorescent chemokines have been used to reveal scavenging activity of ACKR3 in brain slices (Abe et al., 2014) and cell cultures (Kawamura et al., 2014).

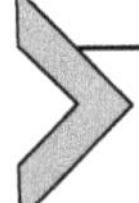

3. MONITORING SCAVENGING WITH RADIOLABELED CHEMOKINES

Iodinated chemokines were initially the standard to measure chemokine receptor expression, binding affinity, and competition. The robust method is also often used to determine the potency and/or efficiency of receptor antagonists. Radiolabeled chemokines are commercially available and tested for functionality. Scavenging activity can be monitored through the degradation of [^{125}I]-labeled chemokines. In essence, chemokines precipitate in the presence of 10% trichloroacetic acid, but peptides generated by lysosomal degradation are resistant and are quantitatively measured in supernatants (Borroni et al., 2009).

The assay is conveniently performed in Transwell® insets (pore size 8 μm) on which endothelial cells or cells of interest expressing scavenging chemokine receptors are grown to confluence. To start the experiment, the culture medium is replaced with Hank's balanced salt solution containing 10 m*M* HEPES and 1% BSA. [^{125}I]-labeled chemokines (0.025 pmol; specific activity 2000 Ci/mmol) are diluted with the appropriate amount of unlabeled chemokine and added on top of the cells (upper compartment) for 3–12 h at 37 °C. To assess degradation or transcytosis of chemokines, the medium on top of the cells and in the bottom chamber is collected separately, cell surface-bound chemokines are removed by the addition of 10 × PBS for 3 min and added to the respective supernatants. To assess intercellular radioactivity, the cells are lysed with 0.4% Triton X-100 in PBS. Trichloroacetic acid (10% final) is added to precipitate intact chemokine from the collect top, bottom, and intracellular medium. Radiolabeled peptides from degraded chemokines will remain in solution. The trichloroacetic acid pellets are dissolved in 2 *M* NaOH and 0.05% (wt/vol) sodium dodecylsulfate (SDS). For quantification, all fractions are measured in a gamma counter. In case of chemokine receptors, which do not recycle the assay becomes saturated with increasing concentrations of chemokine, whereas in case of scavenging receptors that rapidly recycle degradation augments with presented amounts of chemokine (Borroni et al., 2009; Naumann et al., 2010; Pruenster et al., 2009).

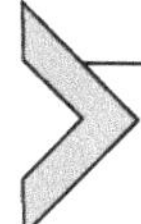

4. MONITORING SCAVENGING WITH FLUORESCENT CHEMOKINES

We first describe two methods to generate fluorescent-labeled chemokines. Part A describes the expression and purification of chemokines fused to fluorescent proteins and expressed in insect cells. Part B delineates the expression of chemokines in bacteria, folding, and purification, followed by enzymatic labeling with phosphopantetheinyl transferases. In the second section, we provide protocols to monitor chemokine uptake by microscopy and flow cytometry (FACS).

4.1 Part A: Expression and Purification of Fluorescent Protein-Tagged Chemokines from Baculovirus-Infected Insect Cells

The use of recombinant baculoviruses provides a rapid and efficient method to express and purify recombinant proteins, including chemokines. Expression in baculovirus-infected insect cells has been successfully applied to produce recombinant active human CC and CXC chemokines from diverse organisms such as *Xenopus*, mouse, and man (Braun et al., 2002;

Heinrich, Ryseck, Macdonald-Bravo, & Bravo, 1993; Imai, 2000; Ishii et al., 1995; Jones, Mulligan, Flory, Ward, & Warren, 1992; Kitaura et al., 1996; Sarris et al., 1993; Ueda et al., 1994; Uguccioni et al., 1996). Additionally, viral encoding chemokines have been expressed in this system (Heo et al., 2015; Paslin, Reykjalin, Tsadik, Schour, & Lucas, 2015). The production in baculovirus-infected insect cells provides the advantage of correct processing, folding, and secretion of the chemokines in the culture supernatants and for ease of purification infected cells are grown without fetal calf serum (FCS) (Braun et al., 2002). The chemokines can easily be purified in their active form from the supernatants applying cation exchange chromatography and size-exclusion chromatography or affinity chromatography using a heparin sulfate matrix (Braun et al., 2002; Volpe et al., 2012). Expression in baculovirus-infected insect cells is a versatile method to produce agonists, antagonists, or chimeric CC and CXC chemokine fusion proteins carrying various fluorescent proteins (Beall, Mahajan, Kuhn, & Kolattukudy, 1996; Brennecke et al., 2013; Chakravarty, Rogers, Quach, Breckenridge, & Kolattukudy, 1998; Heinrich et al., 1993).

4.1.1 Materials

Linearized baculovirus DNAs ProGreen™ or ProEasy™ and transfection buffer Profectin (all AB Vector, San Diego, CA).

4.1.1.1 Insect Culture

Spodoptera frugiperda (*Sf*) 9 cells (Invitrogen, Thermo Fisher Scientific)
Trichoplusia ni High Five™ (H5) (5B1-4, Invitrogen, Thermo Fisher Scientific)
BioWhittaker Insect-Xpress (Lonza)
Gentamicin (GIBCO, Life Technologies, Thermo Fisher Scientific)
Fungizone® Amphotericin B (GIBCO, Life Technologies, Thermo Fisher Scientific)
Pluronic® F-68 (GIBCO, Life Technologies, Thermo Fisher Scientific)
Fetal calf serum (FCS, Sigma-Aldrich)
12.5 and 75 cm^2 cell culture flask (JetBIOFIL® and TPP Techno Plastic Products AG)
Fernbach culture flasks (Schott®, Sigma-Aldrich)

4.1.1.2 Purification

S-Sepharose (GE Healthcare)
Superdex 200 PC 30/3.2 column (GE Healthcare)

HiTrapHeparin 5 mL columns (GE Healthcare)
Automated chromatography system
Certomat BS-1® (B. Braun Melsungen)

4.1.2 Methods

4.1.2.1 Cloning of Fluorescent Chemokine-Encoding DNAs for the Generation of Recombinant Baculoviruses

DNAs encoding CC chemokines or CXC chemokines are amplified from cDNA derived from primary human leukocytes, leukocyte cell lines, or mouse tissue (e.g., liver, spleen). PCR fragments are cloned into the expression vector pcDNA3.1(+) carrying in-frame the cDNA of the fluorescent protein of choice with a short linker sequence at the N-terminus (coding for the amino acids: SGGGGSGGGGSGGGGS). The cDNA encoding for the entire chemokine-fluorescent fusion proteins is excised and ligated into baculovirus transfer vectors pVL1392 or pVL1393 (Volpe et al., 2012).

4.1.2.2 Cell Culture

Spodoptera frugiperda 9 (*Sf*9) cells and *Trichoplusia ni* 5B1–4 cells (H5) are grown at 27 °C in 75 cm^2 cell culture flask in Insect-Xpress supplemented either with heat inactivated 10% FCS and 50 μg/mL gentamicin (*Sf*9) or with 50 μg/mL gentamicin alone (H5).

Cells are passaged after reaching 85–90% confluence (about 3×10^7 cells) every 3 days by diluting the cells 1:3 (*Sf*9) or 1:4 (H5) in fresh medium and transferring them to a new 75 cm cell culture flask.

4.1.3 Production of Recombinant Fluorescent Chemokine-Encoding Baculoviruses

Recombinant baculoviruses are prepared by transfecting *Sf* 9 cells with a 2:1 mixture of the transfer vector (pVL1392 or pVL1393 encoding for fluorescent fusion proteins) and a standard modified linearized Autographa californica multiple nuclear polyhedrosis baculovirus DNA (AcMNPV-baculovirus), which contains a lethal deletion and is rescued by the DNA of the transfer vector. The coexpressed green fluorescent protein (GFP) enables the identification of recombinant viruses. For expression of chemokine-GFP fusion proteins, we recommend the usage of ProEasy™.

4.1.3.1 Procedure

1. Seed *Sf*9 cells to a density of approximately 60% in a 12.5 cm^2 cell culture flask and allow attaching for at least 30 min.

2. Dilute 5 μL Profectin™ transfection reagent 1:10 in sterile deionized water prior transfection as recommended by the provider.
3. Mix the linearized baculovirus ProGreen™ DNA (0.5 μg) and the fluorescent chemokine-encoding transfer vector pVL1393 (1 μg) in a 1.5 mL cup and add drop wise to 50 μl of 10% Profectin™.
4. Incubate the mixture for 20 min at room temperature.
5. Aspirate the culture supernatant of the *Sf9* cells and add 1 mL Insect-Xpress without supplements. Then add the Profectin™/DNA mixture drop wise and incubate the cells for 24 h at 27 °C. To prevent evaporation during the incubation, the *Sf9* cell-containing culture flask is placed into a sterile plastic container containing wet filter paper to increase humidity.
6. After 24 h, add 1 mL of Insect-Xpress supplemented with 10% FCS and 50 μg/mL gentamicin and continue the incubation for 4 days. Harvest the culture supernatants, centrifuge for 5 min, 300 × *g*, at room temperature to remove cells. Collect the virus-containing supernatant (approximately 2 mL).
7. Seed *Sf9* cells at a density of approximately 60% (1.8×10^7 cells) and allow attaching to the cell culture flask (75 cm^2) for 30 min. Aspirate supernatant and add the entire freshly harvested supernatant from step 6.
8. Incubate the cells 1 h at 27 °C, then add 10 mL of Insect-Xpress supplemented with 10% FCS and 50 μg/mL gentamicin. Continue the incubation for 5–7 days (first round of amplification).
9. Harvest the cell culture medium after 5–7 days, centrifuge for 5 min, 300 × *g* at room temperature, and collect the virus-containing supernatant (approx. 12 mL).
10. Seed *Sf9* cells at a density of approximately 60% (3.6×10^7 cells) and allow attaching to the cell culture flask (150 cm^2) for 30 min. Aspirate the medium and add 5 mL of the virus-containing supernatant (step 9) together with 3 mL of Insect-Xpress to the freshly seeded *Sf9* cells. Incubate for 1 h at 27 °C. Then add 22 mL of Insect-Xpress supplemented with 10% FCS and 50 μg/mL gentamicin and continue the incubation for 5–7 days (second round of amplification).
11. High-titer stocks of the baculoviruses are obtained by at least three cycles of amplification in *Sf9* cells.
12. Virus-containing supernatants are ready for protein production in infected H5 insect cells and are stored at 4 °C.

Note

1. Viral titer determination and plaque purification of the recombinant viruses are usually not necessary. Successful infection and production can be monitored by fluorescence microscopy.
2. The often described linearized BaculoGold™ and BaculoGOLD Bright™ baculovirus DNA is no longer marketed by BD Biosciences.
3. Alternative linearized AcMNPV-baculovirus DNA for the generation of recombinant baculoviruses using the transfer vector set pVL1392/1393 are the BacPAK6™ (Takara Clontech) and the Bac-N-Blue™ DNA (Invitrogen, Thermo Fisher Scientific).
4. Alternative methods have been developed to obtain virus stocks without plaque purification for expression of recombinant proteins in infected insect cells. The Bac-to-Bac™ technology (Invitrogen, Thermo Fisher Scientific) avoids homologous recombination in insect cells by using site-specific transposition in *E. coli*. With this technology, recombinant bacmid-DNA is generated that is used to transfect insect cells.

4.1.4 Production of Recombinant Chemokines Tagged with Fluorescent Proteins

For production of recombinant fluorescent chemokine fusion proteins Trichoplusia ni 5B1-4 cells (H5; high five cells) are grown at 27 °C in suspension cultures in Insect-Xpress medium in 1800 mL Fernbach culture flasks. Importantly, the cells grow without FCS facilitating purification of the chemokine from the culture supernatant.

4.1.4.1 Procedure

1. Detach H5 insect cells grown to confluence in six (75 cm^2) or three (150 cm^2) culture flasks (approximately 1.8×10^8 cells) and seed into a Fernbach culture flask in 150 mL H5 Insect-Xpress medium supplemented with 0.2% (v/v) Pluronic F-68, 50 μg/mL gentamicin, and 2.5 μg/mL amphotericin B.
2. H5 cells are kept as suspension culture by incubation at 27 °C on a rotary shaker (shaking amplitude of 25 mm, 80 rpm, Certomat® BS-1) for 2 days.
3. Count the H5 and harvest $\sim 4 \times 10^8$ cells by centrifugation. Resuspend the cell pellet in 50 mL fresh Insect-Xpress medium and transfer back to the Fernbach culture flask.

4. Infect cells with virus-containing cell supernatant (4–8 mL) of the last round of amplification (see previous section) and incubate for 1 h in the Fernbach culture flask on the rotary shaker (80 rpm).
5. After 1 h, add 150 mL Insect-Xpress medium and adjust to (final concentrations) 0.2% (v/v) Pluronic® F-68, 50 μg/mL gentamicin, and 2.5 μg/mL amphotericin B. For chemokine production, incubate the cells in the Fernbach culture flask on the rotary shaker (80 rpm) for 48 h (Fig. 1A).
6. Collect the culture supernatant containing the fluorescent chemokine fusion protein by centrifugation (3000 × *g*, 30 min, 4 °C). Cell-free supernatants can be immediately processed or stored for further use at −80 °C.

Note

1. Nonsecreted chemokines can be easily identified, when fluorescent proteins are found in the pellets of the infected cells, but not in the supernatant.
2. To our experience out of nine human, three murine, and one *Xenopus laevis* CC and CXC chemokines, only one failed to be secreted into the culture supernatant of baculovirus-infected insect cells.

Figure 1 Production and purification of fluorescent chemokines from baculovirus-infected insect cells. (A) Insect cells are grown in Fernbach culture flasks. (B) The culture medium of baculovirus-infected insect cells is applied to affinity chromatography using a HiTrap™ Heparin HP column. (C) Fractions containing the fluorescent chemokine are collected. As shown, all production and purification steps can be visually monitored. (See the color plate.)

4.1.5 Purification of Recombinant Fluorescent Chemokine Fusion Proteins

Recombinant fluorescent chemokine fusion proteins are purified from insect cell culture supernatants by sequential chromatography using (A) ion exchange chromatography and size-exclusion chromatography or (B) affinity chromatography.

4.1.5.1 Procedure

A. Ion-exchange chromatography and size-exclusion chromatography

1. Adjust culture supernatants to pH 7.4 with 0.5 *M* NaOH under stirring.
2. Wash bulk S-Sepharose three times with 20 m*M* NaPi (pH 7.4) and resuspend in the same buffer as a 50% slurry. Add 100 μL of S-Sepharose slurry/100 mL culture medium and incubate the suspension 3 h at 4 °C under continuous agitation (do not use a magnetic stirrer).
3. Collect the S-Sepharose by centrifugation in 50 mL Falcon tubes (5 min, 1000 × *g* at 4 °C). Keep the supernatants to test for residual chemokine.
4. Combine the S-Sepharose pellets in 40 mL of20 m*M* NaPi (pH 7.4) and centrifuge as before. Wash once with 20 mL. Resuspend the S-Sepharose in 1 mL of 80 m*M* NaCl, 20 m*M* NaPi (pH 7.4) and transfer to a 1.5 mL Eppendorf cup and spin down (800 × *g*, 30 s, 4 °C).
5. To elute the chemokines, punch a small hole in the bottom of the tube with a 27-gauge needle and carefully insert the tube in another 1.5 mL Eppendorf cup. Add 200 μL (or 50% of the S-Sepharose bed-volume, depending on the volume) 2 *M* NaCl, 20 m*M* NaPi (pH 7.4) and centrifuge 600 × *g* for 30 s at 4 °C. Readd the eluate from the bottom tube to the S-Sepharose and spin again. Repeat the step twice. For the final spin, use 2000 × *g* centrifugal force to spin out all liquid. The beads should turn almost colorless.
6. Eluates are immediately subjected to size-exclusion chromatography on a Superdex 200 PC 30/3.2 column equilibrated in 80 m*M* NaCl, 20 m*M* NaPi, pH 7.4. Do not leave the fusion proteins for prolonged times in the high salt buffer.
7. Peak fractions containing highly concentrated pure (90%) chemokine (see below) are pooled and stored at −80 °C in small aliquots.

B. Affinity chromatography

1. Fluorescent chemokine fusion proteins are isolated using an automated chromatography system (ÄKTA FPLC or explorer or similar).
2. Load culture supernatants (50–150 mL) on a 5 mL HiTrap™ Heparin HP column equilibrated with buffer A (10 m*M* NaPi, pH 7.3) at a flow rate of 3 mL/min (Fig. 1B).
3. Wash the column with 25 mL buffer A.
4. Elute bound chemokines with 50 mL of a linear gradient of 1 *M* NaCl in buffer A and collect fractions of 2 mL (Fig. 1C).
5. Depending on the chemokine, material of interest elutes between 0.5 and 0.9 *M* NaCl in buffer A.
6. Analyze proteins of the different fractions by SDS-PAGE.

Note:

The purity and concentration of the fluorescent chemokine fusion proteins are calculated as follows:

1. Measure the absorbance at 280 nm and at the fluorescent protein's specific wavelength in the visible light (wavelength can be looked up on the Web, e.g., at http://nic.ucsf.edu/dokuwiki/doku.php?id=fluorescent_proteins).
2. Calculate the protein concentration at each wavelength using the specific extinction coefficients (for the fluorescent protein usually on the same Web page) and for 280 nm *ad hoc* calculated coefficients can be obtained at http://web.expasy.org/protparam/. The ratio conc. fluorescent protein at wavelength/conc. protein 280 nm should be near 1 for the highest purity.
3. Proteins of the main fractions (purity >90%) are additionally analyzed by mass spectrometry.

4.2 Part B: Expression and Purification of Tagged Chemokines from Bacteria for Enzymatic Labeling

The self-induction system in *E. coli*, using the *modified* pET-17B plasmid from which protein expression is induced after a metabolic switch, does not require induction of expression with IPTG and yields several mg of protein from less than a liter of bacterial culture. To facilitate protein isolation, the plasmid encodes for a cleavable histidine tag preceded by an expression-enhancing GroE sequence (Yang et al., 1999), which is introduced after the ATG at the 5′-end. The sequence continues by two codons for the amino acids (LE) arising from a convenient restriction site *Xho*I and by the recognition sequence for enterokinase. The latter allows the silent cleavage of the tag just before the correct N-terminus of any mature chemokine. At the 3′-end, the coding region (without stop codon) is fused in-frame to a *Bsi*WI restriction site followed by the most hydrophilic consensus sequence for phosphopantetheinyl transferase (Sfp) YbbR13 (Zhou et al., 2007) and a stop codon. The two chosen restriction sites, *Xho*I and *Bsi*WI, are absent in the cDNAs of almost all chemokines, thus allowing the easy cassette cloning of PCR fragments encoding for chemokines into these sites.

Unfolded chemokines are isolated from purified inclusion bodies by immobilized-metal ion affinity chromatography (IMAC) in denaturing buffers. After removal of guanidine, the unfolded crude product is precipitated and the pellet is dissolved in a small volume of guanidine in the presence of a strong reductant to cleave any disulfide bridges. The cleared solution is passed over a desalting column equilibrated with a weak buffer at pH 3. Because chemokines possess a basic isoelectric point, the low pH prevents precipitation during guanidine removal. Refolding is obtained by dilution into an arginine-rich buffer with a preset redox potential given by a fixed cysteine–cystine ratio. Folded chemokines are isolated by hydrophobic interaction chromatography. The N-terminus is removed with enterokinase and the mature chemokine with a tag at the C-terminus is purified by reverse-phase HPLC.

4.2.1 Generation of Modified pET-17B

Modified pET-17B is derived from pET-17B (Novagen) as follows. First, the *Xho*I restriction site in pET-17B is removed by site-directed mutagenesis to generate pET-17Bx. Next two primers are designed: (1) a long sense primer starting with an *Nde*I restriction site (including the ATG start codon), then encoding for GroE followed by an eight-His sequence, a LE spacer (*Xho*I restriction), and an enterokinase recognition site (amino acids NH_2-MKDVKHHHHHHHHLE*DDDDK*) followed by 25 bp of any specific chemokine sequence XXXXCAT**ATG**GCTGCTAAAGATGTTAAACATCATCATCATCATCATCATCATCTCGAGGATGATGATGATAAAX_{25}; (2) an antisense primer beginning with an *Eco*R5 restriction site followed by a stop codon and the YbbR13 sequence (Zhou et al., 2007) and an in-frame *Bsi*WI site, and continued with the coding region of any chemokine: RT*DSLEFIASKLA* (amino acids) 5′-XXXGATATC*TTA*TGCAAGTTTACTTGCAATAAACTCAAGACTATCCGTACGX_{25}-3′. Use the primers to amplify a chemokine from appropriate cDNAs (see below). Restrict the plasmid pET-17Bx with *Nde*I (5′) and *Eco*R5 (3′) and ligate the purified *Nde*I/*Eco*R5 restricted PCR fragment. Once the *modified* pET-17B is obtained, all chemokines can easily be cloned in after restriction with *Xho*I and *Bsi*WI.

4.2.1.1 Bacterial Culture

Modified LB (Luria-Bertani) medium for auto-induction: To 1 L LB medium buffered with 50 m*M* NaPi (pH 7.4), add 10.4 mL of 60% glycerol (0.6% final), 5.5 mL of 10% glucose (0.05% final), 26 mL of 8% lactose (0.2% final), and ampicillin for selection (Li, Kessler, van den Heuvel, & Rinas, 2011).

Transform chemical competent *E. coli* (strain BL21(DE3)) with heat shock and plate on LB agar with ampicillin. Pick clones and check for proper insert by PCR. Preculture a single colony in 2 mL modified LB medium for 24 h at 37 °C and control protein expression by SDS-PAGE. Positive clones are expanded in 0.5–1 L modified LB for 24 h yielding ~4–9 g wet bacteria.

4.2.1.2 Purification

Lysozyme (Sigma-Aldrich)
DNAse I (Sigma-Aldrich)
Beckman optima max centrifuge
Ni-agarose for IMAC (Jena Bioscience)
Dialysis tubing (6 kDa cut-off) (Serva, Membra-Cel, MWCO 3500)
Fast desalting column (5 mL, GE Healthcare)
Bulk PorosR2 material (Applied Biosystems, Thermo Fisher Scientific)
Enterokinase (New England BioLabs)
C_{18} reverse-phase HPLC column 25 × 1 cm (length, diameter)

4.2.1.3 Labeling

1 *M* HEPES solution, pH 7.4 (GIBCO, Life Technologies, Thermo Fisher Scientific)
Phosphopantetheinyl transferase (Sfp) (New England BioLabs)
Coenzyme A anhydrous sodium salt (Sigma-Aldrich)
Fluorescent dyes modified with maleimide (e.g., ATTO-labels, Atto-Tec.com; Alexa Fluor, Thermo Fischer Scientific; CG dyes, Sigma-Aldrich; Cyanine Dyes, GE Healthcare)
Alternatively ready to use CoA 488 (New England Biolabs)
C_{18} reverse-phase HPLC columns 25 cm (length)

4.2.2 *Methods*

4.2.2.1 Cloning of Tagged Chemokines-Encoding DNAs for Expression in *E. coli*

Suitable cDNAs encoding for human or mouse chemokines can be obtained by RT-PCR of mRNA from leukocytes or cell lines or from tissue specimens such as tonsil, adenoids (human) or liver, and spleen (mouse). The cDNA of a chemokine is amplified by nested PCR using first a set of primers (25 bp) specific for the leader sequence and the 3′-UTR downstream of the stop codon. For the second PCR design, a sense primer with the restriction site (*Xho*I) upstream of the enterokinase recognition site followed by 25 bp starting in-frame with the first codon of the specific sequence of the mature chemokine (without leader sequence) is as follows: XXXXCTCGAG*GATGATGATGATAAA*X_{25} (underlined *Xho*I, *italic* enterokinase recognition site, X_{25} specific for mature chemokine). The antisense primer matches the last 25–30 bp of the chemokine without stop codon followed immediately by an in-frame *Bsi*WI restriction site and four random nucleotides. The fragment is purified, restricted, and ligated into the *Xho*I/*Bsi*WI restricted *modified* pET-17B vector.

4.2.3 *Purification of Recombinant Tagged Chemokines from Bacteria*

Recombinant tagged chemokines are purified from bacteria inclusion bodies. Proteins are solubilized with guanidine and the His-tagged chemokines purified on an IMAC Ni-affinity column. After dialysis, precipitated protein (mostly chemokine) is collected and solubilized with guanidine in the presence of 60–100 m*M* β-mercaptoethanol (β-ME) at 60 °C. High-speed centrifugation of the solubilized chemokine is recommended before guanidine is removed by passing the solution over a fast desalting column equilibrated in 10 m*M* HCl, 150 m*M* NaCl. Peak fractions without guanidine are subjected to refolding. Refolded chemokines are purified by hydrophobic interaction on PorosR2 resins. After concentration *in vacuo*, the material is digested with enterokinase for 18°h at RT and finally purified by C_{18} reverse-phase HPLC. The two steps (PorosR2 and C_{18} at low pH) efficiently remove residual bacterial-derived LPS.

4.2.3.1 Procedure

1. Resuspend 4–6 g bacterial pellet in 47.5 mL lysis buffer: 50 m*M* Tris–Cl (pH 8.0), 1 m*M* $MgCl_2$, and 0.1% β-ME. Add lysozyme (25 mg) and DNAse (0.5 mg) to the suspension and incubate for 20 min at 4 °C under agitation.

2. Add 2.5 mL of 20% Triton X-100 (1% final) and 300 m*M* NaCl (1.8 g/100 mL) and continue agitation for 20 min.
3. Centrifuge lysate for 20 min at ~40,000 × *g*.
4. Resuspend pellet (inclusion bodies) with a Potter-Elvehjem homogenizer with 2% Triton X-100, 50 m*M* Tris–Cl (pH 8), 2 *M* Urea, 200 m*M* NaCl.
5. Centrifuge for 20 min at ~40,000 × *g*.
6. Freeze (if desired, break point).
7. Dissolve the pellet with a Potter-Elvehjem homogenizer in 10 mL of 50 m*M* NaPi (pH 7.4), 300 m*M* NaCl, 10 m*M* imidazole, 6 *M* guanidine/HCl (28.7 g/50 mL), 10 m*M* β-ME and incubate for 30 min at 50 °C.
8. Freeze (if desired, break point).
9. Thaw and heat (50 °C) up for 20 min.
10. Ultra-centrifuge at 390,000 × *g* for 30 min (e.g., 75,000 rpm MLA-80 rotor in a Beckman Optima Max) to remove any insoluble material.
11. Load supernatant on ~5 mL bed-volume Ni-affinity fast flow column equilibrated in buffer G: (6 *M* guanidine/HCl, 50 m*M* Tris–Cl (pH 8.0), 10 m*M* imidazole, 10 m*M* β-ME).
12. Wash column with 15 mL buffer G.
13. Wash column with 12 mL buffer G containing 30 m*M* imidazole.
14. Elute column with 10 mL buffer G containing 250 m*M* imidazole and collect 0.5 mL fractions.
15. Measure protein at 280 nm against blank containing 250 m*M* imidazole. The molar extinction coefficient of a chemokine with tags can be calculated at http://web.expasy.org/protparam/. Pool peak fractions.
16. Dialyze against PBS (important: use dialysis tube with 3500 Da cut-off).
17. Recover precipitated material by centrifugation in six 2 mL Eppendorf cups.
18. Freeze (if desired, break point).
19. Add to each tube 0.1 mL of 6 *M* guanidine/HCl, 100 m*M* NaCl, 50 m*M* Tris–Cl (pH 8), 50 m*M* β-ME, and heat at 50–60 °C for 20 min. If the pellet does not dissolve (under vigorous vortexing and/or sonication) add more buffer until all is dissolved and the solution is clear.
20. Approximately 1.2–1.8 mL total clear solution (if dark brown add more buffer) is centrifuged for 10 min at 20,000 × *g*.

21. Exchange buffer on 5 mL fast desalting column equilibrated in 10 m*M* HCl, 150 m*M* NaCl.
22. Measure concentration at 280 n*M*. Usually 7–15 mg protein.
23. Dilute to ~0.1 mg/mL in cold (4 °C) folding buffer: 80 m*M* Tris–Cl (pH 8.5), 100 m*M* NaCl, 0.8 *M* arginine, 2 m*M* EDTA, 1 m*M* cysteine, and 0.2 m*M* cystine (from 20 m*M* stock in 0.5 *M* NaOH). 50 mL falcon tubes are most suitable.
24. Gas the tubes with N_2 to remove oxygen and fold under N_2 protection for 18 h at 4 °C under rotation.
25. Pool all tubes and bring to $15\%_{sat}$ $(NH_4)_2SO_4$ (8.2 g/100 mL). Add slowly under continuous stirring to avoid precipitation.
26. Pass the solution through a 0.22 μm filter and load on 3 mL bed-volume Poros R2 1.5 mL/min (equilibrated in $15\%_{sat}$ $(NH_4)_2SO_4$, 10 m*M* Tris–Cl (pH 8), pump A of FPLC), wash with equilibration buffer until UV is at base line. Run linear gradient (5 min) to 100% B: 10 m*M* Tris–Cl, pH 8 (pump B of FPLC). Do not abruptly switch buffer, use gradient. Continue to wash with 5–10 mL with 100% B.
27. Wash pump A of FPLC with 0.1% CH_3COOH (pay attention to wash mixer with pump B before continuing chromatography) and run a 5 min linear gradient to 100% A, continue for 5 min.
28. Wash pump B with 0.1% CH_3COOH, 80% CH_3CN (pay attention to wash mixer with pump A before continuing chromatography) and run a gradient to 100% (B) for 20 min, collect fractions 1/min (Note: 100% CH_3CN can harm the FPLC system.).
29. Collect the symmetric peak fractions, discard late, and trailing (misfolded) fractions.
30. Dry pooled factions under vacuum (SpeedVac™).
31. Dissolve pellet in 1 mL Tris–Cl-buffered saline (TBS), 2 m*M* $CaCl_2$, and measure protein content at 280 nm. Dilute if necessary to 3–4 mg/mL. Measure pH and adjust, if necessary, with 0.5 *M* Tris–Cl (pH 8) until ~pH 7.4.
32. Digest with enterokinase (stock from NEB is 20 ng/μL). Use 10–20 pg/μg protein for 17–18 h at room temperature.
33. Acidify the digest by adding HCOOH (4% final) and load on a 10 mm C_{18} 25 × 1 cm reverse-phase column at a flow rate of 3–4 mL/min. Solvents 0.05% trifluoroacetic acid (TFA) in H_2O (A) and 0.05% TFA in CH_3CN (B), column equilibrated in 20% B. Run linear gradient from 20% to 50% B at 0.5%/min.

34. Chemokines elute between 30% and 45% B. Collect peak fractions and measure protein content at 280 nm.
35. Dry peak *in vacuo* and resuspend the pure chemokine in H_2O.
36. Measure protein concentration and adjust to 150–200 μ*M* chemokine. Analyze by SDS-PAGE and perform mass spectrometry to verify the purity and integrity of the chemokine.

Note

1. Digestion times with enterokinase can vary between lots and should be optimized to avoid star activity. The enzyme works best at high protein concentrations (3–4 mg/mL); measure by SDS-PAGE cleavage efficiency.
2. Confirm correct chemokine expression by mass spectrometry. Calculated monoisotopic mass should be corrected for the two disulfide bridges (mass minus 4).
3. Chemokines are highly stable in H_2O at concentrations above 150 μ*M*. The solution remains acidic due to residual TFA, which also helps to prevent degradation.

4.2.4 Enzymatic Labeling of YbbR13-Tagged Chemokine

Correctly folded chemokines containing a YbbR13 tag (Zhou et al., 2007) at the C-terminus can be labeled with phosphopantetheinyl transferase using fluorescent-labeled CoA. The latter can be obtained by simple thiol-maleimide coupling using fresh CoA and a maleimide-modified fluorescent dye. This easy synthesis gives access to different fluorophores of choice (George et al., 2004; Yin et al., 2006). Alternatively, CoA488 (alexa488 labeled) is commercially available. Labeled chemokines are finally purified by reverse-phase chromatography and stored in H_2O or HEPES-buffered saline.

4.2.4.1 Preparation of Fluorescent-Labeled Coenzyme A

1. Dissolve 10 mg (13.02 μmol) CoA (M_w 767.53 g/mol) in 43 μL of 1 m*M* HCl and store aliquots of 5 μL (1.51 μmol) and 6 μL (1.82 μmol) at −80 °C.
2. Add 5 μL of 100 m*M* Tris–Cl (pH 7.5) + 90 μL dimethylformamide (DMF) to 1.51 μmol CoA.
3. Example dye: Atto565-maleimide (M_w 733 g/mol, 1 mg = 1.36 μmol).
4. Dissolve Atto565-maleimide in 100 μL DMF and add CoA solution (1.51 μmol). Slight excess of CoA helps to modify all fluorescent dye.
5. Let react in the dark at room temperature for 2 h with mixing.
6. Separate on reverse-phase HPLC 25 cm × 1 cm C_{18} column equilibrated with (A) 0.05% $CH_3COOH–NH_3$ (0.083 m*M*) pH 4.35 (for

1 L H_2O use 500 μL CH_3COOH + 200 μL NH_3) and (B) 0.05% CH_3COOH–NH_3 in CH_3CN; equilibrate column at 4 mL/min with 15% B and run a 30 min linear gradient from 15% to 45% at 1%/min; collect fractions (20 s) between 20% and 40% B.

7. Before loading it is important to dilute the reaction mixture to 2 mL with 5% CH_3CN 0.05% CH_3COOH/NH_3.
8. After loading wash column with 15% CH_3CN 0.05% CH_3COOH–NH_3 to remove DMF before starting the gradient.
9. Measure peak fractions at the wavelength of maximum absorption of the fluorophore (datasheet) and at 260 nm. Calculate the contribution of the fluorophore to the absorbance at 260 nm (indicated on the datasheet) and subtract from the value measured at 260 nm. Calculate the CoA concentration ($\varepsilon_{260} = 16{,}400\ M^{-1}\ cm^{-1}$) and of the fluorophore (using extinction coefficient indicated on the datasheet). The molar ratio should be 1:1.
10. Dry peak fraction under vacuum (SpeedVac™).
11. Dissolve content in HEPES-buffered saline conveniently to 0.5–1 m*M* CoA dye and measure concentration.
12. For quality control, run about 5 μL diluted with 90 μL of 0.05% CH_3COOH–NH_3 on an analytical reverse-phase column.

4.2.5 Labeling of Tagged Chemokine

1. Labeled reaction (0.5–1 mL) is prepared as follows:
2. 100 m*M* NaCl (1 *M* stock), 10 m*M* $MgCl_2$ (1 *M* stock), 50 m*M* HEPES (1 *M* stock), 20% glycerol, 5 μ*M* chemokine, and 0.5 μ*M* phosphopantetheinyl transferase (Sfp).
3. Start the reaction by the addition of 12 μ*M* CoA dye and incubate in the dark for 3 h at room temperature.
4. Purify labeled chemokine by reverse-phase HPLC on a 25 cm × 2.1 mm column: solvents 0.05% TFA in H_2O (A) and 0.05% TFA in CH_3CN (B), column equilibrated in 20% B. Run linear gradient from 20% to 50% B at 0.5%/min.
5. Measure peak fractions at the maximum absorption of the fluorophore (datasheet) and at 280 nm. Calculate the contribution of the fluorophore to the absorbance at 280 nm (indicated on the datasheet) and subtract from the value measured at 280 nm. Calculate the chemokine concentration using its molar extinction coefficient (ε_{280} can be calculated at http://web.expasy.org/protparam/) and of the fluorophore

(using extinction coefficient indicated on the datasheet). The molar ratio should be 1:1.

6. Dry peak fractions under vacuum (SpeedVac™).
7. Dissolve labeled chemokines in HEPES-buffered saline at concentrations >100 μ*M* for enhanced stability and store at −20 °C.

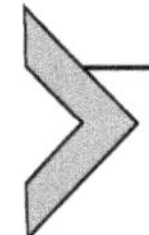

5. MONITOR CHEMOKINE UPTAKE BY MICROSCOPY AND FLOW CYTOMETRY (FACS)

Fluorescent chemokines, when presented to cells expressing cognate receptors, are rapidly internalized and accumulate within 5–30 min in vesicles surrounding/near the nucleus, or in polarized migrating cells in the posterior part. Specificity of the chemokine uptake can be pharmacologically tested with the use of receptor inhibitors/antagonists, such as small molecules or blocking antibodies, or by competition with unlabeled chemokine. To exclude unspecific uptake through micropinocytosis a mixture of differently labeled chemokines is applied, for one of them no receptor is expressed on the cells (Fig. 2).

Measuring uptake by flow cytometry is a fast method and can easily be quantified. Moreover, cells can be separated by cell sorting using multiple parameters. On the other hand, (confocal) microscopy allows the analysis of the mechanism of uptake as well as the tracking of internalized chemokine inside the cell (Fig. 3).

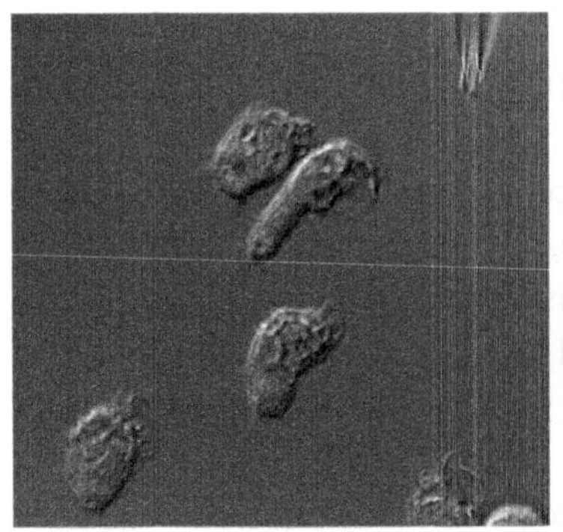

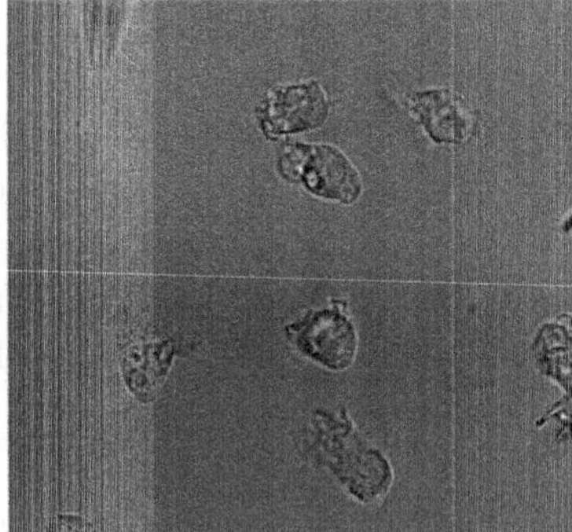

Figure 2 CCL2-mCherry uptake by migrating monocytes. Human monocytes were isolated from peripheral blood (Volpe et al., 2012). Left: 100 n*M* CCL2-mCherry was loaded in the pipette; right: a mixture of 1 μ*M* CCL2 and 1 μ*M* CCL20-venus. Note that monocytes take up only CCL2-mCherry and not CCL20-venus (excluding pinocytosis as mechanism of uptake). CCL2-mCherry is stored in endosomes located in the posterior part of the cells. A selected frame from a time-lapse video is shown. (See the color plate.)

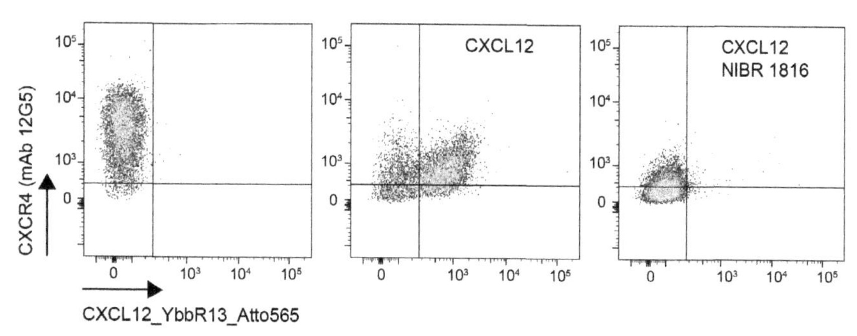

Figure 3 FACS analysis of CXCL12_YbbR13_Atto565 uptake. Mouse pre-B cells 300.19 stably expressing human CXCR4 (10^5/100 μL) were preincubated for 10 min with medium (left, middle) or with medium containing 10 μ*M* NIBR 1816 at 37 °C (right). After centrifugation cells were resuspended in medium (left), medium containing 50 n*M* CXCL12_YbbR13_Atto565 (middle) and 10 μ*M* NIBR 1816 (right) and incubated for 30 min at 37 °C. After a brief acidic wash, cells were kept on ice and stained for CXCR4 using the monoclonal antibody 12G5.

5.1 Materials

Glass bottom dishes (poly-D-lysine coated) (MatTek Corporation)
Protease inhibitors complete EDTA free (Roche, Sigma-Aldrich)
Tissue-Tek® O.C.T. Compound, Sakura® Finetek (VWR)

5.2 Monitoring Scavenging by Flow Cytometry

Cells in suspension (10^6/mL) are incubated in 100 μL culture medium in a 96 round bottom well with 20–70 n*M* fluorescent chemokine at 37 °C for 30 min. Wash cells twice with cold FACS buffer: phosphate-buffered saline (PBS), 2% FCS, and 2 m*M* EDTA and keep them on ice. To remove unwanted chemokine bound to cell-surface receptors or proteoglycans a brief acidic wash is recommended as follows. Resuspend the cell pellet in 50 μL of 100 m*M* NaCl, 50 m*M* glycine/HCl, pH 3 (Infantino, Moepps, & Thelen, 2006) on ice, and dilute after 45–60 s with 200 μL FACS buffer. Wash twice with FACS buffer. If desired, cells can be stained on ice for the expression of surface receptors before analysis by FACS (Fig. 3).

5.3 Monitoring Scavenging by Confocal Microcopy Using Live Cultures or Fixed Samples

Cells seeded and grown on glass bottom culture dishes can be used to continuously monitor chemokine uptake. For this setting is important that the stage of the inverted microscope, if not the entire microscope (for better thermal stability), is heated up to 37 °C to facilitate endocytosis. Cells are seeded on the glass insert in a small volume and are allowed to adhere. Poly-D-lysine coating facilitates adhesion of cells which normally grow in suspension. If the cells need to grow (e.g., for transfection/transduction such as HEK293, HeLa, MDCK, COS), the volume of the medium can be expanded once the cells have colonized the glass surface. The culture medium is removed and replaced with 150 μL of HEPES-buffered colorless medium (w/o phenol red, with protease inhibitors if desired) to form a dome just over the glass bottom (drying the surrounding plastic with a cotton bud helps to constrain the medium over the glass). Z-stacks are then recorded to get a baseline. The medium is carefully replaced with fresh warm medium containing 30–100 n*M* fluorescent chemokine of choice and Z-stacks are recorded over the entire cell volume at selected time points. In case of fluorescent proteins degradation can be prevented by the addition of protease inhibitors (Fig. 4A). Intracellular fluorescence becomes detectable 2–3 min after addition of the fluorescent chemokine and shows a linear increase during at least 20 min (Brennecke et al., 2013). After about 30–45 min, the signal reaches a plateau in the absence of degradation and is stable during several hours.

Alternatively, cells are exposed at 37 °C to fluorescent chemokine in a humidified incubator (with CO_2) for 20–40 min. The reaction is stopped by removing the chemokine solution, washing with PBS, and fixing the sample with 2% paraformaldehyde (PFA) in PBS for 20 min, then remove the PFA and wash with PBS. Specimens can be stored for several days in PBS at 4 °C. Microscopy is performed at room temperature. This approach gives a single endpoint determination (Fig. 5).

For quantitative analysis of the images different software packages can be used. Here we describe a general approach. Files generated by the microscope are imported in the analysis software and the channel of chemokine fluorescence is extracted from all planes of the Z-stacks and from each time point. Only planes through the entire cell body should be considered. The planes from single time points can be summed (collapsed planes). Hence, the fluorescence intensity of endosomes, corresponding to internalized chemokine, is calculated after subtraction of a minimal threshold (background subtraction, same value for all time points, and the set of experiments) using a morphometric tool of the software. The measured parameters should include the area (pixel area) of every single measured structure, its mean fluorescence intensity and total intensity. Filters are applied to define endosomes (e.g., >3 adjacent pixels, depending on the resolution of the images) and to set

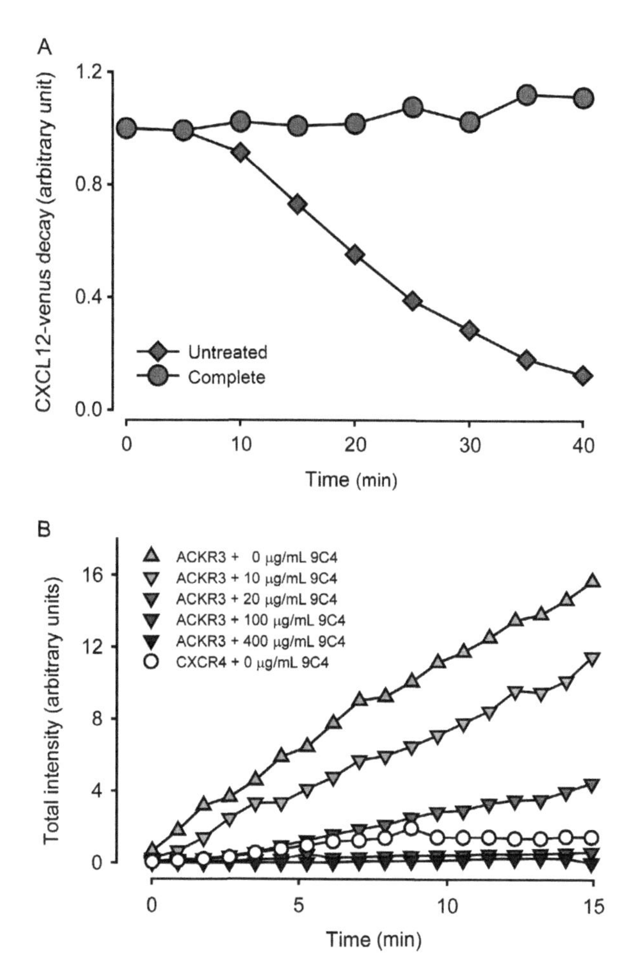

Figure 4 Analysis of CXCL12-venus uptake. MDCK cells stably transfected with human ACKR3 (and CXCR4, bottom) were grown on glass bottom cover slips and exposed to 30 n*M* CXCL12-venus (yellow fluorescent protein). (A) CXCL12-venus degradation. Cells were exposed to CXCL12-venus for 30 min in the presence of protease inhibitor complete (red (dark gray in the print version) circles) or left untreated (green (dark gray in the print version) diamonds). The chemokine-containing medium was removed and replaced with fresh medium. Then Z-stacks were taken at the indicated time points. CXCL12-venus fluorescence decays in the absence of protease inhibitors with an approximately 20 min half life time while in the presence of protease inhibitors the fluorescence remains unchanged. (B) Inhibition of ACKR3 scavenging by a specific monoclonal antibody. At time zero, cells were exposed to CXCL12-venus and uptake was measured every 50 s in the presence of different concentrations of the ACKR3-specific monoclonal antibody 9C4 (Balabanian et al., 2005) (triangles). For comparison, the uptake of CXCL12-venus by the typical chemokine receptor CXCR4 (circles) is shown.

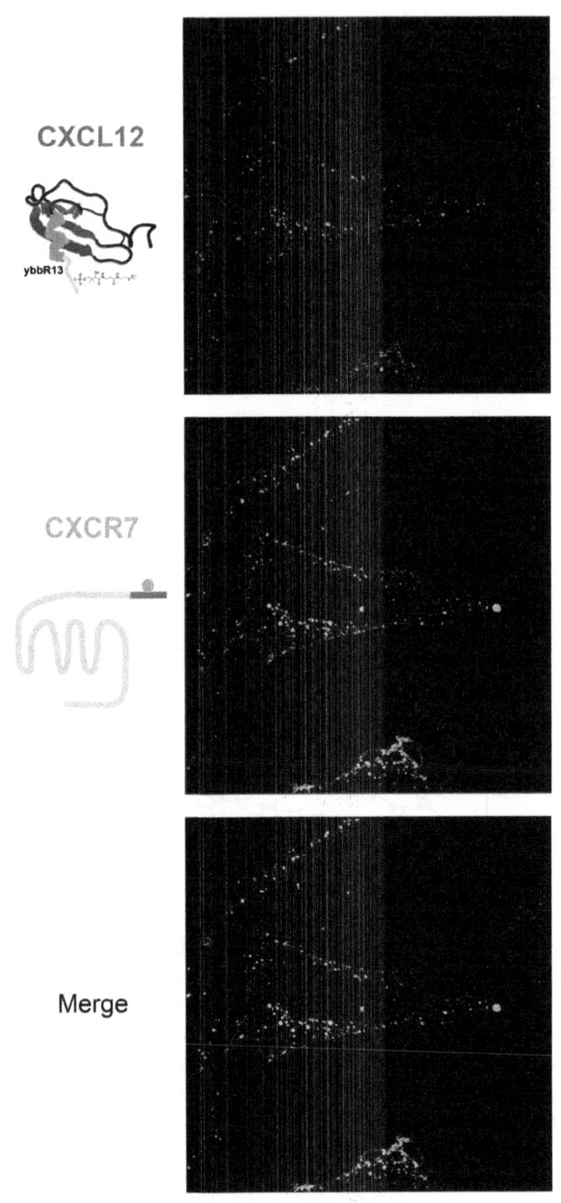

Figure 5 Scavenging of CXCL12_YbbR13_Atto700 by ACKR3. HeLa cells were transiently transfected with human ACKR3 tagged at the N-terminus with the peptidyl carrier protein sequence S6 (Zhou et al., 2007) and grown on glass bottom cover slips as described elsewhere (Humpert et al., 2012). After 48 h, cells were washed and enzymatically labeled at 17 °C with phosphopantetheinyl transferase (Sfp) (Humpert et al., 2012) and CoAAtto565 (Humpert et al., 2012). After labeling, cells were exposed to 50 n*M* CXCL12_YbbR13_Atto700 for 30 min at 37 °C. Cells were fixed with 3.7% PFA and images taken with a confocal laser scanning microscope. (See the color plate.)

a maximum threshold of average fluorescence (fluorescence artifacts not arising from endosomal structures should be excluded). Next, for each collapsed plane the total intensities of all endosomes is summed up resulting in total intensity values/time point and experiment (Fig. 4B).

5.4 Monitoring Scavenging in Whole Tissue

Chemokine scavenging can be measured in freshly isolated tissue specimens, such as human umbilical cord sections or mouse tissue (Naumann et al., 2010). In brief, place sections of 3–5 mm thickness of fresh human umbilical cords (where endothelium is exposed to the medium) in DMEM containing 1% human albumin. For mouse tissues (e.g., heart) expose sites of interest and rinse with D-PBS, 2% FBS, and anticoagulant (Liquemin™), then submerge in DMEM containing 1% bovine albumin. The DMEM medium should also contain 200 n*M* fluorescent chemokine and inhibitors (small molecules or antibodies) or competing unlabeled chemokines to control receptor specificity. After 30 min incubation at 37 °C wash sections with DMEM and mount in O.C.T. for freezing. Cryosections of 20 μm are fixed for 10 min in 4% paraformaldehyde, washed in PBS-0.02% Tween20, and counterstained for 1 h with appropriate antibodies and 1% serum (depending on the antibody host). To stain the nuclei, 4′,6-diamidine-2′-phenylindole dihydrochloride can be added.

5.5 Monitoring Scavenging in Migrating Cells

Leukocytes show a fairly fast (~0.1–0.2 μm/s) amoeboid crawling when migrating along a chemotactic cue (Friedl & Weigelin, 2008). Migration can be monitored continuously when cells are seeded on appropriate extracellular matrix components and are stimulated by a chemoattractant dispensed by a micropipette (Volpe et al., 2012; Volpe, Thelen, Pertel, Lohse, & Thelen, 2010). Here we describe a protocol used to measure chemotaxis and scavenging of fluorescent chemokines by monocytes that can be adapted for other leukocytes. Human peripheral blood monocytes are best isolated by negative selection by magnetic sorting. After isolation, resuspend 10^6 cells in 2 mL D-PBS (containing Ca^{2+} and Mg^{2+} supplemented with 1% FCS, 0.04 m*M* sodium pyruvate, 1 mg/mL fatty acid-free bovine serum albumin, and 1 mg/mL glucose). Seed the cells on poly-D-lysine coated glass bottom dishes, which were preincubated with a 100 μL of 1:80 diluted Matrigel™ (BD Biosciences) solution (added over the glass only) at 4 °C for 30 min. The dish with the cells is placed in a humidified air/CO_2-

controlled incubator which is mounted on a microscope stage inside of a temperature-controlled chamber at 37 °C. Fluorescent labeled chemokines diluted to 50–1000 n*M* in the incubation medium are dispensed with a micropipette (FemptotipII, Eppendorf) controlled by micromanipulator (Eppendorf) at a constant back pressure of 15 hPa (Femtojet, Eppendorf). A laser scan confocal microscope equipped with a temperature-controlled chamber is most suitable for recording time-lapse videos. For some experiments, it might be sufficient to measure single planes otherwise for 3D reconstruction fast Z-stacks should be recorded. Videos showing uptake of fluorescent chemokines are routinely recorded with a 63× objective (Fig. 2).

REFERENCES

Abe, P., Mueller, W., Schutz, D., Mackay, F., Thelen, M., Zhang, P., et al. (2014). CXCR7 prevents excessive CXCL12-mediated downregulation of CXCR4 in migrating cortical interneurons. *Development, 141*, 1857–1863.

Allen, S. J., Hamel, D. J., & Handel, T. M. (2011). A rapid and efficient way to obtain modified chemokines for functional and biophysical studies. *Cytokine, 55*, 168–173.

Ariel, A., Fredman, G., Sun, Y. P., Kantarci, A., Van Dyke, T. E., Luster, A. D., et al. (2006). Apoptotic neutrophils and T cells sequester chemokines during immune response resolution through modulation of CCR5 expression. *Nature Immunology, 7*, 1209–1216.

Bachelerie, F., Ben-Baruch, A., Burkhardt, A. M., Combadiere, C., Farber, J. M., Graham, G. J., et al. (2014). International Union of Pharmacology. LXXXIX. Update on the extended family of chemokine receptors and introducing a new nomenclature for atypical chemokine receptors. *Pharmacological Reviews, 66*, 1–79.

Bachelerie, F., Graham, G. J., Locati, M., Mantovani, A., Murphy, P. M., Nibbs, R., et al. (2014). New nomenclature for atypical chemokine receptors. *Nature Immunology, 15*, 207–208.

Balabanian, K., Lagane, B., Infantino, S., Chow, K. Y., Harriague, J., Moepps, B., et al. (2005). The chemokine SDF-1/CXCL12 binds to and signals through the orphan receptor RDC1 in T lymphocytes. *Journal of Biological Chemistry, 280*, 35760–35766.

Beall, C. J., Mahajan, S., Kuhn, D. E., & Kolattukudy, P. E. (1996). Site-directed mutagenesis of monocyte chemoattractant protein-1 identifies two regions of the polypeptide essential for biological activity. *Biochemical Journal, 313*, 633–640.

Beck-Sickinger, A. G., & Panitz, N. (2014). Semi-synthesis of chemokines. *Current Opinion in Chemical Biology, 22*, 100–107.

Boldajipour, B., Doitsidou, M., Tarbashevich, K., Laguri, C., Yu, S. R., Ries, J., et al. (2011). Cxcl12 evolution—Subfunctionalization of a ligand through altered interaction with the chemokine receptor. *Development, 138*, 2909–2914.

Boldajipour, B., Mahabaleshwar, H., Kardash, E., Reichman-Fried, M., Blaser, H., Minina, S., et al. (2008). Control of chemokine-guided cell migration by ligand sequestration. *Cell, 132*, 463–473.

Bonecchi, R., Borroni, E. M., Anselmo, A., Doni, A., Savino, B., Mirolo, M., et al. (2008). Regulation of D6 chemokine scavenging activity by ligand- and Rab11-dependent surface up-regulation. *Blood, 112*, 493–503.

Borroni, E. M., Buracchi, C., Savino, B., Pasqualini, F., Russo, R. C., Nebuloni, M., et al. (2009). Role of the chemokine scavenger receptor D6 in balancing inflammation and immune activation. *Methods in Enzymology, 460*, 231–243.

Braun, M., Wunderlin, M., Spieth, K., Knochel, W., Gierschik, P., & Moepps, B. (2002). Xenopus laevis Stromal cell-derived factor 1: Conservation of structure and function during vertebrate development. *Journal of Immunology, 168*, 2340–2347.

Brennecke, P., Arlt, M. J., Muff, R., Campanile, C., Gvozdenovic, A., Husmann, K., et al. (2013). Expression of the chemokine receptor CXCR7 in CXCR4-expressing human 143B osteosarcoma cells enhances lung metastasis of intratibial xenografts in SCID mice. *PLoS One, 8*, e74045.

Chakravarty, L., Rogers, L., Quach, T., Breckenridge, S., & Kolattukudy, P. E. (1998). Lysine 58 and histidine 66 at the C-terminal alpha-helix of monocyte chemoattractant protein-1 are essential for glycosaminoglycan binding. *Journal of Biological Chemistry, 273*, 29641–29647.

Clark-Lewis, I. (2000). Synthesis of chemokines. *Methods in Molecular Biology, 138*, 47–63.

Clark-Lewis, I., Kim, K. S., Rajarathnam, K., Gong, J. H., Dewald, B., Moser, B., et al. (1995). Structure–activity relationships of chemokines. *Journal of Leukocyte Biology, 57*, 703–711.

Crick, F. (1970). Diffusion in embryogenesis. *Nature, 225*, 420–422.

Dambly-Chaudiere, C., Cubedo, N., & Ghysen, A. (2007). Control of cell migration in the development of the posterior lateral line: Antagonistic interactions between the chemokine receptors CXCR4 and CXCR7/RDC1. *BMC Developmental Biology*, 7, 23.

Dawson, P. E., Muir, T. W., Clark-Lewis, I., & Kent, S. B. (1994). Synthesis of proteins by native chemical ligation. *Science, 266*, 776–779.

de la Torre, Y. M., Locati, M., Buracchi, C., Dupor, J., Cook, D. N., Bonecchi, R., et al. (2005). Increased inflammation in mice deficient for the chemokine decoy receptor D6. *European Journal of Immunology, 35*, 1342–1346.

Dona, E., Barry, J. D., Valentin, G., Quirin, C., Khmelinskii, A., Kunze, A., et al. (2013). Directional tissue migration through a self-generated chemokine gradient. *Nature, 503*, 285–289.

Edgerton, M. D., Gerlach, L. O., Boesen, T. P., & Allet, B. (2000). Expression of chemokines in Escherichia coli. *Methods in Molecular Biology, 138*, 33–40.

Ford, L. B., Cerovic, V., Milling, S. W., Graham, G. J., Hansell, C. A., & Nibbs, R. J. (2014). Characterization of conventional and atypical receptors for the chemokine CCL2 on mouse leukocytes. *Journal of Immunology, 193*, 400–411.

Ford, L. B., Hansell, C. A., & Nibbs, R. J. (2013). Using fluorescent chemokine uptake to detect chemokine receptors by fluorescent activated cell sorting. *Methods in Molecular Biology, 1013*, 203–214.

Friedl, P., & Weigelin, B. (2008). Interstitial leukocyte migration and immune function. *Nature Immunology, 9*, 960–969.

George, N., Pick, H., Vogel, H., Johnsson, N., & Johnsson, K. (2004). Specific labeling of cell surface proteins with chemically diverse compounds. *Journal of the American Chemical Society, 126*, 8896–8897.

Hansell, C. A., Schiering, C., Kinstrie, R., Ford, L., Bordon, Y., McInnes, I. B., et al. (2011). Universal expression and dual function of the atypical chemokine receptor D6 on innate-like B cells in mice. *Blood, 117*, 5413–5424.

Haraldsen, G., & Rot, A. (2006). Coy decoy with a new ploy: Interceptor controls the levels of homeostatic chemokines. *European Journal of Immunology, 36*, 1659–1661.

Hatse, S., Princen, K., Liekens, S., Vermeire, K., De, C. E., & Schols, D. (2004). Fluorescent CXCL12AF647 as a novel probe for nonradioactive CXCL12/CXCR4 cellular interaction studies. *Cytometry. Part A, 61*, 178–188.

Heinrich, J. N., Ryseck, R. P., Macdonald-Bravo, H., & Bravo, R. (1993). The product of a novel growth factor-activated gene, fic, is a biologically active "C-C"-type cytokine. *Molecular and Cellular Biology*, *13*, 2020–2030.

Heo, J., Dogra, P., Masi, T. J., Pitt, E. A., de, K. P., Smit, M. J., et al. (2015). Novel human cytomegalovirus viral chemokines, vCXCL-1s, display functional selectivity for neutrophil signaling and function. *Journal of Immunology*, *195*, 227–236.

Humpert, M. L., Tzouros, M., Thelen, S., Bignon, A., Levoye, A., Arenzana-Seisdedos, F., et al. (2012). Complementary methods provide evidence for the expression of CXCR7 on human B cells. *Proteomics*, *12*, 1938–1948.

Imai, T. (2000). Chemokine expression in insect cells. *Methods in Molecular Biology*, *138*, 23–32.

Infantino, S., Moepps, B., & Thelen, M. (2006). Expression and regulation of the orphan receptor RDC1 and its putative ligand in human dendritic and B cells. *Journal of Immunology*, *176*, 2197–2207.

Ishii, K., Yamagami, S., Tanaka, H., Motoki, M., Suwa, Y., & Endo, N. (1995). Full active baculovirus-expressed human monocyte chemoattractant protein 1 with the intact N-terminus. *Biochemical and Biophysical Research Communications*, *206*, 955–961.

Jamieson, T., Cook, D. N., Nibbs, R. J., Rot, A., Nixon, C., McLean, P., et al. (2005). The chemokine receptor D6 limits the inflammatory response in vivo. *Nature Immunology*, *6*, 403–411.

Jones, M. L., Mulligan, M. S., Flory, C. M., Ward, P. A., & Warren, J. S. (1992). Potential role of monocyte chemoattractant protein 1/JE in monocyte/macrophage-dependent IgA immune complex alveolitis in the rat. *Journal of Immunology*, *149*, 2147–2154.

Kawamura, T., Stephens, B., Qin, L., Yin, X., Dores, M. R., Smith, T. H., et al. (2014). A general method for site specific fluorescent labeling of recombinant chemokines. *PLoS One*, *9*, e81454.

Kitaura, M., Nakajima, T., Imai, T., Harada, S., Combadiere, C., Tiffany, H. L., et al. (1996). Molecular cloning of human eotaxin, an eosinophil-selective CC chemokine, and identification of a specific eosinophil eotaxin receptor, CC chemokine receptor 3. *Journal of Biological Chemistry*, *271*, 7725–7730.

Leick, M., Catusse, J., Follo, M., Nibbs, R. J., Hartmann, T. N., Veelken, H., et al. (2010). CCL19 is a specific ligand of the constitutively recycling atypical human chemokine receptor CRAM-B. *Immunology*, *129*, 536–546.

Li, Z., Kessler, W., van den Heuvel, J., & Rinas, U. (2011). Simple defined autoinduction medium for high-level recombinant protein production using T7-based Escherichia coli expression systems. *Applied Microbiology and Biotechnology*, *91*, 1203–1213.

Lindley, I., Aschauer, H., Seifert, J. M., Lam, C., Brunowsky, W., Kownatzki, E., et al. (1988). Synthesis and expression in Escherichia coli of the gene encoding monocyte-derived neutrophil-activating factor: Biological equivalence between natural and recombinant neutrophil-activating factor. *Proceedings of the National Academy of Sciences of the United States of America*, *85*, 9199–9203.

Luker, K. E., Lewin, S. A., Mihalko, L. A., Schmidt, B. T., Winkler, J. S., Coggins, N. L., et al. (2012). Scavenging of CXCL12 by CXCR7 promotes tumor growth and metastasis of CXCR4-positive breast cancer cells. *Oncogene*, *31*, 4750–4758.

Luker, K. E., Steele, J. M., Mihalko, L. A., Ray, P., & Luker, G. D. (2010). Constitutive and chemokine-dependent internalization and recycling of CXCR7 in breast cancer cells to degrade chemokine ligands. *Oncogene*, *29*, 4599–4610.

Mahabaleshwar, H., Tarbashevich, K., Nowak, M., Brand, M., & Raz, E. (2012). β-arrestin control of late endosomal sorting facilitates decoy receptor function and chemokine gradient formation. *Development*, *139*, 2897–2902.

Martinez-Martin, N., Viejo-Borbolla, A., Martin, R., Blanco, S., Benovic, J. L., Thelen, M., et al. (2015). Herpes simplex virus enhances chemokine function through modulation of receptor trafficking and oligomerization. *Nature Communications, 6*, 6163.

Naumann, U., Cameroni, E., Pruenster, M., Mahabaleshwar, H., Raz, E., Zerwes, H. G., et al. (2010). CXCR7 functions as a scavenger for CXCL12 and CXCL11. *PLoS One, 5*, e9175.

Paslin, D. A., Reykjalin, E., Tsadik, E., Schour, L., & Lucas, A. (2015). A Molluscum contagiosum fusion protein inhibits CCL1-induced chemotaxis of cells expressing CCR8 and penetrates human neonatal foreskins: Clinical applications proposed. *Archives of Dermatological Research, 307*, 275–280.

Postea, O., Vasina, E. M., Cauwenberghs, S., Projahn, D., Liehn, E. A., Lievens, D., et al. (2012). Contribution of platelet CX(3)CR1 to platelet–monocyte complex formation and vascular recruitment during hyperlipidemia. *Arteriosclerosis, Thrombosis, and Vascular Biology, 32*, 1186–1193.

Proudfoot, A. E., & Borlat, F. (2000). Purification of recombinant chemokines from E. coli. *Methods in Molecular Biology, 138*, 75–87.

Pruenster, M., Mudde, L., Bombosi, P., Dimitrova, S., Zsak, M., Middleton, J., et al. (2009). The duffy antigen receptor for chemokines transports chemokines and supports their promigratory activity. *Nature Immunology, 10*, 101–108.

Sanchez-Alcaniz, J. A., Haege, S., Mueller, W., Pla, R., Mackay, F., Schulz, S., et al. (2011). Cxcr7 controls neuronal migration by regulating chemokine responsiveness. *Neuron, 69*, 77–90.

Sarris, A. H., Broxmeyer, H. E., Wirthmueller, U., Karasavvas, N., Cooper, S., Lu, L., et al. (1993). Human interferon-inducible protein 10: Expression and purification of recombinant protein demonstrate inhibition of early human hematopoietic progenitors. *Journal of Experimental Medicine, 178*, 1127–1132.

Strong, A. E., Thierry, A. C., Cousin, P., Moulon, C., & Demotz, S. (2006). Synthetic chemokines directly labeled with a fluorescent dye as tools for studying chemokine and chemokine receptor interactions. *European Cytokine Network, 17*, 49–59.

Struyf, S., Salogni, L., Burdick, M. D., Vandercappellen, J., Gouwy, M., Noppen, S., et al. (2011). Angiostatic and chemotactic activities of the CXC chemokine CXCL4L1 (platelet factor-4 variant) are mediated by CXCR3. *Blood, 117*, 480–488.

Thelen, M. (2001). Dancing to the tune of chemokines. *Nature Immunology, 2*, 129–134.

Ueda, A., Kawamoto, S., Igarashi, T., Ishigatsubo, Y., Tani, K., Okubo, T., et al. (1994). Human monocyte chemoattractant protein-1 expressed in a baculovirus system. *Gene, 140*, 267–272.

Uguccioni, M., Loetscher, P., Forssmann, U., Dewald, B., Li, H. D., Lima, S. H., et al. (1996). Monocyte chemotactic protein 4 (MCP-4), a novel structural and functional analogue of MCP-3 and eotaxin. *Journal of Experimental Medicine, 183*, 2379–2384.

Ulvmar, M. H., Werth, K., Braun, A., Kelay, P., Hub, E., Eller, K., et al. (2014). The atypical chemokine receptor CCRL1 shapes functional CCL21 gradients in lymph nodes. *Nature Immunology, 15*, 623–630.

Veldkamp, C. T., Peterson, F. C., Hayes, P. L., Mattmiller, J. E., Haugner, J. C., III, de la Cruz, N., et al. (2007). On-column refolding of recombinant chemokines for NMR studies and biological assays. *Protein Expression and Purification, 52*, 202–209.

Venkiteswaran, G., Lewellis, S. W., Wang, J., Reynolds, E., Nicholson, C., & Knaut, H. (2013). Generation and dynamics of an endogenous, self-generated signaling gradient across a migrating tissue. *Cell, 155*, 674–687.

Volpe, S., Cameroni, E., Moepps, B., Thelen, S., Apuzzo, T., & Thelen, M. (2012). CCR2 acts as scavenger for CCL2 during monocyte chemotaxis. *PLoS One, 7*, e37208.

Volpe, S., Thelen, S., Pertel, T., Lohse, M. J., & Thelen, M. (2010). Polarization of migrating monocytic cells is independent of PI 3-kinase activity. *PLoS One, 5*, e10159.

Watts, A. O., Verkaar, F., van der Lee, M. M., Timmerman, C. A., Kuijer, M., van Offenbeek, J., et al. (2013). β-arrestin recruitment and G protein signaling by the atypical human chemokine decoy receptor CCX-CKR. *Journal of Biological Chemistry, 288*, 7169–7181.

Weber, M., Uguccioni, M., Baggiolini, M., Clark-Lewis, I., & Dahinden, C. A. (1996). Deletion of the NH_2-terminal residue converts monocyte chemotactic protein 1 from an activator of basophil mediator release to an eosinophil chemoattractant. *Journal of Experimental Medicine, 183*, 681–685.

Yang, O. O., Swanberg, S. L., Lu, Z., Dziejman, M., McCoy, J., Luster, A. D., et al. (1999). Enhanced inhibition of human immunodeficiency virus type 1 by Met-stromal-derived factor 1beta correlates with down-modulation of CXCR4. *Journal of Virology, 73*, 4582–4589.

Yin, J., Lin, A. J., Golan, D. E., & Walsh, C. T. (2006). Site-specific protein labeling by Sfp phosphopantetheinyl transferase. *Nature Protocols, 1*, 280–285.

Yin, J., Liu, F., Li, X., & Walsh, C. T. (2004). Labeling proteins with small molecules by site-specific posttranslational modification. *Journal of the American Chemical Society, 126*, 7754–7755.

Yin, J., Straight, P. D., McLoughlin, S. M., Zhou, Z., Lin, A. J., Golan, D. E., et al. (2005). Genetically encoded short peptide tag for versatile protein labeling by Sfp phosphopantetheinyl transferase. *Proceedings of the National Academy of Sciences of the United States of America, 102*, 15815–15820.

Zhou, Z., Cironi, P., Lin, A. J., Xu, Y., Hrvatin, S., Golan, D. E., et al. (2007). Genetically encoded short peptide tags for orthogonal protein labeling by Sfp and AcpS phosphopantetheinyl transferases. *ACS Chemical Biology, 2*, 337–346.

CHAPTER SIX

Dual-Color Luciferase Complementation for Chemokine Receptor Signaling

Kathryn E. Luker*[,1], Gary D. Luker*[,†,‡]
*Department of Radiology, Center for Molecular Imaging, University of Michigan, Ann Arbor, Michigan, USA
†Department of Biomedical Engineering, University of Michigan, Ann Arbor, Michigan, USA
‡Department of Microbiology and Immunology, University of Michigan, Ann Arbor, Michigan, USA
[1]Corresponding author: e-mail address: kluker@umich.edu

Contents

1. Introduction 120
2. Methods 121
    2.1 Complementation Reporter Constructs 122
    2.2 Dual-Color Bioluminescence Live-Cell Imaging 125
    2.3 Data Analysis 128
Acknowledgment 128
References 128

Abstract

Chemokine receptors may share common ligands, setting up potential competition for ligand binding, and association of activated receptors with downstream signaling molecules such as β-arrestin. Determining the "winner" of competition for shared effector molecules is essential for understanding integrated functions of chemokine receptor signaling in normal physiology, disease, and response to therapy. We describe a dual-color click beetle luciferase complementation assay for cell-based analysis of interactions of two different chemokine receptors, CXCR4 and ACKR3, with the intracellular scaffolding protein β-arrestin 2. This assay provides real-time quantification of receptor activation and signaling in response to chemokine CXCL12. More broadly, this general imaging strategy can be applied to quantify interactions of any set of two proteins that interact with a common binding partner.

Methods in Enzymology, Volume 570
ISSN 0076-6879
http://dx.doi.org/10.1016/bs.mie.2015.08.024

1. INTRODUCTION

Signaling by chemokine receptors, like most other receptors and signal transduction pathways, relies upon regulated formation and dissociation of protein complexes. A single chemokine may bind to two different chemokine receptors, initiating distinct signaling pathways and biologic outputs. Preferential binding of a chemokine ligand to one of two or more competing receptors can determine activation of specific downstream signaling pathways in addition to magnitude and duration of signaling. Inhibiting chemokine binding to one receptor partner may increase availability of the chemokine ligand to signal through another receptor, potentially altering responses to therapy and contributing to drug resistance. Understanding dynamics of signaling by two different chemokine receptors in response to a common chemokine ligand requires analysis of formation and dissociation of complexes of signaling proteins in physiologic environments. While methods such as immunoprecipitation and immunofluorescence can detect association of multiple proteins, such techniques typically are performed at a limited number of fixed time points, precluding real-time analysis, and quantification of signaling.

Protein fragment complementation assays provide a facile approach to detect and quantify protein interactions in signaling pathways in intact cells and animal models, complementing established biochemical assays (Luker & Luker, 2011). Several different protein fragment complementation assays have been developed, including strategies based on fluorescent proteins, metabolic enzymes, and luciferases (Vidi & Watts, 2009). These assays all are based upon splitting a reporter protein into two inactive fragments (amino (N)- and carboxy (C)-termini) that do not or very minimally reassemble spontaneously. N- and C-terminal reporter fragments then are fused to proteins of interest. When fused proteins of interest interact, N- and C-terminal reporter fragments reconstitute a functional reporter protein. Protein fragment complementation assays based on luciferase enzymes provide a particularly powerful approach to quantify dynamics of protein interactions in chemokine signaling. Unlike fluorescence complementation, luciferase complementation does not require maturation time before producing bioluminescence from interacting proteins, and luciferase complementation also is reversible. Luciferase complementation also provides a large dynamic range of signal with low background activity, and the assay format is compatible with moderate- and high-throughput technologies.

Luciferase complementation assays typically have been used to quantify the magnitude and kinetics of interactions between a single pair of proteins fused to N- and C-terminal fragments of luciferases such as firefly, *Renilla*, or *Gaussia* (Luker et al., 2012, 2004; Paulmurugan & Gambhir, 2003; Remy & Michnick, 2006). However, these strategies cannot analyze two different proteins competing for interaction with a common protein partner as occurs commonly in nodes of signaling pathways. To accomplish this goal, we have leveraged a recently described dual-color luciferase complementation assay based on green and red spectral variants of click beetle luciferase (Coggins et al., 2014; Villalobos et al., 2010). In the dual-color click beetle luciferase complementation assay, N-terminal fragments of click beetle green and red luciferases, respectively, interact with a C-terminal fragment shared by both N-terminal fragments. N-terminal fragments determine the wavelength of bioluminescence produced by complementation. By using emission filters to separate light from complemented green and red luciferases, investigators can quantify interactions of two different proteins with a shared partner in the same population of cells.

In this chapter, we describe methods we have used to quantify interactions between CXCR4 and ACKR3 (formerly designated CXCR7) with the common intracellular scaffolding protein, β-arrestin 2, in cells that coexpress both receptors. Both CXCR4 and ACKR3 share chemokine CXCL12 as a common ligand, although ACKR3 binds CXCL12 with approximately 10-fold higher affinity (Burns et al., 2006). CXCR4 signals through both G protein and β-arrestin pathways, while ACKR3 biases signaling to β-arrestin-mediated outputs (Rajagopal et al., 2010). Using dual-color click beetle complementation, we demonstrated that CXCL12 preferentially signals through ACKR3 in cells that coexpress this receptor with CXCR4, thereby biasing signaling toward β-arrestin 2 (Coggins et al., 2014). While we describe methods for CXCR4 and ACKR3 interacting with β-arrestin 2, the general approach for dual-color click beetle luciferase complementation can be applied readily to other receptors or protein interactions in chemokine signaling pathways.

2. METHODS

The dual-color click beetle luciferase complementation assay we describe is designed to quantify pair-wise interactions between two proteins, such as receptors CXCR4 and ACKR3, with a common binding partner, β-arrestin 2. Intracellular C-termini of CXCR4 and ACKR3 are fused to N-terminal fragments of click beetle green (CBGN) or click beetle red (CBRN) luciferases. The N-terminal fragments of each enzyme determine the color of bioluminescence produced by complementation with the common C-terminal fragment of click beetle luciferase (CLuc), the latter of which is fused to β-arrestin 2. From previous studies, we had determined that fusing the C-terminus of β-arrestin 2 to CLuc produced optimal induction of signaling in response to chemokine CXCL12 (Luker, Gupta, & Luker, 2008). When designing a new complementation assay, we always check all logical orientations of fusions of CBGN, CBRN, and CLuc with proteins of interest.

The overall workflow of this chapter proceeds from generating fusions between proteins of interest and click beetle luciferase fragments and then describes the protocol for cell-based imaging with these constructs to quantify magnitude and kinetics of protein interactions in chemokine signaling.

2.1 Complementation Reporter Constructs

We describe the overall approach for generating click beetle luciferase complementation reporters for live-cell bioluminescence imaging of chemokine receptor signaling. We refer readers to other standard texts, such as Methods in Molecular Biology, for detailed protocols for molecular biology procedures including PCR and ligations.

2.1.1 Required Materials

1. cDNA constructs for interacting proteins of interest.
2. Plasmids for CBGN and CBRN luciferases (Promega) or plasmids with N-terminal (amino acids 1–413) or C-terminal (395–542) fragments of these enzymes. Fragments for click beetle luciferase complementation were selected based on comparable fragments identified for optimal firefly luciferase complementation (Luker et al., 2004; Villalobos et al., 2010). The N-terminal fragment determines the emission spectrum of light from complementation with the common C-terminal fragment.
3. Vectors with constitutive or inducible (optional) promoters for expression in mammalian cells.
4. Lentiviral vectors for expressing gene of interest and packaging proteins (optional).
5. Reagents and equipment for PCR, restriction digests, and DNA ligations.
6. Reagents to validate expression of fusion proteins, such as antibodies for flow cytometry and/or Western blotting.

2.1.2 Generate Constructs for Click Beetle Luciferase Complementation Fusion Proteins

1. Design cloning strategy to fuse intracellular C-termini of chemokine receptors to NLuc fragments of CBGN and red luciferases, respectively, and the common CLuc fragment to β-arrestin 2 (Fig. 1). This cloning strategy is based on quantifying interactions of two different receptors, such as CXCR4 and ACKR3, with a common partner protein, such as β-arrestin 2. NLuc fragments must be fused to the C-termini of receptors for complementation with other intracellular targets and pair-wise determinations of each receptor interacting with β-arrestin 2 based on color of bioluminescence. For new complementation assays, we design and test fusions of luciferase fragments positioned at all spatially relevant N- and C-termini of target proteins. The same general approach can be used to analyze interactions of CBGN and red NLuc fusions of any two proteins of interest with a common binding partner fused to CLuc.
2. We generally add a linker sequence, such as $(GGGGS)_2$, between the protein of interest and luciferase fragments to reduce steric hindrance. A linker is not necessarily required, and other linker sequences and

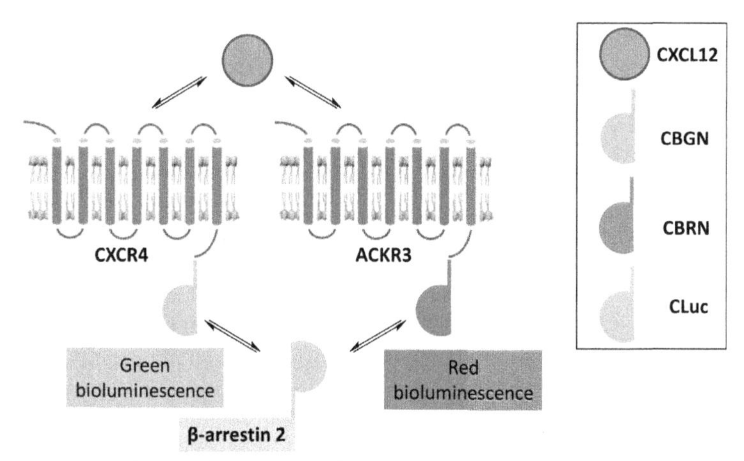

Figure 1 Dual-color luciferase complementation assay for CXCL12 signaling to β-arrestin 2 via CXCR4 or ACKR3. CXCL12 binding to CXCR4 fused to CBGN drives recruitment of β-arrestin 2-CLuc to produce green (light gray in the print version) bioluminescence, while CXCL12 signaling through ACKR3 fused to CBRN produces red (gray in the print version) bioluminescence upon binding to β-arrestin 2. Measuring bioluminescence from green (light gray in the print version) versus red (gray in the print version) click beetle luciferases provides enables real-time analysis of relative signaling through CXCR4 versus ACKR3 in cells coexpressing both receptors. *Figure modified from Wu, Xie, Zhao, Nice, and Huang (2012).*

lengths may be optimal for other protein interaction. Investigators may need to test different linker sequences and fusion constructs to optimize ligand-dependent induction of bioluminescence signal (Chichili, Kumar, & Sivaraman, 2013).

3. Using appropriate molecular biology procedures, generate plasmid constructs with fusion proteins for desired interacting proteins, such as receptors and β-arrestin 2 in this example. We prepare fusion proteins with all logical orientations of NLuc and CLuc fragments on N- and C-termini of target proteins with the goal of optimizing ligand-induced bioluminescence signal above background and negative control interactions (Fig. 2).
4. Produce plasmids with relevant control constructs for background association of CBGN and CBRN fragments with CLuc. Appropriate control constructs should be matched to spatial localization and size of interacting proteins of interest. Possible controls include mismatched

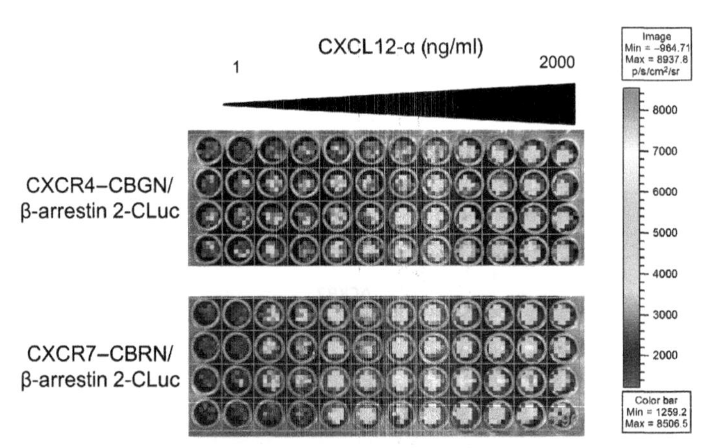

Figure 2 Cell-based click beetle green and red complementation assay for CXCL12-dependent activation of CXCR4–CBGN and CXCR7 (ACKR3)–CBRN association β-arrestin 2-CLuc. MDA-MB-231 breast cancer cells coexpressing CXCR4–CBGN and β-arrestin 2-CLuc or CXCR7 (ACKR3)–CBRN and β-arrestin 2-CLuc were seeded at 1.5×10^4 cells per well in black wall 96-well plates. We incubated cells with vehicle only (far left column) or increasing concentrations of CXCL12-α (1–2000 ng/ml). The figure shows a representative bioluminescence image of green (top) and red (bottom) bioluminescence obtained 12 and 14 min, respectively, after adding CXCL12. The grid overlay is used to quantify photons per well by region-of-interest analysis. Bioluminescence is depicted as a pseudocolor display with red and blue defining high and low values for photon flux. (See the color plate.)

receptors, proteins with mutations in amino acids necessary for protein interaction, and/or other membrane bound and cytosolic proteins.

5. Verify integrity of plasmids with fusion proteins by DNA sequencing.
6. If needed, transfer click beetle luciferase fusion proteins to plasmids for constitutive or inducible (optional) mammalian expression. Inducible expression systems, since as tetracycline regulated promoters, can be useful to investigate how different levels of CBGN or CBRN fusions affect signaling. Using vectors with selectable markers, such as drug-resistance genes or coexpressed fluorescent proteins, facilitate identification of transiently and stably transfected cells. To facilitate repeated experiments using cells with well-characterized expression of reporter constructs, we generate stably transduced cells via lentiviral transduction. We commonly use vectors in which the luciferase complementation reporter is coexpressed via an internal ribosomal entry site or P2A site with a drug-resistance marker or fluorescent protein. This strategy improves the ability to obtain a population in which all cells express desired reporter proteins. We work with batch populations of cells to avoid potential artifacts of selecting individual clones on biologic responses of cells.

2.2 Dual-Color Bioluminescence Live-Cell Imaging

2.2.1 Required Materials

1. Supplies for culturing mammalian cells, including media, sterile culture flasks/dishes, and temperature controlled incubators.
2. HEK-293T cells or other cell line that can be transfected readily for initial testing of constructs and preparation of recombinant viral vectors.
3. Established cell lines or primary cells of interest for specific biologic question(s).
4. Tissue culture treated sterile multiwell plates (96- or 384-well) with black sides, clear bottom, and lid (corning or other vendor). Black sides prevent light generated from one well from contaminating light produced by an adjacent well.
5. Sterile pipette tips (low adherence coating preferred) for dispensing reagents.
6. Standard and multichannel pipettes.
7. Sterile phosphate-buffered saline (PBS) solution.
8. Chemokine ligand for receptor(s) of interest prepared as a stock solution. For chemokine CXCL12 (R&D Systems or other vendor), we

prepare a 50 μg/μl stock in PBS with 0.2% Probumin brand BSA (Millipore) stored at −20 °C or lower. It is important to select a BSA carrier that does not contain proteases that can degrade CXCL12 or other chemokine.

9. D-luciferin (Promega or other vendor) prepared as sterile-filtered stock solution at 15 mg/ml in PBS and stored at −20 °C.
10. Bioluminescence imaging instrument with high sensitivity and software for data quantification and analysis (IVIS, Perkin-Elmer, or similar equipment).

2.2.2 Live-Cell Imaging

1. When developing an assay for new interacting proteins, we first test all orientations (N-terminal and C-terminal fusions as relevant) of CBGN and red NLuc fusions paired with the common interacting CLuc fusion by transient transfection in 293T cells. Investigators may use other cell lines that transfect readily. We also include appropriate control fusion constructs in these tests. The objectives of the initial assays with transient transfections are to (1) identify optimal orientations of fusion proteins for imaging based on greatest ligand-dependent induction of signal above control and (2) verify expression of fusion proteins by means such as Western blotting or flow cytometry. These data inform optimal orientations of NLuc and CLuc fusion proteins to use for stable cell lines. The basic assay protocol detailed in subsequent steps is the same for transiently transfected or stable cell lines.
2. Plate cells at a density of 1×10^4–2×10^4 cells per well in 100 μl complete growth medium with serum in black walled, clear bottom 96-well plates for tissue culture. Since complementation of click beetle luciferase fragments produces less bioluminescence than intact CBGN or red, we typically use 96-well rather than 384-well plates for cell-based assays. This approach allows us to use shorter acquisition times in each emission channel and improves resolution for kinetics of signaling. For 384-well plate assays, we use 3×10^3–5×10^3 cells per well. We typically culture cells for two days in 96-well plates before assays.
3. On the day of the assay, gently aspirate medium from wells and use a multichannel pipette to add 50 μl per well of phenol red free DMEM medium with 0.2% probumin 30 min before imaging. The added medium should be equilibrated to 37 °C in a tissue culture incubator before adding to cells. We use phenol red free DMEM to reduce absorption of emitted bioluminescence. The 0.2% probumin serves as a carrier

for added chemokine, such as CXCL12, and provides a brief period of serum starvation prior to assays. We typically add any inhibitors at desired concentrations during this 30 min incubation. Return cells to the 37 °C incubator.

4. To produce consistent, physiologic signaling responses, it is essential to minimize cooling of cells in 96-well plates during medium exchanges and addition of reagents. When we remove cells from the tissue culture incubator, we place plates on an insulated surface, such as the lid of a Styrofoam box or absorbent bench pad, rather than directly on a counter top. We also work with plates on a counter immediately adjacent to the tissue culture incubator. Adding reagents at room temperature rather than prewarmed to 37 °C slows reaction kinetics and increases variability among experiments.
5. Using a multichannel pipette, add 7 μl per well of D-luciferin from a 15 mg/ml stock and incubate at 37 °C for 5 min. Adding D-luciferin begins the bioluminescence reaction from any preformed protein complexes, so it is necessary to add this and subsequent reagents as rapidly as possible.
6. Immediately before imaging, add 14 μl per well of phenol red free DMEM with 0.2% probumin and desired concentrations of chemokine ligand, such as CXCL12. For CXCL12, we typically use 0–2000 ng/ml. We typically use 4–6 replicate wells per experimental condition. It is important to have a set of wells treated with only vehicle control to use for normalizing imaging data (see Section 2.3).
7. Immediately place plate in bioluminescence imaging instrument at the smallest field of view that will include all samples of interest. For interactions of CXCR4 and ACKR3 with β-arrestin 2, we acquire 20 images with large binning and 2 min exposure, alternating between 530–550 and 690–710 nm emission filters (IVIS Spectrum, Perkin-Elmer). Imaging parameters should be programmed in advance of placing the plate into the instrument to minimize delays in beginning imaging. Acquisition times will vary for other interacting proteins depending on the amount of bioluminescence produced by complementation. Total duration of imaging also will vary based on kinetics of a signaling pathway of interest. As alternatives to wavelengths listed above, filters that collect light <550 nm (green) and >680 nm (red) provide acceptable separation of CBGN and red bioluminescence.
8. Cells may be collected at the end of the assay for subsequent analysis by Western blotting or other biochemical assays as needed.

2.3 Data Analysis

1. Quantify bioluminescence in each well over time by region-of-interest analysis software with the imaging instrument. For IVIS instruments, analyze imaging signal as photon flux rather than counts to correct for any differences in image acquisition time among different experiments. Any wells with saturated pixels cannot be quantified accurately and should be excluded from analysis. Subsequent studies should use shorter image acquisition times to avoid this problem.
2. Normalize ligand-dependent changes in bioluminescence over time (without or with inhibitors) to corresponding signal from untreated controls at the same time point to account for any changes in bioluminescence due to availability of D-luciferin substrate.

ACKNOWLEDGMENT

This work was supported by NIH Grants R01CA170198 and R01CA196018.

REFERENCES

Burns, J., Summers, B., Wang, Y., Melikian, A., Berahovich, R., Miao, Z., et al. (2006). A novel chemokine receptor for SDF-1 and I-TAC involved in cell survival, cell adhesion, and tumor development. *The Journal of Experimental Medicine*, *203*, 2201–2213.

Chichili, V., Kumar, V., & Sivaraman, J. (2013). Linkers in the structural biology of protein–protein interactions. *Protein Science*, *22*, 153–157.

Coggins, N., Trakimas, D., Chang, S., Ehrlich, A., Ray, P., Luker, K., et al. (2014). CXCR7 controls competition for recruitment of beta-arrestin 2 in cells expressing both CXCR4 and CXCR7. *PLoS One*, *9*, e98328.

Luker, K., Gupta, M., & Luker, G. (2008). Imaging CXCR4 signaling with firefly luciferase complementation. *Analytical Chemistry*, *80*, 5565–5573.

Luker, G., & Luker, K. (2011). Luciferase protein complementation assays for bioluminescence imaging of cells and mice. *Methods in Molecular Biology*, *680*, 29–43.

Luker, K., Mihalko, L., Schmidt, B., Lewin, S., Ray, P., Shcherbo, D., et al. (2012). *In vivo* imaging of ligand receptor binding with Gaussia luciferase complementation. *Nature Medicine*, *18*, 172–177.

Luker, K., Smith, M., Luker, G., Gammon, S., Piwnica-Worms, H., & Piwnica-Worms, D. (2004). Kinetics of regulated protein-protein interactions revealed with firefly luciferase complementation imaging in cells and living animals. *Proceedings of the National Academy of Sciences of the United States of America*, *101*, 12288–12293.

Paulmurugan, R., & Gambhir, S. (2003). Monitoring protein-protein interactions using split synthetic renilla luciferase protein-fragment-assisted complementation. *Analytical Chemistry*, *75*, 1584–1589.

Rajagopal, S., Kim, J., Ahn, S., Craig, S., Lam, C., Gerard, N., et al. (2010). Beta-arrestin- but not G protein-mediated signaling by the "decoy" receptor CXCR7. *Proceedings of the National Academy of Sciences of the United States of America*, *107*, 628–632.

Remy, I., & Michnick, S. (2006). A highly sensitive protein-protein interaction assay based on Gaussia luciferase. *Nature Methods*, *3*, 977–979.

Vidi, P., & Watts, V. (2009). Fluorescent and bioluminescent protein-fragment complementation assays in the study of G protein-coupled receptor oligomerization and signaling. *Molecular Pharmacology*, *75*, 733–739.
Villalobos, V., Naik, S., Bruinsma, M., Dothager, R., Pan, M., Samrakindi, M., et al. (2010). Dual-color click beetle luciferase heteroprotein fragment complementation assays. *Chemistry & Biology*, *17*, 1018–1029.
Wu, J., Xie, N., Zhao, X., Nice, E., & Huang, C. (2012). Dissection of aberrant GPCR signaling in tumorigenesis—A systems biology approach. *Cancer Genomics Proteomics*, *9*, 37–50.

CHAPTER SEVEN

Analysis of Arrestin Recruitment to Chemokine Receptors by Bioluminescence Resonance Energy Transfer

J. Bonneterre*, N. Montpas*, C. Boularan†, C. Galés†, N. Heveker*,†,1

*Department of Biochemistry and Research Centre, Sainte-Justine Hospital, Université de Montréal, Montreal, Quebec, Canada

†Institut des Maladies Métaboliques et Cardiovasculaires, Institut National de la Santé et de la Recherche Médicale, U1048, Université Toulouse III Paul Sabatier, Toulouse, France

[1]Corresponding author: e-mail address: nikolaus.heveker@recherche-ste-justine.qc.ca

Contents

1. Introduction 132
2. Methods 133
   2.1 Materials Required 133
   2.2 General Protocol 134
   2.3 Optimizing the Experimental System 136
   2.4 Use of Optimized Systems 140
   2.5 Troubleshooting 141
3. Interpretation and Limitations of BRET Data 145
   3.1 Interpreting $BRET_{max}$ 146
   3.2 Interpreting $BRET_{50}$ 146
   3.3 Comparing Arrestin Recruitment to Different Chemokine Receptors 147
   3.4 Comparing Arrestin Recruitment by Different Ligands to the Same Chemokine Receptor 147
   3.5 Comparing Receptor Mutants 148
4. Perspectives 149
Acknowledgments 150
References 150

Abstract

Chemokine receptors recruit the multifunctional scaffolding protein beta arrestin in response to binding of their chemokine ligands. Given that arrestin recruitment represents a signaling axis that is in part independent from G-protein signaling, it has become a hallmark of G protein-coupled receptor functional selectivity. Therefore, quantification of arrestin recruitment has become a requirement for the delineation of chemokine and drug candidate activity along different signaling axes. Bioluminescence resonance

Methods in Enzymology, Volume 570
ISSN 0076-6879
http://dx.doi.org/10.1016/bs.mie.2015.09.003

energy transfer (BRET) techniques provide methodology for such quantification that can reveal differences between nonredundant chemokines binding the same receptor, and that can be upscaled for high-throughput testing. We here provide protocols for the careful setup of BRET-based arrestin recruitment assays, and examples for the application of such systems in dose-response or time-course experiments. Suggestions are given for troubleshooting, optimizing test systems, and the interpretation of results obtained with BRET-based assays, which indeed yield an intricate blend of quantitative and qualitative information.

1. INTRODUCTION

The multifunctional adaptor protein beta arrestin is recruited from the cytosol to chemokine receptors, and many other seven-transmembrane domain receptors (7TMRs), following ligand binding. Initially found to be involved in the desensitization of G protein-mediated signaling (hence its name) and in receptor internalization, it then emerged that arrestin can also associate with a large number of downstream signaling proteins, and trigger a plethora of arrestin-dependent signaling pathways (Shukla, Xiao, & Lefkowitz, 2011). Arrestin-mediated signaling itself can be both G protein-dependent and -independent. In consequence, G protein-independent arrestin signaling often serves as a paradigm for the concept of functional selectivity. Functional selectivity, or "biased agonism," denominates a concept that gained wide acceptance during the last decade and that describes how agonists of the same receptor may trigger different independent sets of signaling cascades (Urban et al., 2007). It was found that synthetic receptor agonists may be "biased," and selectively activate one or another pathway. This selectivity may be gradual, yielding relative preferences in potency and efficacy of the responses (Galandrin & Bouvier, 2006; Galandrin, Oligny-Longpre, & Bouvier, 2007). It now emerges that functional selectivity may be much more intricate than simply opposing divergent G protein-dependent and -independent responses, but may take place between different G protein-dependent responses (Sauliere et al., 2012) as well as between different arrestin-dependent responses entailed by different arrestin recruitment modalities (Audet et al., 2012; Shukla et al., 2008; Zimmerman et al., 2012; see also below).

Chemokines and their receptors, with their widespread receptor and ligand promiscuity, were critical models to establish that functional selectivity indeed applies to endogenous receptors and ligands, and not only to synthetic drugs (Berchiche, Gravel, Pelletier, St-Onge, & Heveker, 2011;

Colvin, Campanella, Manice, & Luster, 2006; Colvin, Campanella, Sun, & Luster, 2004; Corbisier, Gales, Huszagh, Parmentier, & Springael, 2015; Kohout et al., 2004; Rajagopal et al., 2013). This also definitively dismissed the vision of different chemokine ligands to the same receptor as being "redundant." All chemokine receptors, including the so-called atypical chemokine receptors (with the possible exception of ACKR1/DARC; Chakera, Seeber, John, Eidne, & Greaves, 2008), recruit arrestin. The intriguing case of CXCR7/ACKR3, which generally responds to chemokine stimulation with arrestin recruitment but not G protein-dependent responses (Kalatskaya et al., 2009), has been called a "biased receptor" (Rajagopal et al., 2010) (but see also Odemis et al., 2012).

Historically, arrestin recruitment was first observed by visualizing arrestin relocalization from the cytosol to the plasma membrane using fluorescence-based imaging (Barak, Ferguson, Zhang, & Caron, 1997; Ferguson & Caron, 2004). The advent of bioluminescence resonance energy transfer (BRET) techniques provided a sensitive and quantitative method to measure arrestin recruitment to receptors (Angers et al., 2000). The use of such relatively simple quantitative methodology is also amenable to high-throughput screening and thus facilitates testing of drug candidates on the arrestin pathway (Hamdan, Audet, Garneau, Pelletier, & Bouvier, 2005).

In this chapter, we describe protocols to set up BRET-based arrestin recruitment test systems. We provide suggestions for troubleshooting and identify experimental parameters that can be varied. We also discuss potential pitfalls and, importantly, some limitations of BRET as a quantitatively comparative method. Indeed, BRET is now widely used as a proximity-based assay to detect arrestin recruitment, but it is often overlooked that, besides proximity, the relative orientations of the probes crucially determines BRET efficiency. This somewhat complicates the accurate interpretation of BRET data, but careful setup and combination of different BRET systems hold promise for the study of yet underexplored questions in chemokine receptor signaling.

2. METHODS

2.1 Materials Required

Expression vectors coding for the receptor fused to either the energy donor (luciferase) or acceptor (fluorescent protein), and arrestin fusions to the complementary BRET partner, need to be constructed. Routinely, we give preference to fusing the C-terminal part of the receptor to the energy acceptor, and arrestin (at the N- or C-terminus) to the donor, as the donor expression is kept lower than acceptor expression in BRET experiments. We showed that, when the RLuc-arrestin/receptor-YFP configuration is used, the total RLuc-arrestin levels required for BRET are clearly below endogenous arrestin levels in HEK293 cells (Leduc et al., 2009). This eliminates reservations that the detected receptor:arrestin interactions may depend on abnormally high arrestin levels generated by overexpression of the arrestin fusion construct. Along these lines, receptor levels required for BRET experiments can also be quantified by flow cytometry or cell ELISA, using specific antibodies. Comparison of the obtained values with levels of endogenous receptor in relevant cell types ensures that receptor expression remains in a physiologically relevant range (Percherancier et al., 2005). The energy donor (or acceptor) can be fused to either the C-terminal or N-terminal of arrestin, which may yield different results (Bellot et al., 2015). Depending on the tested receptor and chemokine, either the N-terminal or the C-terminal position of luciferase fused to arrestin may yield optimal results (see also Section 2.5).

2.2 General Protocol

The standard arrestin recruitment BRET[1] protocol is similar for all assay configurations, variations, and setup experiments. We thus give the standard protocol in this section and indicate the parameters that can or must be varied during setup or for specific questions. The standard protocol is also similar between BRET[1] and BRET[2], which use different energy donor and acceptor variants—minor variations between BRET[1] and BRET[2] protocols are directly indicated in the protocol below, while specifics of BRET[2] are dealt with in a separate paragraph below (Section 2.5.2).

In general, BRET arrestin recruitment assays can be performed both on attached cells as well as cells in suspension (Hamdan et al., 2005). We routinely perform and thus describe here the assays on attached cells. Using attached cells facilitates assays in a HTS setting, and cell attachment is required for specific questions such as the determination of the receptor: arrestin dissociation rate, as it most likely better preserves the cellular cytoskeleton organization.

The general protocol for receptor/arrestin BRET experiments is as follows:

Day 1: Plate HEK293 cells in complete DMEM in 6-well plates at a density to reach optimal density for transfection (60–75% confluence) on the next day; let adhere overnight.

Day 2: Transfect the cells. Using the polyethylenimine method, we use a total of 2 μg of DNA per well. Optimal donor and acceptor plasmids quantities should be determined independently (Sections 2.3.1 and 2.3.2). They must always be complemented with empty vector as a DNA carrier to reach 2 μg/well, to equalize transfection efficiency. Always run also a donor-only control transfection, where the energy donor construct is transfected alone, to measure spectral spillover of the donor emission into the acceptor emission channel. Incubate overnight.

Day 3: Trypsin and resuspend the transfected cells in fresh complete medium and seed them into 96-flat-clear-bottom white microplates. Cell adherence to the plate can be increased with precoating with 0.1% poly-D-lysine. Optimal results are achieved with plates with individual clear bottoms surrounded by opaque matrix to increase well-to-well resolution and limit photon spillover, although this is not mandatory. Make sure cells are not clumping, and that equal amounts per well are seeded. Let adhere overnight. Cells should be near confluent, but not overconfluent, the next day.

Day 4:

(i) Change medium to 30 μl/well of BRET buffer (PBS, 0.1% BSA, 0.5 m*M* $MgCl_2$); note that the phenol red in the media interferes with BRET readings. Alternatively, DMEM without phenol red, containing 0.1% BSA but no serum, may be used. With chemokine ligands, we avoid serum in the BRET buffer and prefer the use of bovine serum albumin. Read total fluorescence of all wells using direct acceptor stimulation with a laser and the appropriate filter. This permits to control for the quality of the transfection and the homogenous cell distribution per well, and is required to calculate the [acceptor]/[donor] ratio needed for BRET titration experiments described in Section 2.3.2.

(ii) Then cover the clear microplate bottom with a white sticker. Add 10 μl of chemokine or vehicle to stimulate arrestin recruitment; replace the plate in the incubator for 5 min (or keep in the heated chamber of the plate reader, if available). Different concentrations of chemokines (e.g., in dose-response experiments) are best prepared in a separate 96-well plate and then transferred to the BRET plate with a multichannel pipette. We avoid using injectors for chemokine addition due to the relatively large void volume. After 5 min, add 10 μl of coelenterazine H (5 μ*M* final concentration) for $BRET^1$. Preincubate in the incubator for 10 min. Substrate preincubation allows the luciferase reaction to equilibrate and avoids readings during the initial unstable flash phase of photon emission. For BRET^2, coelenterazine 400 A (DeepBlue C) is used as a substrate—note that no substrate preincubation is required for BRET^2, and should actually be avoided, which somewhat alters the protocol for real-time kinetics readings. The optimal time of addition of the chemokine agonist may vary and can be determined in time-course experiments (the standard protocol given here uses 15 min, a time point where arrestin recruitment—if it occurs—will be clearly visible). For time-course experiments, the chemokine should be added only after the substrate preincubation, followed by immediate plate reading.

(iii) Proceed to BRET readings in the appropriate wavelength channels (in the 460–500 nm (RLuc) and 510–550 nm (YFP) windows for BRET^1). We use filters at 485 ± 14 nm (RLuc) and 530 ± 10 nm (eYFP) for BRET^1. For BRET^2, different filter sets are used (see below). Note that—for BRET^1—the use of different filters and band-passes affects the amplitude of the spillover of the RLuc signal into the YFP emission channel and thus the background signal. The BRET signal (correctly: "BRET ratio") is the ratio of the photon emission in the acceptor channel to the emission in the donor channel. This is converted to BRET$_{net}$ by subtracting the background BRET ratio measured in the absence of cotransfected energy acceptor (energy donor alone). Background BRET derives from detection of photons from the luciferase signal by the acceptor channel and should be a stable, instrument-dependent constant. This constant is also identified during determination of the required donor quantity (Section 2.3.1 and Fig. 1); nevertheless, background (donor-only) controls should also be run as controls within every experiment. BRET$_{net}$ is calculated using the formula:

$$\mathrm{BRET_{net}} = \left(\frac{\text{acceptor}}{\text{donor}}\right)_{\text{sample wells}} - \left(\frac{\text{acceptor}}{\text{donor}}\right)_{\text{"RLuc only" wells}}$$

2.3 Optimizing the Experimental System

2.3.1 Determination of the Optimal Quantity of BRET Donor

The optimal quantity of BRET donor-encoding vector is determined by transfecting increasing amounts of the energy donor plasmid alone. We typically use 5–200 ng of energy donor plasmid. Read luciferase and BRET

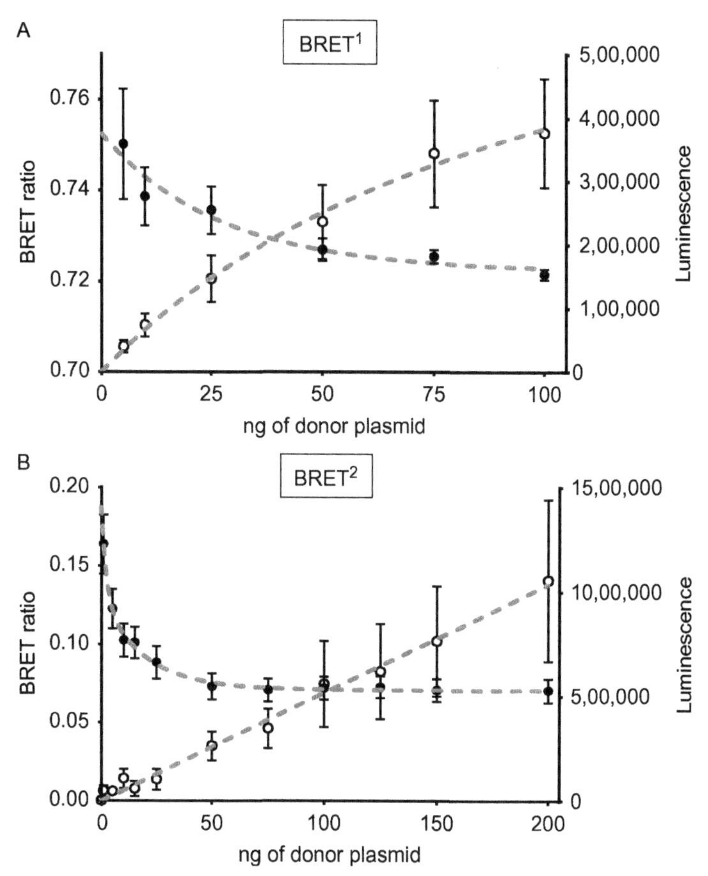

Figure 1 Determination of optimal luciferase quantity and background BRET. Luciferase titrations are plotted as a function of the amount of transfected energy donor DNA, in the absence of energy acceptor (fluorescent protein). The resulting BRET ratio (black circles) is plotted on the left *Y* axis, and the luminescence counts (white circles) are plotted on the right *Y* axis. The detected BRET ratio decreases to reach a background value that is stable over increasing luminescence, and that must be deduced from experimental BRET ratios to obtain $BRET_{net}$. This instrument-dependent stable background BRET is due to bleeding of the donor emission into the acceptor emission channel (note the scale difference between the $BRET^1$ and $BRET^2$ systems). The higher BRET ratios at too low luminescence counts are due to noise detection in the acceptor emission channel.

values (Fig. 1). The optimal amount of luciferase plasmid is the one that yields constant background BRET, but that is small enough to permit BRET saturation upon donor:acceptor cotransfection (next section). In our hands, this quantity is typically around 50 ng/well, but this depends on transfection efficiency. Note that upon donor:acceptor cotransfection, the luciferase expression levels tend to further decrease (up to 50%) due to expression competition between cotransfected expression vectors. It is thus better not to choose the smallest luciferase quantity that generates stable background values if transfected alone. Comparison of the luminescence values in BRET experiments with the donor-only curve will establish if luciferase levels upon donor:acceptor cotransfection remain indeed in the valid range.

2.3.2 BRET Saturation (Also Called BRET Titration Experiments)

BRET saturation experiments determine—among others—the optimal energy acceptor plasmid quantity for specific experiments, which are then executed at a single [acceptor]/[donor] ratio (such as dose-response curves, time-course experiments; see below). BRET donor plasmid is cotransfected at the fixed, optimal concentration with varying amounts of acceptor-encoding vector. Use a wide range of acceptor plasmid concentrations, for example, from 50 to 1900 ng/well. Perform BRET readings, in the absence and presence of saturating agonist concentrations. Plot the calculated $BRET_{net}$ to the measured [acceptor]/[donor] ratios. Note that [acceptor]/[donor] ratios are not represented as respective plasmid concentrations, but derive from actually measured luminescence and fluorescence readings. They are thus strictly instrument-dependent arbitrary units (we also observed slight reproducible differences between identical instrument types from the same manufacturer—these derive from interinstrument variations of the photomultiplier unit). They can thus not be compared between different instruments, or laboratories.

Upon fitting of the results, if the receptor does not constitutively interact with arrestin, $BRET_{net}$ should take the form of a slightly increasing straight line (typically at rather low $BRET_{net}$ values) (Fig. 2A). This so-called bystander BRET derives from donor:acceptor random collisions, and not from specific interactions. Note that even bystander BRET can saturate at high [acceptor]/[donor] ratios—therefore, judgement rather than simply curve fitting alone should be used for interpretation. Ligand-induced arrestin recruitment should yield $BRET_{net}$ values in a nonambiguous hyperbolic function, and most often considerably increased $BRET_{net}$ values compared to bystander BRET (Fig. 2B). Maximal BRET ($BRET_{max}$), and [acceptor]/[donor] ratios that yield half-maximal BRET ($BRET_{50}$) are important comparative values that can be determined by curve fitting.

BRET saturation curves permit to identify the optimal acceptor concentration: the acceptor concentration that—in combination with the optimal

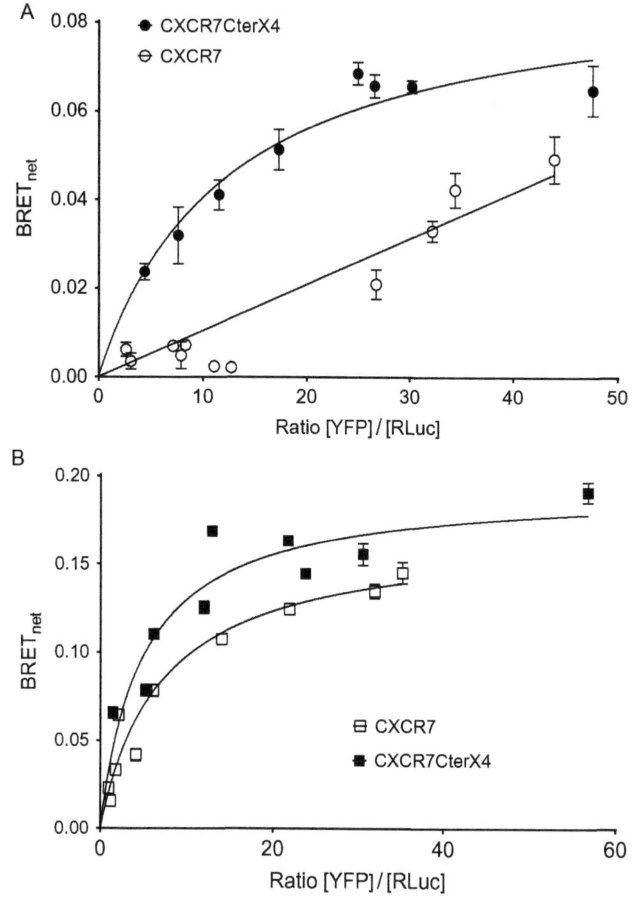

Figure 2 Detection of spontaneous arrestin recruitment by BRET titration experiments. (A) CXCR7-YFP (white circles) and the chimeric receptor CXCR7CterX4-YFP (black circles) were cotransfected at increasing amounts with fixed amounts of beta arrestin 2-RLuc, and $BRET_{net}$ determined *in the absence of agonist.* While CXCR7:arrestin yields a straight line, CXCR7CterX4 yields a clear hyperbola. (B) In presence of agonist, both receptors recruit arrestin and yield hyperbolae, with increased $BRET_{net}$. *Modified from data originally published as supplementary data in Gravel et al. (2010), © the American Society for Biochemistry and Molecular Biology.*

donor concentration—yields near saturation of $BRET_{net}$ ($BRET_{max}$). This acceptor concentration results in optimal sensitivity and interexperiment reproducibility, and is thus to be chosen for experiments that are performed at a single [acceptor]/[donor] ratio. Too low [acceptor]/[donor] ratios yield suboptimal BRET and consequently less sensitivity as well as high interexperiment variation. Too high [acceptor]/[donor] ratios in turn will result in higher bystander BRET and thus will reduce signal-to-noise ratios.

2.4 Use of Optimized Systems

2.4.1 Constitutive Arrestin Recruitment

For the detection of constitutive arrestin recruitment to chemokine receptors, BRET saturation experiments are recommended. In general, arrestin recruitment to chemokine receptors is expected to be ligand dependent, and thus, no specific BRET is anticipated in the absence of chemokine. However, as a special case, some receptors may constitutively associate with arrestin in preassembled complexes. Constitutive recruitment has been reported for the atypical chemokine receptor ACKR2/D6 (McCulloch et al., 2008; Weber et al., 2004). We have shown constitutive arrestin recruitment by a chimeric receptor consisting in CXCR7/ACKR3 with its C-terminal swapped for that of CXCR4 (Gravel et al., 2010; Fig. 2). However, no such constitutive interaction was detected with the reciprocal chimera, or with wild-type CXCR7/ACK3, which both nevertheless recruit arrestin in response to the chemokine CXCL12.

2.4.2 Dose-Response Experiments

The differential activation of signaling pathways along the G protein- or arrestin-dependent signaling axes by different ligands is a hallmark of functional selectivity. Chemokine receptor:arrestin BRET, at the determined optimal donor and acceptor plasmid concentrations, permits the determination of the respective EC_{50}s and maximal BRET responses to different chemokines in dose-response experiments. Relative ligand rank orders of potency can be established that can then be compared to potency rank orders in different readouts such as G-protein activation, informing about potential signaling bias of the respective chemokines. This type of analyses has been reported with a series of chemokine receptors and their multiple chemokine ligands (for example in Berchiche et al., 2011; Corbisier et al., 2015).

Another parameter that may differ significantly between different chemokine ligands at saturating chemokine concentrations is the efficacy of the BRET response, $BRET_{max}$, which is also determined by BRET saturation experiments at saturating ligand concentration. The interpretation of different BRET efficiencies will be dealt with below in detail, but it must be mentioned here that this parameter does describe the efficacy of the BRET response, but—somewhat counter-intuitively—not necessarily the efficacy of arrestin recruitment.

2.4.3 Time-Course Experiments

The dynamics of the receptor:arrestin association and dissociation can be measured by BRET. While the functional correlates of arrestin recruitment dynamics (such as potential shortening of the duration of G-protein signaling) will require further study, the longevity of the receptor:arrestin complex has been implicated in intracellular receptor trafficking and arrestin-dependent downstream signaling (Oakley, Laporte, Holt, Caron, & Barak, 2000; Zimmerman et al., 2012). Initially, the observation of differences in the stability of arrestin-receptor complexes has led to the postulation of "type A" and "type B" receptors, where "type A" receptors engage in stable association with arrestin, and "type B" receptors only transiently recruit the protein (Oakley et al., 2000). However, it turned out that these properties are ligand dependent, rather than receptor dependent (Shukla et al., 2008; Zimmerman et al., 2012). For example, significant differences in CCR2:arrestin dissociation were identified after challenge with the CCR2 ligands CCL2, CCL7, CCL8, and CCL13 (Berchiche et al., 2011; Fig. 3). The dissociation kinetics of receptor:arrestin complexes potentially represents another layer of functional selectivity, and differences in kinetics suggest different modalities of arrestin engagement by the same receptor bound to different ligands.

For arrestin association kinetics, the chemokine ligand is added to the cells after the preincubation with the luciferase substrate, followed immediately by sequential BRET[1] readings. For BRET[2], substrate addition must be matched with each reading, which is best achieved using an automated injector. Therefore, in BRET[2], kinetics readings require the use of multiple wells, each representing one single time point. As arrestin association can be very rapid (*t*½ of few seconds), it can be useful to decelerate the kinetics in order to increase the resolution, by shifting the cells from 37 °C to room temperature. The determination of the receptor:arrestin dissociation kinetics, however, is temperature dependent and requires the cells to be incubated at 37 °C. This can be achieved either by using a heated chamber during the readings or, perhaps more conveniently, by adding the ligand at different time points to cells continuously kept at 37 °C and then recording BRET simultaneously. Moreover, arrestin dissociation kinetics require intact organization of the cytoskeleton and should thus be performed using cells attached to the plate, and not in suspension.

2.5 Troubleshooting

It happens that receptor:arrestin BRET[1] systems do not yield sufficiently strong signals for meaningful analysis. This may occur, for example, if a given

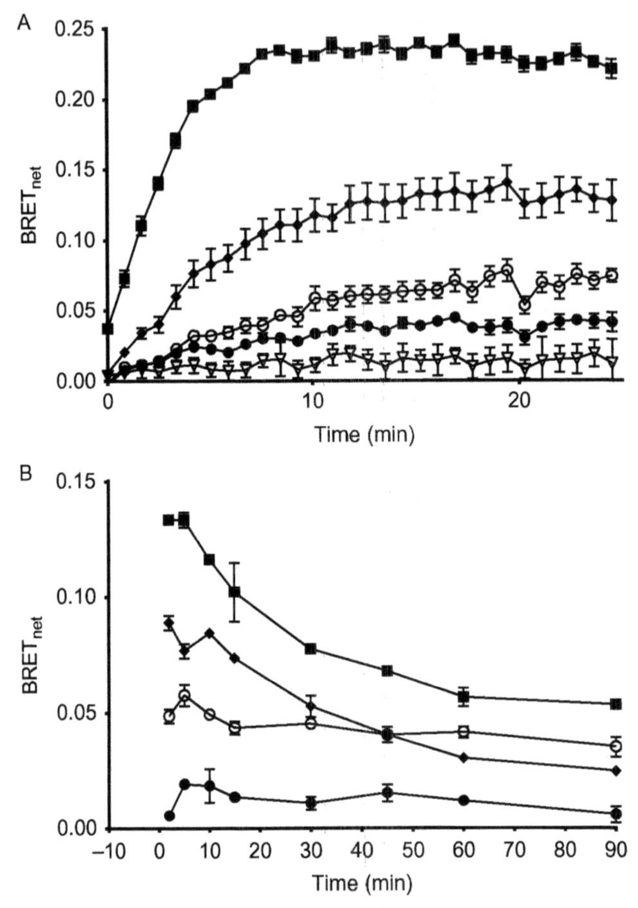

Figure 3 Time-course of arrestin recruitment and dissociation. (A) Association kinetics of arrestin-RLuc to CCR2-YFP after exposure to several of its chemokine ligands (CCL2: black squares; CCL7: black diamonds; CCL8: black circles; CCL13: white circles). (B) Decrease of the arrestin-RLuc:CCR2-YFP association over time after stimulation with the same ligands. *Reprinted with permission from Berchiche et al. (2011).*

system does not bring BRET donor and acceptor in sufficient proximity, or in unfavorable relative orientation to generate adequate BRET signals. Rearrangement of the energy probes or choice of BRET^2 as an alternative detection system may help solve such problems.

2.5.1 Switching the Probe Position

To palliate unfavorable acceptor–donor distance and especially orientation, the fused fluorophores may be switched: either from the arrestin C-terminal to its N-terminal, or between receptor and arrestin (Fig. 4A). Indeed, fusions to either the arrestin N-terminal or C-terminal have been used with success, but not necessarily with identical results (Bellot et al., 2015). We routinely obtained good results using the C-terminally fused arrestin-RLuc constructs with a number of chemokine receptors including CCR2, CXCR3, and CXCR7. While not systematically analyzed, we have found that, at least for CXCR4, the N-terminal arrestin fusion construct RLuc-arrestin does not yield satisfactory results, while the inverse fusion arrestin-RLuc does (Fig. 4B). Alternatively, switching BRET acceptor and donor between receptor and arrestin will also alter the relative donor–acceptor orientation. Probe switching applied to chemokine receptor heterodimers revealed its conformational impact: CXCR4:CCR2 heterodimers responded to CCL2 challenge with BRET increase or decrease, depending on the probe configuration; however, both probe configurations yielded BRET increase in response to CXCL12 (Percherancier et al., 2005). It is also likely that the composition and especially the rigidity of the linker sequence in the fusion proteins affect their ability to yield defined BRET signals, although this issue has not been studied systematically.

2.5.2 Use of BRET^2 to Improve the Signal:Noise Ratio

Alternatively, switching from BRET^1 to BRET^2 (or another generation of donor:acceptor pairs) may increase the intelligibility of the BRET signal (see also Kocan, Dalrymple, Seeber, Feldman, & Pfleger, 2010; Lohse, Nuber, & Hoffmann, 2012). Historically, BRET^1 systems have been developed first and are thus still prevalent in use. However, there is no general reason to favor BRET^1 over BRET^2, if corresponding filter sets are available.

The main difference between BRET^1 and BRET^2 is the use of a different substrate, coelenterazine 400A (DeepBlue C) in combination with a wavelength-shifted fluorescent energy acceptor proteins (GFP2 or GFP10) (Leduc et al., 2009). This leads to much better separation of the respective donor and acceptor emission wavelengths (Denis, Sauliere, Galandrin, Senard, & Gales, 2012) and thus, unlike BRET^1, in little donor emission spillover into the acceptor emission channel. However, this comes at the price of generally lower luminescence signals. This problem can be bypassed by the use of enhanced RLuc variants (RLuc3 and RLuc8) (Loening, Fenn, Wu, & Gambhir, 2006). The improved signal-to-noise ratio of BRET^2 may yield more intelligible signals than BRET^1.

The practical impact of the use of BRET^2 derives from the different emission spectra (365–435 nm (RLuc3) and 505–525 nm (GFP10)

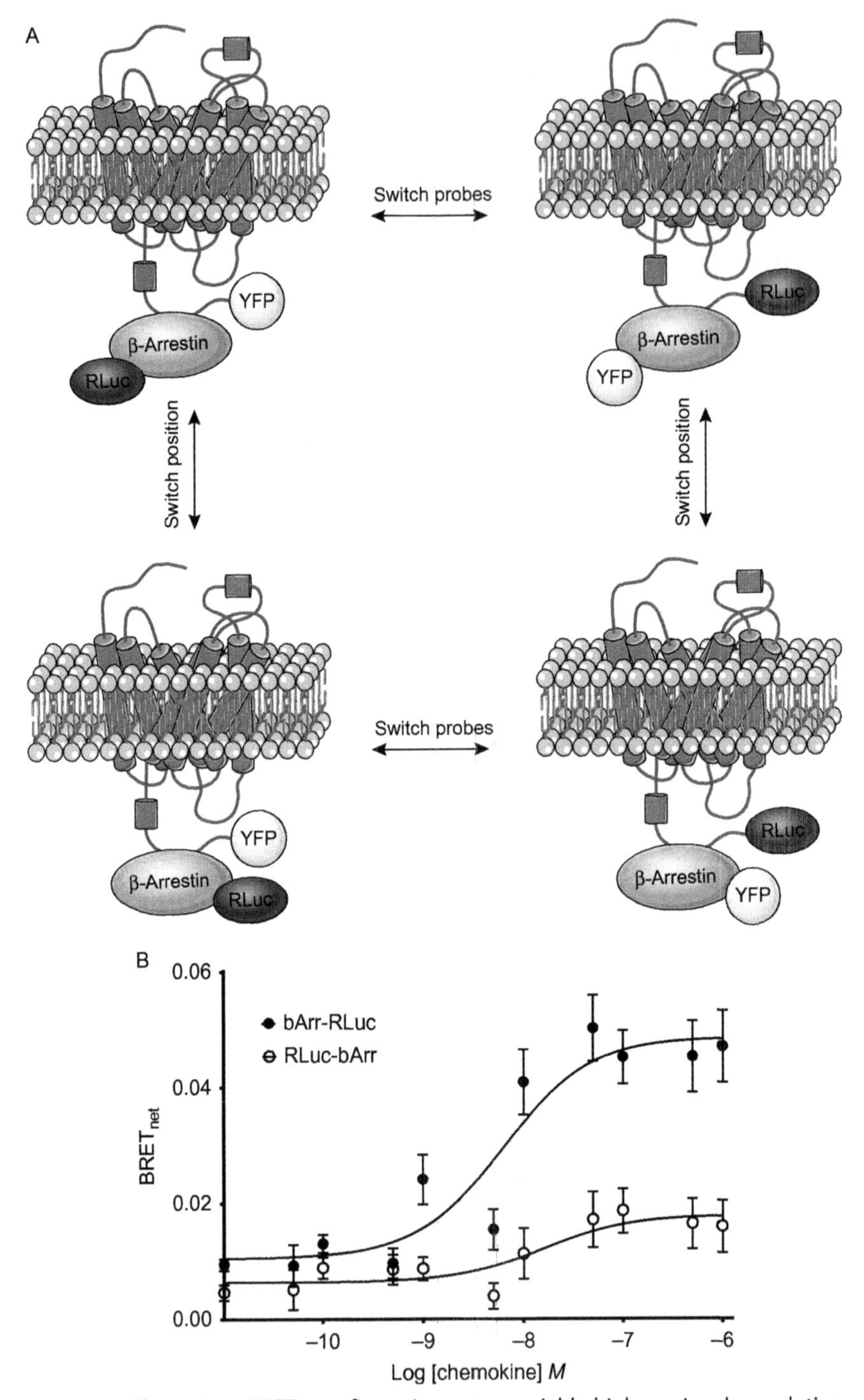

Figure 4 Alternative BRET configurations may yield higher signal resolution. (A) Schematic representation of the different BRET configurations achieved through probe switching or position switching. (B) Increase of resolution to detect arrestin: CXCR4 interaction, through switching from the N-terminally fused RLuc-arrestin (white circles) to the C-terminally fused arrestin-RLuc donor construct (black circles).

windows) that require the use of different filter sets, and from the rather rapid fading of the substrate coelenterazine 400A. For BRET^2, we use filters 390±22 nm (RLuc3) and 510±10 nm (GFP10). Obviously, due to the near absence of energy donor/acceptor spectral overlap, larger bandpasses are possible in BRET^2, and actually recommended to increase signal strength. The rapid fading of the substrate coelenterazine 400A entails that no substrate preincubation is recommended, and that BRET^2 is less suitable for long-term real-time kinetic measurements (unless a separate well is used for each time point, in combination with the use of a substrate injector).

3. INTERPRETATION AND LIMITATIONS OF BRET DATA

To facilitate interpretation of all generated results, it seems useful to recall here that the Förster radius (R_0), which is the distance in which resonance energy transfer occurs between a given donor and acceptor, crucially depends on their relative orientation. Indeed, the orientation factor (k^2) is integral part of the Förster radius and can take values between 0 and 4. Unlike in freely rotating probes, where k^2 can be averaged (Patterson, Piston, & Barisas, 2000), there is no rotational freedom of the probes in experiments with fusion proteins. In fusion proteins, the relative orientation and thus the efficacy of energy transfer between the probes are strongly determined by the conformation of the fused proteins, and of the complex of which they are part (Fig. 5; see also Muller, Galliardt, Schneider, Barisas, & Seidel, 2013). Before dealing with the interpretation of BRET arrestin recruitment data in different settings, we will therefore briefly clarify

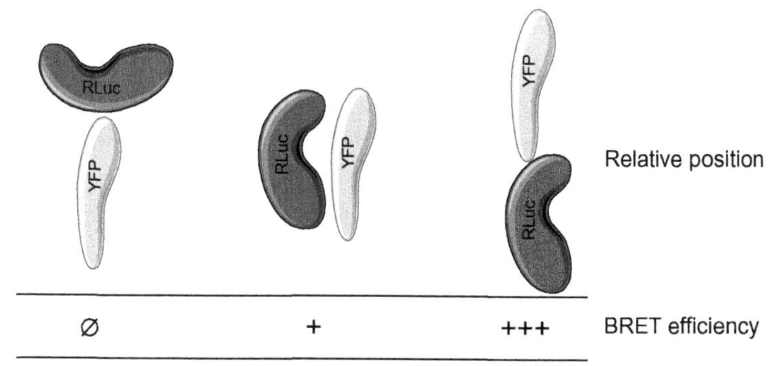

Figure 5 Relative orientation affects energy transfer efficiency. Schematic representation of three different possible relative orientations between donor (RLuc) and acceptor (YFP). Relative orientations may be nonpermissive (left), suboptimal (middle), or optimal (right) for energy transfer.

the meaning of two often used experimentally determined comparative parameters, namely $BRET_{max}$ and $BRET_{50}$.

3.1 Interpreting $BRET_{max}$

Receptor:arrestin $BRET_{max}$ is determined using saturating [acceptor]/[donor] ratios and saturating ligand concentrations. Given that the BRET signal and arrestin recruitment progressively rise with increasing agonist concentrations in dose-response experiments, the amplitude of $BRET_{max}$ is sometimes mistaken as reflecting the efficacy of arrestin recruitment (i.e., high $BRET_{max}$ indicating strong arrestin recruitment and low $BRET_{max}$ representing weak recruitment). This interpretation is not accurate, given the major impact of the BRET partner complex conformation on the orientation factor, and thus Förster radius and BRET efficiency. Rather, different signal intensities at $BRET_{max}$ brought about by different ligands of the same receptor may as well (and probably more often) represent different receptor:arrestin complex conformations. Conformational differences may of course, in addition, also affect the distance between the donor and the acceptor. The conformational aspect of the $BRET_{max}$ amplitude is substantiated by recent experiments where two distinct ligands of the same receptor yielded equal $BRET_{max}$ with RLuc-arrestin, but different $BRET_{max}$ with arrestin-RLuc (Bellot et al., 2015). These differences correlated with distinct intracellular routing of the complex, further suggesting different arrestin recruitment modalities in response to both ligands.

3.2 Interpreting $BRET_{50}$

$BRET_{50}$ depends on the steepness of the BRET saturation curve in [acceptor]/[donor] titration experiments, and thus on the acceptor concentration required to saturate all donor molecules. It is interpreted as reflecting the affinity of the BRET acceptor for the donor (Mercier, Salahpour, Angers, Breit, & Bouvier, 2002). Therefore, $BRET_{50}$ represents a much better parameter than $BRET_{max}$ to address the relative efficacy between arrestin responses generated by different ligands within the same experimental BRET system. Lower $BRET_{50}$ values would result from rapid donor saturation at increasing concentrations of the acceptor, reflecting high propensity of the BRET partners to interact. Relatively higher $BRET_{50}$ in turn would indicate a requirement for higher acceptor concentrations to saturate all donors, and thus lower efficacy of arrestin recruitment.

This being said, such differences in arrestin efficacy may or may not also correlate with differences in arrestin recruitment modalities.

3.3 Comparing Arrestin Recruitment to Different Chemokine Receptors

The dependence of the sensitivity of BRET systems on complex conformation implies that different receptors (for example, two chemokine receptors binding the same chemokine) should be compared with caution. Direct $BRET_{max}$ comparisons are rendered meaningless by conformational differences between different BRET systems. $BRET_{50}$ comparisons, however, are in principle possible. They are best combined with complementary experimental approaches.

3.4 Comparing Arrestin Recruitment by Different Ligands to the Same Chemokine Receptor

Comparison of the effects of different ligands binding to the same receptor is the hallmark of functional selectivity. This concerns both the potency (EC_{50}) and the efficacy (maximal response) of the different signaling axes. Clearly, the EC_{50}s and thus the potency of arrestin responses of different ligands at the same receptor determined by dose-response experiments can be compared (Berchiche et al., 2011; Corbisier et al., 2015), and may reveal signaling bias. Disparate $BRET_{50}$ values may suggest actual differences in the affinity of ligand-bound receptor for arrestin, and thus the efficacy of arrestin recruitment.

As outlined above, the efficacy of the BRET responses ($BRET_{max}$) cannot be equated with the efficacy of arrestin recruitment. If quantitative measures of arrestin recruitment efficacy are sought, this may be better served with the use of reporter complementation assays. These have successfully been used with chemokine receptors (Ikeda, Kumagai, Skach, Sato, & Yanagisawa, 2013; Rajagopal et al., 2013). Note that these assays, while being quantitative, do not permit arrestin dissociation experiments, and that conformational information is lost.

Different $BRET_{max}$, however, is indicative of conformational differences in the receptor:arrestin complex (Fig. 6), which may be of functional importance. Arrestin is a multifunctional protein that can scaffold the assembly (and activation) of a plethora of interacting proteins such as different protein kinases, phosphatases, ubiquitin ligases, and deubiquitinases (Shukla et al., 2011), which differently affect arrestin conformation. Moreover, arrestin can be recruited to receptors as monomeric or oligomeric species,

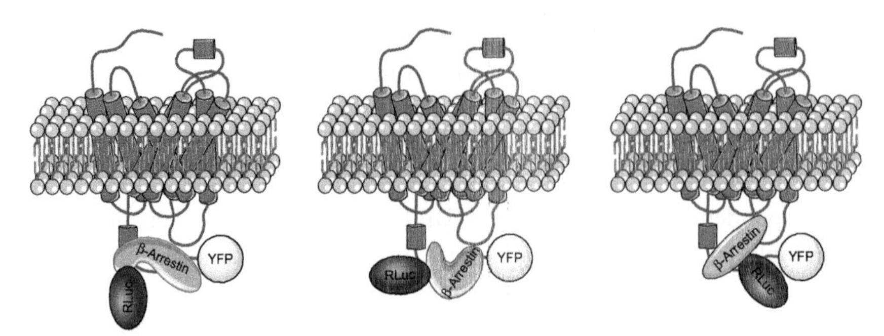

Figure 6 Arrestin recruitment modalities affect BRET efficiency. Schematic representation of three different possible recruitment modalities, which would yield different energy transfer efficiency.

with different functional outcome (Boularan et al., 2007). Such differences in arrestin conformation or "recruitment modalities" (Audet et al., 2012) can be relevant for functional differences between ligands (Audet et al., 2012; Nobles et al., 2011; Shukla et al., 2008; Zimmerman et al., 2012). Therefore, arrestin recruitment modalities may represent an additional layer of functional selectivity that takes also place *within* the arrestin pathway, rather than only between G protein- and arrestin-mediated signaling pathways. For example, this may be the case for CCR2 ligands (Berchiche et al., 2011), where different $BRET_{max}$ induced by different ligands was reported. To conclude on potential differences in arrestin recruitment modalities by ligands, it is recommended to rely on cumulative evidence (such as time-course experiments) or complementary experiments addressing downstream signaling or intracellular receptor transport, rather than on differences in $BRET_{max}$ alone. Arrestin conformation may be also directly assessed using conformational BRET reporters such as dual-brilliance arrestin BRET reporters (RLuc-arrestin-YFP or YFP-arrestin-RLuc) (Charest, Terrillon, & Bouvier, 2005; Nobles et al., 2011; Shukla et al., 2008) that monitor conformational changes within arrestin. While the double-brilliance arrestin probe provides a tool with great potential to evaluate arrestin recruitment by endogenous 7TMRs in primary cells and tissues, for the time being, its practicality is, however, rather limited due to a rather narrow dynamic range.

3.5 Comparing Receptor Mutants

Arrestin recruitment can also be compared between different substitution point mutants of the same receptor. Such comparisons may yield information about key residues for arrestin recruitment, for example, the respective roles of specific phosphorylation sites in the receptor C-terminal (Busillo et al., 2010; de Munnik et al., 2015). Substitution of some of these potential phosphorylation sites, but not others, may result in the absence of detectable arrestin recruitment. One conundrum of such analyses is that amino acid substitutions in the receptor may alter the sensitivity of the BRET system via conformational effects. This is illustrated by an example in which Busillo et al. found that alanine substitutions of phosphorylation sites in the CXCR4 C-terminal could lead to *increased* BRET efficiency—the authors concluded a conformational effect (Busillo et al., 2010). Because these effects may also decrease (or abolish) BRET efficiency for purely technical reasons (and not as an intrinsic property of the mutant receptors), we recommend to gather complementary evidence before concluding on the efficacy of arrestin recruitment. Alternative interpretations such as the occurrence of different arrestin recruitment modalities to different receptor mutants should also be considered. This applies in particular to mutations of phosphorylation sites, as the intricate receptor phosphorylation patterns (bar-codes) by multiple kinases are known to determine the recruited arrestin species (Nobles et al., 2011).

4. PERSPECTIVES

Arrestin recruitment as a main chemokine receptor signaling axis is of significant interest in the light of functional selectivity. The measurement of arrestin recruitment by BRET yields quantitative parameters such as agonist (or antagonist) potency that were challenging to obtain using microscopy. Therefore, BRET has become a standard method for arrestin recruitment. However, it is often underappreciated that results derived from receptor: arrestin BRET systems are not purely quantitative, but also contain conformational information. The inherent blend of quantitative and qualitative information gathered by BRET can be both curse and blessing: while it clearly complicates the interpretation of BRET efficiency, it may also provide first indications for the occurrence of different arrestin recruitment modalities. Besides divergent BRET efficiency between chemokine ligands of the same receptor or receptor mutants, differences in arrestin association and dissociation kinetics suggest different recruitment modalities. To further complicate matters, it cannot be assumed that any given receptor:ligand pair allows only one single recruitment modality: rather, combinations of different recruitment modalities at variable proportions may coexist. For example, the observed biphasic arrestin:receptor dissociation kinetics for the combination CCR2:CCL2, but not CCR2:CCL7 (see also Fig. 3B; Berchiche et al., 2011), may point in this direction. The investigation of mechanisms determining arrestin recruitment modalities—and of the resulting downstream signaling and trafficking effects—is ongoing, and actually just beginning. Application and development of complementary methods will be required for their in-depth investigation.

ACKNOWLEDGMENTS

The authors thank Mathias Plourde for help with the drawings. N.M. was supported by fellowships from the "Fonds de Recherche du Québec–Santé"(FRQS) and the Cole Foundation. N.H. was supported by Canadian Institutes of Health Research grant MOP-123421.

REFERENCES

Angers, S., Salahpour, A., Joly, E., Hilairet, S., Chelsky, D., Dennis, M., et al. (2000). Detection of beta 2-adrenergic receptor dimerization in living cells using bioluminescence resonance energy transfer (BRET). *Proceedings of the National Academy of Sciences of the United States of America, 97*(7), 3684–3689.

Audet, N., Charfi, I., Mnie-Filali, O., Amraei, M., Chabot-Dore, A. J., Millecamps, M., et al. (2012). Differential association of receptor-Gbetagamma complexes with beta-arrestin2 determines recycling bias and potential for tolerance of delta opioid receptor agonists. *The Journal of Neuroscience, 32*(14), 4827–4840.

Barak, L. S., Ferguson, S. S., Zhang, J., & Caron, M. G. (1997). A beta-arrestin/green fluorescent protein biosensor for detecting G protein-coupled receptor activation. *The Journal of Biological Chemistry, 272*(44), 27497–27500.

Bellot, M., Galandrin, S., Boularan, C., Matthies, H. J., Despas, F., Denis, C., et al. (2015). Dual agonist occupancy of AT1-R-alpha2C-AR heterodimers results in atypical Gs-PKA signaling. *Nature Chemical Biology, 11*(4), 271–279.

Berchiche, Y. A., Gravel, S., Pelletier, M. E., St-Onge, G., & Heveker, N. (2011). Different effects of the different natural CC chemokine receptor 2b ligands on {beta}-arrestin recruitment, Gαi signaling, and receptor internalization. *Molecular Pharmacology, 79*(3), 488–498.

Boularan, C., Scott, M. G., Bourougaa, K., Bellal, M., Esteve, E., Thuret, A., et al. (2007). Beta-arrestin 2 oligomerization controls the Mdm2-dependent inhibition of p53. *Proceedings of the National Academy of Sciences of the United States of America, 104*(46), 18061–18066.

Busillo, J. M., Armando, S., Sengupta, R., Meucci, O., Bouvier, M., & Benovic, J. L. (2010). Site-specific phosphorylation of CXCR4 is dynamically regulated by multiple kinases and results in differential modulation of CXCR4 signaling. *The Journal of Biological Chemistry, 285*(10), 7805–7817.

Chakera, A., Seeber, R. M., John, A. E., Eidne, K. A., & Greaves, D. R. (2008). The duffy antigen/receptor for chemokines exists in an oligomeric form in living cells and

functionally antagonizes CCR5 signaling through hetero-oligomerization. *Molecular Pharmacology*, *73*(5), 1362–1370.
Charest, P. G., Terrillon, S., & Bouvier, M. (2005). Monitoring agonist-promoted conformational changes of beta-arrestin in living cells by intramolecular BRET. *EMBO Reports*, *6*(4), 334–340.
Colvin, R. A., Campanella, G. S., Manice, L. A., & Luster, A. D. (2006). CXCR3 requires tyrosine sulfation for ligand binding and a second extracellular loop arginine residue for ligand-induced chemotaxis. *Molecular and Cellular Biology*, *26*(15), 5838–5849.
Colvin, R. A., Campanella, G. S., Sun, J., & Luster, A. D. (2004). Intracellular domains of CXCR3 that mediate CXCL9, CXCL10, and CXCL11 function. *The Journal of Biological Chemistry*, *279*(29), 30219–30227.
Corbisier, J., Gales, C., Huszagh, A., Parmentier, M., & Springael, J. Y. (2015). Biased signaling at chemokine receptors. *The Journal of Biological Chemistry*, *290*(15), 9542–9554.
de Munnik, S. M., Kooistra, A. J., van Offenbeek, J., Nijmeijer, S., de Graaf, C., Smit, M. J., et al. (2015). The viral G protein-coupled receptor ORF74 hijacks beta-arrestins for endocytic trafficking in response to human chemokines. *PLoS One*, *10*(4), e0124486.
Denis, C., Sauliere, A., Galandrin, S., Senard, J. M., & Gales, C. (2012). Probing heterotrimeric G protein activation: Applications to biased ligands. *Current Pharmaceutical Design*, *18*(2), 128–144.
Ferguson, S. S., & Caron, M. G. (2004). Green fluorescent protein-tagged beta-arrestin translocation as a measure of G protein-coupled receptor activation. *Methods in Molecular Biology*, *237*, 121–126.
Galandrin, S., & Bouvier, M. (2006). Distinct signaling profiles of beta1 and beta2 adrenergic receptor ligands toward adenylyl cyclase and mitogen-activated protein kinase reveals the pluridimensionality of efficacy. *Molecular Pharmacology*, *70*(5), 1575–1584.
Galandrin, S., Oligny-Longpre, G., & Bouvier, M. (2007). The evasive nature of drug efficacy: Implications for drug discovery. *Trends in Pharmacological Sciences*, *28*(8), 423–430.
Gravel, S., Malouf, C., Boulais, P. E., Berchiche, Y. A., Oishi, S., Fujii, N., et al. (2010). The peptidomimetic CXCR4 antagonist TC14012 recruits beta-arrestin to CXCR7: Roles of receptor domains. *The Journal of Biological Chemistry*, *285*(49), 37939–37943.
Hamdan, F. F., Audet, M., Garneau, P., Pelletier, J., & Bouvier, M. (2005). High-throughput screening of G protein-coupled receptor antagonists using a bioluminescence resonance energy transfer 1-based beta-arrestin2 recruitment assay. *Journal of Biomolecular Screening*, *10*(5), 463–475.
Ikeda, Y., Kumagai, H., Skach, A., Sato, M., & Yanagisawa, M. (2013). Modulation of circadian glucocorticoid oscillation via adrenal opioid-CXCR7 signaling alters emotional behavior. *Cell*, *155*(6), 1323–1336.
Kalatskaya, I., Berchiche, Y. A., Gravel, S., Limberg, B. J., Rosenbaum, J. S., & Heveker, N. (2009). AMD3100 is a CXCR7 ligand with allosteric agonist properties. *Molecular Pharmacology*, *75*(5), 1240–1247.
Kocan, M., Dalrymple, M. B., Seeber, R. M., Feldman, B. J., & Pfleger, K. D. (2010). Enhanced BRET technology for the monitoring of agonist-induced and agonist-independent interactions between GPCRs and beta-arrestins. *Frontiers in Endocrinology (Lausanne)*, *1*, 12.
Kohout, T. A., Nicholas, S. L., Perry, S. J., Reinhart, G., Junger, S., & Struthers, R. S. (2004). Differential desensitization, receptor phosphorylation, beta-arrestin recruitment, and ERK1/2 activation by the two endogenous ligands for the CC chemokine receptor 7. *The Journal of Biological Chemistry*, *279*(22), 23214–23222.
Leduc, M., Breton, B., Gales, C., Le Gouill, C., Bouvier, M., Chemtob, S., et al. (2009). Functional selectivity of natural and synthetic prostaglandin EP(4) receptor ligands. *The Journal of Pharmacology and Experimental Therapeutics*, *331*(1), 297–307.

Loening, A. M., Fenn, T. D., Wu, A. M., & Gambhir, S. S. (2006). Consensus guided mutagenesis of Renilla luciferase yields enhanced stability and light output. *Protein Engineering, Design & Selection, 19*(9), 391–400.

Lohse, M. J., Nuber, S., & Hoffmann, C. (2012). Fluorescence/bioluminescence resonance energy transfer techniques to study G-protein-coupled receptor activation and signaling. *Pharmacological Reviews, 64*(2), 299–336.

McCulloch, C. V., Morrow, V., Milasta, S., Comerford, I., Milligan, G., Graham, G. J., et al. (2008). Multiple roles for the C-terminal tail of the chemokine scavenger D6. *The Journal of Biological Chemistry, 283*(12), 7972–7982.

Mercier, J. F., Salahpour, A., Angers, S., Breit, A., & Bouvier, M. (2002). Quantitative assessment of beta 1- and beta 2-adrenergic receptor homo- and heterodimerization by bioluminescence resonance energy transfer. *The Journal of Biological Chemistry, 277*(47), 44925–44931.

Muller, S. M., Galliardt, H., Schneider, J., Barisas, B. G., & Seidel, T. (2013). Quantification of Forster resonance energy transfer by monitoring sensitized emission in living plant cells. *Frontiers in Plant Science, 4*, 413.

Nobles, K. N., Xiao, K., Ahn, S., Shukla, A. K., Lam, C. M., Rajagopal, S., et al. (2011). Distinct phosphorylation sites on the {beta}2-adrenergic receptor establish a barcode that encodes differential functions of {beta}-arrestin. *Science Signaling, 4*(185), ra51.

Oakley, R. H., Laporte, S. A., Holt, J. A., Caron, M. G., & Barak, L. S. (2000). Differential affinities of visual arrestin, beta arrestin1, and beta arrestin2 for G protein-coupled receptors delineate two major classes of receptors. *The Journal of Biological Chemistry, 275*(22), 17201–17210.

Odemis, V., Lipfert, J., Kraft, R., Hajek, P., Abraham, G., Hattermann, K., et al. (2012). The presumed atypical chemokine receptor CXCR7 signals through G(i/o) proteins in primary rodent astrocytes and human glioma cells. *Glia, 60*(3), 372–381.

Patterson, G. H., Piston, D. W., & Barisas, B. G. (2000). Forster distances between green fluorescent protein pairs. *Analytical Biochemistry, 284*(2), 438–440.

Percherancier, Y., Berchiche, Y. A., Slight, I., Volkmer-Engert, R., Tamamura, H., Fujii, N., et al. (2005). Bioluminescence resonance energy transfer reveals ligand-induced conformational changes in CXCR4 homo- and heterodimers. *The Journal of Biological Chemistry, 280*(11), 9895–9903.

Rajagopal, S., Bassoni, D. L., Campbell, J. J., Gerard, N. P., Gerard, C., & Wehrman, T. S. (2013). Biased agonism as a mechanism for differential signaling by chemokine receptors. *The Journal of Biological Chemistry, 288*(49), 35039–35048.

Rajagopal, S., Kim, J., Ahn, S., Craig, S., Lam, C. M., Gerard, N. P., et al. (2010). Beta-arrestin- but not G protein-mediated signaling by the "decoy" receptor CXCR7. *Proceedings of the National Academy of Sciences of the United States of America, 107*(2), 628–632.

Sauliere, A., Bellot, M., Paris, H., Denis, C., Finana, F., Hansen, J. T., et al. (2012). Deciphering biased-agonism complexity reveals a new active AT1 receptor entity. *Nature Chemical Biology, 8*(7), 622–630.

Shukla, A. K., Violin, J. D., Whalen, E. J., Gesty-Palmer, D., Shenoy, S. K., & Lefkowitz, R. J. (2008). Distinct conformational changes in beta-arrestin report biased agonism at seven-transmembrane receptors. *Proceedings of the National Academy of Sciences of the United States of America, 105*(29), 9988–9993.

Shukla, A. K., Xiao, K., & Lefkowitz, R. J. (2011). Emerging paradigms of beta-arrestin-dependent seven transmembrane receptor signaling. *Trends in Biochemical Sciences, 36*(9), 457–469.

Urban, J. D., Clarke, W. P., von Zastrow, M., Nichols, D. E., Kobilka, B., Weinstein, H., et al. (2007). Functional selectivity and classical concepts of quantitative pharmacology. *The Journal of Pharmacology and Experimental Therapeutics, 320*(1), 1–13.

Weber, M., Blair, E., Simpson, C. V., O'Hara, M., Blackburn, P. E., Rot, A., et al. (2004). The chemokine receptor D6 constitutively traffics to and from the cell surface to internalize and degrade chemokines. *Molecular Biology of the Cell*, *15*(5), 2492–2508.
Zimmerman, B., Beautrait, A., Aguila, B., Charles, R., Escher, E., Claing, A., et al. (2012). Differential beta-arrestin-dependent conformational signaling and cellular responses revealed by angiotensin analogs. *Science Signaling*, *5*(221), ra33.

CHAPTER EIGHT

Probing Biased Signaling in Chemokine Receptors

Roxana-Maria Amarandi*,†, Gertrud Malene Hjortø*, Mette Marie Rosenkilde*, Stefanie Karlshøj*,[1]

*Laboratory for Molecular Pharmacology, Department of Neuroscience and Pharmacology, Faculty of Health and Medical Sciences, The Panum Institute, University of Copenhagen, Copenhagen, Denmark
†Faculty of Chemistry, Alexandru Ioan Cuza University of Iaşi, Iaşi, Romania
[1]Corresponding author: e-mail address: stefanieth@sund.ku.dk

Contents

1. Introduction 156
2. Types of Bias in the Chemokine System 157
   2.1 Ligand Bias 157
   2.2 Receptor Bias 160
   2.3 Tissue Bias 161
3. Chemokine System-Mediated Intracellular Signaling 162
   3.1 G Protein-Dependent Signaling 162
   3.2 G Protein-Independent Signaling 164
4. Methods 165
   4.1 General 165
   4.2 G Protein Binding ([^{35}S]-GTPγS Binding): Nonselective Assay 166
   4.3 G Protein Signaling (cAMP Assay/SPA-IP$_3$ Assay): Selective Assays 168
   4.4 β-Arrestin Recruitment Assay 171
   4.5 Internalization Assay Using ELISA or Confocal Microscopy 172
   4.6 ERK Phosphorylation 175
   4.7 Chemotaxis 177
Acknowledgments 180
References 181

Abstract

The chemokine system mediates leukocyte migration during homeostatic and inflammatory processes. Traditionally, it is described as redundant and promiscuous, with a single chemokine ligand binding to different receptors and a single receptor having several ligands. Signaling of chemokine receptors occurs via two major routes, G protein- and β-arrestin-dependent, which can be preferentially modulated depending on the ligands or receptors involved, as well as the cell types or tissues in which the signaling event occurs. The preferential activation of a certain signaling pathway to the detriment of others has been termed signaling bias and can accordingly be grouped into ligand bias, receptor bias, and tissue bias. Bias has so far been broadly

Methods in Enzymology, Volume 570
ISSN 0076-6879
http://dx.doi.org/10.1016/bs.mie.2015.09.001

overlooked in the process of drug development. The low number of currently approved drugs targeting the chemokine system, as well as the broad range of failed clinical trials, reflects the need for a better understanding of the chemokine system. Thus, understanding the character, direction, and consequence of biased signaling in the chemokine system may aid the development of new therapeutics. This review describes experiments to assess G protein-dependent and -independent signaling in order to quantify chemokine system bias.

1. INTRODUCTION

The chemokine system mediates leukocyte migration and is therefore crucial for homeostasis and development of the immune system, immune reactions, and defense mechanisms, but also at other occasions requiring directed cell growth and migration, such as embryonic development. Due to these functions, the chemokine system is considered a strategic target for pharmacological intervention in several inflammatory, allergic, and autoimmune diseases (Viola & Luster, 2008), such as rheumatoid arthritis (Szekanecz, Kim, & Koch, 2003), multiple sclerosis (Szczuciński & Losy, 2007), atherosclerosis (Koenen & Weber, 2011), but also HIV infection (Berger, Murphy, & Farber, 1999) and cancer development and progression (angiogenesis and metastasis) (Rosenkilde & Schwartz, 2004). Furthermore, various microorganisms manipulate the endogenous chemokine system in their attempt to establish infection, spread within the body and between host organisms (Rosenkilde, 2005). In particular, viruses encode chemokines, chemokine receptors, and/or chemokine-binding proteins to avoid recognition by the immune system and ensure virus survival (Bachelerie et al., 2014; Rosenkilde & Kledal, 2006; Vischer, Siderius, Leurs, & Smit, 2014). Despite these important roles for the chemokine system, only two small-molecule drugs have reached the market: Maraviroc, a CCR5 antagonist used as HIV-entry inhibitor for R5-tropic strains, and a CXCR4 antagonist, plerixafor, which is used to mobilize hematopoietic stem cells from the bone marrow in patients undergoing chemotherapy (Proudfoot, Power, & Schwarz, 2010; Scholten et al., 2012; Steen, Schwartz, & Rosenkilde, 2009). This highlights the need for a better understanding of the system (O'Hayre, Salanga, Handel, & Hamel, 2010).

The key components of the chemokine system are a group of 8–12 kDa large peptides, known as chemokines, and their cognate 7-transmembrane helix receptors (7TMRs), also called G protein-coupled receptors (GPCRs).

In total, there are ≈24 human endogenous chemokine receptors, including 6 atypical chemokine receptors (ACKR1–6), and about 49 chemokines, which are divided into four groups based on the adjacency of the first two of usually four conserved N-terminal cysteines in their sequences. Thus, there are CC and CXC chemokines (major groups), and one CX_3C and two XC chemokines (minor groups; XC chemokines having only one N-terminal cysteine). Accordingly, there are 10 CC chemokine receptors (CCR1–10), 6 CXC chemokine receptors (CXCR1–6), CX_3CR1, and XCR1 (Allen, Crown, & Handel, 2007; Bachelerie et al., 2014). This complex protein system was commonly described as "redundant" because some chemokines have overlapping *in vivo* functions and chemokine:receptor pairs are usually not exclusive, with chemokines binding to several receptors, and receptors having several ligands (Mantovani, 1999). However, it is increasingly recognized that precise organization, rather than redundancy, is the underlying characteristic of this complex system, with biased signaling being one of the properties currently explored to unravel specific details of chemokine system organization and function (Schall & Proudfoot, 2011; Steen, Larsen, Thiele, & Rosenkilde, 2014; Zweemer, Toraskar, Heitman, & IJzerman, 2014).

Biased signaling, also referred to as functional selectivity or stimulus bias, reflects to the ability of a receptor to preferentially activate particular signaling pathways as a result of differential conformational stabilization by bound ligands (Kenakin, 2011; Kenakin & Christopoulos, 2013; Steen et al., 2014; Zweemer et al., 2014). Although endogenous bias is also found in other 7TM receptors, e.g., the μ-opioid (Rivero et al., 2012; Thompson, Canals, & Poole, 2014) and somatostatin receptors (Zhao et al., 2013), the chemokine system displays by far the largest bias platform, with its numerous ligands and receptors and broad and diverse expression patterns. Overall, biased signaling can be subdivided into three types: ligand bias, receptor bias, and cell/tissue bias, as summarized in Fig. 1 (Steen et al., 2014).

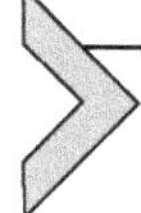

2. TYPES OF BIAS IN THE CHEMOKINE SYSTEM

2.1 Ligand Bias

The term "ligand bias" is used to describe the situation in which the same receptor can activate distinct pathways in response to different ligands (reviewed in Steen et al., 2014; Zweemer et al., 2014). In the chemokine system, ligand bias has been identified in relation to different chemokines acting at the same receptor (Rajagopal et al., 2013), chemokine monomers

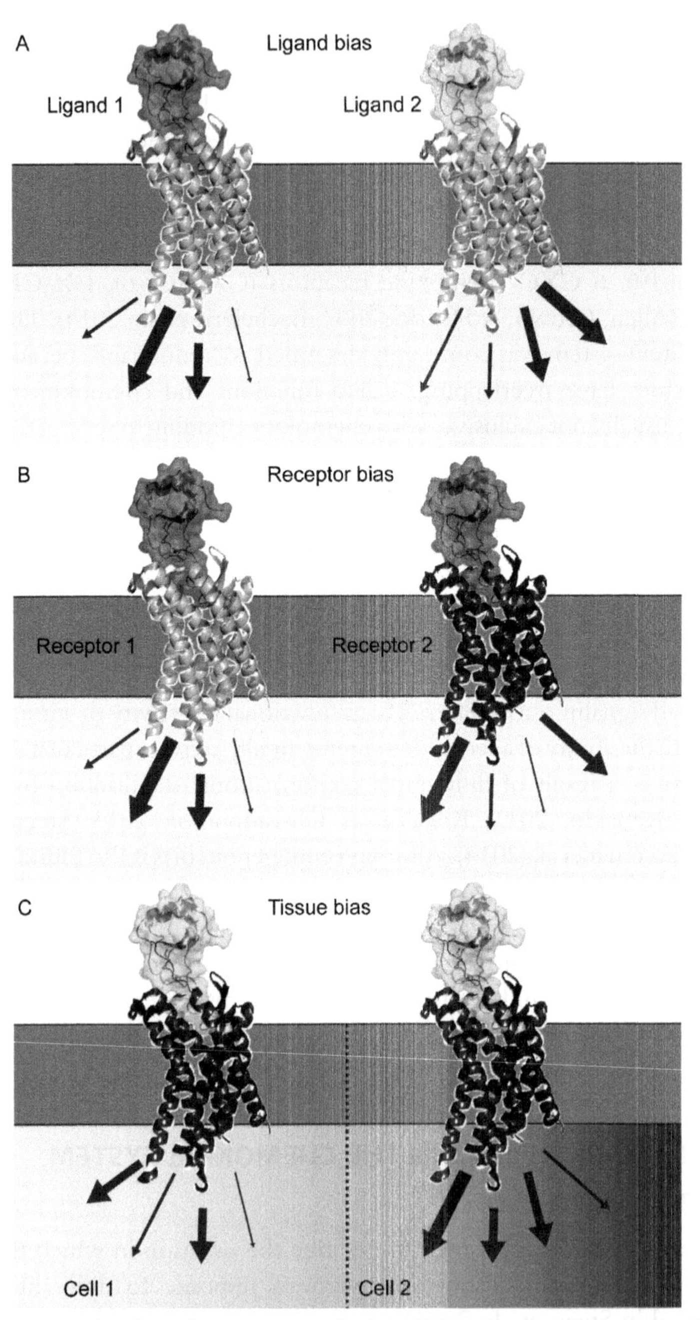

Figure 1 Bias in the chemokine system and other 7TMRs. Biased signaling describes the ability of a receptor to induce different signaling pathways or cellular events. Thereby, different ligands may activate different pathways via the same receptor (ligand bias, A), or the same ligand induces different outcomes at different receptors (receptor bias, B). Also the cell or tissue that "hosts" the ligand:receptor interaction can modulate the induced signaling pathway (tissue bias, C). (See the color plate.)

versus dimers (Drury et al., 2011), or wild-type versus posttranslationally modified chemokines (Savino et al., 2009; Fig. 1A). For example, while both endogenous chemokines of CCR7, CCL19, and CCL21 have the ability to trigger G protein-mediated signaling, only CCL19 is able to induce receptor internalization (Kohout et al., 2004; Zidar, Violin, Whalen, & Lefkowitz, 2009).

The quaternary organization of chemokines can also contribute to biased behavior. For example, the monomeric form of CXCL12 preferentially induces β-arrestin recruitment to CXCR4 and promotes Ca^{2+} flux and chemotaxis through G protein signaling. Dimeric CXCL12, however, loses the ability to induce β-arrestin recruitment and chemotaxis while maintaining its G protein-signaling capacity (Drury et al., 2011).

Posttranslational modifications of chemokines such as N-terminal truncation, glycosylation, or citrullination (Moelants, Mortier, Van Damme, & Proost, 2013; Moelants, Van Damme, & Proost, 2011; Mortier, Van Damme, & Proost, 2008) can also cause differential signaling compared to unmodified ligands. For example, while both full-length isoform CCL14 (1–74) and truncated isoform CCL14(9–74) are able to bind to the atypical chemokine receptor 2 (ACKR2), only the truncated CCL14(9–74) displays β-arrestin bias in comparison to CCL14(1–74) and is able to induce receptor internalization of ACKR2 (Savino et al., 2009).

Ligand bias can be evaluated by plotting two different cellular responses induced by the same concentration of agonist as functions of each other in order to obtain a "bias plot" (Fig. 2; Kenakin & Christopoulos, 2013). This

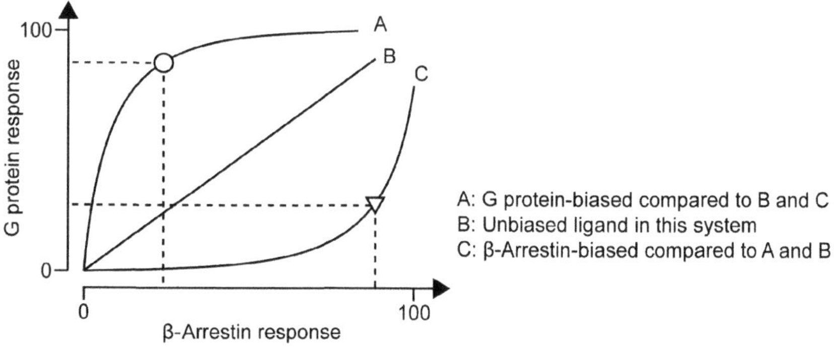

Figure 2 Bias plot. In order to quantify bias properties of one ligand or system compared to another, one can draw a bias plot, where the response of a certain concentration of ligand in one pathway is plotted over its response in another pathway. Note that bias is always comparative, i.e., one ligand has to be described as biased toward one pathway compared to another ligand. This is due to an intrinsic so-called system or observational bias. A standard ligand, e.g., the assumed full agonist endogenous ligand, could be used as this standard ligand.

approach has been successfully applied to identify biased ligands in several receptor systems (Gregory, Hall, Tobin, Sexton, & Christopoulos, 2010; Rajagopal et al., 2011) including the chemokine system (Corbisier, Gales, Huszagh, Parmentier, & Springael, 2015; Rajagopal et al., 2013).

In contrast to endogenous biased behavior, exogenous bias, i.e., bias mediated by nonendogenous ligands, has been extensively reported in both peptide and nonpeptide 7TM receptors and for orthosteric and allosteric ligands (Gregory et al., 2010; Kenakin, 2010; Kenakin & Miller, 2010; Shonberg et al., 2014). Exogenous bias has retrospectively been identified in already marketed drugs, such as the β-adrenergic receptor antagonist carvedilol (Wisler et al., 2007), but has also been successfully explored in the development of next-generation biased drugs targeting 7TM receptors, such as TRV120027 at angiotensin II receptor, which is currently in clinical trials for the treatment of heart failure (Boerrigter et al., 2011; Boerrigter, Soergel, Violin, Lark, & Burnett, 2012; Shonberg et al., 2014).

2.2 Receptor Bias

Receptor bias describes the situation in which the same ligand is able to induce distinct responses at different receptors (Fig. 1B; Steen et al., 2014). In the chemokine system, this type of bias is most obvious in relation to scavenger chemokine receptors, which are able to bind endogenous chemokines, but only elicit β-arrestin-dependent and not G protein-mediated signaling (Bachelerie et al., 2014). CXCL12 binds to CXCR4, its cognate G protein-signaling receptor, but also to ACKR3 (formerly known as CXCR7). While CXCR4 elicits G protein- as well as β-arrestin-dependent signaling in response to CXCL12 (Cheng et al., 2000), ACKR3 functions as a scavenger receptor and exclusively signals in a β-arrestin-mediated fashion (Rajagopal et al., 2010). Receptor bias can also occur in the case of two G protein-signaling chemokine receptors, where the same chemokine will exhibit differentiated behavior. This is the case for CXCL9, -10, and -11, which are endogenous agonists at CXCR3, but are able to antagonize and displace CCL11 from CCR3 (Loetscher et al., 2001). In addition to the human endogenous chemokines and their receptors, viruses often express viral chemokines and chemokine receptors as part of their mimicry and immune system escape mechanisms (Rosenkilde, Waldhoer, Lüttichau, & Schwartz, 2001). Interestingly, most of the viral chemokine receptors display biased signaling when compared to their eukaryotic equivalents in response to the same endogenous chemokine (Rosenkilde, Kledal,

Bräuner-Osborne, & Schwartz, 1999). For example, CX_3CL1 behaves as a full agonist at CX_3CR1, but as a partial inverse agonist at the receptor US28 from human cytomegalovirus—a receptor that also binds several CC chemokines (Casarosa et al., 2001; Kledal, Rosenkilde, & Schwartz, 1998; McLean, Holst, Martini, Schwartz, & Rosenkilde, 2004; Waldhoer et al., 2003). Among the CXC chemokines, CXCL6 is an agonist on CXCR1 and CXCR2 via $G\alpha_i$ activation, but acts as an agonist on the ECRF receptor from herpesvirus saimiri via $G\alpha_q$ activation (Rosenkilde, McLean, Holst, & Schwartz, 2004). Also, CXCL10 and -12 are full agonists in terms of $G\alpha_i$ activity of CXCR3 and CXCR4, respectively, but inverse agonists on the human herpesvirus 8-encoded ORF74 receptor (Rosenkilde et al., 1999; Rosenkilde & Schwartz, 2000).

2.3 Tissue Bias

Tissue bias, sometimes referred to as cell or system bias, reflects the ability of a ligand:receptor pair to differentially activate signaling pathways dependent on the cell, tissue, or species, ultimately reflecting the receptor microenvironment, e.g., availability of G proteins, G protein-coupled receptor kinases (GRKs), β-arrestins, effector or scaffolding proteins (Kenakin & Christopoulos, 2013; Steen et al., 2014; Fig. 1C). The existence of tissue bias in the chemokine system has been acknowledged as early as 1994, when it was coined "cell-dependent signal sorting" (Murphy, 1996). Then, it was found that CCL5 induces distinct signals in eosinophils and basophils (Dahinden et al., 1994); yet, this might really have reflected a case of receptor bias. Even earlier, tissue bias was identified in relation to β-adrenergic receptor activation in the rat atrium (Kenakin, Ambrose, & Irving, 1991). Further examples for the chemokine system include CCL19, which can induce chemotaxis of CCR7-expressing dendritic cells (Ricart, John, Lee, Hunter, & Hammer, 2011), but not T lymphocytes (Nandagopal, Wu, & Lin, 2011). A case of "exogenous" tissue bias has been identified for the small-molecule drug AMD3100. While it acted as antagonist in $G\alpha_i$ activation and Ca^{2+}-flux assays performed in CXCR4-transfected COS-7 and SUP-T1 cells (Bridger et al., 1999; Rosenkilde, Gerlach, et al., 2007), it was found to be a weak partial agonist in CXCR4-transfected CHO and naturally CXCR4-expressing THP-1 cells (Zhang et al., 2002).

These examples emphasize the need to consider tissue bias when planning assays or comparing results. Furthermore, due to the influence of the tissue or cell on the signaling outcome, bias of any ligand or receptor always needs to be looked at in relation to a "benchmark" ligand or receptor in the same cellular system (Kenakin & Christopoulos, 2013). Moreover, while being an important property of the cell, tissue bias can be a hurdle in the drug discovery process, and it is important to evaluate ligands in more than just one assay type and cell line in order to fully characterize their behavior and predict clinical relevance.

The function and signaling of the chemokine system *in vivo* is furthermore influenced by a number of additional factors that can be classified as tissue or rather "situation" bias. These include the up- and downregulation of chemokine receptor expression in response to chemokines. Furthermore, chemokines bind to the very diverse groups of extracellular matrix proteins and glycosaminoglycans (GAGs) with differing affinities, a property that is necessary for the establishment of chemokine gradients. However, specific GAG interactions will influence the composition and availability of chemokines in places of leukocyte trafficking (Johnson, Proudfoot, & Handel, 2005; Kufareva, Salanga, & Handel, 2015). Finally, the immediate history of a receptor influences its response, as uncoupling from G proteins and cross-desensitization might occur in response to a chemokine encounter (reviewed in Zweemer et al., 2014).

3. CHEMOKINE SYSTEM-MEDIATED INTRACELLULAR SIGNALING

Signaling via 7TM receptors can occur through several different pathways (Marinissen & Gutkind, 2001) and is most often described as G protein dependent or G protein independent. The former includes signaling through the Gα or Gβγ subunits of G proteins, while the latter includes signaling induced by β-arrestin recruitment, among others.

3.1 G Protein-Dependent Signaling

G proteins are membrane-bound heterotrimeric GTPases, consisting of an α, β, and γ subunit, and function as the main signal transducers of the cell. In the inactive state, 7TM receptors are usually associated with the inactive GDP-bound G protein. Upon activation, the receptor will promote the exchange of GDP for GTP in the Gα subunit, which induces G protein dissociation from the receptor and its parting into a Gα and Gβγ subunit, the latter of which is an obligate dimer (Lambert, 2008; Oldham & Hamm, 2008).

The Gα subunits can be grouped into four subfamilies: $G\alpha_s$, $G\alpha_{i/o}$, $G\alpha_q$, and $G\alpha_{12/13}$, depending on the downstream effectors activated upon receptor coupling (Milligan & Kostenis, 2006). Endogenous chemokine receptors display a marked preference toward $G\alpha_i$, which leads to inhibition of adenylate cyclase (AC) and reduced cAMP levels, consequently inhibiting protein kinase A (Fig. 3; Bokoch, 1995; Murphy, 1996). $G\alpha_i$-dependent signaling furthermore leads to activation of Src, which induces ERK1/2 phosphorylation through the mitogen-activated protein kinase (MAPK) pathway (Ganju et al., 1998; Kohout et al., 2004). Src also activates phosphatidylinositol-3-kinase (PI3K) to produce phosphatidylinositol-3,4,5-trisphosphate (PIP_3), which induces actin polymerization and formation of protrusions at the leading edge (O'Hayre, Salanga, Handel, & Allen, 2008). Moreover, Gβγ signaling has been tightly linked to chemotaxis and cell migration: PI3K is directly activated by Gβγ and leads to activation of

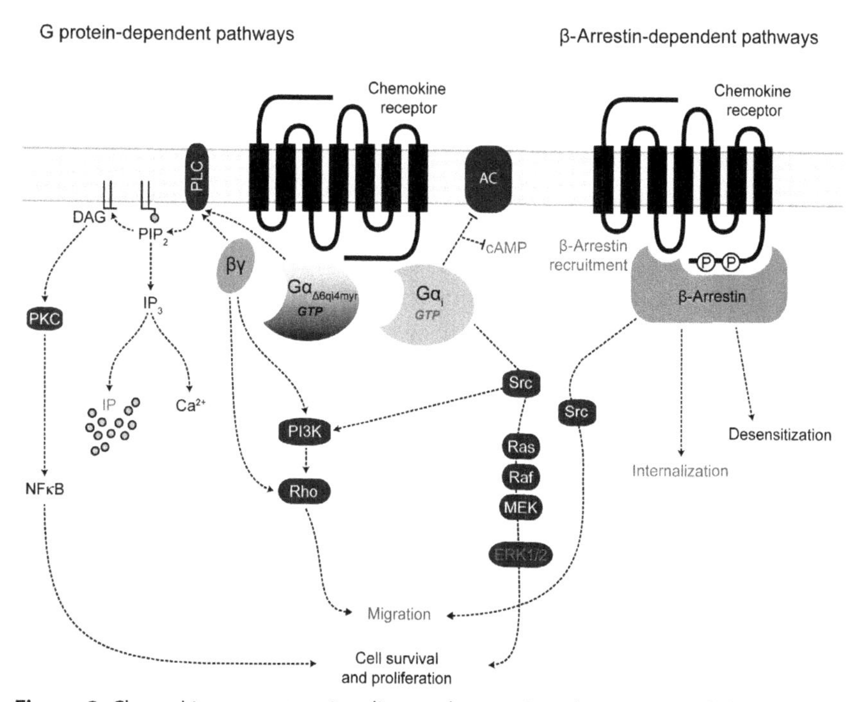

Figure 3 Chemokine receptor signaling pathways. Rough overview of the signaling pathways mediated by chemokine receptors that lead to migration or contribute to cell survival and proliferation. For outcomes highlighted in pink, assays are described in Section 4. The $G\alpha_{\Delta 6qi4myr}$ is an artificial G protein used in an IP_3 turnover assay to measure receptor-mediated activation of $G\alpha_i$. (See the color plate.)

distinct G-nucleotide exchange factors and the small GTPases Rac and Rho at the leading and trailing edge of cell, respectively, inducing actin polymerization at one end and contraction at the other (Rot & von Andrian, 2004). Gβγ also directly activates phospholipase C-β leading to production of inositol-1,4,5-trisphosphate (IP_3) and diacylglycerol from phosphatidylinositol-4,5-bisphosphate (PIP_2). IP_3 then binds to IP_3 receptors in the endoplasmic reticulum and induces Ca^{2+} release and rise in intracellular Ca^{2+} levels. For a detailed discussion of chemokine receptor signaling mechanism leading to cell migration, see Cain & Ridley (2009), O'Hayre et al. (2008), and Rot and von Andrian (2004).

Unlike their eukaryotic counterparts, viral chemokine receptors prefer G_q coupling, thus favoring the accumulation of inositol phosphates (Rosenkilde, 2005). However, some viral chemokine receptors signal through a larger diversity of G proteins, an ability that can be interpreted as host adaptation of the virus. For example, ORF74 can also trigger constitutive signaling through $G\alpha_{i/o}$ (Smit et al., 2002), as well as through $G\alpha_{12}$ or RhoA (Shepard et al., 2001), while US28 can couple in a ligand-dependent manner to $G\alpha_{i/o}$, $G\alpha_q$, and $G\alpha_{12/13}$ subunits, in addition to constitutive signaling through $G\alpha_q$ and Gβγ subunits (Rosenkilde & Kledal, 2006; Rosenkilde et al., 2001).

Finally, the atypical chemokine receptors ACKR1–6 (Bachelerie et al., 2014) are able to recruit chemokines and signal via β-arrestin-dependent pathways, but do not couple to G proteins (Graham, Locati, Mantovani, Rot, & Thelen, 2012). Thus, scavenger or atypical receptors are fully biased toward G protein-independent pathways.

3.2 G Protein-Independent Signaling

Phosphorylation of active-state receptor conformations by GRKs induces β-arrestin recruitment and subsequent displacement of G proteins, uncoupling receptors from their canonical signaling pathways and leading to receptor desensitization and internalization (Luttrell & Lefkowitz, 2002). β-Arrestins can also act as scaffolding proteins for downstream signaling molecules, including the Src family of tyrosine kinases and components of the MAP kinase signaling pathway, thus initiating G protein-independent signaling (DeWire, Ahn, Lefkowitz, & Shenoy, 2007). In the chemokine system, pathways initiated through β-arrestin recruitment regulate chemotaxis (DeFea, 2007; McGovern & DeFea, 2014), apoptosis (Revankar, Vines, Cimino, & Prossnitz, 2004), and other cellular functions

(Lefkowitz & Shenoy, 2005). An important initial step in β-arrestin-mediated signaling is the recruitment of c-Src and other nonreceptor tyrosine kinases, which in turn activate kinases of the MAPK pathway (Luttrell et al., 1999). For CXCR1-expressing neutrophils, this has been found to translate as chemokine-induced granule exocytosis (Barlic et al., 2000). It is important to note, however, that while both G proteins and β-arrestins are able to mediate the activation of MAPKs, the activated enzymes will play differential roles in the cell (Lefkowitz, Rajagopal, & Whalen, 2006). MAPKs activated by G proteins will translocate to the nucleus and phosphorylate transcription factors, thus regulating gene transcription, while β-arrestin-activated MAPKs remain in the cytosol to phosphorylate further substrates likely part of the cytoskeleton and migration machinery (Ahn, Shenoy, Wei, & Lefkowitz, 2004; Luttrell & Gesty-Palmer, 2010; Luttrell et al., 2001; Tohgo, Pierce, Choy, Lefkowitz, & Luttrell, 2002). The role of β-arrestins in cell migration is highlighted by the observation that CXCL12/CXCR4-mediated chemotaxis of lymphocytes is impaired in β-arrestin 2 knockout mice (Fong et al., 2002) and in CXCR4-transfected HEK-293 cells with suppressed β-arrestin 2 expression (Sun, Cheng, Ma, & Pei, 2002). Also, for CXCR1–3- and CCR5-driven chemotaxis, β-arrestin 1, 2 and the MAPK pathway were found to play a role to different extents (reviewed in DeFea, 2007).

In the following, we will describe assays to assess several outcomes of chemokine receptor signaling, starting with G protein activation, secondary messenger production, and β-arrestin recruitment, continuing with further downstream signaling (e.g., ERK phosphorylation), and finally assessing the chemotactic response.

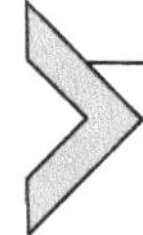

4. METHODS

4.1 General

This review describes methods to experimentally measure chemokine receptor signaling pathways. Standard techniques such as cell culture handling, choice of cells, and transfection methods are not considered here. Please note that the very cheap calcium phosphate transfection method used for transfection of COS-7 cells has been described previously (Jensen & Rosenkilde, 2009). Furthermore, if working with transiently transfected cells using the calcium phosphate transfection method, the assays are carried out 2 days posttransfection, whereas commercial micelle-based transfection methods allow assays to be carried out 1 day after transfection (Benned-Jensen & Rosenkilde, 2010).

4.2 G Protein Binding ([^{35}S]-GTPγS Binding): Nonselective Assay

The initial event of GPCR activation is the exchange of GDP to GTP in the G protein preassociated with the receptor. The GTP-bound G protein is then released from the receptor and dissociates into the Gα and βγ subunits, both of which can trigger activation of effector molecules (Lambert, 2008). As long as the receptor is in its active state, it induces this nucleotide exchange in the Gα subunit and will activate numerous G proteins (Krauss, 2014). This activation of G proteins, i.e., the turnover from the inactive GDP bound to the active GTP-bound form, can be measured by providing [^{35}S]-GTPγS instead of GTP. GTPγS is an analog of GTP, with the third phosphate group exchanged to sulfate, rendering the GTP non-hydrolyzable by the intrinsic GTPase activity of the G protein. Active, GTPγS-bound G proteins will therefore accumulate and can be used as a direct measure of receptor activation. Notably, radioactively labeled [^{35}S]-GTPγS is used, which allows for simple scintillation-based monitoring of active G protein. The described assay is carried out on membrane preparations of the receptor of interest, and not on whole cells.

4.2.1 Required Material

4.2.1.1 Membrane Preparation

PBS (Dulbecco's without calcium, without magnesium; Gibco)
20 m*M* HEPES + 2 m*M* $MgCl_2$, pH 7.4
Protease inhibitor, e.g., Complete, EDTA-free tablets (Roche)
Rubber policemen, dounce homogenizer

4.2.1.2 [^{35}S]-GTPγS Binding

Protease inhibitor, e.g., Complete, EDTA-free tablets (Roche)
500 m*M* HEPES, pH 7.4
100 m*M* HEPES, 0.2% BSA (nuclease-free)
100 m*M* EGTA, pH 7.4–7.6
20 m*M* HEPES + Complete (1 tablet/25 ml)
5 *M* NaCl
1 *M* $MgCl_2$
BSA nuclease-free
Saponin, to be made fresh each time (Sigma)

200 μ*M* GTPγS (Sigma)
1 n*M* [^{35}S]-GTPγS in water (Perkin Elmer)
300 μ*M* GDP (Sigma)
100 mg/ml scintillation proximity assay (SPA) beads WGA-coupled in water (GE Healthcare)
White 96-well plates, clear TopSeals (Perkin Elmer), membrane preparations, ligands, scintillation counter for 96-well plates

4.2.2 Protocol

4.2.2.1 Membrane Preparation

1. Work on ice and with chilled buffers.
2. Wash cells with 10 ml PBS.
3. Add 5–10 ml PBS and harvest cells using a rubber policeman, then transfer cells to dounce homogenizer, and homogenize 10–15 times using pestle B (tighter fitting).
4. Centrifuge for 3 min at 500 rpm (approximately 27 × *g*) at 4 °C, then centrifuge supernatant again at 21,000 × *g* for 30 min at 4 °C; the resulting pellet contains membranes.
5. Resuspend membrane pellet in 250–500 μl per big cell culture flask of 20 m*M* HEPES buffer, 2 m*M* $MgCl_2$.
6. Take out 60 μl for measuring protein concentration using BCA protein assay kit (Pierce, Rockford, IL, USA); the expected protein concentration is approximately 2 mg per big cell culture flask.
7. Dilute 1:2 with 20 m*M* HEPES, 2 m*M* $MgCl_2$ containing Complete protease inhibitor EDTA-free (1 tablet per 25 ml; Roche).
8. Aliquot and freeze at −80 °C.

4.2.2.2 GTPγS-Binding Assay

1. The concentrations of the following components depend on receptor: ligand pairs and have to be optimized: $MgCl_2$ (e.g., 1, 4, or 10 m*M*), NaCl (e.g., 50, 100, or 150 m*M*), GDP (e.g., 2, 5, or 10 μ*M*), and saponin (e.g., 0 or 10 m*M*); example concentrations are given below.
2. Prepare 2 × assay buffer containing 100 m*M* HEPES, 0.2% BSA (nuclease-free), 4 m*M* $MgCl_2$, 100 m*M* NaCl, 2 m*M* EGTA, 2 μ*M* GDP, 10 m*M* saponin; add GDP and saponin fresh on the day of the assay.
3. Make membrane solution (20 μg protein/75 μl/well): dilute membrane preparation to 20 μg per 37.5 μl in 20 m*M* HEPES + Complete (1 tablet/25 ml); then add 37.5 μl 2 × assay buffer.

4. Make [^{35}S]-GTPγS solution (20 μl/well): water:2× assay buffer (1:1), add 1:1000 1 n*M* [^{35}S]-GTPγS.
5. Make SPA-bead solution (50 μl/well): water:2× assay buffer (1:1), add 1:12.5 100 mg/ml SPA beads.
6. Add 75 μl of the membrane solution to each well of a white 96-well plate.
7. Add 5 μl ligand to each well using a 20× standard.
8. Add 5 μl unlabeled 200 μ*M* GTPγS to two additional wells in order to determine unspecific binding.
9. Add 20 μl [^{35}S]-GTPγS solution to each well.
10. Incubate 1 h at 30 °C while shaking softly.
11. Add 50 μl SPA-bead solution (shake well immediately before use) to each well and seal plate.
12. Incubate 30 min at RT while shaking softly.
13. Centrifuge 5 min, 400 × *g* at RT.
14. Count radioactivity bound to SPA beads in a Topcount microplate scintillation and luminescence counter (Packard).

4.3 G Protein Signaling (cAMP Assay/SPA-IP$_3$ Assay): Selective Assays

4.3.1 cAMP Assay

Chemokine receptors induce the activation of Gα_i proteins, which in turn inhibit the activity of AC (O'Hayre et al., 2008). A cell-based assay can be carried out to measure a decrease of cAMP using DiscoveRx HitHunter® kit, an enzyme fragment complementation (EFC) assay. In this assay, the initial cAMP response is elevated by forskolin, which directly stimulates AC (Seamon & Daly, 1981). Following chemokine receptor activation, this basal cAMP level decreases. In EFC, the enzyme β-galactosidase is split into two fragments, enzyme donor (ED) and enzyme acceptor (EA), the smaller of which (ED) is coupled to cAMP. Furthermore, an antibody-binding cAMP is employed. When cAMP-ED is bound to the antibody, it cannot complement the EA and no enzymatic reaction resulting in luminescent signal can occur. Only when the cells produce cAMP as response to AC activation, i.e., at low levels of Gα_i activity, free cAMP can displace cAMP-ED from the antibody, which then can form active β-galactosidase with the EA. The amount of assay signal is therefore reciprocally proportional to the amount of Gα_i activity.

4.3.1.1 Required Material

2 × HBS (280 m*M* NaCl, 50 m*M* HEPES, 1.5 m*M* Na_2HPO_4, pH 7.2)
5 m*M* IBMX (3-isobutyl-1-methylxanthine; stock)
1 m*M* IBMX in 1 × HBS
Forskolin (20 × final concentration of 10–20 μ*M*); receptor dependent, needs to be optimized; avoid repeated freeze–thaw cycles
Ligands (20 × final concentration)
Provided in DiscoveRx HitHunter® cAMP Assay for small molecules or biologics (cat. no. 90-0075):

- Antibody reagent
- ED reagent
- EA reagent
- Galacton-Star
- Emerald II
- Lysis buffer
- From this kit, prepare *solution A* (Galacton-Star®, Emerald II®, lysis buffer (1:5:19)) and *solution A2* (solution A:ED (1:1))

cAMP standard curve ranging from 4×10^{-6} to 2×10^{-10} in PBS
White 96-well plates, clear TopSeals (Perkin Elmer), luminescence plate reader

4.3.1.2 Protocol

1. Wash cells twice with 200 μl 1 × HBS.
2. Add 100 μl prewarmed 1 × HBS containing 1 m*M* IBMX.
3. Preincubate 30 min at 37 °C.
4. Add 5 μl ligands and incubate for 10 min at 37 °C.
5. Add 5 μl forskolin and incubate for further 30 min at 37 °C.
6. Stop on time by aspirating buffer.
7. Add 30 μl PBS and pipet 30 μl of standard cAMP dilutions into separate wells.
8. Add 10 μl antibody reagent.
9. Add 40 μl solution A2.
10. Wrap plates in aluminum foil to protect them from light and shake 1 h at RT with low speed.
11. Add 40 μl EA reagent.
12. Seal plates with TopSeal, wrap in aluminum foil, and shake at low speed for 1 h at RT.
13. Leave plates on bench for 3 h, then count luminescence in Envision reader (Perkin Elmer).

4.3.2 SPA-IP_3 Assay

As alternative to measuring decreasing levels of cAMP in response to $G\alpha_i$ activity and inhibition of AC, an assay using a chimeric, artificial G protein, $G_{\Delta 6qi4myr}$, can be employed to quantify chemokine receptor activation. This $G_{\Delta 6qi4myr}$ protein basically is a $G\alpha_q$ protein, but with the five C-terminal residues substituted with those of $G\alpha_i$. These five $G\alpha_i$ residues are sufficient to provide coupling to chemokine receptors. However, a $G\alpha_q$ response will be elicited upon receptor activation, and an assay measuring the secondary messenger inositol-1,4,5-trisphosphate (IP_3) can be conducted (Conklin, Farfel, Lustig, Julius, & Bourne, 1993; Kostenis, 2001). In order to be able to measure IP_3 levels, cells are grown in the presence of [^{3}H]-myo-inositol, which serves as precursor for the substrate of phospholipase C, the effector molecule of $G\alpha_q$. IP_3 is rapidly degraded intracellularly to inositolmonophosphate (IP), but its final degradation to inositol can be inhibited by the presence LiCl in the assay medium. IP is then quantified using SPA beads that specifically bind IP, but not its precursor [^{3}H]-myo-inositol. This SPA-IP_3 assay is an alternative to the classic IP_3 assay described previously (Jensen & Rosenkilde, 2009). In both assay versions, cells are transfected using calcium phosphate transfection with DNA for receptor:$G_{\Delta 6qi4myr}$ in ratio 2:3. This indirect measurement of $G\alpha_i$ activity has been used in numerous publications within the chemokine system (Hatse et al., 2007; Jensen et al., 2007; Rosenkilde, Andersen, Nygaard, Frimurer, & Schwartz, 2007; Thiele et al., 2012; Thiele, Mungalpara, Steen, Rosenkilde, & Våbenø, 2014).

4.3.2.1 Required Material

[^{3}H]-myo-inositol (Perkin Elmer)
1 *M* LiCl
HBSS (Hank's balanced salt solution; Gibco)
HBSS + 10 m*M* LiCl
10 m*M* Formic acid
12.5 mg/ml YSi poly-D-lysine-coated SPA beads (in water) (Perkin Elmer)
Clear and white 96-well plates, clear TopSeals (Perkin Elmer), scintillation counter for 96-well plates

4.3.2.2 Protocol

1. One day after transfection plate cells into 96-well plates using medium enriched with 5 μl/ml [^{3}H]-myo-inositol; proceed with assay on the following day.

2. Aspirate medium and collect for radioactive waste.
3. Wash twice in 100 μl/well HBSS.
4. Add 100 μl HBSS containing 10 m*M* LiCl, prewarm for 15 min at 37 °C.
5. Add 5 μl/well of 20 × ligand stock (if using antagonists, add them 10–15 min before the agonists).
6. Incubate for 90 min at 37 °C (not shaking).
7. Put the plates on ice and aspirate incubation medium.
8. Add 40 μl 10 m*M* formic acid per well (lysis).
9. Incubate for at least 30 min on ice.
10. Transfer 35 μl of the lysed cells to solid white 96-well plates using a manual multichannel pipette.
11. At this point, the plates could be frozen for later use.
12. Add 80 μl SPA-YSI bead suspension (1 mg/well), shake or stir vigorously before pipetting to avoid sedimentation.
13. Cover the plates with TopSeal stickers.
14. Shake the plates at high speed for 30 min.
15. Spin the plates for 5 min at 400 × *g*.
16. Measure the accumulated IP in a Topcount microplate scintillation and luminescence counter (Packard).

4.4 β-Arrestin Recruitment Assay

Upon activation, chemokine receptors can become phosphorylated by GRKs, which will lead to the recruitment of β-arrestins 2 and/or 1 (DeFea, 2007). This recruitment can be experimentally measured using DiscoveRx PathHunter® Arrestin assays, which employ EFC technique with ED (here also called ProLink, PK) and EA of β-galactosidase fused to the receptor of interest and β-arrestin 1 or 2, respectively. Upon recruitment of the β-arrestin–EA to the activated receptor-ED, active β-galactosidase is formed whose activity can be measured as chemiluminescence after providing substrate. Importantly, any receptor can be fused in frame to ED using the vector pCMV-PK1 (DiscoveRx, cat. no. 93-0491). Alternatively, DiscoveRx offers pCMV-PK2, -ARMS-PK1, and -ARMS-PK2, whereby PK2 displays higher affinity for the EA part fused to β-arrestin, and ARMS is a sequence enhancing arrestin recruitment due to stronger GRK-mediated phosphorylation. We recommend using the tag PK1 since that will provide the least modified response. A cell line equipped with β-arrestin–EA, as well as cell lines carrying both receptor-ED and arrestin–EA, is available from DiscoveRx.

Alternative β-arrestin assays are BRET/FRET-based and have been successfully employed at both chemokine and nonchemokine receptors (Gilliland, Salanga, Kawamura, Trejo, & Handel, 2013; Zhao et al., 2013).

4.4.1 Required Material

Receptor fused to ED (also called ProLink), which is a part of β-galactosidase (DiscoveRx, cat. no. 93-0491)

Cells stably transfected with EA (supplementary part of β-galactosidase)-tagged β-arrestin 1 or 2

Assay medium: cell medium without antibiotics, or OptiMem (Life Technologies)

Provided in DiscoveRx PathHunter® Detection kit (cat. no. 93-0001):

- Galacton-Star
- Emerald II
- From this kit, prepare *detection reagent mix*: Galacton-Star®, Emerald II®, cell assay buffer (1:5:19); protect from light by wrapping the tube with aluminum foil

20 × ligand stock (maximal final DMSO concentration in wells can be 1%)

White 96-well plates, TopSeals (Perkin Elmer), luminescence plate reader

4.4.2 Protocol

1. Aspirate cell culture medium and add 100 μl cell medium.
2. Add 5 μl of 20 × ligand stock.
3. Incubate for 90 min at 37 °C.
4. Add 50 μl detection reagent/well.
5. Seal plates with TopSeal and pack in aluminum foil.
6. Incubate for 60 min on a plate shaker.
7. Count chemiluminescence in Envision reader (PerkinElmer).

4.5 Internalization Assay Using ELISA or Confocal Microscopy

Internalization of a receptor can occur after activation, GRK-mediated phosphorylation, and recruitment of β-arrestins. In consequence, receptors are removed from the cell surface by internalization into endosomes. The further faith of the receptor will either be degradation after sorting to lysosomes, or "rescue" and recycling back to the cell surface. In any case, this internalization event leads to desensitization of the cell toward the receptor's ligands (Marchese, 2014; Marchese, Paing, Temple, & Trejo, 2008; Vroon, Heijnen, & Kavelaars, 2006).

By expressing FLAG-tagged receptors, the amount of receptor on the cell surface in the absence or presence of ligands of interest can be measured by ELISA with anti-FLAG M1 antibodies. A more advanced technique using confocal microscopy and staining of cells prior and post-permeabilization of the cell membrane also allows to visually track the receptor and assess the internalized receptor fraction.

4.5.1 ELISA-Based Internalization Assay

4.5.1.1 Required Material

FLAG-tagged receptor DNA (e.g., in pcDNA3.1+)
Ligands
3.7% Formaldehyde
TBS (0.05 *M* Tris Base, 0.9 (w/v)% NaCl, pH 7.6)
TBS + 2% BSA
2 *M* $CaCl_2$
Primary antibody (1° ab): mouse anti-FLAG M1 antibody (Sigma)
Secondary antibody (2° ab): goat anti-mouse horseradish peroxidase (HRP)-conjugated IgG antibody (Pierce)
TMB Plus (3,3′,5,5′-tetramethylbenzidine; Kem-En-Tec Diagnostics)
0.2 *M* H_2SO_4

4.5.1.2 Protocol

1. Add ligands at desired concentration and over a different time course (e.g., 0, 10, 20, 30 min), and incubate at 37 °C.
2. Aspirate medium and fix cells with 150 μl 3.7% formaldehyde for 10 min; from here work at RT.
3. Wash 3 × with 200 μl TBS.
4. Block in 150 μl TBS + 2% BSA for 30 min.
5. Incubate with 100 μl 1° ab at 2 μg/ml in TBS containing 1% BSA and 1 m*M* $CaCl_2$ for 2 h at RT.
6. Wash 3 × with 200 μl TBS containing 1 m*M* $CaCl_2$.
7. Incubate with 100 μl 2° ab at 0.8 μg/ml in TBS containing 1% BSA and 1 m*M* $CaCl_2$ for 1 h at RT.
8. Wash 3 × with 200 μl TBS containing 1 m*M* $CaCl_2$.
9. Add 150 μl TMB plus, and after 5 min stop reaction by addition of 100 μl 0.2 *M* H_2SO_4.
10. Measure absorbance at 450 nm.

4.5.2 Confocal Microscopy-Based Internalization Assay

4.5.2.1 Required Material

3.7% Paraformaldehyde
TBS (0.05 *M* Tris Base, 0.9 (w/v)% NaCl, pH 7.6)
TBS + 2% BSA
TBS + 1% BSA
Saponin
Primary antibody (1° ab): mouse anti-FLAG M1 antibody (Sigma)
Secondary antibody: goat anti-mouse IgG1, Alexa Fluor 488 conjugate (Molecular Probes)
Secondary antibody: goat anti-mouse IgG1, Alexa Fluor 568 conjugate (Molecular Probes)
SlowFade (Molecular Probes)
Confocal microscope, coverslips, microscope slide, nail polish

4.5.2.2 Protocol

1. Seed and transfect cells on fibronectin-coated coverslips in 6-well plates; cells should be 80% confluent on day 2 posttransfection, where you proceed with the assay.
2. Wash 1 × with cold TBS.
3. Incubate cells in cold cell culture medium containing 1° ab at 2 μg/ml for 1 h at 4 °C.
4. Wash in cold cell culture medium and either fix coverslips immediately in 3.7% paraformaldehyde (negative control) or incubate in prewarmed cell culture medium containing ligand or vehicle at wanted concentration and incubate for different intervals at 37 °C (e.g., 0, 10, 20, 30 min); thereafter wash and fix cells.
5. Block coverslips with TBS containing 2% BSA for 30 min.
6. Detect receptors residing at the cell surface by incubating with goat anti-mouse Alexa Flour 488-conjugated antibody at 2 μg/ml in 1000 μl TBS containing 1% BSA for 30 min.
7. Wash 3 × with approximately 1000 μl TBS containing 1 m*M* $CaCl_2$.
8. Permeabilize cells by adding 1000 μl TBS containing 1% BSA and 0.2 (w/v)% saponin for 20 min.
9. Detect internalized receptors by incubating with goat anti-mouse Alexa Fluor 568-conjugated antibodies at 2 μg/ml in 1000 μl TBS containing 1% BSA for 30 min.
10. Wash 3 × with approximately 1000 μl TBS.

11. Mount coverslips in SlowFade antifade reagent on microscope slide and seal using nail polish.
12. Inspect your samples using confocal microscopy; Alexa Fluor 488 is excited at $\lambda = 495$ nm, and the emission collected at 519 nm; Alexa Flour 568 is excited at $\lambda = 578$ nm, and the emission collected at 603 nm; alternative fluorophores may be used.

4.6 ERK Phosphorylation

The phosphorylation state of various MAPKs regulates many fundamental cellular processes, including chemotaxis. Changes in MAPK phosphorylation states in response to chemokine stimulation of cells can be quantified by traditional Western blotting. Overall, the effect of a chemokine on MAPK phosphorylation state over time is best quantified in cells that have been deprived of growth factors and other signaling molecules before chemokine stimulation. In general, total lysates are made from cells that have been stimulated with chemokines for various time periods after starvation. Protein separation according to size by polyacrylamide gel electrophoresis in the presence of sodium dodecyl sulfate (SDS-PAGE) is followed by protein transfer to a PVDF membrane. Protein membrane immobilization allows for sequential antibody detection of MAPK of interest using primary antibodies directed against a specific kinase followed by incubation with HRP-coupled secondary antibody recognizing the primary antibody. The amount of HRP-coupled antibody bound to the membrane (indirect measure of amount of protein of interest) is detected by luminol-based enhanced chemiluminescence (ECL). In general you detect total MAPK (using an antibody recognizing both native and phosphorylated forms) and after antibody stripping (antibody removal), the membrane can be reprobed with antibodies recognizing only the phosphorylated MAPK form. Finally, you may probe the blot for a housekeeping gene as loading control.

4.6.1 Required Material

Cells
Serum-free cell culture medium
96-Well plates
Chemokines
RIPA buffer (10×; Millipore)
Protease inhibitor cocktail tablets (Mini Complete; Roche)
Phosphatase inhibitor cocktail 2 and 3 (Sigma)
NuPAGE LDS sample buffer (4×; Life Technologies)

NuPAGE Bis–Tris gel (10%; Life Technologies)
NuPAGE MES-SDS running buffer (20 ×; Life Technologies)
Reducing agent DTT
XCell SureLock™ Mini-Cell apparatus and power supply
96% Ethanol
NuPAGE transfer buffer (20 ×; Life Technologies)
Semi-dry blotting apparatus
PVDF membrane (Thermo Scientific)
Blotting paper (Invitrogen)
Blocking buffer (Thermo Scientific)
PBS
Tween-20
Stripping buffer (Thermo Scientific)
Primary antibodies, e.g., p44/42 MAPK (Erk1/2) and Phospho-p44/42 MAPK (Erk1/2, Thr202/Tyr204) (Cell Signaling Technology)
Secondary antibody, e.g., goat anti-rabbit HRP conjugated
ECL substrate (SuperSignal West Pico; Thermo Scientific)
Cooled CCD camera for detection

4.6.2 Protocol

1. Seed equal amounts of cells per well in a 96-well plate (one well per time point, 100,000 cells per well) or bigger format if required (primary leukocytes are used right after seeding, whereas transfected cells are allowed to adhere for 1 day before use).
2. Incubate the cells for 2 h in serum-free medium (high glucose).
3. Stimulate the cells with the relevant chemokines for variable time periods (normally between 10 and 120 min).
4. Transfer the plates to a container with ice and remove medium from the wells.
5. Add 40 μl of RIPA lysis buffer containing protease and phosphatase inhibitors. Lysis is allowed to proceed for 45 min on ice.
6. The lysates are transferred to Eppendorf tubes, and 10 μl SDS sample buffer is added.
7. The lysates are incubated at 95 °C for 5 min and transferred to ice.
8. 30 μl Lysate is loaded on SDS gel, and the gel is allowed to run at 120 V for 2 h.
9. Assemble the transfer apparatus as follows:

- Two blotting papers soaked in transfer buffer (remove air bubbles between each of the following steps by gently rolling a pipette over the assembled layers).
- A PVDF membrane presoaked first in 96% ethanol (2 s), water (2 min), and then transfer buffer (10 min). All membrane handling is done with forceps.
- The gel.
- Two blotting papers soaked in transfer buffer.

10. Place the top of the transfer apparatus onto the layers and gently fasten the screws. Perform transfer for 90 min at 20 V.
11. After protein transfer, immediately incubate the membrane with the protein layer facing upward in blocking buffer for 1 h to block unspecific binding sites on membrane.
12. Incubate in blocking buffer with primary antibody at 37 °C for 2 h or alternatively 4 °C overnight.
13. Remove primary antibody solution and wash membrane twice in PBS with 0.05% Tween-20 (each wash 45 min).
14. Incubate in blocking buffer with secondary antibody at 37 °C for 2 h or 4 °C overnight.
15. Remove secondary antibody solution and wash as before.
16. Remove excess wash buffer from membrane by holding the membrane with forceps allowing one corner to touch a napkin.
17. Place the membrane on a plastic sheet with the protein surface facing upward.
18. Add ECL solution and incubate for 5 min.
19. Add plastic sheet to the top of the membrane and squeeze out excess ECL substrate by running a paper napkin over the plastic sheets several times.
20. Detect the signal using a cooled CCD camera.

4.7 Chemotaxis

4.7.1 Chemotaxis by Classic Transwell Assay

The ultimate cellular response following chemokine receptor activation is chemotactic migration, i.e., chemotaxis of cells (Moser, Wolf, Walz, & Loetscher, 2004). Traditionally, this is measured with an *in vitro* transwell assay, where suspension cells (leukocytes) are placed in droplets on top of a filter, which on its lower side is in contact with medium containing chemoattractant. If the chemoattractant is functional at the investigated receptor, cells will start to migrate through the membrane into the lower compartment. The amount of migrated cells is then quantified using a luminescence-based assay assessing ATP as measure of cell viability. When screening for receptor antagonists, one has to consider that (or if) these should have access to the cells carrying the receptor, i.e., be present both in the cell droplet and the transwell compartment below the filter.

4.7.1.1 Required Material

L1.2 cells stably transfected with receptor, or other suspension cells/ leukocytes
ChemoTx 96-well plates (3.2 mm diameter, 5 μm pore size; Neuroprobe)
Chemotaxis medium (HEPES-modified RPMI 1640, 0.1% BSA)
CellTiter-Glo solution (Promega)
White 96-well plates
Luminescence reader (e.g., Wallac Envision 2104 Multilabel Reader; Perkin Elmer)

4.7.1.2 Protocol

1. Fill lower wells of ChemoTx transwell plates with 31 μl chemotaxis medium containing the desired ligand concentration.
2. Place a 5-μm pore filter on top of the wells and add 2×10^5 cells in a volume of 20 μl chemotaxis medium.
3. Incubate for 5 h at 37 °C in cell incubator, while wrapping the plate in buffer-soaked tissue paper to avoid evaporation of medium.
4. Scrape cells from upper surface of the filter (to avoid cells being soaked through the filter when lifting the membrane) and remove filter.
5. Transfer lower well solution to white 96-well plate.
6. Count cells by the addition of 30 μl CellTiter-Glo dye (Promega) and measure luminescence in appropriate plate reader.

4.7.2 *Ibidi® μ-Slide Chemotaxis 3D*

In comparison to the traditional transwell migration assay, which has an end-point readout in number of migrating cells, Ibidi® μ-Slide Chemotaxis three-dimensions (3D) slides offer a more sophisticated alternative to quantify cell migration, where in addition to estimating the number of cells that perform chemotaxis, one can also get information on speed, directedness, and total migration length of each cell. In addition, migration can be performed in 3D, if the cells are seeded inside a collagen matrix. The time-lapse microscopy recordings in this assay allow precise tracking of cells over time, under controlled temperature and humidity. Time-lapse recordings enable migration analysis of leukocytes or cell lines stably transfected with a chemokine receptor of interest, to elucidate the effect of different chemokines or combinations of chemokine fields in shaping the migration pattern of migrating cells for a period of up to 48 h, i.e., the period for which a stable linear gradient can be maintained inside the channel holding the cells (Fig. 4). In the Ibidi® system, a linear gradient is formed across a narrow channel holding the cells through simple diffusion of the ligand from the source reservoir in contact with the channel on one side, across the channel and to the sink reservoir in contact with the channel on the opposite side. Cells are either seeded on the bottom of the channel (2D) or inside a collagen matrix filling up the channel (3D) (Video 1). After cell attachment or, in the case of 3D studies, collagen polymerization, the sink and source reservoir are filled with medium and chemokine containing medium, respectively (Video 2). This method has been used successfully to study the biased effect of CCL19 and CCL21 on migration of human monocyte-derived dendritic cells (unpublished data), but also in combination with microstructured mazes (Olsen et al., 2013) or channel structures (Hjortø, Olsen, Svane, & Larsen, 2015) fabricated by two-photon polymerization inside the Ibidi® channel to evaluate pore size and channel form on dendritic cell migration.

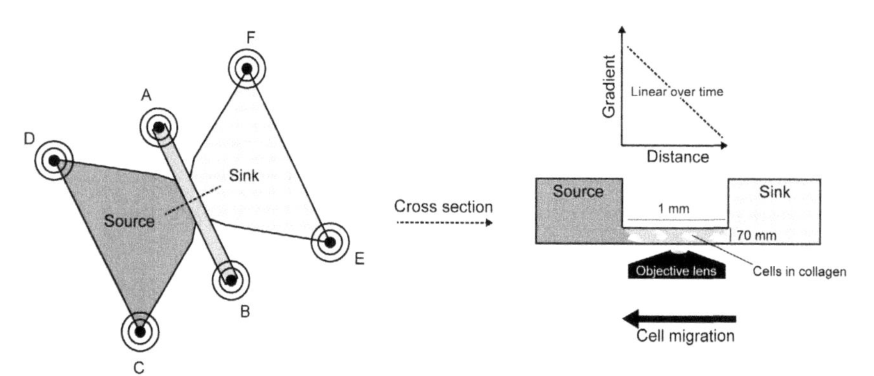

Figure 4 Schematic drawing of the Ibidi® μ-Slide migration chamber. On the left, the chamber is shown from top with inlets A–F. Note that the channel containing cells and collagen is in the center and between inlets A and B. On the right, the cross section of the chamber is shown. Cells will migrate along the short edge of the chamber, which is approximately 1 mm wide. (See the color plate.)

4.7.2.1 Required Material

Cells
Cell culture medium
Collagen I (3 mg/ml, bovine; Advanced BioMatrix)
$NaHCO_3$ (7.5%; Sigma-Aldrich)
10 × MEM (modified Eagle's medium; Sigma-Aldrich)
Chemoattractant
μ-Slide Chemotaxis 3D (Ibidi®)
Time-lapse microscope with temperature- and humidity-controlled incubation chamber
Tracking program (AutoZell or similar)

4.7.2.2 Protocol

1. Close all reservoir inlets (C–F) on Ibidi® Slide by plugs.
2. Prepare cell suspension of $3–9 \times 10^6$ cells/ml.
3. Mix 10 μl of 7.5% $NaHCO_3$ solution with 20 μl of 10 × MEM (avoid bubbles).
4. Add 150 μl of collagen I solution and mix by pipetting (avoid bubbles).
5. Finally add 90 μl cell suspension and mix by pipetting (avoid bubbles).
6. Transfer 6 μl of collagen suspension with cells to Ibidi® channel by placing one drop on the channel inlet (A) and quickly sucking with the same pipette from the channel outlet (B), to draw the collagen suspension into the channel (Video 1).
7. Remove all plugs from all inlets and cover slide with lid, place the slide on a humidified napkin inside a Petri dish in a humidified incubator for 35–40 min to allow for collagen polymerization (Video 1).
8. Close inlets A, B, C, and D with plugs and add 62 μl medium to sink reservoir via inlet E or F (always fill sink before source) (Video 2).
9. Close inlets E and F and add medium with desired chemokine concentration to source reservoir via inlet C or D, then plug inlets C and D (Video 2).
10. Transfer slide to microscope, record at two positions in each channel with a time lapse of 2 (primary immune cells) to 10 min (Video 3).

ACKNOWLEDGMENTS

We thank Mette Simons and Olav Larsen for discussions of some of the presented assay techniques. The scientific work of the authors is supported by Hørslev Foundation, The Danish Research Council, Lundbeck Foundation, Novo Nordisk Foundation, A.P. Møller Foundation, and Aase and Einar Danielsen Foundation.
Conflict of interest: The authors declare no conflicts of interest.

REFERENCES

Ahn, S., Shenoy, S. K., Wei, H., & Lefkowitz, R. J. (2004). Differential kinetic and spatial patterns of beta-arrestin and G protein-mediated ERK activation by the angiotensin II receptor. *The Journal of Biological Chemistry, 279*, 35518–35525.

Allen, S. J., Crown, S. E., & Handel, T. M. (2007). Chemokine: Receptor structure, interactions, and antagonism. *Annual Review of Immunology, 25*, 787–820.

Bachelerie, F., Ben-Baruch, A., Burkhardt, A. M., Combadiere, C., Farber, J. M., Graham, G. J., et al. (2014). International Union of Basic and Clinical Pharmacology. [corrected]. LXXXIX. Update on the extended family of chemokine receptors and introducing a new nomenclature for atypical chemokine receptors. *Pharmacological Reviews, 66*, 1–79.

Barlic, J., Andrews, J. D., Kelvin, A. A., Bosinger, S. E., DeVries, M. E., Xu, L., et al. (2000). Regulation of tyrosine kinase activation and granule release through β-arrestin by CXCR1. *Nature Immunology, 1*, 227–233.

Benned-Jensen, T., & Rosenkilde, M. M. (2010). Distinct expression and ligand-binding profiles of two constitutively active GPR17 splice variants. *British Journal of Pharmacology, 159*, 1092–1105.

Berger, E. A., Murphy, P. M., & Farber, J. M. (1999). Chemokine receptors as HIV-1 coreceptors: Roles in viral entry, tropism, and disease. *Annual Review of Immunology, 17*, 657–700.

Boerrigter, G., Lark, M. W., Whalen, E. J., Soergel, D. G., Violin, J. D., & Burnett, J. C. (2011). Cardiorenal actions of TRV120027, a novel ß-arrestin-biased ligand at the angiotensin II type I receptor, in healthy and heart failure canines: A novel therapeutic strategy for acute heart failure. *Circulation. Heart Failure, 4*, 770–778.

Boerrigter, G., Soergel, D. G., Violin, J. D., Lark, M. W., & Burnett, J. C. (2012). TRV120027, a novel β-arrestin biased ligand at the angiotensin II type I receptor, unloads the heart and maintains renal function when added to furosemide in experimental heart failure. *Circulation. Heart Failure, 5*, 627–634.

Bokoch, G. M. (1995). Chemoattractant signaling and leukocyte activation. *Blood, 86*, 1649–1660.

Bridger, G. J., Skerlj, R. T., Padmanabhan, S., Martellucci, S. A., Henson, G. W., Struyf, S., et al. (1999). Synthesis and structure-activity relationships of phenylenebis(methylene)-linked bis-azamacrocycles that inhibit HIV-1 and HIV-2 replication by antagonism of the chemokine receptor CXCR4. *Journal of Medicinal Chemistry, 42*, 3971–3981.

Cain, R. J., & Ridley, A. J. (2009). Phosphoinositide 3-kinases in cell migration. *Biology of the Cell/Under the Auspices of the European Cell Biology Organization, 101*, 13–29.

Casarosa, P., Bakker, R. A., Verzijl, D., Navis, M., Timmerman, H., Leurs, R., et al. (2001). Constitutive signaling of the human cytomegalovirus-encoded chemokine receptor US28. *The Journal of Biological Chemistry, 276*, 1133–1137.

Cheng, Z. J., Zhao, J., Sun, Y., Hu, W., Wu, Y. L., Cen, B., et al. (2000). β-arrestin differentially regulates the chemokine receptor CXCR4-mediated signaling and receptor internalization, and this implicates multiple interaction sites between beta-arrestin and CXCR4. *The Journal of Biological Chemistry, 275*, 2479–2485.

Conklin, B. R., Farfel, Z., Lustig, K. D., Julius, D., & Bourne, H. R. (1993). Substitution of three amino acids switches receptor specificity of Gq alpha to that of Gi alpha. *Nature, 363*, 274–276.

Corbisier, J., Gales, C., Huszagh, A., Parmentier, M., & Springael, J. Y. (2015). Biased signaling at chemokine receptors. *The Journal of Biological Chemistry, 290*, 9542–9554.

Dahinden, C. A., Geiser, T., Brunner, T., von Tscharner, V., Caput, D., Ferrara, P., et al. (1994). Monocyte chemotactic protein 3 is a most effective basophil- and eosinophil-activating chemokine. *The Journal of Experimental Medicine, 179*, 751–756.

DeFea, K. A. (2007). Stop that cell! Beta-arrestin-dependent chemotaxis: A tale of localized actin assembly and receptor desensitization. *Annual Review of Physiology, 69*, 535–560.

DeWire, S. M., Ahn, S., Lefkowitz, R. J., & Shenoy, S. K. (2007). Beta-arrestins and cell signaling. *Annual Review of Physiology*, *69*, 483–510.

Drury, L. J., Ziarek, J. J., Gravel, S., Veldkamp, C. T., Takekoshi, T., Hwang, S. T., et al. (2011). Monomeric and dimeric CXCL12 inhibit metastasis through distinct CXCR4 interactions and signaling pathways. *Proceedings of the National Academy of Sciences of the United States of America*, *108*, 17655–17660.

Fong, A. M., Premont, R. T., Richardson, R. M., Yu, Y. R., Lefkowitz, R. J., & Patel, D. D. (2002). Defective lymphocyte chemotaxis in beta-arrestin2- and GRK6-deficient mice. *Proceedings of the National Academy of Sciences of the United States of America*, *99*, 7478–7483.

Ganju, R. K., Brubaker, S. A., Meyer, J., Dutt, P., Yang, Y., Qin, S., et al. (1998). The alpha-chemokine, stromal cell-derived factor-1alpha, binds to the transmembrane G-protein-coupled CXCR-4 receptor and activates multiple signal transduction pathways. *The Journal of Biological Chemistry*, *273*, 23169–23175.

Gilliland, C. T., Salanga, C. L., Kawamura, T., Trejo, J., & Handel, T. M. (2013). The chemokine receptor CCR1 is constitutively active, which leads to G protein-independent, beta-arrestin-mediated internalization. *The Journal of Biological Chemistry*, *288*, 32194–32210.

Graham, G. J., Locati, M., Mantovani, A., Rot, A., & Thelen, M. (2012). The biochemistry and biology of the atypical chemokine receptors. *Immunology Letters*, *145*, 30–38.

Gregory, K. J., Hall, N. E., Tobin, A. B., Sexton, P. M., & Christopoulos, A. (2010). Identification of orthosteric and allosteric site mutations in M2 muscarinic acetylcholine receptors that contribute to ligand-selective signaling bias. *The Journal of Biological Chemistry*, *285*, 7459–7474.

Hatse, S., Huskens, D., Princen, K., Vermeire, K., Bridger, G. J., De Clercq, E., et al. (2007). Modest human immunodeficiency virus coreceptor function of CXCR3 is strongly enhanced by mimicking the CXCR4 ligand binding pocket in the CXCR3 receptor. *Journal of Virology*, *81*, 3632–3639.

Hjortø, G. M., Olsen, M. H., Svane, I. M., & Larsen, N. B. (2015). Confinement dependent chemotaxis in two-photon polymerized linear migration constructs with highly definable concentration gradients. *Biomedical Microdevices*, *17*, 30.

Jensen, P. C., Nygaard, R., Thiele, S., Elder, A., Zhu, G., Kolbeck, R., et al. (2007). Molecular interaction of a potent nonpeptide agonist with the chemokine receptor CCR8. *Molecular Pharmacology*, *72*, 327–340.

Jensen, P. C., & Rosenkilde, M. M. (2009). Chapter 8. Activation mechanisms of chemokine receptors. *Methods in Enzymology*, *461*, 171–190.

Johnson, Z., Proudfoot, A. E., & Handel, T. M. (2005). Interaction of chemokines and glycosaminoglycans: A new twist in the regulation of chemokine function with opportunities for therapeutic intervention. *Cytokine & Growth Factor Reviews*, *16*, 625–636.

Kenakin, T. P. (2010). G protein coupled receptors as allosteric proteins and the role of allosteric modulators. *Journal of Receptor and Signal Transduction Research*, *30*, 313–321.

Kenakin, T. P. (2011). Functional selectivity and biased receptor signaling. *The Journal of Pharmacology and Experimental Therapeutics*, *336*, 296–302.

Kenakin, T. P., Ambrose, J. R., & Irving, P. E. (1991). The relative efficiency of beta adrenoceptor coupling to myocardial inotropy and diastolic relaxation: Organ-selective treatment for diastolic dysfunction. *The Journal of Pharmacology and Experimental Therapeutics*, *257*, 1189–1197.

Kenakin, T. P., & Christopoulos, A. (2013). Signalling bias in new drug discovery: Detection, quantification and therapeutic impact. *Nature Reviews. Drug Discovery*, *12*, 205–216.

Kenakin, T. P., & Miller, L. J. (2010). Seven transmembrane receptors as shapeshifting proteins: The impact of allosteric modulation and functional selectivity on new drug discovery. *Pharmacological Reviews*, *62*, 265–304.

Kledal, T. N., Rosenkilde, M. M., & Schwartz, T. W. (1998). Selective recognition of the membrane-bound CX3C chemokine fractalkine, by the human cytomegalovirus-encoded broad-spectrum receptor US28. *FEBS Letters*, *441*(2), 209–214.

Koenen, R. R., & Weber, C. (2011). Chemokines: Established and novel targets in atherosclerosis. *EMBO Molecular Medicine*, *3*, 713–725.

Kohout, T. A., Nicholas, S. L., Perry, S. J., Reinhart, G., Junger, S., & Struthers, R. S. (2004). Differential desensitization, receptor phosphorylation, beta-arrestin recruitment, and ERK1/2 activation by the two endogenous ligands for the CC chemokine receptor 7. *The Journal of Biological Chemistry*, *279*, 23214–23222.

Kostenis, E. (2001). Is Galpha16 the optimal tool for fishing ligands of orphan G-protein-coupled receptors? *Trends in Pharmacological Sciences*, *22*, 560–564.

Krauss, G. (2014). *Biochemistry of signal transduction and regulation* (5th ed.). Weinheim: Wiley-VCH.

Kufareva, I., Salanga, C. L., & Handel, T. M. (2015). Chemokine and chemokine receptor structure and interactions: Implications for therapeutic strategies. *Immunology and Cell Biology*, *93*, 372–383.

Lambert, N. A. (2008). Dissociation of heterotrimeric g proteins in cells. *Science Signaling*, *1*, re5.

Lefkowitz, R. J., Rajagopal, K., & Whalen, E. J. (2006). New roles for beta-arrestins in cell signaling: Not just for seven-transmembrane receptors. *Molecular Cell*, *24*, 643–652.

Lefkowitz, R. J., & Shenoy, S. K. (2005). Transduction of receptor signals by β-arrestins. *Science*, *308*, 512–517.

Loetscher, P., Pellegrino, A., Gong, J. H., Mattioli, I., Loetscher, M., Bardi, G., et al. (2001). The ligands of CXC chemokine receptor 3, I-TAC, Mig, and IP10, are natural antagonists for CCR3. *The Journal of Biological Chemistry*, *276*, 2986–2991.

Luttrell, L. M., Ferguson, S. S., Daaka, Y., Miller, W. E., Maudsley, S., Della Rocca, G. J., et al. (1999). β-arrestin-dependent formation of β2 adrenergic receptor-Src protein kinase complexes. *Science*, *283*, 655–661.

Luttrell, L. M., & Gesty-Palmer, D. (2010). Beyond desensitization: Physiological relevance of arrestin-dependent signaling. *Pharmacological Reviews*, *62*, 305–330.

Luttrell, L. M., & Lefkowitz, R. J. (2002). The role of β-arrestins in the termination and transduction of G-protein-coupled receptor signals. *Journal of Cell Science*, *115*, 455–465.

Luttrell, L. M., Roudabush, F. L., Choy, E. W., Miller, W. E., Field, M. E., Pierce, K. L., et al. (2001). Activation and targeting of extracellular signal-regulated kinases by beta-arrestin scaffolds. *Proceedings of the National Academy of Sciences of the United States of America*, *98*, 2449–2454.

Mantovani, A. (1999). The chemokine system: Redundancy for robust outputs. *Immunology Today*, *20*, 254–257.

Marchese, A. (2014). Endocytic trafficking of chemokine receptors. *Current Opinion in Cell Biology*, *27*, 72–77.

Marchese, A., Paing, M. M., Temple, B. R. S., & Trejo, J. (2008). G protein-coupled receptor sorting to endosomes and lysosomes. *Annual Review of Pharmacology and Toxicology*, *48*, 601–629.

Marinissen, M. J., & Gutkind, J. S. (2001). G-protein-coupled receptors and signaling networks: Emerging paradigms. *Trends in Pharmacological Sciences*, *22*, 368–376.

McGovern, K. W., & DeFea, K. A. (2014). Molecular mechanisms underlying beta-arrestin-dependent chemotaxis and actin-cytoskeletal reorganization. *Handbook of Experimental Pharmacology*, *219*, 341–359.

McLean, K. A., Holst, P. J., Martini, L., Schwartz, T. W., & Rosenkilde, M. M. (2004). Similar activation of signal transduction pathways by the herpesvirus-encoded chemokine receptors US28 and ORF74. *Virology*, *325*, 241–251.

Milligan, G., & Kostenis, E. (2006). Heterotrimeric G-proteins: A short history. *British Journal of Pharmacology*, *147*(Suppl. 1), S46–S55.
Moelants, E. A. V., Mortier, A., Van Damme, J., & Proost, P. (2013). In vivo regulation of chemokine activity by post-translational modification. *Immunology and Cell Biology*, *91*, 402–407.
Moelants, E. A. V., Van Damme, J., & Proost, P. (2011). Detection and quantification of citrullinated chemokines. *PLoS One*, *6*, e28976.
Mortier, A., Van Damme, J., & Proost, P. (2008). Regulation of chemokine activity by post-translational modification. *Pharmacology & Therapeutics*, *120*, 197–217.
Moser, B., Wolf, M., Walz, A., & Loetscher, P. (2004). Chemokines: Multiple levels of leukocyte migration control. *Trends in Immunology*, *25*, 75–84.
Murphy, P. M. (1996). Chemokine receptors: Structure, function and role in microbial pathogenesis. *Cytokine & Growth Factor Reviews*, *7*, 47–64.
Nandagopal, S., Wu, D., & Lin, F. (2011). Combinatorial guidance by CCR7 ligands for T lymphocytes migration in co-existing chemokine fields. *PLoS One*, *6*, e18183.
O'Hayre, M., Salanga, C. L., Handel, T. M., & Allen, S. J. (2008). Chemokines and cancer: Migration, intracellular signalling and intercellular communication in the microenvironment. *The Biochemical Journal*, *409*, 635–649.
O'Hayre, M., Salanga, C. L., Handel, T. M., & Hamel, D. J. (2010). Emerging concepts and approaches for chemokine-receptor drug discovery. *Expert Opinion on Drug Discovery*, *5*, 1109–1122.
Oldham, W. M., & Hamm, H. E. (2008). Heterotrimeric G protein activation by G-protein-coupled receptors. *Nature Reviews. Molecular Cell Biology*, *9*, 60–71.
Olsen, M. H., Hjortø, G. M., Hansen, M., Met, O., Svane, I. M., & Larsen, N. B. (2013). In-chip fabrication of free-form 3D constructs for directed cell migration analysis. *Lab on a Chip*, *13*, 4800–4809.
Proudfoot, A. E. I., Power, C. A., & Schwarz, M. K. (2010). Anti-chemokine small molecule drugs: A promising future? *Expert Opinion on Investigational Drugs*, *19*, 345–355.
Rajagopal, S., Ahn, S., Rominger, D. H., Gowen-MacDonald, W., Lam, C. M., Dewire, S. M., et al. (2011). Quantifying ligand bias at seven-transmembrane receptors. *Molecular Pharmacology*, *80*, 367–377.
Rajagopal, S., Bassoni, D. L., Campbell, J. J., Gerard, N. P., Gerard, C., & Wehrman, T. S. (2013). Biased agonism as a mechanism for differential signaling by chemokine receptors. *The Journal of Biological Chemistry*, *288*, 35039–35048.
Rajagopal, S., Kim, J., Ahn, S., Craig, S., Lam, C. M., Gerard, N. P., et al. (2010). Beta-arrestin- but not G protein-mediated signaling by the "decoy" receptor CXCR7. *Proceedings of the National Academy of Sciences of the United States of America*, *107*, 628–632.
Revankar, C. M., Vines, C. M., Cimino, D. F., & Prossnitz, E. R. (2004). Arrestins block G protein-coupled receptor-mediated apoptosis. *The Journal of Biological Chemistry*, *279*, 24578–24584.
Ricart, B. G., John, B., Lee, D., Hunter, C. A., & Hammer, D. A. (2011). Dendritic cells distinguish individual chemokine signals through CCR7 and CXCR4. *The Journal of Immunology*, *186*, 53–61.
Rivero, G., Llorente, J., McPherson, J., Cooke, A., Mundell, S. J., McArdle, C. A., et al. (2012). Endomorphin-2: A biased agonist at the μ-opioid receptor. *Molecular Pharmacology*, *82*, 178–188.
Rosenkilde, M. M. (2005). Virus-encoded chemokine receptors—Putative novel antiviral drug targets. *Neuropharmacology*, *48*, 1–13.
Rosenkilde, M. M., Andersen, M. B., Nygaard, R., Frimurer, T. M., & Schwartz, T. W. (2007). Activation of the CXCR3 chemokine receptor through anchoring of a small molecule chelator ligand between TM-III, -IV, and -VI. *Molecular Pharmacology*, *71*, 930–941.

Rosenkilde, M. M., Gerlach, L.-O., Hatse, S., Skerlj, R. T., Schols, D., Bridger, G. J., et al. (2007). Molecular mechanism of action of monocyclam versus bicyclam non-peptide antagonists in the CXCR4 chemokine receptor. *The Journal of Biological Chemistry, 282*, 27354–27365.

Rosenkilde, M. M., & Kledal, T. N. (2006). Targeting herpesvirus reliance of the chemokine system. *Current Drug Targets*, 7, 103–118.

Rosenkilde, M. M., Kledal, T. N., Bräuner-Osborne, H., & Schwartz, T. W. (1999). Agonists and inverse agonists for the herpesvirus 8-encoded constitutively active seven-transmembrane oncogene product, ORF-74. *The Journal of Biological Chemistry, 274*, 956–961.

Rosenkilde, M. M., McLean, K. A., Holst, P. J., & Schwartz, T. W. (2004). The CXC chemokine receptor encoded by herpesvirus saimiri, ECRF3, shows ligand-regulated signaling through Gi, Gq, and G12/13 proteins but constitutive signaling only through Gi and G12/13 proteins. *The Journal of Biological Chemistry, 279*, 32524–32533.

Rosenkilde, M. M., & Schwartz, T. W. (2000). Potency of ligands correlates with affinity measured against agonist and inverse agonists but not against neutral ligand in constitutively active chemokine receptor. *Molecular Pharmacology, 57*, 602–609.

Rosenkilde, M. M., & Schwartz, T. W. (2004). The chemokine system—A major regulator of angiogenesis in health and disease. *APMIS: Acta Pathologica, Microbiologica, et Immunologica Scandinavica, 112*, 481–495.

Rosenkilde, M. M., Waldhoer, M., Lüttichau, H. R., & Schwartz, T. W. (2001). Virally encoded 7TM receptors. *Oncogene, 20*, 1582–1593.

Rot, A., & von Andrian, U. H. (2004). Chemokines in innate and adaptive host defense: Basic chemokinese grammar for immune cells. *Annual Review of Immunology, 22*, 891–928.

Savino, B., Borroni, E. M., Torres, N. M., Proost, P., Struyf, S., Mortier, A., et al. (2009). Recognition versus adaptive up-regulation and degradation of CC chemokines by the chemokine decoy receptor D6 are determined by their N-terminal sequence. *The Journal of Biological Chemistry, 284*, 26207–26215.

Schall, T. J., & Proudfoot, A. E. I. (2011). Overcoming hurdles in developing successful drugs targeting chemokine receptors. *Nature Reviews. Immunology, 11*, 355–363.

Scholten, D. J., Canals, M., Maussang, D., Roumen, L., Smit, M. J., Wijtmans, M., et al. (2012). Pharmacological modulation of chemokine receptor function. *British Journal of Pharmacology, 165*, 1617–1643.

Seamon, K. B., & Daly, J. W. (1981). Forskolin: A unique diterpene activator of cyclic AMP-generating systems. *Journal of Cyclic Nucleotide Research*, 7, 201–224.

Shepard, L. W., Yang, M., Xie, P., Browning, D. D., Voyno-Yasenetskaya, T., Kozasa, T., et al. (2001). Constitutive activation of NF-kappa B and secretion of interleukin-8 induced by the G protein-coupled receptor of Kaposi's sarcoma-associated herpesvirus involve G alpha(13) and RhoA. *The Journal of Biological Chemistry, 276*, 45979–45987.

Shonberg, J., Lopez, L., Scammells, P. J., Christopoulos, A., Capuano, B., & Lane, J. R. (2014). Biased agonism at G protein-coupled receptors: The promise and the challenges—A medicinal chemistry perspective. *Medicinal Research Reviews, 34*, 1286–1330.

Smit, M. J., Verzijl, D., Casarosa, P., Navis, M., Timmerman, H., & Leurs, R. (2002). Kaposi's sarcoma-associated herpesvirus-encoded G protein-coupled receptor ORF74 constitutively activates p44/p42 MAPK and Akt via G(i) and phospholipase C-dependent signaling pathways. *Journal of Virology, 76*, 1744–1752.

Steen, A., Larsen, O., Thiele, S., & Rosenkilde, M. M. (2014). Biased and G protein-independent signaling of chemokine receptors. *Frontiers in Immunology, 5*, 277.

Steen, A., Schwartz, T. W., & Rosenkilde, M. M. (2009). Targeting CXCR4 in HIV cell-entry inhibition. *Mini Reviews in Medicinal Chemistry, 9*, 1605–1621.

Sun, Y., Cheng, Z., Ma, L., & Pei, G. (2002). Beta-arrestin2 is critically involved in CXCR4-mediated chemotaxis, and this is mediated by its enhancement of p38 MAPK activation. *The Journal of Biological Chemistry*, *277*, 49212–49219.

Szczuciński, A., & Losy, J. (2007). Chemokines and chemokine receptors in multiple sclerosis. Potential targets for new therapies. *Acta Neurologica Scandinavica*, *115*, 137–146.

Szekanecz, Z., Kim, J., & Koch, A. E. (2003). Chemokines and chemokine receptors in rheumatoid arthritis. *Seminars in Immunology*, *15*, 15–21.

Thiele, S., Malmgaard-Clausen, M., Engel-Andreasen, J., Steen, A., Rummel, P. C., Nielsen, M. C., et al. (2012). Modulation in selectivity and allosteric properties of small-molecule ligands for CC-chemokine receptors. *Journal of Medicinal Chemistry*, *55*, 8164–8177.

Thiele, S., Mungalpara, J., Steen, A., Rosenkilde, M. M., & Våbenø, J. (2014). Determination of the binding mode for the cyclopentapeptide CXCR4 antagonist FC131 using a dual approach of ligand modifications and receptor mutagenesis. *British Journal of Pharmacology*, *171*, 5313–5329.

Thompson, G. L., Canals, M., & Poole, D. P. (2014). Biological redundancy of endogenous GPCR ligands in the gut and the potential for endogenous functional selectivity. *Frontiers in Pharmacology*, *5*, 262.

Tohgo, A., Pierce, K. L., Choy, E. W., Lefkowitz, R. J., & Luttrell, L. M. (2002). β-Arrestin scaffolding of the ERK cascade enhances cytosolic ERK activity but inhibits ERK-mediated transcription following angiotensin AT1a receptor stimulation. *The Journal of Biological Chemistry*, *277*, 9429–9436.

Viola, A., & Luster, A. D. (2008). Chemokines and their receptors: Drug targets in immunity and inflammation. *Annual Review of Pharmacology and Toxicology*, *48*, 171–197.

Vischer, H. F., Siderius, M., Leurs, R., & Smit, M. J. (2014). Herpesvirus-encoded GPCRs: Neglected players in inflammatory and proliferative diseases? *Nature Reviews. Drug Discovery*, *13*, 123–139.

Vroon, A., Heijnen, C. J., & Kavelaars, A. (2006). GRKs and arrestins: Regulators of migration and inflammation. *Journal of Leukocyte Biology*, *80*, 1214–1221.

Waldhoer, M., Casarosa, P., Rosenkilde, M. M., Smit, M. J., Leurs, R., Whistler, J. L., et al. (2003). The carboxyl terminus of human cytomegalovirus-encoded 7 transmembrane receptor US28 camouflages agonism by mediating constitutive endocytosis. *The Journal of Biological Chemistry*, *278*, 19473–19482.

Wisler, J. W., DeWire, S. M., Whalen, E. J., Violin, J. D., Drake, M. T., Ahn, S., et al. (2007). A unique mechanism of beta-blocker action: Carvedilol stimulates beta-arrestin signaling. *Proceedings of the National Academy of Sciences of the United States of America*, *104*, 16657–16662.

Zhang, W.-B., Navenot, J.-M., Haribabu, B., Tamamura, H., Hiramatu, K., Omagari, A., et al. (2002). A point mutation that confers constitutive activity to CXCR4 reveals that T140 is an inverse agonist and that AMD3100 and ALX40-4C are weak partial agonists. *The Journal of Biological Chemistry*, *277*, 24515–24521.

Zhao, P., Canals, M., Murphy, J. E., Klingler, D., Eriksson, E. M., Pelayo, J.-C., et al. (2013). Agonist-biased trafficking of somatostatin receptor 2A in enteric neurons. *The Journal of Biological Chemistry*, *288*, 25689–25700.

Zidar, D. A., Violin, J. D., Whalen, E. J., & Lefkowitz, R. J. (2009). Selective engagement of G protein coupled receptor kinases (GRKs) encodes distinct functions of biased ligands. *Proceedings of the National Academy of Sciences of the United States of America*, *106*, 9649–9654.

Zweemer, A. J. M., Toraskar, J., Heitman, L. H., & IJzerman, A. P. (2014). Bias in chemokine receptor signalling. *Trends in Immunology*, *35*, 243–252.

CHAPTER NINE

Mutagenesis by Phage Display

Pauline Bonvin*,†, Christine Power*,†, Amanda Proudfoot*,†,1, Steven Dunn*,†

*NovImmune S.A., Geneva, Switzerland
†Geneva Research Centre, Merck Serono S.A., Geneva, Switzerland
[1]Corresponding author: e-mail address: amandapf@orange.fr

Contents

1. Introduction 187
2. Investigating the Pharmacophore of Chemokine Binders by Phage Display 188
   2.1 Material 189
   2.2 Cloning of Evasin cDNAs into the pMS101C Vector and Site-Directed Mutagenesis 190
   2.3 Screening of Alanine Mutants by ELISAs 193
3. Phage Display Selection to Modulate the Selectivity of Chemokine Binders 197
   3.1 Construction of Phage Display Libraries 198
   3.2 Phage Display Selections 203
4. Summary 205
References 205

Abstract

Chemokines are small chemoattractant proteins involved in the recruitment of leukocytes to the site of inflammation. Due to their prominent role in the inflammatory process, chemokine inhibitors have been developed by parasites to remain undetected not only by the host immune system but also by various laboratories to develop anti-inflammatory compounds. Taking advantage of the small size of natural chemokine-binding proteins, we report here several methods to facilitate their characterization using phage display to identify the chemokine-binding site and to modulate the selectivity of such inhibitors. Interestingly, these methods could be adapted to display the natural inhibitors of other cytokines or even cytokines on phage surface.

1. INTRODUCTION

Characterization of the regions of a protein that bind to its partner, for example, receptor, signaling molecules, extracellular surface proteins, or proteoglycans, is fundamental to the understanding of the protein's

Methods in Enzymology, Volume 570
ISSN 0076-6879
http://dx.doi.org/10.1016/bs.mie.2015.10.002

mechanism of action. These regions are termed "pharmacophores" and several methods are used for their determination in such structure–function studies. A commonly used method is alanine-scanning mutagenesis, where amino acids presumed to play a role in these bimolecular interactions are systematically replaced with alanine. Such studies, while providing important structure–function information for many proteins, are laborious as they require sequential molecular cloning, protein expression, and purification followed by biological assays.

Phage display is an *in vitro* method commonly used nowadays to isolate monoclonal antibodies but also to characterize the interaction between two proteins (Pande, Szewczyk, & Grover, 2010). This technology relies on the display of an exogenous protein on the surface of a filamentous phage containing the DNA sequence encoding for the protein displayed on its surface, allowing a direct link between the genotype and the phenotype of the protein. Therefore, by amplifying the carrier, both the displayed protein and its sequence are simultaneously amplified. Phage can be rapidly amplified and used in screening protocols to preferentially select those expressing a protein with the desired characteristics. Although filamentous phage have the advantage to tolerate the insertion of exogenous sequences in their genome, the display of large proteins affects phage infectivity and phage display can therefore only be used with small proteins or peptides. Phage display is traditionally used to improve the affinity of a protein for its target or to determine residues critical for the interaction by screening large libraries of mutated proteins and only retrieving the clones with the highest affinity for the target.

Here, we describe two phage display approaches used to investigate the interaction between a chemokine-binding protein (CKBP) and its target(s). The first combines alanine-scanning mutagenesis with phage display, thereby eliminating time-consuming recombinant protein expression and purification and allowing a faster screening of alanine mutants. In the case described, we selected the putative functional amino acids *in silico* using a three-dimensional model of the CKBP complexed with its ligand (Bonvin et al., 2014). The second is the application of phage display to altering the affinity of the binding protein to its ligand.

2. INVESTIGATING THE PHARMACOPHORE OF CHEMOKINE BINDERS BY PHAGE DISPLAY

To determine the amino acids of Evasins, the CKBPs identified in tick salivary glands, involved in the interaction with chemokines, we took advantage of the small size of the Evasin proteins (8–11 kDa) to express them on phage particles. This strategy allows rapid screening of tens of mutants without expressing and purifying them individually and could be adapted to analyze other chemokine/cytokine inhibitors. To allow their display on phage particles, the Evasin cDNAs were cloned into a dedicated phagemid vector, pMS101C. The Evasin open-reading frames were inserted downstream of a repressible lac promoter and of a sequence encoding for a secretion signal peptide. The 3′ sequences of Evasins were fused to a modified c-Myc tag, a seven histidine tag and to the gene encoding for the coat protein gIII of the M13 phage. Evasin and gIII sequences were linked by an amber codon allowing the expression of the full length fusion protein in suppressor strains of *Escherichia coli*. In this context, Evasins could be expressed either as soluble proteins (in presence of IPTG) or as gIII fusion proteins expressed on the surface of phage (following infection with the helper phage M13). We developed dedicated ELISAs to detect the binding of these two forms to chemokines. Phage ELISA (i.e., ELISA performed with the whole phage particle, option 1) allows a strong amplification of the signal but may lead to false positive results due to the tendency of phage to stick to plastic surfaces and to the avidity conferred by the simultaneous expression of several Evasins on the surface of one single phage. On the other hand, results obtained with the soluble ELISA format, performed with the protein secreted in the culture medium or in the bacterial periplasm (option 2), are more reliable due to reduced avidity but signals may be lower or even undetectable if low amounts of the protein are secreted. Depending on the protein analyzed, one format may be more suitable than the other.

We described here the cloning of Evasin open-reading frames into the pMS101C vector, the introduction of alanine mutations into these sequences and the screening of alanine mutants by phage ELISA and soluble ELISA (Figs. 1 and 2).

2.1 Material

Chemokines were produced as previously described (in *E. coli*, either as native proteins or as NusA fusions (Magistrelli et al., 2005; Proudfoot & Borlat, 2000) or purchased from Peprotech (Rocky Hill, NJ). Biotinylated chemokines were obtained from Almac (Craigavon, Northern Ireland) and oligonucleotides from Thermo Fischer Scientific.

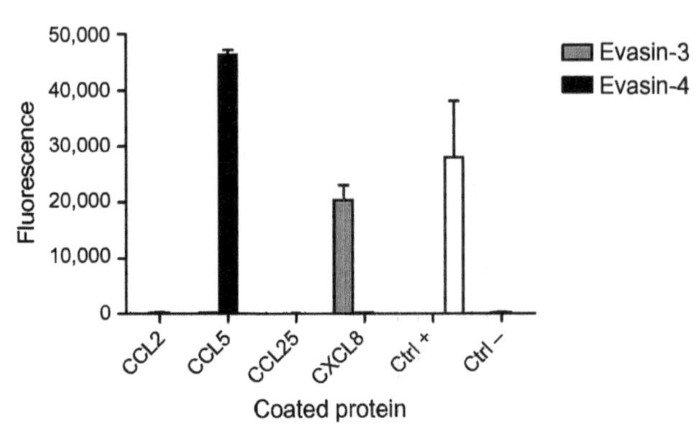

Figure 1 Binding of phage displayed Evasins to chemokines. Evasin-3 and -4 DNA sequences were cloned into pMS101C vector and expressed on the surface of the filamentous phage M13. Phage particles were used in an ELISA assay as explained in Section 2.3. The ability of Evasin-4 to bind CCL5 and of Evasin-3 to bind CXCL8 was retained in this phage format, whereas neither CKBP is able to bind to other chemokines or to the irrelevant protein Tie-2.

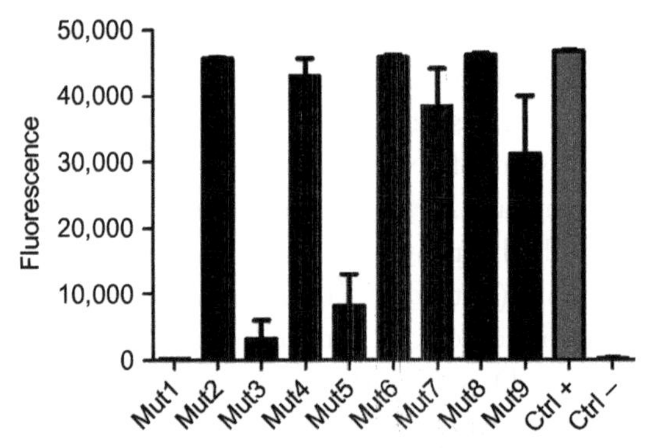

Figure 2 Screening of alanine mutants by phage display. Nine alanine mutants of a CKBP were analyzed by phage ELISA to determine which amino acids are involved in the interaction with the chemokine. Based on these results, mutants 1, 3, and 5 display a clearly reduced binding to the chemokine.

2.2 Cloning of Evasin cDNAs into the pMS101C Vector and Site-Directed Mutagenesis

Required materials

Pwo Master Mix (Roche, cat. no. 03789403001)
Nuclease-free water (Ambion, cat. no. AM9937)
Ultrapure Agarose (Invitrogen, cat. no. 16500100)
SYBER Safe DNA Gel Stain (Invitrogen, cat. no. S33102)

Hyperladder I (Bioline, cat. no. BIO-33026)
Wizard SV Gel and PCR Clean-Up System (Promega, cat. no. A9281)
*Bss*HII restriction enzyme (New England Biolabs, cat. no. R0199L)
*Sal*I-HF restriction enzyme (New England Biolabs, cat. no. R3138L)
CutSmart buffer (New England Biolabs, provided with *Sal*I-HF enzyme)
*Dpn*I restriction enzyme (New England Biolabs, cat. no. R0176L)
T4 DNA Ligase (New England Biolabs, cat. no. M0202L)
Electroporation cuvettes plus, 1 mm gap (BTX, cat. no. 45-0124)
Electrocompetent TG1 *E. coli* (Lucigen, cat. no. 60502)
OneShot Mach1 T1 Phage Resistant Chemically Competent *E. coli* (Invitrogen, cat. no. C862003)
Glucose (Merck Chemicals, cat. no. 346351-1KG)
QIAprep Miniprep Kit (Qiagen, cat. no. 27106)
Appropriate selection antibiotic (e.g., ampicillin, kanamycin)
Tris/borate/EDTA buffer
SOC medium
LB medium
2YT medium

Cloning

1. Amplify the sequence of the gene of interest by polymerase chain reaction (PCR) with the Pwo DNA polymerase using specific primers containing a restriction site for *Bss*HII on 5′ end of the sense primer and a SalI restriction site on the 3′ end of the antisense primer. Mix 10 ng of template DNA with the specific primers to obtain a final concentration of 800 n*M* in a volume of 25 μL. Add 25 μL of Pwo Master Mix and incubate on a thermocycler for 30 cycles of denaturation/annealing/elongation following the manufacturer's instructions.
2. Dilute agarose to 1.1% (w/v) in Tris/borate/EDTA buffer (TBE buffer, 89 m*M* Tris, 89 m*M* boric acid, 2.5 m*M* EDTA). Heat the solution to ensure complete dissolution of agarose. Once the agarose solution has cooled down to 35–40 °C, add SYBER Safe DNA Gel Stain (dilution 1:10,000). Cast a gel using this agarose solution and load the total amount of the PCR product on this gel. After migration, excise the band from the gel and purify the PCR product using the Wizard SV Gel and PCR Clean-Up System or an equivalent system.
3. Digest the PCR product and the destination vector pMS101C with the *Bss*HII and *Sal*I-HF restriction enzymes. For each digestion (PCR product or pMS101C), prepare the following solution:

2.5 μL *Bss*HII (final amount: 10 U)
1 μL *Sal*I-HF (final amount: 20 U)
1 μg DNA (PCR product or pMS101C vector)
5 μL CutSmart Buffer 10×
Nuclease-free H_2O to 50 μL
Incubate for 6 h at 37 °C.

4. Prepare a 1.1% agarose gel as described in step 2 and load the total amount of each digestion product. Excise the band from the gel and purify the PCR product using the Wizard SV Gel and PCR Clean-Up System or an equivalent system. At the end of the DNA purification, elute with 30 μL of nuclease-free H_2O.
5. Ligate the gene of interest into the destination vector using the T4 DNA ligase and following manufacturer's instructions. Incubate overnight at 16 °C. We use a insert:vector ratio of 3:1 but the optimal conditions may vary depending on the vector and the insert used.
6. On the following day, perform the transformation of electrocompetent TG1 bacteria. Place an electroporation cuvette and a microcentrifuge tube on ice. Thaw TG1 bacteria on ice and dilute them 1:6 in sterile H_2O. Transfer 50 μL of diluted bacteria to the chilled microcentrifuge tube and add 2 μL of the ligation product. Mix gently by stirring. Carefully transfer the bacteria/DNA mixture into a chilled electroporation cuvette. Electroporate with a 1700 V pulse of 4 ms (200 Ω, 25 μFD). Immediately add 1 mL of SOC medium to the cuvette and transfer the cells to a sterile culture tube. Incubate at 250 rpm for 1 h at 37 °C.
7. Spread 100 μL of transformed cells on prewarmed 2YT plates containing 2% glucose and the appropriate antibiotic (ampicillin for pMS101C vector). Incubate overnight at 30 °C.
8. Amplify the DNA of five random colonies using a dedicated kit (e.g., QIAprep Miniprep Kit). Sequence the DNA and select the clone with the correct sequence for the next steps.

Alanine scanning by site-directed mutagenesis

1. Design appropriate oligomers to replace the amino acid of interest by an alanine. For each mutation that has to be introduced, prepare one sense primer. The length of these primers should be between 35 and 45 bases and their melting temperature higher than 78 °C. The alanine mutation (modification of up to three nucleotides) should be in the middle of the sequence, with approximately 18–22 bases encoding the new sequence on both sides. For all primers, ensure that the last 3′ base is a G or a C and that the GC content is higher than 40%. Primers that do not fill these criteria may be used but with lower mutation efficiency.

2. Perform a PCR to generate the mutated DNA strand. For each mutation, prepare the following solution:

   100 ng DNA (template vector obtained in previous paragraph)
   125 ng primer
   Nuclease-free H_2O to 15 μL

   Add 15 μL Pwo Master Mix. Incubate on a thermocycler with the following program:
   a. 2 min at 94 °C
   b. 50 s at 94 °C
   c. 8 s at 60 °C
   d. 30 s at 50 °C
   e. 1 min/kb of plasmid length at 68 °C
   f. Repeat steps b–e 18 times
   g. 7 min at 68 °C
3. Add 1 μL of *Dpn*I restriction enzyme (10 U) to each amplification reaction. Mix gently by pipetting up and down several times. Incubate 1 h at 37 °C.
4. Transform chemically competent Mach1 bacteria. For each reaction, transfer 2 μL of *Dpn*I-treated DNA to 50 μL of bacteria. Incubate on ice for 30 min. Heat shock the cells for 30 s at 42 °C and immediately place them on ice for 1 min. Add 250 μL of room temperature SOC medium and incubate at 250 rpm for 1 h at 37 °C. Plate 80–100 μL of each transformation on prewarmed LB plates containing the appropriate antibiotic (ampicillin for pMS101C vector). Incubate overnight at 37 °C.
5. On the following day, randomly pick eight colonies per transformation, amplify them and purify DNA using a standard commercial kit (e.g., QIAprep Miniprep Kit). Sequence the DNA and select the clone with the right sequence for next steps. Electroporate TG1 bacteria with this DNA (as explained in previous paragraph). TG1 cells containing the mutated sequence of the protein of interest can now be used in ELISA assays.

2.3 Screening of Alanine Mutants by ELISAs

Screening ELISAs are classically performed in 96-well microplate. We indicate here the volume of solution corresponding to one ELISA plate. To ensure that results are reproducible, the assay should be performed at least in triplicate, by inoculating three wells with three different individual TG1 colonies containing the same phagemid (step 1).

Required materials

Glucose (Merck Chemicals, cat. no. 346351-1KG)
IPTG (Calbiochem, cat. no. 420322)
Appropriate selection antibiotics (e.g., ampicillin, kanamycin)
2YT medium
TB medium
PBS
M13KO7 helper phage (Invitrogen, cat. no. 18311-019)
96-well maxisorb black plate (Nunc, cat. no. 437111)
NeutrAvidin (Thermo Scientific, cat. no. 31000)
Biotinylated chemokine of interest (Almac Group)
Dry milk (Cell Signaling, cat. no. 9999S)
Polymyxin B sulfate salt (Sigma-Aldricht, cat. no. P1004-1MU)
Guanidine-HCl (Sigma-Aldricht, cat. no. G3272)
Tris pH 7.5 1 *M* (made in-house)
Tween-20 (Merck, cat. no. 8221840500)
Rabbit anti-fd Bacteriophage (Sigma, cat. no. B7786)
Peroxydase-coupled mouse anti-rabbit IgG (Sigma, cat. no. A1949)
Rabbit anti-Evasin (or other specific antibody targeting the protein of interest)
Appropriate secondary antibody coupled to HRP. For the analysis of Evasins, we use an HRP-coupled anti-rabbit IgG (Jackson Immunoresearch, cat. no. 711-036-152)
QuantaBlu Fluorogenic Peroxidase Substrate (Thermo Scientific, cat. no. 15169)

Option 1: phage ELISA

1. Pick individual TG1 colonies containing the mutated pMS101C plasmid from a fresh plate to inoculate 200 μL of 2YT medium containing 2% glucose and ampicillin. Incubate overnight at 30 °C and 750 rpm.
2. On the following morning, inoculate 100 μL of fresh 2YT 2% glucose Amp^+ medium with 3 μL of the overnight culture. Incubate at 30 °C and 750 rpm until the culture medium is turbid (approximately 3.5–4.5 h).
3. Dilute 250 μL of 1×10^{11} M13KO7 phage into 12.5 mL of 2YT 2% glucose Amp^+ medium. Once the culture reaches turbidity, infect by adding 100 μL of 2YT medium containing the M13KO7 helper phage.
4. Leave the plate stationary at 37 °C for 2 h to allow infection and then resuspend the bacteria with a multichannel pipette. Add 5 μL of the

infected culture to 200 μL of fresh 2YT medium containing ampicillin (selection of pMS101C) and kanamycin (selection of infected bacteria) and grow overnight at 30 °C and 750 rpm.

5. On the same day, coat a 96-well maxisorb black plate with 10 μg/mL neutravidin in PBS: add 100 μL/well of the neutravidin solution and incubate the plate stationary overnight at 4 °C. Alternatively, commercially available black plates precoated with streptavidin could be used.
6. On the following day, wash the neutravidin-coated plate three times with a solution of PBS containing 0.1% (v/v) Tween-20. At the end of the washes, remove any remaining wash buffer by inverting the plate and blotting it against paper towels.
7. Add 100 μL/well of 1 μg/mL biotinylated chemokine in PBS and incubate at room temperature for 1 h on a horizontal orbital shaker with gentle agitation (450 rpm).
8. Discard the chemokine solution by inverting the plate. Add 230 μL/well of PBS containing 5% (w/v) nonfat milk and incubate at room temperature for 1 h with gentle agitation (450 rpm). In the meantime, add 100 μL/well of PBS containing 5% milk to plate containing the bacteria. Incubate at room temperature for 30 min with gentle agitation (450 rpm). Do not centrifuge the bacteria plate.
9. Wash the chemokine plate three times as described in the step 6.
10. Add 150 μL/well of the bacteria culture blocked with PBS. This solution contains the phage particles which have been produced by the infected bacteria and which carry the mutated protein at their surface. Incubate for 1 h with gentle agitation (450 rpm).
11. Wash the ELISA plate six times as described in the step 6.
12. Add 100 μL/well of the primary antibody, a rabbit anti-fd Bacteriophage diluted 1:1000 in PBS containing 1% milk. Incubate for 1 h with gentle agitation (450 rpm).
13. Wash the ELISA plate five times as described in the step 6.
14. Add 100 μL/well of the secondary antibody, a peroxidase-coupled mouse anti-rabbit IgG diluted 1:15,000 in PBS containing 1% milk. Incubate for 1 h with gentle agitation.
15. During the incubation, prepare the QuantaBlu Fluorogenic solution by mixing one part of QuantaBlu Peroxide solution with nine parts of QuantaBlu Substrate solution. Incubate without agitation for 1 h protected from light.
16. Wash the ELISA plate five times as described in the step 6.

17. Add 100 μL/well of the QuantaBlu mix and incubate for 1 h protected from light with gentle agitation (450 rpm). Add 100 μL/well of the QuantaBlu Stop solution. Read the fluorescence on a Synergy-H4 or on an equivalent system, with the excitation filter set at 325 nm and the emission filter at 420 nm.

Option 2: ELISA of secreted proteins and/or of proteins present in the periplasm

1. Pick individual TG1 colonies containing the mutated pMS101C plasmid from a fresh plate for inoculation into 200 μL of TB medium containing 2% glucose and ampicillin. Incubate overnight at 30 °C and 750 rpm.
2. On the following morning, inoculate 100 μL of fresh TB 0.1% glucose Amp^+ medium with 3 μL of the overnight culture. Incubate at 30 °C and 750 rpm until the culture medium is turbid.
3. Dilute IPTG to 600 μ*M* in TB 0.1% glucose Amp^+ medium. Once the culture reaches turbidity, induce the expression of single-chain variable fragments by adding 20 μL/well of IPTG 600 μ*M* (final concentration of IPTG: 100 μ*M*). Grow overnight at 30 °C and 750 rpm.
4. On the same day, coat a 96-well maxisorb black plate with 10 μg/mL neutravidin in PBS: add 100 μL/well of the neutravidin solution and incubate the plate without agitation overnight at 4 °C. Alternatively, commercially available black plates precoated with streptavidin could be used.
5. Prepare the periplasmic lysis buffer (S.M.Dunn; previously unpublished): for 50 mL of buffer, mix 0.0138 g of Polymyxin B sulfate salt (final concentration 200 μ*M*), 0.72 g of Guanidine–HCl (final concentration 150 m*M*), 2.5 mL Tris pH 7.5 1 *M* (final concentration 50 m*M*), and 50 μL Tween-20 (final concentration). Agitate until complete dissolution. Sterile filter on a 0.22 μm membrane and keep at 4 °C.
6. On the following day, wash the neutravidin-coated plate three times with a solution of PBS containing 0.1% (v/v) Tween-20. At the end of the washes, remove any remaining wash buffer by inverting the plate and blotting it against paper towels.
7. Add 100 μL/well of 1 μg/mL biotinylated chemokine in PBS and incubate at room temperature for 1 h on a horizontal orbital shaker with gentle agitation (450 rpm).
8. Discard the chemokine solution by inverting the plate. Add 230 μL/well of PBS containing 5% (w/v) nonfat milk and incubate at room temperature for 1 h with gentle agitation (450 rpm).
9. In the meantime, centrifuge the bacterial culture 20 min at 4000 rpm at 4 °C. Transfer the supernatant to a clean 96-well microplate and keep it at 4 °C.

10. Use the pellet to perform the periplasmic extraction: add 50 μL/well of lysis buffer pellet and resuspend the cells by shaking vigorously on an horizontal orbital shaker for 15 min. Incubate for 1 h at 30 °C and 750 rpm. Centrifuge the plate 15 min at 4000 rpm. Transfer the supernatant to a clean 96-well microplate and discard the pellet.
11. Block the periplasmic extract (step 9) and the culture supernatant (step 8) by adding 100 μL/well of PBS containing 5% milk. Incubate for 30 min with gentle agitation (450 rpm) to block the culture.
12. Wash the chemokine plate three times as described in the step 6.
13. Add 150 μL/well of the bacteria culture or of the periplasmic extract. These solutions contain the soluble mutated protein which has been secreted by bacteria. Incubate for 1 h with gentle agitation (450 rpm).
14. Wash the ELISA plate four times as described in step 6.
15. Add 100 μL/well of the primary antibody, which must be a specific antibody that recognizes the mutated protein analyzed in the assay, diluted in PBS containing 1% milk (dilution 1:2000). Incubate for 1 h with gentle agitation (450 rpm).
16. Wash the ELISA plate four times as described in step 6.
17. Add 100 μL/well of the secondary antibody, a peroxidase-coupled antibody which recognizes the species of the primary antibody, diluted in PBS containing 1% milk (dilution 1:15,000). Incubate for 1 h with gentle agitation.
18. During the incubation, prepare the QuantaBlu Fluorogenic solution by mixing one part of QuantaBlu Peroxide solution with nine parts of QuantaBlu Substrate solution. Incubate without agitation for 1 h in the absence of light.
19. Wash the ELISA plate five times as described in step 6.
20. Add 100 μL/well of the QuantaBlu solution and incubate for 1 h in the absence of light with gentle agitation (450 rpm). Add 100 μL/well of the QuantaBlu Stop solution. Read the fluorescence on a Synergy-H4 or on an equivalent system, with the excitation filter set at 325 nm and the emission filter at 420 nm.

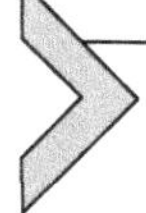

3. PHAGE DISPLAY SELECTION TO MODULATE THE SELECTIVITY OF CHEMOKINE BINDERS

Chemokine inhibitors might need optimization, in order to increase their affinity for a define chemokine or to broaden/restrain their binding profile. In order to optimize the binding profile of Evasin-4, we constructed a phage display library containing randomly mutated DNA sequences encoding for Evasin-4 variants and used this library in phage display selections against various chemokines. We report here the detailed methodology used to perform this work, which might be adapted to isolate new variants of chemokine/cytokine binders.

3.1 Construction of Phage Display Libraries

NB: while constructing the library, always work on ice with DNA.

Required materials

GeneMorph II Random Mutagenesis Kit (Agilent, cat. no. 200550)
Nuclease-free water (Ambion, cat. no. AM9937)
DNA Clean & Concentrator–25 kit (Zymo Research, cat. no. D4033)
*Bss*HII restriction enzyme (New England Biolabs, cat. no. R0199L)
*Sal*I-HF restriction enzyme (New England Biolabs, cat. no. R3138L)
*Dpn*I restriction enzyme (New England Biolabs, cat. no. R0176L)
*Not*I-HF restriction enzyme (New England Biolabs, cat. no. R3189L)
CutSmart Buffer (New England Biolabs, provided with the restriction enzymes)
T4 DNA Ligase (New England Biolabs, cat. no. M0202L)
Electroporation cuvettes plus, 1 mm gap (BTX, cat. no. 45-0124)
Electrocompetent TG1 *E. coli* (Lucigen, cat. no. 60502-2)
Glucose (Merck Chemicals, cat. no. 346351-1KG)
Ampicillin (Calbiochem, cat. no. 171254)
Kanamycin (Calbiochem, cat. no. 420411)
Glycerol (Sigma, cat. no. G5516-1 L)
M13KO7 helper phages (Invitrogen, cat. no. 18311-019)
PEG 6000 (Calbiochem, cat. no. 528877)
Bovine serum albumin (BSA; Merck, cat. no. 1.12018.0100)
PBS (Life technologies, cat. no. 10010-015)
SOC medium
2YT medium
TBE buffer
TB medium
NaCl 5 *M*

Diversification of sequences by random mutagenesis

1. Amplify the sequence of the gene of interest (e.g., Evasin-4) by PCR using specific primers containing a restriction site for *Bss*HII on the 5′ end of the sense primer and a SalI restriction site on the 3′ end of the antisense primer. To introduce random mutations into the sequence,

use an error-prone polymerase (Mutazyme, from the GeneMorph II kit) instead of a standard DNA polymerase. Prepare the following solution in an Eppendorf tube:

1.5 μg of DNA template (e.g., pMS101C-Evasin-4 vector)
2.5 μg of the sense primer
2.5 μg of the antisense primer
20 μL dNTP mix 40 m*M* (final concentration 200 μ*M*)
100 μL 10 × Mutazyme II reaction buffer
Nuclease-free H_2O to 1 mL

Mix well by inverting the tube and add 20 μL of Mutazyme II polymerase. Mix by inversion, centrifuge briefly, and aliquote into 20 PCR tubes (50 μL per tube). Incubate on a thermocycler for 30 cycles of denaturation/annealing/elongation following the manufacturer's instructions.

2. Pool the PCR reactions four by four to obtain five aliquots containing 250 μL each in 2 mL Eppendorf tubes. Add 1250 μL of DNA-binding buffer to each tube and apply on DNA Clean & Concentrator-25 columns (one column per aliquot, i.e., five columns in total). Wash the columns twice, with 300 and 400 μL of DNA Wash Buffer, respectively, and elute with 55 μL H_2O per column. Pool the different aliquots to obtain ca. 275 μL of DNA preparation. Examine 2 μL on a DNA agarose gel to ensure that a single band of the correct size was obtained. Determine the total amount of DNA purified by measuring the DNA concentration.
3. Digest the PCR product and the destination vector pMS101C with the *Bss*HII and *Sal*I-HF restriction enzymes. The amount of each reagent has to be adapted to the amount of DNA purified in step 2. For 10 μg of DNA, prepare the following solutions:

Digestion of insert:

50 μL *Bss*HII (final amount: 200 U)
20 μL *Sal*I-HF (final amount: 400 U)
5 μL *Dpn*I (final amount 100 U)
10 μg DNA (Evasin-4 PCR product)
80 μL CutSmart Buffer 10 ×
Nuclease-free H_2O qsp 800 μL

Digestion of vector:

50 μL *Bss*HII (final amount: 200 U)
20 μL *Sal*I-HF (final amount: 400 U)
10 μL *Not*I-HF (final amount: 200 U)

20 μg DNA (pMS101C)
80 μL CutSmart Buffer 10×
Nuclease-free H_2O to 800 μL
Incubate for 6 h at 37 °C.

4. Process the two digestion reactions through DNA Clean & Concentrator-25 columns (two columns for each reaction), following the manufacturer's instructions. Wash the columns twice, with 300 and 400 μL of DNA Wash Buffer, respectively, and elute with 35 μL of H_2O prewarmed at 65 °C. Pool the eluates to obtain ca. 70 μL of purified DNA for the insert and for the vector. Examine 2 μL on a DNA agarose gel to ensure that single bands of the correct size were obtained. Determine the total amount of DNA purified by measuring DNA concentration.
5. Ligate the gene of interest (e.g., Evasin-4) into the destination vector (pMS101C) using T4 DNA ligase. We use a insert:vector ratio of 6:1 but the optimal conditions may vary depending on the vector and the insert used. To increase the efficiency of the ligation, use the lowest possible reaction volume. Example:

   65 μL digested vector (pMS101C)
   65 μL digested insert (Evasin-4)
   16.1 μL T4 DNA ligase Buffer 10×
   15 μL T4 DNA ligase

   Aliquot the solution into 0.2 mL PCR tubes (20 μL/tube). Incubate on a thermocycler with the following program:
   a. 2 min at 10 °C
   b. 10 s at 10 °C
   c. 30 s at 30 °C
   d. Repeat steps b–c 99 times (or overnight if not limited by the thermocycler software)
6. Pool the reactions and process the sample through a single Clean & Concentrator-25 column using 1 × 300 and 1 × 400 μL washes, followed by elution with 30 μL prewarmed (65 °C) H_2O. Examine 2 μL of the ligation product on a 1% agarose gel.
7. On the following day, perform the transformation of electrocompetent TG1 bacteria. Place 10 electroporation cuvettes and 10 microcentrifuge tubes on ice. Thaw TG1 bacteria on ice. Transfer 50 μL of diluted bacteria to a chilled microcentrifuge tube and add 3 μL of the ligation product. Mix gently by stirring. Carefully transfer the bacteria/DNA mixture into a chilled electroporation cuvette. Electroporate with a

1700 V pulse of 4 ms (200 Ω, 25 μFD). Immediately add 1 mL of SOC medium to the cuvette and transfer the cells to a sterile culture tube. Add again 1 mL of SOC medium to the cuvette to recover all the cells and transfer to the culture tube. Repeat this electroporation nine times.

8. Incubate the ten tubes containing the cells at 250 rpm for 1 h at 37 °C.
9. Pool the cells, measure the total volume of cells (ca. 20 mL) and spread onto five prewarmed 24×24 cm 2YT plates containing 2% glucose and the appropriate selection antibiotic (ampicillin for pMS101C vector). Keep 200 μL of cells for the titration. Incubate overnight at 30 °C.
10. Prepare a serial dilution of bacteria to titrate the number of colonies obtained and determine the size of the library.
    Dilution 10^{-3}: 100 μL cells + 900 μL 2YT medium
    Dilution 10^{-4}: 100 μL of 10^{-3} + 900 μL 2YT medium
    Dilution 10^{-5}: 100 μL of 10^{-4} + 900 μL 2YT medium
    Dilution 10^{-6}: 100 μL of 10^{-5} + 900 μL 2YT medium
    Dilution 10^{-7}: 100 μL of 10^{-6} + 900 μL 2YT medium
    Plate 100 μL of each dilution on 2YT 2% glucose Amp^+ plates and incubate overnight at 30 °C.
11. On the following day, scrape the bacteria using 5 mL 2YT per plate and put them into 50 mL Falcon tubes.
12. Centrifuge the cells for 20 min at 4000 rpm and 4 °C. Decant medium. Resuspend each pellet in 5 mL 2YT medium. Pool the supernatants containing the bacteria and add 0.5 volume of 50% glycerol in H_2O (v/v, sterile filtered on a 0.22 μm membrane) to obtain a final concentration of 16.5% glycerol in 2YT medium. Measure the OD_{600} of the final solution. Prepare 750 μL aliquots and snap freeze them on dry ice prior to store them at −80 °C.
13. In parallel, count the number of colonies on each titration plate to determine the size of the library. Size = average of (dilution factor × number of colonies) × 10 (to have per mL, because we plated 100 μL) × total volume of bacteria (measured in step 9).
14. Randomly pick 20 individual colonies on the titration plates. Amplify their DNA using a dedicated kit (e.g., QIAprep Miniprep Kit). Sequence the DNA to assess the quality of the library.

Production of phage particles (phage rescue)

1. Based on the OD_{600}, determine the concentration of bacteria in your library, knowing that an OD_{600} of 1 corresponds to ca. 8×10^8 cells/mL.

2. Inoculate 200 mL of 2YT 2% glucose Amp^+ with an aliquot of the library to obtain a final concentration around 8×10^7 cells/mL (OD_{600} ca. 0.1). Grow at 30 °C and 180 rpm for 2–3 h until the OD_{600} reaches 0.5 (ca. 4×10^8 cells/mL).
3. Transfer 40 mL of bacteria ($\approx 1.6 \times 10^{10}$ cells) to a clean 50 mL tube. Infect by adding 1.6 mL of M13KO7 helper phage to each tube to have a phage to cells ratio of 10:1 (stock at 10^{11} pfu/mL).
4. Incubate for 1 h at 37 °C and 90 rpm.
5. Centrifuge the cells at 4000 rpm for 20 min. Carefully discard the supernatant and resuspend the pellet in 5 mL 2YT Amp^+ Kan^+. Add to 200 mL of 2YT Amp^+ Kan^+ and incubate overnight at 30 °C and 240 rpm.
6. Prepare the PEG6000/NaCl solution: 125 mL NaCl 5 *M*, 50 g PEG6000, complete with H_2O until 250 mL (final concentration: 2.5 *M* NaCl, 20%, w/v PEG6000). Sterile filter on a 0.22 μm membrane and store at 4 °C.
7. Transfer the overnight rescue culture into four 50 mL tubes (50 mL/tube). Centrifuge at 4 °C at 4000 rpm for 30 min.
8. During the centrifugation, inoculate 50 mL of 2YT 2% glucose medium with an individual TG1 colony from a fresh minimal plate and incubate at 37 °C and 240 rpm.
9. Carefully transfer 40 mL of culture supernatant to a clean 50 mL tube containing 10 mL of PEG 6000/NaCl solution. Repeat this step for the four centrifuged tubes. Mix gently the tubes by inversion and incubate them on ice for 30 min to precipitate the phage.
10. Centrifuge at 4 °C at 4000 rpm for 20 min. Discard the supernatant and place the tubes on ice.
11. Resuspend each pellet in 2 mL PBS with gentle pipetting (some brown-colored material may stay insoluble). Pool 2× two tubes in a 15 mL tube and adjust the volume of each tube to 8 mL.
12. Add 2 mL of PEG/NaCl solution to each tube and mix by inversion. The solution should turn milky white, indicating a high yield of precipitated phage. Place the tubes on ice for 10 min and then centrifuge at 4 °C at 4000 rpm for 20 min.
13. Gently remove all the supernatant from the white phage pellets. Use a P100 pipette to carefully remove any residual liquid from the pellet. Gently resuspend each phage pellet in 4 mL PBS. Pool the two tubes (total volume ca. 8 mL) and aliquote in 8 1.5 mL microcentrifuge tubes (1 mL/tube).

14. Centrifuge at 4 °C at 14,000 rpm for 10 min to pellet any remaining insoluble material. For each tube, carefully transfer 900 μL into a clean 15 mL tube on ice to obtain a final volume of 7.2 mL.
15. Add 800 μL of freshly prepared PBS containing 1.5% BSA (w/v) (filter sterilized on a 0.22 μm membrane) and gently mix by pipetting to obtain 8 mL of phage solution in PBS containing 0.15% BSA.
16. Add 4 mL of 50% glycerol to obtain a final concentration of 16.5%. Aliquote in cryotubes (1 mL/tube) and snap freeze in dry ice. Store at −80 °C. Keep 50 μL of phage for the titration.
17. Prepare a serial dilution to determine the final concentration of phage.

    Dilution 10^{-2}: 10 μL phage + 990 μL 2YT medium
    Dilution 10^{-4}: 10 μL of 10^{-2} + 990 μL 2YT medium
    Dilution 10^{-6}: 10 μL of 10^{-4} + 990 μL 2YT medium
    Dilution 10^{-8}: 10 μL of 10^{-6} + 990 μL 2YT medium
    Dilution 10^{-9}: 100 μL of 10^{-8} + 900 μL 2YT medium
    Dilution 10^{-10}: 100 μL of 10^{-9} + 900 μL 2YT medium
    Dilution 10^{-11}: 100 μL of 10^{-10} + 900 μL 2YT medium

    Measure the OD_{600} of the TG1 bacteria inoculated in step 8. Once it is close to 0.5, mix 100 μL of TG1 bacteria with 100 μL of each dilution from 10^{-8} to 10^{-11}. Incubate for 1 h at 37 °C and 90 rpm.
18. Plate out 100 μL of each dilution on 2YT 2% glucose Amp^{+} plates and incubate overnight at 30 °C.
19. On the following day, count the number of colonies on each titration plate to determine the phage titer. Phage titer (in pfu/mL) = average of (dilution factor × number of colonies) × 10 (to have per mL, because we plated 100 μL) × 2 (dilution with TG1 cells).
20. Randomly pick 20 individual colonies on the titration plates. Amplify their DNA using a dedicated kit (e.g., QIAprep Miniprep Kit). Sequence the DNA to assess the quality of the phage rescue.

3.2 Phage Display Selections

Using a library generated as described above, phage display selections can be performed using standard phage display protocols. As a target, we use biotinylated chemokines captured on magnetic streptavidin beads (Dynabeads M-280 streptavidin, Life Technologies, cat. no. 11205D). An example of selections and the corresponding results is displayed in Table 1.

Table 1 Selections Against CCL3 or CCL5 Using a Library Containing Mutated Evasin-4 Sequences

Target	Selection Round	Input (pfu)	Output (pfu)	Ratio Output/Input	Analyzed Clones	Wild-Type Sequences	Repeated Sequences	Single Sequences	Nonfunctional Sequences
CCL3	1	2×10^{12}	1.22×10^{7}	6.1×10^{-6}	–	–	–	–	–
	2	6×10^{11}	4.40×10^{5}	7.3×10^{-7}	88	58 (66%)	4	18	1
	3	2×10^{10}	3.44×10^{5}	1.7×10^{-5}	80	30 (38%)	10	8	5
CCLS	1	2×10^{12}	3.81×10^{6}	1.9×10^{-6}	–	–	–	–	–
	2	1.7×10^{11}	6.54×10^{5}	3.9×10^{-6}	83	36 (43%)	6	15	6
	3	5×10^{10}	9.85×10^{5}	2.0×10^{-5}	45	16 (36%)	1	5	5

Inputs and outputs are expressed in plaque forming unit (pfu). Repeated sequences: number of sequences present more than once in the sequenced clones; single sequences: number of sequences present only once in the sequenced clones; and nonfunctional sequences: number of sequences containing a stop codon or a frameshift.

4. SUMMARY

Phage display can be adapted to investigate further the interaction between a chemokine and a CKBP such as those used in this study. The adaptation of phage display to the identification of protein pharmacophore is however limited to small proteins but most soluble proteins which play a role as immune messengers, cytokines, chemokines, growth factors, etc., should be susceptible to such an approach.

REFERENCES

Bonvin, P., Dunn, S. M., Rousseau, F., Dyer, D. P., Shaw, J., Power, C. A., et al. (2014). Identification of the pharmacophore of the CC chemokine-binding proteins Evasin-1 and -4 using phage display. *The Journal of Biological Chemistry*, *289*, 31846–31855.

Magistrelli, G., Gueneau, F., Muslmani, M., Ravn, U., Kosco-Vilbois, M., & Fischer, N. (2005). Chemokines derived from soluble fusion proteins expressed in *Escherichia coli* are biologically active. *Biochemical and Biophysical Research Communications*, *334*, 370–375.

Pande, J., Szewczyk, M. M., & Grover, A. K. (2010). Phage display: Concept, innovations, applications and future. *Biotechnology Advances*, *28*, 849–858.

Proudfoot, A. E., & Borlat, F. (2000). Purification of recombinant chemokines from *E. coli*. *Methods in Molecular Biology*, *138*, 75–87.

CHAPTER TEN

Studying Chemokine Control of Neutrophil Migration *In Vivo* in a Murine Model of Inflammatory Arthritis

Yoshishige Miyabe, Nancy D. Kim, Chie Miyabe, Andrew D. Luster[1]
Center for Immunology and Inflammatory Diseases, Division of Rheumatology, Allergy and Immunology, Massachusetts General Hospital, Harvard Medical School, Boston, Massachusetts, USA
[1]Corresponding author: e-mail address: aluster@mgh.harvard.edu

Contents

1. Introduction 208
2. Methods 211
   2.1 K/BxN Serum Transfer Model 211
   2.2 Arthritis Clinical Scoring and Measurement of Paw Thickness 212
   2.3 Arthritis Histological Scoring 213
   2.4 Quantitation of Neutrophil Migration into the Joint 215
   2.5 MP-IVM for Imaging of Joints 221
   2.6 Neutrophil Adoptive Transfer 224
   2.7 Mixed BMC Mice Experiments 228
3. Conclusions 229
References 230

Abstract

Chemokines regulate the migration of cells *in vivo* and dysregulated expression of chemokines and their receptors are implicated in autoimmune and inflammatory diseases. Inflammatory arthritides, such as rheumatoid arthritis (RA), are characterized by the recruitment of inflammatory cells into joints. The K/BxN serum transfer mouse model of inflammatory arthritis shares many similar features with RA. In this autoantibody-induced model of arthritis, neutrophils are the critical immune cells necessary for the development of joint inflammation and damage. In this review, we describe the use of several methods to study the role of chemoattractant receptors, including chemokine receptors, on the recruitment of neutrophils into the joint in the K/BxN model of inflammatory arthritis. This includes both traditional methods, such as flow cytometry, immunohistochemistry, and enzyme assays, as well as multiphoton *in vivo* microscopy that we have adapted to study the role of immune cell trafficking in and around the joint in live mice.

Methods in Enzymology, Volume 570
ISSN 0076-6879
http://dx.doi.org/10.1016/bs.mie.2015.11.002

1. INTRODUCTION

The induction of inflammation requires leukocyte recruitment into the affected tissue. For instance, the ability of the neutrophil to exert its functions relies heavily on its ability to traffic to sites of infection and inflammation (Mayadas, Cullere, & Lowell, 2014). In general, neutrophil migration into inflamed tissue occurs in postcapillary venules and can be described by the adhesion cascade, which begins with the capture of free-flowing leukocytes to the vessel wall, followed by (1) rolling on the vessel wall in the direction of flow, (2) arrest on the endothelium, (3) release from adhesion and crawling in all directions on the vessel to locate a receptive location for, (4) transendothelial migration (TEM), followed then by (5) swarming to a specific location within the tissue (Mayadas et al., 2014). These discrete steps are controlled by a combination of molecular signals, including selectins (rolling), integrins (arrest, crawling), and chemoattractants (arrest, possibly crawling, and TEM).

Chemoattractants regulate the trafficking of immune cells, including neutrophils, between tissues, as well as their positioning and interactions within tissues. Even though chemoattractants consist of a wide range of diverse molecules with differing regulation and biophysical properties, including secreted proteins (e.g., chemokines), proteolytic fragments of serum proteins (e.g., complement component C5a), and bioactive lipids (e.g., leukotrienes), they all induce directed migration by activating seven-transmembrane spanning G protein-coupled receptors (Griffith, Sokol, & Luster, 2014). Differential expression of chemoattractant receptors on cells results in selective recruitment of specific cell types under particular conditions providing appropriate and efficient immune responses tailored to the foreign insult (Griffith et al., 2014). Beyond their pivotal role in the coordinated migration of innate immune cells to inflamed tissue, chemoattractants play important roles in the development of lymphoid tissues, in the maturation of immune cells, and in the generation and delivery of adaptive immune responses (Griffith et al., 2014). Altered expression of chemoattractants and their receptors is implicated in a broad range of human diseases, including cancer, autoimmune, and inflammatory diseases (Chow & Luster, 2014; Griffith et al., 2014).

Once in the tissue, secreted enzymes from leukocytes are important in host defense against invading microbial pathogens, but they are cytotoxic and can also induce host tissue damage (Dufour, 2015; Ji, Ohmura, Mahmood, et al., 2002; Odobasic, Muljadi, O'Sullivan, et al., 2015).

For instance, once neutrophils enter into the inflamed or infected tissue, neutrophils become very activated and release powerful enzymes, such as myeloperoxidase (MPO) and elastase, and form neutrophil extracellular traps (NETs), which disable and kill microbial pathogens. However, when enzymes are released in an unregulated manner, these cytotoxic molecules may damage cells, leading to organ injury and dysfunction. In fact, NETs are seen in the synovial tissue of rheumatoid arthritis (RA; Wright, Moots, & Edwards, 2014), serum levels of MPO increase in patients with active RA (Wang, Jian, Guo, & Ning, 2014), and neutrophil elastase may contribute to the pathogenesis of chronic obstructive pulmonary disease (Hoenderdos & Condliffe, 2013). Thus, active secreted enzymes from leukocytes may contribute to the pathologic processes of inflammatory disease.

Inflammatory arthritis, including RA, is an autoimmune disease where neutrophils are recruited into the diseased joint (Wright et al., 2014). To analyze the mechanism of inflammatory cell recruitment into the joint, and the role of chemoattractants and their receptors in inflammatory arthritis *in vivo*, several animal models of arthritis, including the K/BxN serum transfer model and type II collagen-induced arthritis model, have been studied (Chen, Lam, Kanaoka, et al., 2006; Kim, Chou, Seung, Tager, & Luster, 2006; Miyabe, Miyabe, Iwai, et al., 2013). K/BxN mice are T cell receptor transgenic mice on the NOD background that spontaneously develop arthritis due to the development of autoantibodies to the ubiquitous protein glucose-6-phosphate isomerase (Monach, Mathis, & Benoist, 2008). Transfer of serum from K/BxN mice into wild-type (WT) mice results in the deposition of autoantibody–antigen complexes in joints, which induce neutrophil recruitment and subsequent arthritis (Monach et al., 2008). We and others have previously found that at least four chemoattractant receptors—CCR1, CXCR2, C5aR, and BLT1—contribute to the recruitment of neutrophils into the joint in K/BxN serum transfer model (Chou, Kim, Sadik, et al., 2010; Jacobs et al., 2010; Ji et al., 2002; Kim et al., 2006; Monach, Nigrovic, Chen, et al., 2010; Sadik, Kim, Iwakura, & Luster, 2012). The neutrophil is a key effector cell in this model, and this model can be used to study the role of individual chemoattractants, including chemokines, in neutrophil recruitment into the inflamed joint *in vivo*.

Recent advances in imaging technology have allowed for the direct dynamic observation of biological processes, and provided unprecedented views into immune cell migration in live animals that have deepened our understanding of the molecular control of immune cell trafficking *in vivo*. This technology has recently been applied to study the molecular control of neutrophil trafficking to sites of tissue inflammation, such as thermal liver injury (McDonald, Pittman, Menezes, et al., 2010) and laser skin injury (Lämmermann, Afonso, & Angermann, et al., 2013). The spatial requirements for neutrophil recruitment into the joint have not been described in arthritis as live animal imaging has not been applied to the study of chemoattractant control of immune cell entry into the joint in models of arthritis. However, we recently established a new method of joint imaging using multiphoton intravital microscopy (MP-IVM) in live mice, which allow us to directly visualize and study inflamed joints in live mice. MP-IVM of the joint will provide new insights into the fine control of neutrophil trafficking into the joint and provide new insights into the pathogenesis of arthritis.

Adoptive cell transfer experiments are useful for determining the role of certain molecules, including cytokines, chemokines, and their receptors, on individual immune cell populations, such as neutrophils, T cells, and B cells, in various animal models of diseases. For example, our previous research demonstrated that leukotriene B4 (LTB4) receptor 1 (BLT1)-deficient mice did not develop arthritis, and that adoptive transfer of WT BLT1-expressing neutrophils into BLT1-deficient mice restored arthritis (Kim et al., 2006). Further, mice deficient in 5-lipoxygenase and LTA4 hydrolase, enzymes required to synthesize LTB4, also did not develop arthritis, and adoptive transfer of WT neutrophils into these enzyme-deficient mice also resorted arthritis (Chen et al., 2006). These adoptive transfer studies established that the neutrophil is the critical LTB4-producing and -responding cell type required for neutrophil entry into the joint and for the development of arthritis. Thus, adoptive cell transfer experiments can be used to explore the role of a particular cell type, as well as individual molecules on that cell type, in the activation and migration of individual immune cell populations *in vivo* in models of inflammation and disease.

Labeling cells with cell tracking dyes *ex vivo* prior to adoptive transfer allows for the easy tracking of transferred cells. In addition, cells genetically engineered to express different fluorescent proteins are also commonly used to track cells *in vivo*. The coadoptive transfer of two different cells populations labeled with different fluorescent dyes into the same mouse can be used to compare the trafficking properties of the two cell populations in a short-term competitive cell intrinsic manner. Further, mixed bone marrow chimeric (BMC) mice, generated with BM from different transgenic fluorescent protein expressing mice, allow the study of cell intrinsic effects of chemoattractant receptor deficiency compared to WT cells in the same mice over the entire course of disease development.

Here, we describe how to quantitatively and qualitatively describe neutrophil migratory behavior in a defined peripheral tissue site *in vivo* using multiple complementary methodologies, including immunohistochemistry (IHC), flow cytometry (FCM), enzymatic activity, and MP-IVM in live mice. We also describe the methods of neutrophil isolation from mouse BM, neutrophil labeling with with cell tracker dyes, adoptive neutrophil transfer, and mixed BMC approaches to study the role of chemoattractant receptors in neutrophil recruitment into the joint in a model of inflammatory arthritis.

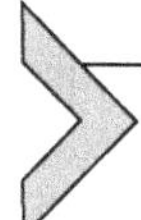

2. METHODS

2.1 K/BxN Serum Transfer Model

The K/BxN serum transfer model is an excellent model to study the effector phase of inflammatory arthritis due to its high penetrance and reproducibility, as well as the ability to easily transfer the disease with serum into genetically deficient strains. It is also an excellent model to study the role of chemoattractants in neutrophil trafficking into the joint as neutrophil entry into the joint is absolutely required for the development of arthritis in this model (Monach et al., 2008; Wipke & Allen, 2001).

2.1.1 Required Materials

2.1.1.1 Isolation of Serum from K/BxN Mice

1. 8- to 10-week-old K/BxN mice. (*Note*: KRN mice are not commercially available and may be requested from the author (Ji et al., 2002). KRN mice are bred with NOD mice (Jackson Laboratory) to generate K/BxN F1 offspring for serum harvest (Ji et al., 2002).)
2. 23-G needles and syringes for cardiac puncture.
3. 1.5 ml Eppendorf tubes.

2.1.1.2 K/BxN Serum Transfer

1. 26 G syringe.
2. K/BxN serum.

2.1.2 K/BxN Serum Transfer Model

2.1.2.1 Isolation of Serum from K/BxN Mice

1. Bleed K/BxN mice by cardiac puncture immediately after CO_2 euthanasia and transfer to Eppendorf tubes. (*Note*: Both male and female K/BxN mice are acceptable donors.)
2. Leave blood at room temperature (RT) for 1 h.

3. Centrifuge at 1,6000 × g for 1 min at RT.
4. Transfer supernatant (serum + blood) into new tube and discard the clotted blood.
5. Centrifuge the supernatant at 850 × g for 10 min at RT.
6. Remove the clarified supernatant (serum) and put into a new tube.
7. Store serum in 1 ml aliquots at −20 °C.

2.1.2.2 K/BxN Serum Transfer

1. Prior to injection of mice, calculate the amount of K/BxN serum required for the experiment. (# mice +1) × 150°μl = volume needed for experiment. Defrost the required amount of K/BxN serum needed for the experiment and combine together prior to injection. K/BxN serum potency may vary from batch to batch; this process will ensure that all mice in a given experiment will receive the same serum.
2. 150 μl K/BxN serum is injected intraperitoneal (i.p.) using a 26-G syringe on day 0 and 2.

2.2 Arthritis Clinical Scoring and Measurement of Paw Thickness

Arthritis clinical score and measurement of paw thickness are the basic measurements of disease activity.

2.2.1 Required Materials

1. Calipers.

2.2.2 Evaluation of Arthritic Clinical Score and Measurement of Paw Thickness

Arthritis clinical scoring and paw thickness are analyzed every 1–3 days after neutrophil adoptive transfer.

1. Arthritis clinical score as described previously (Miyabe, Miyabe, Iwai, et al., 2013). Disease severity for each limb is recorded as follows: 0 = normal; 1 = erythema and swelling of one digit; 2 = erythema and swelling of two digits or erythema and swelling of ankle joint; 3 = erythema and swelling of more than three digits or swelling of two digits and ankle joint; and 4 = erythema and severe swelling of the ankle, foot, and digits with deformity. The clinical arthritis score is defined as the sum of the scores for all four paws of each mouse. (*Note*: A severe swollen ankle is defined as a paw thickness of more than 4 mm.)
2. The thickness of each paw is measured using a pair of digital slide calipers (Fig. 1).

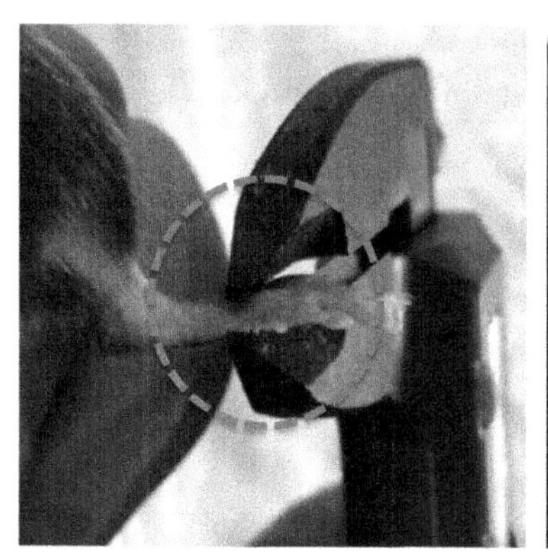

Figure 1 Measurement of paw thickness. Left panel: The red circle shows the measurement of paw thickness of the forefoot. Right panel: The green circle shows measurement of paw thickness from the instep to plantar region in the hind limb. (See the color plate.)

2.3 Arthritis Histological Scoring

Arthritis histological scoring is the standard method to evaluate the extent of inflammation and inflammatory cell infiltration into the joint. This technique, however, is not used describe or quantitate the different types of immune cells that enter the joints.

2.3.1 Required Materials

2.3.1.1 Decalcification

1. 10% Formalin.
2. 70%, 80%, 90%, and 100% ethanol.
3. Mixed reagent (xylene:100% ethanol = 1:3) for removing fat from tissue samples.
4. K-CX decalcification reagent (FALMA).

2.3.1.2 Deparaffinization

1. Xylene.
2. 100% Ethanol.
3. 70% Ethanol.
4. Deionized water (DDW).

2.3.1.3 Hematoxylin and Eosin Staining

1. Hematoxylin and Eosin (H&E) staining kit (Wako Pure Chemical Industries).
2. Mount-Quick (Daido Sangyo) or other solvent-based mounting medium.

2.3.1.4 Dehydration

1. Xylene.
2. 100% Ethanol.
3. 70% Ethanol.

2.3.2 Evaluation of Arthritic Histological Score

1. Harvest mouse ankle joints including the tibiotalar joint to the midfoot.
2. Decalcification.
   - **2.1.** Fix samples with 10% formalin overnight.
   - **2.2.** To remove fat from a sample, place samples into the mixed reagent for at least 30 min. (*Note*: This step is needed so that the KC-X decalcification reagent will sufficiently penetrate through samples. Adequate fat removal will be indicated when the mixed reagent turns a yellow color.)
   - **2.3.** Place each sample into 100%, 90%, 80%, and 70% ethanol, and DDW for 5 min at each step.
   - **2.4.** Place each sample into K-CX decalcification reagent overnight. (*Note*: The decalcification status can be checked by pricking a sample with a needle. The sample will be adequately decalcified if it feels soft.)
   - **2.5.** Wash samples with water for at least 12 h.
   - **2.6.** Samples are embedded with paraffin and cut into 4 μm thick sections on glass slides.
3. Deparaffinization.
   - **3.1.** Paraffinized slides (4-μm-thick sections) are placed into xylene for at least 15 min.
   - **3.2.** Place slides into new xylene for 5 min.
   - **3.3.** Place samples into 100% ethanol for 3 min. This step is performed twice.
   - **3.4.** Place samples into 70% ethanol for 3 min. This step is performed twice.
   - **3.5.** Place samples into DDW for 3 min.
4. H&E staining
   - **4.1.** Place slides into hematoxylin for 5 min.
   - **4.2.** Wash with water for 3 min.

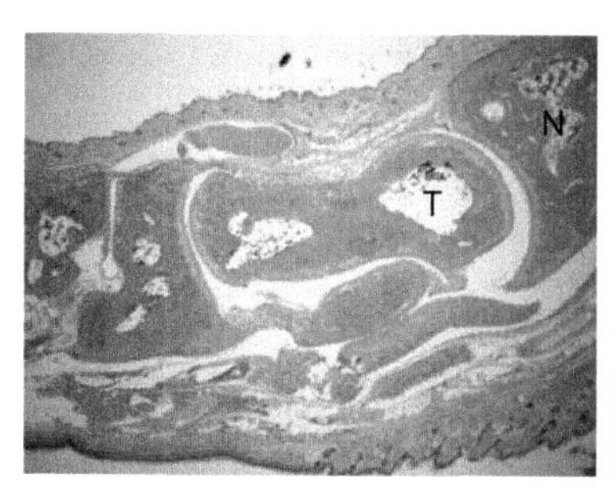

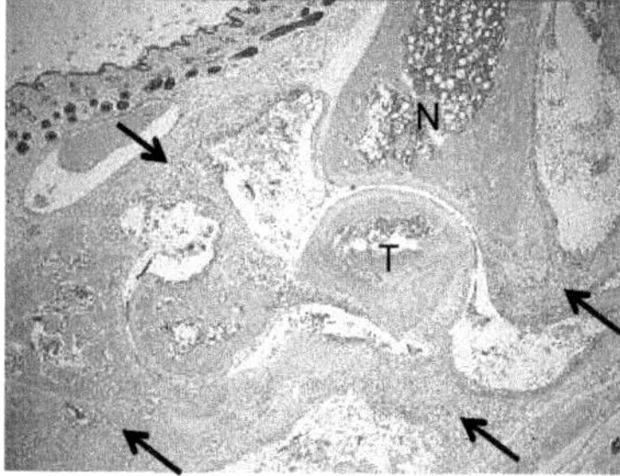

Figure 2 Evaluation of the histological score of joints. H&E stained sections are evaluated histologically and inflammation is scored using the following criteria: 0 = no inflammation, 1 = focal inflammation, and 2 = severe and diffuse inflammatory infiltration. Left panel shows an example of control ankle. Right panel shows an example of diffuse inflammatory infiltration throughout the ankle. Arrow indicates the inflammatory cell infiltration into the joints. T, Taulus; N, Naviculare. (See the color plate.)

  - **4.3.** Place slides into eosin for 5 min.
  - **4.4.** Wash with water for 3 min.
5. Dehydration.
  - **5.1.** Place slides into 75% ethanol for 5 min. This step is performed twice.
  - **5.2.** Place slides into 100% ethanol for 5 min. This step is performed twice.
  - **5.3.** Place slides into xylene for 5 min. This step is performed twice.
  - **5.4.** Mount slides using Mount-Quick.
6. Histological scoring.

Histological scoring of tissue sections uses the following criteria. 0 = no inflammation, 1 = focal inflammatory infiltration, and 2 = severe and diffuse inflammatory infiltration (Fig. 2). (*Note*: It is recommended that two fields per slide, one around the instep area and the other around the Achilles tendon, be used to quantitate arthritis or neutrophil numbers. Severe and diffuse inflammatory infiltration is defined as inflammatory infiltration in the both the instep and Achilles tendon per slide. Focal inflammatory infiltration is defined as inflammatory infiltration in one field per slide.)

2.4 Quantitation of Neutrophil Migration into the Joint

As previously described, H&E staining is an excellent method to evaluate the level of inflammatory cell infiltration into the joint. However, to quantify specific cell types in the joint, additional methods of quantitation are needed. These include IHC and FCM using neutrophil-specific monoclonal antibodies as well enzymatic quantitation for the presence of neutrophil-specific granule proteins (Kim et al., 2006; Miyabe, Miyabe, Miura, et al., 2013; Pulli, Ali, Forghani, et al., 2013).

2.4.1 Immunohistochemistry *(Fig. 3)*

IHC is useful to detect specific cell types in the joint tissue and to determine the location of infiltrated cells in the joint tissue. However, the evaluation of inflammatory cells in the synovial fluid is not amenable to IHC.

2.4.1.1 Required Materials

1. Citrate buffer (Sigma–Aldrich).
2. PBS.
3. Skim milk.
4. Rat purified anti-mouse Ly-6G antibody (clone 1A8, Biolegend).
5. N-Histofine Simple Stain Mouse Max PO (Rat) (Nichirei Biosciences).
6. N-Histofine DAB 3S kit (Nichirei Biosciences).
7. H&E staining (Optional).
8. Antibody Diluent (Dako cytomation).
9. Coplin jars.
10. Slide holders.
11. Humidified slide chamber.

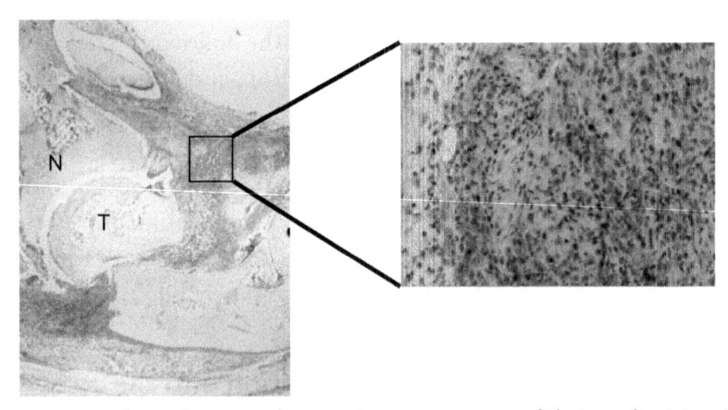

Figure 3 Immunohistochemistry for Ly-6G positive neutrophils into the joint. Brown colored cells are Ly-6G-positive neutrophils in the joint. To quantitate neutrophils in the joint, the number of Ly-6G positive neutrophils in three randomly selected fields per slide are counted. The right photomicrograph is a magnification of the black circle in the left photomicrograph. T, Taulus; N, Naviculare.

2.4.1.2 IHC of Neutrophils in the Arthritic Joints

1. Harvest mouse ankle joints including the tibiotalar joint to the midfoot.
2. Decalcification (see step 2 in Section 2.3.2).
3. Deparaffinization (see step 3 in Section 2.3.2).
4. IHC for neutrophils in the joint.
   - **4.1.** Put slides in Coplin jar filled with 10 m*M* Citrate Buffer for antigen retrieval and place in 65 °C hot water bath for 45 min. (*Note*: Other methods of antigen retrieval, such as microwaving the sample in a water bath, or using 1 m*M* EDTA as a retrieval solution, may also be useful.)
   - **4.2.** Wash. (*Note*: PBS, PBS with Tween 20, and TBS with Tween 20 are appropriate as wash buffers. Each wash step is repeated 3 ×.)
   - **4.3.** Incubate slides in 0.3% H_2O_2 for 30 min to block endogenous peroxidase activity.
   - **4.4.** Wash.
   - **4.5.** Block slides in 1% skim milk for 45 min at RT. (*Note*: 10% normal serum derived from the same species as the detecting antibody or 1% BSA can also be used for blocking.)
   - **4.6.** Wash.
   - **4.7.** Staining: Cover tissue sections with 10 μg/ml purified anti-mouse Ly-6G antibody diluted with antibody diluent and place slides in humidified chamber in 1 h at RT.
   - **4.8.** Wash.
   - **4.9.** Detect antibody binding using a N-Histofine Simple Stain kit according to manufacturer's directions.
   - **4.10.** Wash.
   - **4.11.** H&E staining (Optional) (see step 4 in Section 2.3.2).
5. Dehydration (see step 5 in Section 2.3.2).
6. Evaluation: The number of Ly-6G positive neutrophils in three randomly selected fields per slide are examined under light microscopy. (*Note*: Counting all Ly-6G positive neutrophils in the joint per slide may also be used for quantitation.)

2.4.2 Flow Cytometric Analysis of Neutrophils in Synovial Fluid or Synovial Tissue

FCM is an excellent method to determine the number and type of cells that have infiltrated an inflamed joint or that are present in the synovial fluid. To isolate single cell from the joint tissue, however, enzymatic tissue digestion is needed. Unfortunately, this step can lead to cell death and can also result in the proteolytic cleavage of certain cell surface markers. Thus, this needs to be considered when performing flow cytometric analysis of synovial tissue that has undergone enzymatic digestion to liberate single cells for analysis. This should not be an issue when analyzing synovial fluid with FCM.

2.4.2.1 Required Materials

2.4.2.1.1 Synovial Fluid

1. 10 or 20 μl pipetman and tips.
2. Sterile 1% FCS/PBS.
3. Hemocytometer.
4. Surgical knife.
5. Eppendorf tube.

2.4.2.1.2 Synovial Tissue

1. Surgical knife.
2. Collagenase type IV (Sigma–Aldrich).
3. 2% FCS/PBS.

2.4.2.1.3 Staining with Ly-6G

1. Rat APC-conjugated anti-mouse Ly-6G antibody (clone 1A8, Biolegend).
2. Purified anti-mouse CD16/CD32 antibody (clone 2.4G2, BD Pharmingen) as Fc block.
3. 2% FCS/PBS.
4. 2% Paraformaldehyde.
5. Falcon 5 ml polystyrene round-bottom tubes.

2.4.2.2 Analysis of Neutrophils in Synovial Fluid or Tissue

2.4.2.2.1 Analysis of Neutrophils in Synovial Fluid by FCM (Fig. 4)

1. Euthanize mice and harvest ankle joints, removing skin.
2. Using a surgical knife, make a small incision of about 3–5 mm on one side of the ankle (tibiotalar) joint.
3. Using a 10 or 20 μl pipetman and appropriate tip containing 5 μl sterile 1% FCS/PBS, lavage through the ankle incision and transfer retrieved synovial fluid to an Eppendorf tube. Repeat this step until it appears most cells have been retrieved from the joint space.
4. Synovial fluid cells are counted using a hemocytometer.
5. Resuspend 1×10^6 synovial fluid cells in a volume of 100 μl 1% FCS/PBS, and transfer to a 5 ml polystyrene tube.

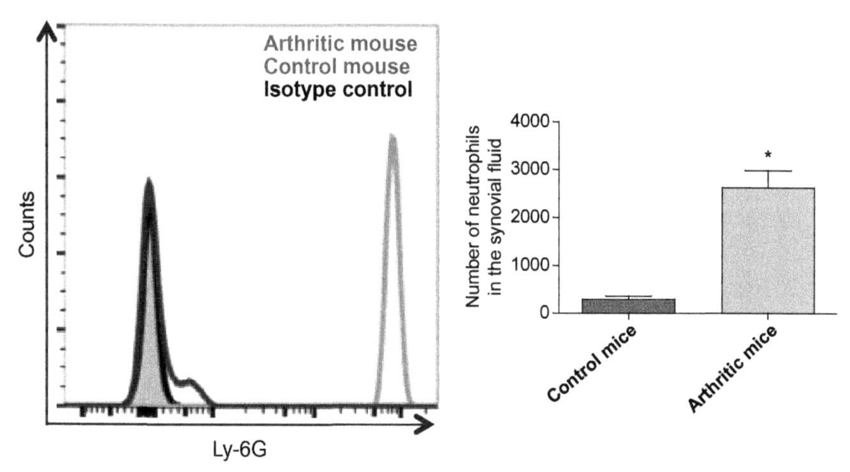

Figure 4 Analysis of neutrophils in synovial fluid by flow cytometry. Synovial fluid was obtained from a control mouse and a mouse with established arthritis (days 7 after K/BxN serum injection). Cells were stained with Ly-6G antibody and analyzed by flow cytometry. Left panel: green line indicates synovial fluid neutrophils from a mouse with established arthritis, red line shows synovial fluid neutrophils from control mouse, and black line reveals staining of established arthritis using an isotype control. The arthritic synovial fluid contains a large population of Ly-6G positive neutrophils while this population was absent in the synovial fluid of a control mouse. Right panel: quantitation of the number of Ly-6G positive cells in control and arthritic joints (control mice: $n=2$, arthritic mice: $n=2$). Data are the mean ± SEM; $^{*}P<0.05$ versus control mice.

6. Incubate synovial fluid cells with 5 μg/ml anti-mouse CD16/CD32 antibody for 5 min at RT to block Fc receptors.
7. Stain with 2 μg/ml anti-mouse Ly-6G antibody for 10 min at RT.
8. Wash with 2% FCS/PBS 3 × and resuspend in 2% paraformaldehyde.
9. Store in the dark at 4 °C until analysis by FCM.

2.4.2.2.2 Analysis of Neutrophils in Synovial Tissue by FCM

1. Euthanize mice and harvest ankle joints, removing skin.
2. Dissect synovial tissue from mouse ankles by a sterile surgical knife.
3. Mince synovial tissue into small pieces using a sterile surgical knife.
4. Incubate synovial tissue with 1.5 mg/ml type IV-collagenase for 1 h at 37 °C (in hot water bath). (Collagenase is usually used for the isolation of single cells from tissue) (Cortez-Retamozo, Etzrodt, Newton, et al., 2012; Deshane, Zmijewski, Luther, et al., 2011. However, collagenase digestion can inadvertently remove the epitope of certain cell surface markers. This should be checked beforehand and if it does, several

modifications can be tried to eliminate the cleavage of the specific epitope, such as changing the type of collagenase, reducing the concentration, and/or reducing the incubation time of the collagenase digestion.)

5. Wash with 2% FCS/PBS and spin down twice.
6. Stain for Ly-6G positive neutrophils and analyze by FCM (see steps 5–9 in Section 2.4.2.2.1).

2.4.3 Enzymatic Quantitation of the Neutrophil-Specific Granule Protein MPO in Synovial Tissue

MPO correlates well with tissue neutrophil content, and can be used as a marker to assess neutrophil infiltration in the tissue (Pulli et al., 2013).

2.4.3.1 Required Materials

1. Mechanical homogenizer.
2. MPO assay kit (Hycult Biotech) containing anti-MPO antibody-coated 96 well plates, dilution buffer, Streptavidin-peroxidase, biotinylated tracer antibody, TMB substrate, and stop solution.
3. PBS.
4. Plate reader.

2.4.3.2 Detection of the Neutrophil-Specific Granule Protein MPO in the Tissue and Synovial Fluid

Neutrophils can be quantitated in synovial tissue and synovial fluid by assaying for the presence of MPO (Nzeusseu Toukap, Delporte, Noyon, et al., 2014; Pulli et al., 2013).

2.4.3.2.1 Synovial Tissue

1. Dissect synovial tissue from mouse ankles (see steps 1–3 in Section 2.4.2.2.2), and homogenize by a mechanical homogenizer. Transfer homogenized tissue to an Eppendorf tube. (*Note*: A sonicator may also be helpful for the homogenization.)
2. Centrifuge synovial tissue homogenate at $15,000 \times g$ for 20 min and remove supernatant, discarding cell pellet.
3. Use supernatant for MPO assays.

2.4.3.2.2 Synovial Fluid

1. Collect synovial fluid from mouse ankles (see steps 1–3 in Section 2.4.2.2.1).
2. Centrifuge the sample at 3000 rpm for 3 min to remove cell.
3. Use supernatant for MPO assays.

2.4.3.2.3 MPO Assay

1. To specifically capture MPO, incubate supernatant in MPO ELISA dilution buffer on anti-mouse MPO antibody-coated 96 well plates for 1 h at RT.
2. Wash assay wells 4 × with washing buffer (PBS with 0.05% Tween 20).
3. Incubate with 100 μl biotinylated tracer antibody for 1 h at RT.
4. Repeat step 5.
5. Incubate with 100 μl Streptavidin peroxidase for 1 h at RT.
6. Repeat step 5.
7. Add 100 μl TMB solution for 30 min at RT.
8. Add 100 μl Stop solution.
9. Analyze assay wells at 450 nm using a microplate reader.

2.5 MP-IVM for Imaging of Joints

Recent advances in imaging technology have provided unprecedented views into immune cell function in live animals, providing entirely new paradigms for immune cell function (Lämmermann et al., 2013; Murooka, Deruaz, Marangoni, et al., 2012). MP-IVM has allowed for the direct observation of leukocyte migration into tissue, and the behavior of leukocytes once within the tissue in live mice. This is a demanding technique that is rather difficult to master and requires many hours of practice.

2.5.1 Required Materials

1. Mouse.
2. Ketamine and Xylazine.
3. Microscissor.
4. Forceps.
5. Agarose.
6. 1 ml syringe.
7. Stage with arm.
8. Coverslip.
9. Slide grass.
10. Crazy glue.
11. Tissue bond.
12. Heater.
13. Flask.
14. Small animal clipper.
15. Upright multiphoton microscope equipped with (a) at least two (ideally three to four) non-descanned PMT detectors (e.g., Ultima V, Prairie

technologies, Middleton, WI; TriMScope, LaVision BioTec, Bielefeld, Germany; TCS SP5, Leica, Wetzlar, Germany, LSM 710 NLO Zeiss, Jena, Germany; or Fluoview FV1000MPE, Olympus, Center Valley, PA), (b) a high numerical aperture objective lens (Olympus XLUMPLFL20XW 20×, 0.95 NA, water immersion, 2 mm working distance), and (c) a femtosecond-pulsed infrared laser (DeepSee, Spectra-Physics/Newport, Mountain View, CA; or Chameleon Ultra II, Coherent, Santa Clara, CA).

2.5.2 MP-IVM of the Joint

To visualize cells, transgenic mice genetically engineered to express a fluorescent protein in cells are often used (Faust, Varas, Kelly, Heck, & Graf, 2000). LysM-GFP mice are commonly used to visualize neutrophils *in vivo* (Devi, Wang, Chew, et al., 2013; Faust et al., 2000; Lämmermann et al., 2013). In these mice, the expression of the green fluorescent protein (GFP) has been engineered to be driven by the lysosomal M promoter, resulting in neutrophils and macrophages that express GFP and are thus green (Faust et al., 2000).

1. Recipient mice are injected with Ketamine (80 mg/kg) and Xylazine (12 mg/kg) i.p. for anesthesia.
2. Shave the mouse leg using clippers and remove additional hair with depilation crème briefly (<30 s), avoiding maceration of the skin. Thoroughly wipe off the crème using moist gauze and repeat if required. All hair should be removed from the observed area, because hair remaining in the imaging field produces undesirable autofluorescence (Fig. 5A left panel).
3. Position the mouse on the microscope stage using tissue bond for fixation of the mouse leg.
4. Using the microscissor, incise the skin over a length of about 1 cm under the surgical scope (Fig. 5A). (*Note*: This step is most important. Surgery may induce neutrophil recruitment, so any surgical damage should be avoided as possible to avoid artifact. In addition, angiogenesis is induced in arthritic joints, compared with control joints, therefore arthritic joints are prone to bleeding. This step will need to be practiced many times for successful imaging.)
5. Melt 1.5% agarose with PBS using a microwave.
6. Cover the imaging field with approximately 1 ml of 1.5% agarose with PBS. (*Note*: Before covering with agarose, the temperature of agarose

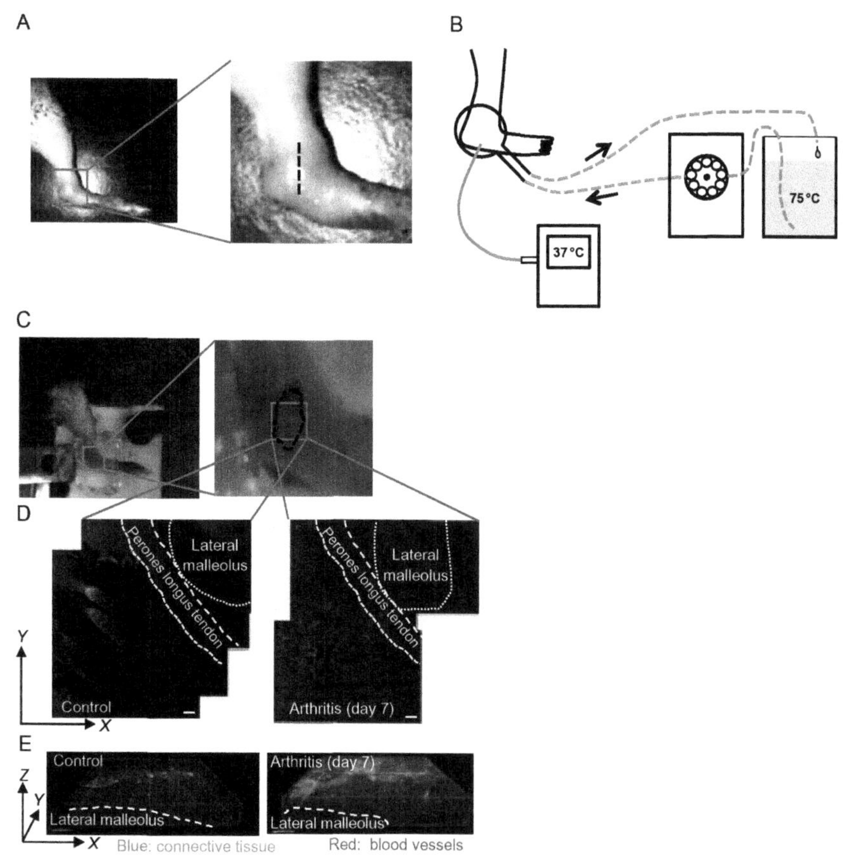

Figure 5 Multiphoton intravital microscopy for imaging joints. (A) Depilated area (left panel). Dashed line suggests position of skin incision (right panel). (B) Schematic of the observed joint and temperature control system. (C) Positioning of experimental animal after surgical preparation (left panel). Dash lined suggests the observed area after surgical preparation (right panel). (D and E) Pictures of the joints in control (left panel) and arthritic mice (right panel). Blue color suggests connective tissue. Red color represents a blood vessel stained with Dextran. Angiogenesis and hyperplasia of synovial tissue is seen in the arthritic joint, compared with control. (See the color plate.)

should be checked. Agarose may be applied to the operators hand for checking. Hot agarose may also induce neutrophil recruitment into the imaging field.)

7. Apply a ring of vacuum grease to the rim of the cover glass and attach it to flexible plastic tubing that will later be connected to the circular water heating system.

8. The stage is transferred to the intravital microscope. The heating element is connected to the roller pump and the water bath and the pump turned on. The thermocoupler is plugged into the digital thermometer (Fig. 5B).
9. Tune the laser source to an appropriate wavelength (e.g., 800 nm for CFSE and CMTMR, 930 nm for EGFP and DsRed).
10. Determine the optimal microscope settings (laser power, PMT gain, offset, etc.). We typically use a pixel resolution of 256 × 256, 16 optical sections, and a cycle time of 15 s, since this provides us with a good compromise between file size, image detail, and the ability to follow individual motile cells for sufficient time to obtain meaningful data. (*Note*: Strong laser power may induce neutrophil recruitment into the imaging area.)
11. Once a suitable location is found, begin image acquisition.
12. Data sets are transformed in Imaris 7.3.1 (Bitplane) to generate maximum intensity projections for export as Quicktime movies (Fig. 6A and B). (*Note*: The neutrophil migration cascade can be observed in the inflamed joint in live mice (Fig. 6C).)

2.6 Neutrophil Adoptive Transfer

Adoptive cell transfer experiments are an excellent experimental approach to determine the role of chemoattractant receptors expressed on neutrophils in K/BxN serum transfer model of arthritis. Different cell tracking dyes can be used to track different populations of transferred cells in short-term homing assays. For instance, WT neutrophils labeled with Cell Tracker Orange CMTMR and chemokine receptor-deficient neutrophils labeled with Cell Tracker Green CMFDA can be coadoptively transferred into the same arthritic mouse to compare the intrinsic function of a specific chemokine receptor on neutrophil function in a competitive setting. Potential issues of cell tracker toxicity can be addressed by switching the dyes between the different cell populations in subsequent experiments. Cell tracker labeling is best used for short-term homing experiments as the dye can leak out of the membrane over time and the cell tracker dye is also diluted with each cell division. For long-term homing experiments, genetically engineered cells that constitutively express fluorescent proteins are more suitable.

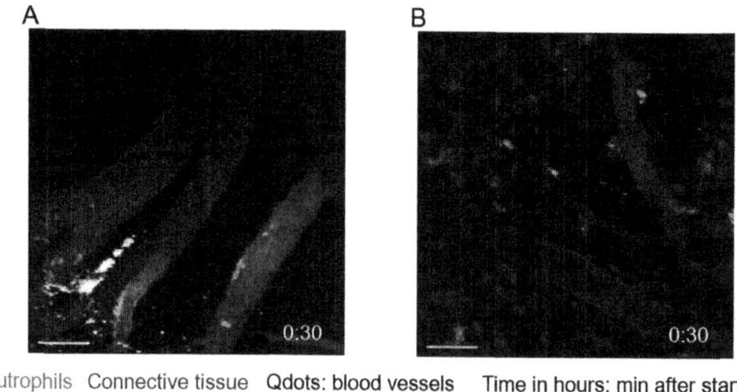

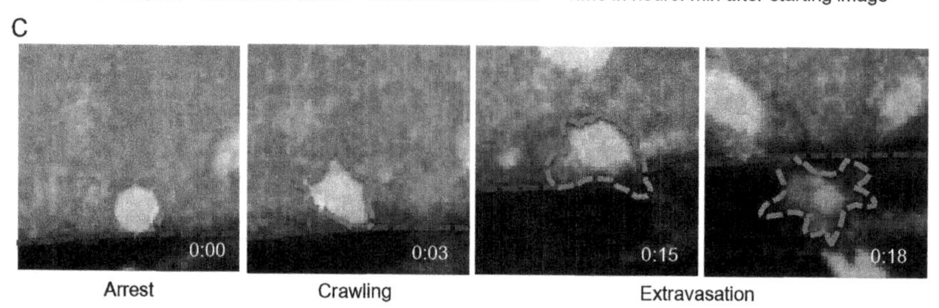

Figure 6 Imaging the joints of live mice. Imaging of the ankle joint of control (A) and arthritic LysM-GFP mice on day 7 after arthritogenic serum injection (B). Green color suggests neutrophils or macrophages. Blue color suggests connective tissue. White color represents blood vessel stained with Q tracker 655 (Qdots). Many extravasated neutrophils are seen in the arthritic joint (B), but not the control joint (A). Imaging depth is typically 100–150 μm below the skin surface. For multiphoton excitation and second harmonic generation, a MaiTai Ti:sapphire laser (Newport/Spectra-Physics) was tuned to between 830 and 920 nm for optimized excitation of the fluorescent probes used. For four-dimensional recordings of cell migration, stacks of 11 optical sections (512×512 pixels) with 4 μm *z*-spacing were acquired every 15 s to provide imaging volumes of 40 μm in depth. Emitted light and second harmonic signals were detected through 455/50 nm, 525/50 nm, 590/50 nm, and 665/65 band-pass filters with non-descanned detectors. (C) Arthritic mice (day 7 after arthritogenic serum injection) were imaged to analyze neutrophils migration cascade in the inflamed joint of live mice. Arrest is characterized by a round shape cell that remains in the same position for at least 30 s. A crawling cell is described as an amoeboid shaped cell and crawls inside of a blood vessel. Transendothelial migration describes a cell that has left the blood vessel lumen and has entered into the tissue. GFP represents WT neutrophils, Qdots blood vessels, and blue suggests connective tissue. Green broken line defines the outline of a neutrophil, and blue broken line suggests the boarder of blood vessels. (See the color plate.)

2.6.1 Required Materials

2.6.1.1 Isolation of BM Neutrophils

1. Bone marrow neutrophil donors: 6- to 8-week-old C57/BL6 male or female mice or desired genetically modified mice, such as CCR1-, CXCR2-, C5aR-, or BLT1-deficient mice (Chou et al., 2010; Jacobs et al., 2010; Kim et al., 2006; Sadik et al., 2012).
2. 1-cc plastic syringes with 23-G needles for i.p. injection of serum from K/BxN mice.
3. 70 μm cell strainers.
4. 1-cc plastic syringes with 27-G needle.
5. EasySep mouse neutrophil enrichment kit (Stemcell technologies).
6. EasySep Magnet (Stemcell technologies).
7. Sterile Cell Isolation Medium: 2% fetal bovine serum (FBS) + PBS without Ca^{++} and Mg^{++} + 1 m*M* EDTA. (*Note*: Other medium, such as 1% FCS/PBS, may also be used for this kit. However, the manufacturer recommends this medium.)
8. 15 and 50 ml Falcon tubes.
9. Falcon 5 ml polystyrene round-bottom tubes.
10. Timer.
11. Forceps.
12. Scissors.
13. 6 cm tissue culture plates.
14. Razor blades.

2.6.1.2 Fluorescently Labeling Neutrophils

1. BM neutrophils.
2. Cell tracker Orange CMTMR, Cell tracker Green CMFDA (Life technologies).
3. 1% FCS/PBS. (*Note*: High serum concentrations, such as 10% FCS/PBS, reduce cell toxicity induced by Cell Tracker, but they reduce staining intensity.)
4. FCS.
5. Water bath.

2.6.1.3 Bone Marrow Neutrophil Adoptive Transfer

1. Ketamine (80 mg/kg) and Xylazine (12 mg/kg) for anesthesia.
2. 1-cc plastic syringe with 23-G needle for i.p. injection of Ketamine and Xylazine.
3. 28-G × ½ insulin syringes for intravenous (i.v.) injection of neutrophils.

4. 1% FBS/PBS (sterile).
5. 1.5 ml Eppendorf tubes.

2.6.2 Isolation of BM Neutrophils

1. Dissect four bones (femurs & tibias) from bone marrow neutrophil donor mice, put in 50 ml tubes containing sterile PBS, and place on ice. [*Note*: One mouse (four bones: femurs & tibias) should yield over 50 million BM cells. If more BM cells from one mouse are desired, two additional bones (humerus) may be also used.]
2. In a sterile hood, cut off the ends of each bone with a clean razor blade. Grasp the bone with forceps and insert a 27-G needle filled with 1 ml Cell Isolation Medium into the marrow space. Flush the marrow out into a 6 cm plate containing 5 ml of Cell Isolation Medium. (*Note*: Repeat this process with the same bone several times until the marrow appears cleared out. Repeat with all bones.)
3. Pass BM through a 27-G needle to break up any clumps.
4. Pass all the BM cells through a 70 μm cell strainer into a 50 ml Falcon tube to remove large pieces of bone and tissue.
5. Add Cell Isolation Medium to the 6 cm plate used for BM harvest to rinse out any remaining cells, and put into a 50 ml Falcon tube through a 70 μm cell strainer.
6. Centrifuge for 5 min, 4 °C, 1500 rpm.
7. The method of neutrophil isolation described below is according to the EasySep manufacturer's protocol (from steps 8 to 17).
8. BM cells are resuspended at a concentration of 1×10^8 cells/ml in Cell Isolation Medium.
9. Add normal Rat Serum at 50 μl/ml of cell suspension for blocking, and EasySep Mouse Neutrophil Enrichment Cocktail at 50 μl/ml of cells. Mix well and incubate on ice for 15 min.
10. Wash cells with Cell Isolation Medium and centrifuge at $300 \times g$ for 10 min.
11. Discard the supernatant, resuspend the cells at 1×10^8 cells/ml in Cell Isolation Medium and put into Falcon 5 ml polystyrene round-bottom tubes.
12. Add EasySep Biotin Selection Cocktail at 50 μl/ml of cells, mix well and incubate on ice for 15 min.
13. Vortex EasySep D Magnetic Particles for 30 s and add these particles at 150 μl/ml of cells, mix well and incubate on ice for 10 min.

14. Bring the cell suspension to total volume of 2.5 ml by adding Cell Isolation Medium.
15. Place the tube into the magnet. Set aside for 3 min.
16. Pour the tube while still in the magnet into a new 10 ml Falcon tube. Neutrophils are in the new tube.
17. Neutrophils derived from BM are resuspended at 1.0×10^8/ml in 100 μl sterile 1% FCS/PBS and stored on ice until use.

2.6.3 Fluorescently Labeling Neutrophils

1. Stain 1×10^7 BM neutrophils from WT or gene-deficient "knockout" (KO) mice in a volume of 1 ml sterile 1% FCS/PBS with Cell Tracker Orange CMTMR Dye (2.5 μ*M*) or Cell Tracker Green CMFDA (2.5 μ*M*) for 15 min at 37 °C (in hot water bath).
2. The staining reaction is stopped by putting cells on ice.
3. Place sample on 2 ml 100% FCS as a cushion in a new 15 ml Falcon tube and then centrifuge at 1500 rpm for 5 min.

2.6.4 Bone Marrow Neutrophil Adoptive Transfer

1. Two different fluorescently labeled neutrophils, for instance, 1×10^7 WT labeled with Cell Tracker Orange and 1×10^7 KO labeled with Cell Tracker Green in 100 μl of sterile 1% FCS/PBS, are coadoptively transferred into arthritic mice.
2. Labeled neutrophil infiltration in the joint can be quantitated by FCM (see Section 2.4.2) and MP-IVM (see Section 2.5).

2.7 Mixed BMC Mice Experiments (Sadik et al., 2012)

Mixed BMC mice are generated by the reconstitution of lethally irradiated WT mice with BM from different strains of mice in a 1:1 ratio. For example, mixed BMC mice can be generated in which GFP-labeled WT BM and RFP-labeled KO BM are transferred into an irradiated WT host. This is an excellent model to compare the function of a specific chemokine receptor in a competitive setting over the entire course of disease development. It is a useful complement to the coadoptive transfer experiments described above, which are more useful for short-term homing assays. However, at least 4 weeks are needed to generate mixed BMC mice. In addition, there is some mortality associated with the preconditioning radiation regime. Also, one needs to ensure that the engraftment occurred in a 1:1 ratio of WT and KO BM. This needs to be confirmed for all mixed BMC mice by FCM before being used for experiments or after the experiment of the cell population of interest is not a circulating leukocyte.

2.7.1 Required Materials

1. Irradiator machine.
2. BM from WT and KO mice.

2.7.2 Mixed BMC Mice Experiments

1. Isolation BM from WT and KO mice labeled with different fluorescently protein (see steps 1–6 in Section 2.6.2).
2. Irradiation of the host mice (10 Gy, 10 min).
3. For a 1:1 ratio, 5×10^6 WT and 5×10^6 KO BM are transferred into irradiated WT host within 24 h after irradiation.
4. Four to six weeks after reconstitution mixed BMC mice were used for experiments.

Prior to use, the efficiency of reconstitution is determined by FCM of the blood. 100 μl of blood is collected into hematocrit tubes containing anticoagulant via the orbit. 1 ml of red blood cell lysing buffer solution is added to each blood sample for 15 min at RT. Cells are stained (see steps 5–9 in Section 2.4.2.2.1) and analyzed by FCM. At the time of the experiment, ≥95% of cells should be of donor origin, and each population of WT and KO cells should be present in roughly equal proportions.

3. CONCLUSIONS

In this chapter, we described IHC, FCM, and MPO assays as complementary methods to quantitate neutrophil migration into the joint. A combination of these techniques is helpful for the accurate quantification of neutrophils in the joint. For instance, IHC is particularly useful to determine the location of neutrophils in the joint. In contrast, FCM is good for the quantification of neutrophils and also for additional phenotyping of the neutrophils in the joint. The MPO assay is a complementary method for quantification of neutrophils infiltration in the joint. However, while these end point studies provide useful information, they do provide information about neutrophil behavior in the joint tissue in live mice.

Recently, imaging technology has provided new insights into immune cell migration in live animals. Therefore, we have been developing new techniques to apply MP-IVM technology to study the arthritic joint in live mice. Our new method of joint MP-IVM allows the analysis of neutrophil migratory behavior in the inflamed joints of live mice. In fact, using this technique, we have begun to determine the specific roles of chemokines on the migration cascade, such as arrest, crawling, and TEM in the joints of live mice. This new technique will provide unprecedented views in chemokine control of immune cell trafficking into the joint *in vivo*. However, joint imaging has several potential caveats that should be considered. For instance, MP-IVM requires an expensive microscope and is time consuming to perform. Also, surgery and the required powerful laser can induce neutrophil recruitment into the joint. Thus, practice and determining the baseline level of neutrophil recruitment is required before starting the experiments.

It is our hope that the addition of MP-IVM to traditional assays will markedly improve our understanding of the molecular control of immune cell recruitment into the joint by chemokine and other chemoattractants and thus the pathogenesis of arthritis and spur the development new therapies for RA and other inflammatory diseases.

REFERENCES

Chen, M., Lam, B. K., Kanaoka, Y., et al. (2006). Neutrophil-derived leukotriene B4 is required for inflammatory arthritis. *Journal of Experimental Medicine, 203*(4), 837–842.

Chou, R. C., Kim, N. D., Sadik, C. D., et al. (2010). Lipid-cytokine-chemokine cascade drives neutrophil recruitment in a murine model of inflammatory arthritis. *Immunity, 33*(2), 266–278.

Chow, M. T., & Luster, A. D. (2014). Chemokines in cancer. *Cancer Immunology Research, 2*(12), 1125–1131.

Cortez-Retamozo, V., Etzrodt, M., Newton, A., et al. (2012). Origins of tumor-associated macrophages and neutrophils. *Proceedings of the National Academy of Sciences of the United States of America, 109*(7), 2491–2496.

Deshane, J., Zmijewski, J. W., Luther, R., et al. (2011). Free radical-producing myeloid-derived regulatory cells: Potent activators and suppressors of lung inflammation and airway hyperresponsiveness. *Mucosal Immunology, 4*(5), 503–518.

Devi, S., Wang, Y., Chew, W. K., et al. (2013). Neutrophil mobilization via plerixafor-mediated CXCR4 inhibition arises from lung demargination and blockade of neutrophil homing to the bone marrow. *Journal of Experimental Medicine, 210*(11), 2321–2336.

Dufour, A. (2015). Degradomics of matrix metalloproteinases in inflammatory diseases. *Frontiers in Bioscience (Scholar Edition)*, 7, 150–167.

Faust, N., Varas, F., Kelly, L. M., Heck, S., & Graf, T. (2000). Insertion of enhanced green fluorescent protein into the lysozyme gene creates mice with green fluorescent granulocytes and macrophages. *Blood, 96*(2), 719–726.

Griffith, J. W., Sokol, C. L., & Luster, A. D. (2014). Chemokines and chemokine receptors: Positioning cells for host defense and immunity. *Annual Review of Immunology, 32*, 659–702.

Hoenderdos, K., & Condliffe, A. (2013). The neutrophil in chronic obstructive pulmonary disease. *American Journal of Respiratory Cell and Molecular Biology, 48*(5), 531–539.

Jacobs, J. P., Ortiz-Lopez, A., Campbell, J. J., Gerard, C. J., Mathis, D., & Benoist, C. (2010). Deficiency of CXCR2, but not other chemokine receptors, attenuates autoantibody-mediated arthritis in a murine model. *Arthritis and Rheumatism, 62*(7), 1921–1932.

Ji, H., Ohmura, K., Mahmood, U., et al. (2002). Arthritis critically dependent on innate immune system players. *Immunity, 16*(2), 157–168.

Kim, N. D., Chou, R. C., Seung, E., Tager, A. M., & Luster, A. D. (2006). A unique requirement for the leukotriene B4 receptor BLT1 for neutrophil recruitment in inflammatory arthritis. *Journal of Experimental Medicine, 203*(4), 829–835.

Lämmermann, T., Afonso, P. V., Angermann, B. R., et al. (2013). Neutrophil swarms require LTB4 and integrins at sites of cell death *in vivo*. *Nature, 498*(7454), 371–375.

Mayadas, T. N., Cullere, X., & Lowell, C. A. (2014). The multifaceted functions of neutrophils. *Annual Review of Pathology, 9*, 181–218.

McDonald, B., Pittman, K., Menezes, G. B., et al. (2010). Intravascular danger signals guide neutrophils to sites of sterile inflammation. *Science, 330*(6002), 362–366.

Miyabe, Y., Miyabe, C., Iwai, Y., et al. (2013). Necessity of lysophosphatidic acid receptor 1 for development of arthritis. *Arthritis and Rheumatism, 65*(8), 2037–2047.

Miyabe, C., Miyabe, Y., Miura, N. N., et al. (2013). Am80, a retinoic acid receptor agonist, ameliorates murine vasculitis through the suppression of neutrophil migration and activation. *Arthritis and Rheumatism, 65*(2), 503–512.

Monach, P. A., Mathis, D., & Benoist, C. (2008). The K/BxN arthritis model. *Current Protocols in Immunology*. Chapter 15:Unit 15.22, Suppl. 81, 15.22.1–15.22.12.

Monach, P. A., Nigrovic, P. A., Chen, M., et al. (2010). Neutrophils in a mouse model of autoantibody-mediated arthritis: Critical producers of Fc receptor gamma, the receptor for C5a, and lymphocyte function-associated antigen 1. *Arthritis and Rheumatism, 62*(3), 753–764.

Murooka, T. T., Deruaz, M., Marangoni, F., et al. (2012). HIV-infected T cells are migratory vehicles for viral dissemination. *Nature, 490*(7419), 283–287.

Nzeusseu Toukap, A., Delporte, C., Noyon, C., et al. (2014). Myeloperoxidase and its products in synovial fluid of patients with treated or untreated rheumatoid arthritis. *Free Radical Research, 48*(4), 461–465.

Odobasic, D., Muljadi, R. C., O'Sullivan, K. M., et al. (2015). Suppression of autoimmunity and renal disease in pristane-induced lupus by myeloperoxidase. *Arthritis and Rheumatism, 67*(7), 1868–1880.

Pulli, B., Ali, M., Forghani, R., et al. (2013). Measuring myeloperoxidase activity in biological samples. *PLoS One, 8*(7), e67976.

Sadik, C. D., Kim, N. D., Iwakura, Y., & Luster, A. D. (2012). Neutrophils orchestrate their own recruitment in murine arthritis through C5aR and FcγR signaling. *Proceedings of the National Academy of Sciences of the United States of America, 109*(46), E3177–E3185.

Wang, W., Jian, Z., Guo, J., & Ning, X. (2014). Increased levels of serum myeloperoxidase in patients with active rheumatoid arthritis. *Life Sciences, 117*(1), 19–23.

Wipke, B. T., & Allen, P. M. (2001). Essential role of neutrophils in the initiation and progression of a murine model of rheumatoid arthritis. *Journal of Immunology, 167*(3), 1601–1608.

Wright, H. L., Moots, R. J., & Edwards, S. W. (2014). The multifactorial role of neutrophils in rheumatoid arthritis. *Nature Reviews. Rheumatology, 10*(10), 593–601.

CHAPTER ELEVEN

Production of Chemokine/Chemokine Receptor Complexes for Structural Biophysical Studies

Martin Gustavsson*, Yi Zheng*, Tracy M. Handel[†,1]
*Skaggs School of Pharmacy and Pharmaceutical Sciences, University of California San Diego, La Jolla, California, USA
[†]Department of Pharmacology, Skaggs School of Pharmacy and Pharmaceutical Sciences, University of California San Diego, La Jolla, California, USA
[1]Corresponding author: e-mail address: thandel@ucsd.edu

Contents

1. Introduction 234
2. Methods 236
   2.1 Design of Constructs 236
   2.2 Baculovirus Production 237
   2.3 Expression 243
   2.4 Purification 245
   2.5 Characterization of Receptor/Chemokine Complex Purity, Homogeneity, and Stability 250
   2.6 Reconstitution 253
3. Summary and Conclusions 257
Acknowledgments 257
References 258

Abstract

The development of methods for expression and purification of seven-transmembrane receptors has led to an increase in structural and biophysical data and greatly improved the understanding of receptor structure and function. For chemokine receptors, this has been highlighted by the determination of crystal structures of CXCR4 and CCR5 in complex with small-molecule antagonists, followed recently by two receptor/chemokine complexes; CXCR4 in complex with vMIP-II and US28 in complex with the CX3CL1. However, these studies cover only a few of the many chemokines and chemokine receptors and production of stable receptor/chemokine complexes remains a challenging task. Here, we present a method for producing purified complexes between chemokine receptors and chemokines by coexpression in Sf9 cells. Using the complex between atypical chemokine receptor 3 and its native chemokine CXCL12 as an example, we describe the virus production, protein expression, and purification process as well as reconstitution into different membrane mimics. This method provides an efficient

Methods in Enzymology, Volume 570
ISSN 0076-6879
http://dx.doi.org/10.1016/bs.mie.2015.10.003

way of producing pure receptor/chemokine complexes and has been used to successfully produce receptor/chemokine complexes for CXC as well as CC receptors.

1. INTRODUCTION

Chemokines are ~10 kDa a signaling proteins that control cell migration in the context of development, immune surveillance, and inflammation by binding to chemokine receptors at the cell membrane (Allen, Crown, & Handel, 2007). Most chemokine receptors belong to the large family of G protein-coupled receptors (GPCRs) that bind extracellular ligands and transmit signals by coupling to G proteins in the cytoplasm (Fredriksson, Lagerstrom, Lundin, & Schioth, 2003; Pierce, Premont, & Lefkowitz, 2002). GPCRs are ubiquitous in human cells and comprise a large fraction of current drug targets (Salon, Lodowski, & Palczewski, 2011). All members of the GPCR family share a common fold with seven-transmembrane helices (TM) and a C-terminal amphipathic helix 8 and are therefore also commonly referred to as 7TM receptors (Katritch, Cherezov, & Stevens, 2013; Venkatakrishnan et al., 2013). Except for rhodopsin, that can be purified in large amounts from a natural source (Palczewski et al., 2000), structural and biophysical studies of GPCRs have long been hampered by the difficulties of producing stable, functional samples in reconstituted systems. In the last few years, the development of new methodology for expression and purification of GPCRs has facilitated the production of purified receptors and enabled studies of GPCRs using a range of structural, biophysical, and biochemical methods that were previously not feasible (Chun et al., 2012; Tate & Schertler, 2009; Venkatakrishnan et al., 2013). Together with advances in data acquisition and sample preparation methods (Caffrey & Cherezov, 2009; Cherezov, Liu, Griffith, Hanson, & Stevens, 2008; Liu et al., 2013; Rosenbaum et al., 2007; Steyaert & Kobilka, 2011), this has greatly improved the understanding of receptor structure and function, and led to a large increase in the number of high-resolution GPCR structures available (Katritch et al., 2013; Venkatakrishnan et al., 2013). The structure of CXCR4, solved both in complex with a small-molecule antagonist and a cyclic antagonist peptide using lipid cubic phase crystallization (Caffrey & Cherezov, 2009), was the first structure of a chemokine receptor (Wu et al., 2010). This structure was followed by an NMR structure of CXCR1 (Park et al., 2012) and a crystal structure of CCR5 solved in complex with the FDA approved small-molecule antagonist maraviroc (Tan et al., 2013), and recently, structures of CXCR4 in complex with the viral chemokine vMIP-II (Qin et al., 2015) and the viral chemokine receptor US28 in complex with CX3CL1 (Burg et al., 2015). In addition, there have been a number of studies where molecular details of recombinantly expressed and purified chemokine receptors and chemokines have been studied using other methods (Kofuku et al., 2009; Kufareva et al., 2014). Still, these studies cover only a few of the more than 20 chemokine receptors and 50 chemokines (Allen et al., 2007) and the production of chemokine receptors, especially in complex with chemokines remains a challenging task. Consequently, there is a great need for development of methods for the production of receptor/chemokine complexes.

Chemokines are small, soluble proteins and can generally be produced using expression in *Escherichia coli* or by chemical synthesis (Allen, Hamel, & Handel, 2011; Veldkamp et al., 2007). However, chemokine receptors, like other 7TM receptors, generally require eukaryotic expression systems for proper folding and membrane insertion (Allen, Ribeiro, Horuk, & Handel, 2009; Burg et al., 2015; Wu et al., 2010). In addition, several chemokine receptors require posttranslational modifications such as tyrosine sulfation and/or glycosylation for efficient ligand binding (Bannert et al., 2001; Fong, Alam, Imai, Haribabu, & Patel, 2002; Veldkamp et al., 2008). Production of stable receptor/chemokine complexes poses additional challenges. For example, receptors and chemokines may prefer different solvent conditions (pH, salt, etc.) for optimal stability *in vitro*. Also, the K_d of chemokines for their receptors varies from the p*M* to high-n*M* range and can be dependent on multiple factors including membrane lipids (Nguyen & Taub, 2002, 2003), and the presence of G proteins bound to the intracellular side of the receptor (Nijmeijer, Leurs, Smit, & Vischer, 2010). Furthermore, purification of 7TM receptors generally requires small-molecule ligands to efficiently extract receptor from membranes and keep them stable during purification (Grisshammer, 2009). This means large amounts of high-affinity small molecules, which may not be commercially available or may be prohibitively expensive, will likely be required. Furthermore, to prepare receptor/chemokine complexes, ligand exchange steps may be needed to replace the small molecule for purified chemokine. The alternative is to produce large quantities of purified chemokine and use them as ligands throughout the extraction and purification process, but this is not practical.

Here, we give detailed instructions for each step involved in production of complexes between chemokine receptors and chemokines through coexpression in Sf9 cells for structural and biophysical studies without the need for small-molecule ligands or separately purified chemokines. Using the production of a complex between atypical chemokine receptor 3,

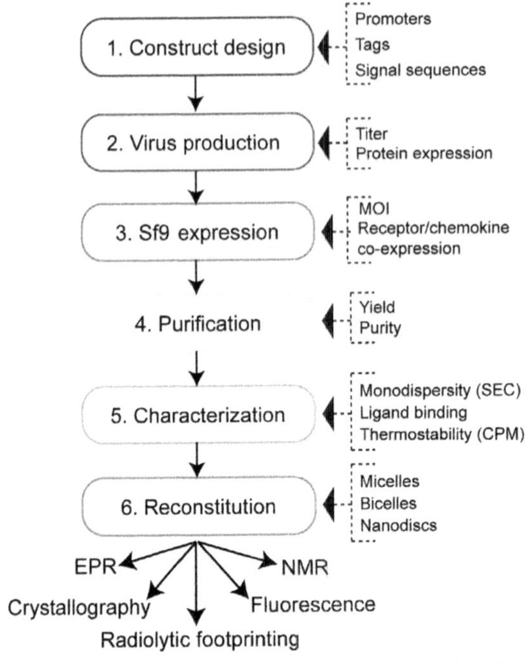

Figure 1 Flow chart for production of complexes. (See the color plate.)

ACKR3 (CXCR7) with CXCL12 (SDF-1) as an example, we show important quality controls and point out potential pitfalls associated with the various steps of the expression and purification process. Figure 1 outlines the major steps involved in the process; and each step is described in detail in the following section.

2. METHODS

2.1 Design of Constructs

2.1.1 Receptor Constructs

Expression constructs were designed with the receptor sequence cloned into a pFastBac vector. A schematic of the receptor construct in the vector is shown in Fig. 2A. This construct utilizes a GP64 or polH promoter to drive receptor expression, an HA signal sequence to enhance localization in the Sf9 cell membrane and a C-terminal Prescission protease (PP) cleavage site followed by FLAG and His tags which permit detection and purification, respectively. Restriction sites before and after the receptor sequence facilitate transfer of sequences between plasmids with different promoters, signal

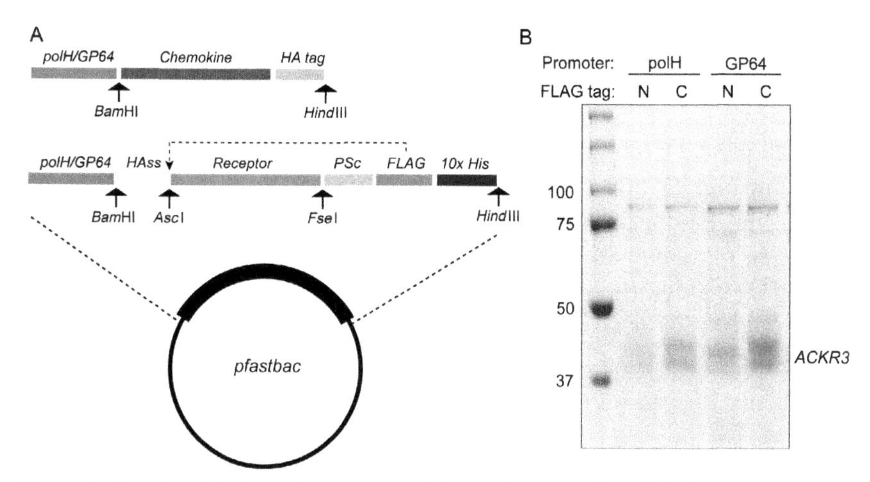

Figure 2 Construct design of receptor and chemokine samples. (A) Schematic overview of chemokine and receptor constructs. (B) SDS-PAGE of purified ACKR3 showing the effect of promoter and tag placement on the final yield of ACKR3 protein. The doubling and fuzziness of the ACKR3 band is due to glycosylation.

sequences, and tags. In our experience, the choice of promoter and the location and identity of the tags can have a major effect on the yield of receptor. For example, Fig. 2B shows the yield of purified ACKR3 with four different combinations of tags and promoters. In this case, a C-terminal FLAG tag gives a significantly higher yield than a construct in which the same tag is placed at the N-terminus of the receptor. Furthermore, a GP64 promoter gives a better yield than a polH promoter. The presence and position of the tag can also sometimes inhibit proteolysis of the receptor termini (data not shown).

2.1.2 Chemokine Constructs

Chemokine constructs with native signal sequences were cloned into pFastBac vectors with a polH promoter and a C-terminal HA tag. A schematic is shown in Fig. 2A, in our experience, the endogenous chemokine signal sequence is sufficient to enable export of the chemokine out of the Sf9 cell but utilizing other signal sequences (HA, other chemokines, etc.) give identical results.

2.2 Baculovirus Production

2.2.1 Equipment

Temperature controlled shaker
Heat block

Centrifuge
Vi-CELL XR Cell Viability Analyzer (Beckman Coulter)
Guava easyCyte 8HT Benchtop Flow Cytometer (Millipore)
Well plate 24 10.4 mL PP individually wrapped (Thomson Instrument Company, 931565-G-1X)
Breathe-Easy sealing membrane (USA Scientific, 9123-6100)
Costar Clear Polystyrene 96-well plates (Corning, 3788)
125 mL polycarbonate Erlenmeyer flask with vent cap (Corning, 431143)
2.2 mL Deep Well Plate, 96 Square Well, PP, 50/Cs, Round Bottom (Phenix Research Products, M-0661)

2.2.2 Reagents

pFastBac vectors with receptor and chemokine of interest
MAX Efficiency DH10Bac Chemically Competent Cells (Thermo Fisher)
SOC medium
LB Agar plates with Gentamicin, Kanamycin, Tetracycline, Bluo-gal, IPTG (Teknova, L1919)
Kanamycin
Gentamicin
Tetracycline
Buffer P1 (Qiagen)
Buffer P2 (Qiagen)
Buffer P3 (Qiagen)
X-tremeGENE HP DNA Transfection Reagent (Roche, 6366244001)
Transfection medium (Expression Systems, 95-020-20)
ESF 921 Insect Cell Culture Medium, Protein-Free (Expression systems, 96-001-01)
Sf9 cells (ATCC® CRL-1711™)
gp64-PE Antibody (Expression Systems, 97-201)
Monoclonal anti-FLAG® M2-FITC antibody produced in mouse (Sigma-Aldrich, F4049)
Monoclonal anti-HA FITC antibody produced in mouse (Sigma-Aldrich, H7411)
7-AAD Viability Staining Solution (eBioscience, 00-6993-50)
Triton X-100
Fungizone® Antimycotic (Thermo Fisher, 15290-018)

2.2.3 Protocol

Baculovirus for expression of chemokines and receptors is produced using the Invitrogen Bac-to-Bac system. For a detailed explanation of virus production using this system, see the manual (https://tools.thermofisher.com/content/sfs/manuals/bactobac_man.pdf). With some exceptions, the steps described below follow the manual.

1. Transformation of pFastBac vectors into DH10 bac cells
   a. Add the pfastbac vector to DH10 bac cells (Thermo Fisher) and incubate on ice for 15 min. The amount of DNA needed is highly dependent on the transformation efficiency of the DH10 bac cells. In our case, 50 ng of DNA is typically added to 20 μL of DH10 bac cells.
   b. Heat shock at 42 °C for 45 s and transfer back to ice for 2 min.
   c. Add 170 μL of SOC media and incubate for 4 h at 37 °C. Simultaneously, incubate LB agar plates with 50 μg/mL kanamycin, 7 μg/mL gentamicin, 10 μg/mL tetracycline, 100 μg/mL Bluo-gal, and 40 μg/mL IPTG (Teknova) at 37 °C.
   d. After 4 h, plate cells onto the LB agar plates. Again, the amount of cells to plate is highly cell dependent and needs to be optimized. In our hands, 15–25 μL is generally sufficient.
   e. Cover the plates from light and incubate at 37 °C for ~40 h or until the color of the colonies (blue/white) can be clearly distinguished. Store plates at 4 °C.
2. Bacmid purification
   a. Prepare overnight cultures by adding antibiotics (50 μg/mL kanamycin, 7 μg/mL gentamicin, and 10 μg/mL tetracycline) to 5 mL of LB media.
   b. Inoculate a single white colony from the LB agar plates in the media and grow over night at 37 °C.
   c. Centrifuge the culture (10 min, 1800 × *g*) and discard the supernatant.
   d. Resuspend the pellet in 300 μL of cold buffer P1 (Qiagen) and transfer to a 2-mL microcentrifuge tube
   e. Add 300 μL of buffer P2 (Qiagen), mix by inverting the tube, and incubate for 5 min at room temperature.
   f. Add 300 μL of cold buffer P3 (Qiagen), incubate on ice for 5 min, and centrifuge for 10 min at 15,000 × *g*.
   g. Transfer 900 μL of supernatant to a new 2-mL microcentrifuge tube.

   h. Add 700 μL of isopropanol, invert gently to mix and incubate for 15 min on ice.
   i. Centrifuge for 15 min at 15,000 × *g*.
   j. Remove the supernatant and add 500 μL of 70% ethanol to wash the pellet
   k. Centrifuge for 5 min 15,000 × *g*.
   l. Carefully remove the supernatant and leave the tube open for ~5 min to allow the remaining ethanol to evaporate. Add 40 μL of TE buffer (10 m*M* TRIS, pH 8, 1 m*M* EDTA), flick the tube gently to solubilize the bacmid and store at 4 °C.
3. P0 virus production
   a. Make a transfection mixture by combining 100 μL of transfection medium (Expression Systems), 3 μL of X-tremeGENE HP DNA Transfection Reagent (Roche), and 5 μL of bacmid (from step 2l) in low-bind DNA microcentrifuge tubes and incubate for 30 min.
   b. Stocks of Sf9 cells are cultured in ESF 921 Insect Cell Culture Medium (Expression systems) at 27 °C shaking at 140 rpm and passaged every 72 h. Transfer 2.5 mL of Sf9 cells at a concentration of 1.2–1.4 cells/mL to a 24-well block (Thomson Instrument Company). Use one well per virus sample and one additional well to use as a control sample of untransfected cells.
   c. Add the transfection mixture (108 μL total) to the cells, cover the block with a sealing membrane (USA Scientific), and incubate at 27 °C for 96 h shaking at 300 rpm.
   d. Transfer 10 μL of cells and 10 μL of anti-GP64 PE conjugated antibody (Expression Systems) to a 96-well plate (Corning) and incubate for 20 min.
   e. While the 96-well plate is incubating, cover the block and centrifuge at 2000 × *g* for 15 min.
   f. Transfer the supernatant to microcentrifuge tubes and save at 4 °C covered from light until use. This is the P0 virus.
   g. Add 180 μL of TBS buffer (50 m*M* TRIS, pH 7.5, 150 m*M* NaCl) to the 96-well plate and read samples using a flow cytometer. Quantify the percent of cells infected by virus by using the untransfected cells as a control. At this point, >80% of the cells should be infected by virus (Fig. 3A).
4. P1 virus production
   a. Prepare 125 mL growth flasks (Corning) with 40 mL of Sf9 cells in early log phase (cell density of 2–3 × 10^6 cells/mL). Add 400 μL of P0 virus and include a control flask with untransfected cells. Incubate at 27 °C shaking at 140 rpm.

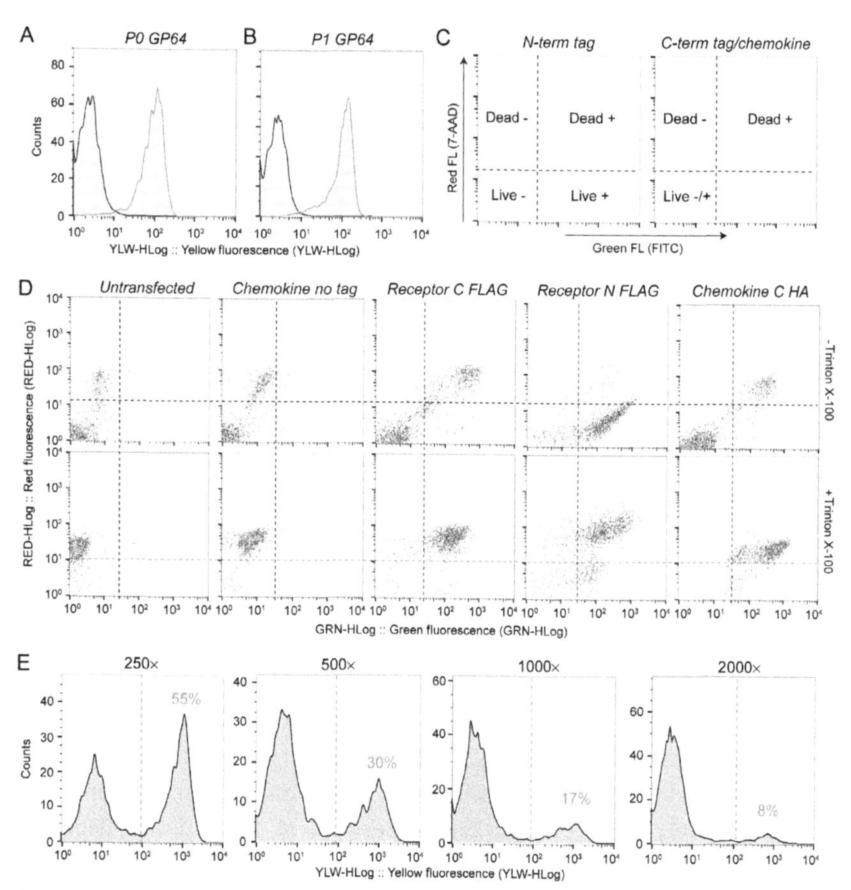

Figure 3 Baculovirus production detected by flow cytometry. (A and B) P0 and P1 virus expression detected by PE conjugated anti-GP64 antibody. (C) Interpretation of FITC assay for expression of receptor and chemokine. (D) FITC assay results for receptor and chemokine samples. Control experiments (columns 1 and 2) were acquired with an anti-FLAG FITC-conjugated antibody, identical results were obtained with anti-HA FITC antibody (data not shown). Note that due to their larger size, cells expressing untagged chemokine have higher nonspecific antibody binding than untransfected cells. Receptor and chemokine experiments (columns 3–5) were acquired with FITC-conjugated FLAG or HA antibodies as indicated in the figure. Samples in the bottom row contained 0.0075% Triton X-100 to permeabilize the cells. (E) Titer of P1 virus using the GP64 assay and serial virus dilutions as indicated above each plot. The % infected cells of each dilution is shown in red (light gray in the print version) above the infected population in each plot.

b. After 20–24 h of incubation, count the cells and measure viability using trypan blue. At this stage, cell growth should be arrested (total cell count $\sim 3 \times 10^6$ cells/mL) and the cell diameter increased by ~20–25% compared to control cells, but the viability should remain above 90%. Measure virus infection using a GP64 assay (see steps 3d and g above). The percent of cells infected at this stage is typically above 90% for successful virus production (Fig. 3B). Transfer flasks back to the shaker and incubate for 24 more hours.

c. After 48 h total incubation, count the cells and measure the viability. At this point, the viability will be lowered (usually 60–80% for chemokines and 50–70% for receptors) and the total cell count is typically $\sim 3 \times 10^6$ cells/mL.

d. Add 10 μL of cells to a 96-well plate (Corning), two wells per virus sample. To the first well, add 10 μL of anti-FLAG-FITC antibody (Sigma-Aldrich) for receptor samples or anti-HA FITC antibody (Sigma-Aldrich) for chemokine samples and 2 μL of 7-AAD staining solution (eBioscience). To the second well, add 10 μL of anti-FLAG antibody (receptor) or anti-HA antibody (chemokine), 2 μL of 7-AAD and 0.0075% (v/v) μL of Triton X-100. Incubate for 20 min.

e. Add 178 μL of TBS buffer to the 96-well plate. Read samples using a flow cytometer. Separate live and dead populations using the 7-AAD dye (red fluorescence) and quantify protein expression using FITC (green fluorescence). Figure 3C shows a summary of the expected FITC results for chemokine samples and for receptor samples with N- and C-terminal tags. In our experience, even if chemokines are secreted from the cell, detection of intracellular depots of chemokines at this stage is important since it confirms that the chemokine is produced by the Sf9 cells. Figure 3D shows experimental data for ACKR3 and CXCL12 samples. Triton X-100 was added to the samples in the bottom row to permeabilize the cells and allow for detection of C-terminally tagged receptors and intracellular chemokines. Note that for chemokines and C-terminally tagged receptors, cell permeabilization is necessary to quantify expression.

f. To harvest the virus, transfer the cells to 50 mL conical tubes and centrifuge for 15 min at 2000 × *g*. Transfer the supernatant to dark conical tubes, add 400 μL of Fungizone (Thermo Fisher) to prevent contamination, and store at 4 °C. This is the P1 virus stock.

5. Titering

a. Prepare 1:10 dilutions of viruses in microcentrifuge tubes by mixing 10 μL of P1 virus with 90 μL ESF 921 insect cell culture medium (Expression Systems).

b. Add 96 μL of medium and 4 μL of the 1:10 virus dilution in column 1 of a 96-well plate (one well per P1 virus sample). Make serial dilutions by mixing 50 μL of column 1 with 50 μL of medium in column 2 and repeat for columns 3 and 4.

c. Transfer 100 μL of Sf9 cells at log phase density (2–3 × 10^6 cells/mL) to a 96-well deep-well block (Phenix Research Products). Copy the configuration of the dilution plate so that all wells that have virus in the dilution plate have cells in the corresponding well of the deep-well block. In addition, add cells to one additional well to use a control.

d. Add 20 μL of the virus dilutions to the cells, seal with a Breathe-Easy sealing membrane (USA Scientific), and incubate at 27 °C, 300 rpm.

e. After 18–24 h measure, the percent of infected cells using a GP64 PE antibody as described in steps 3d and 3g above.

To calculate the virus titer (measured as infectious units (IU) per mL) insert the virus dilution factors (250, 500, 1000, or 2000), number of cells in the well at the time of transfection, percent infected cells from the GP64 assay (%GP64) and the volume of inoculum (0.02 mL) into the equations below. For samples with <30% of the cells infected use:

$$\text{Titer} = \frac{\text{Total cell number } \times \%\text{GP64}/100 \times \text{Viral dilution factor}}{\text{Volume of inoculum}}$$

For samples with >30% of the cells infected, use:

$$\text{Titer} = \frac{\text{Total cell number } \times [-\ln\ (1 - \%\text{GP64}/100)] \times \text{Virus dilution factor}}{\text{Volume of inoculum}}$$

This titering method and calculations are described in more detail by Gueret, Negrete-Virgen, Lyddiatt, and Al-Rubeai (2002) and Li, Ling, Liu, Laus, and Delcayre (2010). Using this method, we will typically obtain titers of $1 \times 10^9 - 4 \times 10^9$ IU/mL with chemokine viruses usually having slightly higher titer than receptor viruses.

2.3 Expression

2.3.1 Equipment

Temperature controlled shaker

125 mL polycarbonate Erlenmeyer flask with vent cap (Corning, 431143)

1 L Polycarbonate Erlenmeyer Flask with Vent Cap (Corning, CLS431147)
Optimum Growth 5 L Flasks (Thomson, 931116)
Vi-CELL XR Cell Viability Analyzer (Beckman Coulter)
Costar Clear Polystyrene 96-Well Plates (Corning, 3788)
Centrifuge
Vi-CELL XR Cell Viability Analyzer (Beckman Coulter)
Guava EasyCyte 8HT Benchtop Flow Cytometer (Millipore)

2.3.2 Reagents and Solutions

ESF 921 Insect Cell Culture Medium, Protein-Free (Expression systems)
Monoclonal anti-FLAG® M2-FITC antibody produced in mouse (Sigma-Aldrich, F4049)
Monoclonal anti-HA FITC antibody produced in mouse (Sigma-Aldrich, H7411)
7-AAD Viability Staining Solution (eBioscience, 00-6993-50)
Triton X-100

2.3.3 Protocol

1. Transfect Sf9 cells when the cells are in log phase (density of $\sim 2.5 \times 10^6\ \text{mL}^{-1}$) by directly adding virus from the receptor and chemokine P1 stocks to the same growth flask. Growth flasks should be chosen based on the volume of expression. We utilize Optimum Growth 5-L Flasks (Thomson) for larger scale expressions (1–2 L), 1 L Polycarbonate Erlenmeyer Flask with Vent Cap (Corning) for intermediate scale expressions (150–350 mL), and 125 mL Polycarbonate Erlenmeyer Flasks with Vent Cap (Corning) for smaller scale expressions (20–40 mL). Calculate the volume of virus to add using:

$$\text{Vol virus (mL)} = \frac{\text{Cell conc (cells/mL)} \times \text{Volume (mL)} \times \text{MOI (IU/cell)}}{\text{Titer (IU/mL)}}$$

The virus multiplicity of infection (MOI) is typically 5–10 but needs to be optimized for each receptor/chemokine pair separately. This can be done through a titration of receptor and chemokine virus MOI in a coexpression experiment (for an example, see Kufareva et al., Chapter 18 of this issue). In the case of ACKR3/CXCL12, the optimal MOI for production of the receptor/chemokine complex was determined to be six for the chemokine when an MOI of six is used for the receptor and higher chemokine MOI suppressed receptor expression. In addition to the coexpressions, transfect control samples, one with receptor virus and one with chemokine virus using the same MOI as for the receptor/chemokine coexpression. The controls are used to confirm receptor and chemokine expression in step 2.

2. 48 h after transfection, count the cells and measure the cell viability by staining with trypan blue. At this point, cell growth should have been arrested, the cell diameter should be ~20–25% larger than for untransfected cells and the cell viability dropped to <80% due to virus infection and receptor/chemokine coexpression. To test for expression, transfer 10 μL of each of the coexpression and control cell cultures into a 96-well assay plate (two wells per sample) for flow cytometry. Harvest the rest of the cells by centrifugation at 2000 × *g* for 15 min and store the pellets at –80 °C. Add anti-HA FITC-conjugated antibody (Sigma-Aldrich) to the first well of the 96-well assay plate to detect chemokine, anti-FLAG FITC-conjugated antibody (Sigma-Aldrich) to the second well to detect receptor and the dye 7-AAD (eBioscience) to both wells to separate live and dead cells. In addition, for cases where the receptor has a C-terminal FLAG tag add Triton X-100 to a final concentration of 0.0075% (v/v) to the second well (calculated based on a final volume of 200 μL) to enable the antibody to access the epitope on the receptor, which is on the cytoplasmic side of the cell membrane. Incubate the plate for 20 min in the dark at 4 °C, dilute to 200 μL with TBS buffer, and read the FITC fluorescence using a flow cytometer. Compare the receptor and chemokine histograms of the coexpressed cells to those of control cells to quantify the expression of the proteins. Detection of the FLAG tag confirms expression of the receptor and detection of the HA-tagged chemokine on the cell surface confirms that the chemokine is expressed and interacting with the receptor. Figure 4 shows a representative example of flow cytometry data for a coexpression of C-terminally tagged ACKR3-FLAG coexpressed with CXCL12-HA and appropriate controls.

2.4 Purification

2.4.1 Equipment

Centrifuge

Kimble-Chase Kontes™ Dounce Tissue Grinders 100, 15, 1 mL (Fisher Scientific, 8853000100, 8853000040, 8853000001)

Poly-Prep Chromatography Columns (Bio-rad, 7311550)

Micro Bio-Spin Columns (Bio-rad, 7326204)

PD-10 desalting column (GE Healthcare, 17-0851-01)

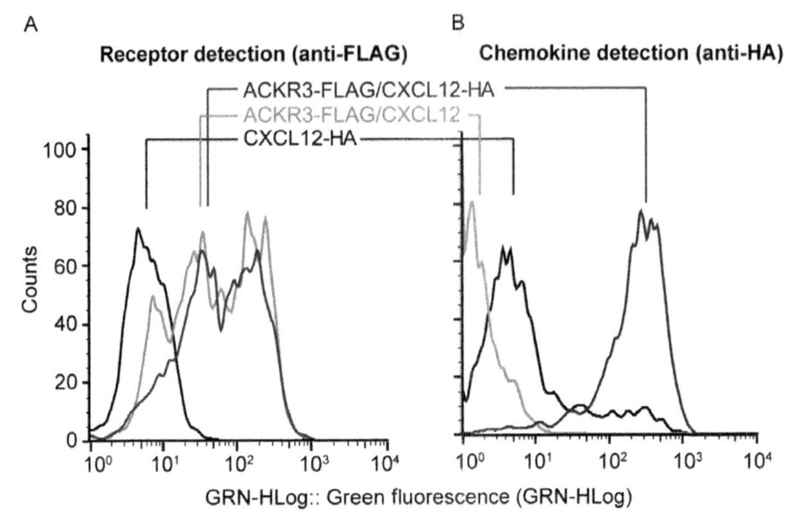

Figure 4 Expression of ACKR3/CXCL12 complex. (A) Detection of total receptor expression in coexpressed samples using anti-FLAG FITC antibody in the presence of Triton X-100 to a final concentration of 0.0075% (v/v). (B) Detection of chemokine binding to receptor at the cell surface using anti-HA FITC antibody. Coexpression with the receptor gives a large signal increase for CXCL12-HA, indicating specific binding of the chemokine to the receptor.

Amicon Ultra-0.5 Centrifugal Filter Unit with Ultracel-100 membrane (EMD Millipore, UFC510096)
Amicon Ultra-4 Centrifugal Filter Unit with Ultracel-100 membrane (EMD Millipore, UFC810096)

2.4.2 Reagents

cOmplete EDTA-free protease inhibitor tablets (Roche, 5056489001)
Iodoacetamide (GE Healthcare, RPN6302)
n-Dodecyl-β-D-maltopyranoside (DDM) (Anatrace, D310)
Cholesteryl hemisuccinate (CHS) (Sigma-Aldrich, C6512)
TALON Superflow Metal Affinity Resin (Clontech, 635507)
HRV 3C Protease (Human Rhinovirus 3C Protease, PreScission Site) (Sino Biological, S3CP01)
PNGaseF (New England Biolabs, P0704S)

2.4.3 Buffers

Hypotonic buffer

10 m*M* HEPES, pH 7.5
10 m*M* $MgCl_2$

20 m*M* KCl
cOmplete EDTA-free protease inhibitor tablets (Roche, 5056489001)

Wash buffer 1
25 m*M* HEPES, pH 7.5
400 m*M* NaCl
10% glycerol
10 m*M* imidazole
0.1/0.02% DDM/CHS

Wash buffer 2
25 m*M* HEPES, pH 7.5
400 m*M* NaCl
10% Glycerol
10 m*M* Imidazole
0.025/0.005% DDM/CHS

Elution buffer
Wash buffer 2
250 m*M* Imidazole

Buffer exchange buffer
25 m*M* HEPES, pH 7.5
400 m*M* NaCl
10% Glycerol
0.025/0.005% DDM/CHS

2.4.4 Protocol

This protocol describes the purification of a receptor/chemokine complex from a 1-L culture of Sf9 cells. Typically, 40 mL of cells is a good starting point for a small-scale expression of a chemokine receptor construct. In the case of a small-scale purification, volumes of buffers and resin can to be scaled down by 25-fold. Throughout the purification all steps should be performed on ice or at 4 °C.

1. Thaw the cell pellets, resuspend in ~120 mL hypotonic buffer, and homogenize with 40 strokes in a dounce homogenizer. Transfer the suspension to 40 mL centrifuge tubes, centrifuge at 50,000 × *g* for 30 min, discard the supernatant, and repeat the process once with hypotonic buffer and three times using high salt buffer (hypotonic buffer + 1 *M* NaCl). For small-scale purifications one round of douncing in hypotonic buffer and two rounds in high salt buffer are usually sufficient. After the last centrifugation, resuspend in hypotonic buffer with 30% glycerol to a final volume of 25 mL, dounce with 40 strokes, transfer to conical tubes,

and flash freeze in liquid nitrogen and store the membranes at –80 °C until the next step of the purification.

2. Thaw the membranes, add protease inhibitors, and dilute with 25 mL of hypotonic buffer. For receptors with exposed cysteine residues, 2 mg/mL iodoacetamide can be added to avoid aggregation during the purification. Incubate for 30 min on a rotisserie.

   Alternative strategy: If large amounts of purified chemokine are available, receptor can be expressed without chemokine in Sf9 cells and chemokine can be added in step 2, at the same time as the protease inhibitors. In this case, the concentration of chemokine used for extraction needs to be optimized. An example of ACKR3 extracted with 3 μ*M* of a CXCL12 mutant (KPV at the N-terminus replaced by LRHQ, LRHQ-CXCL12) (Hanes et al., 2015) or with 100 μ*M* of a small-molecule compound is shown in Fig. 5.

3. Prepare 50 mL of 2 × solubilization buffer (100 m*M* HEPES, pH 7.5, 800 m*M* NaCl, 1.5/0.3% DDM (Anatrace)/CHS (Sigma-Aldrich)). To facilitate the solubilization of DDM and CHS in the buffer, 10/2% (w/v) stocks of DDM/CHS should be prepared in advance by weighing out DDM and CHS, dissolving in 200 m*M* TRIS, pH 8 and filtering the solution to remove insoluble particles.

4. Split the sample into two 50 mL conical tubes, add 25 mL of 2 × solubilization buffer to each tube and incubate for another 3 h. Centrifuge

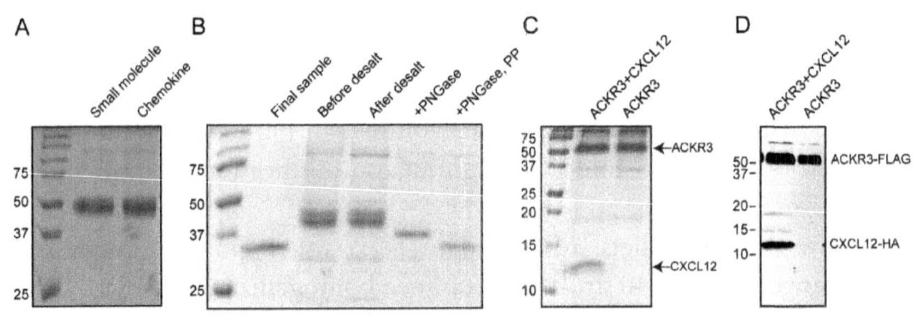

Figure 5 Purification of ACKR3/CXCL12 complex. (A) 10% SDS–PAGE of an bril-ACKR3 fusion protein extracted with the small-molecule CCX777 or with a mutant of CXCL12 where KPV at the N-terminus was replaced with LRHQ (LRHQ-CXCL12) (Hanes et al., 2015). (B) Different stages of purification characterized by SDS-PAGE (10%). (C) 18% SDS-PAGE of a coexpressed and copurified ACKR3/CXCL12 sample (left lane). ACKR3 expressed alone and purified in complex with a small-molecule compound is shown in the right lane. (D) Western blot detecting FLAG-tagged ACKR3 and HA-tagged CXCL12 in samples from (C).

the solubilized samples (50,000 × *g*, 30 min), transfer the supernatant (containing the solubilized receptor/chemokine complex) to two 50 mL conical tubes, and add 1 mL of 50% Talon resin and 10 m*M* of imidazole to each tube. Incubate at 4 °C at a rotisserie over night.

5. Centrifuge the samples for 5 min at 350 × *g*, discard the supernatant, and transfer the resin to Poly-Prep Chromatography Columns (Bio-rad). Wash the resin with 20 mL of wash buffer 1 and 10 mL of wash buffer 2 by adding 2-mL buffer at a time and use gravity flow to run the buffer through the column. After the last wash, elute with 3 mL of elution buffer in 1 mL fractions.
6. Concentrate the eluted sample to 500 μL using Amicon Ultra-4 100 kDa molecular weight cutoff spin concentrators (EMD Millipore).
7. Equilibrate a PD-10 desalting column (GE Healthcare) with 3 × 2 mL of buffer exchange buffer. Apply 500 μL of sample to the column and discard the flow-through. Add 1 mL of buffer and collect the flow-through. At this stage, the sample should be pure enough (as assessed by SDS-PAGE, Fig. 5) to be used directly for basic characterization experiments and for certain sample preparations where the presence of tags is required (i.e., nanodiscs reconstitution, see Section 2.6.5). However, for other applications, it is necessary to cleave off the C-terminal His and FLAG tags on the receptor. This will also serve as an additional purification step.

   Alternative strategy: For small-scale purifications, steps 6 and 7 can be replaced by buffer exchanging samples using Amicon Ultra-0.5 100 kDa molecular weight cutoff spin concentrators (EMD Millipore). In this case, spin down to ~50 μL and bring the volume back to 500 μL three times. After a fourth and final spin, transfer the sample to a new tube and bring the volume to 100 μL.
8. To cleave off the His and FLAG tags, incubate the sample (1 mL) with Prescission Protease (Sino Biological, PP) over night at 4 °C. For glycosylated receptors and chemokines, Pngase F (New England Biolabs) can be added to remove N-linked glycosylation. The amounts of PP and Pngase F needed for cleavage and deglycosylation will vary depending on the receptor and the activity of the enzyme. After overnight incubation, add 500 μL of Talon resin slurry and incubate for 90 min at 4 °C. Transfer the sample to a Micro Bio-Spin column (Bio-rad) and collect the flow-through. Add 500 μL of buffer exchange buffer and again collect the flow-through for a total sample volume of ~1.8 mL. Tag cleavage and deglycosylation can be followed from band shifts using SDS–PAGE (Fig. 5, see next section for experimental details).

9. After step 8, the complex can be concentrated to the desired concentration and characterized for purity, stability, and monodispersity (*see* Section 2.5). In the case of ACKR3/CXCL12, samples can be stored for weeks in detergent or kept frozen at −80 °C for longer periods. However, the sample stability can be highly receptor/chemokine dependent and needs to be tested for each complex individually.

2.5 Characterization of Receptor/Chemokine Complex Purity, Homogeneity, and Stability

2.5.1 Equipment

RotorGene Q 6-plex RT-PCR machine (Qiagen)
HPLC system
Equipment for SDS–PAGE setup (gel chambers, gels, power source)

2.5.2 Reagents and Solutions

Standard solutions for SDS-PAGE (loading buffer, staining/destaining solutions, and running buffers)
Monoclonal anti-FLAG M2 antibody produced in mouse (Sigma-Aldrich, F3165)
Anti-HA High Affinity from rat IgG1 (Roche, 11867423001)
IRDye 800CW goat anti-rat IgG (LI-COR Biosciences, 925-32219)
IRDye 680RT donkey anti-mouse IgG (LI-COR Biosciences, 925-68072)
Sepax SRT-C 300 column, 5 μm, 300 A, 4.6 × 250 mm (Sepax Technologies, 235300-4625)
N-[4-(7-diethylamino-4-methyl-3-coumarinyl) phenyl]maleimide (CPM) dye (Thermo Fisher, D-10251)

2.5.3 Buffers

SEC buffer

25 m*M* HEPES, pH 7.5
400 m*M* NaCl
2.5% glycerol
10 m*M* imidazole
0.025/0.005% DDM/CHS

2.5.4 SDS-PAGE/Western Blot

SDS-PAGE is used to analyze the purity and quantity of the purified complexes while western blots can confirm the presence of chemokine and receptor in the sample. To get sufficient resolution of both receptor and chemokine, it is useful to run a 10% polyacrylamide gel to detect the receptor and an 18% gel to detect the chemokine. Load ~5–10 μg of protein for detection with Coomassie stain and 0.5–1 μg for western blot detection with antibody. Figure 5 shows typical Coomassie-stained gels and western blots for the purified ACKR3/CXCL12 complex. For a pure sample, the receptor and chemokine bands should be easily detectable by SDS-PAGE and be the most prominent bands on the gel. However, if the sample yield is low and/or the sample is not pure, western blots can be used to confirm the presence of receptor and chemokine. For western blots, receptor is detected with a mouse anti-FLAG M2 primary antibody (Sigma-Aldrich) and an IRDye 680 conjugated donkey anti-mouse IgG (LI-COR Biosciences); the presence of chemokine in the complex is confirmed using a rat anti-HA 3F10 primary antibody (Roche) and an IRDye 800 conjugated goat anti-rat IgG (LI-COR Biosciences). The procedure for detecting receptor/chemokine complexes by western blotting is described in more detail in Chapter 18 of this issue (see Kufareva et al.).

2.5.5 Size Exclusion Chromatography

The quality of a size exclusion chromatography (SEC) trace is indicative of the monodispersity of the receptor/chemokine complex since aggregated samples will give broad, asymmetric peaks that elute earlier than well-folded nonaggregated complexes (Fig. 6A). For SEC experiments, equilibrate a Sepax SRT-C 300 column with SEC buffer and inject 5–10 μg of protein onto the column. Measure the absorbance at 280 nm to detect elution of the complex from the column. Figure 6B shows an SEC trace for ACKR3: CXCL12 in DDM/CHS micelles. The elution volume and the sharp symmetric peak indicate that the complex is not aggregated in solution. The elution volume of the complex will be column dependent.

2.5.6 Thermal Unfolding

The midpoint of unfolding or T_m is a useful metric for stability of the protein complex. Unfolding assays utilizing the cysteine-reactive CPM dye is a well-established method to measure the T_m and thereby assess the relative stability of 7TM receptors (Alexandrov, Mileni, Chien, Hanson and Stevens, 2008). The CPM assay has been applied to a large number of receptors, especially in the context of selecting stable constructs and complexes for crystallization. Since chemokines lack free cysteine residues the assay will report on the stability of the receptor, which is highly dependent on the binding of its

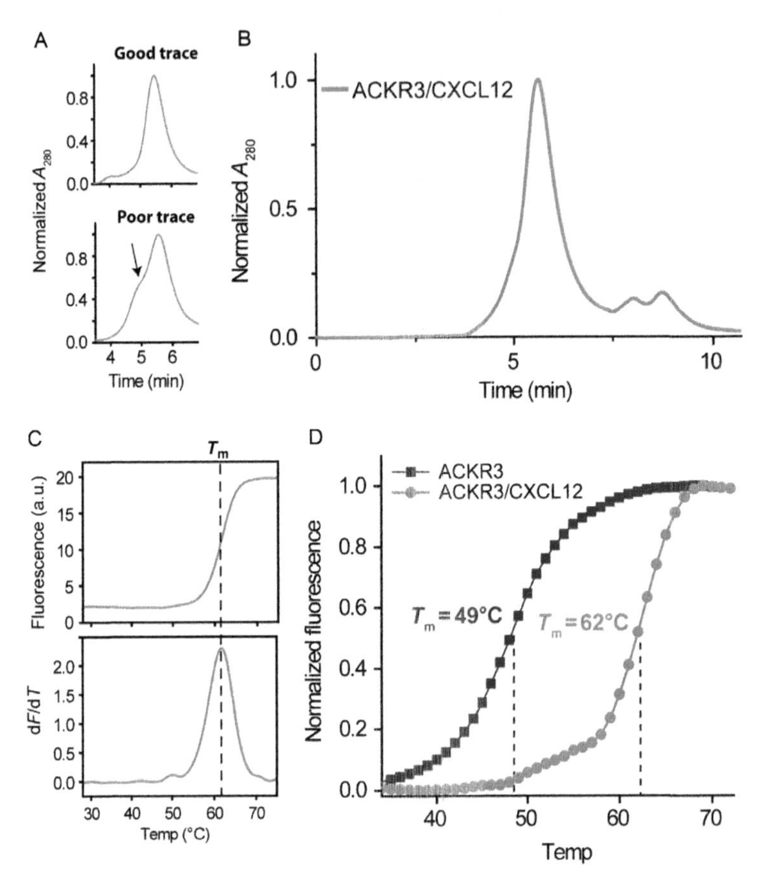

Figure 6 Characterization of receptor/chemokine complexes. (A) Analytical SEC traces of varying quality. The top trace has a sharp, symmetric peak while the peak in the bottom trace has a shoulder (indicated by arrow), which is a sign that the sample is partially aggregated. (B) Analytical SEC trace of the ACKR3/CXCL12 complex. (C) CPM measurements are used to determine the midpoint of thermal unfolding (T_m) of the receptor. Data can be plotted either as CPM fluorescence as a function of temperature (top) or as the derivative of the fluorescence (bottom). (D) CPM experiments with apo ACKR3 and the ACKR3/CXCL12 complex.

chemokine ligand. A high T_m indicates a stable complex that can be used for further studies. In our experience stable complexes between receptors and chemokine have T_m values above ~60 °C and may reach as high as 75 °C in some cases. However, this will be highly dependent on the receptor and chemokine. For CPM experiments, 0.2–0.4 μg of complex is incubated for 20 min with 20 μL of buffer exchange buffer containing 1 μg/mL of CPM dye. Thermal unfolding is measured using a RotorGene Q 6-plex

RT-PCR machine (Qiagen) and the T_m is determined from nonlinear fitting of the unfolding curves (Fig. 6C; Alexandrov et al., 2008). Representative examples of unfolding curves for ACKR3, free and in complex with CXCL12, are shown in Fig. 6D.

2.6 Reconstitution

2.6.1 Equipment

Bath sonicator
Desiccator
FPLC system
Superdex 200 10/300 GL column (GE Healthcare, 17-5175-01)

2.6.2 Reagents and Solutions

1,2-Dimyristoyl-*sn*-glycero-3-phosphocholine (DMPC, 14:0 PC) (Avanti)
Amberlite® XAD®-2 beads (Sigma-Aldrich, 10357)
1,2-Dihexanoyl-*sn*-glycero-3-phosphocholine (DHPC, 6:0 PC) (Avanti)
1-Palmitoyl-2-oleoyl-*sn*-glycero-3-phosphocholine (POPC, 16:0–18:1 PC) (Avanti, 850457)
Sodium cholate (Sigma-Aldrich, C6445)
MSP (~150–200 μ*M* in 25 m*M* HEPES, pH 7.5, 150 m*M* NaCl)

2.6.3 Buffers

Bicelle reconstitution buffer/nanodisc buffer
25 m*M* HEPES, pH 7.5
150 m*M* NaCl

Nanodisc reconstitution buffer
25 m*M* HEPES, pH 7.5
150 m*M* NaCl
200 m*M* Cholate

Nanodisc elution buffer
25 m*M* HEPES, pH 7.5
150 m*M* NaCl
250 m*M* Imidazole

Different biophysical and biochemical methods require reconstitution of the receptor/chemokine complex into different membrane mimics. Many measurements can be done directly in detergent micelles. However, micelles are usually not the ideal environment for receptor stability and are inferior mimics of a native membrane compared to other systems like nanodiscs or bicelles. Reconstitution of receptor/chemokine complexes into these systems may require extensive optimization of reconstitution methods, lipids and buffers. Below are protocols for the reconstitution of ACKR3/CXCL12 into bicelles and nanodiscs.

2.6.4 Bicelles

Bicelles are composed of a mixture of long-chain and short-chain lipids where the long-chain lipid form bilayers with short-chain lipids coating the edges (Whiles, Deems, Vold, & Dennis, 2002). Membrane proteins insert into the bilayer portion and thus bicelles provide a better mimic of a true membrane bilayer than micelles. Bicelles have been used in a number of membrane protein studies, most notably in solid state NMR, where bicelles with a high q (molar ratio of long to short-chain lipid, >3) can be used to align membrane proteins in the magnetic field (Durr, Gildenberg, & Ramamoorthy, 2012), and solution NMR where low q (~0.33–1), isotropic bicelles are utilized (Poget & Girvin, 2007). Several conditions need to be fulfilled for bicelles to form. The long-chain lipids must have saturated chains and a phosphocholine (PC) head group but a portion (up to ~25%) can be exchanged for a lipid with different head group and chain saturation. This is especially relevant for ACKR3 that is stabilized by the presence of negatively charged head groups. The choice of short-chain lipid is less stringent and several detergents can be utilized including 1,2-dihexanoyl-glycero-3-phosphocholine (DHPC), Triton X-100, and 3-[(3-cholamidopropyl)dimethylammonio]-2-hydroxy-1-propanesulfonate (CHAPSO) (Park & Opella, 2010; Whiles et al., 2002). In addition, the concentration of the short-chain lipid needs to be above its CMC and the q value has to be 0.05–1 for the bicelle to have an isotropic behavior in solution as opposed to an aligned bicelle that is formed at q values of ~3 or higher (Whiles et al., 2002). This protocol describes the formation of a 200-μL bicelle sample with $q=0.6$ using DMPC as the long-chain and DHPC as the short-chain lipid. The concentration of DHPC is 20 mM, which is above the CMC (15 mM) and the total lipid concentration is 1.7% (w/v).

1. Weigh out 1.6 mg of DMPC in a glass tube, solubilize in chloroform, and dry away the chloroform under a stream of nitrogen gas until the lipid forms a film at the bottom of the tube. Desiccate over night to remove traces of chloroform.
2. Add 132 μL of reconstitution buffer to the lipids and use vortexing and sonicating (in a bath sonicator) to resuspend the lipid. At this point, the sample should become cloudy and there should be no lipid stuck to the bottom of the tube.

3. Add 40 μL of 10% (w/v) DDM and alternate vortexing (carefully to avoid foam) and sonicating until the solution is clear and the lipid is completely solubilized.
4. Add 10 μL of protein and mix by vortexing (if a different volume of protein is used the volume of reconstitution buffer in step 2 should be adjusted accordingly to get a total sample volume of 182 μL).
5. Add ~70 mg of Amberlite XAD-2 beads (Sigma-Aldrich) and a micro stir bar.
6. Incubate at room temperature on a stir plate for 3–4 h.
7. Remove the sample from biobeads using a gel loading tip. At this time, the solution should be cloudy since the DDM has been absorbed by the biobeads and the sample consists of the receptor/chemokine complex in multilamellar DMPC vesicles.
8. Add 18 μL of D-6-PC from a 10% stock to form the final sample. At this point, the sample should turn completely clear.

2.6.5 Nanodiscs

Nanodiscs (Denisov, Grinkova, Lazarides, & Sligar, 2004) have emerged in the last few years as a membrane mimic that can be used for a number of applications including NMR and fluorescence experiments (Bayburt et al., 2013; Raschle et al., 2009; Whorton et al., 2008). Nanodiscs provide a more native-like membrane mimic than a micelle. Compared to bicelles, nanodiscs avoid the need for detergent in the samples and allow for significantly more flexibility in the choice of lipids. This protocol describes the production of MSP1E3D1 (Bayburt & Sligar, 2010) nanodiscs containing ACKR3 in complex with CXCL12. For simplicity, a single lipid (POPC) is used but similar results are obtained with other lipid mixtures (see Fig. 7A). The protocol is based on methods published by the Sligar lab; see Ritchie et al. (2009), for a protocol describing expression and purification of MSP and a detailed explanation of the reconstitution step.

1. Weigh out 7.6 mg of POPC in a glass tube and add chloroform to dissolve the lipids. Use nitrogen gas to evaporate the chloroform and form a lipid film at the bottom of the tube. Desiccate over night to remove traces of chloroform.
2. Add 120 μL of nanodisc reconstitution buffer. Vortex and sonicate in a bath sonicator (~30 s each) until the solution is clear and the lipid is completely solubilized.
3. Make a reconstitution mixture by combining the following components to get a ACKR3/CXCL12:MSP:POPC molar ratio of 1:10:1250.

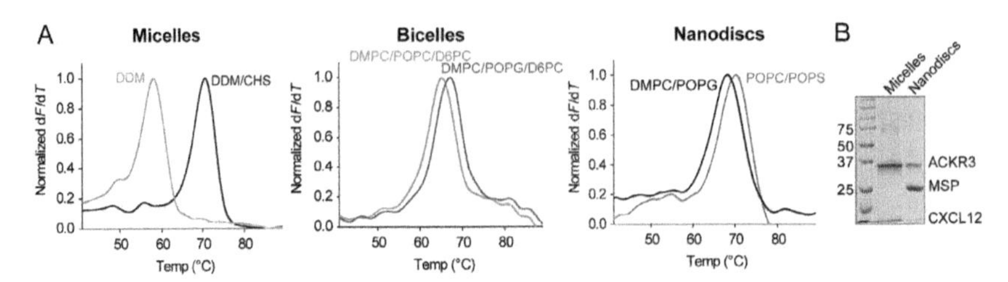

Figure 7 The ACKR3/CXCL12 complex reconstituted into different membrane mimics. (A) The ACKR3/CXCL12 complex is stable in all three membrane mimics as indicated by the sharp peak (derivative of the CPM unfolding curves, see Fig. 6C) and the T_m above 60 °C. (B) SDS-PAGE of ACKR3/CXCL12 sample in nanodiscs and DDM/CHS micelles.

a. 4.5 mg lipid (70 μL of lipid/detergent mixture from step 2)
b. 1.4 mg of MSP1E3D1 (260 μL from a 175 μ*M* solution in 25 m*M* HEPES, 150 m*M* NaCl)
c. 200 μg ACKR3/CXCL12 (120 μL from a 36 μ*M* solution in DDM/CHS)

The volumes and concentrations of MSP and receptor/chemokine stock solutions can be varied. However, if the cholate concentration in the final mixture does not exceed the CMC, additional cholate needs to be added. In this example, the cholate concentration is ~30 m*M*, which is safely above the CMC of ~15 m*M*.

4. Incubate for 90 min on ice
5. Add Amberlite XAD-2 beads (Sigma-Aldrich) and incubate over night at 4 °C stirring.
6. Remove the beads using a gel loading tip
7. Spin down for 10 min at 25,000 × *g* and transfer the supernatant to a new tube
8. Inject the supernatant onto a Superdex 200 10/300 GL column that has been equilibrated with nanodisc buffer and use a fraction collector to collect the elution from the column. In our experience nanodiscs will elute at a volume of ~9–14 mL but this should be confirmed by SDS-PAGE/western blot of the fractions.
9. Pool the fractions containing nanodiscs and add 400 μL of 50% Talon slurry and 10 m*M* Imidazole. Incubate over night at 4 °C in rotisserie. Only receptor-containing nanodiscs should bind to the resin.
10. Centrifuge (350 × *g*, 5 min) and transfer the resin to a Micro Bio-Spin Column (Bio-rad). Wash with 5 × 1 mL of nanodisc buffer using gravity flow.

11. Elute the nanodiscs containing ACKR3/chemokine with 3 × 0.4 mL of nanodisc elution buffer.
12. Buffer exchange the sample using 100 kDa cutoff spin concentrators by concentrating to 50 μL, bringing the volume to 500 μL, and repeating the process three times to get the final sample.
13. Run 10% SDS-PAGE to confirm the nanodisc composition (Fig. 7B). The final sample should consist of only protein-containing nanodiscs, and thus have a 2:1 MSP:receptor molar ratio since each nanodisc is composed of two chains of MSP.

Figure 7A shows unfolding curves of the ACKR3/CXCL12 complex in micelles, bicelles, and nanodiscs acquired using the CPM assay (see Section 2.5.6). In this case, midpoints of unfolding are similar in all three membrane mimics showing that the complex is stable in all three systems. The large increase in T_m after addition of CHS illustrates the importance of the lipid/micelle environment to receptor stability.

3. SUMMARY AND CONCLUSIONS

The methods described in this chapter provide an efficient way of producing receptor/chemokine and has been used in our lab to successfully produce large amounts of receptor/chemokine complexes for CXC as well as CC receptors. Still, it is important to point out that complex formation through coexpression is not a universal method and will not be suitable for all combinations of chemokines and receptors. In the case of ACKR3, coexpression works well for complex formation with CXCL12, which has an affinity of ~0.2 n*M* in cell membranes (Burns et al., 2006). However, complexes between ACKR3 and CXCL11 (ITAC) cannot be produced using the same method due to a ~10-fold lower affinity (Burns et al., 2006), and thus faster off-rate of the chemokine. Another example is the CXCR4/CXCL12, which requires G protein for optimal affinity. In these cases, the protocol above will provide an excellent starting point for modified protocols that implement ligand exchange after purification or coexpression with G proteins that may be needed for successful reconstitution of a receptor/chemokine complex.

ACKNOWLEDGMENTS

This work was supported by grants to T.M.H. (NIAID R01 AI37113, NIGMS U01GM094612, and NIAID R01AI118985) and a Robertson Foundation/Cancer Research Institute Irvington postdoctoral fellowship to M.G.

REFERENCES

Alexandrov, A. I., Mileni, M., Chien, E. Y., Hanson, M. A., & Stevens, R. C. (2008). Microscale fluorescent thermal stability assay for membrane proteins. *Structure*, *16*, 351–359.

Allen, S. J., Crown, S. E., & Handel, T. M. (2007). Chemokine: Receptor structure, interactions, and antagonism. *Annual Review of Immunology*, *25*, 787–820.

Allen, S. J., Hamel, D. J., & Handel, T. M. (2011). A rapid and efficient way to obtain modified chemokines for functional and biophysical studies. *Cytokine*, *55*, 168–173.

Allen, S. J., Ribeiro, S., Horuk, R., & Handel, T. M. (2009). Expression, purification and in vitro functional reconstitution of the chemokine receptor CCR1. *Protein Expression and Purification*, *66*, 73–81.

Bannert, N., Craig, S., Farzan, M., Sogah, D., Santo, N. V., Choe, H., et al. (2001). Sialylated O-glycans and sulfated tyrosines in the NH2-terminal domain of CC chemokine receptor 5 contribute to high affinity binding of chemokines. *Journal of Experimental Medicine*, *194*, 1661–1673.

Bayburt, T. H., & Sligar, S. G. (2010). Membrane protein assembly into nanodiscs. *FEBS Letter*, *584*, 1721–1727.

Bayburt, T. H., Vishnivetskiy, S. A., McLean, M. A., Morizumi, T., Huang, C. C., Tesmer, J. J., et al. (2013). Monomeric rhodopsin is sufficient for normal rhodopsin kinase (GRK1) phosphorylation and arrestin-1 binding. *Journal of Biological Chemistry*, *286*, 1420–1428.

Burg, J. S., Ingram, J. R., Venkatakrishnan, A. J., Jude, K. M., Dukkipati, A., Feinberg, E. N., et al. (2015). Structural biology structural basis for chemokine recognition and activation of a viral G protein-coupled receptor. *Science*, *347*, 1113–1117.

Burns, J. M., Summers, B. C., Wang, Y., Melikian, A., Berahovich, R., Miao, Z., et al. (2006). A novel chemokine receptor for SDF-1 and I-TAC involved in cell survival, cell adhesion, and tumor development. *Journal of Experimental Medicine*, *203*, 2201–2213.

Caffrey, M., & Cherezov, V. (2009). Crystallizing membrane proteins using lipidic mesophases. *Nature Protocols*, *4*, 706–731.

Cherezov, V., Liu, J., Griffith, M., Hanson, M. A., & Stevens, R. C. (2008). LCP-FRAP assay for pre-screening membrane proteins for in meso crystallization. *Crystal Growth and Design*, *8*, 4307–4315.

Chun, E., Thompson, A. A., Liu, W., Roth, C. B., Griffith, M. T., Katritch, V., et al. (2012). Fusion partner toolchest for the stabilization and crystallization of G protein-coupled receptors. *Structure*, *20*, 967–976.

Denisov, I. G., Grinkova, Y. V., Lazarides, A. A., & Sligar, S. G. (2004). Directed self-assembly of monodisperse phospholipid bilayer Nanodiscs with controlled size. *Journal of the American Chemical Society*, *126*, 3477–3487.

Durr, U. H., Gildenberg, M., & Ramamoorthy, A. (2012). The magic of bicelles lights up membrane protein structure. *Chemical Reviews*, *112*, 6054–6074.

Fong, A. M., Alam, S. M., Imai, T., Haribabu, B., & Patel, D. D. (2002). CX3CR1 tyrosine sulfation enhances fractalkine-induced cell adhesion. *Journal of Biological Chemistry*, *277*, 19418–19423.

Fredriksson, R., Lagerstrom, M. C., Lundin, L. G., & Schioth, H. B. (2003). The G-protein-coupled receptors in the human genome form five main families. Phylogenetic analysis, paralogon groups, and fingerprints. *Molecular Pharmacology*, *63*, 1256–1272.

Grisshammer, R. (2009). Purification of recombinant G-protein-coupled receptors. *Methods in Enzymology*, *463*, 631–645.

Gueret, V., Negrete-Virgen, J. A., Lyddiatt, A., & Al-Rubeai, M. (2002). Rapid titration of adenoviral infectivity by flow cytometry in batch culture of infected HEK293 cells. *Cytotechnology*, *38*, 87–97.

Hanes, M. S., Salanga, C. L., Chowdry, A. B., Comerford, I., McColl, S. R., Kufareva, I., et al. (2015). Dual targeting of the chemokine receptors CXCR4 and ACKR3 with novel engineered chemokines. *Journal of Biological Chemistry, 290*, 22385–22397.

Katritch, V., Cherezov, V., & Stevens, R. C. (2013). Structure-function of the G protein-coupled receptor superfamily. *Annual Review of Pharmacology and Toxicology, 53*, 531–556.

Kofuku, Y., Yoshiura, C., Ueda, T., Terasawa, H., Hirai, T., Tominaga, S., et al. (2009). Structural basis of the interaction between chemokine stromal cell-derived factor-1/CXCL12 and its G-protein-coupled receptor CXCR4. *Journal of Biological Chemistry, 284*, 35240–35250.

Kufareva, I., Stephens, B. S., Holden, L. G., Qin, L., Zhao, C., Kawamura, T., et al. (2014). Stoichiometry and geometry of the CXC chemokine receptor 4 complex with CXC ligand 12: Molecular modeling and experimental validation. *Proceedings of the National Academy of Sciences of the United States of America, 111*, E5363–E5372.

Li, Z., Ling, L., Liu, X., Laus, R., & Delcayre, A. (2010). A flow cytometry-based immuno-titration assay for rapid and accurate titer determination of modified vaccinia Ankara virus vectors. *Journal of Virological Methods, 169*, 87–94.

Liu, W., Wacker, D., Gati, C., Han, G. W., James, D., Wang, D., et al. (2013). Serial femtosecond crystallography of G protein-coupled receptors. *Science, 342*, 1521–1524.

Nguyen, D. H., & Taub, D. (2002). Cholesterol is essential for macrophage inflammatory protein 1 beta binding and conformational integrity of CC chemokine receptor 5. *Blood, 99*, 4298–4306.

Nguyen, D. H., & Taub, D. D. (2003). Inhibition of chemokine receptor function by membrane cholesterol oxidation. *Experimental Cell Research, 291*, 36–45.

Nijmeijer, S., Leurs, R., Smit, M. J., & Vischer, H. F. (2010). The Epstein-Barr virus-encoded G protein-coupled receptor BILF1 hetero-oligomerizes with human CXCR4, scavenges Galphai proteins, and constitutively impairs CXCR4 functioning. *Journal of Biological Chemistry, 285*, 29632–29641.

Palczewski, K., Kumasaka, T., Hori, T., Behnke, C. A., Motoshima, H., Fox, B. A., et al. (2000). Crystal structure of rhodopsin: A G protein-coupled receptor. *Science, 289*, 739–745.

Park, S. H., Das, B. B., Casagrande, F., Tian, Y., Nothnagel, H. J., Chu, M., et al. (2012). Structure of the chemokine receptor CXCR1 in phospholipid bilayers. *Nature, 491*, 779–783.

Park, S. H., & Opella, S. J. (2010). Triton X-100 as the "short-chain lipid" improves the magnetic alignment and stability of membrane proteins in phosphatidylcholine bilayers for oriented-sample solid-state NMR spectroscopy. *Journal of the American Chemical Society, 132*, 12552–12553.

Pierce, K. L., Premont, R. T., & Lefkowitz, R. J. (2002). Seven-transmembrane receptors. *Nature Reviews Molecular Cell Biology, 3*, 639–650.

Poget, S. F., & Girvin, M. E. (2007). Solution NMR of membrane proteins in bilayer mimics: Small is beautiful, but sometimes bigger is better. *Biochimica et Biophysica Acta, 1768*, 3098–3106.

Qin, L., Kufareva, I., Holden, L. G., Wang, C., Zheng, Y., Zhao, C., et al. (2015). Structural biology. Crystal structure of the chemokine receptor CXCR4 in complex with a viral chemokine. *Science, 347*, 1117–1122.

Raschle, T., Hiller, S., Yu, T. Y., Rice, A. J., Walz, T., & Wagner, G. (2009). Structural and functional characterization of the integral membrane protein VDAC-1 in lipid bilayer nanodiscs. *Journal of the American Chemical Society, 131*, 17777–17779.

Ritchie, T. K., Grinkova, Y. V., Bayburt, T. H., Denisov, I. G., Zolnerciks, J. K., Atkins, W. M., et al. (2009). Chapter 11 - Reconstitution of membrane proteins in phospholipid bilayer nanodiscs. *Methods in Enzymology, 464*, 211–231.

Rosenbaum, D. M., Cherezov, V., Hanson, M. A., Rasmussen, S. G., Thian, F. S., Kobilka, T. S., et al. (2007). GPCR engineering yields high-resolution structural insights into beta2-adrenergic receptor function. *Science*, *318*, 1266–1273.

Salon, J. A., Lodowski, D. T., & Palczewski, K. (2011). The significance of G protein-coupled receptor crystallography for drug discovery. *Pharmacological Reviews*, *63*, 901–937.

Steyaert, J., & Kobilka, B. K. (2011). Nanobody stabilization of G protein-coupled receptor conformational states. *Current Opinion in Structural Biology*, *21*, 567–572.

Tan, Q., Zhu, Y., Li, J., Chen, Z., Han, G. W., Kufareva, I., et al. (2013). Structure of the CCR5 chemokine receptor-HIV entry inhibitor maraviroc complex. *Science*, *341*, 1387–1390.

Tate, C. G., & Schertler, G. F. (2009). Engineering G protein-coupled receptors to facilitate their structure determination. *Current Opinion in Structural Biology*, *19*, 386–395.

Veldkamp, C. T., Peterson, F. C., Hayes, P. L., Mattmiller, J. E., Haugner, J. C., 3rd, de la Cruz, N., et al. (2007). On-column refolding of recombinant chemokines for NMR studies and biological assays. *Protein Expression and Purification*, *52*, 202–209.

Veldkamp, C. T., Seibert, C., Peterson, F. C., De la Cruz, N. B., Haugner, J. C., Basnet, H., 3rd, et al. (2008). Structural basis of CXCR4 sulfotyrosine recognition by the chemokine SDF-1/CXCL12. *Science Signaling*, *1*, ra4.

Venkatakrishnan, A. J., Deupi, X., Lebon, G., Tate, C. G., Schertler, G. F., & Babu, M. M. (2013). Molecular signatures of G-protein-coupled receptors. *Nature*, *494*, 185–194.

Whiles, J. A., Deems, R., Vold, R. R., & Dennis, E. A. (2002). Bicelles in structure-function studies of membrane-associated proteins. *Bioorganic Chemistry*, *30*, 431–442.

Whorton, M. R., Jastrzebska, B., Park, P. S., Fotiadis, D., Engel, A., Palczewski, K., et al. (2008). Efficient coupling of transducin to monomeric rhodopsin in a phospholipid bilayer. *Journal of Biological Chemistry*, *283*, 4387–4394.

Wu, B., Chien, E. Y., Mol, C. D., Fenalti, G., Liu, W., Katritch, V., et al. (2010). Structures of the CXCR4 chemokine GPCR with small-molecule and cyclic peptide antagonists. *Science*, *330*, 1066–1071.

CHAPTER TWELVE

In Vivo Models to Study Chemokine Biology

F.A. Amaral, D. Boff, M.M. Teixeira[1]
Immunopharmacology, Department of Biochemistry and Immunology, Universidade Federal de Minas Gerais, Belo Horizonte, Brazil
[1]Corresponding author: e-mail address: mmtex@icb.ufmg.br

Contents

1. Introduction 262
2. Methods 264
   2.1 The Use of the Joint to Study the Biology of CXCR2-Dependent Neutrophil Recruitment 264
   2.2 Injection of Chemokines into the Tibiofemoral Joint 265
   2.3 Antigen-Induced Arthritis Model to Study Endogenous Chemokines 266
   2.4 Cell Recovery from the Joint Cavity 269
   2.5 Pleural Cavity as a Model to Study the Recruitment of CCR2-Dependent Monocytes/Macrophages 270
   2.6 Cell Recovery from the Pleural Cavity 272
   2.7 Identification of Cell Types 272
3. Limitations 275
4. Other Perspectives 277
References 277

Abstract

Chemokines are essential mediators of leukocyte movement *in vivo*. *In vitro* assays of leukocyte migration cannot mimic the complex interactions with other cell types and matrix needed for cells to extravasate and migrate into tissues. Therefore, *in vivo* strategies to study the effects and potential relevance of chemokines for the migration of particular leukocyte subsets are necessary. Here, we describe methods to study the effects and endogenous role of chemokine in mice. Advantages and pitfalls of particular models are discussed and we focus on description in model's joint and pleural cavity inflammation and the effects and relevance of CXCR2 and CCR2 ligands on cell migration.

Methods in Enzymology, Volume 570
ISSN 0076-6879
http://dx.doi.org/10.1016/bs.mie.2015.09.015

1. INTRODUCTION

The discovery of the chemokine system has fueled our understanding of the molecular cues used by leukocytes to position themselves *in vivo*. Both under homeostatic and pathological conditions, chemokines and their receptors provide the necessary signals for leukocyte to reach and move within tissues. In addition, chemokines may also activate leukocytes, actions that may be beneficial in the context of infection but may cause tissue damage, as observed in certain chronic inflammatory diseases. The chemokine system is redundant and certainly very complex. For example, a particular chemokine may bind to several receptors (the chemokine CCL5 may bind to CCR1, CCR2, CCR3, and CCR5) and a particular receptor may bind several chemokines (CCR5 binds CCL3, CCL4, CCL5, CCL7, CCL14, and CCL14; Bachelerie et al., 2014). Moreover, there is much genetic control of the expression of chemokines and their receptors in different tissues and cell types in a timely coordinated manner. Because of this complexity, there is much need to comprehend in greater details the effects and endogenous function of chemokines *in vivo*.

In vitro studies are certainly very useful to study chemokine function and are discussed elsewhere in this review series. In the context of leukocyte migration, migration or chemotaxis assays are the most frequently used to infer chemokine function *in vivo*. There are many useful *in vitro* migration assays that go from simple passage via a membrane with a known pore size to more complex systems in which matrix is added to the assay. Regardless of the assay used, none of them can mimic entirely the complexity of leukocyte migration *in vivo*. Indeed, when injected *in vivo* a chemokine will necessarily interact first with resident cells, which may include macrophages, dendritic cells, fibroblasts, endothelial cells, epithelial cells, and other parenchymal cells, depending on the tissue being tested. In addition, the injection will occur in an extravascular compartment, making it necessary that the chemokine moves into the intravascular compartment where it will be presented to rolling leukocytes, likely by interacting with GAGs on endothelial cells (Handel, Johnson, Crown, Lau, & Proudfoot, 2005; Johnson, Proudfoot, & Handel, 2005). Therefore, when a given chemokine is injected *in vivo*, it may attract leukocytes directly by binding to specific receptors on the leukocyte surface. This is similar to *in vitro* studies but with the added complexity of the need to interact with GAGs on endothelial cells (Proudfoot et al., 2003; Wang, Fuster, Sriramarao, & Esko, 2005).

In addition, the injected chemokine may interact with and activate resident cells, such as macrophages, leading to the generation of intermediate chemoattractant molecules that will themselves activate their receptor on the surface of leukocytes (Sokol & Luster, 2015).

There is no single method that is useful to study all chemokines and all chemokine receptors. Here, we will provide two basic strategies to study chemokine-induced leukocytes recruitment *in vivo*. These are just examples on how to exploit *in vivo* models to gain further insight into chemokine function *in vivo*. Initially, we will examine methods that evaluate the effects of the direct injection of the desired chemokine (exogenous chemokine) in a specific compartment of an animal. As a proof of concept for this strategy, we will study the effects of the injection of CXC chemokine CXCL1 and CC chemokine CCL2. This strategy is useful to examine major cell types responding to the given chemokine but there are limitations. In contrast to the situation observed in disease states, a large amount of chemokine is injected at a single time point. Moreover, there are no other mediators or cytokines in tissues that could facilitate chemokine action by, for example, enhancing expression of cell adhesion molecules and GAGs, or enhancing expression of receptors in responding cell types. Some chemokines are rapidly cleaved and inactivated *in vivo* and, hence, may not reach sufficient concentrations to cause leukocyte recruitment when given acutely at a single bolus injection (Mortier, Van Damme, & Proost, 2012). Finally, some leukocyte subsets responding to a particular chemokine may not be abundant in blood under normal conditions and may, hence, not be recruited in sufficient amounts to be detectable in *in vivo* recruitment assays. Indeed, our experience has shown that chemokines that promote the recruitment of neutrophils and eosinophils are easier to study *in vivo* that chemokines that promote preferentially the recruitment of macrophages or lymphocytes.

In the second strategy, we will examine models of endogenous chemokine-dependent leukocyte recruitment *in vivo*. Again focusing on CXCL1 and CCL2, we will examine situations in which these molecules are produced endogenously during a complex inflammatory reaction and drive the recruitment of a given leukocyte. For the latter studies, there is a need for tools, such as antibodies, drugs, or genetically modified animals to study chemokine function.

Mice will be used in all procedures described herewith, because of the general availability of these species for immunology research and the availability of tools, including genetically modified individuals and availability of mouse-specific chemokines tools, i.e., recombinant murine chemokines and specific antibodies. This is a very important aspect as there are many examples of chemokines that do not have cross-species function. Different compartments in mice may be used to investigate the recruitment of leukocytes after injection of a chemokine, including the peritoneum, pleura, joints, and air pouch cavities. The choice of the cavity must be careful as it is influenced by its size (accuracy to inject the chemokine), the basal levels of resident cells (interference in the readout of recruited cells), the option of a technique to identify the cell population (low or high number of recruited cells), and the possibility that a functional readout may be obtained (for instance, the capacity to induce pain when the chemokine is injected in the joint).

2. METHODS

2.1 The Use of the Joint to Study the Biology of CXCR2-Dependent Neutrophil Recruitment

Bone, cartilage, muscle, ligaments, and synovium compose the tibiofemoral joint structure. The synovium is a thin vascularized membrane of resident cells, constituted of synoviocytes, macrophages, and fibroblasts (Khan et al., 2007). These cells, together with the synovial microvasculature, are responsive to different stimuli, including chemokines, creating an inflammatory environment that facilitates the diapedesis of leukocytes. Once migrated, leukocytes can accumulate in the synovial tissue, among the synoviocytes, or into the synovial cavity. One can analyze infiltrated cells in the joint in both compartments (synovial tissue and synovial cavity) by different techniques (in detail below).

Chemokines that bind and activate the receptors CXCR1 and CXCR2 are well-known chemoattractants for neutrophils. These were the first chemokine receptors defined at the molecular level (Holmes, Lee, Kuang, Rice, & Wood, 1991; Murphy & Tiffany, 1991) and can recognize a group of CXC chemokines containing the amino acid motif ELR in the N-terminal domain (Bachelerie et al., 2014). In mice, the best-characterized ELR^+ CXC chemokines are *Cxcl1* (keratinocyte chemoattractant) and *Cxcl2* (macrophage inflammatory protein-2), and they have been extensively studied in a plethora of models of inflammation (Russo, Garcia, Teixeira, & Amaral, 2014).

Targeting Cxcr2 or its canonical ligands reduces recruitment of neutrophils and consequently neutrophil-dependent tissue damage in several models of inflammation (Citro et al., 2015; Coelho et al., 2008; Russo et al., 2009). Several inhibitors of CXCR1 and CXCR2 have been developed and some are undergoing clinical trials for various indications (Dwyer & Yu, 2014; Ha et al., 2015; Moriconi et al., 2007; Rennard et al., 2015). Here, we will detail two protocols to investigate CXCR2-dependent recruitment of neutrophils to the joint by exogenously added or endogenously produced chemokines.

2.2 Injection of Chemokines into the Tibiofemoral Joint

The injection of *Cxcl1* or *Cxcl2* into the tibiofemoral joint of mice leads to rapid migration of significant numbers of neutrophils into the knee. The injection of 100 ng of *Cxcl1* in the joint is optimal to induce recruitment of neutrophils and cells can be easily recovered from the joint cavity 3 h after injection of the chemokine. Injection of human CXCL8 also induces the recruitment of neutrophils, albeit the dose necessary to cause recruitment is higher, approximately 1 μg/joint. Steps for the intra-articular injection of the chemokine and details for the evaluation of cells are given below.

2.2.1 Required Materials

- Insulin syringe (BD Ultra-Fine, 30U, #328322)
- Pipette tip: 0.5–10 μL (white)

2.2.2 Protocol

2.2.2.1 Preparation of Syringe

As the tibiofemoral joint is a very small cavity, it is useful to adapt the syringe in order to facilitate the injection into the joint and to avoid missing the cavity.

- Cut the extremity of a pipette tip and place it around the needle of the syringe in order to keep only the bevel out.
- Use an adhesive tape to fasten the pipette tip to the needle. The pipette tip halts the needle from penetrating too deep and missing the cavity (Fig. 1).

2.2.2.2 Injection into the Joint

- Once anesthetized (according to local ethical committee), remove the fur over the knee (Fig. 2A). It is easily done by pulling the fur using the fingers.
- Slightly flex the knee in order to facilitate the injection (Fig. 2B).
- Use a needle or a nail to find the exact location of the articular cavity (Fig. 2C).

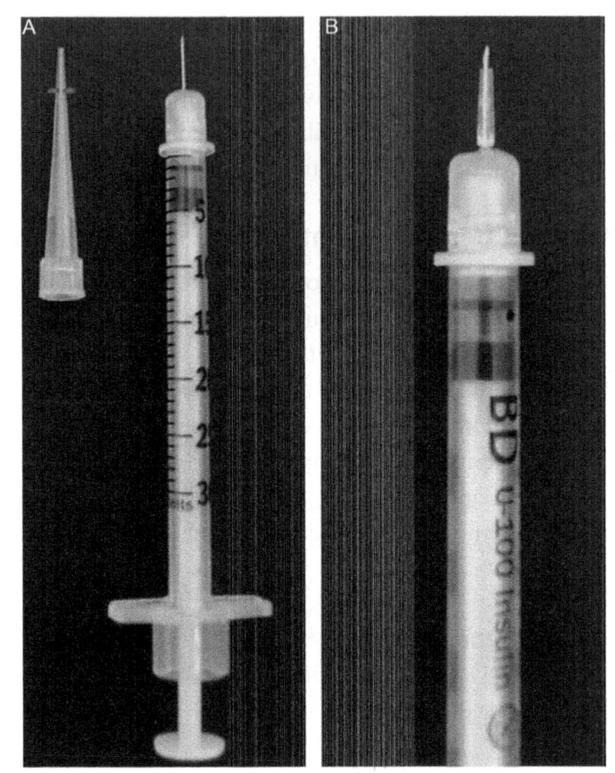

Figure 1 Adaptation of the syringe used to inject in the knee. A pipette tip must be cut (red (gray in the print version) dotted line) (A) and placed around the needle (B).

- Draw 10 μL of the desired chemokine into the syringe. This volume is sufficient to fill all the tibiofemoral cavity of the mouse.
- Inject the volume when the syringe is level perpendicular to the knee (Fig. 2D).

2.3 Antigen-Induced Arthritis Model to Study Endogenous Chemokines

Antigen-induced arthritis (AIA) is a model to induce arthritis in different species, mainly in mice, that was described in 1970s (Brackertz, Mitchell, & Mackay, 1977; Cooke, Hurd, Ziff, & Jasin, 1972), and it is still very useful to investigate mechanisms associated with joint inflammation. AIA may also be used as a proof-of-concept model for the development of new compounds in the preclinical phase. One of the most important features of joint inflammation in AIA is the high number of neutrophils that

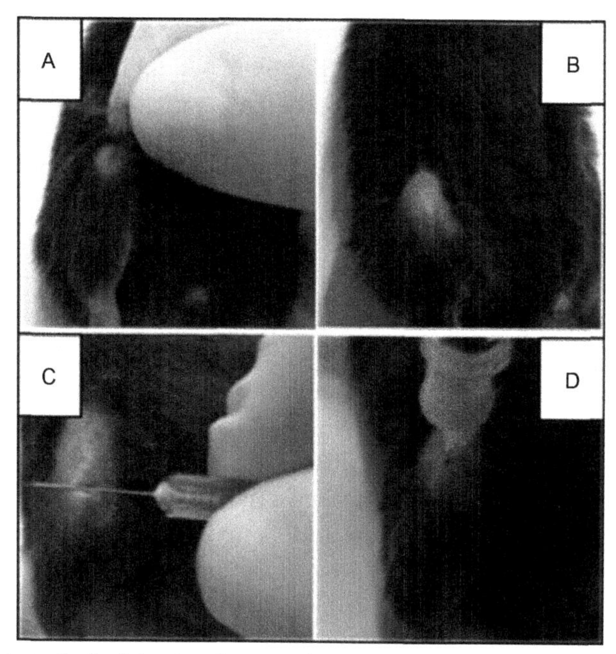

Figure 2 Intra-articular injection. Remove the mouse hair over the knee (A and B), find the articular cavity using a needle (C), and do the intra-articular injection (D).

accumulate in the inflamed joint. Neutrophils contribute significantly to joint damage and dysfunction in this model (Coelho et al., 2008; Grespan et al., 2008). Mechanisms of tissue inflammation in response to the injection of the antigen (methylated bovine serum albumin (mBSA)) rely on the local production of immune complexes and the activation of a Th-1-dependent immune response (Teixeira, Talvani, Tafuri, Lukacs, & Hellewell, 2001). This is followed by production of large amount of cytokines (TNF-α and IFN-γ) and neutrophil-related chemoattractants (CXCL1, LTB4) that mediate the recruitment of neutrophils. Blockade of CXCR2 decreases the recruitment of neutrophils to the joint and consequently reduces joint inflammation and hypernociception caused by AIA (Coelho et al., 2008; Grespan et al., 2008).

In the literature, there are variations in the protocol to induce AIA based on the type of adjuvants, the concentration of the reagents, and the time course of inflammation. Further, different strains of mice and different species of animals may be used to induce AIA (Brackertz et al., 1977). Here, we described a protocol that we have used to induce joint inflammation in C57BL/6 mice.

2.3.1 Required Material

- Complete Freund's adjuvant (CFA; Sigma-Aldrich, #F5881)
- mBSA (Sigma-Aldrich, #A1009)
- Graduated and autoclavable glass syringe with metal locker tip
- Interchangeable plunger
- Insulin syringe (BD Ultra-Fine, 30U, #328322)
- Pipette tip: 0.5–10 μL (white)

2.3.2 Protocol

The first step is the preparation of the emulsion containing the adjuvant and the antigen. For each mouse, 500 μg of mBSA must be dissolved in a final volume of 100 μL (50 μL of CFA + 50 μL of PBS). There are significant losses during the preparation and injection and this must be taken into consideration when deciding the amount of material to be prepared.

- Weigh the required amount of mBSA in a microtube and dissolve it in the required amount of PBS. Avoid foaming of the solution by mixing it gently. Transfer the volume to the glass syringe.
- In another glass syringe, pull the same volume of CFA.
- The next step is to mix both PBS + mBSA and CFA using a connector. Before adapting the connector to both syringes, make sure that no air will be transferred from one to another syringe.
- Start the emulsification. Press the piston of one syringe to transfer the emulsion to the other syringe. Immediately, the emulsion will turn white. Keep a constant movement by approximately 40 × in each hand.
- Initially, the process is smooth but becomes harder at the end of the process.
- To test if the emulsion is ready to be injected, add a drop of the emulsion in a recipient containing PBS. If the drop dissolves, the emulsion is not ready and additional movements with the syringe must be done. Keeping the syringes at 4 °C for 30 min helps emulsification. When the drop of the emulsion is intact in PBS, it is ready to be injected in mice.
- Immunization: in anesthetized mice, remove the fur at the basal of the tail (it is easy to do using the fingers). Inject 100 μL of the emulsion intradermically (Fig. 3). To avoid loss of the emulsion during injection, make sure the needle is deep enough in the dermis. A small glass syringe is required.
- Challenge: 14 days after the immunization, mice should be anesthetized for the intra-articular injection of mBSA (10 μg in 10 μL of PBS). The procedure for i.a. injection was detailed above (see Section 2.2.2.2).

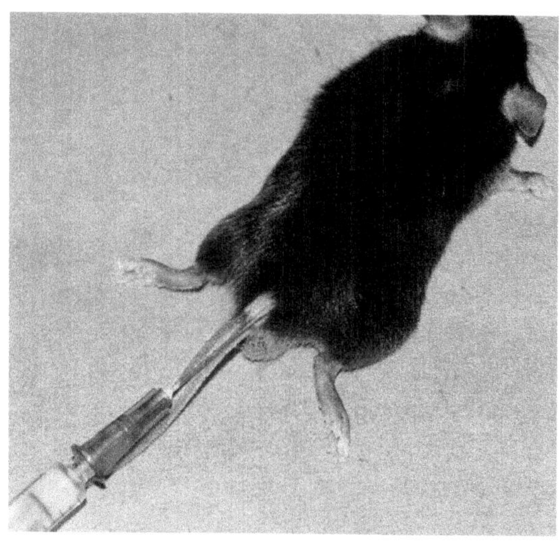

Figure 3 Intradermical injection of the complete Freund's adjuvant + methylated bovine serum albumin in antigen-induced arthritis.

Joint inflammation peaks 24 h after challenge with mBSA in the joint. The main inflammatory parameters that can be evaluated are: infiltration of leukocyte (especially neutrophils) into the joint cavity, production of cytokines and chemokines in periarticular tissue, analysis of the cellular population and activation in lymph node (see Costa et al., 2015 for additional information), histopathology (see Coelho et al., 2008 for additional information), and measurement of joint hypernociception (see Coelho et al., 2008; Sachs et al., 2011 for additional information).

2.4 Cell Recovery from the Joint Cavity

The procedure to recover the cells from the joint cavity can be applied either to AIA and when exogenous chemokines are injected into the joint.

2.4.1 Required Materials

- Micropipette (volume 5 μL, P10)
- Straight micro scissors
- Micro dissecting forceps, straight, delicate
- Bovine serum albumin (BSA) 3% (in PBS solution)

2.4.2 Protocol

After a predetermined time following chemokine injection or in the AIA model, the cells migrated to the articular joint can be recovered.

- Mice should be euthanized according to the guidelines of the local ethics committee.
- Remove the skin from the ankle to the base of the hips in order to visualize the patellar tendon (Fig. 4A).
- Carefully cut the tendon with a scissors (Fig. 4B) and use a dissecting forceps to open the cavity (Fig. 4C).
- Add 90 μL of BSA 3% (in saline or PBS) in an identified tube.
- To recover cells from the joint, wash the cavity with 5 μL of BSA 3% and dispose in the referred tube (Fig. 4D). Repeat it. The final volume will be around 100 μL (a little bit volume could be missed during the wash).
- The periarticular tissue (around 30 mg) can be removed using a scalpel blade and processed to measure chemokines by enzymatic assays (Fig. 4C, dashed line). Keep all the samples on ice during all the steps.

2.5 Pleural Cavity as a Model to Study the Recruitment of CCR2-Dependent Monocytes/Macrophages

The pleural cavity is another interesting compartment to study the recruitment of leukocyte *in vivo*. The pleura is an important mechanical barrier for

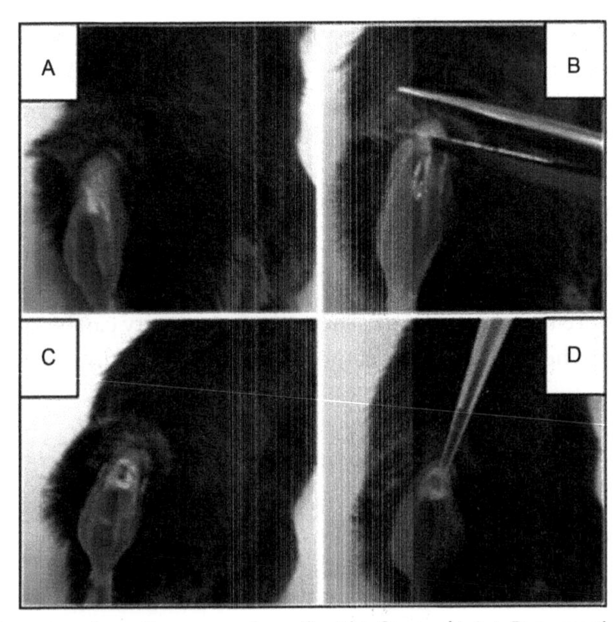

Figure 4 Procedure for cell recovery from the tibiofemoral joint. Remove the skin under the knee (A). Cut the patellar tendon carefully with a delicate scissor (B). Open the articular cavity using tweezers (C). Dashed line can be removed for enzymatic assays (D). Wash the joint cavity with diluent (5 μL, 2×) to recover accumulated cells. (See the color plate.)

visceral protection and is extremely responsive to different stimuli, becoming inflamed and leading to the accumulation of leukocyte into the pleural space (Antony, 2003). The pleural membrane is well vascularized and is composed by mesothelial cells, resident macrophages, mast cells, and lymphocytes in basal conditions (Honma, Abe, & Ito, 1982; Kroegel & Antony, 1997). These cells can produce significant amounts of inflammatory mediators, including chemokines, such as CCL2 and CXCL1 (Antony et al., 1995; Cailhier et al., 2006; Loghmani et al., 2002). Thus, the activation of pleural resident cells during inflammation or due to a single injection of a specific chemoattractant leads to massive infiltration of leukocytes that can be easily recovered from the cavity for cellular identification (Carmo et al., 2014; Klein et al., 2002; Mohammed et al., 1998).

The chemokine CCL2, also known as MCP-1 (monocyte chemotactic protein-1), belongs to CC family of chemokines and is a potent chemoattractant for monocytes by binding to the CCR2 receptor. In inflammatory conditions, different cell types produce CCL2, including endothelial, epithelial, mesothelial cells, and resident macrophages, favoring the recruitment of inflammatory macrophages (Bachelerie et al., 2014). However, since other cells also express CCR2, CCL2 can promote the recruitment of dendritic cells, basophils, NK cells, neutrophils, and T lymphocytes to the tissue (Bachelerie et al., 2014; Talbot et al., 2015). Here, we focus on the recruitment of macrophages to the pleural cavity induced by the injection of CCL2.

2.5.1 Required Materials

- Insulin syringe (BD Ultra-Fine, 30U, #328322)
- Pipette tip: 0.5–10 μL (white)

2.5.2 Protocol

- The syringe should be prepared in the same way as to inject into the knee joint (see Section 2.2.2.1).
- Hold the mouse in the standing position in order to facilitate the injection (Fig. 5).
- Choose the right or left side for the injection. Use your finger to feel the last rib of the mouse. The exact location for the injection is approximately 0.5 cm above the last rib.
- Draw 100 μL of the desired chemokine in the syringe.
- Inject the volume when the syringe is level perpendicular to the right location (Fig. 5).

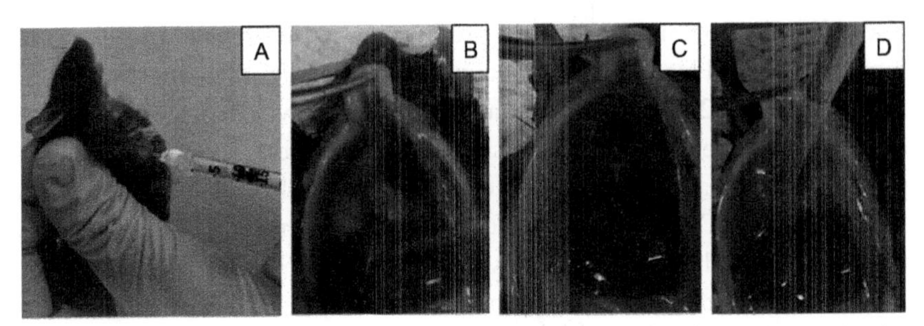

Figure 5 Procedures for the injection and cell recovery from the pleura cavity. Injection in the pleural cavity (A). After laparotomy, the pleura can be easily visualized (B). Open the pleural cavity (C) and wash the cavity for cell recovery (D).

2.6 Cell Recovery from the Pleural Cavity

2.6.1 Required Materials

- Micropipette (volume 100 μL, P100)
- Micro scissors, straight
- Micro dissecting forceps, straight, delicate
- BSA 3% (in PBS solution)

2.6.2 Protocol

After a predetermined time following chemokine injection, the cells migrated to the pleural cavity can be recovered.

- Mice should be euthanized according to the guidelines of the local ethics committee. It is important to mention that cervical dislocation should not be performed after anesthesia. This procedure can cause bleeding and fill the pleural cavity with blood that will compromise the experiment (Fig. 5B).
- Open the pleural cavity with a scissors (Fig. 5C).
- Wash the cavity 2 × with 2 mL of PBS–EDTA (1 m*M*) (Fig. 5D).
- Centrifuge the samples at 1200 rpm for 5 min, remove the supernatant, and resuspend the pellet in 100 μL of BSA 3%.
- Proceed with a cell counting and cytospin as described in Section 2.7.

2.7 Identification of Cell Types

Different protocols can be used to identify the cells that were recruited to the joint and pleural cavities and their activation state. However, the amount of the migrated cells (depending on the stimulus) and the presence of low or high amount of resident cells (depending on the cavity) must be taken into account when deciding the best method for evaluating cell types. Flow cytometry may be ideal to define cell types but greater number of cells are necessary. Here, we will detail a protocol that characterizes the cell types based on the morphology and another protocol that is useful for a more precise identification of the cell population and state of activation.

2.7.1 Identification Based on Morphology (Optical Microscopy)

The cells recovered from the joint or pleural cavity can be counted and differentiated considering their morphology as explained below.

2.7.1.1 Required Materials

- Neubauer chamber
- Microscope slides
- Turk solution
- Staining solution: panoptic or Giemsa's solution
- Optical microscope

2.7.1.2 Protocol

- For total cell counting, prepare a new tube to dilute the samples.
- Dilute the cells recovered from the joint or pleural cavity two or three times in Turk solution (i.e., add 10 or 20 μL of Turk solution, respectively, and 10 μL of your sample). Turk solution lyses the erythrocytes and stains the nuclei of leukocytes, making them easier to be identified under optical microscopy.
- Add 10 μL of the diluted sample in the Neubauer chamber previously covered by an appropriated glass cover.
- Count the total cells present in the four bigger counting grid.
- Calculate the number of total cells using the formula:

$$\text{Total number} = \frac{\text{number of cells} \times 10^4 (\text{correction}) \times \text{dilution}}{\text{number of the big squares counted}}$$

- The remaining sample in the tube (90 μL) is used for differential cell counting using a cytospin centrifuge.
- For a cytospin, place the microscope slides and filters into appropriate slots of the cytospin with the cardboard filters facing the center of the cytospin. Quickly aliquot the sample into the appropriate wells of the cytospin.
- Carefully place the lid of the cytospin over the samples and spin at 400 rpm for 5 min. Remove the filters from the slides without touching the cells (a clear circle filled with cells is evident).
- Allow the microscope slides to dry before the staining.

- Use the panoptic stain method or Giemsa's solution to stain the cells on the microscope slides.
- Allow the microscope slides to dry before the differential counting.
- Use an optical microscope for the differential counting. The cells are counting under oil immersion objective.
- Choose an area where the morphology of the cells is clearly visible. Differentiate the cells based on their morphology.
- Count the cells in a zigzag movement of the slides. Repeat this process in another field of the slide.
- At least, 100 cells (ideally 300) should be counted each time for the calculation of the percentage of cells in your sample.
- The percentage of each type of cell can be converted into number of cells using the total cell counting. For example:

$$\text{Number of the desired cell} = \frac{\text{number of total cells} \times \%\ \text{of desired cell}}{100}$$

2.7.2 *Identification Based on Immunofluorescence: Confocal Microscopy*

2.7.2.1 Material

- Microscope slides
- Cover glass
- Formaldehyde 4% (PBS solution)
- Fc block (CD36)
- Cell permeabilization (Perm/Wash Buffer; BD)
- PBS
- DAPI
- Mounting media

2.7.2.2 Protocol

- Clean the microscope slides and the cover glasses with absolute alcohol and water before use.
- The first step is similar to that described for cytospin. Place the microscope slide (only for support) and the filter into the appropriate slots of the cytospin with the cardboard filters facing the center of the cytospin.
- Attach the cover glass on the exact place where the cells cross the filter's circle. The cells must be fixed on the cover glass. Use an adhesive tape to fasten the cover glass to the filter or to the microscope slide.

- Extract the cells from joint or pleural cavity using only PBS! Do not use BSA to avoid autofluorescence. Count the cells in the Neubauer chamber (see above) and adjust the cell concentration to 3×10^5 cells/mL.
- Proceed to the cytospin centrifugation as described previously.
- Carefully remove the cover glass from the cytospin and place it on a flat surface.
- Fix the cells with formaldehyde 4% for 30 min. Use a Pasteur pipette to cover the circle filled with cells formed during the cytospin centrifugation.
- Wash the cells 3 × with PBS for 5 min each.
- Block the nonspecific binding of the antibodies using 50 μL of Fc block (diluted in PBS) for 30 min in a dark box at room temperature.
- Wash with PBS for 5 min.
- Incubate the cells with the specific antibody (diluted in PBS) for 3 h. Keep the cells in a dark box at room temperature. Use isotypes as negative controls.
- Remove the antibody and wash 4 × with PBS for 5 min.
- Incubate the cells with Perm/Wash Buffer 1 × for 30 min.
- Incubate overnight with 50 μL primary antibody diluted in Perm/Wash Buffer at 4 °C. The cells need stay all time at dark.
- Remove the antibody and wash 4 × with PBS for 5 min.
- Incubate with the 50 μL of secondary antibody diluted in Perm/Wash Buffer for 30 min at dark.
- Wash 5 × with PBS for 5 min.
- Incubate the cells with 100 μL of DAPI (1 mg/mL) diluted in PBS for 10 min at dark.
- Wash 4 × with PBS for 5 min.
- Use a good mounting media in a glass and place the cover glass with the cells side down. Avoid bubbles.
- Wait 30 min to acquire the images or store at 4 °C in a dark box.
- Confocal microscopy: estimate a number of images to be done in each glass and count the cells in the fields using an appropriate program. The result can be expressed as the number of cells in x fields.

3. LIMITATIONS

The use of joint and pleural cavities is a very practical and feasible strategy to study the biology of chemokines *in vivo*. However, several concerns must be taken into account during the choice of the model, including the chemokine under investigation, the time course to evaluate the effect of that chemokine, and the desirable technique for cell identification. Considering the size and the background of resident cells in both cavities, the recovery of cells from the joint is much lower compared to the pleural cavity, especially at baseline (naive or vehicle-injected). In contrast, the low number of cells prevents a more detailed analysis of the cell types and their state of activation. On the other hand, the presence of very few cells at the baseline of the joint cavity facilitates the investigation of weaker chemoattractants, since a smaller increase of the recruitment can be significant.

Another relevant point is that the cellular recruitment induced by the injection of a chemokine may be indirect, mediated by the release of endogenously produced chemoattractants. Indeed, nonhematopoietic cells express chemokine receptors and become activated in the presence of the ligands. Thus, endothelial, epithelial, or mesothelial cells can be a source of different molecules that attract leukocytes (Mancardi et al., 2003). Further, migrated leukocytes can also be the source of different mediators to promote their own recruitment or the recruitment of other cellular types (Griffith, Sokol, & Luster, 2014; Sadik & Luster, 2012). In an experimental model of arthritis, the infiltrated neutrophils into the joint can produce the chemotactic factor LTB4 and the cytokine IL-1β, which is a well-known inducer of other chemoattractants (Sadik, Kim, Iwakura, & Luster, 2012). Thus, the use of specific inhibitors, antagonists, or genetically modified mice that exclude the participation of other mediators could be an alternative to validate the functioning of the chosen chemokine. Moreover, the expression of chemokine receptor on the cellular surface is dynamic, controlled by the presence of the ligands and by different inflammatory mediators (Bachelerie et al., 2014). Thus, during the time course of the experiment, especially at the later time point, the chemokine receptor of interest can be desensitized and others can be positively regulated, bringing complexity into evaluating the results.

Finally, posttranslational modification of chemokines is another level of complexity when trying to understand the effects of chemokines *in vivo*. Recruited leukocytes and the inflamed tissue are a rich source of different enzymes that alter the structure of chemokines, changing their interaction with receptors and glycosaminoglycans and modifying their actions. Proteolysis, glycosylation, citrullination, and nitration are examples of how chemokines can be modified (Mortier et al., 2012). For instance, CD26 (also known as dipeptidylpeptidase 4) is a protease that can remove the first two amino acids from a protein that possesses a proline or alanine in the penultimate N-terminal position (Bongers, Lambros, Ahmad, & Heimer, 1992). The inhibition of CD26 by the treatment with sitagliptin preserves the biologic activity of the chemokine CXCL10. Functionally, mice deficient for CD26 or wild-type mice treated with sitagliptin enhance the recruitment of lymphocytes into tumors in mice, reducing lung metastases (Barreira da Silva et al., 2015). Similarly, mice treated orally with sitagliptin showed an increase of lymphocytes recruitment to the joint cavity when CXCL10 is given intra-articularly (unpublished data).

4. OTHER PERSPECTIVES

Here, we demonstrated some options to study the biology of CXCR2-binding and CCL2 chemokines in *in vivo* models. However, the joint and pleural cavities detailed here can be used beyond the chemokine biology field. For instance, the injection of LPS in the joint or pleural cavities induces an intense local inflammation that can be accurately measured using the techniques described herewith. Mechanistically, there is an option to investigate the steps of leukocyte recruitment along the endothelial cells, including the rolling and adhesion of leukocytes to the microvascular bed of the knee (Coelho et al., 2008). Moreover, the injection of different compounds into the joint can be alternatively used to investigate organ dysfunction, as edema and hypernociception (pain) (Sachs et al., 2011).

REFERENCES

Antony, V. B. (2003). Immunological mechanisms in pleural disease. *The European Respiratory Journal*, *21*(3), 539–544.

Antony, V. B., Hott, J. W., Kunkel, S. L., Godbey, S. W., Burdick, M. D., & Strieter, R. M. (1995). Pleural mesothelial cell expression of C-C (monocyte chemotactic peptide) and C-X-C (interleukin 8) chemokines. *American Journal of Respiratory Cell and Molecular Biology*, *12*(6), 581–588.

Bachelerie, F., Ben-Baruch, A., Burkhardt, A. M., Combadiere, C., Farber, J. M., Graham, G. J., et al. (2014). International Union of Basic and Clinical Pharmacology. [corrected]. LXXXIX. Update on the extended family of chemokine receptors and introducing a new nomenclature for atypical chemokine receptors. *Pharmacological Reviews*, *66*(1), 1–79.

Barreira da Silva, R., Laird, M. E., Yatim, N., Fiette, L., Ingersoll, M. A., & Albert, M. L. (2015). Dipeptidylpeptidase 4 inhibition enhances lymphocyte trafficking, improving both naturally occurring tumor immunity and immunotherapy. *Nature Immunology*, *16*(8), 850–858.

Bongers, J., Lambros, T., Ahmad, M., & Heimer, E. P. (1992). Kinetics of dipeptidyl peptidase IV proteolysis of growth hormone-releasing factor and analogs. *Biochimica et Biophysica Acta*, *1122*(2), 147–153.

Brackertz, D., Mitchell, G. F., & Mackay, I. R. (1977). Antigen-induced arthritis in mice. I. Induction of arthritis in various strains of mice. *Arthritis and Rheumatism*, *20*(3), 841–850.

Cailhier, J. F., Sawatzky, D. A., Kipari, T., Houlberg, K., Walbaum, D., Watson, S., et al. (2006). Resident pleural macrophages are key orchestrators of neutrophil recruitment in pleural inflammation. *American Journal of Respiratory and Critical Care Medicine*, *173*(5), 540–547.

Carmo, A. A., Costa, B. R., Vago, J. P., de Oliveira, L. C., Tavares, L. P., Nogueira, C. R., et al. (2014). Plasmin induces in vivo monocyte recruitment through protease-activated receptor-1-, MEK/ERK-, and CCR2-mediated signaling. *Journal of Immunology*, *193*(7), 3654–3663.

Citro, A., Valle, A., Cantarelli, E., Mercalli, A., Pellegrini, S., Liberati, D., et al. (2015). CXCR1/2 inhibition blocks and reverses type 1 diabetes in mice. *Diabetes*, *64*(4), 1329–1340.

Coelho, F. M., Pinho, V., Amaral, F. A., Sachs, D., Costa, V. V., Rodrigues, D. H., et al. (2008). The chemokine receptors CXCR1/CXCR2 modulate antigen-induced arthritis by regulating adhesion of neutrophils to the synovial microvasculature. *Arthritis and Rheumatism*, *58*(8), 2329–2337.

Cooke, T. D., Hurd, E. R., Ziff, M., & Jasin, H. E. (1972). The pathogenesis of chronic inflammation in experimental antigen-induced arthritis. II. Preferential localization of antigen-antibody complexes to collagenous tissues. *The Journal of Experimental Medicine*, *135*(2), 323–338.

Costa, V. V., Amaral, F. A., Coelho, F. M., Queiroz-Junior, C. M., Malagoli, B. G., Gomes, J. H., et al. (2015). Lithothamnion muelleri treatment ameliorates inflammatory and hypernociceptive responses in antigen-induced arthritis in mice. *PLoS One*, *10*(3), e0118356.

Dwyer, M. P., & Yu, Y. (2014). CXCR2 modulators: A patent review (2009–2013). *Expert Opinion on Therapeutic Patents*, *24*(5), 519–534.

Grespan, R., Fukada, S. Y., Lemos, H. P., Vieira, S. M., Napimoga, M. H., Teixeira, M. M., et al. (2008). CXCR2-specific chemokines mediate leukotriene B4-dependent recruitment of neutrophils to inflamed joints in mice with antigen-induced arthritis. *Arthritis and Rheumatism*, *58*(7), 2030–2040.

Griffith, J. W., Sokol, C. L., & Luster, A. D. (2014). Chemokines and chemokine receptors: Positioning cells for host defense and immunity. *Annual Review of Immunology*, *32*, 659–702.

Ha, H., Debnath, B., Odde, S., Bensman, T., Ho, H., Beringer, P. M., et al. (2015). Discovery of novel CXCR2 inhibitors using ligand-based pharmacophore models. *Journal of Chemical Information and Modeling*, *55*(8), 1720–1738.

Handel, T. M., Johnson, Z., Crown, S. E., Lau, E. K., & Proudfoot, A. E. (2005). Regulation of protein function by glycosaminoglycans—As exemplified by chemokines. *Annual Review of Biochemistry*, *74*, 385–410.

Holmes, W. E., Lee, J., Kuang, W. J., Rice, G. C., & Wood, W. I. (1991). Structure and functional expression of a human interleukin-8 receptor. *Science*, *253*(5025), 1278–1280.

Honma, S., Abe, K., & Ito, T. (1982). Pleural free cells in the mouse: Quantitative and qualitative cell morphology. *Archivum Histologicum Japonicum*, *45*(5), 483–494.

Johnson, Z., Proudfoot, A. E., & Handel, T. M. (2005). Interaction of chemokines and glycosaminoglycans: A new twist in the regulation of chemokine function with opportunities for therapeutic intervention. *Cytokine and Growth Factor Reviews*, *16*(6), 625–636.

Khan, I. M., Redman, S. N., Williams, R., Dowthwaite, G. P., Oldfield, S. F., & Archer, C. W. (2007). The development of synovial joints. *Current Topics in Developmental Biology*, *79*, 1–36.

Klein, A., Pinho, V., Alessandrini, A. L., Shimizu, T., Ishii, S., & Teixeira, M. M. (2002). Platelet-activating factor drives eotaxin production in an allergic pleurisy in mice. *British Journal of Pharmacology*, *135*(5), 1213–1218.

Kroegel, C., & Antony, V. B. (1997). Immunobiology of pleural inflammation: Potential implications for pathogenesis, diagnosis and therapy. *The European Respiratory Journal*, *10*(10), 2411–2418.

Loghmani, F., Mohammed, K. A., Nasreen, N., Van Horn, R. D., Hardwick, J. A., Sanders, K. L., et al. (2002). Inflammatory cytokines mediate C-C (monocyte chemotactic protein 1) and C-X-C (interleukin 8) chemokine expression in human pleural fibroblasts. *Inflammation*, *26*(2), 73–82.

Mancardi, S., Vecile, E., Dusetti, N., Calvo, E., Stanta, G., Burrone, O. R., et al. (2003). Evidence of CXC, CC and C chemokine production by lymphatic endothelial cells. *Immunology*, *108*(4), 523–530.

Mohammed, K. A., Nasreen, N., Ward, M. J., Mubarak, K. K., Rodriguez-Panadero, F., & Antony, V. B. (1998). Mycobacterium-mediated chemokine expression in pleural mesothelial cells: Role of C-C chemokines in tuberculous pleurisy. *The Journal of Infectious Diseases*, *178*(5), 1450–1456.

Moriconi, A., Cesta, M. C., Cervellera, M. N., Aramini, A., Coniglio, S., Colagioia, S., et al. (2007). Design of noncompetitive interleukin-8 inhibitors acting on CXCR1 and CXCR2. *Journal of Medicinal Chemistry*, *50*(17), 3984–4002.

Mortier, A., Van Damme, J., & Proost, P. (2012). Overview of the mechanisms regulating chemokine activity and availability. *Immunology Letters*, *145*(1-2), 2–9.

Murphy, P. M., & Tiffany, H. L. (1991). Cloning of complementary DNA encoding a functional human interleukin-8 receptor. *Science*, *253*(5025), 1280–1283.

Proudfoot, A. E., Handel, T. M., Johnson, Z., Lau, E. K., LiWang, P., Clark-Lewis, I., et al. (2003). Glycosaminoglycan binding and oligomerization are essential for the in vivo activity of certain chemokines. *Proceedings of the National Academy of Sciences of the United States of America*, *100*(4), 1885–1890.

Rennard, S. I., Dale, D. C., Donohue, J. F., Kanniess, F., Magnussen, H., Sutherland, E. R., et al. (2015). CXCR2 antagonist MK-7123. A phase 2 proof-of-concept trial for chronic obstructive pulmonary disease. *American Journal of Respiratory and Critical Care Medicine*, *191*(9), 1001–1011.

Russo, R. C., Garcia, C. C., Teixeira, M. M., & Amaral, F. A. (2014). The CXCL8/IL-8 chemokine family and its receptors in inflammatory diseases. *Expert Review of Clinical Immunology*, *10*(5), 593–619.

Russo, R. C., Guabiraba, R., Garcia, C. C., Barcelos, L. S., Roffe, E., Souza, A. L., et al. (2009). Role of the chemokine receptor CXCR2 in bleomycin-induced pulmonary inflammation and fibrosis. *American Journal of Respiratory Cell and Molecular Biology*, *40*(4), 410–421.

Sachs, D., Coelho, F. M., Costa, V. V., Lopes, F., Pinho, V., Amaral, F. A., et al. (2011). Cooperative role of tumour necrosis factor-alpha, interleukin-1beta and neutrophils in a novel behavioural model that concomitantly demonstrates articular inflammation and hypernociception in mice. *British Journal of Pharmacology*, *162*(1), 72–83.

Sadik, C. D., Kim, N. D., Iwakura, Y., & Luster, A. D. (2012). Neutrophils orchestrate their own recruitment in murine arthritis through C5aR and FcgammaR signaling. *Proceedings of the National Academy of Sciences of the United States of America*, *109*(46), E3177–E3185.

Sadik, C. D., & Luster, A. D. (2012). Lipid-cytokine-chemokine cascades orchestrate leukocyte recruitment in inflammation. *Journal of Leukocyte Biology*, *91*(2), 207–215.

Sokol, C. L., & Luster, A. D. (2015). The chemokine system in innate immunity. *Cold Spring Harbor Perspectives in Biology*, 7(5), a016303.

Talbot, J., Bianchini, F. J., Nascimento, D. C., Oliveira, R. D., Souto, F. O., Pinto, L. G., et al. (2015). CCR2 expression in neutrophils plays a critical role in their migration into the joints in rheumatoid arthritis. *Arthritis & Rheumatology (Hoboken, N.J.)*, *67*(7), 1751–1759.

Teixeira, M. M., Talvani, A., Tafuri, W. L., Lukacs, N. W., & Hellewell, P. G. (2001). Eosinophil recruitment into sites of delayed-type hypersensitivity reactions in mice. *Journal of Leukocyte Biology*, *69*(3), 353–360.

Wang, L., Fuster, M., Sriramarao, P., & Esko, J. D. (2005). Endothelial heparan sulfate deficiency impairs L-selectin- and chemokine-mediated neutrophil trafficking during inflammatory responses. *Nature Immunology*, *6*(9), 902–910.

Monitoring Chemokine Receptor Trafficking by Confocal Immunofluorescence Microscopy

Adriano Marchese[1]

Department of Pharmacology, Loyola University Chicago, Health Sciences Division, Maywood, Illinois, USA
[1]Corresponding author: e-mail address: amarchese@luc.edu

Contents

1. Introduction 282
2. Methods 284
   2.1 Required Materials 284
   2.2 Cell Culture and Transfection 285
   2.3 Passaging Cells onto Cover Slips and Stimulation 286
   2.4 Cell Preparation 288
   2.5 Microscope Image Acquisition 289
3. Concluding Remarks 290
References 291

Abstract

Here, we describe a protocol to detect chemokine receptor CXCR4 by confocal immunofluorescence microscopy in HeLa cells treated with its chemokine ligand CXCL12. Typically, ligand-activated chemokine receptors undergo a multistep process of desensitization and/or internalization from the plasma membrane in order to terminate signaling. Once internalized to endosomes, chemokine receptors readily enter the recycling pathway and return to the cell surface, giving rise to resensitization of signaling. The chemokine receptor CXCR4, when activated by CXCL12 is also internalized to endosomes, but in contrast to many chemokine receptors it is mainly sorted to the degradative pathway, contributing to a loss in the cellular complement of CXCR4 and long-term downregulation of signaling. The trafficking of CXCR4 from early endosomes to lysosomes can be easily detected by confocal immunofluorescence microscopy by immunostaining fixed cells for the receptor and with markers of these vesicular compartments. This approach is advantageous because it can be used to identify factors that regulate the trafficking of CXCR4 from early endosomes to lysosomes. The protocol described here focuses on CXCR4, but it can be easily adapted to other chemokine receptors.

Methods in Enzymology, Volume 570
ISSN 0076-6879
http://dx.doi.org/10.1016/bs.mie.2015.10.004

1. INTRODUCTION

The chemokine ligand CXCL12 and its chemokine receptor CXCR4 have important biological functions (Balkwill, 2004; Domanska et al., 2013; Karpova & Bonig, 2015; Zlotnik & Yoshie, 2012). In addition, CXCR4 has been linked to several diseases, including cancer (Balkwill, 2012). CXCR4 expression is upregulated in at least 20 solid tumors and also in some hematological cancers (Balkwill, 2004; Domanska et al., 2013). CXCR4 expression correlates with poor prognosis, mainly due to its role in metastasis (Muller et al., 2001; Oskarsson, Batlle, & Massague, 2014). In cancer cells, CXCR4 expression is regulated by multiple mechanisms, including at the level of transcription (Vanharanta et al., 2013), translation (Li et al., 2004), and protein (Li et al., 2004). The protocol described here will focus on examining CXCR4 expression at the level of the protein.

CXCR4 expression is directly regulated by several posttranslational modifications, including phosphorylation and ubiquitination (Marchese, 2014). These posttranslational modifications act in concert to target CXCR4 for lysosomal degradation upon activation by CXCL12 (Marchese, 2014). As with most chemokine receptors, CXCR4 is rapidly internalized from the cell surface following agonist binding (Orsini, Parent, Mundell, Marchese, & Benovic, 1999). CXCL12 binding to CXCR4 also promotes rapid phosphorylation of CXCR4 at specific residues within the C-tail, leading to an interaction with the E3 ubiquitin ligase AIP4 (Bhandari, Robia, & Marchese, 2009). AIP4 mediates ubiquitination of the receptor on nearby lysine residues located within the C-tail (Marchese & Benovic, 2001). The ubiquitin moieties serve as a signal for targeting CXCR4 into the degradative pathway (Marchese et al., 2003). This is consistent with fact that upon internalization from the plasma membrane ubiquitinated CXCR4 traffics to early endosomes and then to lysosomes, the terminal degradative compartment (Slagsvold, Marchese, Brech, & Stenmark, 2006). A small fraction of CXCR4 also enters the recycling pathway (Malik & Marchese, 2010), which may be because not all internalized receptors are ubiquitinated or ubiquitin moieties are removed by deubiquitinases thereby facilitating receptor recycling (Marchese, 2014).

Ubiquitinated CXCR4 is targeted to the endosomal sorting complex required for transport (ESCRT) pathway (Marchese et al., 2003). This pathway is comprised of five major protein complexes that act at the limiting membrane of endosomes to selectively target ubiquitinated transmembrane proteins into the lumen of endosomes and multivesicular bodies, which subsequently fuse with lysosomes where the luminal contents are degraded (Henne, Buchkovich, & Emr, 2011). Several factors have been identified that regulate the ability of ESCRTs to target ubiquitinated CXCR4 into lysosomes (Bhandari, Trejo, Benovic, & Marchese, 2007; Holleman & Marchese, 2014; Malik & Marchese, 2010). One of these factors is β-arrestin1, which appears to interact with specific ESCRT proteins on early endosomes to regulate their sorting activity (Malik & Marchese, 2010). In HeLa cells depleted of β-arrestin1 by siRNA, CXCR4 internalizes normally upon ligand binding, but fails to traffic from early endosomes to lysosomes and is therefore not degraded (Bhandari et al., 2007). This defect in trafficking can be easily detected by fixed cell confocal immunofluorescence microscopy. In cells treated with β-arrestin1 siRNA, CXCR4 accumulates on EEA1-positive early endosomes, while in control siRNA treated cells, CXCR4 accumulates on LAMP2-positive late endosomes/lysosomes (Bhandari et al., 2007), suggesting that β-arrestin1 regulates endosome to lysosome trafficking of CXCR4. Although these experiments were performed in HEK293 cells, expressing YFP-tagged CXCR4, similar experiments were performed in HeLa cells that express CXCR4 endogenously (Holleman & Marchese, 2014). Using this approach, it was successfully determined that the E3 ubiquitin ligase DTX3L, also regulates CXCR4 trafficking from early endosomes to late endosomes/lysosomes (Holleman & Marchese, 2014). In the protocol here, we use HeLa cells to examine CXCR4 trafficking from early endosomes to lysosomes in β-arrestin1 depleted cells by fixed cell confocal immunofluorescence microscopy.

Fixed cell immunofluorescence microscopy is advantageous to study receptor trafficking because it provides spatial and temporal resolution of receptor localization to a level that cannot be achieved by other techniques, such as Western blotting. For example, Western blotting clearly shows that in β-arrestin1 depleted cells, agonist promoted degradation of CXCR4 is inhibited as compared to control, confirming that β-arrestin1 is necessary for targeting CXCR4 to the degradative pathway (Bhandari et al., 2007). However, a limitation of the Western blotting approach is that it does not specify where in the trafficking pathway a particular factor actually works. This is an important point because β-arrestins generally mediate internalization of GPCRs from the plasma membrane, and inhibiting internalization will also inhibit receptor degradation, as internalization is a prerequisite for degradation (Marchese & Benovic, 2001). In the case of β-arrestin1, images of cells taken by fixed cell confocal immunofluorescence microscopy showed that β-arrestin1 depletion did not impact agonist promoted internalization of CXCR4, as most of the receptor was found to be internalized to endosomes in cells treated with CXCL12, similar to control (Bhandari et al., 2007). It is important to note that although fixed cell immunofluorescence microscopy was essential to define a role for β-arrestin1 in CXCR4 trafficking from endosomes to lysosomes, follow-up studies were required to understand the mechanism by which this occurs, which culminated in defining a novel noncanonical role for β-arrestin1 in receptor trafficking (Malik & Marchese, 2010).

The protocol here examines intracellular trafficking of CXCR4 in HeLa cells transfected with control siRNA or siRNA targeting β-arrestin1 by fixed cell confocal immunofluoresence microscopy. HeLa cells are an adherent, epithelial cell line derived from cervical cancer and expresses high levels of CXCR4 that can be detected by several techniques using commercially available antibodies (Marchese et al., 2003). The protocol described here can be applied to any adherent cell type that expresses CXCR4, including other cancer cells that express CXCR4 to high levels (Fischer, Nagel, Jacobs, Stumm, & Schulz, 2008).

2. METHODS

2.1 Required Materials

HeLa cells; can be obtained from American Type Culture Collection (ATCC)
DMEM (Hyclone #SH3002201)
FBS (Fisher Scientific #03600511)
0.05% Trypsin–EDTA 1 × (Life Technologies #25300054)
Phosphate-buffered saline (PBS) 1 × (Hyclone #SH30256.01)
Mounting medium with DAPI (Vector Laboratories #H1200; Life Technologies Prolong Gold #P36931)
Microcentrifuge tubes (Dot Scientific #010)
Coverslips, No. 1.5 (Dot Scientific #MC22-15)
Poly-L-lysine (Sigma #P1399)
Formaldehyde (Sigma #X100)
Triton X-100 (Sigma #F1635)
Slides (Fisher Scientific #125442)
10 cm dishes (Falcon #353003)
6-well dishes (Falcon #353046)

Primary antibodies against CXCR4 (BD Transduction Laboratories #BDB551852); LAMP2 (Developmental Studies Hybridoma Bank #H4B4), EEA1 (BD Transduction Laboratories #BDB610456), β-arrestin1 (kindly provided by Jeffrey L. Benovic, Thomas Jefferson University, Philadelphia, PA)
Secondary antibodies (Molecular Probes); goat anti-rat IgG (H + L), $F(ab')_2$ Fragment (Alexa Fluor 649) #A21094; goat anti-mouse IgG (H + L), $F(ab')_2$ Fragment (Alexa Fluor 488 conjugate) #A11029
siRNA (Dharmacon) against Luciferase (P-002099), β-arrestin1 (siGenome SMARTpool #M-011971)
Lipofectamine 3000 Transfection Reagent (Life Technologies #L3000008)
Opti-MEM (Life Technologies #31985062)
Hemacytometer (Fisher Scientific #0267154)
Tissue paper (Fisher Scientific #06666A)
Tweezers
Lens paper
Whatman filter paper
Parafilm
Confocal microscope equipped with 60 × objective

2.2 Cell Culture and Transfection

2.2.1 Passaging and Maintaining HeLa Cells

1. HeLa cells are grown on 10 cm culture dishes at 37 °C, 5% CO_2.
2. Passage cells at 85–95% confluence; approximately every 2 days.
3. Aspirate medium and wash cells 1 × with 10 mL PBS.
4. Add 2 mL trypsin and incubate at 37 °C, 5% CO_2 for 5 min. Cells should be detached from the surface of the culture dish. If many cells are still attached, incubate for an additional minute at 37 °C, 5% CO_2.
5. Add 4 mL DMEM supplemented with 10% FBS.
6. Pipette the entire contents up and down at least 3 ×.
7. Ensure cells are dispersed into a single cell suspension by observing cells under the microscope using a 4 × or 10 × objective.
8. Add 2 mL of equilibrated cell suspension to a 10-cm culture dish containing 10 mL DMEM supplemented with 10% FBS. This is referred to as a 1:3 split.
9. Incubate cells at 37 °C, 5% CO_2 until cells reach 90–95% confluency; this typically takes 48 h.

2.2.2 Cell Counting and Plating

1. Aspirate medium and wash cells 1 × with 10 mL PBS.
2. Add 2 mL trypsin and incubate at 37 °C, 5% CO_2 for 5 min.
3. Add 4 mL DMEM containing 10% FBS.
4. Pipette the entire contents up and down at least 3 ×.
5. Transfer cells to a 50-mL conical tube.
6. Count cells using a Hemacytometer.
7. Seed approximately 1.5×10^5 cells per well of a 6-well culture dish in 2 mL DMEM containing 10% FBS.
8. Incubate cells at 37 °C, 5% CO_2 overnight or approximately for 15–18 h.
9. Cells should be approximately 70–80% confluent and ready for transfection.

2.2.3 Transfection with siRNA

1. Prepare the siRNA/Opti-MEM mixture:
   a. 5 μL of each siRNA (20 μ*M* stock) to a microcentrifuge tube.
   b. 50 μL Opti-MEM.
   c. Incubate at room temperature for 5 min.
2. Prepare transfection reagent/Opti-MEM mixture:
   a. 50 μL Opti-MEM
   b. 6 μL Lipofectamine 3000 transfection reagent
   c. Incubate at room temperature for 5 min
3. Add the transfection reagent/Opti-MEM mixture dropwise to the tube containing the siRNA/Opti-MEM mixture. A single well should be treated with a single siRNA.
4. Tap tube gently to mix.
5. Incubate at room temperature for 20 min.
6. In the mean time, replace media in 6-well culture dish with 2 mL DMEM supplemented with 10% FBS.
7. Add siRNA/transfection reagent mixture to cells dropwise.
8. Incubate cells overnight at 37 °C, 5% CO_2.

2.3 Passaging Cells onto Cover Slips and Stimulation

2.3.1 Coverslip Preparation

1. Dip No. 1.5 coverslips in 100% ethanol and dry by blotting on a clean tissue.
2. Place a single coverslip into a well of a 6-well plate. We typically use 22 × 22 mm coverslips. Other sizes are available.

3. Coat coverslip with 0.1 mg/mL poly-L-lysine (PLL). Pipette approximately 500 μL directly onto the glass surface.
4. Incubate for approximately 1–5 min.
5. Aspirate PLL; save and reuse up to 3 ×. Store at −20 °C.
6. Let coverslips dry in flow hood with lids ajar for approximately 30 min.

2.3.2 Passaging Cells onto Coverslips

1. Aspirate media from each well.
2. Gently wash cells 1 × with 2 mL PBS.
3. Aspirate PBS and add 500 μL trypsin, and incubate at 37 °C, 5% CO_2 for 5 min.
4. Add 2 mL DMEM supplemented with 10% FBS.
5. Pipette entire contents up and down at least 3 × to mix and to disperse cells into a single cell suspension.
6. Pass 1 mL cells onto well with PLL coated coverslip. Each transfection condition should have two parallel wells: one for vehicle and one for agonist treatment. In addition, a third well is recommended to generate a cell lysate to confirm silencing of β-arrestin1 by Western blotting. Ideally, it would be advantageous to costain cells for β-arrestin1, but in our hands we have not found antibodies that can reliably detect endogenous β-arrestin1 by fixed cell immunofluorescence.
7. Incubate overnight at 37 °C, 5% CO_2.

2.3.3 Stimulating Cells

1. The next day, cells should be approximately 80–90% confluent.
2. Aspirate medium from each well and wash 1 × with 2 mL PBS.
3. Add 1 mL complete DMEM in parallel wells containing vehicle or 10 n*M* stromal cell-derived factor (SDF).
   a. Optional: add leupeptin (20 m*M*) and pepstatin A (2 μ*M*) to each well to inhibit lysosomal proteases at least 1 h before and during the stimulation. This is advantageous because it will enhance receptor stability and thereby facilitate detection by immunostaining.
4. Incubate at 37 °C, 5% CO_2 for 3 h. We have observed that at this time point CXCR4 is mainly localized to late endosome/lysosomes. At shorter time points, such as 30 min to 1 h of agonist treatment, CXCR4 is mainly localized to early endosomes in HeLa cells. Time-course studies should be performed for different cell types to determine the time it takes for CXCR4 to accumulate on early endosomes or lysosomes. It is

also important to access the extent to which CXCR4 enters the recycling pathway at each time point.

2.4 Cell Preparation

2.4.1 Fixation and Permeabilization

1. Place plate on ice and wash 2 × with 2 mL ice-cold PBS.
2. To fix cells, add 1 mL 3.7% formaldehyde in PBS, 10 min at room temperature.
3. Rinse cells 3 × with 1 mL PBS at room temperature
4. To permeabilize cells, add 0.1% Triton X-100 in PBS, 10 min at room temperature.

2.4.2 Blocking and Antibody Incubation

1. Incubate cells with 1 mL 5% normal goat serum in 0.1% Triton X-100/PBS, 60 min at 37 °C.
2. Dilute primary antibody in 1% BSA in 0.1% Triton X-100/PBS. Rat anti-CXCR4 (1:100) in combination with mouse anti-EEA1 (1:1000) or anti-LAMP1 (1:1000). EEA1 is used as a marker for early endosomes and LAMP2 is used as a marker for late endosomes/lysosomes. Primary antibody titrations should be performed to identify optimal antibody dilutions to use for staining.
3. Place filter paper inside a sealable container. Moisten filter paper with water in order to maintain a humidified atmosphere during the incubation period.
4. Place a piece of parafilm on top of the moistened filter paper.
5. Add 40 μL antibody solution to parafilm, appropriately spaced to accommodate the desired amount of coverslips.
6. Carefully remove coverslip from the 6-well plate with forceps. We use an 18-gauge needle with a bent tip to help lift the coverslip from the well.
7. Remove excess buffer by touching the edge of the coverslip on a clean tissue.
8. Carefully invert the coverslip and place on drop of antibody solution. Performing the antibody incubations in this manner limits the amount of reagents, especially antibodies, used in an experiment.
9. Seal the container with a lid or parafilm and incubate at 37 °C, 30 min.

2.4.3 Washing

1. Transfer the coverslips back to the 6-well plate. Ensure that the cell side is oriented upward.

2. Rinse cells 3× with 1 mL 0.1% Triton X-100/PBS at room temperature.
3. Dilute secondary antibody in 1% BSA in 0.1% Triton X-100/PBS. Secondary antibody titration experiments should be performed to identify optimal secondary antibody dilutions to use for immunostaining.
4. Place coverslip with cell side onto 50 μL antibody solution.
5. Incubate at 37 °C, 30 min.
6. Wash 3× at room temperature with 1 mL 0.1% Triton X-100/PBS.
7. Leave last wash on as coverslips are being mounted. It is important to ensure that coverslips do not dry during the mounting step.

2.4.4 Mounting

1. Apply four small dots of nail polish to microscope slide and allow to dry. This is meant to preserve the thickness of the sample during confocal sectioning.
2. Place 60 μL mounting medium onto coverslip in the area delimited by the dots.
3. Carefully remove coverslip from the 6-well plate with forceps.
4. Remove excess buffer by touching the edge of the coverslip on a clean tissue.
5. Carefully invert the coverslip and place on drop of mounting medium.
6. If mounting medium seeps from the edges of the cover slip, carefully wipe with tissue.
7. Seal the edges of the coverslip to the slide with nail polish.
8. Samples can be viewed once the nail polish has set.

2.5 Microscope Image Acquisition

2.5.1 Sample Analysis

1. Image fixed samples by confocal microscopy. We mostly use a LSM 510 laser scanning confocal microscope for imaging (Carl Zeiss, Thornwood, NY) and a Plan-Apo 60×/1.4 NA oil lens objective.
2. Acquire images of cells in each channel and of multiple frames of view. We typically image using a 1.4-megapixel cooled extended spectra range RGB digital camera set at 512×512 resolution (Carl Zeiss, Thornwood, NY).
3. Adjust illumination and exposure times accordingly and be careful to remain within the linear range. Keep settings constant during image acquisition of parallel samples within an experiment.
4. Take z-series images. We typically set z-series to 0.2 μ*M*.

5. Analyze images at full bit depth using ImageJ software [(http://imagej.nih.gov/ij/), National Institutes of Health, Bethesda, MD].
6. Carefully perform puncta analysis. Manually scale images using linear transformations to discriminate between true puncta and background noise.
7. Set minimum and maximum threshold values for particle analysis separately for each channel; count particles.
8. Calculate colocalization between proteins using the Pearson coefficient Colocolization Finder plugin in ImageJ.

3. CONCLUDING REMARKS

The protocol described here examines the cellular distribution of the chemokine receptor CXCR4 in the absence or presence of its ligand CXCL12 by fixed cell confocal immunofluorescence microscopy. This protocol can be easily adapted to examine trafficking of other chemokine receptors, depending on the availability of suitable antibodies and cells. As we have shown, when this approach is coupled with gene silencing and other biochemical approaches it can be used to unequivocally identify factors that regulate receptor sorting from endosomes to lysosomes.

Understanding the mechanisms governing endosome to lysosome trafficking is important because this trafficking step can control the amount of receptors on the cell surface and therefore regulate chemokine responsiveness. Although most chemokine receptors are biased toward entering the recycling pathway, CXCR4 in contrast is biased toward entering the degradative pathway (Marchese, 2014). However, a small fraction of CXCR4 can also enter the recycling pathway and return to the plasma membrane, likely contributing to resensitization of signaling. The implication of this is that cellular responsiveness to CXCR4 is likely due in part to biased receptor sorting into the degradative pathway versus the recycling pathway. Dysregulation of this bias could lead to aberrant signaling and have deleterious consequences. For example, in a subset of breast cancer cells, high CXCR4 expression may in part be due to defective sorting of the receptor into the degradative pathway, therefore favoring receptor recycling and enhanced signaling, contributing to tumor metastasis (Li et al., 2004). In this regard, understanding the mechanisms controlling receptor sorting into the degradative pathway is important and may lead to innovative strategies to target receptor sorting therapeutically.

REFERENCES

Balkwill, F. (2004). The significance of cancer cell expression of the chemokine receptor CXCR4. *Seminars in Cancer Biology, 14,* 171–179.

Balkwill, F. R. (2012). The chemokine system and cancer. *Journal of Pathology, 226,* 148–157.

Bhandari, D., Robia, S. L., & Marchese, A. (2009). The E3 ubiquitin ligase atrophin interacting protein 4 binds directly to the chemokine receptor CXCR4 via a novel WW domain-mediated interaction. *Molecular Biology of the Cell, 20,* 1324–1339.

Bhandari, D., Trejo, J., Benovic, J. L., & Marchese, A. (2007). Arrestin-2 interacts with the ubiquitin-protein isopeptide ligase atrophin-interacting protein 4 and mediates endosomal sorting of the chemokine receptor CXCR4. *Journal of Biological Chemistry, 282,* 36971–36979.

Domanska, U. M., Kruizinga, R. C., Nagengast, W. B., Timmer-Bosscha, H., Huls, G., de Vries, E. G., et al. (2013). A review on CXCR4/CXCL12 axis in oncology: No place to hide. *European Journal of Cancer, 49,* 219–230.

Fischer, T., Nagel, F., Jacobs, S., Stumm, R., & Schulz, S. (2008). Reassessment of CXCR4 chemokine receptor expression in human normal and neoplastic tissues using the novel rabbit monoclonal antibody UMB-2. *PLoS One, 3.*

Henne, W. M., Buchkovich, N. J., & Emr, S. D. (2011). The ESCRT pathway. *Developmental Cell, 21,* 77–91.

Holleman, J., & Marchese, A. (2014). The ubiquitin ligase deltex-3 l regulates endosomal sorting of the G protein-coupled receptor CXCR4. *Molecular Biology of the Cell, 25,* 1892–1904.

Karpova, D., & Bonig, H. (2015). Concise review: CXCR4/CXCL12 signaling in immature hematopoiesis—Lessons from pharmacological and genetic models. *Stem Cells, 33,* 2391–2399.

Li, Y. M., Pan, Y., Wei, Y., Cheng, X., Zhou, B. P., Tan, M., et al. (2004). Upregulation of CXCR4 is essential for HER2-mediated tumor metastasis. *Cancer Cell, 6,* 459–469.

Malik, R., & Marchese, A. (2010). Arrestin-2 interacts with the endosomal sorting complex required for transport machinery to modulate endosomal sorting of CXCR4. *Molecular Biology of the Cell, 21,* 2529–2541.

Marchese, A. (2014). Endocytic trafficking of chemokine receptors. *Current Opinion in Cell Biology, 27,* 72–77.

Marchese, A., & Benovic, J. L. (2001). Agonist-promoted ubiquitination of the G protein-coupled receptor CXCR4 mediates lysosomal sorting. *Journal of Biological Chemistry, 276,* 45509–45512.

Marchese, A., Raiborg, C., Santini, F., Keen, J. H., Stenmark, H., & Benovic, J. L. (2003). The E3 ubiquitin ligase AIP4 mediates ubiquitination and sorting of the G protein-coupled receptor CXCR4. *Developmental Cell, 5,* 709–722.

Muller, A., Homey, B., Soto, H., Ge, N., Catron, D., Buchanan, M. E., et al. (2001). Involvement of chemokine receptors in breast cancer metastasis. *Nature, 410,* 50–56.

Orsini, M. J., Parent, J. L., Mundell, S. J., Marchese, A., & Benovic, J. L. (1999). Trafficking of the HIV coreceptor CXCR4. Role of arrestins and identification of residues in the c-terminal tail that mediate receptor internalization. *Journal of Biological Chemistry, 274,* 31076–31086.

Oskarsson, T., Batlle, E., & Massague, J. (2014). Metastatic stem cells: Sources, niches, and vital pathways. *Cell Stem Cell, 14,* 306–321.

Slagsvold, T., Marchese, A., Brech, A., & Stenmark, H. (2006). CISK attenuates degradation of the chemokine receptor CXCR4 via the ubiquitin ligase AIP4. *EMBO Journal, 25,* 3738–3749.

Vanharanta, S., Shu, W., Brenet, F., Hakimi, A. A., Heguy, A., Viale, A., et al. (2013). Epigenetic expansion of VHL-HIF signal output drives multiorgan metastasis in renal cancer. *Nature Medicine, 19*, 50–56.

Zlotnik, A., & Yoshie, O. (2012). The chemokine superfamily revisited. *Immunity, 36*, 705–716.

CHAPTER FOURTEEN

Active Shaping of Chemokine Gradients by Atypical Chemokine Receptors: A 4D Live-Cell Imaging Migration Assay

Kathrin Werth[1], Reinhold Förster[1]
Institute of Immunology, Hannover Medical School, Hannover, Germany
[1]Corresponding authors: e-mail address: werth.kathrin@mh-hannover.de; foerster.reinhold@mh-hannover.de

Contents

1. Background 294
2. Introduction 294
3. Materials and Equipment 296
4. Preparation of Chambers 297
5. Preparation of Cells 299
6. Filling of Chambers 300
7. Time-Lapse Imaging 302
8. Analysis 302
   8.1 Quantitative Analysis of Migrated Cells Using ImageJ 303
   8.2 Semiautomated Cell Tracking of Migrating Cells Using Imaris 305
Acknowledgments 307
References 307

Abstract

Diffusion of chemokines away from their site of production results in the passive formation of chemokine gradients. We have recently shown that chemokine gradients can also be formed in an active manner, namely by atypical chemokine receptors (ACKRs) that scavenge chemokines locally. Here, we describe an advanced method that allows the visualization of leukocyte migration in a three-dimensional environment along a chemokine gradient that is actively established by cells expressing an ACKR. Initially developed to visualize the migration of dendritic cells along gradients of CCL19 or CCL21 that were actively shaped by an ACKR4-expressing cell line, we expect that this chamber system can be exploited to study many other combinations of atypical and conventional chemokine receptor-expressing cells.

Methods in Enzymology, Volume 570
ISSN 0076-6879
http://dx.doi.org/10.1016/bs.mie.2015.09.008

1. BACKGROUND

Directional cell migration relies on the functional interaction of chemokines and their cognate chemokine receptors. Binding of chemokines to these seven transmembrane-spanning receptors leads to heterotrimeric G protein-mediated migration of the cell (Rot & von Andrian, 2004). Besides, chemokines can also bind to other seven-transmembrane-spanning receptors that fail to couple to heterotrimeric G proteins due to a mutation in the DRYLAIV motif at the end of transmembrane domain 3 (Graham, Locati, Mantovani, Rot, & Thelen, 2012; Nibbs & Graham, 2013). Due to their incapability to mediate G protein-mediated signaling, these receptors are referred to as atypical chemokine receptors (ACKRs). Four ACKRs are currently known: ACKR1 (DARC), ACKR2 (D6, also known as CCBP2), ACKR3 (CXCR7, also known as RDC1), and ACKR4 (CCRL1, also known as CCX-CKR and CCR11) (Bachelerie, Ben-Baruch, et al., 2014; Bachelerie, Graham, et al., 2014). Whereas ACKR1 and ACKR2 are rather promiscuous and bind various inflammatory and noninflammatory CC- and, in case of ACKR1, also CXC-chemokines, ACKR3 and ACKR4 show a more specific binding pattern. ACKR3 binds CXCL11 and CXCL12, while ACKR4 binds CCL19, CCL21, CCL25, and in humans also CXCL13 (Ulvmar, Hub, & Rot, 2011). In total, a broad range of chemokines can be bound by ACKRs. After binding their ligands, ACKRs mediate internalization and transport or degradation of the chemokine, which substantially affects chemokine distribution in tissues and enables the shaping of a gradient in an active manner (Dona et al., 2013; Ulvmar et al., 2014).

2. INTRODUCTION

A broad range of systems has been developed to study leukocyte migration *in vitro*. Starting from comparably basic systems such as Boyden chambers (Boyden, 1962), highly sophisticated devices have been described over the years that allow the study of additional parameters like blood or interstitial flow and resulting shear forces (Bonvin, Overney, Shieh, Dixon, & Swartz, 2010; Kwasny et al., 2011). The setup that we describe below is based on a system initially developed by Friedl and colleagues used to investigate leukocyte migration in a three-dimensional environment (Friedl & Brocker, 2004). In this system, coverslips are mounted on glass slides and the space in between is filled with a Collagen I gel containing the cells to be recorded by phase-contrast microscopy. Several modifications allowed us to use this custom-build chamber as a tool for visualization of chemokine gradients that are actively shaped by ACKR4-expressing cells (Ulvmar et al., 2014). First, the use of a flat coverslip as negative chamber template allows the design of consistently flat chambers which improves quality of the recorded videos since only few cells outside the focal plane get recorded. Second, the smaller volume of the chamber allows the active shaping of a chemokine gradient in a reasonable period of time. However, proper polymerization of Collagen I gels is impaired in such flat chambers, which might be due to the increased surface-to-volume-ratio. Therefore, we used Corning® Matrigel® Matrix, a reconstituted basement membrane preparation extracted from the Engelbreth-Holm-Swarm mouse sarcoma (https://www.corning.com/worldwide/en/products/life-sciences/products/surfaces/matrigel-matrix.html). Matrigel mainly consists of the extracellular matrix molecules laminin, collagen IV, and entactin as well as heparan sulfate proteoglycan and can be used in various *in vitro* and *in vivo* applications.

In most of the chamber assays described so far, cells are casted in a gel and subsequently overlaid with chemokine solution (Friedl & Brocker, 2004; Sixt & Lammermann, 2011). Subsequently, the chemokine will diffuse into the gel, thus establishing a chemokine gradient in a passive manner. Video recording allows the analysis of cells migrating along such gradients. In contrast to this setup, the assay described below allows the analysis of cell migration while a chemokine gradient is actively created by cells expressing an ACKR. To this end, the chamber is filled with two gels. In a first step, a gel containing both the consumer cells (expressing an ACKR) and the reporter cells (expressing a conventional chemokine receptor, such as CCR7) as well as the chemokine of interest is casted into one half of the chamber. After polymerization of the first gel, the chamber is filled with a second gel containing no cells but the chemokine at exactly the same concentration as the first gel. The consumer cells in the first half of the chamber will constitutively scavenge the chemokine from the system and as a consequence, the initially homogenous chemokine concentration will be turned into a gradient pointing toward the second half of the chamber. The reporter cells are able to sense these changes in chemokine distribution and will start to migrate toward the second half of the chamber, which initially does not contain any cells. A crucial step of this cell translocation is the crossing of the border between the two gels. Therefore, the casting of this border is presumably the most challenging step of this assay. It has to be rigid enough to prevent mixing of the gels once the second gel is added to the chamber, but at the same time, it has to be soft enough to enable the passage of migrating cells from one part of the chamber to the other.

3. MATERIALS AND EQUIPMENT

For chamber preparation

Glass slides, 75 mm × 25 mm, plain
Coverslips, 24 mm × 60 mm and 18 mm × 18 mm, No. 1 thickness (0.13–0.16 mm)
Paper towels, lint-free
Ethanol 70%
Ethanol 100%
Paraffin wax pellets, e.g., Roti®-Plast, Carl Roth
Vaseline
Glass beaker, 100 ml
Hot plate
Paint brush

For cell preparation

Consumer cells: express an ACKR that continuously scavenges chemokines from the system.
Reporter cells: express at high levels a conventional chemokine receptor for the respective chemokine and are thus capable of sensing changes in chemokine concentration and respond by chemotaxis.

For cell staining (if cells do not encode fluorescent proteins)

TAMRA (*N*,*N*,*N*′,*N*′-tetramethyl-5-(and-6-)-carboxyrhodamine succinimidyl ester; Invitrogen) or a comparable dye
Staining medium: RPMI without phenol red, substituted with 1% FCS and 25 m*M* HEPES, pH 7.4
Chamber medium: RPMI without phenol red, substituted with 10% FCS and 25 m*M* HEPES, pH 7.4

For chamber filling

Corning® Matrigel® Matrix (standard formulation, phenol red free)
Chemokine solution

For live-cell imaging

Live-cell imaging can be performed at any fluorescence microscope equipped with a 5× or 10× objective, camera, heated, and motorized stage, and respective software that allows acquisition of multiple stage positions. Ideally, the microscope is equipped with a mounting frame allowing imaging of multiple slides at the same time.

For our experiments, we use an Axiovert 200M fluorescence microscope (Carl Zeiss), equipped with a heated incubation chamber (Solent Scientific). Images are taken with an AxioCam MRm camera (Carl Zeiss) using a Plan-Neofluar objective 5×/0.15, and processed with AxioVision 4.8 software. In our setup, the four slide mounting frame allows simultaneous acquisition of up to eight chambers.

For analysis

Software, e.g., ImageJ (free software; http://imagej.nih.gov/ij/, National Institutes of Health, Bethesda, MD) or Imaris (commercially available; Bitplane, Switzerland).

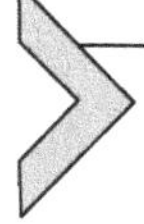

4. PREPARATION OF CHAMBERS

1. Wipe glass slides and coverslips with a lint-free paper towel soaked with 70% ethanol and leave them to dry.
2. Pipette two drops (4 μl each) of 100% ethanol onto the glass slide (Fig. 1A). These drops will temporarily fix your templates without leaving marks and will prevent their unwanted shifting during the following manipulations.
3. Place a small coverslip (18 mm × 18 mm) on each drop and move it to its designated position (Fig. 1B), which is in parallel to the borders of the slide, toward the middle of the slide, leaving an approximately 8-mm broad strip between the two templates. One end of the coverslips should stick out over the border of the glass slide to facilitate removal of the template later on.
4. Firmly press down the coverslip and soak up excessive ethanol with a paper towel. This should be done thoroughly, as excessive ethanol will negatively affect both the stability of the template and the height of the chamber.
5. Mix paraffin and vaseline at a 3:1 ratio by weight in a glass beaker and melt it on a hot plate at approximately 70 °C.
6. Use a paint brush and apply the paraffin mixture onto the three edges of the coverslip that are in contact with the glass slide (Fig. 1C). This has to be done quickly, as the paraffin will become solid instantaneously once it is applied to the colder glass slide. Ideally, one stroke of the brush should cover one edge of the coverslip.

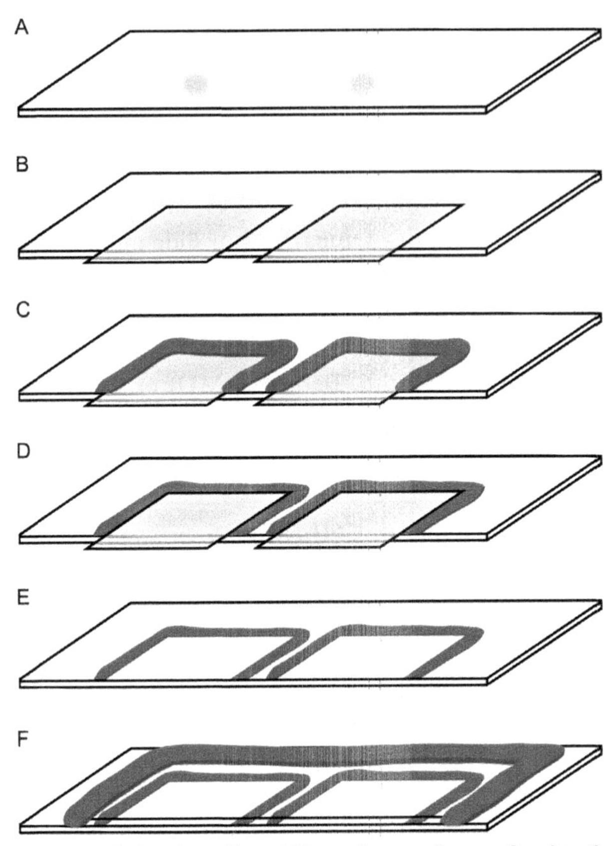

Figure 1 Preparation of chamber: (A and B) small coverslips are fixed to the slide using ethanol, (C) paraffin is applied around the coverslips, (D) excessive paraffin and (E) coverslips are removed, (F) large coverslip is added and fixed with paraffin.

7. Use another glass slide to gently remove excess paraffin above the level of the fixed coverslip. Try to remove the paraffin by scratching parallel to the border of the coverslip or perpendicular toward it rather than away from it as the latter will most likely remove the entire paraffin lane.
8. Remove the small coverslips from the glass slide by lifting the protruding edges. If necessary, remove residual ethanol (Fig. 1E). The glass slide is now equipped with two U-shaped paraffin strips 130–160 μm high.
9. Put a large coverslip (24 mm × 60 mm) on top of the paraffin strips. Leaving some space between the edge of the slide and the edge of the coverslip will ease the subsequent filling of the chamber.

10. Gently press the coverslip on the paraffin and fix it by brushing additional paraffin onto its edges (Fig. 1F). Remember to leave the front side open.
11. The chambers can be prepared in advance and stored in a slide box.

5. PREPARATION OF CELLS

The consumer cells can be primary cells, which are isolated from any organ that contains cells expressing the respective ACKR. Alternatively, a nonmigratory cell line can be genetically modified to express the scavenging receptor. Though the latter option might be more artificial, the use of a cell line bears several advantages. Time consuming isolation of cells is not required, the number of harvested cells is more predictable, and the use of laboratory animals can be diminished. Moreover, the use of a cell line offers the possibility to compare cells expressing an ACKR with control cells transduced with an empty vector control to distinguish the effect of the scavenging receptor from unspecific binding to the cells. In our assays, we use a thymic epithelial cell line which is retrovirally transduced to express ACKR4 and GFP or GFP alone (Ulvmar et al., 2014).

If gradient production is based on ACKR4 and its ligands CCL19 or CCL21, mature dendritic cells are particularly suited as reporter cells as they are known to express high levels of CCR7 upon maturation. Again, these cells can be sorted from primary organs or generated *in vitro*, as described elsewhere (Sixt & Lammermann, 2011). Briefly, bone marrow precursors are isolated under sterile conditions and cultured in the presence of GM-CSF for 8 days. Subsequent treatment with lipopolysaccharide for 24 h will induce maturation and lead to the characteristic phenotype of $CD11c^{+}$ $MHCII^{high}$ $CCR7^{+}$ mature dendritic cells (Ohl et al., 2004). As this method is easy, reliable, and highly reproducible, it is the method of choice in our assays.

Optionally, migratory cells lacking the respective chemokine receptor can be added. These cells—in our case *Ccr7*-deficient dendritic cells—are unable to sense any differences in chemokine concentration but are still able to migrate within the chamber in a random pattern and are therefore perfectly suited as an intrinsic control.

If possible, cells used in the assay should be genetically labeled by the expression of a fluorescent protein such as GFP. If this is not feasible, cells have to be stained with a fluorescent dye. As an example, we next describe labeling of cells using TAMRA.

1. Count cells and adjust the concentration to 2×10^6 cells/ml in staining buffer (1% FCS and 25 m*M* HEPES, pH 7.4, in RRMI without phenol red).
2. Prewarm the cells in a water bath at 37 °C for 10 min.
3. Add TAMRA stock solution to a final concentration of 10 μ*M* and mix immediately to assure homogenous distribution of the dye.
4. Incubate the cells for another 10 min at 37 °C, gently shaking the suspension every 3–4 min.
5. Wash cells twice by adding RPMI and subsequent centrifugation at $300 \times g$ for 5 min.

After harvesting and optional staining, cells are resuspended in chamber medium. Cells are counted and mixed to a final concentration of $4–8 \times 10^6$/ml reporter cells (including intrinsic controls) and $8–16 \times 10^6$/ml consumer cells. In our hands, this concentration is perfectly suited to acquire a moderate cell density that balances the establishment of a gradient and also allows single-cell tracking. Naturally, these numbers may not be ideal in assays using different cell types.

At a later stage of the procedure, the cell suspension is used to dilute the Matrigel (see below). The exact dilution factor depends on the initial protein concentration of the Matrigel used which may vary between different lots. The cell numbers have to be adjusted according to this dilution factor, in order to reach a final concentration in the gel of roughly 2×10^6/ml reporter cells and 4×10^6/ml consumer cells (see example below).

6. FILLING OF CHAMBERS

1. The protein concentration of Matrigel varies between lots ranging from 8 to 12 mg/ml approximately as specified by the manufacturer. We observed optimal conditions for dendritic cell migration in Matrigel matrix at a final protein concentration of 6–7 mg/ml. After calculating the ideal dilution factor for each Matrigel lot, the concentration of cells and the concentration of the chemokine have to be adjusted in both stocks (cell suspension and Matrigel). Chemokine is added to the Matrigel stock. After the Matrigel is mixed with the cell suspension, the final chemokine concentration should be 5–50 n*M*, depending on the applied chemokine. Chemokines with heparan sulfate-binding domains like CCL21 need to be used at higher concentrations.

As an example: If the Matrigel stock is specified by the manufacturer to contain 10 mg/ml protein, an appropriate dilution would be 3:1. Mixing two volumes of Matrigel (containing 7.5 n*M* chemokine) with one volume of cell suspension (containing 6×10^6/ml reporter cells and 12×10^6/ml consumer cells) will result in a final protein concentration of 6.6 mg/ml and a cell concentration of 2×10^6/ml reporter cells and 4×10^6/ml consumer cells as well as 5 n*M* chemokine.

2. Mix cell suspension and Matrigel at the calculated ratio. As Matrigel rapidly aggregates at room temperature, it has to be kept on ice at all times and should only be pipetted with chilled tips.
3. For filling of the chamber, hold the slide in an upright position and place the pipette tip perpendicular to the glass slide at one corner of the open side of the chamber. Add the Matrigel–cell mixture until the chamber is approximately half-full (Fig. 2). To obtain a straight border, it might help to pipette slowly and optionally tilt the slide while filling. Avoid creating air bubbles.
4. Place the chamber in a 50 ml Falcon tube with 3 ml water and close the lid loosely. Incubate it in an upright position at 37 °C and 5% CO_2 for 25–40 min to polymerize the Matrigel and equilibrate the medium.
5. Dilute chemokine containing Matrigel with the chamber medium. It is crucial to dilute the Matrigel at the same ratio in both halves of the chamber, as differences in Matrigel concentrations may hamper or prevent migration of cells across the border created by the subsequent filling step.

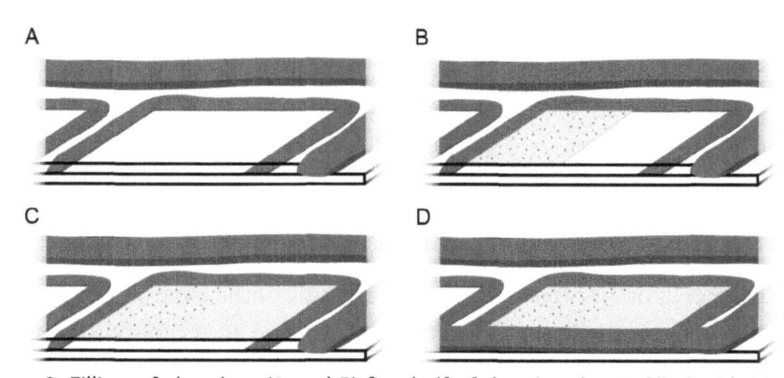

Figure 2 Filling of chamber: (A and B) first half of the chamber is filled with Matrigel matrix containing cells and chemokine, (C) after polymerization of the gel, the second half of the chamber is filled with Matrigel containing chemokine only, and (D) the chamber is sealed with paraffin. (See the color plate.)

6. Fill the second half of the chamber as described in step 3.
7. Seal the open side of the chamber with an additional lane of paraffin and proceed to live-cell imaging.

7. TIME-LAPSE IMAGING

1. Prewarm microscope stage to 37 °C prior to the experiment.
2. Adjust exposure times, stage, and *z*-positions for each chamber to be recorded. As the chambers are pretty flat, acquisition of one *z*-plane might be sufficient in individual cases. However, to improve image quality, three or more planes (30–50 μm step size) should be recorded.
3. Keep in mind that the number of fluorescence channels and *z*-planes collected as well as the number of stage positions determine the duration of each step interval. Ideally, dendritic cells are recorded with one frame per minute. Slower acquisition is possible but might restrict semiautomated cell tracking (see step 2 in Section 8).
4. Depending on the number of consumer cells, the efficiency of chemokine uptake and initial chemokine concentration, directional migration of reporter cells can be seen within the first 2 h. Compared to the classically applied passive diffusion of chemokines, active shaping of the chemokine gradient is a rather slow process. Therefore, we recommend acquisition over a period of at least 10 h.

8. ANALYSIS

We use two approaches to analyze cell migration. The first method is a straightforward and rather basic analysis, based on the measurement of the area covered by cells at the beginning and at the end of each movie using ImageJ (NIH). This procedure resembles an endpoint analysis comparable to that used in a transwell migration assay. However, the migration assay described here compensates the disadvantages of a conventional transwell assay, as moving of the cells is recorded and can be used for further analysis. In case the cells do not translocate from one side of the chamber to the other side, the movies will reveal whether this is due to viability or motility impairments of the cells or due to technical obstacles of the chamber.

The second method describes a more sophisticated procedure based on semiautomated cell tracking using Imaris software (Bitplane). Naturally, many other ways have been described to quantitatively analyze cell migration based on time-lapse imaging, and it should be kept in mind that the methods described here might not be suited to address every possible type of experimental hypothesis.

8.1 Quantitative Analysis of Migrated Cells Using ImageJ

1. Open your file in the ImageJ software and in case the metadata has not been adopted correctly, adjust the scale (Analyze → Set Scale) according to the acquisition settings of your microscope.
2. Use the "Image" → "Stacks" → "Z Project" function to project the movie to a series of 2D images.
3. Only true signals originating from the reporter cells should be visible in the respective channel; if there is spectral overspill from the consumer cells, those false-positive signals have to be removed by subtracting the channels from each other.
4. Define two regions of interest (ROIs; Fig. 3):

   ROI I corresponds to the whole chamber and does not require any further selection.

   ROI II is the right half of the chamber, which initially does not contain cells. This region is defined by the absence of consumer cells in the first frame of the movie and specified using the polygon selection tool.
5. Add ROI II to the ROI manager (ctrl + t).
6. The following steps (7–12) are performed on both, the first and the last frame of every movie.
7. Split channels.
8. Invert the channel that depicts the reporter cells of interest (ctrl + shift + i).
9. Binarize the image using an appropriate threshold (ctrl + shift + t).
10. The command "Edit" → "Selection" → "Create Selection" will select the image area covered by black spots.
11. Pressing "m" starts measurement and the area covered with black pixels will be calculated.
12. For measuring ROI II, the saved area in the ROI manager has to be activated in the frame first. By using "Edit" → "Clear Outside," all spots outside the selected ROI are removed and again, a selection of all spots can be created and the covered area is determined.
13. Repeat steps 7–12 with the last frame of the movie, using the same ROIs.

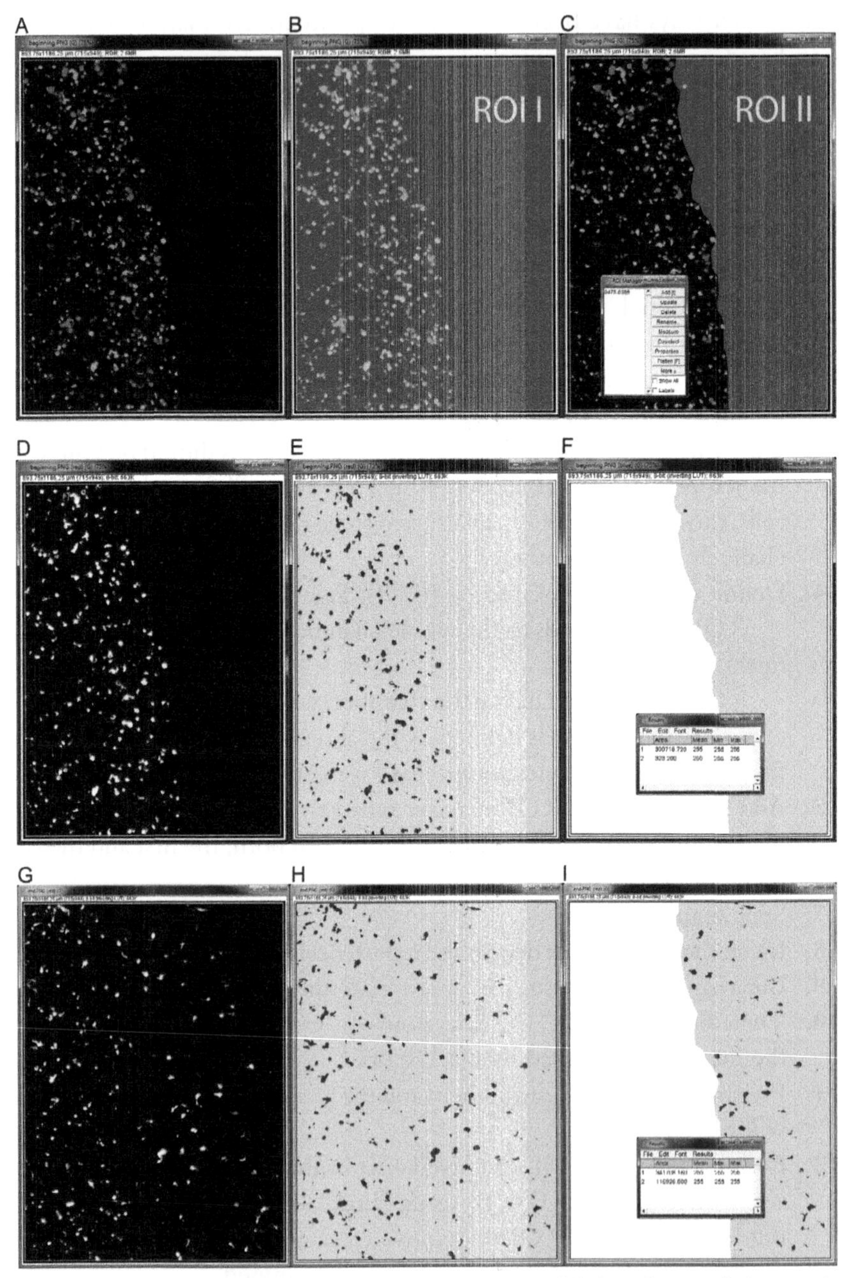

Figure 3 Analysis of cell displacement using ImageJ: (A) first frame of the movie; (B and C) definition of ROI I and ROI II, with the latter being added to the ROI manager; (D) channel depicting the reporter cells is separated from the others using the split channels command; (E) this channel is now inverted and binarized, and all black pixels can be automatically counted in ROI I; (F) after clearing signals outside of ROI II, black pixels can be counted in ROI II; and (G–I) the same operation has to be applied to the last frame of the movie. (See the color plate.)

14. Finally, the ratio of black pixels in the right half of the chamber compared to the whole chamber can be calculated for the beginning and the end of each movie.

8.2 Semiautomated Cell Tracking of Migrating Cells Using Imaris

1. Open your file and check whether existing metadata is adopted correctly. If necessary, change time and scale settings manually.
2. Project the movie to 2D using the respective MatLab plugin.
3. Define a straight border between the two halves of the chamber based on the position of the consumer cells in the first frame of the movie.
4. Rotate the movie around its *z*-axis in order to position this border vertically. The proper turning angle can be found by trial and error. Depending on its size, turning the entire movie might take some time, so it might be advisable to crop the movie down to one frame to find the correct turning angle and then apply this angle to the entire movie.
5. Turning all movies that way not only allows the use of the automatically calculated value "*x*-displacement" (see step 12) but also facilitates direct comparison between movies.
6. Crop the movie to a smaller rectangle with the predefined border in the middle.
7. Any false-positive signals, such as spectral overspill from the consumer cells, have to be removed from the reporter cell channel by channel subtraction (MatLab plugin).
8. Start automated cell tracking by clicking the "spot object button." The appearing spots creation wizard will guide you through the following tracking process.
9. Choose channel of interest.
10. Define the estimated diameter of the cells to be tracked. Spots are automatically detected, however, adjustment of the quality threshold allows you to ensure that cells are included or excluded from the spot selection as desired.
11. Choose an appropriate algorithm and suitable values for "max distance" and "gap size" according to your experiment. Keep in mind that prolonged time intervals between frames require higher values, as the cells might have translocated further between consecutive time points. This inaccuracy in prediction of the cell's position at a subsequent time point may lead to significant errors in cell tracking. Tracks might be

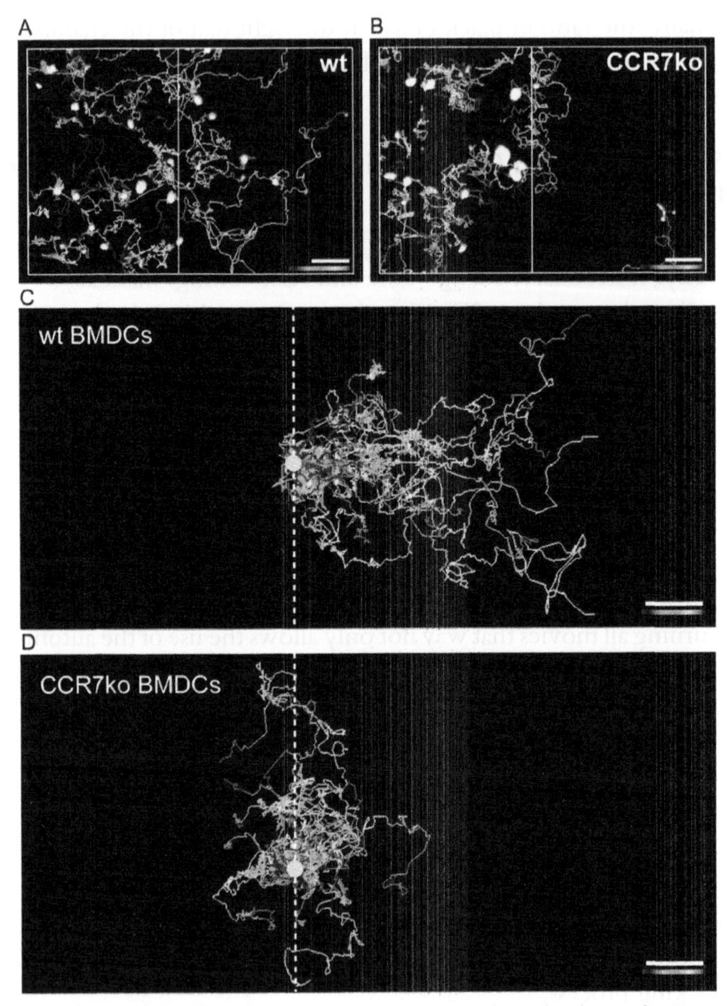

Figure 4 Depiction of cell tracks in Imaris using a time-based color code: (A and B) in original movie, after processing as described in Section 8 step 2; (C and D) after defining a common starting point (white dot). Dotted line indicates *y*-axis at $x=0$. Images show highly directional migration for wild type in contrast to *Ccr7*-deficient bone marrow-derived dendritic cells. Scale bar 100 μm. (See the color plate.)

interrupted or mismatched and therefore need an individual check and manual correction if necessary.

12. Imaris automatically calculates numerous statistics, such as *track duration*, *speed*, *straightness*, and the above mentioned *x-displacement* for each individual track. The latter value reveals whether the cells show a substantial displacement in *x*, which corresponds to the displacement into the

initially cell-free half of the chamber if the movie is processed as described above. All values can be exported into an Excel data sheet for further analysis.

13. For track depiction, statistics- or time-coded coloration of tracks and spots can be applied directly. Alternatively, all tracks can be mathematically processed (MatLab plugin) to overlay the origin of each track to one common starting point (Fig. 4).

ACKNOWLEDGMENTS

We would like to thank Peter Friedl for valuable advices in establishing the system as well as Asolina Braun, Tim Worbs, and Michael Sixt for fruitful discussion and sharing technical expertise. Furthermore, we would like to thank Olga Schulz for critical reading of the manuscript. This work was supported by Deutsche Forschungsgemeinschaft grants (SFB738-B5 and EXC62 to R.F.) ERC advanced grant (322645 to R.F.), the Hannover Biomedical Research School (HBRS), and the Center for Infection Biology (ZIB).

REFERENCES

Bachelerie, F., Ben-Baruch, A., Burkhardt, A. M., Combadiere, C., Farber, J. M., Graham, G. J., et al. (2014). International Union of Basic and Clinical Pharmacology. LXXXIX. Update on the extended family of chemokine receptors and introducing a new nomenclature for atypical chemokine receptors. *Pharmacological Reviews*, *66*(1), 1–79.

Bachelerie, F., Graham, G. J., Locati, M., Mantovani, A., Murphy, P. M., Nibbs, R., et al. (2014). New nomenclature for atypical chemokine receptors. *Nature Immunology*, *15*(3), 207–208.

Bonvin, C., Overney, J., Shieh, A. C., Dixon, J. B., & Swartz, M. A. (2010). A multichamber fluidic device for 3D cultures under interstitial flow with live imaging: Development, characterization, and applications. *Biotechnology and Bioengineering*, *105*(5), 982–991.

Boyden, S. (1962). The chemotactic effect of mixtures of antibody and antigen on polymorphonuclear leucocytes. *Journal of Experimental Medicine*, *115*, 453–466.

Dona, E., Barry, J. D., Valentin, G., Quirin, C., Khmelinskii, A., Kunze, A., et al. (2013). Directional tissue migration through a self-generated chemokine gradient. *Nature*, *503*(7475), 285–289.

Friedl, P., & Brocker, E. B. (2004). Reconstructing leukocyte migration in 3D extracellular matrix by time-lapse videomicroscopy and computer-assisted tracking. *Methods in Molecular Biology*, *239*, 77–90.

Graham, G. J., Locati, M., Mantovani, A., Rot, A., & Thelen, M. (2012). The biochemistry and biology of the atypical chemokine receptors. *Immunology Letters*, *145*(1–2), 30–38.

Kwasny, D., Kiilerich-Pedersen, K., Moresco, J., Dimaki, M., Rozlosnik, N., & Svendsen, W. E. (2011). Microfluidic device to study cell transmigration under physiological shear stress conditions. *Biomedical Microdevices*, *13*(5), 899–907.

Nibbs, R. J., & Graham, G. J. (2013). Immune regulation by atypical chemokine receptors. *Nature Reviews Immunology*, *13*(11), 815–829.

Ohl, L., Mohaupt, M., Czeloth, N., Hintzen, G., Kiafard, Z., Zwirner, J., et al. (2004). CCR7 governs skin dendritic cell migration under inflammatory and steady-state conditions. *Immunity*, *21*(2), 279–288.

Rot, A., & von Andrian, U. H. (2004). Chemokines in innate and adaptive host defense: Basic chemokinese grammar for immune cells. *Annual Review of Immunology*, *22*, 891–928.

Sixt, M., & Lammermann, T. (2011). In vitro analysis of chemotactic leukocyte migration in 3D environments. *Methods in Molecular Biology*, *769*, 149–165.

Ulvmar, M. H., Hub, E., & Rot, A. (2011). Atypical chemokine receptors. *Experimental Cell Research*, *317*(5), 556–568.

Ulvmar, M. H., Werth, K., Braun, A., Kelay, P., Hub, E., Eller, K., et al. (2014). The atypical chemokine receptor CCRL1 shapes functional CCL21 gradients in lymph nodes. *Nature Immunology*, *15*(7), 623–630.

CHAPTER FIFTEEN

The Role of Chemokine and Glycosaminoglycan Interaction in Chemokine-Mediated Migration *In Vitro* and *In Vivo*

Irene del Molino del Barrio, John Kirby, Simi Ali[1]
Institute of Cellular Medicine, Medical Faculty, Newcastle University, Newcastle upon Tyne, United Kingdom
[1]Corresponding author: e-mail address: simi.ali@newcastle.ac.uk

Contents

1. Introduction 310
2. *In Vitro* Chemotaxis 311
   2.1 Diffusion Gradient Chemotaxis of Adherent Cells 312
   2.2 Diffusion-Gradient Chemotaxis of Suspension Cells 314
   2.3 Counting of Migrated Cells 315
   2.4 *In Vitro* Transendothelial Chemotaxis 317
3. *In Vivo* Chemotaxis 318
   3.1 *In Vivo* Leukocyte Recruitment to Murine air Pouch 319
4. Generation of Mammalian Transfectants Expressing Chemokine Receptors 322
   4.1 Chemokine Receptor Cloning 322
   4.2 Transfection of the Vector into the Host Cell Line 323
   4.3 Selection of Stable Transfection Cells 325
5. Discussion 329
Acknowledgments 330
References 331

Abstract

Chemokines have a range of functions, including the activation and promotion of the vectorial migration of leukocytes. They mediate their biological effects by binding to their cognate G-protein-coupled receptors. Upon activation of the heterotrimeric G proteins, the Gα subunit exchanges GDP for GTP and dissociates from the receptor and from the Gβγ subunits, and both G-protein complexes go on to activate other downstream signaling events. In addition, chemokines interact with cell-surface glycosaminoglycans (GAGs). This potential for binding GAG components of proteoglycans on the cell surface or within the extracellular matrix allows the formation of the stable chemokine gradients necessary for leukocyte chemotaxis. In this chapter, we describe techniques for studying chemotaxis both *in vivo* and *in vitro*, as well as the creation of chemokine receptor–expressing cell lines, in order to examine this process in isolation.

Methods in Enzymology, Volume 570
ISSN 0076-6879
http://dx.doi.org/10.1016/bs.mie.2015.09.002

1. INTRODUCTION

Chemokines, *chemo*attractant cyto*kines*, are small (8–14 kDa) proteins that bind to G-protein-coupled receptors composed of seven transmembrane domains (Baggiolini, 1998). The chemokine family encompasses more than 50 members, which can be either homeostatic or inflammatory (Zlotnik & Yoshie, 2000). The former are constitutively expressed in certain tissues and have roles in tissue development, such as angiogenesis or neovascularization, or basal leukocyte migration (Rot & von Andrian, 2004). In the latter, an infection or other proinflammatory stimulus (TNF-α) will cause the release of chemokines that will direct the recruitment of leukocytes (e.g., neutrophils, monocytes, etc.) toward the site of inflammation (Zlotnik, Burkhardt, & Homey, 2011).

During inflammation, several selectins are expressed by the blood vessels' endothelium. Leukocytes can form weak interactions with them as the blood flow pushes the cells forward, making the leukocytes roll along the surface (Lowe, 2003). Meanwhile, the extracellular endothelium is coated with heparan sulfate glycosaminoglycans (GAGs) (Parish, 2006; Salanga, O'Hayre, & Handel, 2009). GAGs immobilize the chemokines, increasing the concentration at the site of production and allowing for the formation of chemokine gradients (Kuschert et al., 1999; Proudfoot et al., 2003; Ulvmar et al., 2014; Weber et al., 2013). When leukocytes come in contact with the presented chemokines, their receptors become activated, causing a signaling cascade to occur. Specifically, receptor activation induces conformational changes and the increased affinity of the leukocytes' integrins for their adhesion molecules (such as ICAM), eliciting adhesion to the endothelium (Ward & Marelli-Berg, 2009). Following arrest, leukocytes extravasate through the endothelium into the tissue (Schenkel, Mamdouh, & Muller, 2004). Most cells follow a paracellular route, migrating through endothelial cell junctions, but it is also possible for them to pass directly through the endothelial cells (transcellular) (Hashimoto et al., 2012). Once cells have crossed the basal lamina and entered the tissue, leukocytes continue to migrate toward the site of inflammation, following the chemokine concentration gradient. Tissue remodeling due to increased collagen networks and the secretion of proteases (e.g., matrix metalloproteinases) helps cells move through the tissue (Lerchenberger et al., 2013). A simplified reproduction of this process can be carried out in the lab through chemotaxis assays, a technique discussed in Section 2. Furthermore, in order to closely mimic the complexities of the process, the migration can be assessed *in vivo*, using mouse models, which we will explain in Section 3.

Chemokine receptors are widely expressed by leukocytes, such as T cells, B cells, or dendritic cells, but they are also present in many cancerous cells (Müller et al., 2001; Scotton, Wilson, Milliken, Stamp, & Balkwill, 2001). In order to facilitate the study of their activation, signaling pathways, and functional consequences, creating a stable cell line that expresses a specific receptor can be useful, especially if the model cell line does not mimic what we see in patients. We will discuss how to create stable transfectants in Section 4.

2. *IN VITRO* CHEMOTAXIS

Chemotaxis assays allow the evaluation of chemokine-receptor–expressing cells' migration in response to their ligand. For these assays, a commercial polycarbonate version of a Boyden chamber (Boyden, 1962) is used. This chamber is composed of an upper and a lower compartment, which are separated by a membrane containing pores of a known size, too small for the migratory cells to simply "fall through"—thus, cells need to undergo morphological and cytoskeletal changes in order to migrate. Generally, an 8-μm pore size is required for mammalian epithelial cells, such as CHO or HEK-293, and cancer cell lines. We tend to use a 5-μm pore size for lymphocytes and monocytes and a 3-μm pore size for neutrophils and leukocytes. A literature search can be performed to determine Boyden-chamber pore sizes and migration times successfully used by other researchers for specific cell lines.

As seen in Fig. 1, two main modalities can be distinguished. Cell migration in diffusion gradient chemotaxis assays is mainly dependent on the strength of the chemokine–receptor interaction, but transendothelial

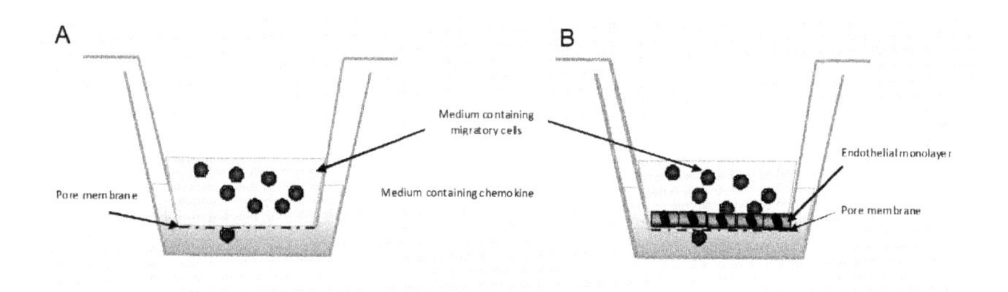

Figure 1 Schematic of a chemotaxis experiment. "A" shows a diffusion gradient assay, and "B" shows a transendothelial chemotaxis assay.

chemotaxis emulates the cell barrier through which cells need to extravasate *in vivo*.

Required materials:

- 24-well companion plate (BD Falcon, Cat. No. 353504)
- Falcon cell-culture inserts (Cat. No. 08-771-4 (3 μm), Cat. No. 08-771-21 (8 μm)).
- DMEM (high glucose, Sigma, Cat. No. D5671)
- RPMI-1640 (Sigma, Cat. No. R8758-500ML)
- Fibronectin (4 μg/ml, Sigma, Cat. No. F1141)
- Gill hematoxylin (Sigma, Cat. No. GHS316)
- DPX mountant (Sigma, Cat. No. 06522)
- Scott's tap-water substitute concentrate (Sigma, Cat. No. S5134)
- Lympholyte-H (CedarLane, Cat. No. CL5015)
- Heparin (Sigma, Cat. No. H3393)
- Bright-Line™ hemacytometer (Sigma, Cat. No. Z359629)
- Anti-human CD45 PE antibody (eBioscience, Cat. No. 12-9459-42)
- CountBright™ absolute counting beads (Life technologies, Cat. No. C36950)

2.1 Diffusion Gradient Chemotaxis of Adherent Cells

This study has been written with chemokine as the ligand, but it is valid for any chemoattractant.

(a) Prepare the cell cultures according to standard cell-culture procedures. For optimal results, starve cells overnight (or for a minimum of 2 h) in serum-free 0.2% BSA/DMEM. These steps should be carried out under sterile conditions, using a tissue culture hood.

(b) Block a 24-well companion plate with 1 ml of 1% BSA/DMEM for 1 h to prevent chemokine depletion due to its binding to the plate.

(c) Upturn and coat the filters from the cell-culture inserts with 150 μl of fibronectin. Incubate for 30 min at room temperature, then remove the excess fibronectin, and allow it to dry for a further 30 min at room temperature before use.

(d) Harvest cells for chemotaxis:

- Remove the medium from the cells and rinse with PBS.
- Add 5 ml PBS-EDTA and incubate for 5 min at 37 °C. Help detach cells by gently tapping flasks and transfer to a labeled universal vessel containing 5 ml of DMEM (serum free).
- Centrifuge at 300 × *g* for 5 min to pellet the cells. Then decant the supernatant.

(e) Resuspend the cells in serum-free 0.2% BSA/DMEM for a final concentration of 4×10^5 cells/ml. Add 500 μl of the cell solution to the cell-culture insert.

The absence of serum is vital,because it contains cytokines and growth factors that will mask the effect of the chemokine added, preventing the formation of a steep chemokine gradient.

(f) Remove 1% BSA/DMEM from the companion plate and replace it with 800 μl of 0.1% BSA/DMEM, with the optimal chemokine concentration. This should be determined beforehand—in our case, 25–50 n*M* of chemokine was selected (Ali, Fritchley, Chaffey, & Kirby, 2002).

Note: Each treatment should be carried out in triplicate, and two controls should be performed. In order to establish background chemotaxis, leave wells without chemokine, and in order to assess chemokinesis, add equal amounts of chemokine in the upper and lower wells.

(g) Carefully lower the cell-culture inserts into the plate wells, allowing for the creation of a chemotactic gradient. Incubate the cell-culture plate for the optimal period in a cell-culture incubator at 37 °C and 5% CO_2.

Note: In transfected CHO and HEK-293 cells, 5–6 h was considered the optimal incubation time. However, for each cell line, the migration time needs to be individually determined.

(h) After incubation, remove the medium from the wells and inserts. Gently wipe the inside of the filter with a cotton bud to remove nonmigrated cells.

(i) Fix filters in methanol at −20 °C overnight (or for a minimum of 1 h).

(j) Remove the methanol with a Pasteur pipette and fill wells and inserts with running tap water. Tap the plate to avoid inserts sticking to the lid, and tip out the water.

(k) Add 1.5 ml of hematoxylin to each well and incubate for 30 min at room temperature. Remove the hematoxylin with a Pasteur pipette and add 1.5 ml of Scott's tap water to blue the stain. Incubate for 10 min at room temperature and tip out the water.

(l) Add 1 ml of 50% ethanol to each well and incubate for 1 min at room temperature. Remove the ethanol and repeat this process with 75%, 90%, and 100% ethanol in order to fully dehydrate the cells. Leave the filters to air-dry for 1 h in a spare companion plate.

(m) Turn the inserts upside down so the filters are pointing up. Using a scalpel, cut the filter around the edge so it is only attached by a thread. Do all filters carefully, so as not to wrinkle them.

(n) Spread a drop of DPX on a microscope slide. With fine tweezers, break free the filter and place it on the DPX, bottom filter up. Place three filters per slide.

(o) Carefully cover the filters with a cover slide. Remove any bubbles formed by pressing firmly with the tweezers and guiding them toward the edges of the slide.

2.2 Diffusion-Gradient Chemotaxis of Suspension Cells

If using peripheral blood mononuclear cells (PBMCs) or suspension cell line, such as THP1 for monocytes or Jurkat cells for T lymphocytes, some changes need to be made to the chemotaxis protocol.

First, to isolate the PBMCs, including lymphocytes, monocytes, and macrophages, complete the following steps:

(a) Collect blood from healthy volunteers and thoroughly mix it with 1 unit of heparin per ml of blood to avoid coagulation. Add an equal volume of serum-free RPMI-1640.

(b) Add 5 ml of lympholyte-H to a 50-ml Falcon tube. Carefully layer the blood over it and centrifuge at $800 \times g$ for 20 min with the brake off to ensure clear layers.

(c) Carefully remove the interface containing the PMBCs with a Pasteur pipette and transfer it to a universal tube. Wash twice by adding 15 ml of serum-free RPMI-1640 and centrifuging at $600 \times g$ for 5 min.

(d) Count the cells, using a Neubauer hemocytometer (see Section 2.3.2)

(e) Resuspend the pellet in serum-free RPMI to a concentration of 1×10^6 cells/ml. Typically, we will have 5–10 times more lymphocytes than monocytes in the suspension. Monocytes will adhere to the underside of the filter, but T cells will migrate through to the lower well.

(f) Serum-starve cells for 1 h prior to use in assays.

For chemotaxis, the following changes apply to Section 2.1 protocol:

- In step (c), there is no need to coat filters with fibronectin, because monocytes will adhere to the filter naturally.
- In step (e), cells need to be at a final concentration of 1×10^6 cells/ml in order to add 5×10^5 cells/insert.
- In step (g), cells only need to be incubated for 90 min.
- In step (h), collect the medium from the wells into 1.5-ml Eppendorf tubes, instead of discarding it. Centrifuge the tubes for $500 \times g$ for 5 min, remove supernatant, and resuspend the pellets. Migrated, nonadhered cells can now be counted using an hemocytometer or flow cytometry (see Section 2.3.3).

2.3 Counting of Migrated Cells

2.3.1 Counting Migrated Cells by Microscopy

This will allow us to count the migrated adherent cells that have stuck to the filter. Using a microscope, count the migrated cells in five high-power fields (e.g., 20 ×) per filter. As seen in Fig. 2, pores appear as randomly scattered round dots, and cells are bigger and purple-colored.

2.3.2 Counting Migrated Cells by Hemocytometer

Cells that have migrated through the filter into the well below can be counted using an improved Neubauer chamber hemocytometer. Following resuspension, pipette 10 μl of the cell suspension beneath the coverslip. Count the cells that settled within the 25-square grid and multiply that number by 10^4 to give the number of cells per milliliter.

2.3.3 Counting Migrated Cells by Beads

Lymphocytes that have fully migrated into the well's media can be counted using flow cytometry. If desired, cells can first be stained for the leukocyte marker CD45.

(a) Resuspend cells in 50 μl of 5% FBS/PBS. Add 2.5 μl of CD45-PE antibody and incubate for at least 30 min at 4 °C.

(b) Add 500 μl of 5% FBS/PBS to wash and then centrifuge at 500 × *g* for 5 min. Remove the supernatant, wash again, and resuspend in 80 μl.

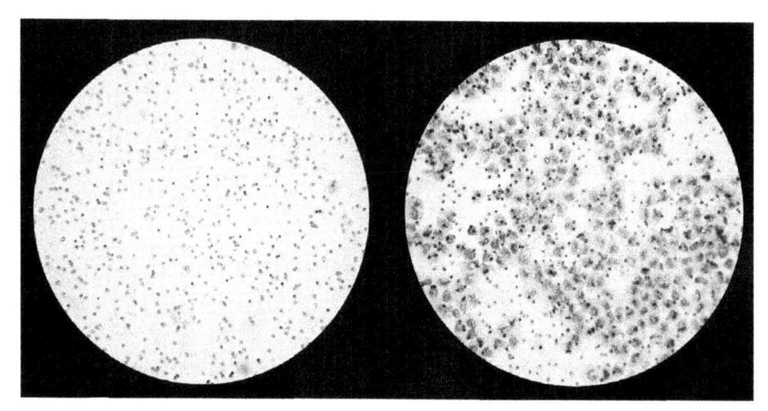

Figure 2 PBMCs were placed in a 3-μm filter above a well that contained the supernatant of untreated macrophages (left) or LPS-stimulated macrophages (right). After a 90 min incubation at 37 °C, filters were fixed and stained with hematoxylin, and migrated monocytes were counted.

To count the cells, the cell solution needs to be “spiked” with a known number of beads—for us, 2000 beads. CountBright™ absolute counting beads contain 5×10^4 beads in 50 μl.

(c) Dilute 50 μl of bead solution in 200 μl of PBS to obtain 200 beads/μl.

(d) Add 10 μl of the counting-bead dilution to 80 μl of cell solution. Transfer to a fluorescence-activated cell sorting (FACS) tube.

(e) Run cells through a flow cytometer to determine the numbers of migrated cells (cell events) and beads (bead events). As seen in Fig. 3, beads appear as a small population on the edge of the plot—to find them, set the side scatter (SSC) as a logarithmic scale.

(f) The number of migrated labeled cells can now be calculated with the following formula:

$$\text{Labeled cells} = \frac{\text{No. of labeled cell events (P1)}}{\text{No. of bead events (P2)}} \times \text{No. of beads added (2000)}$$

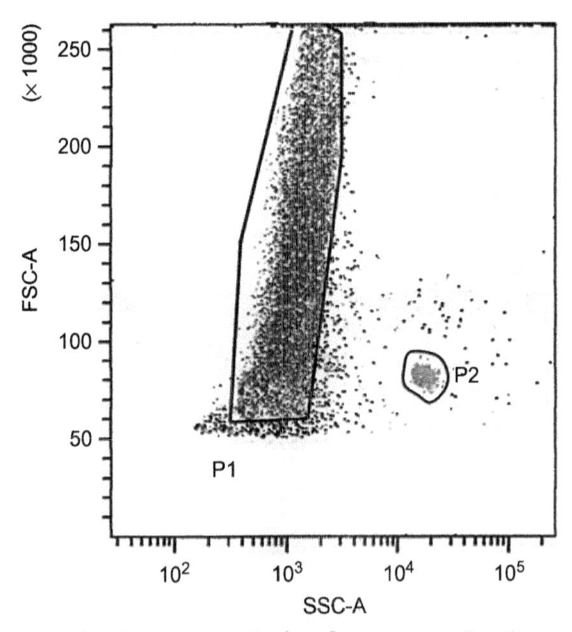

Figure 3 Representative flow-cytometric data from a transmigration assay. The assay was carried out in a BD FACSCANTO II analyzer, and the data was collected using the BD FACSDIVA™ software. The migrated lymphocyte population (P1) was evaluated together with beads (P2) to assess the total cell number as a ratio to a known number of counting beads. Data were then processed using FlowJo single-cell analysis software.

2.4 *In Vitro* Transendothelial Chemotaxis

In order to emulate *in vivo* conditions, the filters can be coated with endothelial cells. This protocol uses EA.hy-926 cells, but others, such as HMEC-1, can be used. For each cell line, the seeding number and media used will vary.

(a) Seventy-two hours prior to the assay, seed the cell-culture inserts with 1×10^5 EA.hy-926 cells and grow them in 500 μl of complete RPMI-1640 at 37 °C in 5% CO_2. The media was only placed in the insert, not in the well underneath, to discourage cells from growing through the filter, forming a "double" monolayer and making final cell-counting more difficult.

(b) When cells are around 80% confluent, activate them by adding 100 units/ml IFNγ and 100 ng/ml TNFα to the culture media overnight (16–24 h). This will induce proinflammatory changes in the endothelial cells and upregulate adhesion molecules such as ICAM and VCAM.

(c) In parallel, treat migratory cells with 300 units/ml of IFNγ in order to induce a more migratory phenotype.

(d) Keep assessing the confluency of the endothelial cell monolayer. This step is critical, given that, if cells are not completely confluent on the day of the assay, migratory cells can pass through the gaps in between the EA.hy-926 cells. And, vice versa, if endothelial cells are over-confluent, motile cells will have a hard time migrating across the monolayer.

(e) On the day of the assay, wash the migratory cells and the endothelial monolayer twice with 0.1% BSA/RPMI.

To the lower chamber, add 800 μl of the chemokine in 0.1% BSA/RPMI.

To the upper chamber, add 1×10^6 migratory cells resuspended in 500 μl of 0.1% BSA/RPMI.

(f) Carefully lower the cell-culture inserts onto the plate wells, allowing for the creation of a chemotactic gradient. Incubate the cell-culture plate for 90 min in a cell culture incubator at 37 °C and 5% CO_2.

(g) Fix and stain the filters as described in Section 2.1 (i–o). As seen in Fig. 4, EA.hy-926 cells will look blurry when the pores are in focus, but monocytes will look more defined and darker in color.

In nonactivated (resting) endothelial monolayers, lymphocytes will barely transmigrate or even adhere to endothelial cells within the assay time period

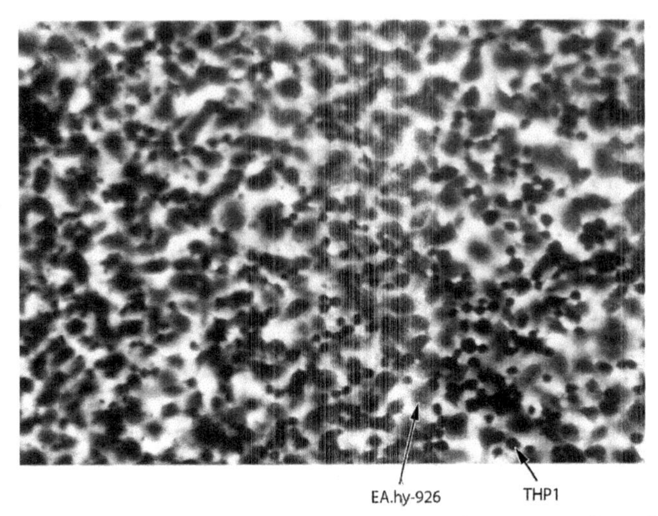

Figure 4 A monolayer of EA.hy-926 cells was grown on the upper surface of a cell culture insert, and THP1 cells were assayed for their ability to migrate. After 2 h incubation at 37 °C, filters were fixed with methanol and stained with hematoxylin, and migrated cells were counted.

(Muller & Weigl, 1992). Thus, only monocyte migration will be assessed. On the other hand, activated endothelial cells support the adhesion of T cells, but their migration will occur at a much slower rate.

3. *IN VIVO* CHEMOTAXIS

Required materials:

- Appropriate chemokines (Peprotech, R&D systems, Almac)
- Phosphate-buffered saline (Sigma, Cat. No. P5493)
- Bright-Line™ hemacytometer (Sigma, Cat. No. Z359629)
- Anti-mouse CD45 PE antibody (eBioscience, Cat. No. 12-0451-81)
- Rat IgG2b K isotype-control PE (eBioscience, Cat. 12-4031-82)
- Anti-mouse CD3 APC antibody (eBioscience, Cat. No. 17-0032-82)
- Rat IgG2b K isotype-control APC (eBioscience, Cat. 17-4031-82)
- CountBright™ absolute counting beads (Life technologies, Cat. No. C36950)
- Methanol (Sigma, Cat. No. 322415)
- Cytospin™ 4 cytocentrifuge (ThermoScientific, Cat. No. A78300003)
- Clipped funnel starter kit (ThermoScientific, Cat. No. 3120110)

3.1 *In Vivo* Leukocyte Recruitment to Murine air Pouch

The generation of an air pouch in mice allows us to easily study *in vivo* the inflammatory response to chemokines by retrieving the infiltrating inflammatory cells. 20–30 g female BALB/c mice (around 8–10 weeks old) should be used to perform these experiments, because, in heavier mice, it is more difficult to correctly form the air pouch (Ali et al., 2005). This study has been written with chemokine as the ligand, but it is valid for any chemoattractant.

(a) In order to generate an air pouch, on day 1, anesthetize the mice via inhalation of isofluorane. Shave and disinfect a patch of skin (about 2 cm^2) on their backs and inject 3 ml of sterile air subcutaneously with a 25-gauge syringe. The injection point should be between the shoulder blades and 1 cm away from the back of the head. Extreme care should be taken not only to minimize the discomfort of the mice, but also to avoid the injected air migrating toward the head or forelimbs and preventing the formation of a round air pouch.

(b) On days 2, 4, and 5, lightly anesthetize the mouse and inject 1 ml of sterile air to "top up" the air pouch. This will produce stable fluid-filled pouches as seen in Fig. 5.

(c) On day 6, inject 10 μg of chemokine in 1 ml of sterile PBS (or PBS only for the negative control) into the air pouch.

If you need to evaluate the ability of a drug (i.e., an antagonist, agonist, or antibody) or a modified chemokine to disrupt or enhance cell

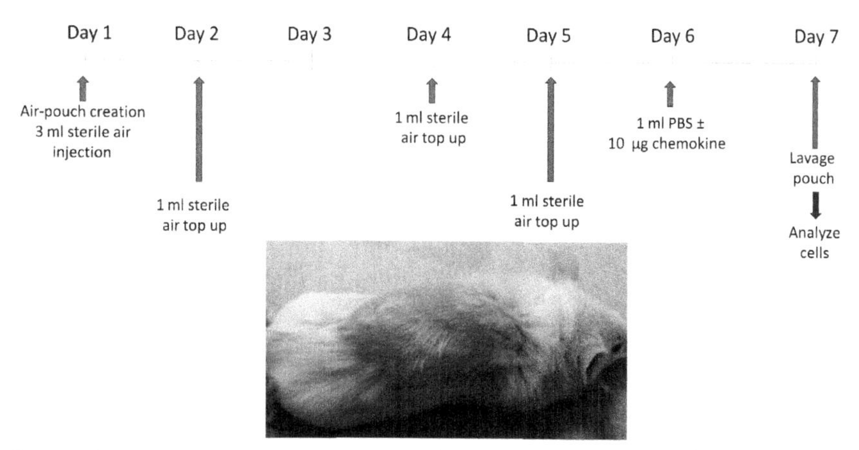

Figure 5 Schematic of the murine air-pouch *in vivo* chemotaxis model. BALB/c mice were anesthetized, their backs were shaved, and the shaved sites were injected with 3 ml of sterile air under the skin. This was repeated on days 2, 4, and 5, and 1 ml of air was injected to create the air pouch.

recruitment to the air pouch, it will need to be intravenously administered. To do so, inject an equimolar ratio of the drug to the chemokine present in the air pouch. Given that the blood volume of the mice is approximately 2 ml and the air pouch volume is 1 ml, for 10 μg of chemokine, 20 μg of the drug will need to be injected in 100 μl of PBS. The sham group will be injected with 100 μl of sterile PBS, so that the inflammatory response can then be evaluated.

(d) Between six and twenty-four hours after the chemokine injection (day 7), euthanize the mice by asphyxiation with carbon dioxide. Whenever possible, avoid breaking their necks because it may cause bleeding into the air pouch. The necessary time will depend on the chemokine being evaluated—neutrophil-attractant chemokines, such as CXCL8, will see recruitment after 6–8 h, but monocyte recruitment with chemokines, such as CCL2, will take 16–24 h.

Completely shave the air pouch and recover the recruited cells by lavaging the pouch twice with 0.75 ml of 3 m*M* EDTA/PBS.

(e) Centrifuge the exudates at 500 × *g* for 5 min and resuspend the cell pellets in 200 μl of 1% FBS/PBS. Cells can now be analyzed and counted by flow cytometry or using Cytospin.

3.1.1 Cell Staining and Counting Using Flow Cytometry

The process followed is very similar to the one described in Section 2.3.3.

(a) Count cells with a hemocytometer, as described in Section 2.3.2. Cell number can be extremely variable, but numbers between 10 and 200×10^4 cells/ml are expected.

(b) Prepare three FACS tubes—one for unstained cells, one for the isotype control, and one to label with the antibody. If enough cells are retrieved, resuspend 200,000 cells/tube in 50 μl of 5% FBS/PBS. Otherwise, divide the cells equally in three tubes. Add 1 μl of Fc block in the isotype and antibody tubes and incubate for 30 min on ice.

(c) Add 2.5 μl of isotype or murine CD45-PE and CD3-APC antibodies and incubate for at least 30 min at 4 °C. This will allow us to differentiate T lymphocytes, leukocytes, and monocytes.

As these two fluorochromes have nonoverlapping emission/excitation spectra, there is no need to have extra tubes with just one of the antibodies for compensation.

Add 700 μl of 5% FBS/PBS to wash and centrifuge at 500 × *g* for 5 min. Remove the supernatant and repeat. Resuspend cells in 200 μl of 5% FBS/PBS.

(d) To count cells, first dilute 50 μl of CountBright™ absolute counting beads in 200 μl of PBS to obtain 200 beads/μl. Add 10 μl of the counting-bead dilution to the "unstained" FACS tube.

(e) Run cells through a flow cytometer to determine the number of migrated cells as a ratio to beads. In the stained-tube count, $CD45^{+}$ $CD3^{+}$ cells correspond to T lymphocytes, $CD45^{+}$ cells to leukocytes, and $CD45^{+}$ SSC^{hi} to monocytes.

3.1.2 Cell Staining and Counting Using Cytospin

(a) Count cells with the hemocytometer, as described in Section 2.3.1. Spin the cells at 2000 rpm for 3 min, Decant the supernatant and resuspend them in 200 μl of PBS. The supernatant can be stored and used in the future for cytokine estimation.

(b) To prepare the samples, place the cytoslide in the stainless steel clip, with the writing facing upward. Align on top the desired filter card and the funnel (opening facing upward) and clamp the construct tightly with the cytoclip. The clamp should also be facing upward when closed.

(c) Load 200 μl of the cell suspension into the funnel and place it in the centrifuge, ensuring that the lid is closed. Spin at 1000 rpm for 3 min.

(d) Unclip the sample and remove the filter and funnel. Cells should have attached to the slide.

(e) Place slides in a glass chamber filled with methanol to the top and fix overnight in −20 °C.

(f) Air-dry the slide and observe under the microscope. Cells can be stained using hematoxylin and eosin (H&E) or with a specific cell marker (see examples in Fig. 6).

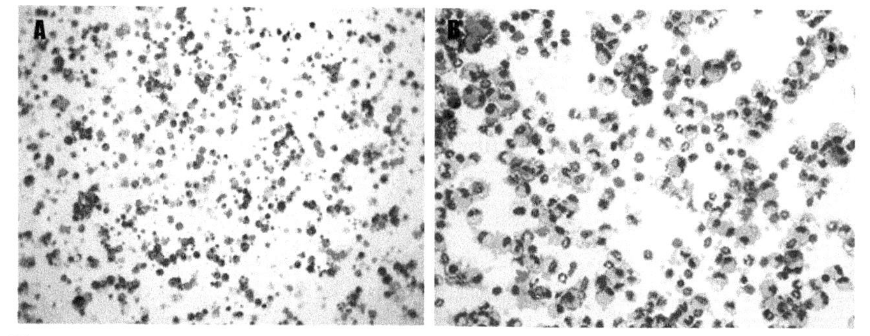

Figure 6 *In vivo* leukocyte recruitment to CCL7. The air pouch in the mouse was injected with 10 μg CCL7, and the mice were sacrificed 16 h later. The air pouch lavage was spun in the Cytospin4 cytocentrifuge and stained. Panel (A) shows staining with the mouse macrophage marker F4/80, and panel (B) shows hematoxylin and eosin (H&E) staining of the total cell population.

4. GENERATION OF MAMMALIAN TRANSFECTANTS EXPRESSING CHEMOKINE RECEPTORS

Detailed understanding of chemokine action requires the molecular cloning of chemokine receptors. The transfectant cells can be used to characterize the functional role of the chemokine receptor in isolation, facilitating its study (Harvey et al., 2007).

Required materials:

- Vector
  - pcDNA™3.1 (+) mammalian expression vector (Life technologies, Cat. No. V790-20), containing the G418 selectable marker
  - *Alternative*: pcDNA™3.1/Zeo (+) mammalian expression vector (Life technologies, Cat. No. V860-20)
- Subcloning Efficiency™ DH5α™ competent cells (Life technologies, Cat. No. 18265-017)
- LB broth (Miller, Sigma, Cat. No. L3522)
- LB broth with agar (Lennox, Sigma, Cat. No. L2897)
- QIAprep spin miniprep kit (Qiagen, Cat. No. 27104)
- Nucleofector™ 2b device (Lonza, Cat. No. AAB-1001)
- Amaxa™ Nucleofector™ solution V (Lonza, Cat. No. VCA-1003)
- Effectene transfection reagent (1 ml) (Qiagen, Cat. No. 301425)
- Cloning cylinders, glass (Sigma, Cat. No. C1059-1EA)
- Trypsin-EDTA solution (Sigma, Cat. No. T4049)

4.1 Chemokine Receptor Cloning

In order to create a transfectant cell line, the desired sequence must first be cloned into an appropriate expression vector. The vector must have both a bacterial selection marker (most commonly ampicillin) and a mammalian-cell selection marker (usually either geneticin (G418) or zeocin). One of the most common vectors, and the one used for this study, is the pcDNA3.1 vector, which can be purchased with either G418 or zeocin resistance.

Once the plasmid has been created using conventional molecular cloning methods (Green & Sambrook, 2012), it can be amplified by transforming competent *Escherichia coli*.

(a) First, prepare LB broth (liquid) and LB agar. After autoclaving, cool the agar to approximately 50 °C before pouring to avoid antibiotic denaturation. Add 100 μg/ml of ampicillin to the LB agar and pour the plates (thickness should be around 5 mm). Leave it to set for around half an hour.

(b) Recover a vial of chemically competent *E. coli* from −80 °C and thaw on ice (approximately 20–30 min).

(c) Mix your plasmid with 50 μl of competent cells in an Eppendorf. The optimal quantity of the plasmid will vary, depending on the vector, and thus, a titration from 1 to 5 μl of the ligation reaction should be carried out. In our hands, 2 ng of plasmid was found to be optimal.

(d) Place the competent cell/DNA mixture on ice for 30 min.

(e) Heat-shock the transformation tube by placing it into a 42 °C water bath for 20 s. This allows for the formation of pores in the *E. coli* membrane through which the plasmid can enter the cell.

(f) Place the tubes back on ice for 2 min. This will cause the pores to close, trapping the plasmid inside the cell.

(g) Add the (now transformed) cells to 1 ml of LB media and incubate for 1–2 h at 37 °C on a shaker (225 rpm) at 37 °C to recover the antibiotic resistance.

(h) Spread 100 and 200 μl of the cell mixture onto the preprepared LB plates with ampicillin. Store the remaining cells at 4 degrees in case the *E. coli* is too confluent and needs to be plated again.

(i) Incubate at 37 °C overnight. Incubate also a negative control dish with no cells plated.

(j) On the next day, pick up individual colonies and grow them overnight in universal tubes containing 10 ml of LB with antibiotic on a shaker at 37 °C.

4.2 Transfection of the Vector into the Host Cell Line

Once the transformed *E. coli* have been generated, the plasmid can be extracted using a kit and following the manufacturer's protocol. In our hands, the QIAprep spin miniprep kit produced good plasmid yields. The plasmid can now be sent for sequencing to confirm correct ligation.

4.2.1 Dose–Response Curve

First, a killing curve should be carried out to optimize the lowest antibiotic concentration that will kill untransfected cells.

(a) For each antibiotic concentration to be assessed, plate 1×10^5 cells per well in a 6-well plate and grow them overnight. It is important to bear in mind that the kill curve can be influenced by cell density, so, ideally, cells should be about 40% confluent the next day.

(b) The next day, change to selective media containing the desired antibiotic concentration. Typically, zeocin sensitivity should be assessed around 0–400 μg/ml and G418 around 0–1200 μg/ml.

(c) Assess confluence daily for 8 days. Replace the media with fresh antibiotic every 3 days. The chosen antibiotic concentration should ideally kill all cells during the given time period.

4.2.2 Transfection with Lipid Reagents

For most cell lines, lipid-based reagents allow for efficient transfection using the minimal amount of plasmid. In our group, we have used Effectene with good results, but many other reagents have a similar mechanism of action. This protocol has been optimized for CHO-K1 cells, and thus, the quantities of plasmid and Effectene can be altered depending on the cell line used.

(a) The day before transfection, seed 6×10^5 cells in a 6-cm dish in 5 ml of complete media. The number should be optimized for each cell line, so that they are about 80% confluent on the next day. Incubate overnight at 37 °C and 5% CO_2.

(b) The next day, dilute 2 μg of plasmid (it must be at a concentration equal or higher than 0.1 μg/μl) in buffer EC to a total volume of 150 μl and add 16 μl of enhancer. This will allow for the condensation of DNA, so that it is easier to pack. Mix by vortexing for 1 s and incubate at room temperature for 3 min.

(c) Add 20 μl of Effectene to the plasmid mixture and vortex for 10 s. Incubate samples for 8 min at room temperature, so that the Effectene micelles can surround the plasmid. In the meantime, remove the media from the 6-cm dish, wash the cells with 5 ml of PBS, and add 4 ml of fresh media.

(d) Mix 1 ml of media with the Effectene–DNA complex mixture and pipette it drop-wise onto the 6-cm plate. Swirl the dish to ensure the mixture reaches all the cells. The complexes will now undergo endocytosis, and the plasmid will be released from the endosome into the cytoplasm, from where it will migrate to the nucleus.

(e) Incubate cells overnight or for a minimum of 6 h. The next day, replace the media withg fresh material containing the selection antibiotic.

(f) Transfection efficiency can then be assessed using flow cytometry or RT-PCR.

4.2.3 Transfection with Amaxa

For hard-to-transfect cells, electroporation can be used to introduce the plasmid inside the cell. This method has a higher transfection rate, but also a higher cell death due to the electrical pulses required to open pores in the cells.

(a) Harvest cells for transfection by completing the following steps:
- Remove the medium from the cells and rinse them with PBS.
- Add 5 ml of trypsin and incubate for 5 min at 37 °C. Help detach the cells by gently tapping flasks and transfer them to a labeled universal tube containing 5 ml of media.
- Centrifuge the culture at 300 × *g* for 5 min to pellet the cells, and then decant the supernatant.

(b) Resuspend 1×10^6 cells in 100 μl of Amaxa™ Nucleofector™ solution. The solution will vary depending on the cell transfected—for instance, solution V can be used to transfect HEK-293 and CHO-K1 cells. A comprehensive list is available on the Lonza website.

(c) Add 5 μg of the transformed vector and pipette gently to mix.

(d) Transfer the cell solution to an Amaxa™ Nucleofector™ electroporation cuvette and insert in the machine.

(e) Run the appropriate program in the Amaxa machine (at 50 μFarads capacitance, between 300 and 500 V). The complete list of programs is available on the website—for instance, run program A-23 for HEK-293 or X-13 for MDA-MB-231 cells.

(f) Place the cuvette on ice for 5 min.

(g) Grow electroporated cells in a T-25 flask with complete media overnight at 37 °C. Given that the selection of transfectants is a long process, growth in flasks reduces the risk of infection, as opposed to growing them in Petri dishes. On the other hand, the lack of a removable lid makes placing cloning cylinders a more complex process (see Section 4.3.1).

(h) The next day, replace with fresh media containing the selection antibiotic. Transfection efficiency can be assessed at 24 and 48 h post-transfection, using flow cytometry.

4.3 Selection of Stable Transfection Cells

4.3.1 Colony-Picking

The isolation of transfectant colonies can be carried out in about 2–3 weeks, when transformant colonies of medium size have formed (such as the ones seen in Fig. 7). These colonies can be observed under the microscope, but they should also be visible to the naked eye when looking from below into the flask.

Figure 7 Example of a medium-sized colony formed by stably transfected CHO-K1 cells.

(a) The previous day, autoclave the following:
- Silicone grease. Ideally, sterilize the grease by placing a small amount in the bottom a glass Petri dish (never plastic; it will melt) and autoclaving it. Otherwise, it can be autoclaved in an Eppendorf tube inside a glass bottle and spread afterward in a sterile plastic Petri dish.
- Forceps (or thick tweezers).
- Cloning rings/cylinders. These can be purchased or created by cutting the wider-edge 1000-μl tips or Eppendorfs. In order to obtain a clean cut, the blade should be heated under the flame and then quickly used to cut the plastic. Autoclave the rings or cylinders in a paper bag.

(b) While looking at the flask from below against the light, draw a circle around the colonies on the bottom of the plate using a marker pen (they will look like small white dots). Be careful not to mark across the colony, because this will make it harder to see the cells under the microscope. Double-check that the colonies marked are alive and well, using the microscope.

The following steps should be done in a class-2 safety cabinet to ensure sterility:

(c) Prepare 24-well plates with 1 ml of media containing the selection antibiotic. Each marked colony should be placed in a separate well.

(d) If cells were cultured in a T-75, the lid will need to be cut using a hot blade. Simply follow the upper edge of the flask (the blade will need to be heated and cleaned from the melted plastic several times) until the upper plastic cover is almost totally cut. Use a pair of sterile tweezers to break the last bit and lift the "lid."

(e) Remove and discard the growth medium. Rinse the plate with PBS to remove any floating cells.

The following steps should be done as quickly as possible to avoid cells drying out:

(f) Using sterile forceps, pick up a cloning cylinder. Gently press the flat bottom (if they are self-made, the cut side will be ragged.) into the smooth silicone grease and remove with a sudden vertical motion to ensure an even distribution. Set the cylinder over a marked colony and press down evenly with the forceps. Try not to slide the cylinder across the colony, because this will smear the silicone grease over the cells and prevent the trypsin from contacting them.

(g) If a Petri dish was used, check under the microscope to ensure there is only one colony inside the cylinder.

(h) Add 200 μl of prewarmed trypsin to the cloning cylinders and incubate at 37 °C for 2–5 min. Alternatively, let the trypsin work at room temperature for 5–10 min.

(i) Examine the cells under the microscope to confirm that they have detached from the flask. Normally you cannot see the cells floating, given the small surface, but cells will appear rounded and with a surrounding glow if they have detached.

(j) Pipette the trypsin up and down the cloning cylinder and add it to one of the wells of the prepared 24-well plate(s). Rinse the cylinder with 200 μl of the well's media to transfer any remaining cells. Repeat for each marked colony.

(k) Incubate the cells at 37 °C and 5% CO_2 until colonies have grown (around 7–10 days). When cells are confluent in a 24-well plate, transfer them to a 6-well plate and then to a T-75.

(l) Expression of the gene can then be assessed using flow cytometry (see Fig. 8A) or RT-PCR (see Fig. 8B).

4.3.2 Single-Cell Dilution

Single-cell dilution is a more effective method for ensuring monoclonality but it has the downside of being a slow and inefficient process, because colonies do not always grow from a single cell. It is thus recommended for cells that grow quickly and easily.

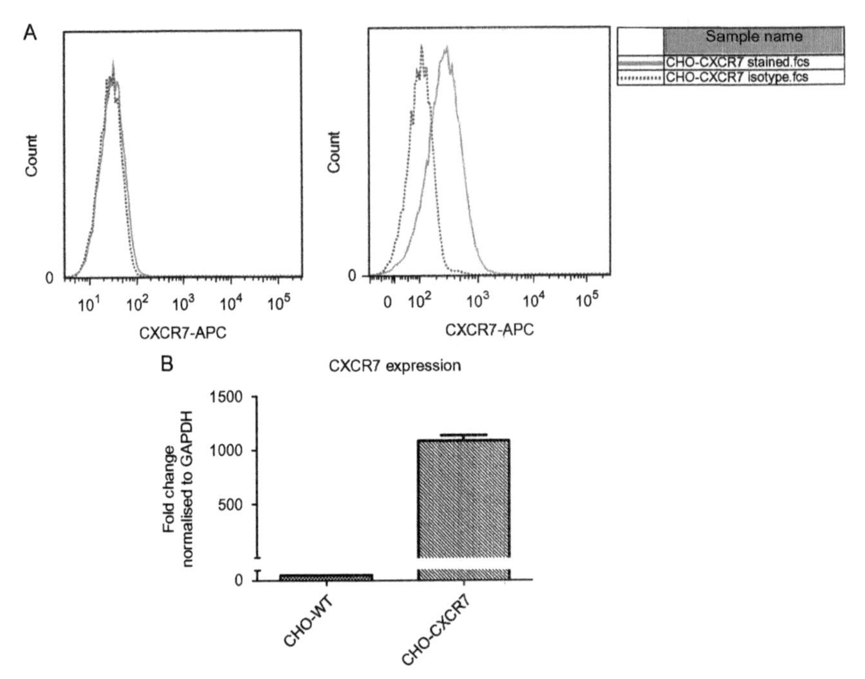

Figure 8 CHO cells were stably transfected with the CXCR7-pcDNA3 plasmid using Effectene and stably selected using 800 µg/ml of G418. Several colonies were isolated using cloning rings and expanded. CXCR7 expression was assessed using flow cytometry (A) and RT-PCR (B), and the levels were normalized to the housekeeping gene GAPDH, before being compared to CHO-WT.

(a) Discard the growth medium from your flask or plate. Rinse the plate with PBS to remove any floating cells and add 5 ml of trypsin. Incubate for 5 min at 37 °C until the cells are detached, and transfer them to a labeled universal tube containing 5 ml of media.

(b) Count the cells and centrifuge them at 300 × *g* for 5 min to pellet the cells. Then decant the supernatant. Resuspend the cells in selection media until you only have 100 cells/ml.

(c) Prepare three 96-well plates with 100 µl of selective media on columns 2–12.

(d) Add 200 µl of the cell solution to well A1. Mix well and pipette 100 µl into well A2. Mix well again and pipette 100 µl into well A3. Continue the serial dilution until well A12. If done correctly, single-cell colonies should form on wells A5 and A6.

(e) Repeat the process on rows B–H and on the second and third plates. The remaining cells can be placed in a 10-cm dish in case colonies do not grow, and the process needs to be started anew.
(f) Check on the microscope to confirm the presence of cells on the first rows.
(g) Incubate at 37 °C for 3–4 days. Check which wells have only one colony growing and mark them.
(h) Incubate for 1 week or until colonies have grown enough. Transfer colonies to a well of a 24-well plate, then a 6-well plate, and to a T-75 at last.
(i) Transfection efficiency can then be assessed using flow cytometry or RT-PCR.

5. DISCUSSION

Although the process is not perfect, Boyden chambers allow us to reproduce *in vitro* the chemotaxis process. This chemotaxis assay allows for the formation of a stable chemokine gradient that permits cell migration on the basis of chemokine–chemokine receptor strength interaction. Thus, there is complete disregard for the effect of cell–cell interaction, and cell motility is solely assessed (Cukierman, Pankov, & Yamada, 2002). This can be of particular importance if the effect of a chemotaxis inhibitor, such as a natural compound (Kobayashi et al., 2007), a truncated chemokine (Proost et al., 1998), or a blocking antibody, is to be assessed. Similarly, agonists can be used to enhance (Campbell et al., 1998) or impair chemotaxis through desensitization (Armstrong, 1995; Sogawa, Ohyama, Maeda, & Hirahara, 2011). Furthermore, its quickness and ease of use make it ideal for eliminating other confounding factors such as proliferation (Lukas, Lukas, & Bartek, 2004), and it is easily reproducible for inexperienced researchers.

Although the system contemplates the distinction of chemotaxis and chemokinesis, it does not account for the complexities of the extravasation process. To emulate these, the filter can be coated in advance. Even though some studies use components of the extracellular matrix such as collagen, we have found that the culture of endothelial cells on top of the filter optimally mimics *in vivo* conditions. Similarly, other chemotaxis assays are available in the market, most of them based on a two-chamber model. For instance, the Zigmond chamber and its improvement, the Dunn chamber allow for the visualization of the cell migration toward the chemoattractant, but they do not allow for a more *in vivo* simulation (Sackmann, Fulton, & Beebe, 2014).

Another option is the Cellix system, which reproduces the blood flow in which migrating cells roll and adhere to the endothelium (Blow, 2007; Naemi, Carter, Kirby, & Ali, 2013). This system allows the visualization of cells and closely mimics the *in vivo* system thanks to its sheath flow, but it requires the purchase of separate equipment and specific microchips.

All of these methods can only reproduce part of the process that occurs *in vivo*, therefore, a robust *in vivo* model is needed. Chemotaxis can be reproduced *in vivo* using intraperitoneal injection instead of creating an air pouch with similar results (Gaspar, Sakai, & De Gaspari, 2014). However, the peritoneal cavity does not allow for the formation of an enclosed cavity, which allows for a more reproducible environment (Edwards, Sedgwick, & Willoughby, 1981). This is especially useful when studying an anti-inflammatory compound, because the changes in the recruited cells can be closely monitored. Furthermore, the air pouch is an excellent model for the synovial lining, making it the best option when investigating arthritis (da Silva Guerra et al., 2011). The inflammatory microenvironment can be reproduced by injecting the cavity with LPS/Carrageenan. Air pouch experiments have also been carried out in humanized mouse models (O'Boyle et al., 2012), which have an advantage when evaluating drug candidates for potential human treatment. Additionally, the air pouch lining can be preserved and stained using immunohistochemical techniques. Other alternatives, such as intratracheal injections, are more invasive for the mice and require sutures, but they can be used when assessing acute lung inflammation (Ulich et al., 1991).

Lastly, single-cell dilution is a more-effective selection technique than cloning cylinders in ensuring monoclonality when picking stable transfectants, but it has the downside of longer cell-growth time or no growth. A technique that encompasses both is cell-sorting using flow cytometry (FACS). Cells are stained for the cloned chemokine receptor and injected in the flow machine. The single stream of cells is then hit by a laser, the cell fluorescence is measured, and positive cells are charged and selected using deflection plates (Bonner, Hulett, Sweet, & Herzenberg, 1972). This allows for high-chemokine receptor-expressing cells to be selected in large numbers, but it has the disadvantage of possible sample contamination and the requirement for additional equipment.

ACKNOWLEDGMENTS

The authors would like to thank Catriona Barker for the picture in Fig. 5, Yvonne Jenkins for the diagram in Fig. 1 and the graph in Fig. 3, and Helen Lawrence for the picture in Fig. 2.

This work was supported by grants from FP7-MCITN (POSAT, 606979), British Heart Foundation, and the Women's Cancer Detection Society.

REFERENCES

Ali, S., Fritchley, S. J., Chaffey, B. T., & Kirby, J. A. (2002). Contribution of the putative heparan sulfate-binding motif BBXB of RANTES to transendothelial migration. *Glycobiology, 12*(9), 535–543.

Ali, S., Robertson, H., Wain, J. H., Isaacs, J. D., Malik, G., & Kirby, J. A. (2005). A non-glycosaminoglycan-binding variant of CC chemokine ligand 7 (monocyte chemoattractant protein-3) antagonizes chemokine-mediated inflammation. *The Journal of Immunology, 175*(2), 1257–1266.

Armstrong, R. A. (1995). Investigation of the inhibitory effects of PGE2 and selective EP agonists on chemotaxis of human neutrophils. *British Journal of Pharmacology, 116*(7), 2903–2908.

Baggiolini, M. (1998). Chemokines and leukocyte traffic. *Nature, 392*(6676), 565–568.

Blow, N. (2007). Microfluidics: In search of a killer application. *Nature Methods, 4*(8), 665–670.

Bonner, W. A., Hulett, H. R., Sweet, R. G., & Herzenberg, L. A. (1972). Fluorescence activated cell sorting. *Review of Scientific Instruments, 43*(3), 404–409.

Boyden, S. (1962). The chemotactic effect of mixtures of antibody and antigen on polymorphonuclear leucocytes. *The Journal of Experimental Medicine, 115*(3), 453–466.

Campbell, J. J., Bowman, E. P., Murphy, K., Youngman, K. R., Siani, M. A., Thompson, D. A., et al. (1998). 6-C-kine (SLC), a lymphocyte adhesion-triggering chemokine expressed by high endothelium, is an agonist for the MIP-3β receptor CCR7. *The Journal of Cell Biology, 141*(4), 1053–1059.

Cukierman, E., Pankov, R., & Yamada, K. M. (2002). Cell interactions with three-dimensional matrices. *Current Opinion in Cell Biology, 14*(5), 633–640.

da Silva Guerra, A. S. H., do Nascimento Malta, D. J., Laranjeira, L. P. M., Maia, M. B. S., Colaço, N. C., de Lima, M. do. C., et al. (2011). Anti-inflammatory and antinociceptive activities of indole–imidazolidine derivatives. *International Immunopharmacology, 11*(11), 1816–1822.

Edwards, J. C. W., Sedgwick, A. D., & Willoughby, D. A. (1981). The formation of a structure with the features of synovial lining by subcutaneous injection of air: An in vivo tissue culture system. *The Journal of Pathology, 134*(2), 147–156.

Gaspar, E. B., Sakai, Y. I., & De Gaspari, E. (2014). A mouse air pouch model for evaluating the immune response to Taenia crassiceps infection. *Experimental Parasitology, 137*, 66–73.

Green, M. R., & Sambrook, J. (Eds.), (2012). *Plasmid vectors. Molecular cloning: A laboratory manual: Vol. 1* (pp. 1.3–1.105). New York: Cold Spring Harbor Laboratory Press.

Harvey, J. R., Mellor, P., Eldaly, H., Lennard, T. W. J., Kirby, J. A., & Ali, S. (2007). Inhibition of CXCR4-mediated breast cancer metastasis: A potential role for heparinoids? *Clinical Cancer Research, 13*(5), 1562–1570.

Hashimoto, K., Kataoka, N., Nakamura, E., Hagihara, K., Okamoto, T., Kanouchi, H., et al. (2012). Live-cell visualization of the trans-cellular mode of monocyte transmigration across the vascular endothelium, and its relationship with endothelial PECAM-1. *The Journal of Physiological Sciences, 62*(1), 63–69.

Kobayashi, H., Kitamura, K., Nagai, K., Nakao, Y., Fusetani, N., van Soest, R. W. M., et al. (2007). Carteramine A, an inhibitor of neutrophil chemotaxis, from the marine sponge Stylissa carteri. *Tetrahedron Letters, 48*(12), 2127–2129.

Kuschert, G. S. V., Coulin, F., Power, C. A., Proudfoot, A. E. I., Hubbard, R. E., Hoogewerf, A. J., et al. (1999). Glycosaminoglycans interact selectively with chemokines and modulate receptor binding and cellular responses. *Biochemistry, 38*(39), 12959–12968.

Lerchenberger, M., Uhl, B., Stark, K., Zuchtriegel, G., Eckart, A., Miller, M., et al. (2013). Matrix metalloproteinases modulate ameboid-like migration of neutrophils through inflamed interstitial tissue. *Blood*, *122*(5), 770–780.

Lowe, J. B. (2003). Glycan-dependent leukocyte adhesion and recruitment in inflammation. *Current Opinion in Cell Biology*, *15*(5), 531–538.

Lukas, J., Lukas, C., & Bartek, J. (2004). Mammalian cell cycle checkpoints: Signalling pathways and their organization in space and time. *DNA Repair*, *3*(8), 997–1007.

Müller, A., Homey, B., Soto, H., Ge, N., Catron, D., Buchanan, M. E., et al. (2001). Involvement of chemokine receptors in breast cancer metastasis. *Nature*, *410*(6824), 50–56.

Muller, W. A., & Weigl, S. A. (1992). Monocyte-selective transendothelial migration: Dissection of the binding and transmigration phases by an in vitro assay. *The Journal of Experimental Medicine*, *176*(3), 819–828.

Naemi, F. M. A., Carter, V., Kirby, J. A., & Ali, S. (2013). Anti-donor HLA class I antibodies: Pathways to endothelial cell activation and cell-mediated allograft rejection. *Transplantation*, *96*(3), 258–266.

O'Boyle, G., Fox, C. R. J., Walden, H. R., Willet, J. D. P., Mavin, E. R., Hine, D. W., et al. (2012). Chemokine receptor CXCR3 agonist prevents human T-cell migration in a humanized model of arthritic inflammation. *Proceedings of the National Academy of Sciences of the United States of America*, *109*(12), 4598–4603.

Parish, C. R. (2006). The role of heparan sulphate in inflammation. *Nature Reviews. Immunology*, *6*(9), 633–643.

Proost, P., De Meester, I., Schols, D., Struyf, S., Lambeir, A.-M., Wuyts, A., et al. (1998). Amino-terminal Truncation of Chemokines by CD26/dipeptidyl-peptidase IV conversion of RANTES into a potent inhibitor of monocyte chemotaxis and HIV-1-Infection. *The Journal of Biological Chemistry*, *273*(13), 7222–7227.

Proudfoot, A. E. I., Handel, T. M., Johnson, Z., Lau, E. K., LiWang, P., Clark-Lewis, I., et al. (2003). Glycosaminoglycan binding and oligomerization are essential for the in vivo activity of certain chemokines. *Proceedings of the National Academy of Sciences of the United States of America*, *100*(4), 1885–1890.

Rot, A., & von Andrian, U. H. (2004). Chemokines in innate and adaptive host defense: Basic chemokinese grammar for immune cells. *Annual Review of Immunology*, *22*, 891–928.

Sackmann, E. K., Fulton, A. L., & Beebe, D. J. (2014). The present and future role of microfluidics in biomedical research. *Nature*, *507*(7491), 181–189.

Salanga, C. L., O'Hayre, M., & Handel, T. (2009). Modulation of chemokine receptor activity through dimerization and crosstalk. *Cellular and Molecular Life Sciences*, *66*(8), 1370–1386.

Schenkel, A. R., Mamdouh, Z., & Muller, W. A. (2004). Locomotion of monocytes on endothelium is a critical step during extravasation. *Nature Immunology*, *5*(4), 393–400.

Scotton, C. J., Wilson, J. L., Milliken, D., Stamp, G., & Balkwill, F. R. (2001). Epithelial cancer cell migration a role for chemokine receptors? *Cancer Research*, *61*(13), 4961–4965.

Sogawa, Y., Ohyama, T., Maeda, H., & Hirahara, K. (2011). Formyl peptide receptor 1 and 2 dual agonist inhibits human neutrophil chemotaxis by the induction of chemoattractant receptor cross-desensitization. *Journal of Pharmacological Sciences*, *115*(1), 63–68.

Ulich, T. R., Yin, S., Guo, K., Yi, E. S., Remick, D., & Del Castillo, J. (1991). Intratracheal injection of endotoxin and cytokines. II. Interleukin-6 and transforming growth factor beta inhibit acute inflammation. *The American Journal of Pathology*, *138*(5), 1097.

Ulvmar, M. H., Werth, K., Braun, A., Kelay, P., Hub, E., Eller, K., et al. (2014). The atypical chemokine receptor CCRL1 shapes functional CCL21 gradients in lymph nodes. *Nature Immunology*, *15*(7), 623–630.

Ward, S., & Marelli-Berg, F. (2009). Mechanisms of chemokine and antigen-dependent T-lymphocyte navigation. *The Biochemical Journal, 418*, 13–27.
Weber, M., Hauschild, R., Schwarz, J., Moussion, C., de Vries, I., Legler, D. F., et al. (2013). Interstitial dendritic cell guidance by haptotactic chemokine gradients. *Science, 339*(6117), 328–332.
Zlotnik, A., Burkhardt, A. M., & Homey, B. (2011). Homeostatic chemokine receptors and organ-specific metastasis. *Nature Reviews. Immunology, 11*(9), 597–606.
Zlotnik, A., & Yoshie, O. (2000). Chemokines: A new classification system and their role in immunity. *Immunity, 12*(2), 121–127.

CHAPTER SIXTEEN

Examining Roles of Glycans in Chemokine-Mediated Dendritic–Endothelial Cell Interactions

Xin Yin[*,†,‡,1], Scott C. Johns[†,‡,1], Roland El Ghazal[†,‡], Catherina L. Salanga[§], Tracy M. Handel[§], Mark M. Fuster[†,‡,2]

[*]Jiangsu Key Laboratory of Marine Pharmaceutical Compound Screening, School of Pharmacy, Huaihai Institute of Technology, Lianyungang, China
[†]VA San Diego Healthcare System, Medical and Research Sections, La Jolla, California, USA
[‡]Department of Medicine, Division of Pulmonary and Critical Care, University of California San Diego, La Jolla, California, USA
[§]Department of Pharmacology, Skaggs School of Pharmacy and Pharmaceutical Sciences, University of California San Diego, La Jolla, California, USA
[2]Corresponding author: e-mail address: mfuster@ucsd.edu

Contents

1. Introduction 336
2. Methods 339
   2.1 DC Migration Toward LEC as Mediated by HS–Chemokine Interaction 341
   2.2 DC Adhesion to and Transmigration Across LECs as Mediated by HS–Chemokine Interaction 343
   2.3 Visualizing Chemokine–Chemokine Receptor Interaction *In Situ* 347
3. Perspectives 350
Acknowledgments 353
References 353

Abstract

Interactions between glycosaminoglycans (GAGs) and chemokines play a critical role in multiple physiological and pathological processes, including tumor metastasis and immune-cell trafficking. During our studies examining the genetic importance of the GAG subtype known as heparan sulfate (HS) on lymphatic endothelial cells (LECs), we established a repertoire of methods to assess how HS affects chemokine-mediated cell–cell interactions. In this chapter, we describe methods for monitoring migration and adhesion interactions of dendritic cells (DCs), the most potent antigen-presenting cells, with LECs. We will also report a methodology to assess chemokine–receptor interactions while incorporating approaches to target HS in the system. This includes *in situ* methods to visualize and quantify direct interactions between chemokines and chemokine receptors on DC surfaces, and how targeting HS produced by LECs or even DCs

[1] These authors contributed equally to this manuscript.

Methods in Enzymology, Volume 570
ISSN 0076-6879
http://dx.doi.org/10.1016/bs.mie.2015.09.017

affects these interactions. These methods enable the mechanistic and functional characterization of GAG–chemokine interactions in cell-based studies that model physiologic interactions *ex vivo*. They may also be used to obtain novel insights into GAG-mediated biological processes.

1. INTRODUCTION

Dendritic cells (DCs) are master regulators of immunity. They are a heterogeneous population of professional antigen-presenting cells that can prime naïve T cells to induce either protective immune responses or immune tolerance. Upon exposure to pathogens, DCs undergo rapid maturation, migrate to draining lymph nodes (DLN), and activate protective T cell responses. In some settings such as cancer, however, a predominance of "tolerogenic" DCs may arrive at the DLN, induce anergy in antigen-specific T cells as well as the generation of Foxp3+ regulatory T (Treg) cells. In either setting, "downstream" DC functions critically rely on DC trafficking from the periphery to the DLN.

Trafficking of DCs from the periphery through afferent lymphatic vessels to the DLN is a multistep process involving local chemotaxis within the extracellular matrix (ECM), adhesion to lymphatic endothelium, trans-endothelial migration, intralumenal transit following lymph flow, and adhesion to, as well as extravasation from, lymphatic vessels into the LN parenchyma (Fig. 1).

This process depends on many chemokine–receptor interactions, among which the CCR7–CCL21/CCL19 axis is the most potent and best characterized. The LN-homing chemokine receptor CCR7 is upregulated in DC during DC maturation. Through interactions with its cognate ligands CCL19 and CCL21, which are produced by stromal cells in LN and lymphatic endothelial cells (LECs), CCR7 essentially guides the directional migration of antigen-presenting DCs toward the DLN (Forster, Davalos-Misslitz, & Rot, 2008; Randolph, Angeli, & Swartz, 2005). In addition to CCL19 and CCL21, other chemokines, including CCL3, CCL4, CCL5, CCL7, CCL13, CCL15, CCL20, CCL25, CXCL12, CX3CL1, and their corresponding receptors, are also suggested to regulate DC trafficking *in vivo* (Dieu-Nosjean, Vicari, Lebecque, & Caux, 1999; Johnson & Jackson, 2013). Despite the biological significance of chemokine–receptor interactions, little is known about the molecular mechanisms that control DC trafficking at various steps en route toward the DLN.

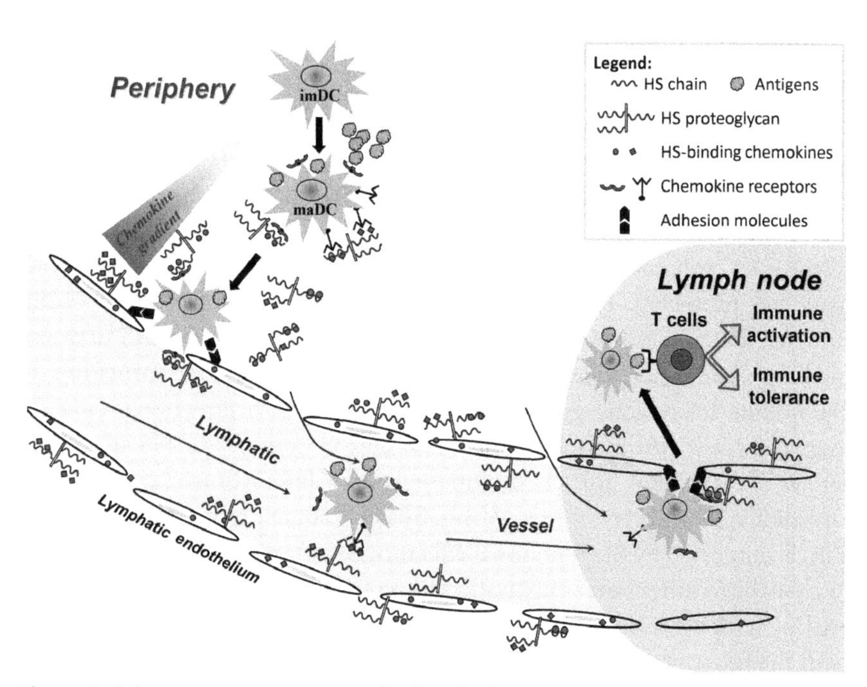

Figure 1 Schematic representation of DC traffic from the periphery to the draining LN (DLN) and the potential mechanisms for HS in this process. Immature DCs (imDC) pick up antigen in the periphery, undergo maturation to become mature DCs (maDC), and are associated with the upregulation of certain chemokine receptors, such as CCR7. Following HS-mediated chemokine gradients, lymphatic flow, and adhesion events that may also be mediated by lymphatic endothelial HS, antigen-loaded DCs travel from the periphery to the DLN, where they present antigen to T cells and modulate the balance between immune activation and immune tolerance. Importantly, HS on soluble proteoglycans may also mediate chemokine–receptor interactions *in trans*. (See the color plate.)

Heparan sulfate (HS) is a negatively charged, linear polysaccharide that is ubiquitously expressed on cell surfaces, in the ECM, and in basement membranes via covalent linkage to a small number of core proteins to form HS proteoglycans. Structurally, HS is a linear chain of 10–200 disaccharide units of *N*-acetyl-D-glucosamine and D-glucuronic acid, with occasional epimerization of the latter to iduronic acid at discrete points along the glycan chain (Rabenstein, 2002). The disaccharide repeat unit is modified by different HS biosynthetic enzymes to include *N*- and *O*-sulfation (6-*O*- and 3-*O*-sulfation of the glucosamine and 2-*O*-sulfation of the uronic acid) and epimerization of β-D-glucuronic acid to α-L-iduronic acid (Rabenstein, 2002; Sarrazin, Lamanna, & Esko, 2011). These modifications confer structural and subsequent functional diversity to HS. This glycan regulates many cellular processes including proliferation, adhesion, migration, endocytosis, and signal transduction mainly through interactions with a variety of protein ligands, including growth factors such as basic fibroblast growth factor (FGF-2), splice variants of vascular endothelial growth factor A (VEGF-A), VEGF-C, and chemokines such as interleukin 8 (IL-8/CXCL8), CCL21, and stroma-derived factor 1α (SDF-1α/CXCL12) (Capila & Linhardt, 2002). Through interactions with HS, chemokines may be protected from proteolytic degradation (Rot, 2010; Sadir, Imberty, Baleux, & Lortat-Jacob, 2004; Wagner et al., 1998); HS may also mediate chemokine oligomerization, which facilitates the formation of chemokine haptotactic gradients that provide critical directional cues for migrating cells (Proudfoot et al., 2003) and in some cases HS may even act as a coreceptor for chemokine–receptor interactions (Lortat-Jacob, Grosdidier, & Imberty, 2002; Wang, Sharp, Handel, & Prestegard, 2013). This is supported by the functional effects of genetically altering HS sulfation or chain biosynthesis on either trafficking cells or endothelial beds while correlating with tissue or *ex vivo* demonstration of altered interactions of CCL21 with the relevant cell surfaces (where signaling by cognate CCR7 is altered as a result) (Bao et al., 2010; Tsuboi et al., 2013; Yin et al., 2014).

We have been interested in understanding the effects of HS on distinct steps of chemokine-mediated cell trafficking. Our recent studies demonstrate that lymphatic endothelial HS modulates the optimal presentation of chemokine CCL21 both *in cis* (on either CCR7+ lymphatic endothelium or in more recent studies, on DCs) and *in trans* (on CCR7+ migrating cells), while regulating the formation of chemokine gradients and mediating migration signaling (Yin et al., 2014, 2010). Functionally, disrupting the biosynthesis of lymphatic endothelial HS, by targeting either HS chain initiation or chain sulfation, may significantly reduce chemokine-dependent tumor cell metastasis and DC trafficking in the lymphatic system (Yin et al., 2014, 2010). While performing these genetic validation studies, we have established a repertoire of experimental methods to examine chemokine-mediated cell–cell interactions. We have also initiated studies to probe the HS-dependent engagement of key lymphatic chemokines with their cognate receptors. In this chapter, we describe three main protocols for examining roles of HS on chemokine-mediated DC–LEC interactions, namely: (1) *ex vivo* DC migration toward cultured primary LECs, (2) DC adhesion with and transmigration across LEC monolayers under shear flow, and (3) interactions between chemokines secreted from LECs with chemokine receptors on the DC surface. The focus of this chapter is not on a specific chemokine–receptor pair, although the CCL21–CCR7 interaction happens to serve as a functionally useful model system, but on how to optimally modify and apply available methodology to address how targeting key modulators (e.g., HS) affects chemokine-receptor-dependent phenomena such as the trafficking behaviors of cells.

2. METHODS

All methods described in this chapter involve the use of two main cell types, DCs and LECs. The proper culture and treatment of these two cell types ensures accuracy and consistency while addressing various questions in working assays. With specific regard to DCs, since different DC maturation stages (as well as different DC subtypes at the same stage of culture) may present distinct expression profiles of chemokine receptors (thus responding to different chemokines; McColl, 2002), it is important to use the proper DC subtype/stage expressing the chemokine receptor of interest. In our studies, major interest lies in the CCL21–CCR7 axis; therefore, we use mature myeloid DCs that have been demonstrated to express CCR7 (Hansson et al., 2006; McColl, 2002). For this purpose, we differentiate DCs from mouse bone marrow precursor cells in the presence of granulocyte/macrophage colony-stimulating factor (GM-CSF), as described elsewhere (Madaan, Verma, Singh, Jain, & Jaggi, 2014). In our hands, cells collected on Day 7 or 8 after GM-CSF induction are approximately 90% positive for CD11c (Fig. 2A) and variably positive (depending on maturation induction by lipopolysaccharides, LPS) for CD86 and MHCII (biomarkers for mature DCs, Fig. 2B).

We have used these cells in various biological assays. To alter HS expression or its fine structure in DCs, we isolated DCs from mice bearing whole-body or myeloid-specific knockouts in HS biosynthesis (Sarrazin et al., 2011).

For LECs, we initiated our studies using the human lung lymphatic microvascular endothelial cells (HMVEC-LLy) that are commercially available from Lonza (CC-2814; Basel, Switzerland). These are primary cells isolated from microvascular lymphatic vessels in distal lymphatic endothelium. These cells provide an optimal model to study DC trafficking behaviors because the distal lymphatic endothelium are where DC adhesion, transmigration, and entry occur *in vivo* during the initial steps of DC trafficking from the lung. They are cultured in endothelium cell growth medium (EGM medium; Lonza) that is appropriately supplemented (2MV BulletKit;

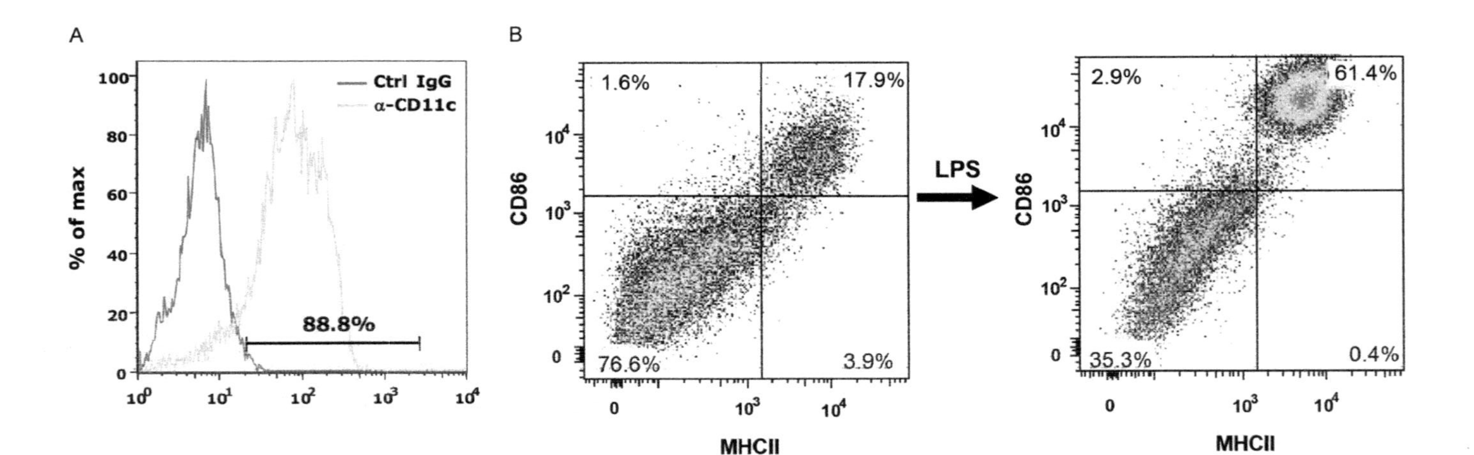

Figure 2 Granulocyte/macrophage colony-stimulating factor (GM-CSF) drives differentiation of bone marrow cells into mature myeloid DCs. On day 8 after GM-CSF differentiation, cells were analyzed by flow cytometry for DC marker (% positive for CD11c, A), as well as maturation markers (i.e., CD11c+cells that are both positive for MHCII and CD86, B). The treatment of LPS significantly promotes the maturation of DCs. The percentage of MHCII+CD86+ DCs increased from 18% without LPS treatment to 61% after treatment (B).

Lonza); and cells propagated until a total of 12 population doublings (from initial tissue harvest) are used in our biological assays. At third passage from commercial receipt (i.e., total of 5–6 population doublings), the cells are confirmed in our own hands to be >99% pure by Prox-1 nuclear staining (Yin et al., 2011). In addition, we have begun employing mouse primary LECs from other commercial sources (e.g., LN primary lymphatic-derived LECs; Cell Biologics) in order to target and/or employ mouse-specific lines in some of our assays. As an RNAi-based method to alter the expression or HS fine structure in cultured LECs, we transiently transfect the cells with a mixture of three siRNA duplexes (IDT, Coralville, IA) targeting a specific HS biosynthetic enzyme. In our experience, between 48 and 72 h after transfection, we typically achieve approximately 70–80% knockdown on the expression of a specific target HS biosynthetic enzyme (data not shown).

2.1 DC Migration Toward LEC as Mediated by HS–Chemokine Interaction

In this protocol, we use the transwell chamber to examine the migration of DCs toward LEC monolayers in the presence of neutralizing antibodies against chemokines or cognate chemokine receptors. To assess the importance of HS in this migration process, we isolate DCs or LECs from mice bearing genetic alterations in HS biosynthesis targeted to either tissue type. Alternatively, we transiently transfect LECs or DCs with siRNA targeting a particular HS biosynthetic enzyme. DCs are generally nonadherent in culture, although adherence properties may change if induced to mature with exogenous cytokines. Thus, when used directly from culture as immature/unstimulated cells, the DCs will not firmly attach to the reverse side of the transwell insert, but will migrate into the medium toward the bottom-well LEC monolayer. In order to accurately quantify the number of migrated DCs, we label DCs with a fluorescent dye such as calcein AM (acetoxymethyl ester form), which is a cell-permeant dye that is hydrolyzed to a strong green-fluorescent form by cellular esterases once the molecule enters live cells (Braut-Boucher et al., 1995). Alternatively, we have used GFP-labeled DCs isolated from GFP-expressing mice.

2.1.1 Required Materials

- Cells: Log-phase LECs and DCs cultured and treated accordingly
- Transwell insert (3.0 μm pore size; Corning; Cat. No. 3415)
- 24-well tissue culture plate (Corning; Cat. No. 3526)
- Calcein AM (10 m*M* in DMSO; eBioscience; Cat. No. 68-0853-39)

- 0.25% trypsin/EDTA (Life Technologies; 25200056)
- Cell scraper (Corning; Cat. No. 3010)
- 72-well Terasaki plate (Sigma; Cat. No. M5812)
- Neutralizing antibodies for chemokine or chemokine receptor: anti-CCR7 (R&D; Cat. No. MAB197), anti-CCL21 (R&D; Cat. No. AF366), anti-CCL19 (R&D; Cat. No. AF361), anti-CXCR4 (R&D; Cat. No. MAB172); anti-CXCL12 (R&D; Cat. No. AF-310-NA); anti-CCL5 (eBioscience; Cat. No. 14-7993-81).
- Fluorescence microscope (Nikon; Eclipse 80i)
- Serum-free EGM media (Lonza; Cat. No. CC-3156)
- EGM-2MV BulletKit (Lonza; Cat. No. CC-3202)

2.1.2 Procedures

1. Seed LECs (1×10^5) into each well of 24-well plate in triplicate for each condition to be tested. Leave three blank wells as a no-cell (NC) negative control.
2. After 24 h, LECs grow to approximately 80–90% confluency. Treat LECs accordingly as described below.
   a. For siRNA transfection, transfect the cells in OptiMEM using Lipofectamine following the manufacturer's instructions for 6 h. Then remove the supernatant and directly add 500 μL of prewarmed complete LEC medium to each well (no washing after removing the transfection supernatant). Incubate the cells at 37 °C, 5% CO_2 for a further 42 h.
   b. For treatment with neutralizing antibodies, wash the cells once with PBS, add serum-free EGM medium with anti-CCR7 (100 ×), anti-CXCR4 (100 ×), anti-CCL5 (200 ×), anti-CXCL12 (50 ×), anti-CCL19 (50 ×), or anti-CCL21 (50 ×) (500 μL/well), and incubate at 37 °C, 5% CO_2. Leave all neutralizing antibodies in the well for the whole migration period.
3. Label DCs with calcein AM: add calcein AM directly into the tissue culture plate containing DCs to a final concentration of 2 μ*M*. Incubate at 37 °C, 5% CO_2 for 1–2 h. DCs successfully labeled with calcein AM should present a green-yellow color.
4. Place transwell insert into 24-well plate. Lift calcein AM-labeled DCs with cell scraper. Count DCs and load 1 to 5×10^5 cells in 100 μL of serum-free EGM medium into the top well of the insert.
5. Let migration proceed at 37 °C, 5% CO_2 for 3–6 h.

Note: The migration time frame depends mainly on the migration velocity of target cells, which needs to be determined in preliminary experiments. Allowing too long of a period may saturate DC migration in a relatively early time frame and may reduce the true signal-to-noise ratio thereby masking the difference between different experimental conditions.

6. After the migration period, remove the plate. All of the following steps should be done on a lab bench with minimal exposure to light:
7. Lift the transwell insert out of the well, and after discarding medium from the top well of the insert, transfer all medium from the bottom well to a clean eppendorf tube. This step collects DCs that migrated into the medium of the bottom well.
8. Place the transwell insert back into the well. Add 500 μL of 0.25% trypsin into the bottom well and rock the plate gently at 37 °C for 5 min. The trypsin will help to lift DCs attached to the bottom side of the insert and cells (both LECs and DCs) from the bottom well.
9. Remove the transwell insert and transfer all trypsin solution from the bottom well to the same eppendorf tube from step 7.
10. Centrifuge the cells in the eppendorf tube from step 9 at $500 \times g$ for 10 min, discard the supernatant, resuspend the cell pellet in 15–20 μL PBS, and load 2 μL onto a 72-well Terasaki (microtiter) plate, and image the cells under fluorescence microscope (40 ×).
11. The images are analyzed using Image J software, with the areas of fluorescent calcein signal quantified and averaged from triplicates.

2.1.3 Results

A schematic illustration of the migration assay is shown in Fig. 3A, with a representative image of migrated fluorescent DCs in the Terasaki plate-well in Fig. 3B, along with quantification of DC migration toward LEC monolayers as affected by either neutralizing antibodies or siRNA targeting HS biosynthetic enzymes in Fig. 3C.

2.2 DC Adhesion to and Transmigration Across LECs as Mediated by HS–Chemokine Interaction

In this protocol, we examine DC adhesion to and transmigration across LECs under shear flow, which, for example, mimics DC behavior when exiting the lymphatic vessels into the LN. This assay is performed using a BioFlux system, an automated system for live cell assays under shear flow

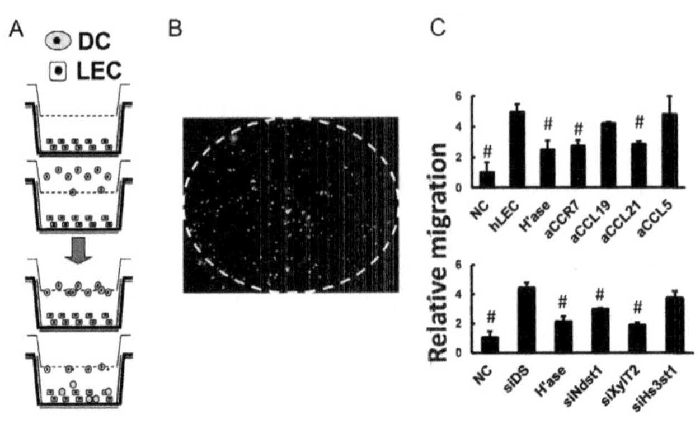

Figure 3 Altering LEC-associated HS reduces LEC-driven directional migration of DCs across liquid medium. (A) A scheme for the transwell migration assay using DCs prelabeled with calcein and LECs. (B) A representative microscope image (100×) of migrated calcein-labeled DCs in Terasaki plate (well edges marked by the dashed line). (C) Transwell migration of DCs into wells containing no cells (NC) or LECs treated as indicated was quantified and normalized to NC. αCCR7, αCCL19, αCCL21, or αCCL5, blocking antibodies to CCR7, CCL19, CCL21, or CCL5, respectively; H'ase, LECs pretreated with heparin lyases; siDS, LECs transfected with control (scrambled duplex) RNA; siNdst1, siXylT2, or siHs3st1, LECs transfected with siRNA targeting corresponding HS biosynthetic enzymes. Data presented represent at least three independent experiments. $^{*}P<0.05$, $^{\#}P<0.01$, as compared to LEC control group in (C) (upper panel), and siDS control group in (C) (lower panel).

developed by Fluxion Biosciences. Since the adhesion and transmigration experiment is performed on a microscope platform outside of the incubator environment, it is important to include HEPES (final concentration: 10 m*M*) in the cell culture medium to buffer the pH. The significance of chemokine–receptor interactions and HS is assessed by using the same approaches described in Protocol 1.2; i.e., the neutralizing antibodies for chemokines or chemokine receptors, and siRNA targeting specific HS biosynthetic enzymes, respectively.

2.2.1 Required Materials

- Cells: Log-phase LECs and DCs cultured and treated as described above
- Bioflux 200 system (Fluxion Biosciences)
- Bioflux 48-well plate (Fluxion Biosciences; Cat. No. 900-0014)
- Matrigel (BD Bioscience; Cat. No. 356230)
- Fluorescence microscope (Perkin Elmer; UltraView Vox Spinning Disk Confocal Microscope)

2.2.2 Procedures

2.2.2.1 Preparation of LEC Monolayer in the Flow Channel of a Bioflux Plate

1. Prime/coat the channels: Add diluted matrigel (1:50 in PBS, 100 μL/well) into the outlet wells. Connect the interface and apply a shear force of 5 dyn/cm^2 for 5 min to push PBS into the inlet wells. Incubate the plate at 37 °C for 1 h.

   Note: Before using the BioFlux Plates, it is important to prime the channels that will be used in the experiment. The purpose of priming is to introduce fluid all the way through the channel and to prevent air bubbles from getting into the system. The purpose of coating the channel is to promote cell adhesion. The selection of the coating reagent depends on the cell type. For LECs, our preliminary trial suggests matrigel is a better coating reagent than collagen or gelatin.
2. Add 500 μL of prewarmed LEC growth medium to the outlet well and wash coated channels at 5 dyn/cm^2 for 10 min.
3. Remove all excess liquid from upper and lower reservoir of both inlet and outlet wells, being extremely careful not to remove liquid from the channel entrance, which can be seen as a small bead in the lower reservoir.
4. Balance the liquid level in the inlet and outlet wells by first adding 50 μL of complete LEC medium into the inlet well followed by 50 μL of medium containing LECs that are well suspended as single cells (to avoid clogging of the channels) into the outlet well.

   Note 1: The purpose of this step is to prevent gravity flow of cells between the inlet and outlet wells after seeding LECs into the flow channel. It is important to finish this whole seeding process within 5 min in small groups of no more than six channels at a time.

   Note 2: In order to achieve a confluent monolayer of LEC cells the second day, the cell density in the outlet wells needs to be predetermined according to cell type, generally between 2×10^6–10^7 cells/mL. For LECs, the optimal density of seeding is approximately 1×10^7 cells/mL. At this point, either primary LECs or LECs that have been transfected with siRNA can be seeded into the channel.
5. Perfuse the cells into the channels at a shear force of 2 dyn/cm^2 and monitor this process on microscope. Stop the shear flow (generally within 5–7 s) when you see the whole channel is evenly covered with cell suspension.
6. Remove the interface and place the tissue culture lid on the plate. Incubate for 1–2 h in the tissue culture incubator to allow initial cell

attachment to the coated channel. Visually inspect channels under the microscope. At least 20–30% of the cells should start to attach to the channel; if not, extend the attachment time. In our hands, 2 h is sufficient for the initial attachment of LECs.

7. Add 200 μL of complete LEC growth medium into both the inlet and outlet wells. Incubate in the tissue culture incubator for another 16–18 h.

2.2.2.2 Examine DC Adhesion to and Transmigration Across LECs

1. The next day, label DCs with calcein AM: add calcein AM directly into the tissue culture plate containing DCs to a final concentration of 2 μ*M*. Incubate at 37 °C, 5% CO_2 for 1–2 h. DCs successfully labeled with calcein AM should present a green-yellow color.
2. Meanwhile, treat LECs in the flow channel with neutralizing antibodies for 1 h by replacing the medium in both the inlet and outlet wells with fresh serum-free EGM medium containing the corresponding neutralizing antibody.
3. Lift calcein AM-labeled DCs using the cell scraper, wash the cells once with PBS, and resuspend them at a concentration of 2×10^6 cells/mL in serum-free EGM medium. The neutralizing antibodies could also be added to the DCs at this step.

   Note: To minimize the interference of active components from the serum, such as growth factors or complex carbohydrates, we use serum-free medium during the experiment. In addition, it is important to maintain both cell types in a healthy state during the adhesion/transmigration experiment. Since LECs are more fragile and sensitive than DCs in response to media change, we use basal LEC media, i.e., EGM, instead of RPMI (basal DC medium) to resuspend the DCs.
4. Add 100 μL of DC suspension to the inlet well and perfuse at 2 dyn/cm^2 for 1 min or until they are visible in the viewing window of the microscope. Adjust the shear force to promote attachment of the DCs, which should be determined based on the endothelial cell type and the physiological or pathological process under investigation. To mimic the low shear force within terminal lymphatic vessels, we choose 0.14 dyn/cm^2 and proceed with a 15-min adhesion interval.
5. Upon the completion of the adhesion interval, remove unused cells from the inlet well, replenish the inlet well with 500 μL of serum-free EGM media, and initiate flow at 1 dyn/cm^2 for 5 min to wash out any loosely attached DCs. At the end of the wash period, image the flow channel for

quantitative measurement of adhesion events and also for an initial image for transmigration at time 0 (T_0).

6. Place the plate on a transparent 37 °C heating adaptor, the permissive temperature for transmigration, reset the shear flow to 0.14 dyn/cm^2, and allow transmigration to proceed at 37 °C for 30 min. To monitor the transmigration process, capture an end-point image in the same field after 30 min (T_{30}). The percentage transmigration is calculated as [(No. of adherent DCs at T_0 − No. of adherent DC at T_{30})/No. of adherent DCs at T_0] × 100%.

2.2.3 Results

Representative images of adherent DCs before and after transmigration are shown in Fig. 4A and the quantification of the adhesion events and percentage transmigration as affected by targeting different chemokines or lymphatic endothelial HS is shown in Fig. 4B.

2.3 Visualizing Chemokine–Chemokine Receptor Interaction *In Situ*

In this protocol, we examine the interaction between the chemokine CCL21 and its receptor CCR7 within the mouse Lewis Lung Carcinoma (LLC) tumor xenograft though the use of a proximity ligation assay (PLA). The PLA assay detects the proximity of two targets (with the maximum distance between two epitopes being 30–40 nm) through the use of primary antibodies specific for the two targets and a pair of oligonucleotide-labeled secondary antibodies (Zatloukal et al., 2014). These two oligonucleotide probes are then ligated together and fluorescently visualized after a rolling-circle amplification. Detecting the proximity of these two targets in a tumor microenvironment is valuable for subsequent studies on how HS can affect this interaction. This method could prove useful as a building block for further studies that involve targeting HS by in transgenic models wherein HS biosynthesis is targeted.

2.3.1 Required Materials

- Formalin-fixed paraffin-embedded tissue slides
- Hemo-De (Fisher; Cat. No. HD150A)
- 100% Ethanol
- 10 m*M* Citrate solution pH 6.0
- TTBS solution (100 m*M* NaCl, 50 m*M* Tris, and 0.05% Tween)
- 20 × SSC (Life Technologies; Cat. No. 15557-044)

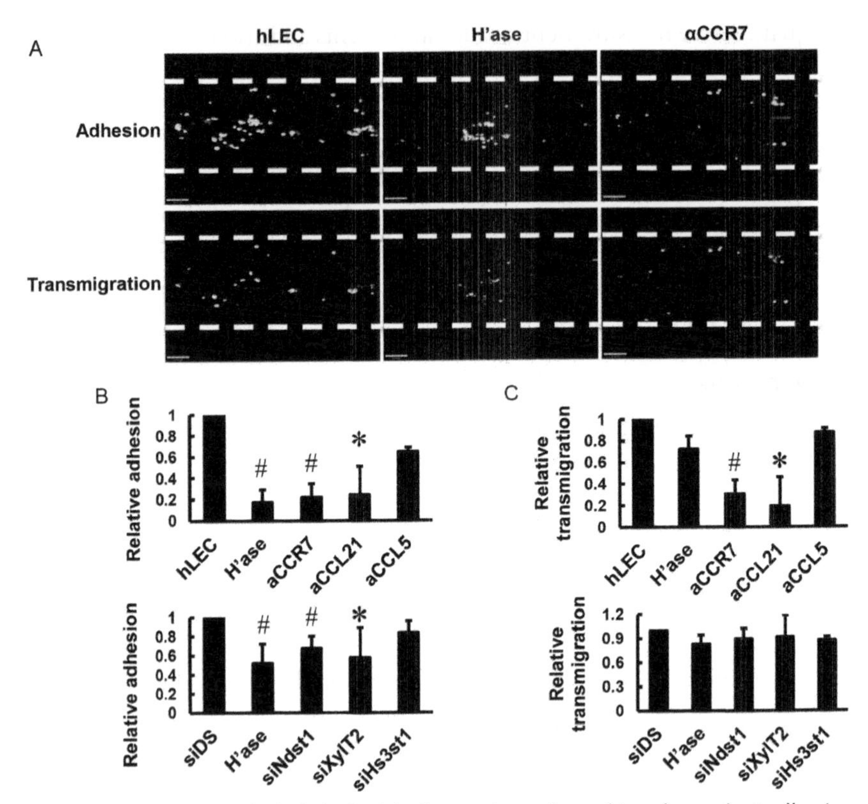

Figure 4 Lymphatic endothelial HS critically regulates chemokine-dependent adhesion of DCs under low-level physiologic shear flow. LECs were seeded into a BioFlux flow chamber (edges marked by dashed lines) and pretreated with neutralizing antibodies (against CCL21, CCR7, or CCL5) or siRNA targeting the indicated HS biosynthetic enzymes (siDS). DCs were labeled with calcein AM and introduced into the flow chamber at a constant shear stress of 0.14 dyn/cm^2 at 37 °C. After 15 min, nonadherent cells were washed off and the adherent DCs were imaged using fluorescence microscopy [(A) upper panels, fluorescence in green (gray in the print version) over black flow channel background], quantified with Metamorph software, and normalized to the LEC control group [upper graph in (B)] and to the siDS control group for siRNA-targeting experiments [lower graph in (B)]. The transmigration of adherent DCs was recorded following an additional 30-min cell-free flow period (from T_0 to T_{30}), imaged under a fluorescence microscope [lower panels in (A)], and quantified using Metamorph software, with the percentage transmigration calculated as [(No. of adherent DC at T_0 − No. of adherent DC at T_{30})/No. of adherent DC at T_0] × 100%. All transmigration data were normalized to LEC control group [upper graph in (C)] and to the siDS control group for siRNA experiments [lower graph in (C)]. $^*P<0.05$, $^{\#}P<0.01$, as compared to LEC control group in (B) and (C) upper graphs, and to siDS control group in (B) and (C) lower graphs.

- PAP Pen (RPI: Cat. No. 195506)
- PLA Probe anti-rabbit Plus (Sigma; Cat. No. DUO92002)
- PLA Probe anti-mouse Minus (Sigma; Cat. No. DUO92004)
- Detection Reagents Orange (Sigma; Cat. No. DUO92007)
- Mounting medium with DAPI (Sigma; Cat. No. DUO82040)
- 0.5 mg/mL Mouse IgG (Vector; Cat. No. I-2000)
- 0.5 mg/mL Rabbit IgG (Vector; Cat. No. I-1000)
- 0.5 mg/mL Mouse anti-CCL21 (R&D Systems; Cat. No. MAB3661)
- 0.5 mg/mL Rabbit anti-CCR7 (Novus; Cat. No. NB110-55680)
- Fluorescence microscope (Nikon; Eclipse 80i)

2.3.2 Procedures

1. Deparrafinze mouse LLC tissue by adding slide to Hemo-De for 2× 15 min.
2. Rehydrate tissue through a stepwise ethanol dilution: 2×10 min each of 100%, 95%, and 70% ethanol.
3. For antigen retrieval, incubate tissue slides in a preheated 10 m*M* citrate solution in a 95 °C water bath for 20 min.
4. Remove the slides from the citrate solution and allow them to cool down at room temperature for 20 min.
5. Wash the slides briefly with TTBS.
6. Encircle the tissue on the slides with a PAP pen.
7. Add blocking solution from the PLA Probe kit.

   Note 1: The amount of solution added to the tissue block in this and all subsequent steps depends on the size of the tissue sample. It is important to ensure that the tissue is completely covered.

   Note 2: In our experience, the blocking solution provided with the PLA Probe kit generates a lower nonspecific staining background than 1% BSA solution prepared in PBS.
8. Incubate at 37 °C for 30 min in a humidified chamber.
9. Wash the slides 3 × 1 min in TTBS.
10. Add diluted mouse anti-CCL21 and rabbit anti-CCR7 antibodies (in the Antibody Diluent provided with the PLA Probe kit) to the tissue. As negative controls, isotype-matched control IgG antibodies were added at same dilution.
11. Incubate overnight in a humidified chamber at 4 °C with gentle agitation.
12. Wash slides 3× 1 min in TTBS.

13. Dilute anti-rabbit plus and anti-mouse minus PLA probes (1:10) in Antibody Diluent, add to tissues and incubate for 1 h at 37 °C in a humidified chamber.
14. Wash slides 3 × 1 min in TTBS.
15. Add ligation solution (prepared following the recipe provided in PLA Probe kit manual) to the tissue and incubate for 1 h at 37 °C in a humidified chamber.
16. Wash slides 3 × 1 min in TTBS.
17. Add amplification solution (prepared following the recipe provided in PLA Probe kit manual) to the tissue and incubate for 2 h at 37 °C in a humidified chamber.
18. Wash slides in a series of increasingly diluted SSC solutions: 1 × 2 min in each of 2 ×, 1 ×, 0.2 ×, 0.02 × SSC.

    Note: This final wash method was found to provide superior PLA signal as compared to the wash method recommended in the PLA Probe kit manual.
19. Wash the slides 1 × 2 min in 70% ethanol.
20. Air dry slides in the dark.
21. Mount the cover slip on the slides using fluorescent mounting media with DAPI.
22. Visualize the PLA product on a fluorescent microscope with wavelengths set to λ_{ex} 554 nm; λ_{em} 576 nm.

2.3.3 Results

A representative image of PLA signal from tumor tissues labeled with either isotype-matched control IgGs or specific CCR7 and CCL21 antibodies are shown in Fig. 5. Both images were from the same region on two sequential slides of the same LLC tumor tissue.

3. PERSPECTIVES

In this chapter, we provide three methods for examining DC–LEC interactions driven by chemokines. The first two allow for phenotypic characterization of DC–LEC interactions, namely static chemotactic migration and adhesion/transmigration under shear flow, as affected by chemokine–receptor interactions or HS targeting. The third allows us to visualize direct interactions between chemokines relevant to the lymphatic microenvironment and their cognate receptors on the surface of trafficking cells (in this case LLC tumor cells). Based on these methods, we were able to

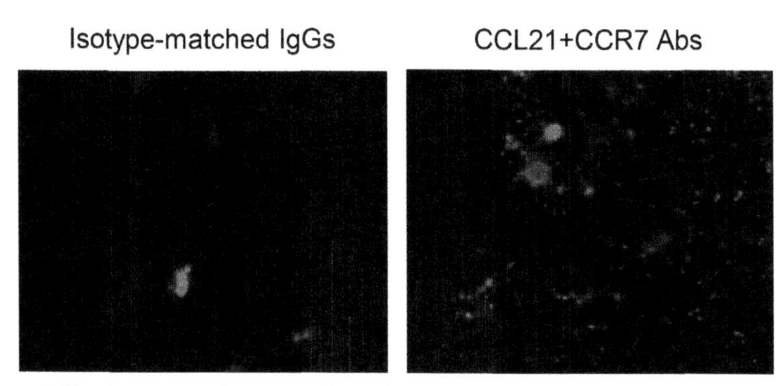

Figure 5 The interaction between chemokine CCL21 and its cognate receptor CCR7 is detectable by proximity ligation assay (PLA) in mouse xenograft tumor tissues. Mouse xenograft tumors derived from Lewis Lung Carcinoma (LLC) cells were examined by PLA assay for the interaction/proximity between CCL21 and CCR7. As a negative control, the isotype-matched control IgGs were used. Representative images from the same field of sequential slides stained with either control IgGs (left panel) or specific CCL21 + CCR7 antibodies (Abs, right panel) are shown. The positive PLA signal presents as small red (gray in the print version) fluorescent dots on the tissue.

examine the biological significance of HS in chemokine-mediated cell migratory behavior within the lymphatic microenvironment. In particular, we illustrate methods to characterize the coreceptor role of HS in CCR7–CCL21 interactions and its importance in facilitating DC adhesion, migration, and/or invasion toward LECs. In earlier studies, we also demonstrated parallel behavior by tumor cells during LEC-driven migration, which appear to usurp the same HS-sensitive chemokine-receptor systems (Yin et al., 2014, 2010).

The transwell migration assay and related methods have been widely applied to study the migratory and/or invasive behavior of different cell types in response to various molecules such as chemokines, growth factors, lipids, nucleotides; and used with various other cell types. Although a static system is not capable of replicating the effects of a complex microenvironment that mediates *in vivo* cell migration or invasion, the transwell assay does provide a simple and highly reproducible way to measure the migratory capability of target cells under various controlled test conditions. In our system, by separating DCs from LECs via liquid medium, we were able to assess DC migration in response to variously altered glycans secreted (on matrix-proteoglycans) from LECs. By introducing neutralizing antibodies to specific chemokines or chemokine receptors, or targeting HS biosynthesis by LECs, we could examine the importance of chemokine signaling and the effects of LEC-HS on unique chemoattractants that drive this migration process. Several critical factors need to be predetermined for the transwell migration assay: first, the specific time frame for migration that is dependent mainly on the migration velocity of target cells (in our system, we found that 3–6 h is optimal for examining DC migration while 12–18 h is needed for tumor cells); second, inclusion of proper controls, including negative control with no LECs in the bottom well, positive control with chemokine alone in the bottom well, and specificity control with neutralizing antibodies targeting different chemokines or chemokine receptors; and third, selection of a proper transwell pore size for the target cells to migrate across (where 3 μm works well for DCs while 5 μm is optimal for many carcinoma cells).

To complement the transwell migration assay and more closely mimic DC–LEC interactions within the lymphatic vessel, we adopted the high-throughput Bioflux system to examine DC adhesion to and transmigration across LEC layers under low-level shear flow to mimic the conditions that might occur during low-pressure flow over lymphatic endothelium. This system allows replicated set-up of each experimental condition and simultaneous examination of multiple conditions, minimizing variations between experiments and facilitating statistical analysis. As detailed in Protocol 2 provided above, the establishment of a confluent LEC monolayer is critical for examining DC adhesion and subsequent transmigration. The labeling of DCs with fluorescence greatly simplifies data quantification and analysis for Protocols 1 and 2.

The PLA is a highly sensitive method to visualize whether two molecules are proximal to each other (within 30–40 nm). It works with oligonucleotide-tagged antibodies against the molecules of interest, with proximity leading to hybridization, rolling-circle amplification, and detection with nucleic-acid specific fluorescent probes. More generally, PLA has been successfully applied to detect a single protein (Fredriksson et al., 2002; Gullberg et al., 2004), protein–protein interactions (Soderberg et al., 2006), and posttranslational modification of a target protein (Jarvius et al., 2007). PLA presents several advantages over conventional methods for protein–protein interactions such as coimmunoprecipitation and colocalization analysis. First, PLA enables *in situ* detection of transient and endogenous protein–protein interactions; second, it confers high sensitivity for detecting interactions with single-molecule resolution; third, it can be used to detect interactions on the cell surface, within subcellular compartments, or within tissues. In our laboratory, we successfully applied PLA to examine HS effects on chemokine–chemokine receptor interactions on the surface of cultured cells (Yin et al., 2014, 2011, 2010); and here show its proof-of-principle success in mouse tissues. In our experience, the success of PLA highly depends on the selection of optimal antibodies for each target, titration of antibody concentrations, and inclusion of proper negative controls to minimize nonspecific signals and positive controls to optimize various experimental parameters.

In summary, there are multiple options for exploring the molecular mechanisms and characterizing the functional consequences of HS in regulating chemokine-mediated cell–cell interactions. Proper selection of an *in vitro* assay system should help to address mechanism(s) involved in a specific biological question or physiological state, which ideally should be validated *in vivo*. The findings may also guide rational applications or targeting *in vivo* to further validate new hypotheses. While carrying out these assays, one should carefully consider each component in the assay system and optimize various parameters to ensure the collection of meaningful experimental data.

ACKNOWLEDGMENTS

We acknowledge funding support from NIH/NHLBI (R01-HL107652 to M.M.F.) and NIH/NIAID (R01 AI37113-13 to T.M.H.). Figures 2A, 3, and 4 are adapted or reproduced from our previously published open-access study entitled "Lymphatic specific disruption in the fine structure of heparan sulfate inhibits dendritic cell traffic and functional T cell responses in the lymph node" (*J. Immunol.* 2014 Mar 1;192(5):2133-42. doi: 10.4049/jimmunol.1301286).

REFERENCES

Bao, X., Moseman, E. A., Saito, H., Petryniak, B., Thiriot, A., Hatakeyama, S., et al. (2010). Endothelial heparan sulfate controls chemokine presentation in recruitment of lymphocytes and dendritic cells to lymph nodes. *Immunity*, *33*(5), 817–829. http://dx.doi.org/10.1016/j.immuni.2010.10.018.

Braut-Boucher, F., Pichon, J., Rat, P., Adolphe, M., Aubery, M., & Font, J. (1995). A nonisotopic, highly sensitive, fluorimetric, cell-cell adhesion microplate assay using calcein AM-labeled lymphocytes. *Journal of Immunological Methods*, *178*(1), 41–51. Retrieved from, http://www.ncbi.nlm.nih.gov/pubmed/7829864.

Capila, I., & Linhardt, R. J. (2002). Heparin-protein interactions. *Angewandte Chemie International Edition in English*, *41*(3), 391–412. Retrieved from, http://www.ncbi.nlm.nih.gov/entrez/query.fcgi?cmd=Retrieve&db=PubMed&dopt=Citation&list_uids=12491369.

Dieu-Nosjean, M. C., Vicari, A., Lebecque, S., & Caux, C. (1999). Regulation of dendritic cell trafficking: A process that involves the participation of selective chemokines. *Journal of Leukocyte Biology*, *66*(2), 252–262. Retrieved from, http://www.ncbi.nlm.nih.gov/pubmed/10449163.

Forster, R., Davalos-Misslitz, A. C., & Rot, A. (2008). CCR7 and its ligands: Balancing immunity and tolerance. *Nature Reviews. Immunology, 8*(5), 362–371. http://dx.doi.org/10.1038/nri2297.

Fredriksson, S., Gullberg, M., Jarvius, J., Olsson, C., Pietras, K., Gustafsdottir, S. M., et al. (2002). Protein detection using proximity-dependent DNA ligation assays. *Nature Biotechnology, 20*(5), 473–477. http://dx.doi.org/10.1038/nbt0502-473.

Gullberg, M., Gustafsdottir, S. M., Schallmeiner, E., Jarvius, J., Bjarnegard, M., Betsholtz, C., et al. (2004). Cytokine detection by antibody-based proximity ligation. *Proceedings of the National Academy of Sciences of the United States of America, 101*(22), 8420–8424. http://dx.doi.org/10.1073/pnas.0400552101.

Hansson, M., Lundgren, A., Elgbratt, K., Quiding-Jarbrink, M., Svennerholm, A. M., & Johansson, E. L. (2006). Dendritic cells express CCR7 and migrate in response to CCL19 (MIP-3beta) after exposure to Helicobacter pylori. *Microbes and Infection, 8*(3), 841–850. http://dx.doi.org/10.1016/j.micinf.2005.10.007.

Jarvius, M., Paulsson, J., Weibrecht, I., Leuchowius, K. J., Andersson, A. C., Wahlby, C., et al. (2007). In situ detection of phosphorylated platelet-derived growth factor receptor beta using a generalized proximity ligation method. *Molecular & Cellular Proteomics, 6*(9), 1500–1509. http://dx.doi.org/10.1074/mcp.M700166-MCP200.

Johnson, L. A., & Jackson, D. G. (2013). The chemokine CX3CL1 promotes trafficking of dendritic cells through inflamed lymphatics. *Journal of Cell Science, 126*(Pt. 22), 5259–5270. http://dx.doi.org/10.1242/jcs.135343.

Lortat-Jacob, H., Grosdidier, A., & Imberty, A. (2002). Structural diversity of heparan sulfate binding domains in chemokines. *Proceedings of the National Academy of Sciences of the United States of America, 99*(3), 1229–1234. http://dx.doi.org/10.1073/pnas.032497699.

Madaan, A., Verma, R., Singh, A. T., Jain, S. K., & Jaggi, M. (2014). A stepwise procedure for isolation of murine bone marrow and generation of dendritic cells. *Journal of Biological Methods, 1*(1), e1. http://dx.doi.org/10.14440/jbm.2014.12.

McColl, S. R. (2002). Chemokines and dendritic cells: A crucial alliance. *Immunology and Cell Biology, 80*(5), 489–496. http://dx.doi.org/10.1046/j.1440-1711.2002.01113.x.

Proudfoot, A. E., Handel, T. M., Johnson, Z., Lau, E. K., LiWang, P., Clark-Lewis, I., et al. (2003). Glycosaminoglycan binding and oligomerization are essential for the in vivo activity of certain chemokines. *Proceedings of the National Academy of Sciences of the United States of America, 100*(4), 1885–1890. http://dx.doi.org/10.1073/pnas.0334864100.

Rabenstein, D. L. (2002). Heparin and heparan sulfate: Structure and function. *Natural Product Reports, 19*(3), 312–331. Retrieved from, http://www.ncbi.nlm.nih.gov/pubmed/12137280.

Randolph, G. J., Angeli, V., & Swartz, M. A. (2005). Dendritic-cell trafficking to lymph nodes through lymphatic vessels. *Nature Reviews. Immunology, 5*(8), 617–628. http://dx.doi.org/10.1038/nri1670.

Rot, A. (2010). Chemokine patterning by glycosaminoglycans and interceptors. *Frontiers in Bioscience (Landmark Ed), 15*, 645–660. Retrieved from, http://www.ncbi.nlm.nih.gov/pubmed/20036838.

Sadir, R., Imberty, A., Baleux, F., & Lortat-Jacob, H. (2004). Heparan sulfate/heparin oligosaccharides protect stromal cell-derived factor-1 (SDF-1)/CXCL12 against proteolysis induced by CD26/dipeptidyl peptidase IV. *The Journal of Biological Chemistry, 279*(42), 43854–43860. http://dx.doi.org/10.1074/jbc.M405392200.

Sarrazin, S., Lamanna, W. C., & Esko, J. D. (2011). Heparan sulfate proteoglycans. *Cold Spring Harbor Perspectives in Biology. 3*(7). http://dx.doi.org/10.1101/cshperspect.a004952.

Soderberg, O., Gullberg, M., Jarvius, M., Ridderstrale, K., Leuchowius, K. J., Jarvius, J., et al. (2006). Direct observation of individual endogenous protein complexes in situ by proximity ligation. *Nature Methods*, *3*(12), 995–1000. http://dx.doi.org/10.1038/nmeth947.

Tsuboi, K., Hirakawa, J., Seki, E., Imai, Y., Yamaguchi, Y., Fukuda, M., et al. (2013). Role of high endothelial venule-expressed heparan sulfate in chemokine presentation and lymphocyte homing. *Journal of Immunology*, *191*(1), 448–455. http://dx.doi.org/10.4049/jimmunol.1203061.

Wagner, L., Yang, O. O., Garcia-Zepeda, E. A., Ge, Y., Kalams, S. A., Walker, B. D., et al. (1998). Beta-chemokines are released from HIV-1-specific cytolytic T-cell granules complexed to proteoglycans. *Nature*, *391*(6670), 908–911. http://dx.doi.org/10.1038/36129.

Wang, X., Sharp, J. S., Handel, T. M., & Prestegard, J. H. (2013). Chemokine oligomerization in cell signaling and migration. *Progress in Molecular Biology and Translational Science*, *117*, 531–578. http://dx.doi.org/10.1016/B978-0-12-386931-9.00020-9.

Yin, X., Johns, S. C., Kim, D., Mikulski, Z., Salanga, C. L., Handel, T. M., et al. (2014). Lymphatic specific disruption in the fine structure of heparan sulfate inhibits dendritic cell traffic and functional T cell responses in the lymph node. *Journal of Immunology*, *192*(5), 2133–2142. http://dx.doi.org/10.4049/jimmunol.1301286.

Yin, X., Johns, S. C., Lawrence, R., Xu, D., Reddi, K., Bishop, J. R., et al. (2011). Lymphatic endothelial heparan sulfate deficiency results in altered growth responses to vascular endothelial growth factor-C (VEGF-C). *The Journal of Biological Chemistry*, *286*(17), 14952–14962. http://dx.doi.org/10.1074/jbc.M110.206664.

Yin, X., Truty, J., Lawrence, R., Johns, S. C., Srinivasan, R. S., Handel, T. M., et al. (2010). A critical role for lymphatic endothelial heparan sulfate in lymph node metastasis. *Molecular Cancer*, *9*, 316. http://dx.doi.org/10.1186/1476-4598-9-316.

Zatloukal, B., Kufferath, I., Thueringer, A., Landegren, U., Zatloukal, K., & Haybaeck, J. (2014). Sensitivity and specificity of in situ proximity ligation for protein interaction analysis in a model of steatohepatitis with Mallory-Denk bodies. *PLoS One*, *9*(5). http://dx.doi.org/10.1371/journal.pone.0096690.

CHAPTER SEVENTEEN

Preparation and Analysis of N-Terminal Chemokine Receptor Sulfopeptides Using Tyrosylprotein Sulfotransferase Enzymes

Christoph Seibert*, Anthony Sanfiz*,[1], Thomas P. Sakmar*, Christopher T. Veldkamp†,[2]

*Laboratory of Chemical Biology and Signal Transduction, The Rockefeller University, New York, USA
†Department of Chemistry, University of Wisconsin–Whitewater, Whitewater, Wisconsin, USA
[2]Corresponding author: e-mail address: veldkamc@uww.edu

Contents

1. Introduction 358
2. Methods 361
 2.1 Expression and Purification of TPST-1 and -2 from Mammalian Cells 361
 2.2 Expression and Refolding of Functional TPST-1 from *E. coli* 363
 2.3 Analysis and Purification of PAPS 368
 2.4 *In Vitro* Sulfation of N-Terminal Chemokine Receptor Peptides Using TPST Enzymes 370
 2.5 Reversed-Phase HPLC of Sulfopeptides 371
 2.6 Mass Spectrometry of Sulfotyrosine Peptides 373
 2.7 Characterization of Sulfopeptides by Protein NMR 378
3. Caveats and Limitations 380
4. Perspectives 381
Acknowledgments 385
References 385

Abstract

In most chemokine receptors, one or multiple tyrosine residues have been identified within the receptor N-terminal domain that are, at least partially, modified by posttranslational tyrosine sulfation. For example, tyrosine sulfation has been demonstrated for Tyr-3, -10, -14, and -15 of CCR5, for Tyr-3, -14, and -15 of CCR8, and for Tyr-7, -12, and -21 of CXCR4. While there is evidence for several chemokine receptors that tyrosine

---

[1] Current address: Department of Microbiology, New York University Medical Center, New York, NY, USA.

ISSN 0076-6879
http://dx.doi.org/10.1016/bs.mie.2015.09.004

sulfation is required for optimal interaction with the chemokine ligands, the precise role of tyrosine sulfation for chemokine receptor function remains unclear. Furthermore, the function of the chemokine receptor N-terminal domain in chemokine binding and receptor activation is also not well understood. Sulfotyrosine peptides corresponding to the chemokine receptor N-termini are valuable tools to address these important questions both in structural and functional studies. However, due to the lability of the sulfotyrosine modification, these peptides are difficult to obtain using standard peptide chemistry methods. In this chapter, we provide methods to prepare sulfotyrosine peptides by enzymatic *in vitro* sulfation of peptides using purified recombinant tyrosylprotein sulfotransferase (TPST) enzymes. In addition, we also discuss alternative approaches for the generation of sulfotyrosine peptides and methods for sulfopeptide analysis.

1. INTRODUCTION

With up to 1% of eukaryotic proteins potentially containing sulfotyrosine residues, sulfation of tyrosines is a common posttranslational modification whose biological impact has only just started to be elucidated (Ludeman & Stone, 2014; Moore, 2003; Seibert et al., 2008). In humans and most mammals, there are two isoforms of the enzyme responsible for tyrosine sulfation; these enzymes are tyrosylprotein sulfotransferase 1 and 2 (TPST-1 and TPST-2) (Moore, 2003; Seibert & Sakmar, 2008). TPST-1 and TPST-2 are located in the *trans*-Golgi network and this limits tyrosine sulfation to secreted or membrane proteins (Moore, 2003; Seibert & Sakmar, 2008). Both enzymes utilize the cosubstrate PAPS, or 3′-phosphoadenosine-5′-phosphosulfate, as the sulfate donor to catalyze the sulfation of a tyrosine's phenolic hydroxyl in a substrate protein or peptide as seen in Fig. 1 (Moore, 2003; Seibert & Sakmar, 2008).

Early studies of sulfotyrosine-containing proteins involved in blood coagulation made it clear that sulfotyrosine posttranslational modifications

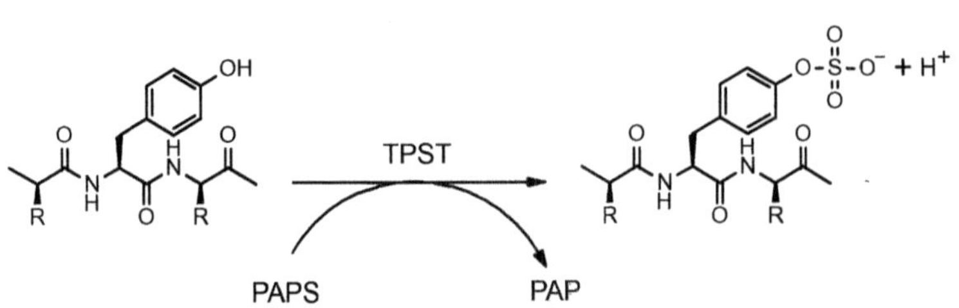

Figure 1 Reaction catalyzed by TPST enzymes.

influence protein–protein interactions (Moore, 2003; Seibert & Sakmar, 2008). Coagulation factor VIII, or antihemophilic factor, contains a sulfotyrosine at position 1680 that is required for a strong interaction of the inactive form of factor VIII with von Willebrand factor. Mutation of tyrosine 1680 in factor VIII to phenylalanine leads to the loss of sulfotyrosine at this position and results in a form of hemophilia A (Leyte et al., 1991). Hirudin, the well-known anticoagulant from the medicinal leech *Hirudo medicinalis*, is an inhibitor of thrombin that binds 10-fold more tightly to thrombin when it contains a sulfotyrosine at amino acid position 63 (Stone & Hofsteenge, 1986).

Just as coagulation and the coagulation cascade involve numerous protein–protein interactions that include sulfotyrosine residues, leukocyte chemoattraction, including leukocyte rolling, tight adhesion, and trans-endothelial migration, does as well. For example, P-selectin glycoprotein ligand-1 (PSGL-1)'s binding to selectins is essential for leukocyte rolling and tethering, the initial step in the leukocyte extravasation cascade (Carlow et al., 2009). This PSGL-1 selectin interaction is, in part, mediated by sulfotyrosine residues located in the N-terminus of mature PSGL-1 (Rodgers, Camphausen, & Hammer, 2001; Wilkins, Moore, McEver, & Cummings, 1995). Resting T-cells expressing PSGL-1 show enhanced chemotaxis toward the homeostatic chemokines CCL19 and CCL21 and increased recruitment to secondary lymphoid organs (Veerman et al., 2007). However, these enhancements are not a result of the canonical PSGL-1 selectin interactions, but result from the N-terminus of PSGL-1 binding directly to CCL19 or CCL21 (Veerman et al., 2007; Veldkamp et al., 2015). It is not surprising that the acidic sulfotyrosine containing N-terminus of PSGL-1 binds CCL19 or CCL21. The N-termini of chemokine receptors are acidic and many contain or are predicted to contain sulfotyrosine residues that enhance affinity for chemokine ligands (Farzan, Babcock, et al., 2002; Farzan et al., 2000; Ludeman & Stone, 2014; Seibert & Sakmar, 2008).

Farzan and colleagues were the first to show that specific tyrosine residues in the N-termini of the chemokine receptors CCR5 and CXCR4 were sulfated and that tyrosine sulfation increased the affinity for these receptors' chemokine ligands (Farzan, Babcock, et al., 2002; Farzan, Chung, et al., 2002; Farzan et al., 1999). Furthermore, they have also shown that tyrosine sulfation of CCR5, which is a major coreceptor for HIV-1, is required for viral entry into host cells (Farzan et al., 1999). Sulfation of CXCR4 in contrast, which also acts as a major HIV-1 coreceptor, is not required for coreceptor function (Farzan, Babcock, et al., 2002).

Chemokines are hypothesized to activate their receptors through a two-site, two-step binding and activation model in which the chemokine receptor N-terminus binds to the chemokine domain (site one) followed by the chemokine N-terminus binding to a second site on the receptor leading to receptor activation (Crump et al., 1997; Kufareva, Salanga, & Handel, 2015). Farzan and colleagues illustrated the importance of sulfotyrosines for the site one interaction using an inactive CCR5 mutant lacking N-terminal residues 2–17 (Bannert et al., 2001; Farzan, Chung, et al., 2002). CCL3 could only activate a CCR5 Δ2–17-induced intracellular calcium flux when the receptor was rescued by the presence of synthetic CCR5 N-terminal peptides containing sulfated tyrosines but not unsulfated counterparts (Bannert et al., 2001; Farzan, Chung, et al., 2002). Subsequently, many researchers have used protein NMR and sulfotyrosine-containing peptides corresponding to a chemokine receptor N-terminus to mimic and study the site one interaction between a chemokine receptor and its chemokine ligand (Duma, Haussinger, Rogowski, Lusso, & Grzesiek, 2007; Millard et al., 2014; Simpson, Zhu, Widlanski, & Stone, 2009; Veldkamp et al., 2008; Veldkamp, Seibert, Peterson, Sakmar, & Volkman, 2006).

While several investigators have used chemically synthesized sulfotyrosine-containing peptides to mimic a chemokine receptor's N-terminus (Bannert et al., 2001; Cormier et al., 2000; Duma et al., 2007; Farzan, Chung, et al., 2002; Ludeman & Stone, 2014; Millard et al., 2014; Simpson et al., 2009; Tan, Ludeman, et al., 2013; Tan, Zhu, et al., 2013), our approach differs in that we used recombinant TPSTs to enzymatically sulfate peptides. Moore and colleagues cloned human TPST-1 and TPST-2 and showed recombinant TPSTs from mammalian expression systems had sulfotransferase activity in the presence of the cosubstrate PAPS (Moore, 2003; Ouyang, Lane, & Moore, 1998; Ouyang & Moore, 1998). We and others have used TPST-1 and TPST-2 to characterize the sulfation of N-terminal peptides from CCR5 (Jen, Moore, & Leary, 2009; Seibert, Cadene, Sanfiz, Chait, & Sakmar, 2002). We have also used TPST enzymes to characterize the enzymatic sulfation of CXCR4 (Seibert et al., 2008), a chemokine receptor that plays significant roles in cancer metastasis (Ben-Baruch, 2008; Muller et al., 2001). Using sulfotyrosine-containing CXCR4 N-terminal peptides, the structural basis for the site one interaction between the CXCR4 N-terminus and CXCL12 was also probed (Seibert et al., 2008; Veldkamp et al., 2008, 2006).

Here, we describe approaches for utilizing recombinant TPST-1 and TPST-2 to enzymatically sulfate N-terminal chemokine receptor peptides and for the characterization of these sulfopeptides. For a more expansive introduction to TPST enzymes and proteins containing sulfotyrosine post-translational modifications, see the following references: Ludeman and Stone (2014), Moore (2003, 2009), Seibert and Sakmar (2008), and Stone, Chuang, Hou, Shoham, and Zhu (2009).

2. METHODS

2.1 Expression and Purification of TPST-1 and -2 from Mammalian Cells

TPSTs are type II transmembrane proteins with a single α-helical transmembrane segment that anchors the catalytic domain in the Golgi lumen (Moore, 2003, 2009). To improve heterologous expression and purification, soluble recombinant variants of human TPST-1 and TPST-2 have been engineered that are suitable for *in vitro* sulfation of peptide substrates (Ouyang & Moore, 1998). These TPST variants lack the cytoplasmic N-terminus and the transmembrane domain, which are not required for enzymatic activity. Furthermore, an N-terminal transferring signal peptide followed by a protein C epitope was N-terminally fused to the catalytic domain of both TPSTs (residues 25–370 and 25–377 of full-length TPST-1 and TPST-2, respectively) to aid in expression and purification of the enzymes.

In this section, we provide methods for the expression and purification of recombinant engineered TPST-1 and TPST-2 using transiently transfected HEK293-T cells. With these methods, which were adapted from published procedures (Ouyang & Moore, 1998; Seibert et al., 2002), 100 μg quantities of purified TPST-1 and TPST-2 can be obtained that are sufficient for analytical- and semipreparative-scale peptide sulfation reactions (Seibert et al., 2002, 2008; Veldkamp et al., 2008, 2006).

2.1.1 Required Materials

HEK293-T cells (ATCC)
pMSH1TH and pMSH2TH expression vectors (Dr. Kevin L. Moore, University of Oklahoma Health Sciences Center)
Cell culture media and supplements (Gibco)
Disposable cell culture materials (Corning or Falcon)
Plasmid purification kit (Qiagen)
LipofectAMINE Plus (Invitrogen)

Complete™ protease inhibitor mixture (Roche)
Anti-protein C resin (Roche)
Disposable plastic columns with stopcocks (BioRad)
Centricon YM10 concentrators (Millipore)
SDS–PAGE gels (10%), electrophoresis buffers, staining solution, and electrophoresis system (BioRad)
Analytical-grade reagents (Sigma-Aldrich or Fisher Scientific)
Standard cell culture equipment (cell culture incubator, sterile cell culture hood, centrifuge, etc.)

2.1.2 Cell Culture and Transfection

1. Prepare transfection-grade pMSH1TH and/or pMSH2TH DNA using a plasmid purification kit according to the manufacturer's protocol.
2. Culture HEK293-T cells in DMEM supplemented with 10% FBS using 100-mm cell culture plates and maintain cells at 37 °C with 5% CO_2.
3. Expand HEK293-T cells into as many 100-mm cell culture plates as required, assuming a yield of approximately 4 μg purified TPST-1 or 8 μg purified TPST-2 per plate of transfected HEK293-T cells.
4. Transfect HEK293-T cells with pMSH1TH or pMSH2TH plasmid (4 μg plasmid per plate) using LipofectAMINE Plus™ (20 μl Plus reagent and 30 μl LipofectAMINE reagent per plate) according to the manufacturer's protocol.
5. After 48 h, wash plates with 5 ml PBS each and harvest transfected HEK293-T cells in 2 ml per plate of ice-cold PBS supplemented with Complete™ protease inhibitor mixture (without EDTA).
6. Collect cells by centrifugation (15 min at 10,000 × *g*) and proceed with anti-protein C immunoaffinity purification (Section 2.1.3).

2.1.3 Anti-protein C Immunoaffinity Purification of TPSTs

1. Resuspend transfected HEK293-T cells in 0.5 ml per plate of ice-cold solubilization buffer (SB) (20 m*M* TAPS, pH 9.0, 100 m*M* NaCl, 1% Triton X-100, Complete™ protease inhibitor mixture without EDTA) and incubate for 1–2 h at 5 °C (on a Nutator).
2. Centrifuge (20 min at 10,000 × *g*) to separate solubilized proteins from cell debris. Collect supernatant and continue with step 3 or store at −80 °C.
3. Wash anti-protein C resin (use 30–50 μl of resin per plate of transfected HEK293-T cells) with equilibration buffer (50 m*M* MOPS, pH 7.5, 100 m*M* NaCl, 5 m*M* $CaCl_2$, 1% (v/v) Triton X-100).

4. Add 50 m*M* MOPS, pH 7.5, 5 m*M* $CaCl_2$, and 10% glycerol to supernatant and incubate with washed anti-protein C resin over night at 4 °C (on a Nutator).
5. Pour anti-protein C resin slurry into a disposable column.
6. Wash anti-protein C resin with 10 column volumes of ice-cold wash buffer 1 (20 m*M* MOPS, pH 7.5, 2 *M* NaCl, 2 m*M* $CaCl_2$, 0.1% Triton X-100) followed by 10 column volumes of ice-cold wash buffer 2 (20 m*M* MOPS, pH 7.5, 150 m*M* NaCl, 2 m*M* $CaCl_2$, 0.1% Triton X-100).
7. Elute bound TPST protein with 10 column volumes of ice-cold elution buffer (20 m*M* MOPS, pH 7.5, 150 m*M* NaCl, 10 m*M* EDTA, 0.1% Triton X-100, 10% glycerol) and collect eluent fractions.
8. Analyze TPST eluent fractions by SDS–PAGE on 10% gels under reducing conditions and staining with Coomassie brilliant blue R-250 according to standard protein chemistry protocols.
9. Pool TPST-containing fractions, concentrate with Centricon YM10 concentrator to yield a final TPST concentration between 0.3 and 0.8 μg/μl, and store aliquots at -80 °C.
10. Determine concentration of purified TPST preparations by densitometry of Coomassie brilliant blue R-250-stained SDS–PAGE gels with BSA as an internal standard.
11. Test the enzymatic activity of purified TPST preparations by tyrosine sulfation assays (see Section 2.4) using the PSGL-1 1–15 peptide (QATEYEYLDYDFLPE-NH_2) as a standard substrate.

2.2 Expression and Refolding of Functional TPST-1 from *E. coli*

For certain structural and functional studies, in particular NMR spectroscopy studies, rather large quantities of sulfotyrosine peptides in the 10-mg range are required. Due to the low tyrosine sulfation activities of soluble TPST enzymes for peptide substrates, enzymatic peptide sulfation at this scale requires quantities of TPST enzymes that are difficult and expensive to obtain from transfected HEK293 cells.

In this section, we describe methods to produce 100 mg quantities of a soluble recombinant variant of human TPST-1 by using an *E. coli* expression system. Expression of this TPST-1 variant in *E. coli* results in incorrectly folded protein that accumulates in inclusion bodies. Active enzyme is obtained by solubilization of inclusion bodies followed by functional refolding of TPST-1.

Due to the overlapping substrate specificities of TPST enzymes (Seibert et al., 2008), it should be possible to efficiently sulfate most peptide substrates with TPST-1 at sufficiently high enzyme concentrations. However, if TPST-2 is required for sulfation of a specific peptide, a protocol for *E. coli* expression and functional refolding of TPST-2 can be found elsewhere (Teramoto et al., 2013).

2.2.1 Required Materials

pMSH1TH expression vectors (Dr. Kevin L. Moore, University of Oklahoma Health Sciences Center) and pET28a(+) expression vector (Novagen) or pET28a-TPST1 expression vector

Transformation-competent *E. coli* BL21 (DE3) cells (Novagen)

Luria Broth (LB) (Gibco)

Complete™ protease inhibitor cocktail (Roche)

Guanidine hydrochloride (GndHCl) (Sigma-Aldrich)

n-Dodecyl-β-D-maltoside (Anatrace)

Reduced glutathione (GSH) and oxidized glutathione (GSSG) (Sigma-Aldrich)

15-ml Polycarbonate tubes with conical bottom (Falcon, Greiner, or Sarstedt)

Parafilm

YM10 ultrafiltration membranes and ultrafiltration device (Amicon Millipore)

Bradford protein assay (BioRad)

SDS–PAGE gels (10%), electrophoresis buffers, staining solution, and electrophoresis system (BioRad)

Analytical-grade reagents (Sigma-Aldrich or Fisher Scientific)

French press pressure cell

FPLC system with Superdex 200 26/60 size-exclusion column (GE Healthcare)

Peristaltic pump with 0.01 in. ID silicon tubing (Rainin)

Standard laboratory equipment (bacterial incubator, centrifuge, ultracentrifuge, UV–Vis spectrometer, Nutator, magnetic stirrer, etc.)

2.2.2 Transformation and Expression of TPST-1 as Inclusion Bodies in E. coli

1. Construction of the pET28-TPST1 plasmid for expression of recombinant engineered human TPST-1 in *E. coli*: The cDNA encoding the catalytic domain and stem region of human TPST-1 (amino acids 25–370)

was amplified by polymerase chain reaction (PCR) using the eukaryotic expression vector pMSH1TH as a template and introducing 5′ *Nhe*I and 3′ *Xho*I restriction sites. After *Nhe*I and *Xho*I cleavage, the PCR fragment was spliced into a pET28a(+) vector, which was modified to introduce a hexa-histidine (H_6) tag followed by a PreScission Protease cleavage site (LEVLFQ/GP) at the N-terminus of the TPST-1 construct. Due to the subcloning strategy, a total of 26 residues (MGSSHHHHHHSSGLEVLFQ/GPHMASM) were fused to Gly-25 of TPST-1. The pET28-TPST1 plasmid can be obtained from the authors (after obtaining permission from Dr. Kevin L. Moore) or constructed according to the described strategy.

2. Transform competent *E. coli* BL21 (DE3) cells with the pET28a-TPST1 expression plasmid and grow transformed *E. coli* at 37 °C in LB medium supplemented with 50 mg/ml of kanamycin.
3. Inoculate 1.6-l cultures with 1.6 ml each of a 100 ml overnight starter culture and incubate at 37 °C on a shaker platform.
4. At a cell density corresponding to an OD_{600nm} of approximately 0.4, add IPTG at a final concentration of 0.4 m*M* to induce TPST-1 expression, and incubate cultures for 3 h at 37 °C.
5. Harvest *E. coli* cells by centrifugation (25 min, 3600 rpm, 4 °C) and resuspend in ice-cold lysis buffer (50 m*M* Tris–HCl, pH 8.0; 10 m*M* $MgCl_2$; 1 m*M* DTT; 1 m*M* PMSF; 1 tablet per 25 ml of Complete™ protease inhibitor cocktail without EDTA) using 3 ml of lysis buffer for each gram of *E. coli* cell pellet. Freeze aliquots of resuspended *E. coli* cells with liquid nitrogen and store at -80 °C.

2.2.3 Solubilization of TPST-1 from Inclusion Bodies

1. Thaw *E. coli* cells containing TPST-1 inclusion bodies (from Section 2.2.2) on ice and lyse cells by three passages through a French press pressure cell at 15,000 psi.
2. Collect inclusion bodies by centrifugation for 1 h at 10,000 × *g* and 4 °C and remove the yellow layer of membrane fragments from on top of the white inclusion body pellet.
3. Wash inclusion bodies with ice-cold WB1 (50 m*M* Tris–HCl, pH 8.0; 100 m*M* NaCl; 10 m*M* EDTA; 1% (w/v) Triton X-100; 1 m*M* PMSF; 1 tablet per 25 ml of Complete™ protease inhibitor cocktail without EDTA).
4. Wash inclusion bodies with ice-cold WB2 (20 m*M* Tris–HCl, pH 8.0; 200 m*M* NaCl; 1 m*M* EDTA; 1 m*M* PMSF).

5. Resuspend inclusion bodies in SB (100 m*M* Tris–HCl, pH 8.0; 6 *M* GndHCl; 5 m*M* EDTA; 10 m*M* DTT) using 10 ml SB for 1 g of inclusion bodies and incubate at 5 °C with constant agitation using a Nutator. After 12 h, add fresh DTT (10 m*M*) and continue incubation for 2 h at room temperature.
6. Remove insoluble material by centrifugation (30 min, 125,000 × *g*, 4 °C) and concentrate supernatant containing solubilized unfolded TPST-1 by ultrafiltration using a 10-kDa cutoff membrane (YM10, Amicon Millipore).
7. Dilute concentrated TPST-1 solution 10-fold with buffer A (100 m*M* Na-acetate, pH 4.5; 6 *M* GndHCl; 10 m*M* DTT), concentrate by ultrafiltration using a 10-kDa cutoff membrane (YM10, Amicon Millipore) to about 30 mg/ml, and store aliquots of this TPST-1 stock solution at -80 °C.
8. Determine protein concentrations by the Bradford method with bovine serum albumin (BSA) as a standard.

2.2.4 Purification of TPST-1 by Size-Exclusion Chromatography Under Denaturing Conditions (Optional)

Optionally, unfolded TPST-1 can be purified by size-exclusion chromatography under denaturing conditions to separate monomeric from aggregated TPST-1.

1. Equilibrate a Superdex 200 26/60 size-exclusion column with buffer B (100 m*M* Na-acetate, pH 4.5; 6 *M* GndHCl; 5 m*M* DTT) at a flow rate of 2 ml/min and with UV detection at 280 nm.
2. Load 500 μl of concentrated TPST-1 stock solution in buffer A to the column and collect eluent in 5-ml fractions.
3. Pool fractions corresponding to monomeric TPST-1, concentrate by ultrafiltration using a 10-kDa cutoff membrane (YM10, Amicon Millipore) to a final concentration of about 30 mg/ml, and store aliquots at -80 °C.
4. Determine protein concentrations by the Bradford method with BSA as a standard.

2.2.5 TPST-1 Refolding

TPST-1 refolding is achieved by diluting a concentrated (30 mg/ml) stock solution of unfolded TPST-1 in buffer A into a large (200-fold) volume of the refolding buffer. The refolding conditions were first optimized in small-scale (1 ml) refolding trials using a fractional factorial refolding screen

(Chen & Gouaux, 1997). Conditions were then adjusted and further optimized for large-scale refolding reactions by successively increasing the reaction volume from 100 to 2000 ml.

1. For a 1000-ml scale refolding reaction, transfer approximately 5 ml of concentrated TPST-1 stock solution in buffer A (containing approximately 150 mg of unfolded TPST-1) into a 15-ml polycarbonate tube with conical bottom that is kept on ice.
2. Prepare 1000 ml of filtered (0.22 μm) 50 m*M* Tris–HCl, pH 8.5, 500 m*M* GndHCl, 10 m*M* NaCl, 0.4 m*M* KCl, 1 m*M* EDTA buffer, transfer to a 1500-ml beaker, add a magnetic stir bar, and refrigerate to 5 °C. Finalize the refolding buffer by flushing with argon and adding 0.14 m*M* of DDM, 5 m*M* of GSH, and 2.5 m*M* of GSSG. Cover the beaker with Parafilm and place on a magnetic stirrer.
3. Assemble the refolding setup in a cold room or a large refrigerator kept at 5 °C. To slowly add the TPST-1 stock solution to the refolding buffer, use a peristaltic pump with 0.01 in. ID silicon tubing. Place one end of the silicon tubing into the TPST-1 stock solution so that it reaches to the bottom of the conical tube. Place the other end of the silicon tubing so that it just touches the surface of the refolding buffer (use P10 tip) (keep covered with Parafilm).
4. To start the refolding reaction, first switch on the magnetic stirrer and adjust the stir speed so that a small vortex forms, which just reaches the bottom of the beaker. Next, start the peristaltic pump and set it to a flow rate between 10 and 20 μl/min. After TPST-1 addition is complete, stop the magnetic stirrer and incubate the refolding mixture for 20 h at 4 °C.
5. Centrifuge (30 min, 125,000 × *g*, 4 °C) the refolding mixture to remove precipitated protein.
6. Concentrate the supernatant containing refolded TPST-1 by ultrafiltration using a membrane with 10 kDa cutoff (YM10, Amicon Millipore).
7. Dialyze the concentrated refolding mixture against storage buffer (20 m*M* MOPS, pH 7.5; 150 m*M* NaCl; 10 m*M* EDTA; 0.1% Triton X-100; 10% glycerol). Store aliquots at -80 °C.
8. Determine protein concentrations by the Bradford method using BSA as a standard.
9. Analyze TPST-1 preparations by SDS–PAGE on 10% gels and by TPST-1 sulfation assay using PSGL-1 peptide (QATEYEYLDYDFLPE-NH_2) as a standard (Section 2.4) (on average, 80% yield of TPST-1 protein with 30–50% specific activity compared to TPST-1 from eukaryotic expression).

2.3 Analysis and Purification of PAPS

In our experience, PAPS from commercial sources was generally about 80% pure. However, significantly lower purities have been observed in some cases. A major impurity in any PAPS preparation is 3′-phosphoadenosine-5′-phosphate (PAP), which is formed by spontaneous hydrolysis of PAPS. As a by-product of the TPST-catalyzed sulfotransfer reaction, PAP acts as a product inhibitor of tyrosine sulfation (Danan, Yu, Hoffhines, Moore, & Leary, 2008; Danan et al., 2010), which is one reason why sulfation rates of *in vitro* sulfation reactions decrease over time. Thus, for efficient *in vitro* sulfation of peptide substrates, in particular, if multiple sulfation sites are present, high-purity (≥80%) PAPS preparations with low (<10%) PAP content are to be used. For these reasons, we found it necessary to routinely analyze all PAPS preparations by ion-pair HPLC, and to repurify low-purity PAPS preparations by Mono Q anion-exchange chromatography. Furthermore, to minimize the formation of PAP, PAPS preparations should be kept at neutral pH and at very low temperatures (-80 °C).

2.3.1 Required Materials

PAPS and PAP can be obtained from various commercial sources (Calbiochem/EMD Biosciences, Sigma-Aldrich, Fluka, R & D Systems)
HPLC-grade solvents (Pierce or Fisher Scientific)
Tetrabutylammonium phosphate (TBAP) (Alltech)
Dowex 50WX8/H^+ resin (400–200 mesh) p.a. (Fluka)
Analytical-grade reagents (Sigma-Aldrich or Fisher Scientific)
HPLC system with UV–Vis detector
LiChrospher 100 RP-18 endcapped (5 μm, 250 mm × 4.6 mm) (EMD Millipore) or similar analytical RP-HPLC column
FPLC system with Mono Q HR5/5 column (GE Healthcare)
Standard laboratory equipment (SpeedVac™, lyophilizer, UV–Vis spectrometer, etc.)

2.3.2 Preparation of PAPS Stock Solutions

1. Dissolve lyophilized PAPS in HPLC-grade water at 3–4 m*M*.
2. Measure pH and neutralize by adding approximately 10 μl of 1 *M* Tris–HCl, pH 7.5 to 1 ml of PAPS solution if pH is below 7.0.
3. Precisely determine PAPS concentration by UV spectroscopy using a molar extinction coefficient of 15,400 M^{-1} cm^{-1} (at 259 nm and pH 7.0).
4. Analyze PAPS stock solution by ion-pair RP-HPLC (Section 2.3.3).
5. Store aliquots at -80 °C.

2.3.3 Analysis of PAPS by Ion-Pair RP-HPLC

This method was modified after a published procedure by Pennings and van Kempen (1979).

1. Prepare eluent A (10 m*M* $NH_4H_2PO_4$, pH 5.5, 5 m*M* TBAP, in water) and eluent B (100% acetonitrile), filter, and degas.
2. Equilibrate an analytical LiChrospher 100 RP-18 HPLC column with 20% B at a flow rate of 1.5 ml/min. For each HPLC analysis, inject a 20-μl sample onto the column and apply an eluent gradient from 20% to 50% B in 20 min at a flow rate of 1.5 ml/min with UV detection at 260 nm.
3. To test and calibrate the HPLC system, first perform a blank analysis by injecting a water sample. Next analyze a PAP reference, which should elute at ca. 12 min followed by a PAPS reference, which should elute at ca. 16 min.
4. Analyze the PAPS sample and calculate PAPS purity from peak integration data of the HPLC.

2.3.4 Purification of PAPS by Mono Q Anion-Exchange Chromatography (Optional)

Optionally, if PAPS purity is less than 80% or PAP content exceeds 10%, PAPS can be purified by Mono Q anion-exchange chromatography (Burkart, Izumi, Chapman, Lin, & Wong, 2000).

1. Prepare eluent A (water) and eluent B (1 *M* NH_4HCO_3, in water), filter, and degas.
2. Equilibrate a Mono Q HR5/5 column with 100% A at a flow rate of 1 ml/min.
3. Inject 250-μl sample of PAPS stock solution onto the column and apply an eluent profile of 0% B for 5 min followed by 0–100% B in 55 min, and 100% B for 5 min at a flow rate of 1 ml/min with UV detection at 245 nm.
4. Collect 0.5-ml eluent fractions and pool fractions corresponding to PAPS (PAP should elute at ca. 26 min and PAPS at ca. 40 min).
5. Using a vacuum filtration device with a sintered glass filter, wash Dowex 50WX8/H^+ resin with water followed by methanol and ethanol, and air-dry resin.
6. Neutralize pooled PAPS fractions (from step 4) by stepwise adding washed Dowex 50WX8/H^+ resin and carefully monitoring the pH to avoid acidification.
7. Filter the PAPS solution, freeze with liquid nitrogen, and lyophylize.

8. To convert PAPS to its tetralithium salt, prepare Dowex 50WX8/Li^+ resin: Treat Dowex 50WX8/H^+ resin stepwise with 1 *M* LiCl until the pH is neutral, wash with water followed by ethanol, and air-dry resin.
9. Dissolve PAPS (from step 7) with HPLC-grade water (at 3–4 m*M*) and treat with Dowex 50WX8/Li^+ resin. Filter the PAPS solution and neutralize with approximately 10 μl of 1 *M* Tris–HCl, pH 7.5 for 1 ml of PAPS solution.
10. Analyze PAPS solution by ion-pair RP-HPLC (Section 2.3.3), determine PAPS concentration by UV spectroscopy (Section 2.3.2), and store aliquots at -80 °C.

2.4 *In Vitro* Sulfation of N-Terminal Chemokine Receptor Peptides Using TPST Enzymes

We have not found fluoride, manganese, or 5′-AMP to be essential for TPST-1 or TPST-2 activity. Dithiothreitol (DTT) is also optional as precaution for preventing possible methionine oxidation if peptides contain methionine(s).

2.4.1 Required Materials

400 m*M* PIPES, pH 6.8
10% Triton X-100
5 *M* NaCl
1 *M* DTT (optional)
3–4 m*M* PAPS (>80%, see Section 2.3)
0.5 mg/ml TPST-1 or TPST-2
Concentrated stock of N-terminal peptide dissolved in water, pH 6.8

2.4.2 Conditions for Enzymatic Sulfation of Peptides

1. Combine the above reagents with a volume of HPLC-grade water to produce final concentrations of 40 μ*M* PIPES, 100 m*M* NaCl, 0.10% Triton X-100, 10 m*M* DTT (optional), 400 μ*M* PAPS, 50–100 μ*M* peptide, and 0.05 mg/ml TPST-1 or TPST-2. Control reactions can leave out the PAPS cosubstrate or TPST enzyme.
2. Incubate at 16 °C.
3. The reaction can be monitored using RP-HPLC and mass spectrometry by removing 50 μl of the reaction every 12–24 h. See RP-HPLC and mass spectrometry sections for protocols. New peak(s) with increasing intensity in the RP-HPLCs along with a decreased peak intensity for the unsulfated peptide peak correlate with tyrosine sulfation (see Fig. 2).

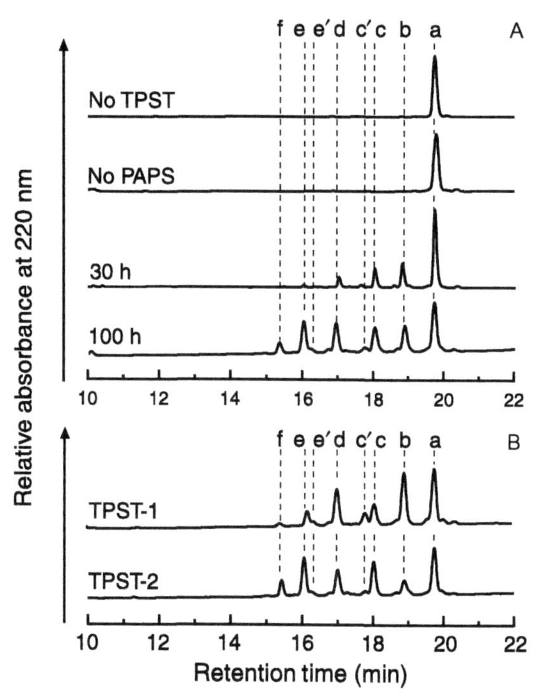

Figure 2 RP-HPLC analysis of CCR5 2–18 sulfation products. (A) Characterization of the *in vitro* sulfation reaction. Peptide CCR5 2–18 (0.1 mg/ml, ~50 μ*M*) was incubated with a mixture of TPST-1 and TPST-2 (20 μg/ml each) and in the presence of the sulfation cosubstrate PAPS (400 μ*M*). After 30 or 100 h at 16 °C, 60-μl aliquots were analyzed by RP-HPLC. In negative-control experiments (100-h incubation time), either the TPST mixture (no TPST) or the PAPS (no PAPS) was omitted. Peaks were labeled a–f in increasing order of hydrophilicity. (B) Comparison of TPST-1 and TPST-2. CCR5 2–18 (0.1 mg/ml, ~50 μ*M*) was incubated for 100 h with TPST-1 (40 μg/ml) or TPST-2 (40 μg/ml) in the presence of PAPS (400 μ*M*). Peak a corresponds to CCR5 2–18, peak b to CCR5 2–18 sY14, peak c to CCR5 2–18 sY15, peak c′ to CCR5 2–18 sY10/sY14, peak d to CCR5 2–18 sY14/sY15, peak e′ to CCR5 2–18 sY3/sY10/sY15, peak e to CCR5 2–18 sY10/sY14/sY15, and peak f to CCR5 2–18 sY3/sY10/sY14/sY15. *Reproduced from Seibert et al. (2002). Copyright 2002 National Academy of Sciences, USA.*

2.5 Reversed-Phase HPLC of Sulfopeptides

While successful separation of sulfotyrosine peptides has been described with standard RP-HPLC solvent systems containing 0.1% trifluoroacetic acid, we found that separation at pH 6.5 using 20 m*M* ammonium acetate as a buffer gave best results. Under these conditions, the potential loss of the sulfotyrosine modification caused by acid-catalyzed hydrolysis is minimized (Seibert & Sakmar, 2008; Stone & Payne, 2015). Furthermore, at pH 6.5

resolution of closely related sulfotyrosine peptides was greatly improved compared to resolution in the standard acidic solvent system.

2.5.1 Required Materials

2:1 Chloroform:methanol solution
Bioselect™ C18 SPE columns (218SPE3000, Grace Vydac)
Analytical RP-HPLC column (218TP54, Grace Vydac)
Semipreparative RP-HPLC column (218TP510, Grace Vydac)
SpeedVac™ with heater disabled or lyophilizer
RP-HPLC with UV–Vis detector
HPLC buffer A: 20 m*M* ammonium acetate, pH 6.5 in water
HPLC buffer B: 20 m*M* ammonium acetate, pH 6.5, 70% acetonitrile
0.22-μm Spin filter

2.5.2 Analytical RP-HPLC

RP-HPLC can be used to follow sulfation reaction (monitoring absorbance at 220 nm) for the appearance of new sulfopeptide peaks and the loss of the unsulfated peak (Seibert et al., 2002, 2008). Peaks can be collected to purify individual sulfopeptides.

2.5.2.1 Sample Clean Up

1. Extract the sulfation reaction or portion thereof with 3.5–4 times the reaction volume of 2:1 chloroform:methanol solution. Save the aqueous layer.
2. Equilibrate the SPE column(s) (1 ml HPLC buffer A, followed by 1 ml HPLC buffer B, followed by 3 ml HPLC buffer A).
3. Load the aqueous layer onto the SPE column. Do not exceed 5 mg of total peptide; use additional SPE columns.
4. Wash the SPE column with 1 ml buffer A.
5. Elute the SPE column with 1.5 ml of a 50:50 mixture of HPLC buffers A and B.
6. Strip the SPE column with 100% acetonitrile.
7. SpeedVac™ or lyophilize the elution.

2.5.2.2 RP-HPLC

1. Dissolve the cleaned up, dried sulfation reaction in a volume of water equivalent to the reaction volume from which the sample came. For example, if a 50-μl aliquot of a sulfation reaction was prepared for HPLC, dissolve the cleaned up, dried sample in 50 μl of water.

2. Spin filter the dissolved sulfation reaction.
3. Analyze 50 μl of the dissolved, filtered sulfation reaction using the analytical RP-HPLC column with a linear gradient of HPLC buffer B from 5% to 60% over 40 min. If using an HPLC with a detector limited to one wavelength, monitor at 220 nm.
4. Collect eluting fractions corresponding to peptide peaks.
5. Pool peaks and SpeedVac™ or lyophilize.

2.5.3 Semipreparative RP-HPLC

Purification of large amounts of sulfopeptides requires scaling up the RP-HPLC procedure.

1. Semipreparative RP-HPLC follows the same general procedure as the analytical RP-HPLC with some exceptions. A semipreparative column is used with a flow rate of 3 ml/min and with a larger 1, 5, or 10 ml injection loop.
2. The gradient may also need to be optimized to obtain baseline separation of peaks when purifying large quantities of sulfopeptides (0.5–1 mg of total peptide). Start by injecting 50 μl and incrementally increasing the percent HPLC buffer B at the start of the gradient while decreasing the percent HPLC buffer B at the end of the gradient until peptide peaks are very well resolved. Then incrementally increase the injected volume or total peptide injected. Stop increasing when it appears baseline separation will be lost or column capacity may be exceeded. For example, using a linear gradient from 19% to 30% HPLC buffer B over 40 min up to 1 mg of total peptide from a sulfation reaction using the CXCR4 N-terminus as substrate could be separated into the individual resulting sulfopeptides (Seibert et al., 2008) (see Fig. 3).

2.6 Mass Spectrometry of Sulfotyrosine Peptides

Analysis of sulfotyrosine peptides and localization of sulfotyrosine positions in the presence of multiple potential sulfation sites can be a challenging task, in particular, if multiple sulfotyrosines are present in a single peptide chain, which is the case with most N-terminal chemokine receptor peptides. While mass spectrometry analysis of protein phosphorylation on a proteomics scale is well established, this is not the case for protein tyrosine sulfation. Due to the inherent lability of the sulfotyrosine sulfoester bond, partial or complete loss of the sulfotyrosine modification is generally observed as a neutral loss of SO_3 (ΔM_r = -80 Da) under standard mass spectrometry conditions. In particular, irrespective of the desorption/ionization method employed, positive

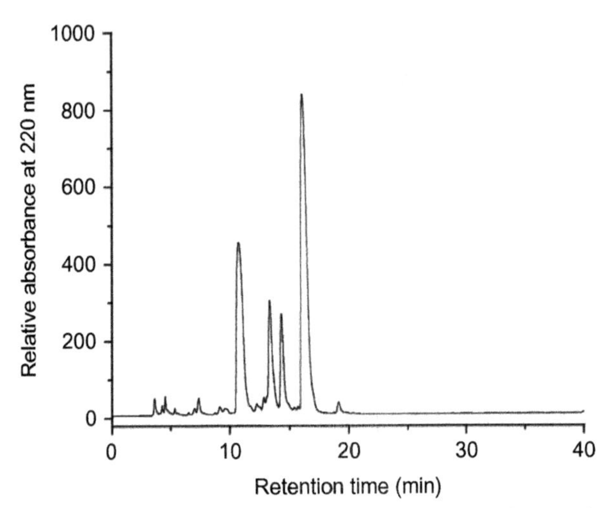

Figure 3 Representative RP-HPLC for purification of CXCR4 sulfopeptides on a semi-preparative scale (see Section 2.5.3).

ion mode mass spectrometry methods generally lead to complete loss of the sulfotyrosine modification. In negative ion mode mass spectrometry, in contrast, using optimized experimental conditions, a high degree of sulfotyrosine retention has been observed (see Seibert & Sakmar, 2008 for review). For example, analyzing an N-terminal CXCR4 peptide with three sulfotyrosines in negative ion mode electrospray ionization (ESI) mass spectrometry, we observed complete retention of all three sulfotyrosine modifications. Analysis of the same peptide using matrix-assisted laser desorption/ionization-time-of-flight (MALDI-TOF) mass spectrometry in negative ion linear mode, on the other hand, resulted in a major peak corresponding to triple sulfated peptide with additional peaks corresponding to the neutral loss of one or two SO_3 (Seibert et al., 2008).

The biggest obstacle for sulfotyrosine analysis to date is the lack of a robust and reliable mass spectrometry fragmentation method for sequencing sulfotyrosine peptides and localizing the sulfation sites by tandem mass spectrometry (MS/MS). In particular, collision-induced dissociation (CID), which in the past has been the workhorse method for fragment ion generation in MS/MS and therefore is widely available in mass spectrometry labs and core facilities, generally results in extensive loss of sulfate in the observed fragment ions (see Seibert & Sakmar, 2008 for review). In recent years, novel fragmentation methods such as electron capture dissociation, electron transfer dissociation, electron detachment fragmentation, metastable atom-activated dissociation, and ultraviolet photodissociation have been developed, which favor peptide backbone fragmentation over fragmentation of labile side chain modifications. Using these fragmentation methods, it was demonstrated that sulfated fragment ions sufficient for sulfotyrosine localization can be produced, in particular, if these methods were adopted for negative ion mode (see Robinson, Moore, & Brodbelt, 2014; Seibert & Sakmar, 2008 and references therein). While these results are promising, most of these methods have so far been used in pilot studies only, and therefore, the available data are limited to a small number of model peptides with one or two sulfotyrosines. It remains to be seen which of these methods are capable of analyzing more challenging peptides and proteins containing multiple sulfotyrosines in highly acidic sequences and at a proteomics scale.

These novel fragmentation methods require highly specialized equipment that is currently not readily available at many institutions. Therefore, we provide alternative methods for sulfotyrosine localization that are based on the generation of peptide fragments by proteolytic cleavage of sulfotyrosine peptides. While this approach can be laborious and is not suited for sulfotyrosine analysis at a proteomics scale, it might still be the best option, if a small number of closely related sulfotyrosine peptides of known sequence are to be analyzed, as is the case with identifying the products of an *in vitro* sulfation reaction (Seibert et al., 2002, 2008).

Another robust method for sulfotyrosine localization, which does not rely on sulfotyrosine retention in fragment ions and should be considered as an alternative to the method described by us, was developed by Leary and coworkers (Yu, Hoffhines, Moore, & Leary, 2007). This method is based on the acetylation of unsulfated tyrosine hydroxyl groups using sulfosuccinimidyl acetate, leaving sulfated tyrosine residues unmodified. MS/MS analysis of the derivatized peptides in positive ion mode using CID as the fragmentation method results in complete loss of tyrosine sulfation but leaves tyrosine acetylation intact. Thus, sulfotyrosine residues are identified as unmodified tyrosines, whereas unsulfated tyrosines are observed as acetyltyrosines in the fragment ions. Using this method, Leary and coworkers were able to localize sulfotyrosine residues in numerous peptides, including several N-terminal chemokine receptor peptides with multiple sulfation sites (Jen et al., 2009; Yu et al., 2007).

2.6.1 Required Materials

Sequencing-grade proteases (Roche)

α-Cyano-4-hydroxycinnamic acid (4HCCA) (Sigma)

HPLC-grade solvents (Pierce or Fisher Scientific)

RP-C18 ZipTips (Millipore)
Analytical-grade reagents (Sigma-Aldrich or Fisher Scientific)
Mass spectrometry instrumentation for MALDI-TOF MS and/or ESI MS

2.6.2 Proteolytic Cleavage of Sulfotyrosine Peptides

To localize sulfotyrosine residues in peptides with multiple potential sulfation sites, peptides can be cleaved with specific proteases to generate fragments containing subsets of the sulfation sites (Seibert et al., 2002, 2008). If sulfation sites in fragments are ambiguous, the cleavage process can be repeated for the fragments using proteases with different specificities. In our experience, endoproteinases Asp-N or Glu-C are most useful for fragmenting sulfotyrosine peptides, because sulfotyrosine peptides generally contain multiple acidic residues in proximity of the sulfation sites. However, for peptides with complex tyrosine sulfation patterns, specific cleavage with additional proteases such as chymotrypsin, or generating a peptide ladder by using time-dependent carboxypeptidase Y digestion might be required.

1. Determine the number of sulfotyrosine residues in the uncleaved (RP-HPLC-purified) peptide by mass spectrometry. If the sulfation sites are ambiguous, proceed with step 2.
2. Identify proteolytic cleavage sites in the peptide and design a strategy for fragment generation.
3. Digest peptide (0.1–1 μg) with sequencing-grade proteases according to the manufacturer's instructions and take (5 μl) samples after different incubation times.
4. Optionally, to separate cleavage fragments subject samples to analytical RP-HPLC (see Section 2.5.2) and collect peak fractions.
5. Dry samples using a SpeedVac™, dissolve in a small volume (3 μl) of saturated 4HCCA matrix solution, and analyze by MALDI-TOF MS (see Section 2.6.3). Alternatively, desalt samples using RP C18 ZipTips and analyze by ESI MS.
6. If sulfation sites in fragments are ambiguous, repeat steps 2–5 for (RP-HPLC separated) fragments or uncleaved peptide using different protease.

2.6.3 MALDI-TOF Mass Spectrometry

In our experience, an ultra-thin layer sample preparation method (Cadene & Chait, 2000) using ammonium acetate as an additive gave best results for MALDI-TOF MS of sulfotyrosine peptides (Seibert et al., 2002, 2008).

Under optimized measurement conditions, with low laser power in linear negative ion mode, complete sulfate retention was routinely observed in the most abundant ion. However, some loss of sulfate was generally observed, giving rise to additional signals that could be mistaken for contaminating peptide species. Hence, it is crucial to use RP-HPLC-purified sulfotyrosine peptides for MALDI-TOF MS analysis. For a more detailed discussion of the characteristic loss-of-sulfate peak patterns observed in MALDI-TOF mass spectra of sulfotyrosine peptides, we refer to the following references: Seibert et al. (2002, 2008) and Seibert and Sakmar (2008).

1. Purify the MALDI matrix 4HCCA by HCl precipitation.
2. Prepare the ultra-thin layer solution by diluting a saturated solution of 4HCCA in TWA (0.1% TFA in H_2O/ACN 2:1) 1:3 with isopropanol.
3. Prepare a saturated 4HCCA matrix solution in a 2:1 (v/v) mixture of water and acetonitrile with 10 m*M* ammonium acetate as an additive.
4. Wash sample plate with methanol, followed by water, followed by methanol and let it dry thoroughly.
5. To create an ultra-thin layer of the 4HCCA matrix, evenly distribute 20 μl of the ultra-thin layer solution over the sample plate with the flat side of a 200-μl pipette tip. Wait until approximately two-third of the liquid has dried and use a Kimwipe paper to distribute the remaining liquid. Carefully remove white material from the sample plate with a Kimwipe paper wrapped around a finger, leaving a yellowish ultra-thin layer of 4HCCA on the surface.
6. Dilute peptide sample 1:10 in matrix solution and spot a small aliquot (0.5–1 μl) of peptide–matrix solution onto the ultra-thin layer-coated sample plate. Immediately when crystals form, remove excess liquid using a 10-μl pipette tip attached to a vacuum line.
7. Perform MALDI-TOF mass measurements in negative linear, delayed extraction mode. The precise measurement parameters will depend on the specific instrumentation used and require some optimization. As a guideline, the instrument settings for a Voyager DE-STR instrument (Applied Biosystems) can be found elsewhere (Seibert et al., 2002, 2008).

2.6.4 ESI Mass Spectrometry

Alternatively, sulfotyrosine peptides can be analyzed using ESI MS in negative ion mode. In our experience, using optimized measurement conditions, virtually no loss of sulfate was observed, and therefore, interpretation of negative ion mode ESI mass spectra of sulfotyrosine peptides was straight forward (Seibert et al., 2008). If sodium adduct formation is observed, the use of ammonium acetate as an additive should be considered to improve signal intensities.

1. Desalt sulfotyrosine peptide samples by binding to RP-C18 ZipTips according to the manufacturer's protocol. After washing with HPLC-grade water, elute peptides with 70% acetonitrile.
2. Perform ESI mass measurements in negative ion mode. The precise measurement parameters will depend on the specific instrumentation used and require some optimization. As a guideline, the instrument settings for a QSTAR XL hybrid electrospray quadrupole–quadrupole time-of-flight mass spectrometer (Applied Biosystems) equipped with a nano-ESI source can be found elsewhere (Seibert et al., 2008).

2.7 Characterization of Sulfopeptides by Protein NMR

If uniformly isotopically labeled (^{15}N or $^{15}N/^{13}C$) peptides are used in TPST-driven sulfation reactions and purified by RP-HPLC, protein NMR, in addition to mass spectrometry, can be used to confirm the location of tyrosine sulfation.

2.7.1 Required Material

High-field nuclear magnetic resonance spectrometer ($\geq$500 MHz)
RP-HPLC-purified [U-^{15}N or U-$^{15}N/^{13}C$] peptide
RP-HPLC-purified [U-^{15}N or U-$^{15}N/^{13}C$] sulfopeptide
Chemical shift assignments for [U-$^{13}C/^{15}N$] peptide determined using standard assignment strategies (Markley, Ulrich, Westler, & Volkman, 2003)
^{15}N-^{1}H heteronuclear single-quantum coherence pulse program such as hsqcf3gpph19 from the Bruker sequence library

2.7.2 ^{15}N-^{1}H HSQC Spectroscopy

1. Prepare separate NMR samples consisting of purified [U-^{15}N or U-$^{15}N/^{13}C$] unsulfated peptide and [U-^{15}N or U-$^{15}N/^{13}C$] sulfopeptides in an appropriate NMR buffer. For example, 25 m*M* deuterated MES, 10% D_2O, 0.2% NaN_3, pH 6.8.
2. Collect ^{15}N-^{1}H HSQC spectra and overlay each sulfopeptide spectra with that of the unsulfated peptide.
3. Compare the spectra paying particular attention to tyrosine residues and residues proximal to tyrosines. Similarities in chemical shift correlate with the absence of sulfation while differences in chemical shift are indicative of tyrosine sulfation (see Fig. 4).

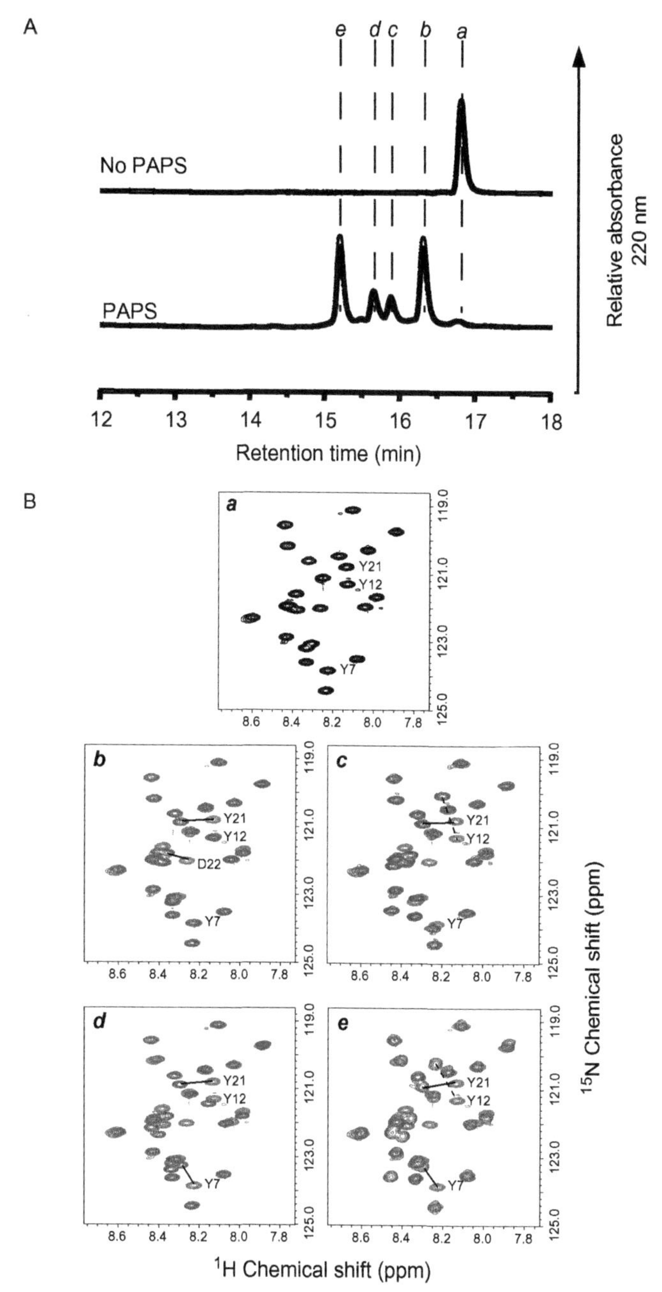

Figure 4 Protein NMR analysis of CXCR4 sulfopeptides. (A) Analytical-scale RP-HPLCs of TPST-1-catalyzed sulfation reactions of 50 μ*M* CXCR4 1–38 (C24A) without and with 400 μ*M* PAPS. Peak a corresponds to CXCR4 1–38 (C24A), peak b to CXCR4 1–38 (C24A) sY21, peak c to CXCR4 (C24A) sY12/sY21, peak d to CXCR4 (C24A) sY7/sY21, and peak e to CXCR4 (C24A) sY7/sY12/sY21. (B) ^{15}N-^{1}H HSQC spectra of peak a and overlays of ^{15}N-^{1}H HSQC spectra of peak b, c, d, or e (red (dark gray in the print version)) onto peak a (gray).

3. CAVEATS AND LIMITATIONS

While making chemokine receptor sulfopeptides enzymatically is one approach, Payne and colleagues have turned what were pitiful methods for chemically synthesizing sulfopeptides into a usable approach (Stone & Payne, 2015). A purported strength of the chemical synthesis approach allows for control of where sulfotyrosines are placed in a peptide or the pattern of sulfotyrosines in a peptide containing more than one. Hence, a possible limitation, or strength depending on perspective, to using TPSTs to sulfate chemokine receptor peptides is that one is limited to the tyrosine sulfation pattern resulting from enzymatic sulfation versus the control afforded through chemical synthesis. For example, one could not use TPSTs to produce a CXCR4 N-terminal peptide, which contains tyrosines at positions 7, 12, and 21, sulfated at only Y7 as analysis of CXCR4 sulfation kinetics indicates Y21 is sulfated first followed by either 7 or 12 with the remaining tyrosine sulfated thereafter (Seibert et al., 2008). Hence, the impact of only sY7 on the interaction of a CXCR4 N-terminal peptide with CXCL12 could not be addressed using this system. However, this limitation can be viewed as a strength because receptor peptides with more physiologically relevant sulfation patterns will be investigated. It is unlikely that it is just a coincidence that TPSTs sulfate an N-terminal CXCR4 peptide at Y21 first and that, in the context of intact CXCR4, sY21 has the most impact on CXCL12 binding (Farzan, Babcock, et al., 2002).

While sulfation of the CXCR4 N-terminus has little impact on its role as an HIV-1 coreceptor, sulfotyrosines in the N-terminus of CCR5 greatly enhance HIV-1 infection or affinity for chemokine ligands (Farzan, Babcock, et al., 2002; Farzan, Chung, et al., 2002; Farzan et al., 1999). A peptide corresponding to residues 2–18 of the CCR5 N-terminus, which contains tyrosines at positions 3, 10, 14, and 15, is sulfated by TPST-1 or TPST-2 first at tyrosines 14 and 15 followed by sulfation at position 10 and lastly at position 3 (Jen et al., 2009; Seibert et al., 2002). But, a CCR5 1–18 peptide is sulfated at Y3 first followed by Y14/Y15 and finally Y10 (Jen et al., 2009). While CCR5 Y3 is sulfated first, sY3 appears least important among other sY's in full-length CCR5 for HIV-1 infection but sY3 does have a small impact on chemokine binding (Bannert et al., 2001; Farzan et al., 1999). While a CCR5 sY10/sY14 N-terminal peptide is sufficient to rescue chemokine signaling through CCR5 Δ2–17 (Farzan, Chung, et al., 2002) or reconstitute HIV-1 infection in cells expressing

CCR5 Δ2–17 (Liu et al., 2014), neither study investigated sY3 in the context of an N-terminal CCR5 peptide. As one would expect, knowledge regarding the order of tyrosine sulfation by TPSTs, which in some instances appears to correlate with functional importance, along with the ability to control the location and pattern of sulfotyrosines chemical synthesis will provide the broadest picture of functional importance for individual sulfotyrosine residues.

While lack of control over the location of sulfotyrosines in the sulfopeptides produced enzymatically may be a limitation or perhaps a strength, there is a much more important caveat to mention. A major caveat to using recombinant TPST enzymes to generate sulfotyrosine containing chemokine receptor peptides or other peptides is the requirement for highly pure PAPS cosubstrate, generally greater than 80% pure. This is because PAPS can spontaneously degrade to sulfate and PAP at a rate that increases with increasing temperature. In our experience, PAPS is relatively stable over a period of a couple months at -80 °C, but will degrade to PAP at 37 °C in a matter of a day. Additionally, highly pure PAPS has not always been commercially available and when it is shipping time or conditions can have an impact on PAPS purity. If PAPS purity is less than 80% or PAP content exceeds 10%, PAPS can be purified by Mono Q anion-exchange chromatography as described above (Burkart et al., 2000).

4. PERSPECTIVES

Native sources of TPSTs have been used to enzymatically sulfate peptides, but producing enzymatically sulfated peptides on a large scale really became realistic upon the cloning and purification of recombinant TPSTs from mammalian cell lines by Moore and colleagues (Ouyang et al., 1998; Ouyang & Moore, 1998). Here, we provide a protocol that closely follows Moore and colleague's method for preparing recombinant human TPSTs from mammalian cells and our methods for sulfation of N-terminal chemokine receptor peptide (Ouyang et al., 1998; Ouyang & Moore, 1998). We also present a method for expression and functional refolding of recombinant, enzymatically active TPST-1 from *E. coli*, which we have also used to sulfate N-terminal chemokine receptor peptides and PSGL-1 peptides. The structure of human TPST-2 was recently solved using recombinant TPST-2 from *E. coli* and has provided structural information on the TPST's mechanism (Teramoto et al., 2013). The availability of bacterial-produced enzymes makes large-scale preparation of sulfopeptides using TPSTs even more practical. An added benefit of using TPSTs to produce sulfopeptides is that the reactions can be monitored over time and, if multiple tyrosines are present, the order of tyrosine sulfation can be determined. This can be done simply by quantifying absorbance of RP-HPLC peaks (Seibert et al., 2002, 2008) or through monitoring the reactions using more complex mass spectrometry approaches pioneered by Leary and colleagues (Danan et al., 2008, 2010; Jen et al., 2009). Knowledge of tyrosine sulfation order provides information of which sulfopeptides are most physiologically relevant for study and such kinetic data can also provide information on the enzymatic mechanism of TPSTs (Danan et al., 2008, 2010; Jen et al., 2009).

We have used TPSTs to make large quantities of sulfopeptides for structural studies. For example, TPSTs were used to sulfate milligram quantities of CXCR4 N-terminal peptides of natural isotopic abundance or labeled uniformly with N-15 or N-15 and C-13. Because CXCR4 N-terminal peptides promote CXCL12 dimer formation, we used a covalently locked CXCL12 dimer to solve the structure of the CXCL12 dimer with various N-terminal CXCR4 peptides that were either unsulfated, sulfated at position 21, or sulfated at tyrosines 7, 12, and 21 (Veldkamp et al., 2008, 2006; Ziarek et al., 2013). Utilizing this structure, small molecule inhibitors that target the CXCR4 sulfotyrosine 21 binding site on CXCL12 have been developed (Smith et al., 2014; Veldkamp, Ziarek, Peterson, Chen, & Volkman, 2010). The availability of isotopically labeled sulfopeptides, which is possible when producing sulfopeptides enzymatically but not synthetically, aided in solving a high-quality NMR structure of a locked CXCL12 dimer with CXCR4 sulfopeptides. However, Stone and colleagues have shown through their structure of CCL11 bound to a synthetic CCR3 sulfopeptide that isotopic labeling of sulfopeptides is not required for structure determination (Millard et al., 2014). Stone and colleagues provide a comparison of their CCL11/CCR3 sulfopeptide structure to that of our locked CXCL12 dimer with CXCR4 sulfopeptides (Millard et al., 2014), while Handel and colleagues provide analysis of the locked CXCL12 dimer with CXCR4 sulfopeptide structure in the context of their recently determined structure of intact CXCR4 and the viral chemokine vMIP-II (Kufareva et al., 2015; Qin et al., 2015).

The classical view of sulfotyrosine is that it is a posttranslational modification that increases affinity between binding partners. We have shown this to be the case for CXCL12 and enzymatically produced CXCR4 N-terminal peptides with affinity increasing with increasing sulfation

(Seibert et al., 2008). Others have shown similar increases in affinity between chemokines and synthetic sulfotyrosine containing receptor peptides from CCR5 (Duma et al., 2007), CCR3 (Simpson et al., 2009; Zhu et al., 2011), and CCR2 (Tan, Ludeman, et al., 2013). However, a more significant role for sulfotyrosine posttranslational modifications in some chemokine receptors likely involves the role of sulfotyrosines in ligand bias or biased agonism. Many chemokine receptors have two or more chemokine ligands and are excellent examples of ligand bias or biased agonism with CCR7 and its ligands CCL19 and CCL21 being a protypical example (Zidar, Violin, Whalen, & Lefkowitz, 2009). Different sulfation patterns affect the selectivity of synthetic N-terminal peptides from CCR3 for CCL11/eotaxin-1, CCL24/eotaxin-2, and CCL26/eotaxin-3 and are hypothesized to contribute to ligand-biased signaling in intact CCR3 (Zhu et al., 2011). Unlike CCR3 or CCR7, which have multiple and in most cases monomeric ligands (Love et al., 2012; Millard et al., 2014; Veldkamp et al., 2015), CXCR4 has only one natural ligand CXCL12. Peptides and sulfopeptides corresponding to the N-terminus of CXCR4 promote CXCL12 dimer formation and a covalently locked CXCL12 dimer with a nearly identical structure to the wild-type CXCL12 dimer signals through CXCR4 as a partial agonist (Drury et al., 2011; Veldkamp et al., 2008, 2006; Ziarek et al., 2013). Similarly to CCR3, it could be hypothesized that sulfotyrosines in CXCR4 are contributing to bias signaling, but instead of conferring selectivity for one ligand over another the sulfotyrosines are affecting selectivity for an oligomeric state and thereby influencing bias signaling.

Some CC chemokines form dimers; however, these dimers are not thought to activate, even partially, or bind with significant affinity to their chemokine receptors as exemplified by a covalently locked CCL2 dimer (Tan, Ludeman, et al., 2013) or CCL4 (Jin, Shen, Baggett, Kong, & LiWang, 2007). CC chemokine monomers are believed to be full receptor agonists; for example, an obligate CCL2 monomer is a CCR2 agonist (Tan, Ludeman, et al., 2013). Yet, CCR2 N-terminal sulfopeptides bind more tightly to the covalently locked CCL2 dimer and the obligate monomer than unsulfated peptides (Tan, Ludeman, et al., 2013). In the context of wild-type CCL2, these same sulfopeptides promote a shift in the CCL2 monomer–dimer equilibrium toward monomer (Ludeman & Stone, 2014; Tan, Ludeman, et al., 2013). Stone and colleagues conclude from these results that for CC chemokines that dimerize receptor sulfotyrosines promote formation of the receptor agonist or the CC chemokine monomer

(Ludeman & Stone, 2014; Stone & Payne, 2015; Tan, Ludeman, et al., 2013). Recall Farzan and colleagues used CCR5 N-terminal sulfopeptides to rescue CCL3 activation of a CCR5 Δ2–17 receptor (Bannert et al., 2001; Farzan, Chung, et al., 2002). Both CCL3 and CCL4 form tight dimers (Czaplewski et al., 1999; Lodi et al., 1994). Interestingly, the results from Stone and colleagues may suggest another interpretation of Farzan and colleagues' results. The CCR5 N-terminal sulfopeptides may not necessarily be rescuing an inactive CCR5 Δ2–17 receptor so much as these sulfopeptides are promoting formation of a receptor agonist, a monomeric chemokine.

While recombinant TPSTs allow for the study of the enzymes themselves and for preparation of sulfopeptides useful as reagents, other approaches to making sulfopeptides or proteins exist. For example, chemical synthesis approaches pioneered by Stone and Payne (2015), Liu and Shultz's incorporation of sulfotyrosine in recombinantly expressed proteins through amber codon suppression (Liu & Schultz, 2006), or simply recombinant expression in mammalian cell lines containing endogenous TPSTs (references in Seibert & Sakmar, 2008) have all been reported. Still others have coexpressed both a protein of interest and a TPST in mammalian cells. For example, Farzan and colleagues used this coexpression approach to generate a fusion of CD4-Ig with a small C-terminal CCR5 sulfopeptide mimetic and showed this fusion to be more potent at preventing HIV-1 infection than the best broadly neutralizing HIV-1 antibodies (Gardner et al., 2015). They also used a coexpression approach in rhesus macaque vaccine studies by using adeno-associated viruses coding for production of the CD4-Ig with a small C-terminal CCR5 sulfopeptide fusion and TPST-2 (Gardner et al., 2015). We believe the potential exists for using recombinant TPSTs to sulfate purified membrane or secreted proteins through *in vitro* sulfation reactions similar to those used for sulfation of peptides. For example, Handel and colleagues used a disulfide trap to preform a CXCR4/vMIP-II complex for their structure of CXCR4 with the viral chemokine in which the majority of the CXCR4 N-terminus is absent (Qin et al., 2015; Wu et al., 2010). This involved coexpression of both CXCR4 and vMIP-II with each containing an additional cysteine for formation of the disulfide-trapped complexes in insect cells (Qin et al., 2015; Wu et al., 2010). While coexpression or coinfection of a third gene of interest, a TPST, for sulfation of the CXCR4 N-terminus might generate high enough affinity for the chemokine that the receptor N-terminus adopts an observable conformation, asking insect cells to heterologously express three different genes at the same time may be a daunting task (Sokolenko et al., 2012).

An alternative approach, which might prove to be equally daunting, would be to utilize recombinant TPSTs to sulfate tyrosines in purified receptors like CXCR4 or CCR5.

ACKNOWLEDGMENTS

We apologize to anyone whose works we have failed to discuss or cite either through inadvertent omission or due to space constraints. This work was supported by NIH grant 1R15CA159202-01 to C.T.V. We wish to thank our collaborators Martine Cadene, Brian T. Chait, Francis C. Peterson, and Brian F. Volkman. We also thank Kevin L. Moore for providing TPST constructs.

REFERENCES

Bannert, N., Craig, S., Farzan, M., Sogah, D., Santo, N. V., Choe, H., et al. (2001). Sialylated O-glycans and sulfated tyrosines in the NH_2-terminal domain of CC chemokine receptor 5 contribute to high affinity binding of chemokines. *The Journal of Experimental Medicine, 194*(11), 1661–1673.

Ben-Baruch, A. (2008). Organ selectivity in metastasis: Regulation by chemokines and their receptors. *Clinical & Experimental Metastasis, 25*(4), 345–356.

Burkart, M. D., Izumi, M., Chapman, E., Lin, C. H., & Wong, C. H. (2000). Regeneration of PAPS for the enzymatic synthesis of sulfated oligosaccharides. *The Journal of Organic Chemistry, 65*(18), 5565–5574.

Cadene, M., & Chait, B. T. (2000). A robust, detergent-friendly method for mass spectrometric analysis of integral membrane proteins. *Analytical Chemistry, 72*(22), 5655–5658.

Carlow, D. A., Gossens, K., Naus, S., Veerman, K. M., Seo, W., & Ziltener, H. J. (2009). PSGL-1 function in immunity and steady state homeostasis. *Immunological Reviews, 230*(1), 75–96.

Chen, G. Q., & Gouaux, E. (1997). Overexpression of a glutamate receptor (GluR2) ligand binding domain in *Escherichia coli*: Application of a novel protein folding screen. *Proceedings of the National Academy of Sciences of the United States of America, 94*(25), 13431–13436.

Cormier, E. G., Persuh, M., Thompson, D. A., Lin, S. W., Sakmar, T. P., Olson, W. C., et al. (2000). Specific interaction of CCR5 amino-terminal domain peptides containing sulfotyrosines with HIV-1 envelope glycoprotein gp120. *Proceedings of the National Academy of Sciences of the United States of America, 97*(11), 5762–5767.

Crump, M. P., Gong, J. H., Loetscher, P., Rajarathnam, K., Amara, A., Arenzana-Seisdedos, F., et al. (1997). Solution structure and basis for functional activity of stromal cell-derived factor-1; dissociation of CXCR4 activation from binding and inhibition of HIV-1. *The EMBO Journal, 16*(23), 6996–7007.

Czaplewski, L. G., McKeating, J., Craven, C. J., Higgins, L. D., Appay, V., Brown, A., et al. (1999). Identification of amino acid residues critical for aggregation of human CC chemokines macrophage inflammatory protein (MIP)-1alpha, MIP-1beta, and RANTES. Characterization of active disaggregated chemokine variants. *The Journal of Biological Chemistry, 274*(23), 16077–16084.

Danan, L. M., Yu, Z., Hoffhines, A. J., Moore, K. L., & Leary, J. A. (2008). Mass spectrometric kinetic analysis of human tyrosylprotein sulfotransferase-1 and -2. *Journal of the American Society for Mass Spectrometry, 19*(10), 1459–1466.

Danan, L. M., Yu, Z., Ludden, P. J., Jia, W., Moore, K. L., & Leary, J. A. (2010). Catalytic mechanism of Golgi-resident human tyrosylprotein sulfotransferase-2: A mass spectrometry approach. *Journal of the American Society for Mass Spectrometry, 21*(9), 1633–1642.

Drury, L. J., Ziarek, J. J., Gravel, S., Veldkamp, C. T., Takekoshi, T., Hwang, S. T., et al. (2011). Monomeric and dimeric CXCL12 inhibit metastasis through distinct CXCR4 interactions and signaling pathways. *Proceedings of the National Academy of Sciences of the United States of America*, *108*(43), 17655–17660.

Duma, L., Haussinger, D., Rogowski, M., Lusso, P., & Grzesiek, S. (2007). Recognition of RANTES by extracellular parts of the CCR5 receptor. *Journal of Molecular Biology*, *365*(4), 1063–1075.

Farzan, M., Babcock, G. J., Vasilieva, N., Wright, P. L., Kiprilov, E., Mirzabekov, T., et al. (2002). The role of post-translational modifications of the CXCR4 amino terminus in stromal-derived factor 1 alpha association and HIV-1 entry. *The Journal of Biological Chemistry*, *277*(33), 29484–29489.

Farzan, M., Chung, S., Li, W., Vasilieva, N., Wright, P. L., Schnitzler, C. E., et al. (2002). Tyrosine-sulfated peptides functionally reconstitute a CCR5 variant lacking a critical amino-terminal region. *The Journal of Biological Chemistry*, *277*(43), 40397–40402.

Farzan, M., Mirzabekov, T., Kolchinsky, P., Wyatt, R., Cayabyab, M., Gerard, N. P., et al. (1999). Tyrosine sulfation of the amino terminus of CCR5 facilitates HIV-1 entry. *Cell*, *96*(5), 667–676.

Farzan, M., Vasilieva, N., Schnitzler, C. E., Chung, S., Robinson, J., Gerard, N. P., et al. (2000). A tyrosine-sulfated peptide based on the N terminus of CCR5 interacts with a CD4-enhanced epitope of the HIV-1 gp120 envelope glycoprotein and inhibits HIV-1 entry. *The Journal of Biological Chemistry*, *275*(43), 33516–33521.

Gardner, M. R., Kattenhorn, L. M., Kondur, H. R., von Schaewen, M., Dorfman, T., Chiang, J. J., et al. (2015). AAV-expressed eCD4-Ig provides durable protection from multiple SHIV challenges. *Nature*, *519*(7541), 87–91.

Jen, C. H., Moore, K. L., & Leary, J. A. (2009). Pattern and temporal sequence of sulfation of CCR5 N-terminal peptides by tyrosylprotein sulfotransferase-2: An assessment of the effects of N-terminal residues. *Biochemistry*, *48*(23), 5332–5338.

Jin, H., Shen, X., Baggett, B. R., Kong, X., & LiWang, P. J. (2007). The human CC chemokine MIP-1beta dimer is not competent to bind to the CCR5 receptor. *The Journal of Biological Chemistry*, *282*(38), 27976–27983.

Kufareva, I., Salanga, C. L., & Handel, T. M. (2015). Chemokine and chemokine receptor structure and interactions: Implications for therapeutic strategies. *Immunology and Cell Biology*, *93*(4), 372–383.

Leyte, A., van Schijndel, H. B., Niehrs, C., Huttner, W. B., Verbeet, M. P., Mertens, K., et al. (1991). Sulfation of Tyr1680 of human blood coagulation factor VIII is essential for the interaction of factor VIII with von Willebrand factor. *The Journal of Biological Chemistry*, *266*(2), 740–746.

Liu, X., Malins, L. R., Roche, M., Sterjovski, J., Duncan, R., Garcia, M. L., et al. (2014). Site-selective solid-phase synthesis of a CCR5 sulfopeptide library to interrogate HIV binding and entry. *ACS Chemical Biology*, *9*(9), 2074–2081.

Liu, C. C., & Schultz, P. G. (2006). Recombinant expression of selectively sulfated proteins in *Escherichia coli*. *Nature Biotechnology*, *24*(11), 1436–1440.

Lodi, P. J., Garrett, D. S., Kuszewski, J., Tsang, M. L., Weatherbee, J. A., Leonard, W. J., et al. (1994). High-resolution solution structure of the beta chemokine hMIP-1 beta by multidimensional NMR. *Science*, *263*(5154), 1762–1767.

Love, M., Sandberg, J. L., Ziarek, J. J., Gerarden, K. P., Rode, R. R., Jensen, D. R., et al. (2012). Solution structure of CCL21 and identification of a putative CCR7 binding site. *Biochemistry*, *51*(3), 733–735.

Ludeman, J. P., & Stone, M. J. (2014). The structural role of receptor tyrosine sulfation in chemokine recognition. *British Journal of Pharmacology*, *171*(5), 1167–1179.

Markley, J. L., Ulrich, E. L., Westler, W. M., & Volkman, B. F. (2003). Macromolecular structure determination by NMR spectroscopy. *Methods of Biochemical Analysis*, *44*, 89–113.

Millard, C. J., Ludeman, J. P., Canals, M., Bridgford, J. L., Hinds, M. G., Clayton, D. J., et al. (2014). Structural basis of receptor sulfotyrosine recognition by a CC chemokine: The N-terminal region of CCR3 bound to CCL11/eotaxin-1. *Structure*, *22*(11), 1571–1581.

Moore, K. L. (2003). The biology and enzymology of protein tyrosine O-sulfation. *The Journal of Biological Chemistry*, *278*(27), 24243–24246.

Moore, K. L. (2009). Protein tyrosine sulfation: A critical posttranslation modification in plants and animals. *Proceedings of the National Academy of Sciences of the United States of America*, *106*(35), 14741–14742.

Muller, A., Homey, B., Soto, H., Ge, N., Catron, D., Buchanan, M. E., et al. (2001). Involvement of chemokine receptors in breast cancer metastasis. *Nature*, *410*(6824), 50–56.

Ouyang, Y., Lane, W. S., & Moore, K. L. (1998). Tyrosylprotein sulfotransferase: Purification and molecular cloning of an enzyme that catalyzes tyrosine O-sulfation, a common posttranslational modification of eukaryotic proteins. *Proceedings of the National Academy of Sciences of the United States of America*, *95*(6), 2896–2901.

Ouyang, Y. B., & Moore, K. L. (1998). Molecular cloning and expression of human and mouse tyrosylprotein sulfotransferase-2 and a tyrosylprotein sulfotransferase homologue in *Caenorhabditis elegans*. *The Journal of Biological Chemistry*, *273*(38), 24770–24774.

Pennings, E. J. M., & van Kempen, G. M. J. (1979). Analysis of 3′-phosphoadenylylsulphate and related compounds by paired-ion high-performance liquid chromatography. *Journal of Chromatography*, *176*, 478–479.

Qin, L., Kufareva, I., Holden, L. G., Wang, C., Zheng, Y., Zhao, C., et al. (2015). Structural biology. Crystal structure of the chemokine receptor CXCR4 in complex with a viral chemokine. *Science*, *347*(6226), 1117–1122.

Robinson, M. R., Moore, K. L., & Brodbelt, J. S. (2014). Direct identification of tyrosine sulfation by using ultraviolet photodissociation mass spectrometry. *Journal of the American Society for Mass Spectrometry*, *25*(8), 1461–1471.

Rodgers, S. D., Camphausen, R. T., & Hammer, D. A. (2001). Tyrosine sulfation enhances but is not required for PSGL-1 rolling adhesion on P-selectin. *Biophysical Journal*, *81*(4), 2001–2009.

Seibert, C., Cadene, M., Sanfiz, A., Chait, B. T., & Sakmar, T. P. (2002). Tyrosine sulfation of CCR5 N-terminal peptide by tyrosylprotein sulfotransferases 1 and 2 follows a discrete pattern and temporal sequence. *Proceedings of the National Academy of Sciences of the United States of America*, *99*(17), 11031–11036.

Seibert, C., & Sakmar, T. P. (2008). Toward a framework for sulfoproteomics: Synthesis and characterization of sulfotyrosine-containing peptides. *Biopolymers*, *90*(3), 459–477.

Seibert, C., Veldkamp, C. T., Peterson, F. C., Chait, B. T., Volkman, B. F., & Sakmar, T. P. (2008). Sequential tyrosine sulfation of CXCR4 by tyrosylprotein sulfotransferases. *Biochemistry*, *47*(43), 11251–11262.

Simpson, L. S., Zhu, J. Z., Widlanski, T. S., & Stone, M. J. (2009). Regulation of chemokine recognition by site-specific tyrosine sulfation of receptor peptides. *Chemistry & Biology*, *16*(2), 153–161.

Smith, E. W., Liu, Y., Getschman, A. E., Peterson, F. C., Ziarek, J. J., Li, R., et al. (2014). Structural analysis of a novel small molecule ligand bound to the CXCL12 chemokine. *Journal of Medicinal Chemistry*, *57*(22), 9693–9699.

Sokolenko, S., George, S., Wagner, A., Tuladhar, A., Andrich, J. M., & Aucoin, M. G. (2012). Co-expression vs. co-infection using baculovirus expression vectors in insect cell culture: Benefits and drawbacks. *Biotechnology Advances*, *30*(3), 766–781.

Stone, M. J., Chuang, S., Hou, X., Shoham, M., & Zhu, J. Z. (2009). Tyrosine sulfation: An increasingly recognised post-translational modification of secreted proteins. *New Biotechnology*, *25*(5), 299–317.

Stone, S. R., & Hofsteenge, J. (1986). Kinetics of the inhibition of thrombin by hirudin. *Biochemistry, 25*(16), 4622–4628.

Stone, M. J., & Payne, R. J. (2015). Homogeneous sulfopeptides and sulfoproteins: Synthetic approaches and applications to characterize the effects of tyrosine sulfation on biochemical function. *Accounts of Chemical Research, 48*, 2251–2261.

Tan, J. H., Ludeman, J. P., Wedderburn, J., Canals, M., Hall, P., Butler, S. J., et al. (2013). Tyrosine sulfation of chemokine receptor CCR2 enhances interactions with both monomeric and dimeric forms of the chemokine monocyte chemoattractant protein-1 (MCP-1). *The Journal of Biological Chemistry, 288*(14), 10024–10034.

Tan, Q., Zhu, Y., Li, J., Chen, Z., Han, G. W., Kufareva, I., et al. (2013). Structure of the CCR5 chemokine receptor-HIV entry inhibitor maraviroc complex. *Science, 341*(6152), 1387–1390.

Teramoto, T., Fujikawa, Y., Kawaguchi, Y., Kurogi, K., Soejima, M., Adachi, R., et al. (2013). Crystal structure of human tyrosylprotein sulfotransferase-2 reveals the mechanism of protein tyrosine sulfation reaction. *Nature Communications, 4*, 1572.

Veerman, K. M., Williams, M. J., Uchimura, K., Singer, M. S., Merzaban, J. S., Naus, S., et al. (2007). Interaction of the selectin ligand PSGL-1 with chemokines CCL21 and CCL19 facilitates efficient homing of T cells to secondary lymphoid organs. *Nature Immunology, 8*(5), 532–539.

Veldkamp, C. T., Kiermaier, E., Gabel-Eissens, S. J., Gillitzer, M. L., Lippner, D. R., DiSilvio, F. A., et al. (2015). Solution structure of CCL19 and identification of overlapping CCR7 and PSGL-1 binding sites. *Biochemistry, 54*(27), 4163–4166.

Veldkamp, C. T., Seibert, C., Peterson, F. C., De la Cruz, N. B., Haugner, J. C., III, Basnet, H., et al. (2008). Structural basis of CXCR4 sulfotyrosine recognition by the chemokine SDF-1/CXCL12. *Science Signaling, 1*(37), ra4.

Veldkamp, C. T., Seibert, C., Peterson, F. C., Sakmar, T. P., & Volkman, B. F. (2006). Recognition of a CXCR4 sulfotyrosine by the chemokine stromal cell-derived factor-1alpha (SDF-1alpha/CXCL12). *Journal of Molecular Biology, 359*(5), 1400–1409.

Veldkamp, C. T., Ziarek, J. J., Peterson, F. C., Chen, Y., & Volkman, B. F. (2010). Targeting SDF-1/CXCL12 with a ligand that prevents activation of CXCR4 through structure-based drug design. *Journal of the American Chemical Society, 132*(21), 7242–7243.

Wilkins, P. P., Moore, K. L., McEver, R. P., & Cummings, R. D. (1995). Tyrosine sulfation of P-selectin glycoprotein ligand-1 is required for high affinity binding to P-selectin. *The Journal of Biological Chemistry, 270*(39), 22677–22680.

Wu, B., Chien, E. Y., Mol, C. D., Fenalti, G., Liu, W., Katritch, V., et al. (2010). Structures of the CXCR4 chemokine GPCR with small-molecule and cyclic peptide antagonists. *Science, 330*(6007), 1066–1071.

Yu, Y., Hoffhines, A. J., Moore, K. L., & Leary, J. A. (2007). Determination of the sites of tyrosine O-sulfation in peptides and proteins. *Nature Methods, 4*(7), 583–588.

Zhu, J. Z., Millard, C. J., Ludeman, J. P., Simpson, L. S., Clayton, D. J., Payne, R. J., et al. (2011). Tyrosine sulfation influences the chemokine binding selectivity of peptides derived from chemokine receptor CCR3. *Biochemistry, 50*(9), 1524–1534.

Ziarek, J. J., Getschman, A. E., Butler, S. J., Taleski, D., Stephens, B., Kufareva, I., et al. (2013). Sulfopeptide probes of the CXCR4/CXCL12 interface reveal oligomer-specific contacts and chemokine allostery. *ACS Chemical Biology, 8*(9), 1955–1963.

Zidar, D. A., Violin, J. D., Whalen, E. J., & Lefkowitz, R. J. (2009). Selective engagement of G protein coupled receptor kinases (GRKs) encodes distinct functions of biased ligands. *Proceedings of the National Academy of Sciences of the United States of America, 106*(24), 9649–9654.

CHAPTER EIGHTEEN

Disulfide Trapping for Modeling and Structure Determination of Receptor:Chemokine Complexes

Irina Kufareva*, Martin Gustavsson*, Lauren G. Holden*, Ling Qin*, Yi Zheng*, Tracy M. Handel†,1

*Skaggs School of Pharmacy and Pharmaceutical Sciences, University of California San Diego, La Jolla, California, USA

†Department of Pharmacology, Skaggs School of Pharmacy and Pharmaceutical Sciences, University of California San Diego, La Jolla, California, USA

[1]Corresponding author: e-mail address: thandel@ucsd.edu

Contents

1. Introduction 390
2. Architecture of Receptor:Chemokine Interfaces 392
3. Cysteine as a Natural Crosslinking Agent 395
4. Disulfide Trapping 396
   4.1 Overall Strategy 396
   4.2 Selection of Candidate Residue Pairs 396
   4.3 Generation of Disulfide-Trapped Receptor:Chemokine Complexes 400
   4.4 Characterization of Disulfide-Trapped Receptor:Chemokine Complexes 403
5. Conclusion and Perspectives 416
Acknowledgments 417
References 417

Abstract

Despite the recent breakthrough advances in GPCR crystallography, structure determination of protein–protein complexes involving chemokine receptors and their endogenous chemokine ligands remains challenging. Here, we describe disulfide trapping, a methodology for generating irreversible covalent binary protein complexes from unbound protein partners by introducing two cysteine residues, one per interaction partner, at selected positions within their interaction interface. Disulfide trapping can serve at least two distinct purposes: (i) stabilization of the complex to assist structural studies and/or (ii) determination of pairwise residue proximities to guide molecular modeling. Methods for characterization of disulfide-trapped complexes are described and evaluated in terms of throughput, sensitivity, and specificity toward the most energetically favorable crosslinks. Due to abundance of native disulfide bonds at receptor: chemokine interfaces, disulfide trapping of their complexes can be associated with intramolecular disulfide shuffling and result in misfolding of the component proteins;

Methods in Enzymology, Volume 570
ISSN 0076-6879
http://dx.doi.org/10.1016/bs.mie.2015.12.001

because of this, evidence from several experiments is typically needed to firmly establish a positive disulfide crosslink. An optimal pipeline that maximizes throughput and minimizes time and costs by early triage of unsuccessful candidate constructs is proposed.

1. INTRODUCTION

Chemokines promote cell migration in the context of development, immunity, inflammation, and many other pathological and physiological processes (Baggiolini, 1998; Charo & Ransohoff, 2006; Gerard & Rollins, 2001; Griffith, Sokol, & Luster, 2014; Murdoch & Finn, 2000; Ransohoff, 2009). They do so by the virtue of binding to and activating seven-transmembrane (7TM) receptors on the surface of migrating cells. In humans, there are approximately 45 chemokines that, based on the pattern of the conserved cysteine motif in their N-terminus, are divided into CC, CXC, CX_3C, or XC families (Bachelerie et al.). The 22 chemokine receptors that are expressed in human tissues exhibit remarkable specificity in their recognition of the chemokines of different families, e.g., some receptors exclusively bind and are activated by CC chemokines while others strictly prefer CXC chemokines; based on this observation, the receptors are also classified into the same four subfamilies. Some receptors interact with multiple chemokines within their subfamily, while others have but a single endogenous chemokine ligand. Finally, several members of the *Herpesviridae* (herpesvirus) family encode chemokines and/or chemokine receptors in their genomes (Montaner, Kufareva, Abagyan, & Gutkind, 2013); these viral proteins interact with human receptors or chemokines, respectively, frequently demonstrate broad specificity spanning both CC and CXC families, and hijack chemokine receptor signaling cascades in host cells for the replicative advantage of the virus.

Knowledge of the structural basis of the high affinity, specificity, and pharmacology of receptor:chemokine interactions is extremely important, both from the standpoint of understanding the biology and for the development of therapeutics. Yet crystallography of chemokine receptors and especially their complexes with chemokines have proved to be quite challenging. As most members of the 7TM receptor family, chemokine receptors are unstable outside their native membrane environment and conformationally heterogeneous; they also lack hydrophilic surfaces for crystal formation. Due to advances in protein engineering, screening, and crystallization (Bill et al.,

2011; Ghosh, Kumari, Jaiman, & Shukla, 2015; Liu, Wacker, Wang, Abola, & Cherezov, 2014; Moraes, Evans, Sanchez-Weatherby, Newstead, & Stewart, 2014), the last few years were marked by dramatic progress in structure determination of 7TM receptors. However, even with engineered receptor constructs and with novel crystallization techniques, structure determination of protein–protein *complexes* involving chemokine receptors and their endogenous chemokine ligands remains difficult. The binding affinity of chemokines to detergent-solubilized receptors may be reduced in comparison to that observed in cell membranes, contributing to lower stability of the complexes. Further, some chemokines bind with high affinity only to select conformational (e.g., G protein-coupled, active) states of their receptors (Nijmeijer, Leurs, Smit, & Vischer, 2010) and these states are challenging to reproduce in detergent-solubilized conditions and in the absence of intracellular effectors and scaffolding proteins. Finally, crystallization of a 7TM receptor with any ligand frequently relies on slow complex dissociation kinetics (Zhang, Stevens, & Xu, 2015); such kinetics may be an inherent property of some receptor:chemokine pairs (e.g., the virally encoded receptor US28 and human CX_3CL1/fractalkine; Burg et al., 2015), but not others.

Here, we describe *disulfide trapping* (also called *cysteine trapping* or *disulfide crosslinking*)—an experimental methodology for generating irreversible covalent binary protein complexes from unbound protein partners by introducing two cysteine residues (one per interaction partner) at strategically selected positions within their interaction interface. Disulfide trapping can serve at least two distinct purposes: (i) stabilization of the complex to prevent spontaneous dissociation and to trap a specific conformation, both of which can facilitate structural studies, and/or (ii) evaluation of residue proximities that, in conjunction with molecular modeling, can provide insight into the structural basis of the interaction even if the complex does not yield to crystallization efforts.

With respect to the second application (determining residue proximities), the disulfide trapping approach has been successfully applied to complexes of several receptors with small molecules and peptides (Buck & Wells, 2005; Dong et al., 2012; Hagemann, Miller, Klco, Nikiforovich, & Baranski, 2008; Kufareva et al., 2014; Monaghan et al., 2008). Residue proximities are established on the basis of crosslinks of varying strength and specificity and used to guide molecular modeling and interface mapping. This is in contrast to disulfide trapping for structure determination where only the strongest, energetically favorable crosslinks help stabilization and crystallization of the complexes. The first application of disulfide trapping for crystallization of a GPCR was achieved in 2011 by Rasmussen et al. (2011) when they crystallized an irreversible complex between a small-molecule agonist and a nanobody stabilized β_2 adrenergic receptor. Last year was marked by a successful application of the methodology to a complex between a receptor and a chemokine, yielding the first high-resolution structural insight into chemokine recognition by an intact receptor (Qin et al., 2015).

Disulfide trapping has a number of advantages compared to other types of intermolecular crosslinking. First, unlike unnatural amino acids (Grunbeck et al., 2012; Grunbeck & Sakmar, 2013; Huber, Naganathan, Tian, Ye, & Sakmar, 2013; Kim, Axup, & Schultz, 2013), cysteine can be incorporated into recombinant proteins using straightforward cloning techniques and a wide range of expression systems. Next, when implemented this way, the incorporation is 100% efficient. Finally, unlike bulky moieties used in photoaffinity labeling (Chen, Pinon, Miller, & Dong, 2009, 2010; Coin et al., 2013; Dong et al., 2007, 2011; Miller et al., 2011; Pham & Sexton, 2004; Wittelsberger et al., 2006), a cysteine residue side chain is small and does not tend to induce artificial distortions of complex geometry.

Although a powerful approach, disulfide trapping is prone to artifacts and false positives, especially when cysteines are introduced at interfaces that are rich in native intramolecular disulfide bonds as in chemokine receptor complexes. Because of that, combined evidence from several assays is typically required to establish a strong positive intramolecular disulfide crosslink. This chapter provides an account of methods for generating disulfide crosslinked complexes and for experimentally characterizing the efficiency and quality of the crosslinks. A strategy for streamlining the approaches into a cost-effective pipeline for screening multiple candidates and quickly triaging unsuccessful pairs is also discussed.

2. ARCHITECTURE OF RECEPTOR:CHEMOKINE INTERFACES

All chemokines share a conserved topology that consists of a flexible N-terminus, a double-cysteine motif, a three-strand β-sheet, and a C-terminal helix (Fig. 1). The cysteines in the double-cysteine motif may be adjacent to one another (in the CC chemokine family), separated by one residue (in the CXC family), or separated by three residues (in the

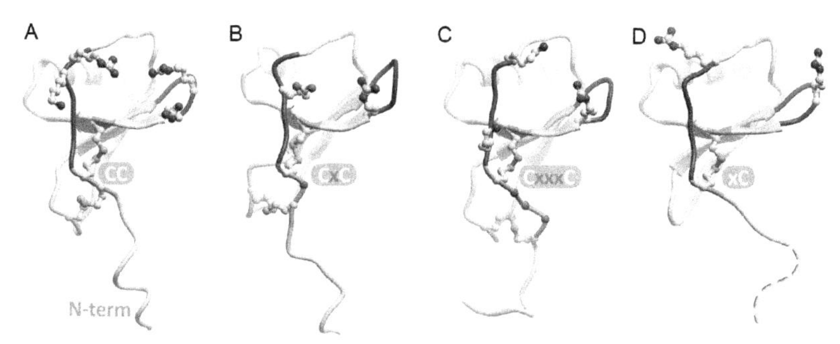

Figure 1 Representative chemokines of CC (A), CXC(B), CX_3C (C), and XC (D) families. The conserved cysteine pattern is shown in yellow (gray in the print version) sticks; residues separating the two N-terminal cysteines are shown as magenta balls (B and C). Loops believed to be involved in coordinating the N-terminus of the receptor (N-loop and 40s-loop) are colored blue (dark gray in the print version) and charged residues in these regions are shown as sticks.

CX_3C family that only has one member, CX_3CL1/fractalkine). In two human chemokines, XCL1/lymphotactin and XCL2/SCM-1, only one cysteine is present in this region. The cysteines are involved in intramolecular disulfide bonds connecting the N-terminus to the loop between the first and the second β-strands (in CC, CXC, and CX_3C families) and to the third β-strand (in all chemokines) (Fig. 1). Chemokine receptors, like all members of the 7TM receptor superfamily, share a conserved topology with seven membrane-spanning helices and a disulfide bond between TM3 and the extracellular loop (ECL) 2. All chemokine receptors except CXCR6 also have a disulfide bond connecting their flexible N-terminus to ECL3 (Fig. 2).

From biochemical, biophysical, mutagenic, modeling, and structural studies of the chemokine receptor system to-date (Brelot, Heveker, Montes, & Alizon, 2000; Burg et al., 2015; Gupta, Pillarisetti, Thomas, & Aiyar, 2001; Kofuku et al., 2009; Kufareva et al., 2014; Qin et al., 2015; Saini et al., 2011; Zhou & Tai, 2000) it is clear that the interaction between the receptor and the chemokine involves two distinct epitopes (Kufareva, Salanga, & Handel, 2015). In the so-called chemokine recognition site 1 (CRS1; Scholten et al., 2011), the extended N-terminus of the receptor binds to the globular core of the chemokine. In CRS2, the flexible N-terminus of the chemokine reaches into the TM domain binding pocket of the receptor. For most chemokines, their distal N-termini are recognized as critical signaling domains that directly induce the activation-related conformational changes in the TM domains of the receptors via CRS2 interaction (Chevigne, Fievez, Schmit, & Deroo, 2011; Proost et al., 2008; Van

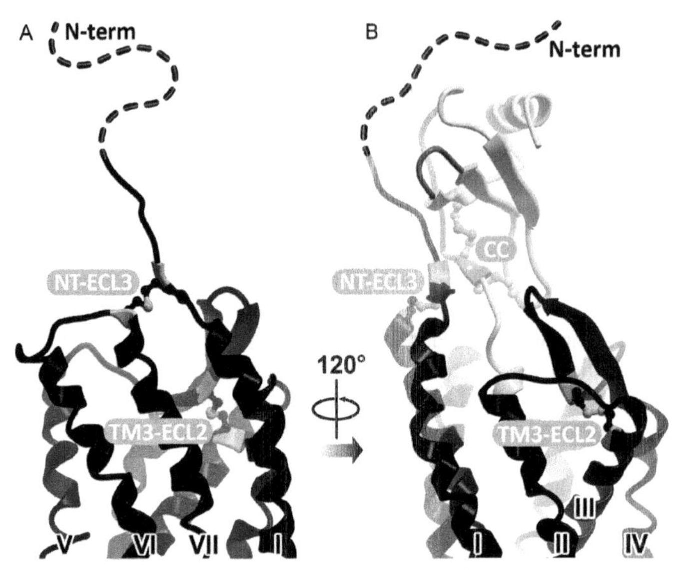

Figure 2 (A) Conserved intramolecular disulfide bonds in the extracellular part of CXCR4. (B) Both disulfides are a part of the receptor:chemokine interface, with one of the disulfides found in direct proximity to the conserved N-terminal disulfide pattern of the chemokine.

Damme et al., 1989). The first two structures of the receptor:chemokine complexes (Burg et al., 2015; Qin et al., 2015) demonstrated that the two epitopes are joined by an intermediate region that we refer to as CRS1.5 and that brings into close proximity the conserved N-terminal cysteine of the receptor with the conserved cysteine motif of the chemokine (Fig. 2). This understanding is especially important in the context of disulfide trapping, because introducing additional artificial cysteines at disulfide-rich interfaces has high potential of shuffling the native disulfide connectivity and misfolding the proteins while still exhibiting, at least in some assays, the behavior of a positive crosslink.

While retaining the conserved overall architecture, the multiple receptor:chemokine pairs appear to utilize distinct structural determinants and conformational mechanisms to achieve specificity of binding and signaling (Burg et al., 2015; Choi et al., 2005; Duchesnes, Murphy, Williams, & Pease, 2006; Pakianathan, Kuta, Artis, Skelton, & Hebert, 1997; Qin et al., 2015; Saini et al., 2011). This diversity hinders transfer of structural knowledge between different complexes by homology and emphasizes the importance of detailed studies of individual receptor:chemokine pairs; and while this task may be cost-prohibitive for X-ray crystallography, it is more tractable by other approaches such as molecular modeling in conjunction with disulfide trapping.

3. CYSTEINE AS A NATURAL CROSSLINKING AGENT

Cysteine is the most chemically reactive of the 20 natural amino acids: its thiol side chain is easily oxidized due to its nucleophilic nature. Oxidation of two spatially proximal cysteine residues leads to formation of disulfide bonds that play a major role in protein folding and stability. The dissociation energy of a typical disulfide bond is about 50–60 kcal/mol, which makes it 10–20 times stronger than the strongest hydrogen bond, but still much weaker than a typical covalent bond (e.g., carbon–carbon). Disulfide bonds are susceptible to breakage in the presence of other nucleophiles, but thiol-disulfide exchange is inhibited at physiological and acidic pH.

An ideal disulfide bond has a S_γ–S_γ distance of 2.04 ± 0.07 Å and a specific relative orientation of remaining atoms within the two cysteines (C_β–S_γ–S_γ–C_β dihedral angle of $90 \pm 12°$) (Pellequer & Chen, 2006). Bonds with dihedral angles of 0°–180° occur in protein structures, but are significantly weaker than those with ideal geometry.

Most cellular compartments are rich in glutathione and thus represent a reducing environment in which disulfide bonds are not stable. Consequently, cysteine residues are usually present in their free form in soluble cytosolic and nuclear proteins. However, the oxidizing environment in the extracellular space, in the lumen of the rough endoplasmic reticulum, and in the mitochondrial intermembrane space favors formation of disulfide bonds. As a result, intramolecular disulfide bonds are common in secreted proteins, with chemokines being a perfect example (Fig. 1). Intramolecular disulfide bonds are also frequently found in the extracellular domains of transmembrane proteins, which, as illustrated in Fig. 2, includes the chemokine receptors. Contrary to the extracellular fragments, cysteine residues deeper in the TM domain of the receptor or in intracellular loops appear to be reduced and not involved in disulfide formation. In live cells, the interaction between chemokines and their receptors occurs in the extracellular oxidizing environment and thus additional disulfide bonds introduced at their interface are expected to form and remain stable when the cysteines are in the correct proximity and orientation. However, when receptors are extracted from their natural cellular context, there is a danger of transmembrane and especially intracellular cysteine residues, which are normally reduced, to become oxidized and lead to the formation of nonspecific covalent adducts with chemokines or each other.

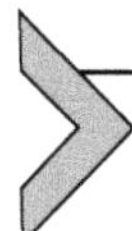

4. DISULFIDE TRAPPING

4.1 Overall Strategy

Disulfide trapping serves the goal of generating an irreversible covalently bound receptor:chemokine complex from otherwise dissociable protein partners. To accomplish this goal, a single nonnative cysteine residue is introduced into the receptor and into the chemokine. Depending on the method used to generate the binding partners (e.g., bacterial expression, eukaryotic cell expression, or chemical synthesis), the mutated proteins can be either expressed separately and mixed, or coexpressed in the same cells. In both approaches, the main assumption is that if formation of the native complex positions the two engineered cysteine residues in favorable proximity and geometry, a disulfide bond will spontaneously form thus locking the complex and preventing its dissociation. The resulting irreversible complexes can be detected using nonreducing SDS-PAGE and/or Western blotting and further characterized in terms of overall yield, crosslinking efficiency, stability, and monodispersity. In our own applications for generating receptor:chemokine complexes, we had the best success with coexpressing the mutated receptor and the mutated chemokine in Sf9 insect cells. Among other advantages, this strategy avoids the need to make purified chemokine, which, due to an extra free cysteine residue, may be complicated by intramolecular disulfide shuffling and formation of covalent chemokine oligomers.

4.2 Selection of Candidate Residue Pairs

Finding optimal position pairs for introducing receptor:chemokine disulfide crosslinks might at first seem like an impossible endeavor; however, in our implementation, the disulfide trapping method involves 3D model-guided "cherry-picking" of residue pairs for cysteine mutagenesis. Starting with a model of the complex, we assign a numeric score to each pair of receptor:chemokine residues that reflects the likelihood of them spontaneously forming a disulfide bond when mutated to cysteine. In a hypothetical situation when a high-resolution structure or model is already available, the residues are mutated *in silico* to cysteines, a disulfide bond is imposed, the side-chains of the two bonded residues and the neighboring residues are optimized by conformational sampling, and the disulfide bond energy is evaluated by

$$E_{SS} = 10 \times \left(d_{S:S}^2 - 2.05^2\right)^2 + 5 \times \left(d_{S_1:C_{\beta 2}}^2 - 3.051^2\right)^2 + 5 \times \left(d_{S_2:C_{\beta 1}}^2 - 3.051^2\right)^2 + 9 \times \left(d_{C_\beta:C_\beta}^2 - 3.855^2\right)^2,$$

where d_{SS}, $d_{S_i:C_{\beta j}}$, and $d_{C_\beta:C_\beta}$ stand for pairwise distances between the indicated atoms in the disulfide-bonded residues. This energy term is combined with the overall energy involving steric compatibility of the *in silico* introduced disulfide bond with its immediate environment and used for scoring and ranking of the residue pairs. In typical real-life situations where only an approximate model is available, relaxed criteria should be used, mostly based on appropriate Cα–Cα and Cβ–Cβ distances between the residues in question. In general, due to the size and geometry of cysteine residues, disulfide crosslinking works best for residue pairs that are involved in backbone–backbone contacts; residues with only side chain–side chain contacts are usually poor crosslinking candidates.

Receptor:chemokine residue pairs are rank ordered by decreasing score. After that, penalties are introduced to account for possible liabilities of introducing a reactive cysteine at each candidate location. For example, a penalty is imposed if one of the designed cysteine residues is located in a fragment without a defined secondary structure and is adjacent or proximal (within two to three residue positions) to a native cystine (disulfide-bonded cysteine) in the sequence. Introducing an artificial cysteine in such positions increases the probability of protein misfolding by formation of an unwanted disulfide bond between the counterpart of the native and the newly introduced cysteine (shuffling, Fig. 3A). An immediately adjacent native cystine within a helix or a β-strand may be free of such liabilities because (i) these regions are expected to have less flexibility and (ii) adjacent residues in these secondary structures point in different directions (Fig. 3C, in the case of an α-helix) or strictly opposite directions (in the case of a β-strand, Fig. 3B). On the other hand, taking into account residue geometry within α-helices and β-strand (Fig. 3D and E), we usually penalize an artificial cysteine introduced two residues up or down from a native cysteine in a β-strand, and three to four residues up and down from a native cysteine in an α-helix.

Our early attempts in designing a disulfide crosslink between CXCR4 and CXCL12 were guided by the NMR structure (Veldkamp et al., 2008) of a CXCL12 dimer in complex with a 38-residue peptide isolated

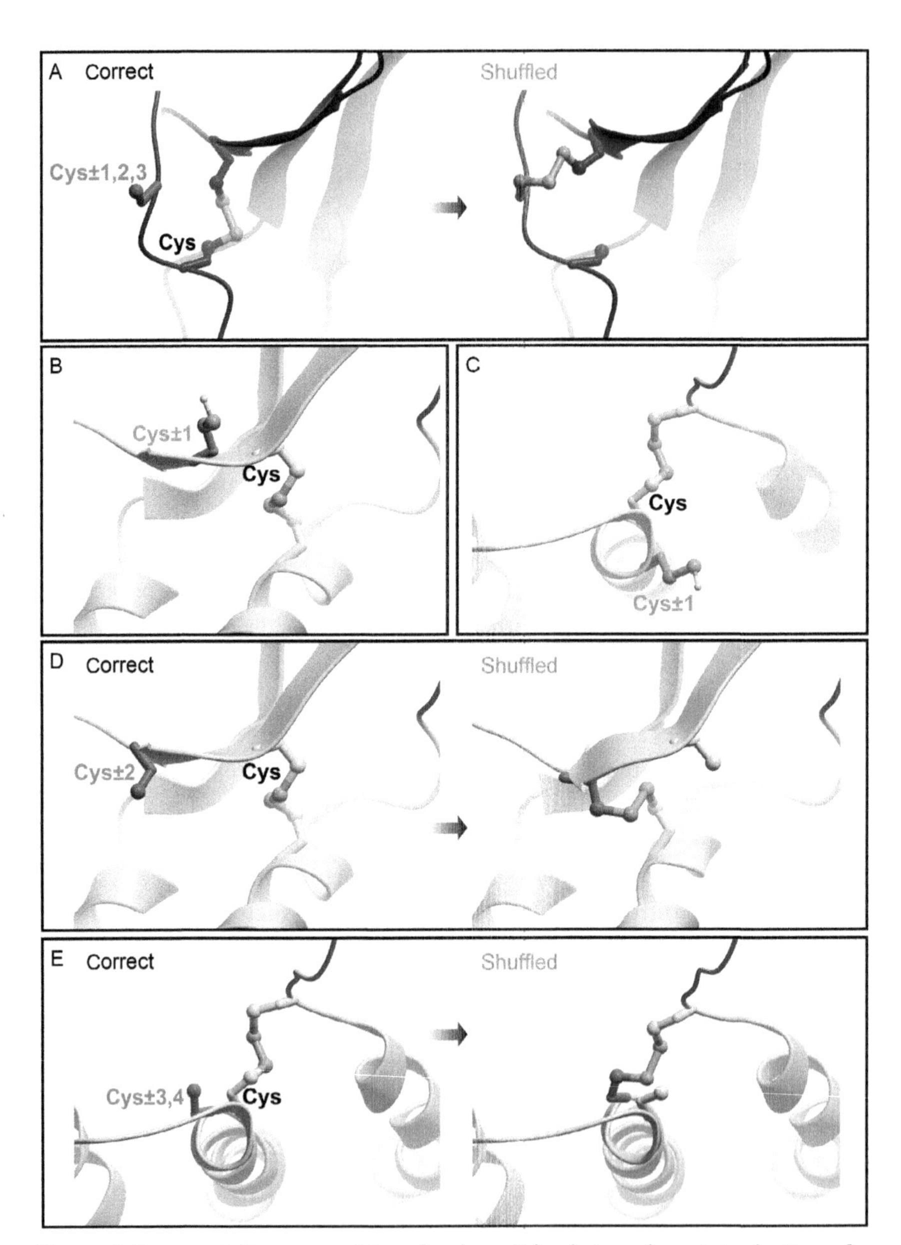

Figure 3 Representative geometries of polypeptide chains where introduction of a cysteine poses high risk of intramolecular disulfide shuffling. (A) Within 1–3 residues from a native cysteine in a loop without defined secondary structure. (B and C) Within an α-helix and a β-strand, positions proximal to a native cysteine are generally safe. Instead, positions two residues up or down from a native cysteine within a β-strand (D) or 3–4 residues up or down from a native cysteine within an α-helix (E) are subject to disulfide shuffling liability.

from the CXCR4 N-terminus (CRS1). The *in silico* disulfide bonds introduced between all possible proximal CXCR4:CXCL12 residue pairs were ranked by both the disulfide bond energy, E_{SS}, according to the equation above, and by the overall energy of the disulfide in the context of the complex; the latter included van der Waals, electrostatics, hydrogen bonding, hydrophobicity, and polar surface area terms (Kufareva et al., 2014). Figure 4A illustrates the scatter plot of E_{SS} and overall energy values for the 11 residue pairs selected for experimental validation. Mutant pair characterization performed as described in the next section confirmed that the pair predicted to form the best crosslink, CXCR4(K25C):CXCL12 (S16C), did indeed crosslink very efficiently (Fig. 4B), while the remaining

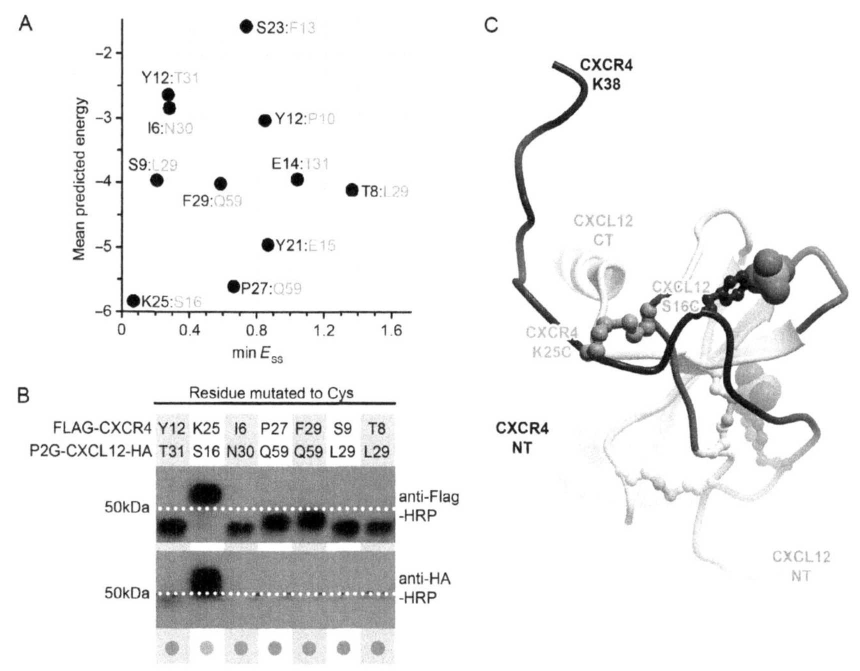

Figure 4 Identification of the first CRS1 crosslink between CXCR4 and CXCL12 from the CXCR4:CXCL12 NMR structure (PDB 2 k05). (A) 11 crosslinks were designed *in silico* and ranked according to the overall predicted energy (vertical) and disulfide bond energy (horizontal). (B) Following coexpression of the mutant pairs in insect cells, the best ranking pair, CXCR4(K25C):CXCL12(S16C) demonstrated efficient crosslinking in Western blotting as evidenced by the double (antireceptor and antichemokine) staining and the molecular weight shift of the material in the respective lane. (C) The identified crosslink (green; dark gray in the print version) is shown in the context of the NMR structure from which it was generated: CXCL12 is in white ribbon and the N-terminal peptide of CXCR4 is in black ribbon. The sulfate groups on CXCR4 sulfotyrosines sTyr7, sTyr12, and sTyr21 are shown as CPK. *Panel (B) is adapted from Kufareva et al. (2014).*

pairs showed no or ambiguous crosslinking. This single CRS1 crosslink (Fig. 4C) was later used in molecular docking to build and refine the model of the complex between the full-length receptor and the chemokine (Kufareva, Handel, & Abagyan, 2015; Kufareva et al., 2014).

4.3 Generation of Disulfide-Trapped Receptor:Chemokine Complexes

The methodology for cloning, expressing, and purifying disulfide crosslinked receptor:chemokine complexes is no different from that for noncovalent complexes, which is described in detail by Gustavsson et al. in chapter "Production of Chemokine/Chemokine Receptor Complexes for Structural and Biophysical Studies" of this Methods of Enzymology volume. Below we provide a brief summary of the major steps while stressing the nuances important in the context of disulfide crosslinking.

4.3.1 Cloning

Baculovirus infected *Spodoptera frugiperda* (Sf9) insect cells represent an appropriate expression system for production of membrane proteins and complexes in many applications including biophysical studies and structure determination. The Bac-to-Bac Baculovirus Expression System (Invitrogen) is used to generate bacmids and baculovirus.

First, the receptors are cloned into a pFastBac™ vector (Invitrogen) under either a gp64 or a polH promoter, and are N-terminally or C-terminally tagged with a FLAG tag (DYKDDDD, for detection). The placement of the tag has to be optimized for each individual receptor as it may differentially affect yield and trafficking even for homologous receptors (e.g., CXCR4 tolerates the tag on the N-terminus while ACKR3 constructs express substantially better when the tag is placed at the C-terminus). Additionally, a His_{10} tag is placed at the C-terminus for metal affinity purification. The chemokine constructs include the native signal sequences, are C-terminally tagged with an HA-tag (YPYDVPDYA, for detection) and are cloned into a pFastBac™ vector under a polH promoter. Codon optimization reflecting the inherent bias of the expression system may provide an additional way to optimize the expression constructs. Cysteine mutations are introduced at the desired positions in the receptor and chemokine constructs using standard QuikChange site-directed mutagenesis protocols.

Recombinant bacmids separately carrying the genes for the receptor and the chemokine are generated by transforming the pFastBac™ constructs into DH10Bac™ competent *Escherichia coli* cells (Invitrogen) according to the manufacturer's instructions. Propagation of a single selected white colony with subsequent bacmid purification is described in detail by Gustavsson et al. in Chapter 11.

4.3.2 Generation of Baculovirus Stocks

To generate high-titer (>10^9 viral particles per mL) recombinant baculovirus, recombinant bacmids (5 μL of the purified material at the concentration obtained through purification) containing the target genes are transfected *separately* into *Sf9* cells (2.5 mL at a density of 1.2×10^6 cells/mL) using 3 μL of Xtreme Gene Transfection Reagent (Roche) and 100 μL of Transfection Medium (Expression Systems). Cell suspensions are incubated for 96 h with shaking at 27 °C. P0 viral stocks are isolated and used to infect larger Sf9 cultures for generation of P1 viral stock. Typically, 400 μL of P0 virus is used to infect 40 mL of culture at this step. Virus titer can be quantified by flow cytometry following cell staining with PE-conjugated anti-gp64 antibody (Expression Systems).

4.3.3 Sf9 Coexpression

Coexpression of disulfide-trapped complexes between receptors and chemokines in insect Sf9 cells is similar to coexpression of noncovalent complexes and is described in detail by Gustavsson et al in Chapter 11. Briefly, the P1 stocks of the two types of particles (those carrying receptor and those carrying chemokine) are used to coinfect Sf9 cells at a density of $2\text{–}2.6 \times 10^6$ cells/mL. Even if the multiplicity of infection (MOI) has been previously separately optimized for each vector, it frequently needs to be reoptimized in the context of the coexpression experiment, likely because the cell expression machinery gets taxed to a different degree when producing individual components versus the complex (Fig. 5).

The cells are allowed to grow while expressing the two proteins for a period that may also need optimization, but typically for 44–48 h. Biomass is harvested by centrifugation and stored at –80 °C until further use.

It is unknown how and when the receptor:chemokine complexes form when the two proteins are coexpressed in Sf9 cells. One possibility is that the chemokine molecules are secreted in the cell culture medium and from there, bind to the receptors that are expressed and trafficked to the plasma membrane. Alternatively, the proteins may bind one another at the time of folding in the endoplasmic reticulum and be trafficked to the cell surface together as a complex. In either case, the interaction interface is in an oxidizing environment which promotes disulfide bond formation for those

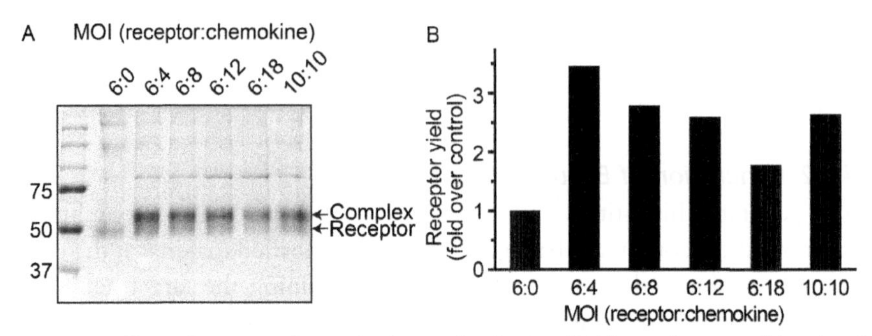

Figure 5 Effect of varying the multiplicity of infection (MOI) on the overall levels of expression and the efficiency of crosslinking between a pair of single-cysteine mutants of ACKR3 and CXCL12. (A) Nonreducing 10% SDS-PAGE of the samples following metal affinity purification by the tag on the receptor. The lower and the upper bands indicate non-crosslinked receptor and crosslinked complex, respectively. (B) Band densitometry for quantification of receptor expression levels. Receptor yield is expressed as fold increase over the control sample (receptor only, no chemokine). Although the calculated fraction of crosslinked complex is close to 80% across all samples, the receptor yield is maximal at the MOI of 6:4 (receptor:chemokine).

cysteine residue pairs that are in spatial proximity and favorable orientation. As a result, covalently trapped complexes may be harvested from the cell membranes along with noncovalent complexes and uncomplexed receptors.

4.3.4 Membrane Preparation

Harvesting and metal affinity purification of disulfide-trapped receptor: chemokine complexes is performed in the same way as purification of isolated receptors or noncovalent complexes, as described in detail by Gustavsson et al in Chapter 11. Briefly, cell biomass is thawed, lysed in hypotonic buffer (10 m*M* HEPES, pH 7.5, 10 m*M* $MgCl_2$, 20 m*M* KCl, and EDTA-free protease inhibitor cocktail (Roche)), and membranes are purified by repeated (3 ×) Dounce homogenization (40 strokes per round) followed by centrifugation at 50,000 × *g* at 4 °C for 30 min. Membrane pellets are resuspended in the above hypotonic buffer during round 1 and in a high salt buffer (10 m*M* HEPES, pH 7.5, 10 m*M* $MgCl_2$, 20 m*M* KCl, 1 *M* NaCl, and EDTA-free protease inhibitor cocktail) during rounds 2–3. Following the last centrifugation, membranes are resuspended and homogenized in hypotonic buffer supplemented with 30% glycerol (v/v) and flash-frozen for storage at –80 °C until further use.

4.3.5 *Purification of Receptor Complexes*

Purified membranes are thawed on ice, homogenized with an equal volume of hypotonic buffer, mixed with an equal volume of 2 × solubilization buffer (100 m*M* HEPES, pH 7.5, 800 m*M* NaCl, 1.5% *n*-dodecyl-ß-D-maltopyranoside (DDM, Anatrace), 0.3% cholesteryl hemisuccinate (CHS, Sigma)), incubated for 3 h at 4 °C, and then centrifuged at 25,000 × *g* for 30 min. The supernatant is incubated overnight at 4 °C with TALON IMAC resin (Clontech) and 20 m*M* imidazole. After binding, the resin is washed with 20 column volumes of wash buffer (25 m*M* HEPES, pH 7.5, 400 m*M* NaCl, 0.025% DDM, 0.005% CHS, 10% glycerol). Resin is then resuspended in 6 column volumes of wash buffer and incubated with 10 μL of PreScission Protease at 15 mg/mL (purified in house in our case) for 3 h at 4 °C. Purified protein is collected as flow through.

The above composition of solubilization, wash, and elution buffers is such that target proteins remain folded, and tightly bound slow dissociating or covalent complexes remain intact. In case modifications to the buffer compositions are necessary, it is important that nucleophiles and reducing agents are avoided to ensure stability of any intra- or intermolecular disulfide bonds through the purification process.

4.4 Characterization of Disulfide-Trapped Receptor: Chemokine Complexes

Identification of an efficient and energetically favorable disulfide crosslink can be performed using multiple experimental approaches. Furthermore, because of the complexities of the disulfide-rich receptor:chemokine interfaces, and because of general and implementation-specific variations in assay throughput, material requirements, sensitivity, and specificity, no single assay is ideal and crosslink characterization typically requires combined evidence from several assays. We were successful in crosslink characterization using the following methods:

- flow cytometry, to detect complexes directly on cells
- denaturing, but nonreducing SDS-PAGE of the purified protein material, to evaluate the presence and relative abundance of the covalently linked complexes
- Western blotting, to confirm the nature of the observed bands
- differential scanning fluorimetry (DSF) using cysteine-reactive CPM dye (CPM-DSF) to characterize thermal stability of the purified complexes
- size-exclusion chromatography (SEC), to study monodispersity of the complexes

Table 1 summarizes time and material requirements for the five assays as implemented in our hands, and provides an estimate of their throughput as well as specificity and sensitivity in separating strong energetically favorable disulfide crosslinks from weak nonspecific ones. According to these parameters, the assays can be prioritized in the workflow in order to quickly triage unsuccessful pairs. Notably, SDS-PAGE and Western blotting appear to be the central and indispensable experiments in crosslink characterization despite their moderate throughput, sensitivity, or specificity. This is because they can provide proper context for interpreting results of other experiments, especially when these results are ambiguous or inconclusive.

4.4.1 Detection of Covalent Complexes on Cells Using Flow Cytometry

Flow cytometry allows the initial detection of disulfide-trapped complexes directly in the cells, prior to purification of receptor material. It requires staining of complex-expressing cells with fluorescent antibodies against a tag on the chemokine. At least with some chemokines (including CXCL11, CXCL12, and CXCL14), C-terminal placement of the tag on the chemokine prevents interference between the receptor:chemokine and chemokine:antibody binding and ensures that the receptor-bound chemokine is detectable on the cell surface with anti-tag antibodies. This may or may not be the case for other chemokines; for example, so far, we have been unsuccessful in detection of cell surface-bound vMIP-II using this method.

4.4.1.1 Experiment

For flow cytometry experiments, Sf9 cells coexpressing the disulfide-trapped complexes are washed twice in TBS + 0.5% BSA and stained in 1:100 dilution of FITC-conjugated anti-tag antibody. We use mouse monoclonal anti-FLAG® M2-FITC antibody and mouse monoclonal anti-HA-FITC antibodies, both from Sigma-Aldrich (item numbers F4049 and H7411, respectively). Following 20 min incubation on ice in the dark, the cells are diluted in TBS and analyzed using a Guava bench top miniflow cytometer (Millipore). Controls must include untransfected cells in order to quantify the levels of nonspecific antibody binding; in cases where nonspecific binding is nonnegligible, repeated washing of the stained cells with TBS + 0.5 BSA may help. Controls should also include cells coexpressing WT receptor and WT chemokine that are presumed to form efficient but dissociable complexes (Fig. 6).

For some receptor:chemokine complexes (e.g., ACKR3:CXCL12), the affinity and dissociation kinetics is such that large amounts of cell

Table 1 Comparison of Experimental Approaches for Characterization of Disulfide-Trapped Receptor:Chemokine Complexes

	Amount of Sample Needed (μg)	Optimal protein Concentration (μg/ul)	Preparation Time (h)[a]	Throughput[b]	Specificity[c]	Sensitivity[c]
Flow cytometry	n/a (performed on cells)		~1	+++	–	+++
SDS-PAGE	5–10	0.3–0.5	~1	++	++	++
Western blotting	0.5–1	0.1–0.5	~4–6	++	++	+++
CPM-DSF	0.2–0.4	0.2–0.4	~1	+++	+++	–
SEC	5–10	0.3–0.5	~0.5–1	+	+++	–

[a]The estimated preparation time does not include protein purification. Protein purification is required for all experiments except flow cytometry and, in some implementations, Western blotting, and typically takes at least 3 days.
[b]Number of samples that can be reasonably tested in parallel in a single experiment, including controls: up to 5–6 (+), up to 12–15 (++), up to 48–96 (+++).
[c]A qualitative estimate of specificity and sensitivity of each assay towards strong crosslinks favorably compatible with the native complex geometry.

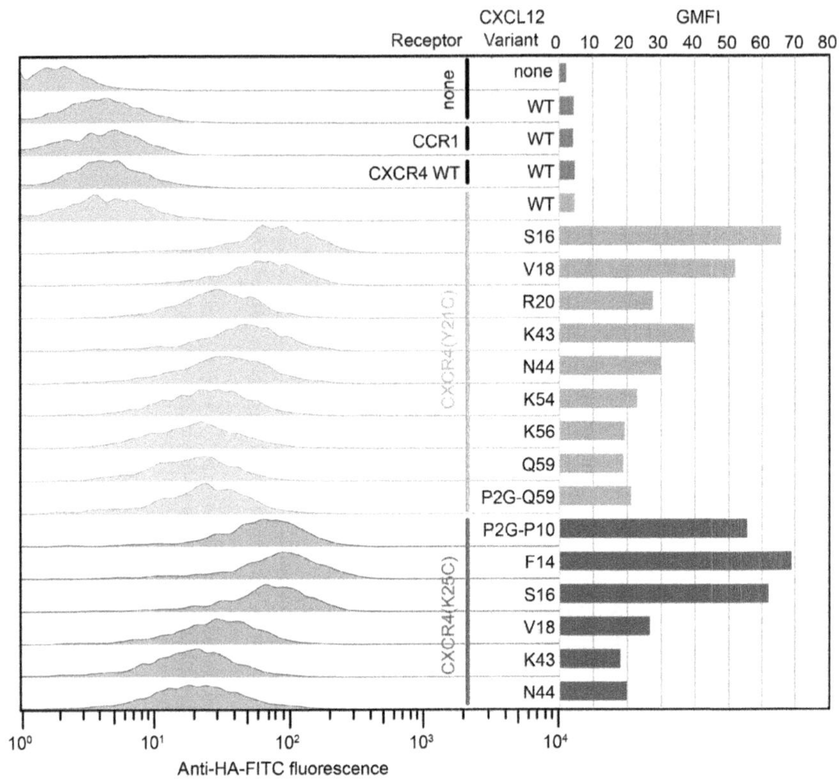

Figure 6 Detection of crosslinked CXCR4:CXCL12 and CXCR4:P2G-CXCL12 complexes by flow cytometry. Each of two receptor variants (CXCR4(Y21C), cyan (light gray in the print version) or CXCR4(K25C), magenta (gray in the print version)) was coexpressed with multiple chemokine variants. Left: chemokine on the cell surface is detected by fluorescently labeled antibody against a tag on its C-terminus. Right: geometric mean fluorescence intensity is plotted for each receptor:chemokine pair. Notably, for CXCR4:CXCL12 complexes, staining of non-crosslinked complexes (CXCR4 WT:CXCL12 WT, gray and CXCR4 (Y21C):CXCL12 WT, top cyan (light gray in the print version)) is low and comparable to staining of the control sample that does not have the receptor (sample 2); this indicates lower complex affinity and/or fast dissociation and makes the experiment possible without unlabeled competitor. This is in contrast to Fig. 7 showing a similar experiment for a slower dissociating complex of ACKR3:CXCL12.

surface-bound chemokine are present even if the covalent disulfide crosslink between the receptor and the chemokine is not formed. To separate disulfide-trapped complexes from long-lived noncovalent complexes, prolonged incubation of samples with excess unlabeled competitor (a chemokine or a small molecule) is helpful. Such preincubation dissociates the

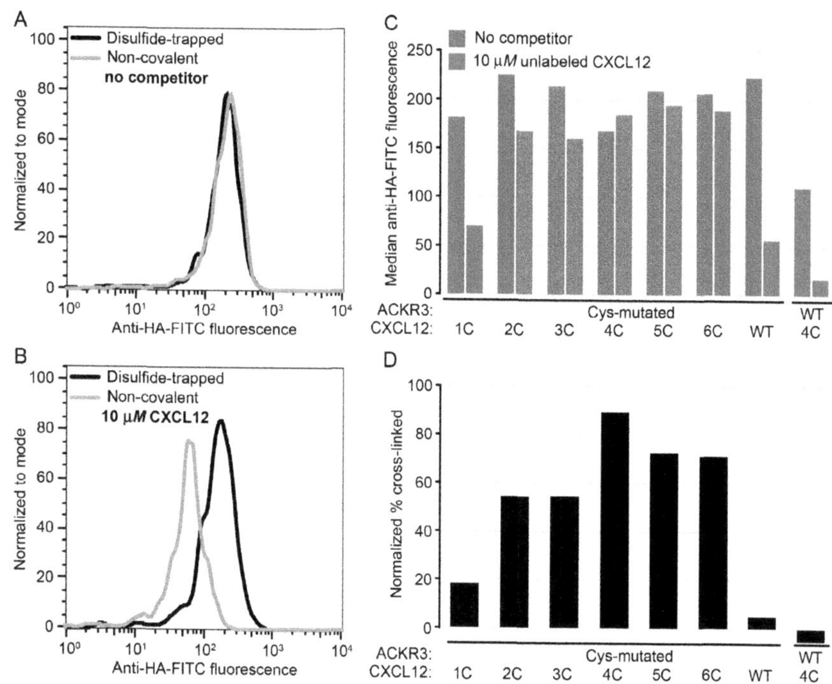

Figure 7 Detection of crosslinked ACKR3:CXCL12 complexes by flow cytometry following incubation with an unlabeled competitor. (A) Due to high affinity and/or slow dissociation of the chemokine, antichemokine staining is indistinguishable between the noncovalent (green; gray in the print version) and disulfide-trapped (black) complexes, (B) Prolonged incubation with excess unlabeled CXCL12 dissociates noncovalent (green; gray in the print version) but not disulfide-trapped (black) complexes, effectively reducing surface staining. (C) Flow cytometry screening of multiple disulfide-trapped candidate complexes side-by-side with controls (ACKR3(Cys):CXCL12 WT and ACKR3 WT: CXCL12(4C)). (D) Percent crosslinked for each complex was estimated by dividing median anti-HA-FITC fluorescence in the presence of competitor by that in the absence of competitor and subtracting nonspecific signal from control samples.

noncovalent complexes, but has no effect on the covalent disulfide-trapped complexes (Fig. 7).

4.4.1.2 Interpretation

The result of a flow cytometry experiment is a histogram of cell distribution by the total amount of surface-bound chemokine, with or without an unlabeled competitor ligand. This assay provides an easy way to quickly screen numerous candidate disulfide-trapped construct combinations *without the complexities of sample purification*; however, the utility of this method is limited to chemokines that are detectable on the cell surface by a

C-terminal tag. Furthermore, even for chemokines that are efficiently detected, the average levels of specific anti-tag antibody fluorescence on cells may vary significantly as a result of varying surface expression of the *receptor*; therefore, flow cytometry at best allows one to qualitatively *estimate* the presence of disulfide-trapped complexes or to *rank* on a relative scale several chemokine mutants that are coexpressed with a single-receptor mutant as a part of the same experiment. In this sense, the assay has relatively low specificity. On the other hand, as long as antibody detection of the surface-bound chemokine is established, the assay has high sensitivity and thus rarely produces false negatives: the absence of antichemokine staining in these cases signals an unsuccessful crosslink that can be eliminated from further consideration.

4.4.2 Quantification and Characterization of Disulfide-Trapped Complexes by SDS-PAGE and Western Blotting

Following metal affinity purification, promising candidate complexes are analyzed for yield and crosslinking efficiency by nonreducing SDS-PAGE and Western blotting. These assays appear central in crosslink characterization as they provide the proper context for interpretation of all other experiments.

4.4.2.1 Experiment

For nonreducing SDS-PAGE analysis, 5–10 μg of purified protein is mixed with Laemmli buffer containing no reducing agents and loaded onto a 10% polyacrylamide gel. With some membrane receptors, it is important that the SDS-denatured samples are *not* heated prior to loading. Room temperature is a reasonable tradeoff between avoiding protein aggregation (that occurs when samples are heated) and maintaining the solubility of SDS (that precipitates if samples are kept on ice). Gels are stained with Coomassie stain, and destained 3 × with a 50%/40%/10% (v/v/v) mixture of H_2O, methanol, and acetic acid. Molecular weight shift and the relative band intensity are used as indicators of the presence and relative abundance of the disulfide-trapped complex, respectively. In a separate control experiment, the same samples are run using standard reducing Laemmli buffer.

Western blot detection is used to specifically identify Flag-tagged receptor and HA-tagged chemokine bands. In this case, nonreducing SDS-PAGE is performed as above and, without staining, transferred to a nitrocellulose membrane at 100 V for 1 h. The membrane is incubated in 10 mL TBS-T (1 × TBS, pH 7.4, 0.1% Tween (v/v)) supplemented with 5% milk for 1 h at room temperature. 1 μL of primary antibodies (mouse anti-Flag M2 primary antibody (Sigma, for receptor) and rat anti-HA 3 F10 primary antibody (Roche, for chemokine)) is added to 10 mL of fresh TBS-T with 5% milk and again incubated at room temperature for 1 h. 0.5 μL of secondary antibodies (IRDye 680 donkey anti-mouse IgG and IRDye 800 goat anti-rat IgG (LI-COR Biosciences)) is added to 10 mL of TBS-T with 5% BSA and incubated for 1 h at room temperature. Following incubation, the membrane is washed 3 × with 10 mL of fresh TBS-T for 10 min. The membrane is finally rinsed with 1 × sterile PBS and imaged using the Odyssey IR imaging system (LI-COR Bioscience). Membranes can also be dried overnight between layers of filter paper, and then kept and imaged dry for up to 2 weeks with minimal loss in signal provided they are protected from light.

We usually conduct Western blot analysis on purified samples and interpret it in conjunction with other assays described herein: SDS-PAGE, SEC, and DSF. However, unlike these other assays, Western blotting does not *require* the samples to be purified and thus can be conducted on whole cell lysates as an alternative first line of analysis to quickly screen for promising disulfide-trapped constructs.

4.4.2.2 Interpretation

Following a nonreducing SDS-PAGE, most frequently observed bands are located at the following molecular weights (from smallest to largest):

- Uncomplexed chemokine monomer
- Uncomplexed chemokine dimer
- Uncomplexed receptor monomer (Fig. 8A)
- Receptor:chemokine complex (Fig. 8A)

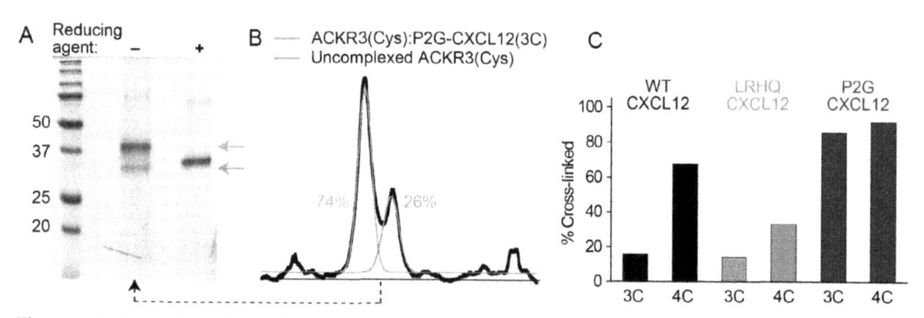

Figure 8 SDS-PAGE analysis of a crosslinked complex between ACKR3(Cys) and N-terminal mutants of CXCL12. (A) Relative intensity of the bands on a nonreducing and a reducing SDS-PAGE for P2G-CXCL12 (3C), (B) Band intensity from (A) quantified by densitometry. (C) The crosslinking efficiency with ACKR3(Cys) is quantified for cysteine residues introduced at two proximal positions of CXCL12, each in the context of three N-terminally modified chemokine variants; the observed difference in crosslinking efficiency of homologous mutants reflects conformational variations between the chemokine variants.

- Receptor:chemokine dimer complex
- Uncomplexed receptor dimer
- Receptor dimer:chemokine complex
- Aggregated protein material

Additional proximal bands may appear as a result of proteolytic degradation of receptor (effectively lowering its MW) or heterogeneous posttranslation modifications (increasing MW). In all cases, it is important to include the following controls on the same gel:

- A homologous well-behaving insect cell expressed receptor construct, if available
- A non-crosslinked insect cell expressed receptor:chemokine pair (e.g., including a nonmutated receptor and/or nonmutated chemokine)
- A well-behaving disulfide-trapped pair, if available.

By ensuring that an equal amount of protein is loaded into each lane of the gel, and by providing adequate controls, relative ranking of the crosslink quality becomes possible.

An ideal crosslinked complex shows as a bright band on a nonreducing SDS-PAGE and is possibly accompanied by a much dimmer (if present at all) band corresponding to uncomplexed monomeric receptor, in the faint or absent background of other irrelevant bands (Fig. 8A). Relative brightness of the two bands may serve as a measure of crosslinking efficiency (the ratio of crosslinked to non-crosslinked receptor, Fig. 8B and C), although in many cases, it may overestimate the fraction of crosslinked receptor due to poor membrane extraction of non-crosslinked species. Western blot analysis of an ideal crosslinked pair will show clear and weak (if present at all) receptor-only staining of the lower MW band with clear and strong costaining of the disulfide-trapped higher MW complex band (Fig. 9). Further confirmation of the successful crosslink may be obtained by repeating the SDS-PAGE and Western blotting analysis under reducing conditions, which is expected to completely dissociate the higher MW complex band, to increase the density of the receptor-only band (Fig. 8A), and to bring about the lower molecular weight chemokine-only band (assuming the gel percentage is sufficient to detect low MW proteins in the sample).

We consider successful clear-cut intermolecular crosslinks as evidence of spatial proximity and favorable orientation of the mutated residues in the context of the native complex geometry, even if these crosslinks do not withstand the scrutiny of the more stringent SEC and DSF assays described below. As such, they can be used for validation and refinement of 3D models of the complex (Kufareva, Handel, & Abagyan, 2015), at least in the form of

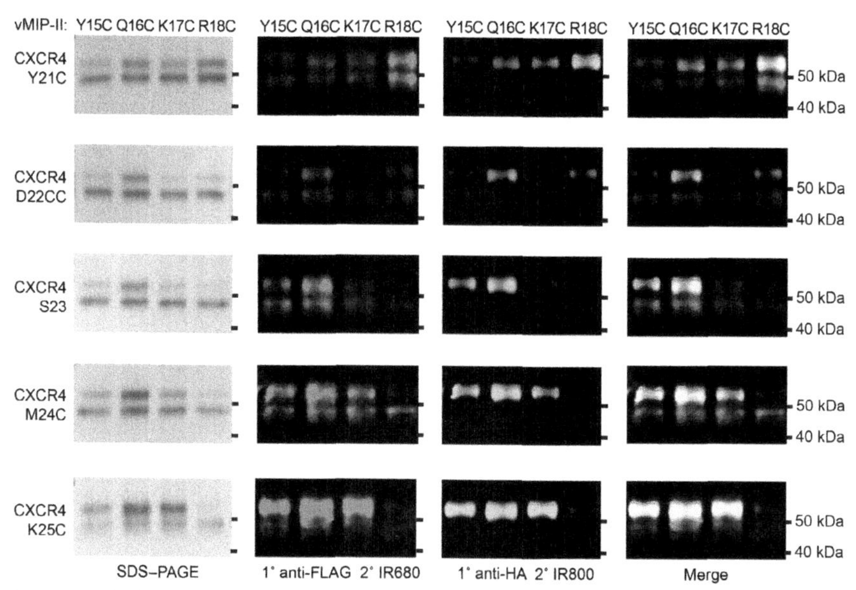

Figure 9 SDS-PAGE and Western blotting analysis of CRS1 disulfide crosslinked complexes of CXCR4 and vMIP-II. Left column: Coomassie stained SDS-PAGE gels of five different CXCR4 Cys mutants with four different vMIP-II Cys mutants. Second column: detection of the same SDS-PAGE gel by Western with an anti-FLAG antibody against a tag on the receptor. Third column: detection with an anti-HA antibody against a tag on the chemokine. Fourth column: merge of the anti-Flag and anti-HA Westerns. *Figure adapted from Qin et al. (2015).* (See the color plate.)

interresidue distance restraints. Stronger evidence may be required in order to integrate the crosslink in the modeling procedure in the form of an explicit intermolecular disulfide bond.

Unfortunately, ideal crosslinks are rare, and in most cases, the relative density of the two bands is nonoptimal and additional irrelevant bands are present. Such bands may indicate, for example, formation of covalent homodimers or higher order homooligomers from the mutated and possibly misfolded chemokines or receptors, or formation of covalent receptor:chemokine complexes of noncanonical stoichiometry, all of which can be deduced from Western blotting staining against the receptor and the chemokine. The prevalence of such bands indicates that the artificially introduced cysteine residue pair interferes with proper folding of the individual proteins or with their interactions, and warrants caution in application of the crosslink in modeling procedures.

Given the prevalence and the role of native intramolecular disulfide bonds at the receptor:chemokine interface, any artificially introduced cysteine can potentially result in disulfide shuffling. This scenario may exhibit the behavior of an efficient crosslink on SDS-PAGE or a Western blot, although in reality, the artificially introduced cysteines are not proximal. Consequently, interpretation of the crosslinks in the context of the aminoacid sequence and the nascent 3D model of the complex may be instrumental in separating true hits from results of disulfide shuffling.

4.4.3 Characterization of Monodispersity and Thermal Stability of Disulfide-Trapped Complexes with SEC and CPM-DSF

SEC and CPM-DSF represent the most stringent and specific assays for characterization of disulfide-trapped complexes. This is because even the most clear-cut crosslinks may introduce minor distortions to the native favorable positions and orientations of the receptor and the chemokine in the complex; such distortions are detectable by lowered thermal stability or increased aggregation of the purified complexes as compared to an ideal geometry crosslink.

4.4.3.1 SEC and CPM-DSF Experimental Protocols

Complex monodispersity is analyzed by analytical SEC using a Sepax SRT-C 300 column. Details of the assay are described in Chapter 11 by Gustavsson et al. Absorbance at 280 nm (A_{280}) is recorded and plotted as a function of elution time to obtain the SEC trace. As implemented in our hands, the samples are analyzed one-by-one with each run taking 15–20 min (assuming the column is properly equilibrated), i.e., it is lower throughput than other experiments described herein (Table 1).

Thermostabilities of purified complexes are analyzed by a DSF assay adapted from previous publications (Alexandrov, Mileni, Chien, Hanson, & Stevens, 2008). Receptor:chemokine samples are mixed with a thiol-reactive 7-diethylamino-3-(4′-maleimidylphenyl)-4-methylcoumarin (CPM) dye that increases in fluorescence upon interaction with internal (not disulfide-bonded) cysteines of the receptor that gradually become exposed and accessible for labeling as a result of thermal denaturation. The melting curve is obtained by plotting CPM fluorescence as a function of temperature (Figs. 10A and 11A); in the canonical case, this curve has a clear sigmoidal shape and its first derivative has a clear sharp peak whose x-coordinate is interpreted as the melting temperature (T_m) of the sample (selected curves in Figs. 10B and 11B). We perform this assay using a Rotor-Gene Q 6-plex RT-PCR machine (Qiagen), which allows miniaturization and relatively high-throughput (up to 72 samples can be analyzed at a time and sample

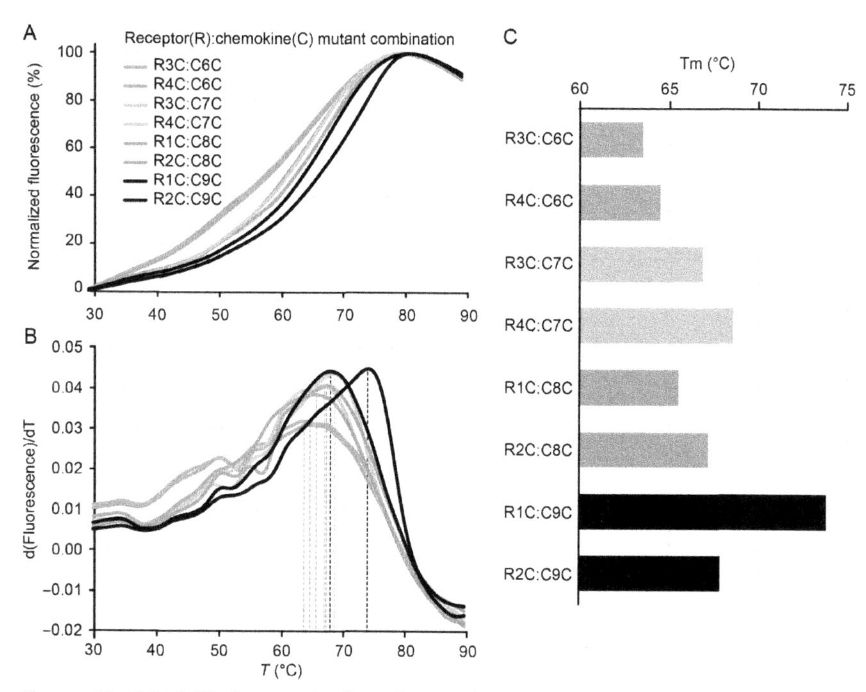

Figure 10 CPM-DSF characterization of several cysteine mutant pairs for a CC chemokine and its receptor. (A) Melting curves plotting the increasing fluorescence of the CPM dye as a function of temperature. (B) First derivative curves of (A) demonstrate the positions of the peaks used to calculate T_m. (C) Melting temperatures are calculated from the derivative curves. One mutant combination stands out as having a significantly higher T_m. (See the color plate.)

volume can be as low as 20 μL). Briefly, 0.2–0.5 μg of protein is mixed with 5 μ*M* CPM dye in 25 m*M* HEPES, pH 7.5, 400 m*M* NaCl, 0.025% DDM, 0.005% CHS, and 10% glycerol to a final volume of 20 μL; samples are incubated for 5 min at room temperature and then heated gradually from 28 °C to 90 °C at a rate of 1.5 °C/min, with CPM fluorescence (excitation 365 nm, emission 460 nm) recorded at every 1 °C increase. Rotor-Gene Q—Pure Detection software is used to extract the T_m from the first derivative plot of the denaturation curve.

4.4.3.2 Interpretation

An ideal SEC curve has a narrow sharp peak at the elution time that represents the hydrodynamic volume of the protein or complex in question. Protein:protein complexes typically elute earlier than the isolated components due to higher hydrodynamic volume. An ideal peak is tall (compared to

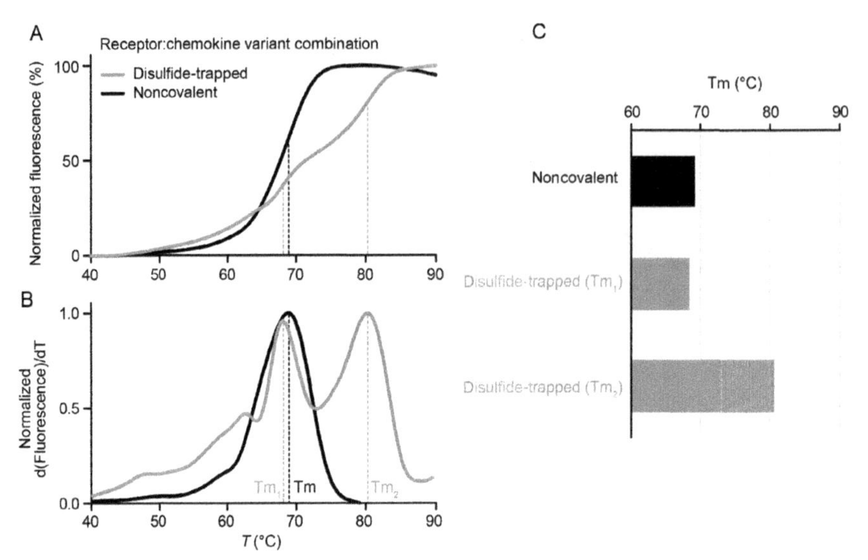

Figure 11 CPM-DSF characterization of noncovalent and disulfide-trapped complexes of a CXC chemokine and its receptor. (A) Melting curves plotting the increasing fluorescence of the CPM dye as a function of temperature. A biphasic transition is observed for the disulfide-trapped sample, likely due to the presence of both noncovalent and disulfide-trapped subpopulations. (B) First derivative curves of (A) demonstrate the positions of the peaks used to calculate T_m. A double-peak is clearly present for the disulfide-trapped sample. (C) Melting temperatures are calculated from the derivative curves. The disulfide-trapped sample shows two T_ms one of which is indistinguishable from the noncovalent sample T_m.

controls), narrow, and sharp with no shoulders on either side, indicating high concentration and high degree of monodispersity (homogeneity) of the sample (e.g., Fig. 12C). A widened peak may result from conformational heterogeneity or from formation of dimers and higher order oligomers. A shoulder on the left of the main peak is typically due to protein aggregates and is undesirable. The heights and shapes of SEC peaks obtained in parallel for several disulfide-trapped complex candidates (with proper controls) can be used to rank the candidates in terms of both protein concentration (or, with uniform treatment, yield) and monodispersity.

A canonical CPM-DSF curve has a clear sigmoidal shape and its first derivative has a single peak, indicating a monophasic melting transition of the sample. The height of the derivative peak is *not* necessarily representative of the sample yield; instead, it is a characteristic of the sharpness of the melting transition. Poor (flat) transitions, with short and wide derivative peaks

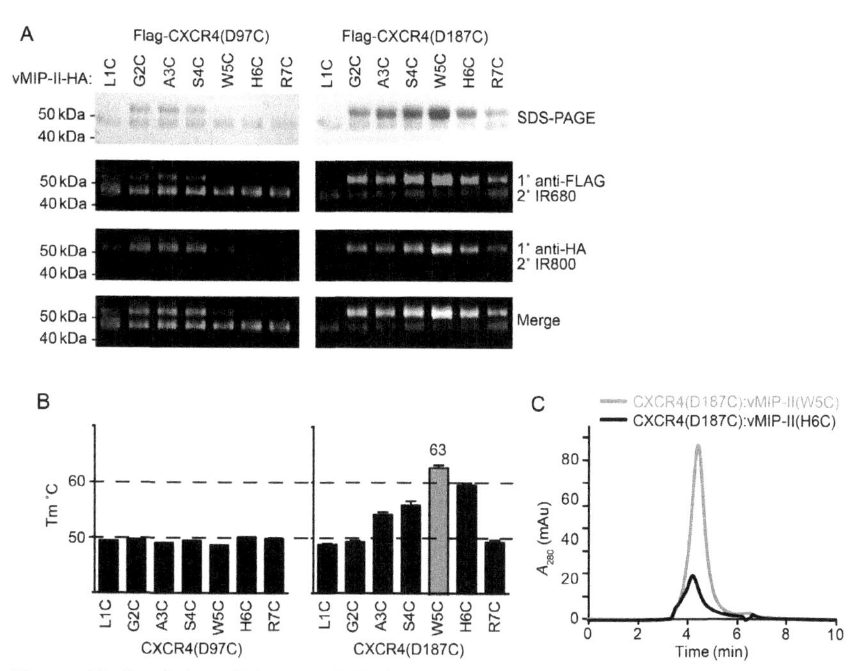

Figure 12 Candidate CRS2 crosslinks between CXCR4 and vMIP-II identified by SDS-PAGE and Western blotting (A) were further subjected to CPM-DSF (B) and SEC (C). Although spatially close and similar in both yield and crosslinking efficiency to the neighboring residues (A), pairs of CXCR4(D187C) with vMIP-II(W5C) and vMIP-II(H6C) stand out as having higher thermal stability (B). Furthermore, the complex with vMIP-II(W5C) has a much better peak shape and height than the complex with vMIP-II (H6C), indicating a higher yield and higher degree of monodispersity than vMIP-II(H6C) and other candidate pairs. *Figure adapted from Qin et al. (2015).*

(Fig. 10A, pink curves), may result not only from insufficient sample concentrations but also from high degree of heterogeneity or the overall low quality and stability of the sample; they may also be an inherent property of the protein in question (e.g., due to a very small number of buried cysteine residues that can potentially become exposed and accessible for CPM labeling). In high-quality disulfide-trapped samples, two distinct transitions are sometimes discernible: one corresponding to melting of the non-crosslinked population and another (at a higher T_m) corresponding to the crosslinked population (Fig. 11A and B). As for individual samples above, the relative height of the derivative peaks should not be interpreted as relative abundance of one population over another, or as fraction crosslinked, but rather simply as the sharpness of the melting transition.

Compared to SDS-PAGE and Western blotting, SEC and DSF typically provide much better resolution in distinguishing optimal from neighboring suboptimal crosslinks. Depending on implementation, DSF and SEC may also be more or less time- and resource-consuming than SDS-PAGE and Western blotting. In our hands, the Rotor-Gene Q 6-plex RT-PCR instrument allows one to run up to 72 samples in parallel, with each of them taking only about 5% of the protein that is typically used for SDS-PAGE, thus far exceeding both SDS-PAGE and Western blotting in throughput and miniaturization. In view of this, when complex stabilization (and not simply determination of residue proximities) is the primary goal, high-throughput DSF can be used as a first line of screening of the candidate pairs, with only the complexes showing clear melting transitions and potentially increased melting temperatures being subject to SDS-PAGE.

Figure 12 illustrates an example where multiple promising crosslinks (all spatially close) were identified in a receptor:chemokine complex by gel electrophoresis and Western blotting (Fig. 12A), but only two of them resulted in a significant increase in the complex melting temperature (Fig. 12B), and only one of them stood out in terms of complex yield and monodispersity as demonstrated by the height and the quality of the SEC trace (Fig. 12C). Following these stringent validation steps, the most promising crosslinks may serve as tools for structure determination of the complex, as was the case for a CXCR4:vMIP-II complex by Qin et al. (2015).

5. CONCLUSION AND PERSPECTIVES

Although a useful and relatively straightforward methodology for both modeling and structure determination of protein:protein complexes, disulfide trapping has numerous caveats, especially when applied to interfaces rich in native intramolecular disulfide bonds such as those of chemokines with their receptors. Combined evidence from several biophysical experiments as well as critical assessment of results in the context of 3D models enable one to separate false positives from energetically favorable crosslinks that stabilize the native geometry of the complex. Prioritizing high-throughput high-sensitivity assays helps optimize the pipeline by early triage of unsuccessful candidates. Iterative application of the methodology in conjunction with molecular modeling is a viable strategy for elucidating structural determinants of receptor:chemokine interactions.

ACKNOWLEDGMENTS

Authors thank Dr. Ruben Abagyan (UCSD) for useful discussions. This work was partially supported by NIH grants U01 GM094612, U54 GM094618, R01 GM071872, R01 AI118985, and R01 AI37113.

REFERENCES

Alexandrov, A. I., Mileni, M., Chien, E. Y. T., Hanson, M. A., & Stevens, R. C. (2008). Microscale fluorescent thermal stability assay for membrane proteins. *Structure, 16*(3), 351–359. Retrieved from, http://www.sciencedirect.com/science/article/B6VSR-4S1K22J-8/2/541b65b93c931f875a192deee2cf6575.

Bachelerie, F., Ben-Baruch, A., Combadiere, C., Farber, J. M., Graham, G. J., Horuk, R., et al. (2015). *Chemokine receptors, introduction. IUPHAR/BPS guide to PHARMACOLOGY.* Retrieved from, http://www.guidetopharmacology.org/GRAC/FamilyIntroductionForward?familyId=14.

Baggiolini, M. (1998). Chemokines and leukocyte traffic. *Nature, 392*(6676), 565–568. Retrieved from, http://dx.doi.org/10.1038/33340.

Bill, R. M., Henderson, P. J. F., Iwata, S., Kunji, E. R. S., Michel, H., Neutze, R., et al. (2011). Overcoming barriers to membrane protein structure determination. *Nature Biotechnology, 29*(4), 335–340. Retrieved from, http://dx.doi.org/10.1038/nbt.1833.

Brelot, A., Heveker, N., Montes, M., & Alizon, M. (2000). Identification of residues of CXCR4 critical for human immunodeficiency virus coreceptor and chemokine receptor activities. *Journal of Biological Chemistry, 275*(31), 23736–23744. Retrieved from, http://www.hubmed.org/display.cgi?uids=10825158.

Buck, E., & Wells, J. A. (2005). Disulfide trapping to localize small-molecule agonists and antagonists for a G protein-coupled receptor. *Proceedings of the National Academy of Sciences of the United States of America, 102*(8), 2719–2724. http://dx.doi.org/10.1073/pnas.0500016102.

Burg, J. S., Ingram, J. R., Venkatakrishnan, A. J., Jude, K. M., Dukkipati, A., Feinberg, E. N., et al. (2015). Structural basis for chemokine recognition and activation of a viral G protein–coupled receptor. *Science, 347*(6226), 1113–1117. http://dx.doi.org/10.1126/science.aaa5026.

Charo, I. F., & Ransohoff, R. M. (2006). The many roles of Chemokines and Chemokine receptors in inflammation. *New England Journal of Medicine, 354*(6), 610–621. http://dx.doi.org/10.1056/NEJMra052723.

Chen, Q., Pinon, D. I., Miller, L. J., & Dong, M. (2009). Molecular basis of glucagon-like peptide 1 docking to its intact receptor studied with carboxyl-terminal photolabile probes. *Journal of Biological Chemistry, 284*(49), 34135–34144. http://dx.doi.org/10.1074/jbc.M109.038109.

Chen, Q., Pinon, D. I., Miller, L. J., & Dong, M. (2010). SPATIAL Approximations Between Residues 6 And 12 In The Amino-Terminal Region Of Glucagon-Like Peptide 1 And Its Receptor: A region critical for biological activity. *Journal of Biological Chemistry, 285*(32), 24508–24518. http://dx.doi.org/10.1074/jbc.M110.135749.

Chevigne, A., Fievez, V., Schmit, J.-C., & Deroo, S. (2011). Engineering and screening the N-terminus of chemokines for drug discovery. *Biochemical Pharmacology, 82*(10), 1438–1456. http://dx.doi.org/10.1016/j.bcp.2011.07.091.

Choi, W.-T., Tian, S., Dong, C.-Z., Kumar, S., Liu, D., Madani, N., et al. (2005). Unique ligand binding sites on CXCR4 probed by a chemical biology approach: Implications for the design of selective human immunodeficiency virus type 1 inhibitors. *Journal of Virology, 79*(24), 15398–15404. Retrieved from, http://www.hubmed.org/display.cgi?uids=16306611.

Coin, I., Katritch, V., Sun, T., Xiang, Z., Siu, F. Y., Beyermann, M., et al. (2013). Genetically encoded chemical probes in cells reveal THE binding path OF urocortin-I TO Crf class B gpcr. *Cell, 155*(6), 1258–1269. http://dx.doi.org/10.1016/j.cell.2013.11.008.

Dong, M., Lam, P. C. H., Gao, F., Hosohata, K., Pinon, D. I., Sexton, P. M., et al. (2007). Molecular approximations between residues 21 and 23 of secretin and its receptor: Development of a model for peptide docking with the amino terminus of the secretin receptor. *Molecular Pharmacology, 72*(2), 280–290. Retrieved from, http://www.hubmed.org/display.cgi?uids=17475809.

Dong, M., Lam, P. C. H., Pinon, D. I., Hosohata, K., Orry, A., Sexton, P. M., et al. (2011). Molecular basis of secretin docking to its intact receptor using multiple photolabile probes distributed throughout the pharmacophore. *Journal of Biological Chemistry, 286*(27), 23888–23899. Retrieved from, http://www.hubmed.org/display.cgi?uids=21566140.

Dong, M., Xu, X., Ball, A. M., Makhoul, J. A., Lam, P. C. H., Pinon, D. I., et al. (2012). Mapping spatial approximations between the amino terminus of secretin and each of the extracellular loops of its receptor using cysteine trapping. *The FASEB Journal, 26*(12), 5092–5105. Retrieved from, http://www.hubmed.org/display.cgi?uids=22964305.

Duchesnes, C. E., Murphy, P. M., Williams, T. J., & Pease, J. E. (2006). Alanine scanning mutagenesis of the chemokine receptor CCR3 reveals distinct extracellular residues involved in recognition of the eotaxin family of chemokines. *Molecular Immunology, 43*(8), 1221–1231. Retrieved from, http://www.sciencedirect.com/science/article/B6T9R-4GWC0KX-2/2/4257c4d48a2d0ae36c4d6d45012b06d4.

Gerard, C., & Rollins, B. J. (2001). Chemokines and disease. *Nature Immunology, 2*(2), 108–115. Retrieved from, http://dx.doi.org/10.1038/84209.

Ghosh, E., Kumari, P., Jaiman, D., & Shukla, A. K. (2015). Methodological advances: The unsung heroes of the GPCR structural revolution. *Nature Reviews. Molecular Cell Biology, 16*(2), 69–81. http://dx.doi.org/10.1038/nrm3933. http://www.nature.com/nrm/journal/v16/n2/abs/nrm3933.html#supplementary-information.

Griffith, J. W., Sokol, C. L., & Luster, A. D. (2014). Chemokines and chemokine receptors: Positioning cells for host defense and immunity. *Annual Review of Immunology, 32*(1), 659–702. http://dx.doi.org/10.1146/annurev-immunol-032713-120145.

Grunbeck, A., Huber, T., Abrol, R., Trzaskowski, B., Goddard, W. A., & Sakmar, T. P. (2012). Genetically encoded photo-cross-linkers Map the binding site of an allosteric drug on a G protein-coupled receptor. *ACS Chemical Biology*, 7(6), 967–972. http://dx.doi.org/10.1021/cb300059z.

Grunbeck, A., & Sakmar, T. P. (2013). Probing G protein-coupled receptor—Ligand interactions with targeted photoactivatable cross-linkers. *Biochemistry, 52*(48), 8625–8632. http://dx.doi.org/10.1021/bi401300y.

Gupta, S. K., Pillarisetti, K., Thomas, R. A., & Aiyar, N. (2001). Pharmacological evidence for complex and multiple site interaction of CXCR4 with SDF-1alpha: Implications for development of selective CXCR4 antagonists. *Immunology Letters, 78*(1), 29–34. http://dx.doi.org/10.1016/S0165-2478(01)00228-0.

Hagemann, I. S., Miller, D. L., Klco, J. M., Nikiforovich, G. V., & Baranski, T. J. (2008). Structure of the complement factor 5a receptor-ligand complex studied by disulfide trapping and molecular modeling. *Journal of Biological Chemistry, 283*(12), 7763–7775. http://dx.doi.org/10.1074/jbc.M709467200.

Huber, T., Naganathan, S., Tian, H., Ye, S., & Sakmar, T. P. (2013). *Unnatural amino acid mutagenesis of GPCRs using amber codon suppression and bioorthogonal labeling. In P. M. Conn (Ed.), Methods in Enzymology: Vol. 520* (pp. 281–305): Academic Press. http://www.sciencedirect.com/science/bookseries/00766879/520/supp/C.

Kim, C. H., Axup, J. Y., & Schultz, P. G. (2013). Protein conjugation with genetically encoded unnatural amino acids. *Current Opinion in Chemical Biology, 17*(3), 412–419. http://dx.doi.org/10.1016/j.cbpa.2013.04.017.

Kofuku, Y., Yoshiura, C., Ueda, T., Terasawa, H., Hirai, T., Tominaga, S., et al. (2009). Structural basis of the interaction between Chemokine stromal cell-derived factor-1/CXCL12 and its G-protein-coupled receptor CXCR4. *Journal of Biological Chemistry, 284*(50), 35240–35250. http://dx.doi.org/10.1074/jbc.M109.024851.

Kufareva, I., Handel, T. M., & Abagyan, R. (2015). Experiment-guided molecular modeling of protein–protein complexes involving GPCRs. In M. Filizola (Ed.), *G protein-coupled receptors in drug discovery: Vol. 1335* (pp. 295–311). New York, NY: Springer.

Kufareva, I., Salanga, C. L., & Handel, T. M. (2015). Chemokine and chemokine receptor structure and interactions: Implications for therapeutic strategies. *Immunology & Cell Biology, 93*(4), 372–383. http://dx.doi.org/10.1038/icb.2015.15.

Kufareva, I., Stephens, B. S., Holden, L. G., Qin, L., Zhao, C., Kawamura, T., et al. (2014). Stoichiometry and geometry of the CXC chemokine receptor 4 complex with CXC ligand 12: Molecular modeling and experimental validation. *Proceedings of the National Academy of Sciences of the United States of America, 111*(50), E5363–E5372. http://dx.doi.org/10.1073/pnas.1417037111.

Liu, W., Wacker, D., Wang, C., Abola, E., & Cherezov, V. (2014). Femtosecond crystallography of membrane proteins in the lipidic cubic phase. *Philosophical Transactions of the Royal Society of London. Series B: Biological Sciences, 369*(1647), 20130314. http://dx.doi.org/10.1098/rstb.2013.0314.

Miller, L. J., Chen, Q., Lam, P. C.-H., Pinon, D. I., Sexton, P. M., Abagyan, R., et al. (2011). Refinement of glucagon-like peptide 1 docking to its intact receptor using Mid-region photolabile probes and molecular modeling. *Journal of Biological Chemistry, 286*(18), 15895–15907. http://dx.doi.org/10.1074/jbc.M110.217901.

Monaghan, P., Thomas, B. E., Woznica, I., Wittelsberger, A., Mierke, D. F., & Rosenblatt, M. (2008). Mapping peptide hormone – receptor interactions using a disulfide-trapping approach†. *Biochemistry, 47*(22), 5889–5895. http://dx.doi.org/10.1021/bi800122f.

Montaner, S., Kufareva, I., Abagyan, R., & Gutkind, J. S. (2013). Molecular mechanisms deployed by virally encoded G protein-coupled receptors in human diseases. *Annual Review of Pharmacology and Toxicology, 53*(1), 331–354. http://dx.doi.org/10.1146/annurev-pharmtox-010510-100608.

Moraes, I., Evans, G., Sanchez-Weatherby, J., Newstead, S., & Stewart, P. D. S. (2014). Membrane protein structure determination—The next generation. *Biochimica et Biophysica Acta (BBA) - Biomembranes, 1838*(1, Pt. A), 78–87. http://dx.doi.org/10.1016/j.bbamem.2013.07.010.

Murdoch, C., & Finn, A. (2000). Chemokine receptors and their role in inflammation and infectious diseases. *Blood, 95*(10), 3032–3043. Retrieved from, http://www.bloodjournal.org/bloodjournal/95/10/3032.full.pdf.

Nijmeijer, S., Leurs, R., Smit, M. J., & Vischer, H. F. (2010). The Epstein-Barr virus-encoded G protein-coupled receptor BILF1 hetero-oligomerizes with human CXCR4, scavenges G{alpha}i proteins, and constitutively impairs CXCR4 functioning. *Journal of Biological Chemistry, 285*(38), 29632–29641. Retrieved from, http://www.hubmed.org/display.cgi?uids=20622011.

Pakianathan, D. R., Kuta, E. G., Artis, D. R., Skelton, N. J., & Hebert, C. A. (1997). Distinct but overlapping epitopes for the interaction of a CC-chemokine with CCR1, CCR3 and CCR5. *Biochemistry, 36*(32), 9642–9648. Retrieved from, http://www.hubmed.org/display.cgi?uids=9289016.

Pellequer, J.-L., & Chen, S.-w. W. (2006). Multi-template approach to modeling engineered disulfide bonds. *Proteins: Structure, Function, and Bioinformatics, 65*(1), 192–202. http://dx.doi.org/10.1002/prot.21059.

Pham, V., & Sexton, P. M. (2004). Photoaffinity scanning in the mapping of the peptide receptor interface of class II G protein—Coupled receptors. *Journal of Peptide Science, 10*(4), 179–203. http://dx.doi.org/10.1002/psc.541.

Proost, P., Loos, T., Mortier, A., Schutyser, E., Gouwy, M., Noppen, S., et al. (2008). Citrullination of CXCL8 by peptidylarginine deiminase alters receptor usage, prevents proteolysis, and dampens tissue inflammation. *The Journal of Experimental Medicine*, *205*(9), 2085–2097. http://dx.doi.org/10.1084/jem.20080305.

Qin, L., Kufareva, I., Holden, L. G., Wang, C., Zheng, Y., Zhao, C., et al. (2015). Crystal structure of the chemokine receptor CXCR4 in complex with a viral chemokine. *Science*, *347*(6226), 1117–1122. http://dx.doi.org/10.1126/science.1261064.

Ransohoff, R. M. (2009). Chemokines and Chemokine receptors: Standing at the crossroads of immunobiology and neurobiology. *Immunity*, *31*(5), 711–721. http://dx.doi.org/10.1016/j.immuni.2009.09.010.

Rasmussen, S. G. F., Choi, H.-J., Fung, J. J., Pardon, E., Casarosa, P., Chae, P. S., et al. (2011). Structure of a nanobody-stabilized active state of the b2 adrenoceptor. *Nature*, *469*(7329), 175–180. Retrieved from, http://dx.doi.org/10.1038/nature09648, http://www.nature.com/nature/journal/v469/n7329/abs/nature09648.html#supplementary-information.

Saini, V., Staren, D. M., Ziarek, J. J., Nashaat, Z. N., Campbell, E. M., Volkman, B. F., et al. (2011). The CXC Chemokine receptor 4 ligands ubiquitin and stromal cell-derived factor-1alpha function through distinct receptor interactions. *Journal of Biological Chemistry*, *286*(38), 33466–33477. http://dx.doi.org/10.1074/jbc.M111.233742.

Scholten, D. J., Canals, M., Maussang, D., Roumen, L., Smit, M. J., Wijtmans, M., et al. (2011). Pharmacological modulation of chemokine receptor function. *British Journal of Pharmacology*, *165*(6), 1617–1643. http://dx.doi.org/10.1111/j.1476-5381.2011.01551.x.

Van Damme, J., Decock, B., Lenaerts, J.-P., Conings, R., Bertini, R., Mantovani, A., et al. (1989). Identification by sequence analysis of chemotactic factors for monocytes produced by normal and transformed cells stimulated with virus, double-stranded RNA or cytokine. *European Journal of Immunology*, *19*(12), 2367–2373. http://dx.doi.org/10.1002/eji.1830191228.

Veldkamp, C. T., Seibert, C., Peterson, F. C., De la Cruz, N. B., Haugner, J. C., III, Basnet, H., et al. (2008). Structural basis of CXCR4 sulfotyrosine recognition by the chemokine SDF-1/CXCL12. *Science Signaling*, *1*(37), ra4. http://dx.doi.org/10.1126/scisignal.1160755. http://stke.sciencemag.org/content/1/37/ra4.

Wittelsberger, A., Corich, M., Thomas, B. E., Lee, B.-K., Barazza, A., Czodrowski, P., et al. (2006). The Mid-region of parathyroid hormone (1 − 34) serves as a functional docking domain in receptor activation†. *Biochemistry*, *45*(7), 2027–2034. http://dx.doi.org/10.1021/bi051833a.

Zhang, X., Stevens, R. C., & Xu, F. (2015). The importance of ligands for G protein-coupled receptor stability. *Trends in Biochemical Sciences*, *40*(2), 79–87. http://dx.doi.org/10.1016/j.tibs.2014.12.005.

Zhou, H., & Tai, H. H. (2000). Expression and functional characterization of mutant human CXCR4 in insect cells: Role of cysteinyl and negatively charged residues in ligand binding. *Archives of Biochemistry and Biophysics*, *373*(1), 211–217. Retrieved from, http://www.hubmed.org/display.cgi?uids=10620340.

CHAPTER NINETEEN

Analysis of G Protein and β-Arrestin Activation in Chemokine Receptors Signaling

Alessandro Vacchini*,†, Marta Busnelli‡, Bice Chini‡, Massimo Locati*,†,1, Elena Monica Borroni*,†

*Department of Medical Biotechnologies and Translational Medicine, Università degli Studi di Milano, Milano, Italy
†Humanitas Clinical and Research Center, Rozzano, Italy
‡CNR Institute of Neuroscience, Milan, Italy
[1]Corresponding author: e-mail address: massimo.locati@humanitasresearch.it

Contents

1. Introduction 422
2. G Proteins Signaling 423
   2.1 Detection of Gα Protein Activation by Measurement of Second Messengers 424
   2.2 Detection of Gα Protein Activation by G Proteins Conformational Changes 428
3. β-Arrestins Signaling: Detection of β-Arrestins Recruitment by Protein Conformational Changes 431
   3.1 Required Materials 431
   3.2 Measurement of Ligand-Dependent β-Arrestin Recruitment 431
   3.3 Measurement of Constitutive β-Arrestin Recruitment 433
4. Calculation of Biased Signaling 434
5. Summary 436
References 437

Abstract

Chemokines are key regulators of leukocyte migration and play fundamental roles in immune responses. The chemokine system includes a set of over 40 ligands which engage in a promiscuous fashion a panel of over 25 receptors belonging to a distinct family of 7 transmembrane-domain receptors (7TM) widely expressed on a variety of cells. Although responses evoked by chemokine receptors have long been considered the result of balanced activation of the G protein- and β-arrestin-dependent signaling modules, evidence is accumulating showing that these receptors are capable, as other 7TMs, to activate different signaling modules in a ligand- and cell/tissue-specific manner. This biased signaling, or functional selectivity, confers a hitherto largely uncharacterized level of complexity to the chemokine system and challenges our present understanding of its redundancy. At the same time, it also provides new insights of

Methods in Enzymology, Volume 570
ISSN 0076-6879
http://dx.doi.org/10.1016/bs.mie.2015.09.016

relevance for chemokine receptors targeting drug development plans. Here, we provide current methods to study biased signaling of chemokine receptors by dissecting G proteins and β-arrestins activation upon chemokine stimulation.

1. INTRODUCTION

Chemokines orchestrate leukocyte trafficking and play fundamental roles both in physiological and pathological immune responses and in inflammatory diseases (Charo & Ransohoff, 2006). They act via chemokine receptors, which belong to the family of 7TMs and are differentially expressed by all leukocytes and many nonhematopoietic cells, including cancer cells (Bonecchi et al., 2009). As for other 7TMs, responses evoked by chemokine receptors rely on multiple signal transduction modules, as the G protein-dependent signaling pathways which promote cell migration are tightly integrated with β-arrestins, which on the one hand functionally uncouple G proteins and on the other function as signalosome adaptor/scaffolding proteins activating signaling pathways involved in the control of different cellular functions (Bonecchi et al., 2009).

Presently, 22 chemokine receptors and 48 chemokines have been identified in the human proteome (Bachelerie et al., 2014). In this complex biological system, several receptors are activated by multiple ligands, and a specific ligand often has the ability to activate multiple receptors (Zlotnik & Yoshie, 2012). This high degree of promiscuity makes the chemokine system particularly prone to biased signaling, also known as functional selectivity, a relatively new emerging phenomenon in 7TMs biology where by one ligand may mainly activate G proteins whereas another preferentially activates β-arrestins (Zweemer, Toraskar, Heitman, & IJzerman, 2014). Biased signaling does not simply comprise distinct signaling via either G proteins or β-arrestins, but also includes more subtle differences in the activation of other downstream signaling proteins, as different ligands can preferentially activate different subtypes of G proteins and β-arrestins isoforms, or differently affect signaling events such as ERK activation or Ca^{2+} mobilization (Zweemer et al., 2014). Evidence of biased signaling in the chemokine system is rapidly accumulating (Corbisier, Gales, Huszagh, Parmentier, & Springael, 2015; Steen, Larsen, Thiele, & Rosenkilde, 2014; Zidar, 2011). CCR7 is activated by CCL19 and CCL21, which are equivalent agonists for G protein signaling but differ in their ability to activate GRK and β-arrestins (Byers et al., 2008; Kohout et al., 2004); CCR2 activates G protein/β-arrestin balanced signaling when engaged by all ligands except CCL8, which was found biased for signaling to β-arrestin2 versus β-arrestin1 (Berchiche, Gravel, Pelletier, St-Onge, & Heveker, 2011); CXCR3 is also subjected to biased signaling when engaged by CXCL9 but the extent and the nature are determined by the amount of receptor expression and the cellular assay used (Rajagopal et al., 2013; Watts, Scholten, Heitman, Vischer, & Leurs, 2012). Evidence for biased signaling appears to be even more relevant for chemokine receptors with a large number of ligands. As an example, CCR1 has partial agonists (CCL14, CCL15, CCL23) becoming fully active after processing of their extended N-terminal domain (Tian et al., 2004), β-arrestin-biased ligands (CCL3 and CCL15; Rajagopal et al., 2013), and G protein-biased agonists (CCL5 and CCL23; Rajagopal et al., 2013). A second type of bias occurring in the chemokine system refers to cases where the same chemokine activates different signaling pathways from a given receptor in a tissue/cell-specific manner or in a species-specific manner (Rajagopal et al., 2013; Steen et al., 2014). As an example, CCL19 has different roles in inducing chemotaxis depending on the cell type where CCR7 is expressed on (Nandagopal, Wu, & Lin, 2011; Ricart, John, Lee, Hunter, & Hammer, 2011). Besides these evidences of ligand bias observed for conventional chemokine receptors, atypical chemokine receptors (ACKRs) represent an example of a bias form operating at the receptor level. ACKRs bind a variety of chemokines but no G protein-dependent signaling has been recorded upon binding of any of them (Bonecchi, Savino, Borroni, Mantovani, & Locati, 2010; Cancellieri, Vacchini, Locati, Bonecchi, & Borroni, 2013). On the contrary, ACKRs are particularly prone to recruit and elicit signaling through β-arrestins (Borroni et al., 2013). In conclusion, the existence of different forms of biased signaling in the chemokine system challenges our present understanding of redundancy and suggests that signaling properties of chemokine receptors should be taken into consideration by immunologists when deciphering the biology of the system as well as pharmacologists interested in targeting it. Here, we are providing methods to study chemokine receptors signaling by dissecting G protein and β-arrestin activation upon stimulation of one representative member of this family, namely CCR5.

2. G PROTEINS SIGNALING

One critical component for chemokine receptors signaling is the heterotrimeric G protein complex that directly associates with the chemokine receptor and transduces signals to downstream intracellular signaling molecules (Carman & Benovic, 1998; Lefkowitz, 1998). In most cases, chemokine receptors are coupled to Gαi, thus reducing intracellular cyclic AMP (cAMP) levels and PKA activation. This Gαi-dependent pertussis toxin (PTX)-sensitive pathway is of key relevance for the ability of chemokines to induce cell migration and coordinate leukocyte recruitment. However, emerging evidence suggests that an alternative Gαq-dependent pathway is engaged by a subset of chemokines and is critically required for regulating chemokine receptors-mediated inositol trisphosphate (IP3) generation and calcium release (Shi et al., 2007). In this section, we describe procedures for studying Gαi and Gαq activation by direct and indirect energy transfer-based approaches which evaluate G proteins conformational changes (BRET technology) or downstream second messenger levels (AlphaScreen and HTRF technologies). Procedures have been performed in Human Embryonic Kidney (HEK293 and HEK293T) or Chinese Hamster Ovary (CHO) cell lines grown in complete DMEM or DMEM/F12 supplemented with 10% fetal bovine serum (FBS), 25 m*M* HEPES (Lonza), 100 U/ml of penicillin/streptomycin (P/S; Lonza).

2.1 Detection of Gα Protein Activation by Measurement of Second Messengers

Recently, a number of nonradioactive methods to measure second messengers have been made commercially available that require short compound and agonist incubation times (Gabriel et al., 2003; Zhang & Xie, 2012). Among these methods, the Amplified Luminescent Proximity Homogeneous Assay (Alpha) Screen and the Homogenous Time-Resolved Fluorescence (HTRF) are the most high-throughput screening feasible techniques which share advantages of broad linear range, high signal-to-background, miniaturization, limited number of dispensing steps, and are particularly recommended for cells expressing low levels of 7TMs because of their higher sensitivities.

2.1.1 Required Materials

HEK293, HEK293T, or CHO-K1 cells lines can be purchased from various sources (purchased by American Type Culture Collection (ATCC)) and stably expressing CCR5 described previously (Borroni et al., 2013)

DMEM (Lonza; Cat. No. BE12-604F/U1)

DMEM/F12 (Lonza; Cat. No. BE04-687F/U1)

FBS (Sigma–Aldrich; Cat. No. F7524)

HEPES (Lonza; Cat. No. 17-737)
Penicillin/Streptomycin solution (10,000 U/ml) (P/S, Lonza; Cat. No. 17-602)
Versene (PBS + 0.5 m*M* EDTA, Gibco/Thermo Fisher Scientific; Cat. No. 15040-033)
AlphaScreen cAMP Detection Kit (Perkin Elmer; Cat. No. 6760635D)
PTX (Calbiochem/MerckMillipore; Cat. No. 516560)
Hanks Balanced Salt Solution (HBSS, Lonza; Cat. No. BE10-527F)
3-isobutyl-1-methylxanthine (IBMX, Sigma–Aldrich; Cat. No. I5879)
Bovine Serum Album (BSA, Sigma–Aldrich; Cat. No. A7030)
Forskolin (Sigma–Aldrich; Cat. No. F6886)
Recombinant human CCL3, CCL3L1, CXCL8 (R&D Systems; Cat. No. 270-LD, 509-MI and 208-IL, respectively)
IP-One Tb kit (Cisbio Bioassays; Cat. No. 62IPAPEB)
Polylysine (Sigma–Aldrich; Cat. No. P4707)
White opaque 384-well microplate (OptiPlate-384, PerkinElmer; Cat. No. 6007290)
96-well black plates (Greiner Bio-One; Cat. No. 655 077)

2.1.2 Measurement of Gαi Protein Activation by AlphaScreen Technology

AlphaScreen is a bead-based technology used to study biomolecular interactions in a microplate format. Binding of molecules captured on the beads leads to an energy transfer from one bead to the other, ultimately producing a luminescent signal. Among many available applications, AlphaScreen allows to quantify cAMP based on the energy transfer occurring inside a molecular complex consisting of streptavidin-coated donor beads (DB), biotinylated cAMP, and acceptor beads (AB) containing thioxene derivatives and coated with an antibody against cAMP. Sample irradiation at 680 nm induces the generation of singlet oxygen molecules from DB that diffuse within 200 nm and therefore interact only with AB in close proximity, resulting in a cascade of chemical reactions culminating in light production between 520 and 620 nm referred to as AlphaScreen signal (Eglen et al., 2008). The ability of the energy transfer to occur only when AB bind biotinylated cAMP and in turn streptavidin on DB can be exploited to measure the amount of endogenous cAMP as reduction of chemiluminescent emission due to the displacement of biotinylated cAMP from AB and the subsequent inhibition of streptavidin binding on DB. Differently from Gαs, Gαi proteins activation is measured following sample stimulation with forskolin and results in the inhibition of the forskolin-induced adenylyl cyclase activation and subsequent increased intracellular cAMP levels. We set up a protocol to quantify Gαi protein activation upon ligand stimulation in HEK293/CCR5 cells using the AlphaScreen cAMP Detection kit (Fig. 1).

1. 8×10^5 cells are seeded into P6 plate and 24 h after are treated with PTX (100 ng/ml) or vehicle in complete medium without P/S.
2. After 18 h, cells are washed twice with PBS, detached with Versene and resuspended at 10^7 cells/ml in stimulation buffer (StimB: HBSS containing 0.5 m*M* IBMX, 0.1% w/v BSA, and 5 m*M* HEPES, pH 7.4).
3. 5 μl of cells/anti-cAMP AB mix solution containing StimB (30%) + 10^7 vehicle- or PTX-treated cells/ml (20%, corresponding to 10^4 cells final) + anti-cAMP AB working solution (50%, prior 1 to 25 dilution of AB stock solution in StimB) are plated in triplicate on a white opaque 384-well microplate.
4. Chemokine serial dilutions (10^{-6} to 10^{-10} *M*) are prepared 2× concentrated in StimB containing 2× concentrated forskolin (20 μ*M*) and finally 5 μl of each dilution is added in triplicate to the wells. In addition, vehicle is performed in triplicate by adding to the cells 5 μl of StimB mixed 1 to 2 with 20 μ*M* forskolin.
5. Control points are performed in triplicate as follows: 5 μl cells/anti-cAMP AB mix with 5 μl StimB without forskolin and chemokine to

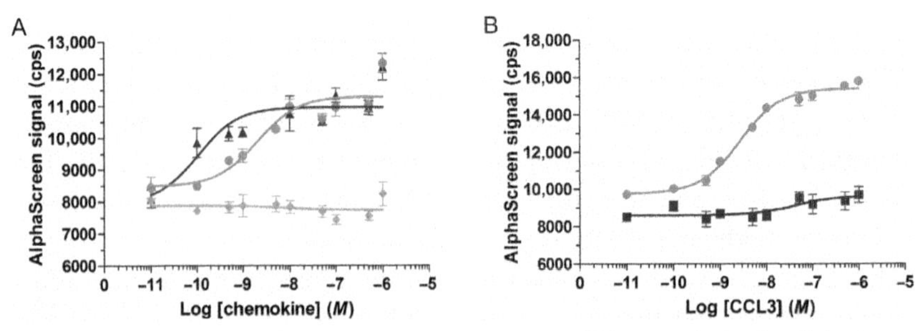

Figure 1 Quantification of Gαi protein activation upon CCR5 stimulation by AlphaScreen technology. HEK293/CCR5 cells are stimulated for 30 min with forskolin (10 μ*M*) and the indicated concentrations of (A) CCL3L1 (blue (dark gray in the print version), ▲), CCL3 (red (gray in the print version), •) or CXCL8 (green (light gray in the print version), ◆) or alternatively with (B) CCL3 (red (gray in the print version), •) after 18 h pretreatment with PTX (blue (dark gray in the print version), ■; 100 ng/ml). Graphs show the amount of AlphaScreen signal expressed as counts per seconds (cps). Results refer to one representative experiment performed in triplicate. Data are analyzed with nonlinear curve fitting equations and the Median Effective Concentration (EC_{50}) is calculated for each CCR5 ligand ($EC_{50_{CCL3L1}} = 0.187$ n*M*; $EC_{50_{CCL3}} = 2.03$ n*M*; and $EC_{50_{CXCL8}}$ = not determined).

measure the maximal recordable AlphaScreen signal in resting cells, 5 μl of a 1 to 2 dilution of anti-cAMP AB working solution in StimB with either 5 μl of StimB or 5 μl of cAMP (5 μ*M*, prior 1 to 10 dilution of 50 μ*M* cAMP stock solution in StimB), to detect the maximal and the minimal AlphaScreen signal, respectively.

6. Plate is incubated 30 min at room temperature (RT) in the dark.
7. Streptavidin DB/biotinylated cAMP detection mix is prepared in lysis/detection buffer (prior 1 to 10 dilution of 10 × stock solution in MilliQ H_2O) by 1 to 150 dilution of the Streptavidin DB stock and 1 to 24 dilution of the biotinylated cAMP solution (1 μ*M*, prior 1 to 10 dilution of 10 μ*M* biotinylated cAMP stock solution in PBS). This solution has to be prepared 30 min before use and incubated at RT in the dark to allow biotinylated cAMP binding to streptavidin DB.
8. Chemokine stimulation is blocked by adding to each well 15 μl of Streptavidin DB/biotinylated cAMP detection mix to each well. Plate is incubated for at least 1 h at RT in the dark.
9. The measurement of AlphaScreen signal is performed by a Synergy H4 Microplate Reader (BioTek) equipped with the following filter setting: excitation is provided by a tungsten lamp through a 680/30 nm filter placed on the excitation wheel and plug on emission wheel, whereas emission arising from the sample is subsequently acquired after filter switching through a 570/100 nm filter placed on the emission wheel with a plug on excitation wheel. Dose–response curves can be generated plotting the AlphaScreen signal measured against agonist concentration expressed as logarithm of molarity and analyzed with nonlinear curve fitting equations (GraphPad Prism 5 software). Statistics can be obtained by ANOVA and Bonferroni post-hoc test.

2.1.3 Measurement of Gαq Protein Activation by HTRF Technology

HTRF technique combines standard FRET technology, which is based on the transfer of energy between two fluorophores in close proximity (donor: europium/terbium cryptate (Em: 620 nm); acceptor: d2/XL665 (Em: 665 nm), with time-resolved measurement of fluorescence, thus allowing elimination of short-lived background fluorescence (Degorce et al., 2009; Norskov-Lauritsen, Thomsen, & Brauner-Osborne, 2014). Differently from AlphaScreen, HTRF provides assay to measure inositol-1-phosphate (IP1) and it is particularly suited to quantify Gαq stimulation, as it is known to induce phospholipase C activation and triggers the inositol cascade following receptor activation. IP1, a downstream metabolite of IP3, accumulates in cells and is stable in the presence of lithium chloride (LiCl), making it ideal for Gαq functional assays. To measure Gαq protein activation, HTRF assays have been developed for the direct quantitative determination of IP1 based on a competitive immunoassay that uses terbium cryptate-labeled anti-IP1 monoclonal antibody and d2-labeled IP1. We set up a protocol to quantify Gαq protein activation upon ligand stimulation in CHO-K1/CCR5 cells using IP-One Tb cell-based assay in adherent conditions described as follows (Borroni et al., 2013).

1. Cells are seeded in 100 μl of complete DMEM/F12 at a density of 5×10^4 into 96-well black plates previously coated with 100 μl/well of polylysine for 30 min at 37 °C.
2. After 18 h at 37 °C, cells are washed twice with 100 μl/well of stimulation buffer (StimB) provided by the kit and 35 μl of StimB are finally added to each well containing 200 n*M* of chemokine (Borroni et al., 2013) or PBS as vehicle in triplicate. 35 μl of StimB alone is added in duplicate to perform negative control. Meanwhile, IP1 standard curve (average range of 0.011–11 μ*M*) is prepared according to manufacturer's instructions and added to cell-free wells in duplicate.
3. After 1 h of incubation at 37°C, HTRF reagents (1st distribution: IP1-d2 then 2nd distribution: Ab-cryptate) are diluted at indicated working concentration in the lysis buffer provided by the kit and 15 μl are added to each wells (samples and standards) except in the negative control wells where only 15 μl of lysis buffer instead of IP1-d2 are added. Plate is covered by lid and incubated for 1 h at RT.
4. Plate is read on the compatible HTRF reader Mithras LB 940 (Berthold Technologies). Results are calculated from the 665/620 nm ratio*10^4 (*R*) and expressed in Delta *F*% obtained by the following equation: (standard or sample $R-$negative *R*)/(negative *R*)*100. Standard curve is drawn up by plotting Delta *F*% versus IP1 concentration and Delta *F*% obtained for samples can be reported on the standard curve to deduce respective IP1 concentrations. Statistical significant differences between cells stimulated with the agonist and vehicle is evaluated using Student's *t*-test (GraphPad Prism 5 software).

2.2 Detection of Gα Protein Activation by G Proteins Conformational Changes

The bioluminescent resonance energy transfer (BRET) technique is based on the energy transfer between a luciferase enzyme acting as donor and a fluorescent protein acting as acceptor (Pfleger & Eidne, 2006). The classical

BRET assay ($BRET^1$) uses as donor the *Renilla Reniformis* luciferase (Rluc) that catalyzes the transformation of the substrate Coelenterazine h into coelenteramide with concomitant light emission, peaking at 480 nm. The energy emitted is absorbed by the acceptor, a variant of the jellyfish *Aequorea Victoria* Green Fluorescent Protein (YFP or Venus), and results in an emission of light at 530 nm. Energy transfer efficiency, referred to as BRET signal, is determined ratiometrically and reflects the proximity between the donor and the acceptor. $BRET^1$ is characterized by a poor spectral resolution (50 nm), with a partial overlap of the Rluc on the YFP emission spectra that contributes to increase the YFP background and to decrease the signal to background ratio. The new generation $BRET^2$ uses GFP^2/GFP^{10}(Ex: 400 nm; Em: 511 nm) as acceptor and a modified Coelenterazine h substrate, DeepBlue C, which shifts the Rluc peak emission from 480 to 400 nm, thus providing a broader spectral resolution (115 nm) and improved quality of the detection (Bertrand et al., 2002). Of note, as DeepBlue C has 100-fold lower quantum yield and decays threefold faster compared to Coelenterazine h, $BRET^2$ assays should be preferred when subtle changes in the BRET signal are investigated due to their higher signal-to-background ratio, whereas $BRET^1$ assays are more indicated for real-time long kinetics measurements (Hamdan, Audet, Garneau, Pelletier, & Bouvier, 2005).

Taking advantage of $BRET^2$ properties, G proteins BRET biosensors have been developed to quantify biased signaling in 7TMs instead of the classical GTPγS-binding assays (Bidlack & Parkhill, 2004; Strange, 2010), as they represent a dynamic model to detect early signaling events that occur shortly after ligand stimulation and also enable the identification of G protein subtypes activated by the receptors (Namkung et al., 2015). Recently, these biosensors have been used to test the ability of chemokine receptors to activate specific G protein subunits (Corbisier et al., 2015). They have been generated by Galés and Bouvier's groups (Gales et al., 2006; Sauliere et al., 2012) through Rluc insertion within the Gα subunit amino acid sequence and GFP^{10} fusion at Gγ2 subunit N-terminal. When the receptor is activated, G protein heterotrimeric complex undergoes a conformational rearrangement upon GDP/GTP exchange, which determines a lower BRET energy transfer from the donor to the acceptor as a consequence of Gα subunit separation from Gγ. Here, we provide a brief description of the protocol set up and published by Galés and collaborators (Gales et al., 2006; Sauliere et al., 2012) optimized by themselves to quantify Gα subtypes activation upon CCR5 stimulation in HEK293T cells (Corbisier et al., 2015).

2.2.1 Required Materials

G protein biosensors-encoding plasmids (Gα-Rluc, GFP^{10}-Gγ2, and Gβ1 subunits; Corbisier et al., 2015)
hCCR5-encoding pcDNA3 (Corbisier et al., 2015)
Versene (Gibco/Thermo Fisher Scientific; Cat. No. 15040-033)
Dulbecco's phosphate-buffered saline (PBS, Gibco/Thermo Fisher Scientific; Cat. No. 14190-144)
BRET assay buffer (PBS, pH 7.4 + 0.5 m*M* $MgCl_2$ + 0.1% (w/v) glucose)
DC Protein Assay kit (Bio-Rad; Cat. No. 500-0112)
Coelenterazine 400a/DeepBlue C (Biotium; Cat. No. 10125)
Black/White Isoplate-96 Black Frame White Well (Perkin Elmer; Cat. No. 6005030)

2.2.2 Detection of Gα Protein Activation

1. HEK293T cells are transiently transfected by calcium phosphate method with plasmids encoding the G protein biosensors and CCR5-pcDNA3.
2. 48 h after transfection, cells are washed twice with PBS, detached with Versene, and resuspended in BRET assay buffer (PBS, pH 7.4 + 0.5 m*M* $MgCl_2$ + 0.1% (w/v) glucose).
3. Cells are quantified as protein content with DC Protein Assay, as per manufacturer's protocol, and 80 μg of proteins/well are distributed in a 96-wells Black/White Isoplate.
4. Cells are incubated with 100 n*M* of chemokine (see Section 2.1.1), previously diluted in PBS, or PBS as vehicle.
5. After 1 min of agonist stimulation, Coelenterazine 400a (5 μ*M*) is added to each well and immediately $BRET^2$ readings are collected using an Infinite F200 reader (Tecan), which allows the sequential integration of signals detected with Rluc filter (370–450 nm) and GFP^{10} filter (510–540 nm). $BRET^2$ ratio is calculated as follows: $EmGFP^{10}/EmRluc$. Changes in BRET induced by the ligands are expressed on graphs as Δ BRET using the formula: $\left(EmGFP^{10}{}_{ligand}/EmRluc_{ligand}\right) - \left(EmGFP^{10}{}_{PBS}/EmRluc_{PBS}\right)$ and reflect the amount of G protein activation. Statistical significant differences between cells stimulated with the agonist and vehicle is evaluated using one-way analysis of variance with Dunnet's post-hoc test or where appropriate using Student's *t*-test (GraphPad Prism 5 software).

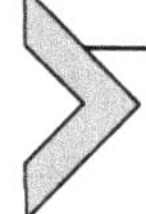

3. β-ARRESTINS SIGNALING: DETECTION OF β-ARRESTINS RECRUITMENT BY PROTEIN CONFORMATIONAL CHANGES

Activated chemokine receptors also engage a parallel set of regulatory mechanisms initiated by recruitment of kinases which phosphorylate agonist-occupied receptors at key residues in the C-terminal domain (Pitcher, Freedman, & Lefkowitz, 1998). Following phosphorylation, the receptor associates at plasma membrane with β-arrestins, scaffolding proteins expressed as two isoforms (β-arrestin1 and β-arrestin2) which not only sterically impede the receptor from further coupling to G proteins and promote receptor internalization/desensitization, but also function as signalosome by recruiting other signaling proteins to the receptor involved in controlling several cellular functions (DeWire, Ahn, Lefkowitz, & Shenoy, 2007). In this section, we are providing procedures for detecting β-arrestins recruitment by protein conformational changes through BRET.

3.1 Required Materials

Refer to Sections 2.1.1 and 2.2.1, with the addition of:

Plasmids encoding β-arrestin1 or 2-YFP or GFP^2, generated by cloning β-arrestins cDNA in pEYFP-N1 (Clontech; Cat. No. 6006-1) or $pGFP^2$-N3 (Biosignal Packard/Perkin Elmer; Cat. No. 6310240) with the PPVAT amino acid linker.

Plasmid encoding hCCR5 cloned in pRluc-N3 (CCR5-Rluc, Biosignal Packard/Perkin Elmer; Cat. No. 6310220), fusing CCR5 C-terminus to Rluc with the amino acid linker GDPRVPVAT.

Lipofectamine 2000 (Invitrogen/Thermo Fisher Scientific; Cat. No. 11668-019)

Coelenterazine h (Biotium; Cat. No. 10111)

3.2 Measurement of Ligand-Dependent β-Arrestin Recruitment

The ligand-induced translocation of β-arrestins to the 7TM C-terminal tail and the technical advantages of the BRET technology applied to the evaluation of protein–protein interactions have made the measurement of β-arrestins recruitment to 7TMs by BRET, an approach complementary to conventional biochemical and imaging techniques (Ferguson & Caron, 2004) frequently used in desensitization studies of several 7TMs (Namkung et al., 2015), including chemokine receptors (Corbisier et al., 2015; Hamdan et al., 2005). In this assay, the receptor is modified at the C-terminal tail with the donor Rluc and coexpressed with a fusion protein between Venus/YFP and β-arrestins (β-arrestin1 or β-arrestin2). Variations in the BRET signal can be monitored kinetically or in concentration-dependent manner after ligand addition. Here, we provide a protocol to measure β-arrestins recruitment to CCR5 upon ligand stimulation in HEK293T cells by a BRET[1] assay (Fig. 2).

1. Cells are seeded in 10 ml of complete DMEM at a density of 3×10^6 into 100-mm tissue culture dish and after 18 h at 37 °C, cells are transiently transfected with Lipofectamine 2000 following manufacturer's instruction, with plasmids encoding CCR5-Rluc and β-arrestin1 or 2-YFP.
2. 24 h after transfection, cells are treated with PTX (100 ng/ml) or vehicle in complete medium without P/S.
3. After 18 h, cells are washed twice with PBS, detached with Versene, and resuspended in BRET assay buffer.
4. Cells are quantified as protein content by DC Protein Assay and 80 μg of proteins/well are distributed in a 96-wells Black/White Isoplate.

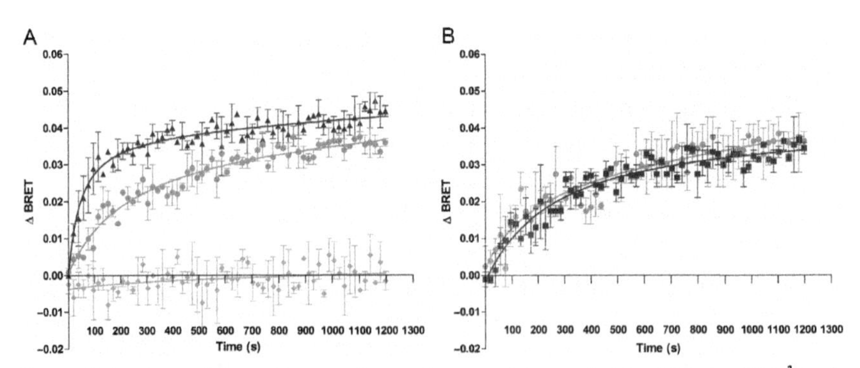

Figure 2 Analysis of ligand-induced recruitment of β-arrestin2 to CCR5 by BRET[1] technique. Real time measurement of BRET signal in HEK293T cells expressing CCR5-Rluc and β-arrestin2-YFP and stimulated for 20 min with 100 n*M* of (A) CCL3L1 (blue (dark gray in the print version), ▲), CCL3 (red (gray in the print version), •) or CXCL8 (green (light gray in the print version), ◆) or alternatively with (B) CCL3 (red (gray in the print version), •) after 18 h pretreatment with PTX (blue (dark gray in the print version), ■; 100 ng/ml). Results are expressed as Δ BRET, corresponding to the difference in BRET ratio between stimulated and PBS-treated cells. Data represent the means ± SEM of two independent experiments.

5. Cells are incubated with 5 μ*M* of Coelenterazine h for 8 min before stimulation with 100 n*M* of chemokines or PBS as vehicle. All reagents are previously diluted in PBS. Alternatively, dose–response curves can be generated by incubating cells with various concentrations of chemokines and after 20 min Coelenterazine h is added for 8 additional min.
6. $BRET^1$ signal between Rluc and YFP is measured using a Synergy H4 reader (BioTek) that allows the sequential integration of signals detected with Rluc filter (440/40 nm) and YFP filter (540/35 nm). Calculate $BRET^1$ ratio as follows: Em_{YFP}/Em_{Rluc}. The changes in BRET induced by the ligands reflect the amount of β-arrestins recruitment to the receptor. These are monitored in real-time up to 20 min and expressed on graphs as Δ BRET using the formula: $\left(EmYFP_{ligand}/EmRluc_{ligand}\right) - \left(EmYFP_{PBS}/EmRluc_{PBS}\right)$. Kinetic curves can be generated using nonlinear regression fitting equations and BRET max and half/time parameters are determined using GraphPad Prism 5 software.

3.3 Measurement of Constitutive β-Arrestin Recruitment

Recently, several evidences suggested that some 7TMs, including chemokine receptors (Gilliland, Salanga, Kawamura, Trejo, & Handel, 2013), are able to adopt active conformations also in the absence of agonist binding. This results in constitutive G protein activation and/or β-arrestins recruitment to the receptor, leading to the initiation of signal transduction pathways and receptor down-modulation (Seifert & Wenzel-Seifert, 2002). The functional relevance of chemokine receptors constitutive activity deserves deeper investigation, especially taking into account its impact on pharmacology. Taking advantage of $BRET^2$ properties, here we provide a protocol to evaluate constitutive recruitment of β-arrestins to CCR5 in HEK293T cells by donor saturation assay.

1. Cells are seeded in 1 ml of complete DMEM at a density of 4×10^5 into P12 plate.
2. After 18 h at 37 °C, cells are transiently transfected with 200 ng of CCR5-Rluc and 10 serial 1 to 2 dilutions starting from 1 μg of empty $pGFP^2$ vector, β-arrestin1-GFP^2 and β-arrestin2-GFP^2, comprising of a blank sample transfected with donor plasmid only. To each transfection sufficient amount of pcDNA3 or any empty vector is added to bring the final DNA amount to 1.2 μg.
3. 48 h after transfection, cells are washed twice with PBS, detached with Versene, and resuspended in BRET assay buffer.

4. Cells are quantified as protein content by DC Protein Assay, resuspended at 1 mg/ml and 80 μg of proteins/well are distributed in a 96-wells Black/White Isoplate.
5. Using a Tecan F500 reader, fluorescence is measured through filters with excitation peak at 340/35 nm and excitation at 535/25 nm and refers to the amount of GFP^2(β-arrestins) transfected. Subsequently, add Coelenterazine h to the final 5 μ*M* concentration, incubate 8 min in the dark, and detect luminescence to obtain luciferase quantification.
6. 80 μg of cells are plated in each well, Coelenterazine 400a (5 μ*M*) is added and immediately $BRET^2$ readings are collected using the same reader. $BRET^2$ ratio is calculated as previously described (see Section 2.2).
7. Background fluorescence measured from the blank sample is subtracted from all the samples, and fluorescence of each transfectant is normalized over its own luminescence to obtain a fluorescence/luminescence ratio (acceptor/donor). The measured $BRET^2$ is plotted as a function of the acceptor/donor ratio for each donor-acceptor couple, resulting in a linear trend in the case of nonspecific interactions and in a hyperbolic curve indicating specific association of the donor molecule with the acceptor.

Several observations clearly indicate that β-arrestins can adopt multiple different conformations that are dependent on the ligand that stimulates the 7TM (Shukla et al., 2008). In order to provide new insights into the structural rearrangements incurred by β-arrestins upon its recruitment to the engaged receptor, Charest and collaborators developed a protocol based on a single molecule BRET biosensor termed double brilliance, in which β-arrestin2 is tagged with Rluc and YFP at the N- and C-terminus, respectively, and extended it also to monitor β-arrestins rearrangements upon chemokine receptors stimulation (Charest, Terrillon, & Bouvier, 2005). Finally, thanks to these recent advances in physics, an improved $BRET^1$ strategy called imaging or visual BRET, has been developed by Perroy and collaborators to image and quantify dynamics of protein–protein interactions, including 7TM and β-arrestins, at the subcellular level (nucleus, plasma membrane, endocytic vesicles) in the tens-of-seconds to tens-of-minutes time frame (Coulon et al., 2008; Perroy, 2010).

4. CALCULATION OF BIASED SIGNALING

The ability to experimentally quantify functional selectivity has required the development of methods to calculate bias (Kenakin &

Christopoulos, 2013), which have been also recently adapted to the chemokine receptor system (Zweemer et al., 2014). In general, bias can be visualized if two cellular responses (referring to pathway A and B) to a defined ligand are plotted as functions of each other: that is, the amount of signal produced in B is plotted as a function of signal produced in A in response to equivalent concentrations of a given ligand (Kenakin & Miller, 2010). This results in a *bias plot* as shown in Fig. 3A. Unless it results in a straight line, the bias plot indicates that the ligand produces a relatively more efficient response to one pathway compared with the other, and is useful for graphically expressing the differential activation of two pathways by the same ligand. However, it is optimal to quantify bias in terms of a single parameter, as this allows the application of statistical methods to assess true differences in ligand activity. Using potency (EC_{50}) to characterize and quantify ligand activity is inadequate for ligands that produce different maximal responses (efficacy, E_{max}), such as partial agonists. Conversely, using only efficacy fails to differentiate between ligands that behave as full agonists, which can produce a stimulus that exceeds the signaling capabilities of the system, yielding for all ligands a uniform maximal response (Rajagopal et al., 2011). Concerning CCR5, two methods have been recently proposed to obtain a single parameter that incorporate the minimal elements to describe ligand activity

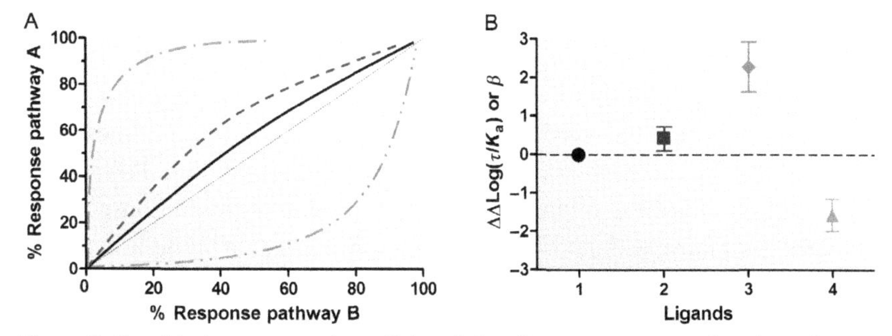

Figure 3 Graphical representation of biased signaling. Responses to four ligand (numbered 1–4) in two assay systems (defined pathway A and B) are shown as graphs representing (A) bias plot and (B) bias factor. Graphs show the response in assay A as a function of the corresponding response in assay B for all ligands compared to ligand 1 (black, •) that is chosen as the reference ligand. Ligand 2 (blue (dark gray in the print version), — —, ■) produces comparable responses both in assay A and B, indicating a balanced response. Ligand 3 (red (gray in the print version), – ▪ –, ◆) produces considerably greater responses in assay A for defined responses in assay B, indicating a bias toward response A. Ligand 4 (green (light gray in the print version), – ▪▪ –, ▲) produces a considerably lower response in assay A as a function of its responses in assay B, indicating a bias toward response B.

and generate *bias factor* as shown in Fig. 3B. In the first method (Kenakin, Watson, Muniz-Medina, Christopoulos, & Novick, 2012; van der Westhuizen, Breton, Christopoulos, & Bouvier, 2014), the single parameter is the Transduction Coefficient (TC) $=\log(\tau/K_A)$, where the term τ incorporates ligand efficacy, receptor density, and coupling within the system, and the dissociation constant K_A is the reciprocal of the conditional affinity of the ligand in the functional system. TC value is calculated for each ligand by fitting dose–response curves obtained for pathway A and B with the Black-Leff operational model (Black & Leff, 1983). TC values of tested ligands undergo the subtraction of the one generated by the reference ligand chosen arbitrarily (usually a balanced ligand) to obtain a relative effectiveness of the ligand for each signaling pathway $[\Delta\log(\tau/K_a)=\log(\tau/K_a)_{\text{ligand}}-\log(\tau/K_a)_{\text{reference}}]$ that can be subsequently compared across the signaling pathways analyzed $[\Delta\Delta\log(\tau/K_a)=\Delta\log(\tau/K_a)_A-\Delta\log(\tau/K_a)_B]$. Statistical analysis is performed using a two-way unpaired Student's *t*-test on the $\Delta\log(\tau/K_a)$ to make pairwise comparisons between two pathways activated by a given ligand, where $p<0.05$ is considered statistically significant. This method allows to calculate bias even if the affinity between ligand, receptor, and signaling complexes changes between assay systems, but unfortunately it does not fit well in the cases of weak partial agonists. In the second method (Corbisier et al., 2015; Ehlert, 2005), the single parameter is the Bias Index $(\beta)=\log(\text{RA}_{\text{AB,ligand}}/\text{RA}_{\text{AB,reference}})$, where Relative Activity (RA) values $=E_{max}/\text{EC}_{50}$ denotes the relative efficacy of a test ligand through pathway A and B relative to a reference ligand, allowing to estimate bias in conditions where ligands display different maximal potencies, in example of partial agonists activity or allosteric modulation of signaling. RA value is calculated for test and reference ligands by fitting dose–response curves obtained for pathway A and B with nonlinear regression using a sigmoidal dose–response model (GraphPad Prism software), yielding EC_{50} and E_{max} which are used as follow: $\text{RA}_{\text{AB}}=(E\text{max}_A/E_{50_A})(E_{50_B}/E\text{max}_B)$. Statistical analysis is performed using one-way ANOVA analysis of variance with Dunnett's post-hoc test, and statistical significance related to the reference ligand is taken as $p<0.05$.

5. SUMMARY

Chemokine receptors have then long been considered to function as "light switches," being either off/inactive or on/activated after chemokine engagement. In this scenario, chemokines were considered as only being able to always activate both G proteins and β-arrestins signaling pathways of the receptor in a balanced fashion, and thus pharmacological selectivity of given chemokine receptor inhibitor was only defined by its ability to block one or another specific receptor, but not to modulate its specific signaling properties. Conversely, as consequence of the redundancy intrinsic of the chemokine system, evidences are accumulating showing that chemokine receptors are capable of much more than simple binary signals, being now considered able to occupy a number of discrete conformations, each potentially associated with distinct signaling mechanisms (Zweemer et al., 2014). This biased signaling, also called functional selectivity, refers to the ability of different chemokine to activate a certain intracellular signaling pathway over another on a given receptor (Rajagopal et al., 2013) and, up to now, has not been taken into account upon developing chemokine receptors small molecule antagonists and have been put forward to explain the failure to develop chemokine receptor-targeted medicines in inflammation (Schall & Proudfoot, 2011). Based on these observations, the identification and pharmacological exploitation of biased agonists is presently considered a promising approach to move from receptor-selective to signaling pathway-selective therapies in the chemokine field (Whalen, Rajagopal, & Lefkowitz, 2011). Therefore, the efforts in developing increasingly accurate approaches aimed at fine-dissecting G proteins and β-arrestins signaling pathways evoked upon chemokine engagement are required to improve understanding of chemokine receptors signaling mechanisms in order to provide a framework for more effective drug discovery that have selective efficacy and fewer side effects.

REFERENCES

Bachelerie, F., Ben-Baruch, A., Burkhardt, A. M., Combadiere, C., Farber, J. M., Graham, G. J., et al. (2014). International Union of Basic and Clinical Pharmacology. [corrected]. LXXXIX. Update on the extended family of chemokine receptors and introducing a new nomenclature for atypical chemokine receptors. *Pharmacological Reviews*, *66*(1), 1–79.

Berchiche, Y. A., Gravel, S., Pelletier, M. E., St-Onge, G., & Heveker, N. (2011). Different effects of the different natural CC chemokine receptor 2b ligands on beta-arrestin recruitment, Gαi signaling, and receptor internalization. *Molecular Pharmacology*, *79*(3), 488–498.

Bertrand, L., Parent, S., Caron, M., Legault, M., Joly, E., Angers, S., et al. (2002). The BRET2/arrestin assay in stable recombinant cells: A platform to screen for compounds that interact with G protein-coupled receptors (GPCRS). *Journal of Receptor and Signal Transduction Research*, *22*(1–4), 533–541.

Bidlack, J. M., & Parkhill, A. L. (2004). Assay of G protein-coupled receptor activation of G proteins in native cell membranes using [35S]GTP gamma S binding. *Methods in Molecular Biology*, *237*, 135–143.

Black, J. W., & Leff, P. (1983). Operational models of pharmacological agonism. *Proceedings of the Royal Society of London B: Biological Sciences*, *220*(1219), 141–162.
Bonecchi, R., Galliera, E., Borroni, E. M., Corsi, M. M., Locati, M., & Mantovani, A. (2009). Chemokines and chemokine receptors: An overview. *Frontiers in Bioscience (Landmark Ed)*, *14*, 540–551.
Bonecchi, R., Savino, B., Borroni, E. M., Mantovani, A., & Locati, M. (2010). Chemokine decoy receptors: Structure-function and biological properties. *Current Topics in Microbiology and Immunology*, *341*, 15–36.
Borroni, E. M., Cancellieri, C., Vacchini, A., Benureau, Y., Lagane, B., Bachelerie, F., et al. (2013). β-arrestin-dependent activation of the cofilin pathway is required for the scavenging activity of the atypical chemokine receptor D6. *Science Signaling*, *6*(273), ra30 31-11, S31-33.
Byers, M. A., Calloway, P. A., Shannon, L., Cunningham, H. D., Smith, S., Li, F., et al. (2008). Arrestin 3 mediates endocytosis of CCR7 following ligation of CCL19 but not CCL21. *The Journal of Immunology*, *181*(7), 4723–4732.
Cancellieri, C., Vacchini, A., Locati, M., Bonecchi, R., & Borroni, E. M. (2013). Atypical chemokine receptors: From silence to sound. *Biochemical Society Transactions*, *41*(1), 231–236.
Carman, C. V., & Benovic, J. L. (1998). G-protein-coupled receptors: Turn-ons and turn-offs. *Current Opinion in Neurobiology*, *8*(3), 335–344.
Charest, P. G., Terrillon, S., & Bouvier, M. (2005). Monitoring agonist-promoted conformational changes of beta-arrestin in living cells by intramolecular BRET. *EMBO Reports*, *6*(4), 334–340.
Charo, I. F., & Ransohoff, R. M. (2006). The many roles of chemokines and chemokine receptors in inflammation. *The New England Journal of Medicine*, *354*(6), 610–621.
Corbisier, J., Gales, C., Huszagh, A., Parmentier, M., & Springael, J. Y. (2015). Biased signaling at chemokine receptors. *The Journal of Biological Chemistry*, *290*(15), 9542–9554.
Coulon, V., Audet, M., Homburger, V., Bockaert, J., Fagni, L., Bouvier, M., et al. (2008). Subcellular imaging of dynamic protein interactions by bioluminescence resonance energy transfer. *Biophysical Journal*, *94*(3), 1001–1009.
Degorce, F., Card, A., Soh, S., Trinquet, E., Knapik, G. P., & Xie, B. (2009). HTRF: A technology tailored for drug discovery—A review of theoretical aspects and recent applications. *Current Chemical Genomics*, *3*, 22–32.
DeWire, S. M., Ahn, S., Lefkowitz, R. J., & Shenoy, S. K. (2007). Beta-arrestins and cell signaling. *Annual Review of Physiology*, *69*, 483–510.
Eglen, R. M., Reisine, T., Roby, P., Rouleau, N., Illy, C., Bosse, R., et al. (2008). The use of AlphaScreen technology in HTS: Current status. *Current Chemical Genomics*, *1*, 2–10.
Ehlert, F. J. (2005). Analysis of allosterism in functional assays. *The Journal of Pharmacology and Experimental Therapeutics*, *315*(2), 740–754.
Ferguson, S. S., & Caron, M. G. (2004). Green fluorescent protein-tagged beta-arrestin translocation as a measure of G protein-coupled receptor activation. *Methods in Molecular Biology*, *237*, 121–126.
Gabriel, D., Vernier, M., Pfeifer, M. J., Dasen, B., Tenaillon, L., & Bouhelal, R. (2003). High throughput screening technologies for direct cyclic AMP measurement. *Assay and Drug Development Technologies*, *1*(2), 291–303.
Gales, C., Van Durm, J. J., Schaak, S., Pontier, S., Percherancier, Y., Audet, M., et al. (2006). Probing the activation-promoted structural rearrangements in preassembled receptor-G protein complexes. *Nature Structural & Molecular Biology*, *13*(9), 778–786.
Gilliland, C. T., Salanga, C. L., Kawamura, T., Trejo, J., & Handel, T. M. (2013). The chemokine receptor CCR1 is constitutively active, which leads to G protein-independent, beta-arrestin-mediated internalization. *The Journal of Biological Chemistry*, *288*(45), 32194–32210.

Hamdan, F. F., Audet, M., Garneau, P., Pelletier, J., & Bouvier, M. (2005). High-throughput screening of G protein-coupled receptor antagonists using a bioluminescence resonance energy transfer 1-based beta-arrestin2 recruitment assay. *Journal of Biomolecular Screening*, *10*(5), 463–475.

Kenakin, T., & Christopoulos, A. (2013). Signalling bias in new drug discovery: Detection, quantification and therapeutic impact. *Nature Reviews. Drug Discovery*, *12*(3), 205–216.

Kenakin, T., & Miller, L. J. (2010). Seven transmembrane receptors as shapeshifting proteins: The impact of allosteric modulation and functional selectivity on new drug discovery. *Pharmacological Reviews*, *62*(2), 265–304.

Kenakin, T., Watson, C., Muniz-Medina, V., Christopoulos, A., & Novick, S. (2012). A simple method for quantifying functional selectivity and agonist bias. *ACS Chemical Neuroscience*, *3*(3), 193–203.

Kohout, T. A., Nicholas, S. L., Perry, S. J., Reinhart, G., Junger, S., & Struthers, R. S. (2004). Differential desensitization, receptor phosphorylation, beta-arrestin recruitment, and ERK1/2 activation by the two endogenous ligands for the CC chemokine receptor 7. *The Journal of Biological Chemistry*, *279*(22), 23214–23222.

Lefkowitz, R. J. (1998). G protein-coupled receptors. III. New roles for receptor kinases and beta-arrestins in receptor signaling and desensitization. *The Journal of Biological Chemistry*, *273*(30), 18677–18680.

Namkung, Y., Radresa, O., Armando, S., Devost, D., Beautrait, A., Le Gouill, C., et al. (2015). Quantifying biased signaling in GPCRs using BRET-based biosensors. *Methods*. http://dx.doi.org/10.1016/j.ymeth.2015.04.010. [Epub ahead of print].

Nandagopal, S., Wu, D., & Lin, F. (2011). Combinatorial guidance by CCR7 ligands for T lymphocytes migration in co-existing chemokine fields. *PLoS One*, *6*(3), e18183.

Norskov-Lauritsen, L., Thomsen, A. R., & Brauner-Osborne, H. (2014). G protein-coupled receptor signaling analysis using homogenous time-resolved Forster resonance energy transfer (HTRF(R)) technology. *International Journal of Molecular Sciences*, *15*(2), 2554–2572.

Perroy, J. (2010). Subcellular dynamic imaging of protein-protein interactions in live cells by bioluminescence resonance energy transfer. *Methods in Molecular Biology*, *591*, 325–333.

Pfleger, K. D., & Eidne, K. A. (2006). Illuminating insights into protein-protein interactions using bioluminescence resonance energy transfer (BRET). *Nature Methods*, *3*(3), 165–174.

Pitcher, J. A., Freedman, N. J., & Lefkowitz, R. J. (1998). G protein-coupled receptor kinases. *Annual Review of Biochemistry*, *67*, 653–692.

Rajagopal, S., Ahn, S., Rominger, D. H., Gowen-MacDonald, W., Lam, C. M., Dewire, S. M., et al. (2011). Quantifying ligand bias at seven-transmembrane receptors. *Molecular Pharmacology*, *80*(3), 367–377.

Rajagopal, S., Bassoni, D. L., Campbell, J. J., Gerard, N. P., Gerard, C., & Wehrman, T. S. (2013). Biased agonism as a mechanism for differential signaling by chemokine receptors. *The Journal of Biological Chemistry*, *288*(49), 35039–35048.

Ricart, B. G., John, B., Lee, D., Hunter, C. A., & Hammer, D. A. (2011). Dendritic cells distinguish individual chemokine signals through CCR7 and CXCR4. *The Journal of Immunology*, *186*(1), 53–61.

Sauliere, A., Bellot, M., Paris, H., Denis, C., Finana, F., Hansen, J. T., et al. (2012). Deciphering biased-agonism complexity reveals a new active AT1 receptor entity. *Nature Chemical Biology*, *8*(7), 622–630.

Schall, T. J., & Proudfoot, A. E. (2011). Overcoming hurdles in developing successful drugs targeting chemokine receptors. *Nature Reviews. Immunology*, *11*(5), 355–363.

Seifert, R., & Wenzel-Seifert, K. (2002). Constitutive activity of G-protein-coupled receptors: Cause of disease and common property of wild-type receptors. *Naunyn-Schmiedeberg's Archives of Pharmacology*, *366*(5), 381–416.

Shi, G., Partida-Sanchez, S., Misra, R. S., Tighe, M., Borchers, M. T., Lee, J. J., et al. (2007). Identification of an alternative G{alpha}q-dependent chemokine receptor signal transduction pathway in dendritic cells and granulocytes. *The Journal of Experimental Medicine, 204*(11), 2705–2718.

Shukla, A. K., Violin, J. D., Whalen, E. J., Gesty-Palmer, D., Shenoy, S. K., & Lefkowitz, R. J. (2008). Distinct conformational changes in beta-arrestin report biased agonism at seven-transmembrane receptors. *Proceedings of the National Academy of Sciences of the United States of America, 105*(29), 9988–9993.

Steen, A., Larsen, O., Thiele, S., & Rosenkilde, M. M. (2014). Biased and g protein-independent signaling of chemokine receptors. *Frontiers in Immunology, 5*, 277.

Strange, P. G. (2010). Use of the GTPγS ([35S]GTPγS and Eu-GTPγS) binding assay for analysis of ligand potency and efficacy at G protein-coupled receptors. *British Journal of Pharmacology, 161*(6), 1238–1249.

Tian, Y., New, D. C., Yung, L. Y., Allen, R. A., Slocombe, P. M., Twomey, B. M., et al. (2004). Differential chemokine activation of CC chemokine receptor 1-regulated pathways: Ligand selective activation of Galpha 14-coupled pathways. *European Journal of Immunology, 34*(3), 785–795.

van der Westhuizen, E. T., Breton, B., Christopoulos, A., & Bouvier, M. (2014). Quantification of ligand bias for clinically relevant beta2-adrenergic receptor ligands: Implications for drug taxonomy. *Molecular Pharmacology, 85*(3), 492–509.

Watts, A. O., Scholten, D. J., Heitman, L. H., Vischer, H. F., & Leurs, R. (2012). Label-free impedance responses of endogenous and synthetic chemokine receptor CXCR3 agonists correlate with Gi-protein pathway activation. *Biochemical and Biophysical Research Communications, 419*(2), 412–418.

Whalen, E. J., Rajagopal, S., & Lefkowitz, R. J. (2011). Therapeutic potential of beta-arrestin- and G protein-biased agonists. *Trends in Molecular Medicine, 17*(3), 126–139.

Zhang, R., & Xie, X. (2012). Tools for GPCR drug discovery. *Acta Pharmacologica Sinica, 33*(3), 372–384.

Zidar, D. A. (2011). Endogenous ligand bias by chemokines: Implications at the front lines of infection and leukocyte trafficking. *Endocrine, Metabolic & Immune Disorders: Drug Targets, 11*(2), 120–131.

Zlotnik, A., & Yoshie, O. (2012). The chemokine superfamily revisited. *Immunity, 36*(5), 705–716.

Zweemer, A. J., Toraskar, J., Heitman, L. H., & IJzerman, A. P. (2014). Bias in chemokine receptor signalling. *Trends in Immunology, 35*(6), 243–252.

CHAPTER TWENTY

Flow Cytometry Detection of Chemokine Receptors for the Identification of Murine Monocyte and Neutrophil Subsets

Ornella Bonavita*,†, Matteo Massara*,†, Achille Anselmo*, Paolo Somma*, Hilke Brühl‡, Matthias Mack‡, Massimo Locati*,†, Raffaella Bonecchi*,§,1

*Humanitas Clinical and Research Center, Rozzano, Italy
†Department of Medical Biotechnologies and Translational Medicine, Università degli Studi di Milano, Milano, Italy
‡Universitätsklinikum Regensburg, Regensburg, Germany
§Department of Biomedical Sciences, Humanitas University, Rozzano, Italy
[1]Corresponding author: e-mail address: raffaella.bonecchi@humanitasresearch.it

Contents

1. Introduction 442
2. Blood Collection and Preparation of Blood Cells from Mice 444
   2.1 Cardiac Puncture 444
   2.2 Retro-Orbital Blood Collection 444
   2.3 Preparation of Blood Cells for Flow Cytometry Analysis 445
3. Staining of Cell Surface Markers and Chemokine Receptors 445
4. Flow Cytometry Analysis 449
   4.1 Quality and Reproducibility Controls 449
   4.2 Compensation Procedures 450
   4.3 Data Acquisition 451
   4.4 Gating Strategy 451
   4.5 Data Presentation and Analysis 451
5. Discussion 453
6. Concluding Remarks 454
Acknowledgments 455
References 455

Abstract

Chemokine receptors are differentially expressed on leukocyte subpopulations dictating their ability to migrate both in physiological and pathological conditions. Their expression is modulated during leukocyte differentiation and maturation and they can be used as markers to identify and characterize the frequency and the activation state of

Methods in Enzymology, Volume 570
ISSN 0076-6879
http://dx.doi.org/10.1016/bs.mie.2015.09.011

leukocytes present in a tissue. Here, we will describe flow cytometry approaches to detect chemokine receptors identifying subpopulations of circulating monocytes and neutrophils.

1. INTRODUCTION

The chemokine system is the main regulator of leukocyte migration during homeostatic and inflammatory conditions. The system is composed by 45 chemokines, small cytokines that can be functionally distinguished as inflammatory or homeostatic, the former being produced by any cell type and recruiting leukocytes during inflammatory process, the latter being produced by specific tissues and cell types and regulating leukocyte traffic inside tissues in homeostatic conditions (Bonecchi et al., 2009; Griffith, Sokol, & Luster, 2014). Chemokines exert their biological function through the interaction with receptors belonging to a distinct subfamily of the large seven transmembrane domain receptor family. Chemokine receptors include 18 canonical receptors, which induce leukocyte migration along their cognate ligand gradient, and a small subgroup of four atypical chemokine receptors, which are unable to support cell migration but play a role in the establishment and shaping of the chemokine gradients in tissues (Bonecchi, Savino, Borroni, Mantovani, & Locati, 2010; Nibbs & Graham, 2013).

Different leukocyte subpopulations express distinct pattern of chemokine receptors, and changes in the chemokine receptor expression profile distinguishes distinct activation states of a given leukocyte subpopulation. This concept has been particularly investigated in T cells, where chemokine receptors differential expression was exploited in order to identify distinct T cell subsets. Chemokine receptors are now used as selective markers for the identification of these cells. For instance, Th1 lymphocytes express CXCR3, Th2 express CCR4 (Bonecchi et al., 1998), Th17 express CCR6, and naïve T cells express CCR7 (Islam & Luster, 2012). The differential expression of chemokine receptors was also found for other leukocyte subsets, such as dendritic cells (Sozzani et al., 1998) and B lymphocytes (Breitfeld et al., 2000).

It is emerging that also circulating monocytes and polymorphonuclear neutrophils (PMN) are represented by heterogeneous populations. In mouse, at least two different monocyte subpopulations exist, which can be distinguished for differential expression of the antigen Ly6C. Ly6C^{high} are the classical or proinflammatory monocytes that are recruited to inflammatory sites where they produce high levels of inflammatory cytokines becoming tissue macrophages (Geissmann, Jung, & Littman, 2003), while Ly6C^{low} are the nonclassical, patrolling, or alternative monocytes, considered to be important in the reparative process through the production of IL-10 and their differentiation in proresolving macrophages (Auffray et al., 2007). Whether monocyte subsets originate as distinct populations from a common precursor or by transition from Ly6C^{high} to Ly6C^{low} as well as their respective role in pathology are still debated issues. Conversely, it is clear that monocyte subsets have different homing capabilities, which are supported by distinct chemokine receptors, with Ly6C^{high} monocytes being CCR2^{high}/CX3CR1^{low} and Ly6C^{low} monocytes CCR2^{low}/CX3CR1^{high}. In the last few years, the existence of subsets was also described for PMN (Beyrau, Bodkin, & Nourshargh, 2012). In the mouse, most of circulating PMN are Ly6G^{high}/CD62L^{pos}/CXCR2^{high}. When an inflammatory situation occurs, a fraction of circulating PMN downregulate L-selectin (CD62L) and CXCR2, and upregulate ICAM-1/CD54, CCR1, and CCR2. These circulating PMN are activated and have increased killing ability due to reactive oxygen species production (Sionov, Fridlender, & Granot, 2014). Heterogeneity of circulating myeloid cells has also been reported in the human setting, and despite the fact they are labeled with different markers, the different subsets have functions and chemokine receptor expression similar to the murine setting. Inflammatory or classical monocytes are CD14^{high}/CD16^{neg} and CCR2^{high}/CX3CR1^{low}, while the nonclassical or patrolling ones are CD14^{low}/CD16^{pos} and CCR2^{low}/CX3CR1^{high} (Tacke & Randolph, 2006). Referring to circulating PMN, a single CD16^{high}/CD62L^{high}/CXCR2^{high} population is detected in healthy conditions, while under inflammatory conditions CD16^{high}/CD62L^{dim} PMN expressing both CCR1 and CCR2 are found in the circulation (Pillay et al., 2012). These data indicate that analysis of chemokine receptors expression on circulating myeloid cells identify distinct subpopulations with different functions in homeostatic and pathologic conditions.

Multicolor flow cytometry allows simultaneous detection of several antigens and represents the best technique available for the identification and quantification of these populations. In this review, we outline the methodologies for flow cytometric detection of chemokine receptors on murine circulating monocyte and PMN subsets.

2. BLOOD COLLECTION AND PREPARATION OF BLOOD CELLS FROM MICE

Flow cytometry analysis of murine myeloid cells was performed on blood cells of C57BL6 mice; however, the protocol that here we describe is appropriate for all mouse strains. Blood can be collected either by cardiac puncture or from the retro-orbital sinus, as here described.

2.1 Cardiac Puncture

1. Anesthetize mouse accordingly to approved procedures (e.g., ketamine/xylazine).
2. Place the animal on the back.
3. Disinfect the skin using 70% ethanol.
4. Insert a needle mounted on a 1-ml 25GA X5/8 in syringe (BD Plastipak, #301359) slightly left under sternum and then into the heart. Needle and syringe should be held 20–30° off horizontal.
5. Apply a negative pressure on the syringe and withdraw blood. Exert a moderate negative pressure to prevent heart chamber collapse.
6. To prevent blood cells rupture by sheering forces, remove the needle before transferring blood in a K2-EDTA spray-coated tube (tube size 13 × 75, draw volume 3.0 ml, BD Bioscience, #368856). It is recommended to keep the sample on ice.
7. Perform euthanasia.

2.2 Retro-Orbital Blood Collection

1. Manually restrain mouse and apply one drop of topical ophthalmic anesthetic, such as proparacaine or tetracaine, on the eye. An alternative to topical anesthesia is general anesthesia (e.g., ketamine/xylazine).
2. Wait a few seconds and remove the excess with a sterile gauze.
3. Place a sterile 100-μl microcapillary pipette at the canthus of eye.
4. With a gentle rotating motion, insert tube through membrane.
5. Continue rotating tube on back of orbit until blood flows.
6. If excessive bleeding occurs, apply gentle pressure with a gauze.
7. Transfer blood in a K2-EDTA spray-coated tube (tube size 13 × 75, draw volume 3.0 ml, BD Bioscience, #368856). It is recommended to keep the sample on ice.

8. Perform euthanasia. Retro-orbital collection can be considered a non-terminal procedure only if less of 10% of the total blood (about 200 μl) is collected.

2.3 Preparation of Blood Cells for Flow Cytometry Analysis

1. Add 100 μl blood to each tube (5 ml 12 × 75 mm polystyrene round bottom test tubes for flow cytometry, BD Falcon, #352052).
2. Add 500 μl FACS buffer (1 × PBS, 1% BSA, 0.1% NaN_3; pH 7.2–7.4).
3. Centrifuge at 1500 rpm/450 × *g* for 5 min.
4. Discard supernatant.
5. Lyse red blood cells by adding 1.2 ml of ammonium chloride–potassium (ACK) lysis buffer (0.15 *M* NH_4Cl, 10 m*M* $KHCO_3$, 0.1 m*M* EDTA; pH 7.2–7.4).
6. Resuspend cells by gentle agitation.
7. Incubate on ice for 5 min.
8. Add 2 ml FACS buffer and centrifuge at 1500 rpm for 5 min at 4 °C.
9. Proceed to step 10 if a ring of red blood cells surrounding the white cellular pellet will be evident, indicating that lysis was successful. If lysis has not been successful, repeat steps 5–9.
10. Carefully aspirate supernatant with a p200 tip attached to the aspirator.

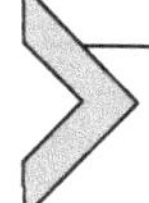

3. STAINING OF CELL SURFACE MARKERS AND CHEMOKINE RECEPTORS

In this section, we describe the strategy used to identify and analyze myeloid cell subpopulations, in particular monocytes and PMN, based on different expression of surface markers (CD45, CD115, Ly6C, and CD11b for monocyte subpopulations; CD45, Ly6G, CD11b, and ICAM-1/CD54 for PMN subpopulations) and chemokine receptors. The correct combination of fluorochrome conjugates to the antibody chosen for flow cytometry analysis is an important step to obtain the best signal and to reduce inconvenience related to the compensation. In particular, in order to optimize the signal, fluorochromes with the highest staining index should be reserved for rare cellular events. Signals from antigens with low expression, including chemokine receptors, can be enhanced using amplification approaches, a multistep staining procedure that requires purified primary antibodies and antispecies secondary antibodies directly conjugated or biotinylated. This staining approach improves the fluorescence signal promoting the exposure of the fluorophore to the laser beam. The appropriate fluorochrome combinations required to analyze cell populations of interest in four different mix and their respective experimental controls, represented by fluorescence minus one (FMO) and control isotypes, reported in Table 1 (Roederer, 2002).

The following procedure describes the staining of blood myeloid cells. The staining can also be performed in full EDTA blood and red cells can also be lysed after staining.

Table 1 Experimental Controls (FMO and Isotype) Used for Monocytes and PMN Staining

Fluorescence Minus One (FMO) and Isotype Controls

	BV605	BV421	FITC	PerCP-Cy5.5	PE	PE-Cy7	APC
Staining 1							
Monocytes CCR2							
FMO BV421 + Isotype	CD45	Isotype	Ly6C	CD11b	CD115	X	X
FMO BV422	CD45	None	Ly6C	CD11b	CD115	X	X
Staining 2							
Monocytes CX3CR1							
FMO APC	CD45	X	Ly6C	CD11b	CD115	X	None
Staining 3							
PMN CCR2/CCR1							
FMO APC	CD45	CCR2	X	CD11b	CCR1	Ly6G	None
FMO BV421 + Isotype	CD45	Isotype	X	CD11b	CCR1	Ly6G	ICAM
FMO PE + Isotype	CD45	CCR2	X	CD11b	Isotype	Ly6G	ICAM
FMO BV421	CD45	None	X	CD11b	CCR1	Ly6G	ICAM
FMO PE	CD45	CCR2	X	CD11b	None	Ly6G	ICAM
Staining 4							
PMN CXCR2							
FMO APC	CD45	X	X	CD11b	CXCR2	Ly6G	None
FMO PE + Isotype	CD45	X	X	CD11b	Isotype	Ly6G	ICAM
FMO PE	CD45	X	X	CD11b	None	Ly6G	ICAM

1. Add 100 μl of Fc blocker (purified anti-mouse CD16/32 antibody; BioLegend, clone 93, #101302) diluted at 1:200 ratio in ice-cold FACS buffer. Fc block contributes to avoid nonspecific binding of antibody to Fc receptors and reduces background fluorescence.
2. Incubate on ice for 15 min.
3. Add 500 μl of FACS buffer and centrifuge at 1500 rpm for 5 min at 4 °C.
4. Discard supernatant.
5. Incubate with the anti-CCR2 MC-21 generated by Dr. Mack et al. (2001) at 2.7 μg/ml and anti-CX_3CR1 (rabbit anti-mouse, Abcam, #ab8021) at 2 μg/ml in 100 μl of FACS buffer, or relative isotype control (rat IgG2b,k) at 5 μg/ml for 1 h on ice.
6. Wash 3 × in 500 μl FACS buffer at 1500 rpm for 5 min at 4 °C.
7. For CCR2 staining, add monoclonal biotin-labeled mouse anti-rat IgG2b (BD Biosciences, #553898) at 5 μg/ml in 100 μl of FACS buffer.
8. Incubate the samples for 30 min on ice and wash 3 × in 500 μl FACS buffer at 1500 rpm for 5 min at 4 °C.
9. Incubate with fluorochrome-labeled antibodies (Table 2) diluted in FACS buffer as described below:
   - **(a)** For monocytes staining, incubate cells in 100 μl FACS buffer with anti-CD45 (4 μg/ml, BD Bioscience, #563053), anti-CD11b (4 μg/ml, BD Bioscience, #550993), anti-CD115 (4 μg/ml, eBioscience, #12-1152-82), and anti-Ly6C (10 μg/ml, BD Bioscience, #553104).
   - **(b)** For PMN staining, incubate cells in 100 μl FACS buffer with anti-CD45 (4 μg/ml, BD Bioscience, #563053), anti-CD11b (4 μg/ml, BD Bioscience, #550993), anti-Ly6G (2 μg/ml, BD Bioscience, # 560601), and anti-ICAM-1/CD54 (2 μg/ml, BioLegend, #116119).
   - **(c)** For CX3CR1 staining, add fluorochrome-labeled secondary antibody at 4 μg/ml (goat anti-rabbit indicated in Table 2).
   - **(d)** For CCR2 staining, add fluorochrome-labeled streptavidin at 2 μg/ml (BioLegend, #405226).
10. Incubate for 30 min at 4 °C in the dark.
11. Wash the cells 3 × by centrifugation at 1500 rpm for 5 min at 4 °C.
12. Aspirate the supernatant.
13. Resuspend cells in 500 μl FACS buffer.
14. Store the cell suspension immediately at 4 °C in the dark. Acquire samples as soon as possible. If you need to wait longer than 1 h before the acquisition, a fixation procedure is required.

Table 2 Antibodies Combination for Chemokine Receptors Staining in Monocytes and PMN

Antibody	Fluorochrome	Cat No.	Supplier
Staining monocytes CCR2/CX3CR1			
CD45	BV 605	563053	BD
CD11b	PerCP-Cy5.5	550993	BD
Ly6C	FITC	553104	BD
CD115	PE	12-1152-82	eBioscience
CCR2	Purified	X	Homemade
Mouse anti-rat IgG2b	X	553898	BD
Streptavidin	BV 421	405226	BioLegend
CCR2 isotype control	Purified	555846	BD
Rabbit anti-mouse CX3CR1	X	ab8021	Abcam
Goat anti-rabbit	Alexa 647	A-21244	Life Technologies
Staining PMN CCR2/CCR1/CXCR2			
CD45	BV 605	563053	BD
CD11b	PerCP-Cy5.5	550993	BD
Ly6G	PeCy7	560601	BD
ICAM	APC	116119	BioLegend
CCR2	Purified	X	Homemade
Mouse anti-rat IgG2b	X	553898	BD
Streptavidin	BV 421	405226	BioLegend
CCR2 isotype control	Purified	555846	BD
CCR1	PE	FAB5986P	R&D
CCR1 isotype control	PE	IC013P	R&D
CXCR2	PE	FAB2164P	R&D
CXCR2 isotype control	PE	IC006P	R&D

Fixation procedure:

15. Resuspend cells in 300 μl FACS fix (1× PBS, 1% formaldehyde; pH 7.2–7.4).
16. Incubate for 30 min at room temperature.

17. Centrifuge at 1500 rpm for 5 min at 4 °C and resuspend in 500 μl FACS buffer.
18. Store the cell suspension immediately at 4 °C in the dark.

4. FLOW CYTOMETRY ANALYSIS

Experiments have been analyzed using a LSR Fortessa cytometer (BD Biosciences) equipped with four lasers (405-nm violet laser, 488-nm blue laser, 561-nm yellow laser, 640-nm red laser) and configured for fluorochromes detection as indicated in Table 3. A series of control and calibration procedures has to be performed before sample acquisition.

4.1 Quality and Reproducibility Controls

Quality and reproducibility control allows the characterization of the fluorescence detectors and the entire optical configuration of the instrument. For quality and reproducibility controls, we use CS&T Research beads (BD immunocytometry system, #655051), which are dim/mid/brightly dyed polystyrene beads specifically produced for the automated procedure of cytometer setup, baseline definition, and performance tracking (BD FACSDiva™ software), as described below:

1. Baseline definition

   Whenever CS&T beads lot or instrument configurations are changed, new performance values, called target values, must be automatically calculated, running CS&T beads in order to determine new reference baseline values.

2. Performance tracking

   The comparison between the baseline and the daily CS&T beads values settles the stability of the instrument (CV range and the size of PhotoMultiplier Voltage adjustment (ΔPMTV)), pointing out the state

Table 3 LSR Fortessa Cytometer Configuration for Fluorochromes Detection

		Laser Setting						
		Fluorochromes						
		BV605	**BV421**	**FITC**	**PerCP-Cy5.5**	**PE**	**PE-Cy7**	**APC**
Filters	BP	610/20	450/50	530/30	710/50	586/15	780/60	670/14
	LP	600LP		505LP	685LP		750LP	
Laser	(nm)	405	405	488	488	561	561	640

of quality and reproducibility of the experiment. Furthermore, the performance check procedure sets up instrument parameters as PMTVs values and laser delay in order to obtain optimal fluorescence detection.

4.2 Compensation Procedures

In order to minimize spectral overlap during a multicolor flow cytometry analysis, a compensation procedure before acquiring data is required. Use the FACSDiva software package to automatically calculate a compensation matrix from single-stained controls (CompBeads, BD immunocytometry system, #552843; see Table 4). Below, a brief guide for getting started:

1. Adjust side scatter (SSC) and forward scatter (FSC) on an unstained beads control to put the population of interest on scale.
2. Acquire all single-stained compensation controls. If necessary, adjust the PMTV gains for each detector starting from those defined by CS&T procedure.

Table 4 Single-Stained Controls Used for Compensation Procedure

	BV605	BV421	FITC	PerCP-Cy5.5	PE	PE-Cy7	APC
COMPBEADS-BV605	CD45						
COMPBEADS-BV421		CCR2					
COMPBEADS-FITC			Ly6C				
COMPBEADS-PERCP-Cy5.5				CD11b			
COMPBEADS-PE					CCR1		
COMPBEADS-PE					CXCR2		
COMPBEADS-PE					CD115		
COMPBEADS-PE-Cy7						Ly6G	
COMPBEADS-APC							ICAM
COMPBEADS-APC							CX3CR1

3. Automatically calculate compensation values across all included detectors and apply them to all the experimental controls and samples of interest.

4.3 Data Acquisition

1. Set FSC and SSC values in order to detect all the leukocyte subpopulations present in the sample.
2. Modify the FSC threshold value in order to remove debris from the acquisition.
3. Acquire and save 100,000 events from the samples of interest. A higher number of events is required to generate statistically relevant data for the determination of chemokine receptors expression levels in rare populations.

4.4 Gating Strategy

The gating strategy to analyze murine monocyte subpopulations from blood-derived leukocytes is described in Fig. 1A. The evaluation of CCR2 and CX3CR1 expression levels on the surface of LY6C^{high} and LY6C^{low} monocytes is defined as shown in Fig. 1B.

The morphological identification and doublet discrimination of total leukocytes derived from murine blood is performed as previously described in Fig. 1A. PMN are identified as described in Fig. 2A on CD45 positive cells (not shown). The evaluation of CCR2, CCR1, and CXCR2 expression levels in ICAM-1/CD54^{neg} and ICAM-1/CD54^{pos} PMN subsets is defined as shown in Fig. 2B.

4.5 Data Presentation and Analysis

Data are analyzed by flow cytometry analysis software (BD FACS Diva, BD Biosciences; FlowJo, TreeStar). Data presentation using the frequency of positive events is appropriate only in presence of a bimodal distribution of the emitted fluorescence. In this case, there is no particular preference for the use of the histogram or the dot plot to present data. In case of a non-bimodal distribution of the emitted fluorescence, data should be reported as relative MFI (i.e., the ratio between the MFI values of the sample stained with all the experimental markers and the MFI values of the negative control sample) and the histogram layout should be preferred. Differently from frequency data, this analytical approach provides "relative" quantitative information of the chemokine receptor expression levels on the surface of each

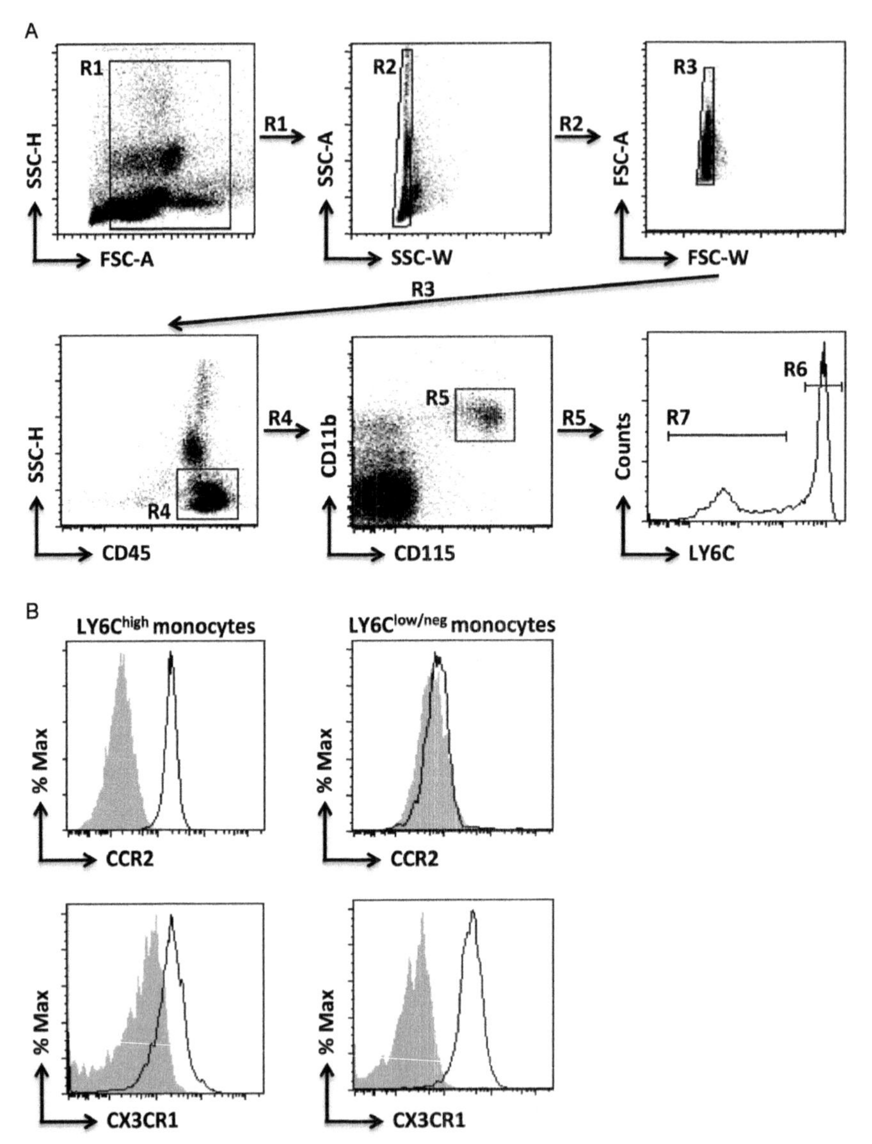

Figure 1 Flow cytometry analysis of murine monocytes from whole blood. (A) After morphological gating (R1) and doublets discrimination (R2 and R3), total mononuclear cells were gated as CD45^{pos}/SSC^{low} events (R4). Total monocytes were identified as CD11b^{pos}/CD115^{pos} events (R5) and subsequently Ly6C^{high} (R6) and Ly6C$^{low/neg}$ (R7) monocyte subpopulations were gated. (B) Surface CCR2 and CX3CR1 expression levels (black lines) were compared with the relative negative control (FMO plus isotype control for CCR2 and FMO for CX3CR1; gray fill) in each monocyte subpopulation (R6 and R7) according to the protocol described herein.

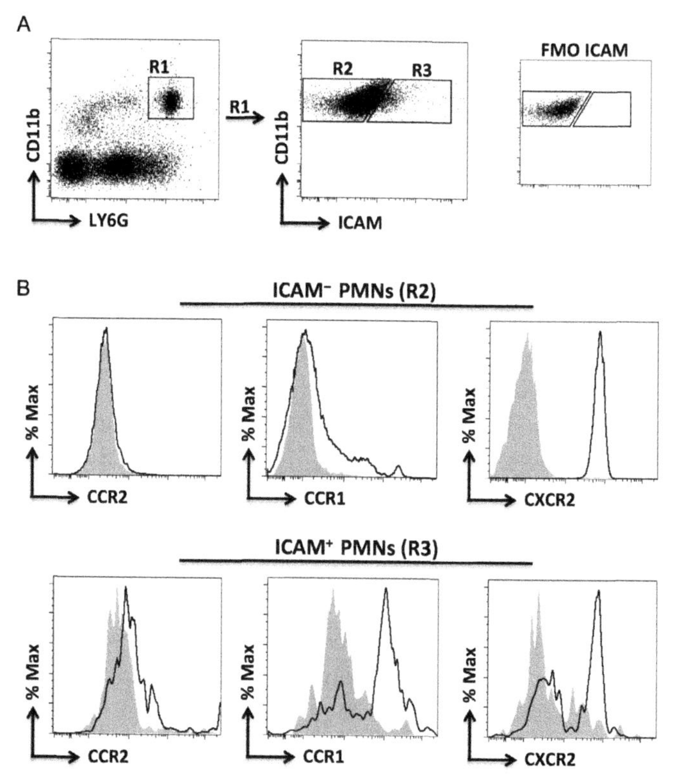

Figure 2 Flow cytometry analysis of PMN from whole blood of tumor bearing mice. (A) Total PMN were identified as CD11b^{pos}/Ly6G^{pos} events (R1) and subsequently ICAM-1/CD54^{neg} (R2) and ICAM-1/CD54^{pos} (R3) PMN cell subsets were gated. R3 gate was defined using FMO as negative control. (B) Surface CCR2, CCR1, and CXCR2 expression levels (black lines) were compared with the relative negative controls (FMO plus isotype control; gray fill) in each PMN subpopulation (R2 and R3) according to the protocol described herein.

cell subset analyzed. In this context, the presentation of the negative control allows a more accurate interpretation of the presented results.

5. DISCUSSION

In the last years, flow cytometry has become the technique of choice to detect chemokine receptors expressed by leukocytes. Added values of this technique include the possibility to analyze chemokine receptors expression also on rare cell populations with no need of purification procedures when combinations of different antibodies identifying leukocyte subpopulations are used, and the generation of quantitative data, including the percentage and absolute number of cells expressing a chemokine receptor of interest and, when combined with fluorochrome-specific beads, the number of molecules on the cell surface (Chan, Jilani, Chang, & Albitar, 2007). However, the detection and measurement of chemokine receptors by FACS analysis on myeloid cells poses some technical challenges which in some cases have led to irreproducible results. As chemokine receptors are in most cases expressed at low levels, and myeloid cells are highly autofluorescent and express high levels of Fc receptors, most difficulties rely on the need of an appropriate noise-to-signal ratio. To overcome this issue, myeloid cells must be preincubated with FcR-blocking antibodies and then labeled with antibodies conjugated with fluorophores with an emission wavelength far from the visible spectrum (i.e., PECy-7, allophycocyanin (APC), Alexa-647), or antibodies specifically mutated in the Fc region and unable to bind to Fc receptors should be used (REAfinity, Miltenyi). Some chemokine receptors also require to improve the fluorophore-versus-protein (FTP) ratio. This goal can be achieved by signal amplification approaches, such as biotin-conjugated primary antibodies followed by staining with a streptavidin-conjugated brilliant fluorophore. To further increase the FTP ratio, a three steps labeling strategy can also be adopted, using an unconjugated primary antibody followed by a biotinylated secondary antibody and a streptavidin-conjugated brilliant fluorophore, as we described here for CCR2. In addition, new techniques have been developed based on the selective labeling of the Fab fragments or on the amplification signal emitted by the fluorophore-conjugated antibodies, such as the commercially available fluorescence amplification by sequential employment of reagents technology (Anselmo et al., 2014). Finally, it has to be considered that chemokine receptors can also be localized inside the cell and receptor detection in intracellular districts requires labeling procedures including cell permeabilization steps.

6. CONCLUDING REMARKS

Definition of heterogeneity in myeloid populations is a complex field of intense investigation. The relevance of a precise definition of identity and underlying functions of distinct myeloid cell subsets is being increasingly appreciated in several fields, starting from tumor immunology where circulating and intratumoral myeloid populations, long been considered homogeneously suppressive, have been recently appreciated to be highly diversified. Combined with single-cell transcriptional profiling, multicolor flow cytometry has driven our growing understanding of complexity and holds even more ambitious promises based on technological advances such as spectral unmixing image analysis techniques. Being key in differential recruitment of distinct leucocyte subsets, chemokine receptors have always been regarded as promising subset markers, as their established role in T and dendritic cell subset phenotyping testifies. Here, we summarized evidence for their role also in myeloid subsets phenotyping and provided a practical for their effective detection.

ACKNOWLEDGMENTS

Research activities in the lab are supported by Ministero dell'Istruzione dell'Università e della Ricerca (FIRB projects), the Italian Association for Cancer Research (AIRC), and the European Community's Seventh Framework Programme (FP7-2007-2013) under grant agreement HEALTH-F4-2011-281608 (TIMER).

REFERENCES

Anselmo, A., Mazzon, C., Borroni, E. M., Bonecchi, R., Graham, G. J., & Locati, M. (2014). Flow cytometry applications for the analysis of chemokine receptor expression and function. *Cytometry A*, *85*(4), 292–301.

Auffray, C., Fogg, D., Garfa, M., Elain, G., Join-Lambert, O., Kayal, S., et al. (2007). Monitoring of blood vessels and tissues by a population of monocytes with patrolling behavior. *Science*, *317*(5838), 666–670.

Beyrau, M., Bodkin, J. V., & Nourshargh, S. (2012). Neutrophil heterogeneity in health and disease: A revitalized avenue in inflammation and immunity. *Open Biology*, *2*(11), 120134.

Bonecchi, R., Bianchi, G., Bordignon, P. P., D'Ambrosio, D., Lang, R., Borsatti, A., et al. (1998). Differential expression of chemokine receptors and chemotactic responsiveness of type 1 T helper cells (Th1s) and Th2s. *The Journal of Experimental Medicine*, *187*(1), 129–134.

Bonecchi, R., Galliera, E., Borroni, E. M., Corsi, M. M., Locati, M., & Mantovani, A. (2009). Chemokines and chemokine receptors: An overview. *Frontiers in Bioscience (Landmark Edition)*, *14*, 540–551.

Bonecchi, R., Savino, B., Borroni, E. M., Mantovani, A., & Locati, M. (2010). Chemokine decoy receptors: Structure-function and biological properties. *Current Topics in Microbiology and Immunology*, *341*, 15–36.

Breitfeld, D., Ohl, L., Kremmer, E., Ellwart, J., Sallusto, F., Lipp, M., et al. (2000). Follicular B helper T cells express CXC chemokine receptor 5, localize to B cell follicles, and support immunoglobulin production. *The Journal of Experimental Medicine*, *192*(11), 1545–1552.

Chan, H. E., Jilani, I., Chang, R., & Albitar, M. (2007). Detection of chromosome translocations by bead-based flow cytometry. *Methods in Molecular Biology*, *378*, 167–174.

Geissmann, F., Jung, S., & Littman, D. R. (2003). Blood monocytes consist of two principal subsets with distinct migratory properties. *Immunity*, *19*(1), 71–82.

Griffith, J. W., Sokol, C. L., & Luster, A. D. (2014). Chemokines and chemokine receptors: Positioning cells for host defense and immunity. *Annual Review of Immunology*, *32*, 659–702.

Islam, S. A., & Luster, A. D. (2012). T cell homing to epithelial barriers in allergic disease. *Nature Medicine*, *18*(5), 705–715.

Mack, M., Cihak, J., Simonis, C., Luckow, B., Proudfoot, A. E., Plachy, J., et al. (2001). Expression and characterization of the chemokine receptors CCR2 and CCR5 in mice. *Journal of Immunology*, *166*(7), 4697–4704.

Nibbs, R. J., & Graham, G. J. (2013). Immune regulation by atypical chemokine receptors. *Nature Reviews. Immunology*, *13*(11), 815–829.

Pillay, J., Kamp, V. M., van Hoffen, E., Visser, T., Tak, T., Lammers, J. W., et al. (2012). A subset of neutrophils in human systemic inflammation inhibits T cell responses through Mac-1. *The Journal of Clinical Investigation*, *122*(1), 327–336.

Roederer, M. (2002). Compensation in flow cytometry. *Current Protocols in Cytometry*. Chapter 1, Unit 1.14. http://dx.doi.org/10.1002/0471142956.cy0114s22.

Sionov, R. V., Fridlender, Z. G., & Granot, Z. (2014). The multifaceted roles neutrophils play in the tumor microenvironment. *Cancer Microenvironment*.

Sozzani, S., Allavena, P., D'Amico, G., Luini, W., Bianchi, G., Kataura, M., et al. (1998). Differential regulation of chemokine receptors during dendritic cell maturation: A model for their trafficking properties. *Journal of Immunology*, *161*(3), 1083–1086.

Tacke, F., & Randolph, G. J. (2006). Migratory fate and differentiation of blood monocyte subsets. *Immunobiology*, *211*(6–8), 609–618.

CHAPTER TWENTY-ONE

Molecular Pharmacology of Chemokine Receptors

Raymond H. de Wit[2], Sabrina M. de Munnik[2], Rob Leurs, Henry F. Vischer[3], Martine J. Smit[1,3]
Amsterdam Institute for Molecules Medicines and Systems (AIMMS), Division of Medicinal Chemistry, Vrije Universiteit, Amsterdam, The Netherlands
[1]Corresponding author: e-mail address: mj.smit@vu.nl

Contents

1. Introduction 458
   1.1 Chemokines and Their Receptors 458
   1.2 Chemokine Receptor Therapeutics 459
2. Pharmacological Quantification of Chemokine Receptor Binding 462
   2.1 Binding Theory 462
   2.2 Orthosteric Versus Allosteric Ligand Binding 463
   2.3 Binding Experiments 465
3. Pharmacological Quantification of Chemokine Receptor Signaling 482
   3.1 Determining the Mode of Action of Chemokine Receptor Ligands in Functional Studies 484
   3.2 Functional Assays to Measure Chemokine Receptor Signaling 488
4. Conclusions and Future Perspectives 505
Acknowledgments 507
References 507

Abstract

Chemokine receptors are involved in various pathologies such as inflammatory diseases, cancer, and HIV infection. Small molecule and antibody-based antagonists have been developed to inhibit chemokine-induced receptor activity. Currently two small molecule inhibitors targeting CXCR4 and CCR5 are on the market for stem cell mobilization and the treatment of HIV infection, respectively. Antibody fragments (e.g., nanobodies) targeting chemokine receptors are primarily orthosteric ligands, competing for the chemokine binding site. This is opposed by most small molecules, which act as allosteric modulators and bind to the receptor at a topographically distinct site as compared to chemokines. Allosteric modulators can be distinguished from orthosteric ligands by unique features, such as a saturable effect and probe dependency. For successful

[2] These authors contributed equally to this work.
[3] Co-senior authors.

Methods in Enzymology, Volume 570
ISSN 0076-6879
http://dx.doi.org/10.1016/bs.mie.2015.12.002

drug development, it is essential to determine pharmacological parameters (i.e., affinity, potency, and efficacy) and the mode of action of potential drugs during early stages of research in order to predict the biological effect of chemokine receptor targeting drugs in the clinic. This chapter explains how the pharmacological profile of chemokine receptor targeting ligands can be determined and quantified using binding and functional experiments.

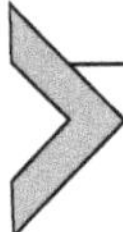

1. INTRODUCTION

1.1 Chemokines and Their Receptors

Chemokines and their receptors play a crucial role in the immune system. Chemokines are soluble proteins (8–12 kDa) that mainly direct migration (i.e., chemotaxis) of immune cells between the bone marrow, blood, and peripheral tissues during homeostatic leukocyte homing or to sites of inflammation (Griffith, Sokol, & Luster, 2014) by forming a chemotactic gradient through the binding to glycosaminoglycans (GAGs) located at the surface of endothelial cells (Handel, Johnson, Crown, Lau, & Proudfoot, 2005). In addition, chemokines and their receptors are also involved in embryonic development, tissue repair, and angiogenesis (Dimberg, 2010; Martins-Green, Petreaca, & Wang, 2013; Wang & Knaut, 2014). To date, 43 human chemokines and 23 human chemokine receptors (CKRs) have been identified. Chemokines are divided into four classes (C, CC, CXC, and CX3C), based on the number and arrangement of conserved cysteine residues. The CKR nomenclature is based on the chemokine subclass that they bind. Some chemokines are able to bind to multiple CKR subtypes, whereas others bind exclusively to a single CKR subtype (Bachelerie et al., 2014; de Munnik, Smit, Leurs, & Vischer, 2015).

CKRs belong to the class A (rhodopsin) family of G protein-coupled receptors (GPCRs) that activate intracellular heterotrimeric G proteins. Activated CKRs recruit β-arrestins, which inhibit further G protein activation (desensitization) and induce receptor internalization (Patel, Channon, & McNeill, 2013). An exception is formed by the atypical CKRs ACKR1 (DARC), ACKR2 (D6), ACKR3 (CXCR7), ACKR4 (CCXCKR), and ACKR5 (CCRL2) that bind chemokines, but fail to induce classical G protein-dependent signaling and chemotaxis. However, the ACKRs recruit β-arrestins and internalize upon chemokine binding, probably to regulate chemokine availability and/or to facilitate chemokine transport from the basolateral site to the luminal side of vascular endothelial cells (Ulvmar, Hub, & Rot, 2011). In addition to its role in receptor regulation, the scaffolding protein β-arrestin is able to function as signaling hub, interacting with cytosolic proteins, thus G protein-independently activating signaling cascades (e.g., MAPK and Akt) (Rajagopal et al., 2010; Wang et al., 2008).

Chemokines and their receptors play an important role in several diseases (Raman, Sobolik-Delmaire, & Richmond, 2011). CCR5 and CXCR4 act as coreceptors for HIV to facility viral entry in macrophages and $CD4^+$ T cells, respectively (Deng et al., 1996). Inappropriate expression of chemokines and/or CKRs might lead to excessive leukocyte recruitment to inflammatory sites and underlies diseases such as asthma, chronic obstructive pulmonary disease (COPD), inflammatory bowel disease, rheumatoid arthritis, atherosclerosis, multiple sclerosis, and psoriasis (Koelink et al., 2012). Many cancer types show upregulated expression of chemokines and/or CKRs. Aberrant CKR signaling contributes to tumorigenesis, invasion, and cancer metastasis. For example, CXCR4 is often overexpressed in breast cancer cells, which can metastasize to distant organs that express its ligand CXCL12 (e.g., lung) (Wang & Knaut, 2014). Furthermore, infiltrating leukocytes promote tumorigenesis (Raman et al., 2011).

Some herpesviruses encode chemokines and CKRs that have probably been hijacked from the human genome during evolution (Vischer, Siderius, Leurs, & Smit, 2014). The virally encoded CKRs are expressed in host cells after infection, causing a dysregulation of endogenous signaling networks. In contrast to their human counterparts, most of these viral CKRs can signal in a constitutively active manner (i.e., signaling in the absence of an agonist) and promiscuously couple to multiple G protein subtypes (e.g., $G\alpha_{i/o}$, $G\alpha_{q/11}$, $G\alpha_{12/13}$) to activate multiple signaling pathways. Furthermore, most viral CKRs bind a broad range of chemokines from different subclasses that modify the constitutive signaling. Human cytomegalovirus (HCMV)-encoded US28 has been linked to tumorigenesis and vascular diseases and several small molecules have been developed that inhibit chemokine binding and constitutive signaling of US28 (Casarosa et al., 2003; Hulshof et al., 2005; Kralj, Kurt, Tschammer, & Heinrich, 2014; Vischer et al., 2010). Kaposi's sarcoma-associated herpesvirus (KSHV)-encoded ORF74 plays a key role in the angioproliferative disease Kaposi's sarcoma (Montaner, Kufareva, Abagyan, & Gutkind, 2013; Vischer et al., 2014).

1.2 Chemokine Receptor Therapeutics

Considering the role of chemokines and their receptors in various pathologies, this system is an appealing target for pharmacological intervention.

The majority of drug development efforts aim to antagonize the action of chemokines through their cognate receptors. Although this can be achieved by targeting chemokines (Daubeuf et al., 2013; Klarenbeek et al., 2012; Koenen & Weber, 2010) or the chemokine–GAG binding interface (Adage et al., 2012; Johnson et al., 2004; Suffee et al., 2012), this chapter will focus on ligands that target CKRs. These receptors are promising drug targets since they are characterized by cell type-specific expression profiles, which allows tissue-specific targeting (Schall & Proudfoot, 2011). Furthermore, CKRs belong to the GPCR receptor family, the most successfully targeted family of proteins, indicating that the receptors are tractable for drug development.

The orthosteric ligand binding site of CKRs is defined as the binding site of chemokines, the endogenous agonists. Chemokines are considered to activate receptors via a two-step binding model (Allen, Crown, & Handel, 2007), which is supported by the recently reported US28-CX3CL1 crystal structure (Burg et al., 2015). During the first step, the positively charged chemokine core binds to the negatively charged N-terminal region and extracellular loops of the receptor through primarily ionic interactions, mainly contributing to high-affinity binding between chemokine and receptor. After this recognition step, the flexible N-terminus of the chemokine interacts with additional more buried extracellular or transmembrane residues, stabilizing an active CKR conformation (Allen et al., 2007; Scholten, Canals, Maussanget al., 2012; Thiele & Rosenkilde, 2014). That chemokine N-termini are essential for receptor activation is demonstrated by N-terminal CC and CXC truncation variants (e.g., CCL5, CCL14a, and CXCL11) which are able to bind CKRs with high affinity, yet fail to elicit agonistic effects (Clark-Lewis, Mattioli, Gong, & Loetscher, 2003; Gong, Uguccioni, Dewald, Baggiolini, & Clark-Lewis, 1996; Richter et al., 2009). Interestingly, this canonical two-step mechanism does not appear to be entirely conserved for all chemokines, since CXCL12 truncations show impaired affinity for CXCR4 (Richter et al., 2014).

CKRs initially received attention as drug targets due to the role of CXCR4 and CCR5 in HIV-1 entry, resulting in the first clinically approved small molecule drugs AMD3100 and maraviroc (Fig. 1), respectively. Maraviroc inhibits binding of HIV-1-encoded gp120 (which is essential for viral entry) to CCR5 (Kuritzkes, Kar, & Kirkpatrick, 2008), resulting in a decrease of viral load in CCR5-tropic HIV-1 patients (Fatkenheuer et al., 2005; Watson, Jenkinson, Kazmierski, & Kenakin, 2005). The CXCR4 competitive antagonist AMD3100/plerixafor was initially

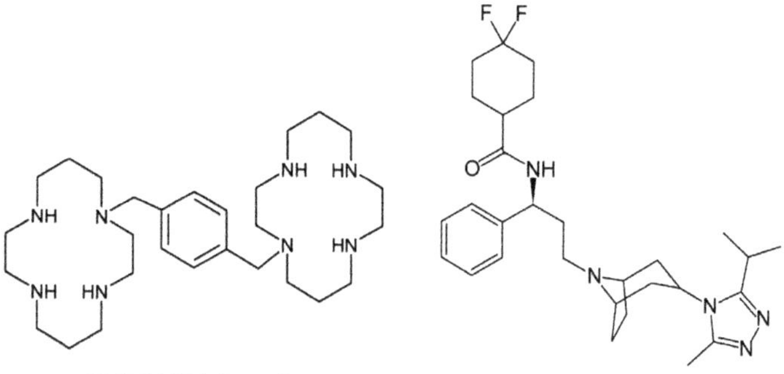

AMD3100/plerixafor Maraviroc

Figure 1 Molecular structures of chemokine receptor therapeutics AMD3100/plerixafor and maraviroc.

developed as an anti-HIV drug, but is currently FDA-approved for stem cell mobilization in non-Hodgkin's lymphoma and multiple myeloma by inhibiting CXCL12 binding (De Clercq, 2009; Wijtmans, Scholten, de Esch, Smit, & Leurs, 2012). Maraviroc and AMD3100 provide a clear proof of concept of successfully targeting CKRs with small molecules. In addition to maraviroc and AMD3100, many other small molecules have been developed for therapeutic targeting CKRs; however, no novel small molecule entities have reached the market since 2008. A number of candidates were evaluated in human clinical trials for treatment of malignancies and inflammatory diseases (e.g., Crohn's disease, asthma, COPD, rheumatoid arthritis, and multiple sclerosis); still the drug candidates failed to meet the expected end points in clinical trials (Proudfoot, Power, & Schwarz, 2010). Better understanding of disease mechanisms and advancements in small molecule drug development campaigns may result in successful CKR-targeting drugs. The availability of ligand-bound crystal structures of CCR5 (Tan et al., 2013), CXCR4 (Qin et al., 2015; Wu et al., 2010), and US28 (Burg et al., 2015) enables structure-based virtual screening and ligand design that may result in the identification of more effective small molecule CKR modulators.

During the last decade, the development of antibody-based biologics targeting CKRs has gained momentum, which in part can be attributed to improved immunization strategies, selection tools, GPCR expression, and purification methods (Klarenbeek et al., 2012; Mujic-Delic, de Wit, Verkaar, & Smit, 2014; Vela, Aris, Llorente, Garcia-Sanz, & Kremer, 2015). In 2012, Mogamulizumab (Poteligeo®) a glyco-engineered CCR4-targeting antibody was approved for clinical use in Japan for adult

T-cell leukemia and is currently the only CKR-targeting biologic available on the market (Beck & Reichert, 2012; Ishida et al., 2012). Especially in the field of (immuno)oncology, CKR antibodies have great potential due to their high affinity, selectivity, and on-target additional immune responses through antibody-dependent cellular cytotoxicity and complement-dependent cytotoxicity (Kubota et al., 2009). The selectivity limits off-target effects, whereas the additional immune response induces cell death of oncogenic CKR-expressing cells. It is therefore not surprising that a number of CCR2, CCR4, CCR5, and CXCR4 antibodies are currently evaluated in clinical trials (Klarenbeek et al., 2012; Vela et al., 2015). Several antibody-based formats have been developed (e.g., single-chain Fv, Fab, and nanobody/VHH) (Mujic-Delic et al., 2014; Muyldermans, 2013). The nanobody/VHH platform has shown great promise (decreased size and complexity, efficient molecular cavity binding) in the effective and selective targeting of CKRs and has been extensively reviewed elsewhere (Mujic-Delic et al., 2014). Recently, nanobodies have been described that potently inhibit cellular signaling of CXCR2, CXCR4, and CXCR7 (Bradley et al., 2015; Jahnichen et al., 2010; Maussang et al., 2013).

The aim of this chapter is to describe how binding and functional parameters of CKR ligands can be quantified. Furthermore, this chapter describes the different mechanisms of action of CKR ligands and how to distinguish between orthosteric and allosteric ligands. Knowledge on the mechanism of action may improve the understanding of drug behavior in a therapeutic setting.

2. PHARMACOLOGICAL QUANTIFICATION OF CHEMOKINE RECEPTOR BINDING

2.1 Binding Theory

A ligand can elicit biological effects upon binding to its target receptor. In ligand–receptor binding, a reversible two-state model is generally assumed (Eq. 1) wherein the ligand (A)–receptor (R) pair is either free (A + R) or constitutes a complex (AR).

$$[A] + [R] \underset{k_2}{\overset{k_1}{\rightleftarrows}} [AR] \tag{1}$$

Binding equilibrium is reached dependent on concentrations of reactants ([A] + [R]) and the association (k_1) and dissociation (k_2) rate constants of the AR complex. In equilibrium, the association and dissociation rates are equal (Eq. 2); thus, the equilibrium dissociation rate constant (K_d) is used as a pharmacological parameter representing the affinity of the ligand for the receptor (Eq. 3):

$$k_1[A][R] = k_2[AR] \tag{2}$$

$$\frac{k_2}{k_1} = \frac{[A][R]}{[AR]} = K_d \tag{3}$$

The Langmuir adsorption isotherm can be applied to receptor binding equilibrium, thus yielding the fraction of receptor occupancy (Kenakin, 2009b; Eq. 4). The equation dictates that a ligand concentration equal to K_d occupies 50% ($\rho_{AR} = 0.5$) of the receptors. Furthermore, receptor occupancy reaches a horizontal asymptote toward unity reflecting the finite number of receptors (R_{total}) in any biological system.

$$\rho_{AR} = \frac{[AR]}{[R_{total}]} = \frac{[A]}{[A] + K_d} \tag{4}$$

Ligands can prevent binding of the endogenous agonist through binding and sterically hinder the orthosteric site, which is termed competitive binding. In the presence of two competitive ligands (A and B), the receptor can be in three distinct states: unbound (R), ligand A-bound (AR), or ligand B-bound (BR) (Eq. 5).

$$[A] + [B] + [R] \rightleftarrows [AR] + [B] \rightleftarrows [A] + [BR] \tag{5}$$

In the presence of the competitor B, the fractional occupancy of the ligand A ($\rho_{AR,B}$) is dependent on the concentrations and affinities of the individual ligands (Eq. 6):

$$\rho_{AR,B} = \frac{[AR]}{R_{total}} = \frac{[AR]}{[AR] + [BR] + [R]} = \frac{\frac{[A]}{K_{d,A}}}{\frac{[A]}{K_{d,A}} + \frac{[B]}{K_{d,B}} + 1} \tag{6}$$

2.2 Orthosteric Versus Allosteric Ligand Binding

Orthosteric ligands bind to the chemokine binding site and consequently impair chemokine function through steric hindrance. In contrast, allosteric modulators bind sites that are distinct from the orthosteric binding site. Allosteric modulators can alter the receptor conformation and thus affect binding and/or efficacy (see Section 3) of the orthosteric ligand and vice versa (Kenakin, 2010). Consequently, small allosteric modulators (500–600 Da) can affect the pharmacological profile of much larger protein ligands, such as chemokines (8–12 kDa). Maraviroc is an example of a allosteric modulator. Whereas competitive ligands do not affect ligand dissociation, it was revealed from kinetic binding studies that maraviroc accelerates the dissociation rate of the orthosteric chemokine CCL3 from CCR5, illustrating that binding of the allosteric modulator affects binding characteristics of the orthosteric ligand (Garcia-Perez et al., 2011). The cooperativity factor α denotes the modulation of orthosteric ligand affinity by the allosteric modulator. In case of a positive allosteric modulator (PAM) $\alpha > 1$, indicating an increased observed affinity of the probe ligand (K_{obs}), whereas a negative allosteric modulator (NAM) is characterized by $\alpha < 1$ and consequently decreased observed affinity. An important distinction between orthosteric and allosteric binding is that allosterism and thus changes in K_{obs} are saturable (i.e., when the allosteric sites are fully occupied). Proportional (fractional) receptor occupancy by the orthosteric ligand (A) (e.g., chemokine) in the presence of an allosteric modulator (M) can be described by Eq. (7), which illustrates that the effect on observed orthosteric ligand affinity is saturable (i.e., K_d is multiplied by a certain factor) and depends on the concentration [M] and affinity (K_M) of the allosteric modulator and cooperativity factor α:

$$\rho_{AR} = \frac{[A]}{[A] + K_{obs}} \text{ with } K_{obs} = K_d \left(\frac{\left(1 + \frac{[M]}{K_M}\right)}{\left(1 + \frac{\alpha[M]}{K_M}\right)} \right) \tag{7}$$

The conformational change of the receptor induced by the allosteric modulator might differentially affect the binding characteristics of distinct orthosteric ligands and as a consequence α is probe dependent. As most CKRs respond to multiple chemokines, it is important to test an allosteric modulator in combination with different chemokines (Bernat, Brox, Heinrich, Auberson, & Tschammer, 2015; Sachpatzidis et al., 2003). For example, the allosteric modulator aplaviroc blocks binding of the HIV-1-encoded gp120 to CCR5 and subsequently suppresses the infectivity and replication of different HIV-1 strains in peripheral blood mononuclear cells (Maeda, Nakata, et al., 2004) and in an AIDS mouse model (Nakata et al.,

2005). Interestingly, this compound preserves CCL4 and CCL5 binding to CCR5 and consequent chemotaxis and receptor internalization (Maeda, Nakata, et al., 2004), but inhibits CCL3 binding (Watson et al., 2005). Furthermore, the presence of different chemokine receptor conformational (sub)populations is also involved in probe dependency since different chemokines may recognize distinct chemokine receptor conformations. This phenomenon is demonstrated for ORF74, where CXCL1, CXCL8, and CXCL10 bind to different ensembles of receptor conformations, which is shown by diverging observed receptor expression levels and/or partial heterologous radioligand displacement patterns when different probe ligands are applied in binding experiments (Verzijl et al., 2006).

High-affinity chemokine binding is frequently dependent on G protein coupling (Maeda, Kuroki, Haase, Michel, & Reilander, 2004; Springael et al., 2006; Vischer, Watts, Nijmeijer, & Leurs, 2011). In fact, G proteins act as PAMs on chemokine binding to their receptors. This is illustrated by CXCL12–CXCR4 binding which can be abrogated through $G\alpha_i$ protein scavenging (Nijmeijer, Leurs, Smit, & Vischer, 2010). In a similar fashion, uncoupling G proteins from CXCR2 and CXCR3 by the treatment with GTPγS or PTX eliminates CXCL8 and CXCL10 binding, respectively (Cox et al., 2001; de Kruijf et al., 2009). β-Arrestin also stabilizes a high-affinity state of the angiotensin II type 1A (AT_{1A}) receptor, the glucagon-like peptide-1 (GLP-1) receptor, the neurokinin NK1 receptor, and the thyrotropin-releasing hormone receptor (Gurevich, Pals-Rylaarsdam, Benovic, Hosey, & Onorato, 1997; Jorgensen, Martini, Schwartz, & Elling, 2005; Martini et al., 2002; Sanni et al., 2010), but this has not yet been described for CKRs.

2.3 Binding Experiments

Binding experiments using radioligands as probes have proven to be very successful in characterizing ligand–receptor interactions. Radioligand-binding experiments are applicable to virtually all chemokine-receptor pairs and have resulted in accurate and robust quantification of ligand-binding parameters, such as affinity and association/dissociation rate constants.

In addition to the specific on-target binding, there is always nonspecific binding of the radioligand to free adsorption sites on the assay plate and/or nonspecific protein/membrane sites. Nonspecific binding is increased linearly with ligand concentration and the nonspecific binding constant (k). Thus, the observed total binding in biological assays is the sum of specific and nonspecific binding (Eq. 8):

$$\begin{aligned}\text{total binding} &= \text{nonspecific binding} + \text{specific binding}\\ &= k[\mathrm{A}] + \frac{[\mathrm{A}]}{[\mathrm{A}] + K_\mathrm{d}}\end{aligned} \tag{8}$$

While performing binding experiments, one should consider that CKRs exist as an ensemble of native conformations on the cell surface (Kenakin & Miller, 2010). Cellular background or assay conditions can significantly shift the conformational equilibrium and consequently affect results. Membrane isolation could affect receptor interaction with G proteins and/or other intracellular molecules that allosterically affect chemokine binding to the receptor. This potentially explains differences in observed affinities and probe-dependent results in membrane versus intact cell experiments. In our lab, we frequently perform membrane and intact cell-binding experiments in parallel. In general, initial ligand and/or target characterization experiments are performed to evaluate the presence of experimental setup-dependent effects. When the results are comparable, we usually continue with membrane-binding experiments since this format allows high-throughput screening (e.g., 96-well format harvesting and washing) and is more flexible (e.g., membrane content per well can be varied to optimize assay, less stringent temperature, and buffer conditions) than intact cell binding. Below the different binding experiments (kinetic, saturation, and displacement) are described using CKR membranes.

2.3.1 Materials

Membrane isolation

- *Mammalian cells*, for general purposes we use HEK293T or COS-7 cells because of high transfection efficiency and CKR expression levels and ability to test mutant CKRs. If possible, physiologically relevant cell lines should be considered since binding characteristics are prone to be context dependent. Yet, the expression levels of CKRs in these cells are often low.
- *Phosphate-buffered saline (PBS) without Ca^{2+}/Mg^{2+}*, 137 m*M* NaCl, 2.7 m*M* KCl, 1.5 m*M* KH_2PO_4, and 8 m*M* Na_2HPO_4 (pH 7.4).
- *PBS/EDTA*, PBS without Ca^{2+}/Mg^{2+} supplemented with 1 m*M* EDTA.

- *Membrane buffer*, 15 m*M* Tris, 0.3 m*M* EDTA, 2 m*M* $MgCl_2$ (pH 7.5 at 4 °C). The addition of a protease inhibitor cocktail is recommended to prevent receptor degradation (cOmplete, Roche, 1 tablet/50 ml of membrane buffer).
- *Tris–Sucrose buffer*, 20 m*M* Tris, 250 m*M* sucrose (pH 7.4 at 4 °C).
- *Teflon pestle glass homogenizer (potter) with rotor (1000–1500 rpm)* for efficient and homogeneous cell disruption.
- *Liquid* N_2 for snap freezing of membrane suspensions.
- *Ultracentrifuge and polycarbonate tubes*, minimum speed 40,000 × *g* at 4 °C.
- *Syringe with 18-27G needle*, for homogenization of membrane pellet in Tris–sucrose buffer.

Binding experiments

- *Biological target*, isolated membranes from mammalian cells transiently or stably expressing CKR of interest.
- *HEPES binding buffer*, 50 m*M* HEPES–HCl, pH 7.4, 1 m*M* $CaCl_2$, 5 m*M* $MgCl_2$, 0.1 *M* NaCl. In case of chemokine binding, binding buffer should be supplemented with 0.5% (w/v) bovine serum albumin (BSA) to prevent nonspecific binding (e.g., adsorbent surface on assay plate) of the "sticky" chemokines.
- *High-affinity radioligand*, chemokines containing tyrosine residues can be ^{125}I-labeled in house with straightforward and efficient Iodogen® technique (de Munnik, Kooistra, et al., 2015; Gruijthuijsen et al., 2002). Many ^{125}I-labeled chemokines are commercially available from PerkinElmer, the ligands are labeled using lactoperoxidase or Bolton–Hunter techniques and subsequently purified ensuring a maximal specific activity of 2200 Ci/mmol. Since radioactivity decreases in time, it is important to determine the specific activity (Ci/mmol) and radioactive concentration (mCi/ml) at the time of experiment. This is essential since exact radioligand concentrations are required to determine K_d and calculate K_i. The universal decay calculator (PerkinElmer) is a convenient tool to determine specific activity and concentrations on the date of intended use.
- *Cold (unlabeled) ligands*: Chemokines (Peprotech, R&D systems), nanobodies, or small molecule ligands to displace the radioligand probe.
- *Unifilter-GF/C filter plates*, glass fiber membrane filter 96-well plates with a 1.2 μ*M* pore size (PerkinElmer) to collect radioligand-bound membranes.
- *0.5% Polyethylenimine (PEI) solution*, diluted with dH_2O from 50% PEI (Sigma-Aldrich) stock solution. Used to reduce nonspecific binding of radioligands to glass fiber.

- *HEPES wash buffer*, HEPES binding buffer with additional NaCl for washing of cells/membranes to remove residual unbound radioligand. 50 m*M* HEPES–HCl, pH 7.4, 1 m*M* $CaCl_2$, 5 m*M* $MgCl_2$, 0.5 *M* NaCl.
- *Filtration harvester* for collection of ligand-bound material. For radioligand membrane-binding experiments, a unifilter-96 harvester (PerkinElmer) is used to separate the radioligand-bound membranes from free radioligand on Unifilter-GF/C filter plates.
- *Scintillation fluid* (MicroScint™-O, PerkinElmer) for nonaqueous solutions is used in membrane-binding experiments to detect radioactivity.
- *Topseals and Backseals* (PerkinElmer), seals to prevent leaking of scintillation fluid.
- *Measurement.* In radioligand membrane-binding experiments, Microbeta liquid scintillation counter is used to quantify radioactivity of bound radioligands.

^{125}I-radiolabeling

- *Recombinant chemokine or nanobody (5–25 μg)*, protein sequence should contain available tyrosine residues for iodogen ^{125}I-labeling. Since tyrosine conjugation occurs randomly, radiolabeling can affect ligand affinity if residues are important for target recognition.
- *Na-^{125}I radionuclide (0.5 mCi)* (PerkinElmer), 100 mCi/ml (pH 8–11).
- *Pierce iodination-labeling tubes* (Thermo Fisher Scientific), glass tubes coated with iodogen reagent, the labeling reaction takes place at the surface of the water-insoluble oxidant.
- *Labeling buffer*, 125 m*M* Tris–HCl (pH 6.8), 150 m*M* NaCl.
- *Elution buffer*, 125 m*M* Tris–HCl (pH 7.4), 150 m*M* NaCl, freshly supplemented with 0.5% (w/v) BSA to prevent nonspecific binding of radiolabeled protein.
- *PD-10 desalting columns* (GE Healthcare), separates radiolabeled protein from free radionuclide.
- *L-Tyrosine* (Sigma-Aldrich), quenches labeling reaction by scavenging free radionuclide. 4 mg is dissolved in 1 ml labeling buffer. After vigorous vortexing, the solution is cleared by filtration (0.45 μm pore size) to remove residual undissolved L-tyrosine.
- *50% Trichloroacetic acid solution* (Sigma-Aldrich), used to precipitate protein.
- *Dose calibrator and gamma counter* to measure radioactivity.

2.3.2 Methods

Membrane isolation

- Cells are grown in 10 cm cell culture dishes. For transiently transfected cells, the optimal expression level is obtained 24–48 h posttransfection.
- In order to prevent protein degradation all further steps should be performed on ice.
- Cell culture medium is aspirated, and cells are washed with 5 ml/dish cold (4 °C) PBS without Ca^{2+}/Mg^{2+}.
- Cold (4 °C) PBS/EDTA 5 ml/dish is added and incubated on ice until the cells start to detach (10–20 min).
  - Note: For cells that detach easily (e.g., HEK293T cells), it is not necessary to use PBS/EDTA and cells can be detached using cold PBS and collected by resuspension.
- The cells are collected using a cell scraper and pooled in conical polypropylene centrifugation tubes.
- Centrifuge cell suspension at 1500 × *g* and 4 °C for 10 min. Remove supernatant, resuspend cells in 2.5 ml/dish PBS without Ca^{2+}/Mg^{2+}, and centrifuge again at 1500 × *g* and 4 °C for 10 min. During this step residual PBS/EDTA is removed.
- Discard supernatant and resuspend cells in 0.5 ml/dish membrane buffer.
- Transfer cell suspension to glass potter tube and homogenize using a rotor-connected Teflon pestle (1100–1200 rpm) by 10 strokes (up and down). Cell lysate is transferred to conical polypropylene tubes.
- Two freeze–thaw cycles of the cell lysates are performed using liquid N_2.
- Cell lysate is transferred to polycarbonate ultracentrifuge tubes and tared using membrane buffer. Ultracentrifugation is performed at 40,000 × *g* and 4 °C for 30 min to pellet the membranes.
- The supernatant is aspirated, and the membrane pellet is carefully washed using 1 ml Tris–sucrose buffer to remove residual membrane buffer.
- Tris–sucrose buffer (150 μl/dish) is added and the membrane suspension is homogenized by syringe-needle resuspension.
- The membrane suspension is aliquoted and snap-frozen using liquid N_2. Membranes aliquots are stored at −80 °C.
  - Note: CKR membranes are generally stable at −80 °C, and the aliquots can be stored for 1–2 years. Multiple freeze–thaw cycles greatly affect membrane stability and thus should be prevented.

- Membrane protein content can be determined using a protein estimation assay (e.g., Pierce BCA assay). The membrane protein content is necessary to estimate receptor expression levels (B_{max}) from binding data.

^{125}I-radiolabeling

- All experimental procedures involving radioactivity are performed in a protected environment. The radionuclide is continuously stored behind a lead screen in a fume hood. The labeling reaction is carried out at room temperature. Before and after labeling, the protein solutions are stored at −20 °C.
- The iodonation tube is washed with 100 μl labeling buffer.
- 35 μl labeling buffer, 10 μl (0.5 μg/μl chemokine or 1.0–2.5 μg/μl nanobody) protein solution, and 5 μl radionuclide are carefully added in consecutive order to the iodination tube. The solution should be exposed to the iodogen coating since the labeling reaction is restricted to the oxidant surface.
- The reaction mixture is incubated for 12 min, and every 2 min the tube is carefully tapped to expose the reagents to the oxidant surface. The total amount of added radioactivity [S] is measured by a dose calibrator.
- The labeling reaction is quenched by addition of 450 μl (4 mg/ml) L-tyrosine solution and incubated for 5 min to scavenge free reactive radionuclide.
- 10 μl reaction mix sample is added to 990 μl H_2O. This solution is split into two tubes containing 400 μl fractions (total + TCA precipitation). 100 μl 5% BSA solution and 100 μl 50% TCA are added to one tube to precipitate the labeled protein. The tubes are vortexed and incubated on ice for 30 min. The supernatant is added to counting tubes and radioactivity is determined using a gamma counter.
- The radiolabeled protein is purified from the reaction mixture using a PD-10 desalting column. The column is equilibrated by 20 ml elution buffer prior to loading of the reaction mixture. The purified material is collected in 1 ml fractions, and radioactivity is measured by a dose calibrator. Labeled proteins are usually eluted after 3–4 ml, whereas the free radionuclide is eluted much later (9–12 ml). The purified radiolabeled protein is aliquoted and stored at −20 °C. A radioligand sample (10 μl) is measured using a gamma counter.
- ^{125}I incorporation = (total − TCA)/total.
- Activity/protein (Ci/g) = ([S]·incorporation)/amount of protein.
- Specific activity (Ci/mmol) = activity/protein (Ci/g)/molecular weight (mmol/g)
- Concentration (mCi/ml) = (activity purified radiolabeled protein (cpm/ml)/specific activity (cpm/mmol)) · specific activity (mCi/mmol).

Association binding

- Radioligand (final concentration 10–100 p*M* for ^{125}I-chemokines and 50–500 p*M* for ^{125}I-nanobodies) in HEPES binding buffer supplemented with 0.5% BSA (HBB/BSA), 50 µl/well, is added to a 96-well plate. A fixed concentration of radioligand is used per association time curve. In order to correct for nonspecific binding, a saturating concentration of cold competitive ligand is incubated with the radioligand for each radioligand concentration and all time points.
- At different time points membranes in HBB/BSA are added (0.1–10 µg in 50 µl/well) to start radioligand association. Membranes are incubated at designated temperature (in general between 4 and 37 °C) on a plate shaker.
- Harvesting and measurement of binding reaction (see "harvesting and measurement" section below).

Dissociation binding

- Radioligand (final concentration 10–100 p*M* for ^{125}I-chemokines and 50–500 p*M* for ^{125}I-nanobodies) in HEPES binding buffer supplemented with 0.5% BSA (HBB/BSA), 50 µl/well, is added to a 96-well plate. A fixed concentration of radioligand is used per dissociation time curve.
- Membranes in HBB/BSA are added (0.1–10 µg in 50 µl/well) to start radioligand association. Membranes are incubated at designated temperature (in general between 4 and 37 °C) on a plate shaker until binding equilibrium is reached.
- At different time points, a saturating concentration of displacer (cold orthosteric ligand) is added to the binding reaction to initiate radioligand dissociation.
- Harvesting and measurement of binding reaction (see "harvesting and measurement" section below).

Saturation binding

- Increasing concentrations of radioligand (final concentrations ranging from 1 p*M* to 100 n*M* for ^{125}I-chemokines and ^{125}I-nanobodies) in HEPES binding buffer supplemented with 0.5% BSA (HBB/BSA), 25 µl/well, is added to 96-well plate. At least two curves should be pipetted, for total binding and nonspecific binding.
- For nonspecific binding of radioligand, a saturating concentration of cold competitive ligand in HBB/BSA (25 µl/well) should be added to each radioligand concentration.
- Binding reaction is initiated by adding membranes in HBB/BSA to the wells (0.1–10 µg/well in 50 µl), incubation at designated temperature on a plate shaker until binding equilibrium is reached.

- Harvesting and measurement of binding reaction (see "harvesting and measurement" section below).

Displacement binding

- Increasing concentrations of cold ligand in HEPES binding buffer supplemented with 0.5% BSA (HBB/BSA), 25 μl/well, is added to a 96-well assay plate.
- Radioligand (final concentration 10–100 p*M* for ^{125}I-chemokines and 50–500 p*M* for ^{125}I-nanobodies), 25 μl/well, is added to the plate. A fixed concentration of radioligand is used in displacement binding experiments.
- Binding reaction is initiated by adding membranes in HBB/BSA to the wells (0.1–10 μg/well in 50 μl), and incubation at designated temperature on a plate shaker until binding equilibrium is reached. If ligand kinetics are unknown, we usually perform a 2 h incubation. In most cases, this is more than sufficient to reach equilibrium and most chemokines and nanobodies reach equilibrium within 30 min of binding.
- Harvesting and measurement of binding reaction (see "harvesting and measurement" section below).

Harvesting and measurement

- Membrane binding incubations are terminated using the following procedures.
- GF/C filter plates are presoaked with 0.5% PEI solution in H_2O for at least 30 min at 22 °C prior to harvesting. PEI is used to minimize binding of free radioligand to glass fibers.
- Upon completion of time course, radioligand-bound membranes are collected on GF/C filter plates using a harvester. Membranes are washed, three times with ice-cold wash buffer to remove residual unbound radioligand.
  - Note: Since binding equilibrium is disrupted during harvesting, it is crucial that the washing steps are performed quickly and with ice-cold buffer. Time and temperature are important factors in ligand dissociation, so should be considered during harvesting procedures.
- GF/C filter plates are dried in an oven at 55 °C until dry; usually 30–45 min is sufficient. Add a backseal (supplied with GF/C plate) to the filter plate.
  - Note: The plates can also be dried O/N at room temperature and overdrying does not affect the results.
- Scintillation fluid (MicroScint™-O) is added (25 μl/well) using a multichannel or multidrop reagent dispenser.

- Add Topseal to filter plates to prevent leaking of scintillation fluid.
- Count radioactive decay of radioligand-bound membranes on GF/C filter plates using a liquid scintillation counter (Microbeta, PerkinElmer).
  - Note: After scintillation fluid addition, we normally include a delay of at least 3 h prior to counting of the plate. The delay results in decreased variation in the data.

2.3.3 Kinetic Binding

In ligand-binding experiments, it is usually a good first step to start with kinetic binding since this results in information regarding the independent association and dissociation rates, and the resultant time to reach equilibrium. The acquired data can be applied to efficiently design equilibrium (saturation and/or displacement) experiments. It is important to notice that equilibrium is reached more rapidly when ligand concentration is increased (Eq. 9). Upon equilibrium, the free radioligand is removed (e.g., by filtration of centrifugation) or a saturating concentration of cold (unlabeled) orthosteric ligand can be added to disrupt the equilibrium. From this point onward, the radioligand will dissociate in time until it is fully displaced from the target (Fig. 2).

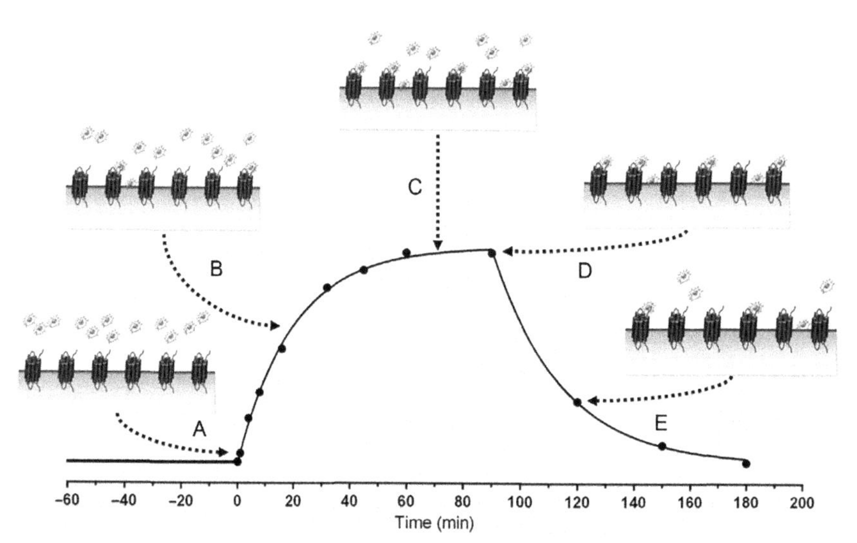

Figure 2 Time course of radioligand binding kinetics. Radioligand is introduced to the system (A) and allowed to bind (B) until equilibrium is reached (C). Free radioligand is removed or outcompeted by an excess of orthosteric cold ligand (D) which results in a time-dependent dissociation of radioligand (E).

Traditionally in drug development, affinity was considered the primary parameter for drug potential; however, recent studies have indicated that kinetics could play a prominent role in drug efficacy in the clinic. For this reason, drug target-residence time (RT) is frequently applied as a key pharmacological parameter as indication for therapeutic potential (Hoffmann et al., 2015; Planaguma et al., 2015; Swinney et al., 2014).

Important parameters in kinetic binding studies are the association and dissociation half-lives, which can be used to determine association and dissociation rate constants (k_1 and k_2) and thus also K_d (k_2/k_1) in an indirect manner. The dissociation half-life can be directly determined from the raw dissociation binding curve data when nonlinear regression is applied, since dissociation is only dependent on the dissociation rate constant (k_2) (Eq. 10). RT is the reciprocal of the dissociation rate and represents the time that a ligand resides on the receptor (Eq. 11). In contrast to k_2, an observed association rate constant (k_{obs}) is obtained from association rate curves, due to the fact that the association rate constant (k_1) is dependent on multiple variables, namely the association rate constant (k_1), radioligand concentration ([A]), and the dissociation rate constant (k_2) (Eq. 9). This is exemplified by the impact of association and dissociation curves when different radioligand concentrations are used. Higher radioligand concentrations result in a faster onset of the equilibrium, while the dissociation curve is unaffected by radioligand concentration (Fig. 3).

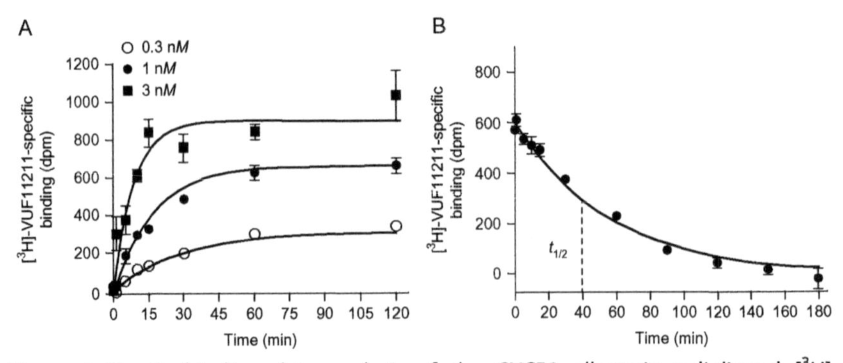

Figure 3 Kinetic binding data analysis of the CXCR3 allosteric radioligand [^{3}H]-VUF11211. (A) CXCR3 association binding with [^{3}H]-VUF11211 radioligand. With increasing radioligand concentration, equilibrium is reached more rapidly and results in more specific binding. (B) CXCR3 dissociation binding with [^{3}H]-VUF11211 radioligand. The dissociation half-life is only dependent on the dissociation rate constant (k_2) and independent of radioligand concentration. *Data adapted from Scholten et al. (2015).*

$$\text{association } t_{1/2} = \frac{\ln(2)}{k_{\text{obs}}} \text{ with } k_{\text{obs}} = k_1[\text{A}] + k_2 \tag{9}$$

$$\text{dissociation } t_{1/2} = \frac{\ln(2)}{k_2} \tag{10}$$

$$RT = \frac{1}{k_2} \tag{11}$$

Kinetic binding experiments also allow the distinction between orthosteric and allosteric binders, which is a challenge in displacement studies since allosteric modulators with small cooperativity factors ($\alpha < 0.1$) behave very similar to competitive ligands. However in kinetic dissociation studies the distinction is easily made. In the presence of an orthosteric ligand, tracer ligand dissociation should be entirely unaffected since this is only dependent on the tracer dissociation rate constant (Eq. 10). However upon introduction of an allosteric modulator, the receptor conformation is changed which may affect the affinity and thus also k_2 of the tracer ligand, which can be clearly observed from kinetic binding data. This is highlighted by a pyrazinyl-sulfonamide allosteric modulator targeting CCR4, through kinetic experiments it was demonstrated that this small molecule increases the dissociation rate constant of the orthosteric agonist CCL22 (Andrews, Jones, & Wreggett, 2008).

2.3.4 Saturation Binding

Saturation binding experiments are performed under equilibrium conditions with increasing concentrations of radioligand in the absence and presence of saturating concentration unlabeled competitor to determine total and nonspecific binding, respectively (Fig. 4).

Specific and saturable radioligand binding to the receptor is obtained by subtraction of nonspecific from total binding. The resultant data can be used to directly derive K_d and B_{max} values. Although exact knowledge of kinetic ligand parameters is valuable for experimental design of equilibrium experiments such as saturation binding, this information is not crucial, since a longer incubation period ensures equilibrium. It should be kept in mind that the incubation period should be based on the lowest concentration of the slowest ligand, since this condition is slowest in reaching equilibrium.

Dependent on the radioligand, diverging receptor expression levels (B_{max} values) can be detected, since different radioligands potentially recognize distinct ensembles of receptor conformations (see Section 2.2). Hypothetical data for saturation binding are depicted in Fig. 5, which shows a

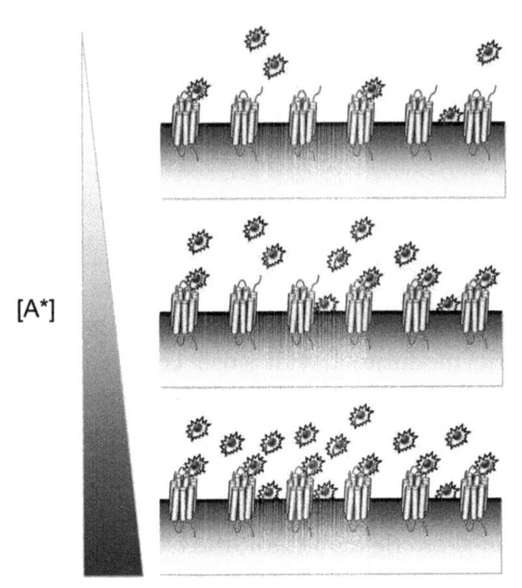

Figure 4 Saturation binding. Increasing radioligand concentrations results in saturation of specific binding sites.

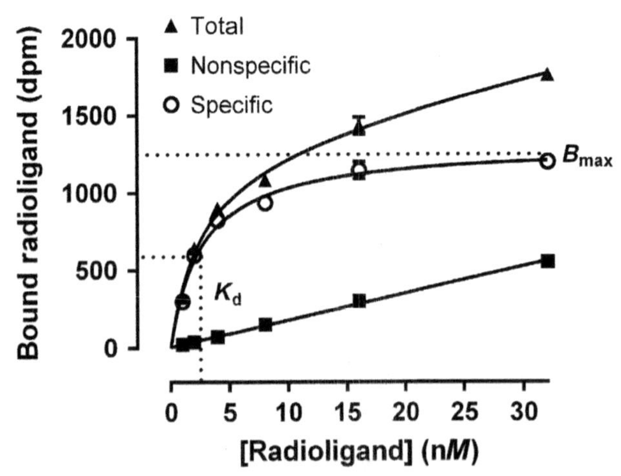

Figure 5 Saturation binding data analysis. Data of a saturation binding experiment depicting total, nonspecific, and specific binding. Specific binding is obtained by subtracting linear nonspecific binding from the total binding curve. The ligand concentration resulting in 50% of maximal specific binding is the K_d value, a parameter for affinity, whereas the horizontal asymptote is equal to the total number of receptor binding sites (B_{max}) in the experimental biological system.

total, nonspecific, and resultant-specific binding curve. The pharmacological parameters K_d and B_{max} can be obtained from the specific binding curve. The specific binding data can be transformed using a Scatchard plot. By dividing the specific binding (AR) with the radioligand concentration ([A]) on the y-axis and plotted against specific binding (AR) on the x-axis, the data are linearized to visualize tendencies (Eq. 12). In case of one-site binding, the trend line should be linear with a slope equal to K_d. If the trend is not linear, this is an indication for multiple independent binding sites. Furthermore, the linear x-axis intercept is equal to B_{max}. Caution should be taken with interpretation of Scatchard plot data, since the transformation violates the principles of linear regression (i.e., independent vs. dependent variables plot) and is prone to error.

$$\frac{[AR]}{[A]} = \frac{B_{max}}{K_d} - \frac{[AR]}{K_d} \tag{12}$$

2.3.5 Displacement Binding

Displacement or competition experiments are performed with a fixed radioligand concentration, usually in the range of 0.2–0.5 K_d, which is displaced with increasing concentrations of cold ligand (Fig. 6). In line with saturation binding, displacement studies are performed under equilibrium conditions. Displacement binding allows pharmacological characterization of nontractable ligands. This is a major advantage since a single radioligand can be used to determine the affinity of many unlabeled orthosteric ligands through competition for the same receptor binding site. Since a fixed (low) radioligand concentration is used in displacement studies, the required amounts of radioligand are diminished and nonspecific binding is reduced, while identical pharmacological parameters ($K_{d/i}$ and B_{max}) are obtained from saturation binding experiments, albeit in an indirect manner. The displacer can be chemically identical to the radioligand (homologous displacement) or be an entirely different ligand (heterologous displacement). Whereas the affinity of the radioligand is represented by equilibrium dissociation constant (K_d), the obtained affinity parameter of a competitive ligand is termed the inhibition constant (K_i) due to radioligand competition/inhibition.

By performing nonlinear regression $\left(Y = \text{bottom} + (\text{top} - \text{bottom}) / \left(1 + 10^{(X - \log \text{IC}_{50})}\right)\right)$ on displacement data, a negative sigmoidal curve

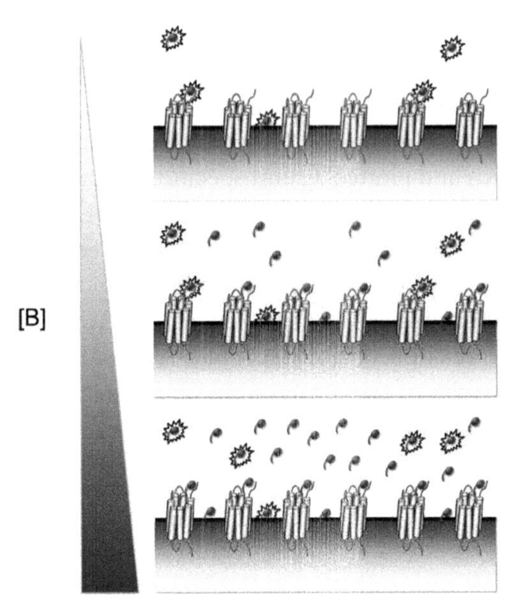

Figure 6 Displacement binding. Radioligand and cold ligand compete for receptor binding. A single radioligand concentration is displaced by increasing concentrations of competitive ligand.

should be obtained. An important point in this curve is the IC_{50} value, which reflects the concentration of cold ligand displacing 50% of maximal displacement by this ligand. It is important to realize that the IC_{50} value is dependent on the concentration and affinity of the radioligand. In competition binding, the equation for fractional receptor occupancy in the presence of a competitor (Eq. 6) can be applied, which dictates that when IC_{50} is equal to [B], $\rho_{AR,B} = 0.5 \times \rho_{AR}$. Rearrangement of fractional occupancy under these conditions yields the Cheng–Prusoff equation (Eq. 13) that is used to calculate affinity (K_i) from competitive displacement curve data (Cheng & Prusoff, 1973). Figure 7 depicts typical CKR binding results of homologous (Fig. 7A) and heterologous (Fig. 7B) competitive displacement assays.

$$K_i = \frac{IC_{50}}{1 + \left(\frac{[A]}{K_d}\right)} \tag{13}$$

The pharmacological parameter K_i is identical to K_d, a measure for intrinsic ligand affinity. However, since K_i values are calculated from radioligand displacement data, the interpretation of obtained values can

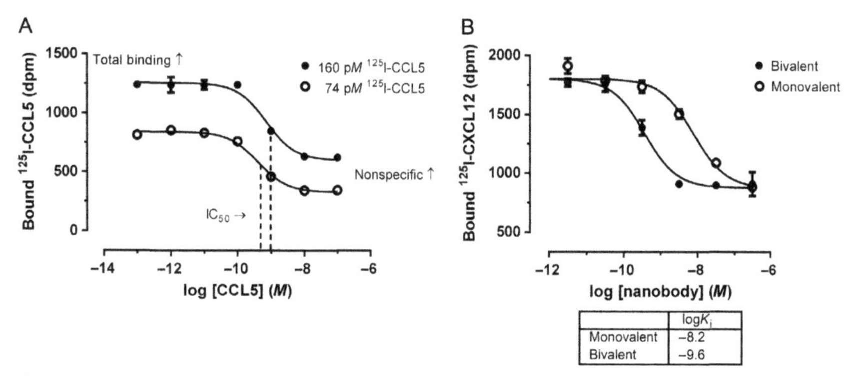

	logK_i
Monovalent	−8.2
Bivalent	−9.6

Figure 7 Competitive binding. (A) Homologous competitive binding of CCL5 to HCMV-encoded US28. Increasing the radioligand concentration results in increased total and nonspecific binding, and a higher IC_{50} (rightward shift of the curve) as more cold ligand is required to displace the radioligand. (B) Heterologous competitive binding of CXCR4-specific nanobodies. The affinity (K_i) of several cold ligands is determined by displacement of a single radioligand (^{125}I-CXCL12). The displacement curves indicate that bivalent nanobodies have a higher affinity for CXCR4 than their monovalent counterparts.

be ambiguous in comparison to K_d values that are determined from saturation binding. The determination of this parameter is not entirely independent (in contrast to K_d from saturation binding), since K_i is calculated from IC_{50} and radioligand concentration and affinity (Eq. 13). Therefore, caution is warranted in K_d determination from displacement data. Radioligand depletion (>10% of added radioligand is bound) also contributes to ambiguous results; if a large proportion of radioligand is bound, this results in a significant decrease of free radioligand that competes with unlabeled ligand for free binding sites, thus affecting the binding equilibrium and obtained IC_{50} values.

Competitive binding to the orthosteric binding site is characterized by an indefinite rightward shift of IC_{50} with increasing radioligand concentration (Fig. 7A). In contrast, noncompetitive binding of an unlabeled ligand, as a consequence of slow off rate, results in identical IC_{50} values in the presence of different concentrations radioligand (Fig. 8A). An example of a noncompetitive compound is JNJ-27141491, a CCR2 antagonist that blocks CCL2 chemokine function regardless of chemokine concentration (Buntinx et al., 2008). Displacement binding curves might also detect allosteric interaction between radioligand and unlabeled ligand. This is in particular relevant for CKR drug development since most small molecules

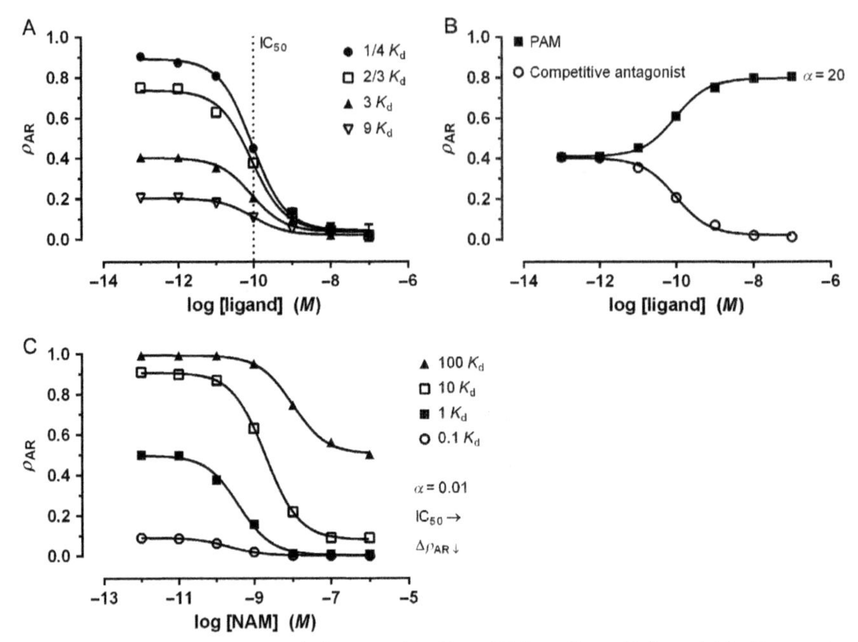

Figure 8 Noncompetitive and allosteric binding. (A) Specific radioligand binding to the receptor is inhibited by increasing concentrations of noncompetitive ligand. Radioligand concentration does not affect the IC_{50} of a noncompetitive binder. (B) Comparison of competitive ligand and positive allosteric modulator binding. A single concentration of radioligand is coincubated with increasing concentrations of competitive ligand (white circles) or positive allosteric modulator (black squares). Whereas the competitive ligand competes for the orthosteric site and thus displaces the radioligand, the positive allosteric modulator binds to a distinct site and induces a different receptor conformation, thus increasing radioligand affinity (20-fold increase at saturating PAM concentration, $\alpha = 20$), which results in a PAM concentration-dependent increase of radioligand binding. (C) Radioligand displacement by a negative allosteric modulator. Radioligand (four concentrations ranging from 0.1 to 100 K_d) is displaced by a curve of NAM concentrations. Binding of the NAM changes receptor conformation which results in a decrease of radioligand affinity (100-fold decrease at saturating NAM concentration, $\alpha = 0.01$) and thus a concentration-dependent displacement of radioligand. The IC_{50} is dependent on radioligand concentration. The allosteric modulation is saturable; consequently, this results in partial displacement of the radioligand, which is more apparent at higher radioligand concentrations.

targeting CKRs are allosteric modulators that bind to a distinct binding site as compared to chemokines (Scholten, Canals, Maussang, et al., 2012). PAM increases specific binding of the radioligand, resulting in sigmoidal binding curves contrasting the inverse sigmoidal curves of competitive binding (Fig. 8B). A number of structurally similar small molecule metal ion chelators increase specific binding of ^{125}I-CCL3 to CCR5-expressing cells and are thus characterized as PAMs. Interestingly, the same small molecules displace ^{125}I-CCL5 from CCR5, demonstrating probe dependency of allosteric modulation (Thiele et al., 2011). In contrast to PAMs, characterization of NAMs is less trivial because displacement curves of NAMs with a very small α (<0.1) may appear strikingly similar to orthosteric ligands when competition is performed against a single radioligand concentration. However, allosterism is saturable, and with increasing radioligand concentration, this might result in partial radioligand displacement with increased IC_{50} (Fig. 8C).

On the other hand, this partial radioligand displacement might also be the consequence of conformational selection. For instance, the cognate chemokines CCL17 and CCL22 recognize distinct CCR4 populations and can consequently not fully displace each other from the chemokine binding site (Viney et al., 2014).

Since allosteric ligands do not interact in a competitive manner with orthosteric radioligands, the observed IC_{50} values cannot be converted in assay-independent K_i values using the Cheng–Prusoff equation. Therefore, in case of allosteric modulation, IC_{50} values can be reported as a quantitative measure for affinity. Nevertheless, it should be considered that the IC_{50} value is not an independent variable since radioligand concentration affects this parameter. As a result, IC_{50} values can only be compared if identical assay conditions are applied.

Development of allosteric radioligands that bind the same (allosteric) binding site as unlabeled allosteric modulators allows quantification of K_i values of the latter, as shown using the radiolabeled allosteric inverse agonist VUF11211 (Scholten et al., 2015) and allosteric modulator RAMX3 on CXCR3 (Bernat, Heinrich, Baumeister, Buschauer, & Tschammer, 2012), and the allosteric antagonist CCR2-RA on CCR2 (Zweemer et al., 2013).

2.3.6 Binding Assay Optimization and Quality Control

The reversible two-state model (Eq. 1) assumes that free radioligand concentration is not significantly affected by target binding; however in practice, (free) ligand depletion may occur, thereby affecting the binding equilibrium and should be prevented. Generally up to 10% radioligand depletion is accepted for affinity determinations and can be reduced by increasing the assay volume without adjusting radioligand assay concentration and maintaining the same amount of biological content.

In equilibrium saturation and competition binding, the incubation period should be long enough to be certain of equilibrium and the conditions can be evaluated in kinetic binding experiments. The equilibrium time can be shortened through an increase of temperature; however, this may also affect membrane stability. Radioligand concentration also affects equilibrium time since the ligand association rate is linearly related to ligand concentration (Eq. 9); an increase in radioligand concentration should shorten radioligand equilibrium time.

In all radioligand experiments, it is critical to minimize nonspecific binding. Due to the hydrophobic nature of many proteins (e.g., membranes and chemokines) and small molecules, the assay buffer can be supplemented with nonspecific protein (e.g., BSA) or mild detergents (e.g., Tween, CHAPS) to saturate nonspecific adsorption sites or to solubilize membranes while preserving structural integrity, respectively; both supplements minimize nonspecific binding.

Finally, in all experimental setups, there should be a good signal-to-noise ratio (this also applies to functional experiments as described in Section 3). In case of radioligand binding, there should be a large assay window (i.e., signal-to-noise ratio) between total and nonspecific binding and minimal variation between the replicate data points. The Z'-factor can be used as measure for assay quality (Zhang, Chung, & Oldenburg, 1999). This statistical parameter reflects both signal-to-noise ratio and signal variation (Eq. 14). In our lab, we aim for Z' factors >0.5 since this ensures excellent assay quality. The equation requires the absolute values and standard deviations of the positive control (total binding) and negative control (nonspecific binding). If the Z' factor <0.5, this can be improved by increasing the amount of biological material (i.e., higher membranes content) to increase the number of specific binding sites:

$$Z' = 1 - \left[\frac{(3\text{SD total binding}) + (3\text{SD nonspecific binding})}{|\text{total binding} - \text{nonspecific binding}|}\right] \tag{14}$$

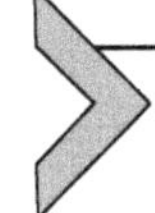

3. PHARMACOLOGICAL QUANTIFICATION OF CHEMOKINE RECEPTOR SIGNALING

In recent years, many functional assays have been developed to quantify the effect of ligands on CKR function. Concentration–response curves, in which the functional response of a ligand is plotted as a function of its concentration (Fig. 9A), provide information on ligand efficacy and

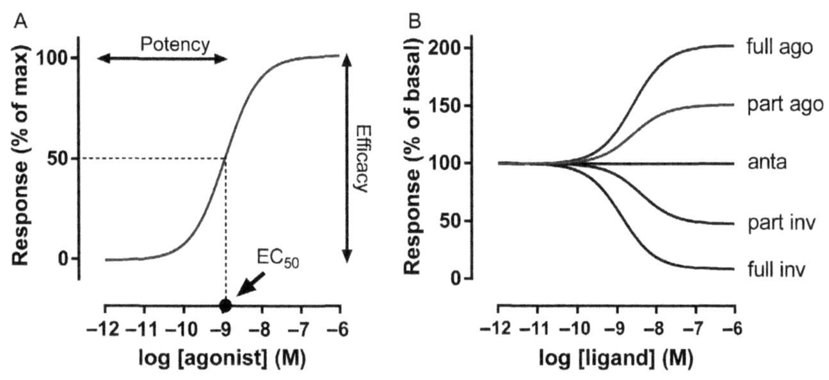

Figure 9 The potency and efficacy of ligands can be determined from concentration–response curves. (A) The potency of an agonist defines the location of the concentration–response curve along the concentration axis, whereas the efficacy of a ligand defines the location of the concentration–response curve along the response axis (but efficacy might also affect the ligand potency). Potency is often expressed as the EC_{50} value (indicated by an arrow). (B) Ligands can be divided in full (full ago) or partial agonists (part ago) (red (gray in the print version) curves), antagonists (anta) (black curve), or full (full inv) or partial inverse agonists (part inv) (blue (dark gray in the print version) curves). A partial agonist has a lower efficacy as compared to a full agonist, whereas inverse agonists have a negative efficacy. Antagonists have no efficacy or potency.

potency. The efficacy of a ligand determines its response at full receptor occupancy (Fig. 9A) (Kenakin, 2013). Ligands can have multiple efficacies, dependent on the response that is studied. This phenomenon is called biased signaling and is further discussed in more detail in chapter "Mutagenesis by Phage Display" by Bonvin et al. Potency is the propensity of a ligand to produce a response and is often expressed as an EC_{50} value, which is the ligand concentration that produces half of its maximal response (Fig. 9A). The potency of a ligand is dependent on its affinity but also on cell-based properties such as the efficiency by which the activated receptor couples to the downstream signaling cascade and receptor numbers. In efficiently coupled signaling pathways, not all receptors have to be occupied to produce the maximum response (i.e., receptor reserve). This produces a left-shift in potency as compared to receptor occupancy (affinity) (Kenakin, 2009b). The potency of an agonist might thus vary in different cell types or tissues.

Functional assays are able to discriminate between different classes of ligands (i.e., agonists, antagonists, and inverse agonists). Agonists elicit a response and full agonists can be distinguished from partial agonists as the latter produce a submaximal response at full receptor occupancy and thus have a lower efficacy in comparison to full agonists (Fig. 9B). Reciprocally, inverse agonists decrease basal signaling of constitutively active receptors. Antagonists bind to the receptor but elicit no response and consequently have no potency and efficacy (Fig. 9B). However, an antagonist can block (inverse) agonist-induced responses, which might be beneficial in situation of excessive agonism. How antagonists affect agonist EC_{50} values and/or efficacy depends on their binding site (orthosteric vs. allosteric) and kinetics, which can be determined in functional assays.

3.1 Determining the Mode of Action of Chemokine Receptor Ligands in Functional Studies

The first step to functionally characterize (new) CKR ligands is to determine the effect of the ligand on CKR signaling to identify potential (partial) agonists or inverse agonists. Next, the effect of these ligands on the response of a fixed concentration of agonist is determined. Orthosteric antagonists will inhibit the agonist-induced response till baseline at full receptor occupancy, whereas a NAM does not completely inhibit agonistic responses and consequently does not reach the baseline due to the saturation of the allosteric effect (unless α is very small, see Section 2.3.5) (Fig. 10). However, a similar elevation in baseline is observed when an orthosteric partial agonist at full receptor occupancy competes with a full agonist due to the (low level of) efficacy of the partial agonist and the two modes of action cannot be distinguished from each other using this functional analysis. Reciprocally, when the ligand is a competitive inverse agonist for a constitutively active

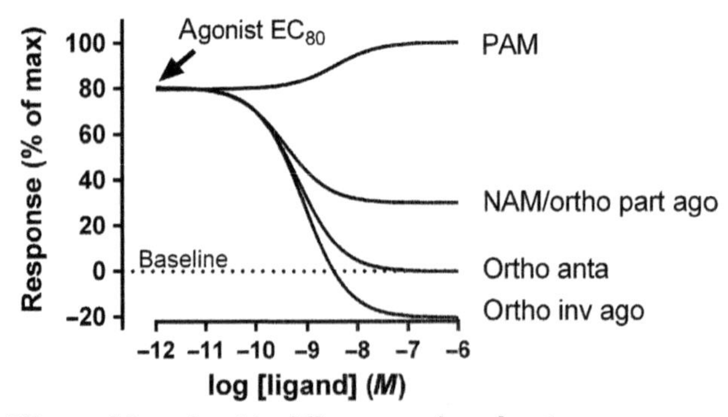

Figure 10 Effects of ligands with different modes of action on an agonist-induced response. By comparing the level of inhibition at full receptor occupancy with the baseline, a distinction can be made between an orthosteric antagonist (ortho anta), orthosteric inverse agonist (ortho inv ago), negative allosteric modulator (NAM), and positive allosteric modulator (PAM).

receptor, a response below baseline is observed at full receptor occupancy (Fig. 10). Potentiation of the agonistic response can only be explained by a PAM (Fig. 10). To distinguish these different modes of action, it is important to use an optimal agonist concentration (e.g., the EC_{80}). If the agonist concentration is too high, an increased response mediated by a PAM is not detectable. On the other hand, if the agonist concentration is too low, it is difficult to distinguish a competitive antagonist from a NAM, as also the NAM will approach the baseline (see also Fig. 8C).

To further elucidate the mode of action of a ligand, agonist concentration–response curves are generated in the absence or presence of increasing antagonist/allosteric modulator concentrations. An orthosteric antagonist produces a parallel rightward shifts of the agonist concentration–response curve (larger EC_{50} values, Fig. 11A), as higher concentrations of agonists are required to compete with the antagonist and obtain the same response as compared to the situation without antagonist. A characteristic of competitive antagonism is that the maximum response remains unaltered as a high concentration of agonist would always overcome the action of a competitive antagonist (if tested at equilibrium). Furthermore, the rightward shift of the agonist concentration–response curve is theoretically unsaturable (but is limited by issues such as solubility or toxic effects) due to the competitive nature of the antagonist. The magnitude of the rightward shift of the agonist concentration–response curve is quantified by equiactive dose ratios (DR) of agonist in the absence or presence of antagonist ($DR = EC_{50}[\text{presence of antagonist}]/EC_{50}[\text{absence of antagonist}]$). DR is solely dependent on the concentration ([B]) and the affinity (K_B) of the antagonist and is described by the "Schild equation" (Wyllie & Chen, 2007; Eq. 15):

$$\log(DR - 1) = \log[B] - \log K_B \tag{15}$$

When log(DR − 1) is plotted as a function of log[B] in a "Schild plot," a linear correlation between log(DR − 1) and log[B] with a slope of unity is observed for competitive antagonists (Fig. 11B; Wyllie & Chen, 2007), as observed for the inhibition of CXCL12-induced CXCR4 signaling by competitive nanobodies (Jahnichen et al., 2010).

A saturable decrease in orthosteric agonist affinity by increasing concentrations of allosteric modulator ($\alpha < 1$) is reflected by decreasing DRs (Fig. 11C) and consequently a nonlinear Schild plot with a downward curvature (Fig. 11D). This allows to distinguish orthosteric antagonists from

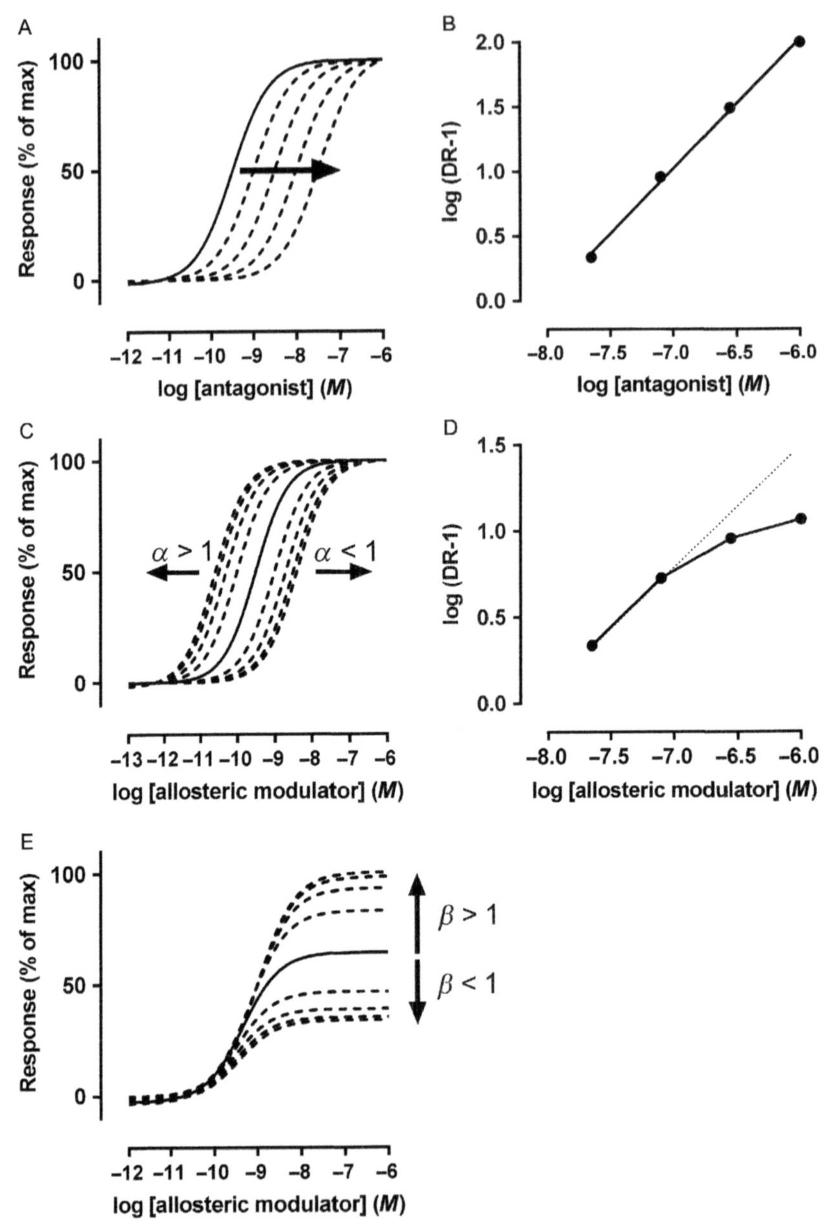

Figure 11 Effects of orthosteric antagonists and allosteric modulators on agonist concentration–response curves. (A) Orthosteric antagonists cause an unsaturable parallel rightward shift of the agonist concentration–response curve and consequently show a linear Schild plot with a slope of unity (B). Allosteric modulators can cause a saturated left- or rightward shift of the agonist concentration–response curve, quantified by cooperativity factor α (C), resulting in a curved Schild plot (D). Allosteric modulators might also increase or decrease the maximal response (quantified by cooperativity factor β) (E).

allosteric antagonists. When α is very small and a large allosteric modulator concentration range is required for the saturation effect to become evident (see Section 2.3.5), the use of a highly potent agonist is required to create enough "space" for the rightward shift of the agonist concentration–response curve. An upward curved Schild plot (reflecting increasing DRs with increasing concentrations antagonist/modulator) is often explained by toxic effects (Kenakin, 2009a).

Allosteric modulators can also change the efficacy of orthosteric ligands independent of the change in affinity, which is quantified by cooperativity factor β. For example, aplaviroc has no significant effects on CCL5 binding to CCR5, but inhibits CCL5-induced Ca^{2+} responses (Watson et al., 2005). If $\beta < 1$, the allosteric modulator decreases the maximal response of the agonist (Fig. 11E), which might be preceded by a rightward shift of the full agonist concentration–response curve in the presence of spare receptors. On the other hand, allosteric modulators with a value of $\beta > 1$ increase the maximal agonist response. Importantly, the effect of the allosteric modulator on the efficacy of the orthosteric ligand is also saturable (Fig. 11E; Slack et al., 2013).

On the other hand, a decreased maximal response might also be explained by noncompetitive (pseudo-irreversible or insurmountable) orthosteric antagonism, which occurs when a slow dissociating competitive antagonist cannot be displaced by an agonist within the time frame of the functional response (i.e., reequilibration). This results in a reduced number of receptors available for the agonist and a subsequent depression of the maximal response without changing the EC_{50} value, which cannot be reversed by high concentrations of agonist. Detection of noncompetitive orthosteric antagonism might be dependent on the presence of spare receptors.

When the allosteric modulator is difficult to distinguish from noncompetitive orthosteric antagonists (due to a very small value of β), additional unique allosteric effects such as probe dependency (see Section 2.2) should be identified. Alternatively, mutagenesis studies could be performed to determine the antagonist binding pocket. It is important to note that a lack of an allosteric effect could never be used as evidence against allosterism. For example, the CXCR3 antagonist NBI-74330 inhibited the maximal response of CXCL11-induced [^{35}S]-GTPγS binding (see Section 3.2.1) and Ca^{2+} mobilization (see Section 3.2.3) without any saturable effect. Moreover, NBI-74330 did not affect the dissociation rate of ^{125}I-CXCL11. Yet, different mutations in the transmembrane domain of CXCR3 affected

NBI-74330 binding without changing the affinity for CXCL11. This suggests that NBI-74330 and CXCL11 do not bind to the same binding site and that NBI-74330 acts as an allosteric CXCR3 ligand (Scholten et al., 2014). Since most of the classical pharmacological models are described for orthosteric interactions, it is challenging to analyze data of allosteric modulators. In order to study allosterism, it is prudent to perform multiple complementary experiments, and combining these data should give an indication for potential allosteric mode of action.

3.2 Functional Assays to Measure Chemokine Receptor Signaling

Most human CKRs preferentially activate $G\alpha_{i/o}$ proteins, resulting in the inhibition of cyclic adenosine monophosphate (cAMP) production by the enzyme adenylyl cyclase (AC). cAMP activates exchange proteins activated by cAMP (Epac) and protein kinase A (PKA) and the latter subsequently activates the transcription factor cAMP-responsive element (CRE). The released Gβγ subunits activate phospholipase C-β (PLC-β), resulting in the formation of inositol 1,4,5-triphosphate (IP_3) and diacylglycerol (DAG) from phosphatidylinositol-4,5-bisphosphate (PIP_2). IP_3 subsequently increases intracellular Ca^{2+} release from endoplasmic reticulum (ER) stores, which results in the activation of protein kinase C (PKC) and the transcription factor nuclear factor of activated T cells (NFAT) (Neptune & Bourne, 1997; Patel et al., 2013). Furthermore, CKRs are phosphorylated on their intracellular domains by G protein-coupled receptor kinases (GRKs) and subsequently recruit β-arrestins (Fig. 12).

This section describes several assays to measure G protein-dependent signaling of CKRs at different levels of the signal transduction pathway (from activation of G protein to accumulation of second messengers and activation of transcription factors), β-arrestin recruitment, downstream responses (e.g., chemotaxis), or integrated responses such as changes in impedance. Protocols, required materials, and points of attention are described for each assay. Mock-transfected cells (lacking expression of the CKR of interest) and/or receptor-specific NAMs or orthosteric antagonists should be taken along as a control for the specificity of the agonist-induced signal. Likewise, small molecule inhibitors (or activators) should be validated to exclude off-target effects (especially when high concentrations are used). See for further details chapter "Preparation and Characterization of Glycosaminoglycan Chemokine Co-Receptors" by Kitic et al.

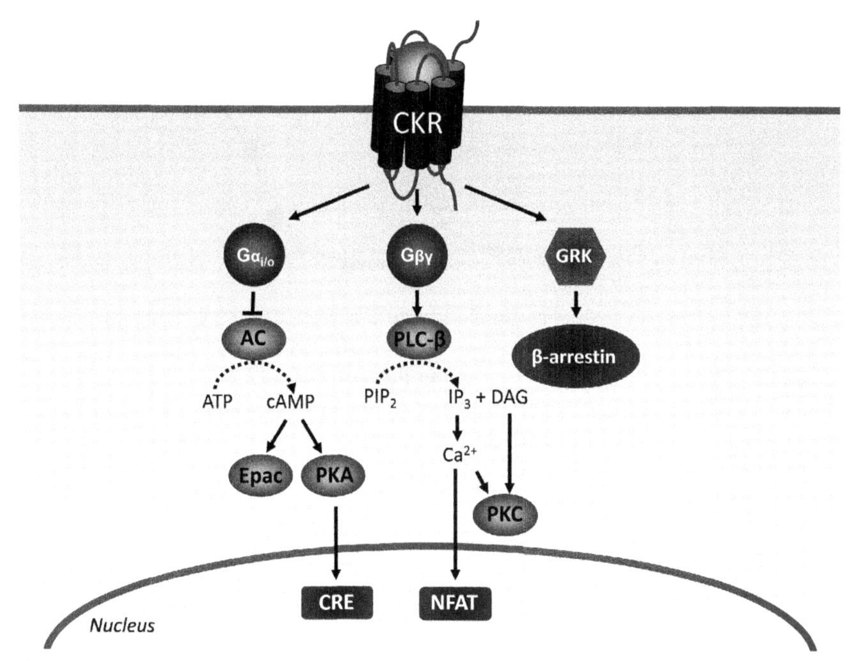

Figure 12 CKR signaling. CKRs preferentially signal via $G\alpha_{i/o}$ proteins and recruit β-arrestin upon receptor activation. Abbreviations: AC, adenylyl cyclase; ATP, adenosine triphosphate; cAMP, cyclic adenosine monophosphate; CRE, cAMP-responsive element; DAG, diacylglycerol; Epac, exchange proteins activated by cAMP; GRK, G protein-coupled receptor kinase; IP_3, inositol 1,4,5-triphosphate; NFAT, nuclear factor of activated T-cells; PIP_2, phosphatidylinositol-4,5-bisphosphate; PKA, protein kinase A; PKC, protein kinase C; PLC-β, phospholipase C-β.

3.2.1 G Protein Activation: GTPγS-Binding Assay

Activation of CKRs leads to the release of GDP from the Gα subunit and the formation of a high-affinity complex between the activated receptor and G protein. This is followed by binding of GTP to the Gα subunit, the dissociation of the Gα and Gβγ subunits, and the subsequent activation of downstream effector proteins (Oldham & Hamm, 2008). By using a radiolabeled and poorly hydrolysable analog of GTP ($[^{35}S]$GTPγS), G protein activation can be measured by counting $[^{35}S]$-incorporation. This assay is performed on isolated membranes and the method is quite similar to the binding protocol described in Section 2.3.2. For example, this assay has been used to determine CXCR3 activation in response to CXCL11 and the small molecule VUF10661 (Fig. 13).

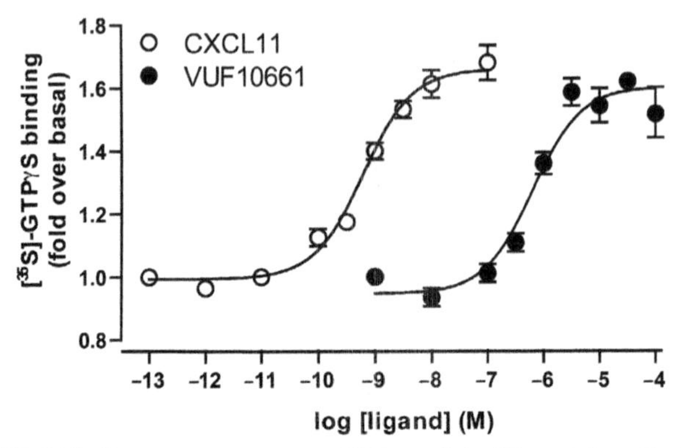

Figure 13 GTPγS binding assay to measure CKR-mediated G protein activation. CXCL11 and the small molecule VUF10661 dose-dependently increase [^{35}S]-GTPγS binding at membranes from HEK293-CXCR3 stable cells. *Figure was adapted from Scholten, Canals, Wijtmans, et al. (2012).*

Materials

- *Biological target*, isolated membranes (see Section 2.3.2) from mammalian cells transiently or stably expressing the CKR of interest.
- *Assay buffer*, 50 m*M* HEPES, 10 m*M* $MgCl_2$, 100 m*M* NaCl (pH 7.2) supplemented with 5 μg/well saponin, 3 μ*M* GDP.
- *Wash buffer*, 50 m*M* Tris–HCl, pH 7.4.
- *[^{35}S]-GTPγS* (PerkinElmer).
- *GF/B UniFilter plate* (PerkinElmer), to separate bound from unbound [^{35}S]-GTPγS.
- *Filtration harvester, scintillation liquid, topseals and backseals, and Microbeta liquid scintillation counter*, as described in Section 2.3.1.

Methods

- 500 p*M* [^{35}S]-GTPγS (25 μl/well) in assay buffer is added to a 96-well plate (Greiner).
- Increasing concentrations ligand (25 μl/well) are added to the plate.
- Membranes (2–5 μg/well, 50 μl/well in assay buffer) are added to the plate to initiate the binding reaction and incubated for 1 h at 22 °C.
- [^{35}S]-GTPγS-bound membranes are collected by filtration through a GF/B filter plate using a harvester. Membranes are washed quickly three times with ice-cold wash buffer to remove free [^{35}S]-GTPγS.
- GF/B filter plates are dried at 55 °C for approximately 1 h and a backseal is added, prior to the addition of scintillation liquid (25 μl/well).
- Bound [^{35}S]-GTPγS binding is quantified by a Microbeta scintillation counter after including a 3 h delay.

GDP concentrations (ranging from 0 to 10 μ*M*), [^{35}S]-GTPγS concentrations, and the amount of membranes might need to be optimized for each receptor of interest. This assay is especially suited for $G\alpha_{i/o}$-coupled receptors such as CKRs, as $G\alpha_{i/o}$ proteins are generally more abundant than other G proteins (Harrison & Traynor, 2003). However, the GTPγS binding assay can also be performed for $G\alpha_s$- or $G\alpha_q$-coupled receptors such as ORF74 in combination with immunoprecipitation of [^{35}S]-GTPγS-bound Gα (Liu, Sandford, Fei, & Nicholas, 2004). The advantage of this assay is that it measures CKR activation proximal to the receptor, resulting in signals that are less subjected to downstream modulations (e.g., by cross talk with other signaling pathway) (Harrison & Traynor, 2003). A disadvantage is that cytosolic proteins allosterically modulating the receptor (see Section 2.2) might be removed during membrane isolation (Strange, 2010).

3.2.2 cAMP Production

CKR-mediated inhibition of cAMP production can be quantified by an Epac-based bioluminescence energy transfer (BRET) biosensor (CAMYEL), in which a BRET donor (Renilla luciferase 8, Rluc8) and an acceptor (enhanced yellow fluorescent protein, eYFP) are both fused to the cAMP target protein Epac. In the absence of cAMP, Rluc8 and eYFP are in close proximity and BRET can be measured upon adding the Rluc8 substrate coelenterazine-h. However, Epac changes conformation upon cAMP binding, resulting in an increased distance between Rluc8 and eYFP and consequently a decrease in BRET (Fig. 14A). Importantly, forskolin (FSK) is added for the direct activation of adenylyl cyclase to increase basal cAMP levels and to create a window for measuring inhibition of cAMP formation by $G\alpha_i$-coupled CKRs, such as CXCR3 (Fig. 14B). This results in an inhibitory concentration–response curve, when expressed as percentage of the FSK-induced response (Fig. 14B). Furthermore, the phosphodiesterase (PDE) inhibitor isobutylmethylxanthine (IBMX) is often added in cAMP assays to inhibit cAMP degradation.

PKA-based cAMP sensors with the BRET donor and acceptor fused to the regulatory and catalytic subunits of PKA, respectively, have the disadvantage that PKA forms a heterotetrameric complex as compared to the single-chain Epac sensor. Hence, Rluc8- or eYFP-tagged PKA subunits might interact with endogenous PKA subunits and not all BRET proteins will form a functional BRET sensor. The Epac-based sensors show increased BRET efficiency, a better signal-to-noise ratio, faster activation kinetics,

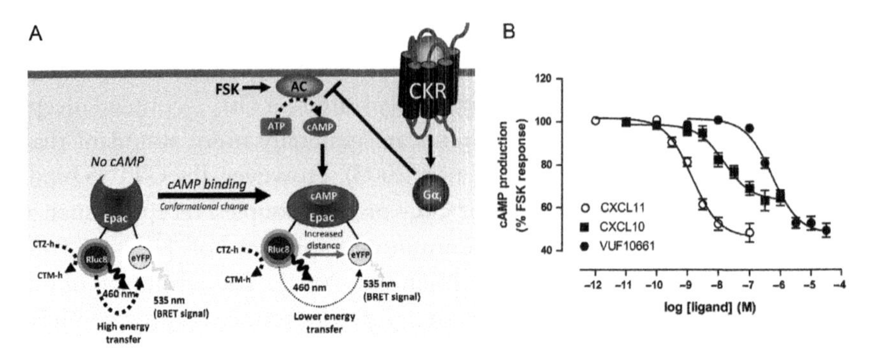

Figure 14 BRET-based Epac sensor to measure CKR-mediated inhibition of cAMP production. (A) In the absence of a CKR agonist, energy is transferred from Rluc8 to eYFP upon oxidation of the Rluc8 substrate coelenterazine-h (CTZ-h) to coelenteramide-h (CTM-h). Forskolin (FSK)-induced activation of adenylyl cyclase (AC) results in the production of cAMP, which binds to Epac and subsequently induces a conformational change, resulting in a decreased BRET. CKR activation inhibits the production of cAMP and consequently increases BRET between Rluc8 and eYFP. (B) The chemokines CXCL10 and CXCL11 and the small molecule VUF10661 inhibit FSK-induced cAMP accumulation in a dose-dependent manner in CXCR3-expressing HEK293 cells. *Figure was adapted from Scholten, Canals, Wijtmans, et al. (2012).* (See the color plate.)

and an extended dynamic range as compared to PKA-based sensors (Salahpour et al., 2012; Sprenger & Nikolaev, 2013; Willoughby & Cooper, 2008).

Materials

- *Biological target*, we often use HEK293T cells transiently transfected with 250 ng/10^6 cells CKR DNA in combination with 1000 ng/10^6 cells CAMYEL using PEI (see de Munnik, Kooistra, et al., 2015 for transfection protocol). However, these amounts may be optimized for each receptor.
- *Hank's Balanced Salt Solution (HBSS)* (Thermo Fisher Scientific).
- *Coelenterazine-h* (Promega), the Rluc8 substrate.
- *Forskolin*, directly activates AC and subsequently increases basal cAMP levels.
- *IBMX*, a PDE inhibitor which increases cAMP levels (for example from Sigma-Aldrich).
- *Multilabel plate reader*, suitable to measure BRET (emission 535 nm) and Rluc8 (emission 480 nm), such as Victor3 (PerkinElmer), PheraStar (BMG Labtech), or Mithras (Berthold Technologies).

Methods

- HEK293T cells expressing CKR and CAMYEL are seeded (50,000 cells/well) in poly-L-lysine-coated white 96-well plates (Greiner) 24 h posttransfection.
- 48 h posttransfection, cells are washed with HBSS (100 μl/well) to remove medium prior to the addition of 40 μl/well HBSS (or 20 μl when antagonists are added).
- Cells are subsequently incubated with 5 μ*M* (final concentration) coelenterazine-h and 40 μ*M* (final concentration) of the PDE inhibitor IBMX in HBSS (20 μl/well) for 5 min at 37 °C. Real-time BRET measurements to study temporal cAMP dynamics should be conducted in the absence of IBMX.
  - Antagonists might be preincubated for 30–60 min (depending on their binding kinetics) prior to the addition of coelenterazine-h (20 μl/well).
- Next, increasing concentrations of agonist in HBSS (20 μl/well) are added and cells are incubated at 37 °C. Incubation time should be optimized for different CKR subtypes and depends on binding kinetics of agonist and antagonists (if used). When incubations are longer than 30 min, coelenterazine-h might be added 5 min before the read-out to prevent substrate depletion. Alternatively, the long-acting substrate EnduRen (Promega) might be used. As an example, CXCL11 was incubated for 10 min to measure CXCR3-mediated inhibition of cAMP formation.
- For $G\alpha_i$-coupled receptors, cells are then incubated with 1 μ*M* FSK (final concentration) for 5 min at 37 °C (20 μl/well). The amount of FSK might be optimized as the EC_{50} value for FSK might vary for different cell types (Hill, Williams, & May, 2010). Too high concentrations of FSK are difficult to overcome by $G\alpha_i$-coupled receptors. Adding FSK before the agonist has bound to the receptor might lead to high-preformed cAMP that cannot be inhibited by CKR signaling in the presence of a PDE inhibitor.
- BRET (emission 535 nm) and Rluc8 luminescence (emission 480 nm) are measured for 0.5–1 s/well using a multilabel plate reader, such as Victor3 (PerkinElmer), PheraStar (BMG Labtech), or Mithras (Berthold Technologies). The BRET ratio (535 nm/480 nm) is calculated to correct for (minor) variation in expression of the Epac-based cAMP biosensor.

Prior to the dilution of (expensive) chemokines and coelenterazine-h, expression of the Epac biosensor could easily be confirmed by fluorescent microscopy or quantified in a plate reader. Since coelenterazine-h is light sensitive, assay plates should be protected from light during the incubation steps.

Homogeneous time resolved fluorescence-based kits from Cisbio and PerkinElmer (Lance kit) are commercially available to measure cAMP accumulation after 30 min of agonist stimulation. In the Cisbio kit, d2-labeled cAMP competes with endogenous cAMP for the binding to a Eu^{3+} cryptate-labeled anti-cAMP antibody. Energy transfer between d2 and Eu^{3+} is thus inversely proportional to the endogenously formed cAMP concentration and the signal increases in response to the activation of $G\alpha_i$-coupled receptors. The same assay principle applies to the Lance kit from PerkinElmer that uses an Eu^{3+} chelate-labeled cAMP tracer and a cAMP-specific antibody labeled with the ULight™ dye.

Importantly, the signal-to-noise window for $G\alpha_i$-coupled receptors might be negatively influenced if only a fraction of the cells express CKR (e.g., transient transfections), whereas FSK increases cAMP in all cells.

3.2.3 Reporter Gene Assay

In a reporter gene assay, CKR-mediated activation of transcription factors can be quantified. A plasmid encoding the CKR of interest is cotransfected with a plasmid encoding a reporter protein of which expression can easily be detected, such as firefly luciferase (Fluc) (or another bioluminescent protein or enzyme such as Gaussia luciferase (Gluc) or β-galactosidase). Luciferase transcription is controlled by an inducible promoter containing multiple response elements for a specific transcription factor. Hence, activation (or inhibition) of this transcription factor in response to CKR activation leads to the expression of luciferase (Fig. 15A).

CRE-luciferase reporter gene assay is used to measure inhibitory signaling of $G\alpha_i$-coupled CKRs to AC in the presence of FSK (Fig. 15A). For example, this CRE-luciferase reporter gene assay was used to measure the CXCR4 signaling in response to CXCL12 (Fig. 15B). The small molecule AMD3100 inhibits the CXCL12 response, resulting in an increase in luminescence (Fig. 15C).

Materials

- *Biological target*, we often use HEK293T cells transiently transfected with 500 ng/10^6 cells CRE-luciferase DNA in combination with 100–500 ng/10^6 cells CKR DNA using PEI (the amount of CKR DNA should be optimized for each receptor; see below).

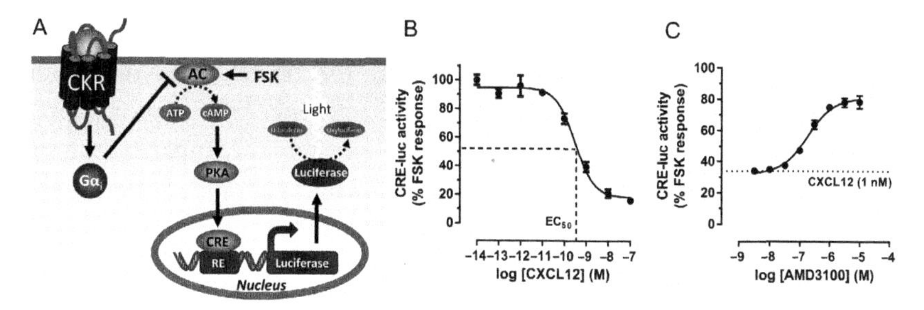

Figure 15 Reporter gene assay to measure CKR-induced inhibition of CRE activity. (A) Forskolin (FSK) activates adenylyl cyclase (AC) to increase basal cAMP-responsive element (CRE) activation and to create a window for measuring inhibition of CRE activation. Activated CRE translocates to the nucleus where it binds to the CRE response element (RE) in the promoter region that regulates luciferase expression. The substrate D-luciferin is converted to oxyluciferin by luciferase and luminescence can be measured to quantify CRE activation. CKR activation inhibits AC and consequently inhibits CRE activation. (B) CXCL12 dose-dependently inhibits FSK-induced CRE activity in CXCR4-expressing HEK293T cells. (C) The small molecule AMD3100 dose-dependently inhibits CXCL12 (1 n*M*)-induced activation of CXCR4, resulting in an increase in CRE activity. Abbreviations: ATP, adenosine triphosphate; cAMP, cyclic adenosine monophosphate; PKA, protein kinase A.

- *Forskolin*, directly activates AC and subsequently increases basal CRE activation.
- *Luciferase assay reagent*, 0.82 m*M* ATP, 230 μg/ml beetle luciferin (Promega, Madison, WI), 18.4 m*M* $MgCl_2$, 77 μ*M* $Na_2H_2P_2O_7$, 23.7 m*M* Tris–H_3PO (pH 7.8), 23.7% (v/v) glycerol, 1.6% (v/v) Triton X-100, and 527 μ*M* dithiothreitol.
- *Multilabel plate reader* (suitable to measure luminescence), such as Victor3 (PerkinElmer), PheraStar (BMG Labtechnologies), or Mithras (Berthold Technologies).

Methods

- HEK293T cells expressing CKR are seeded (50,000 cells/well) in poly-L-lysine-coated white 96-well plates 24 h posttransfection.
- 24 h after seeding the cells, growth medium is replaced by low-serum (0–1% serum) medium supplemented with 0.05% BSA, 1 μ*M* FSK, and agonist in the absence or presence of an antagonist and cells are incubated for 5–6 h at 37 °C/5% CO_2.
- After incubation, stimulation medium is removed and cells are incubated for 30 min with 25 μl/well of luciferase assay reagent (protected from light).
- CRE activity is quantified by measuring luminescence in a multilabel plate reader.

Transfection amounts might be optimized depending on the CKR. Too high CKR expression levels might saturate the translational machinery at the expense of luciferase expression. The influence of receptor expression on luciferase expression may give problems when comparing the signaling of different receptors (for example, wild type and mutants) that show differences in expression levels. A dual luciferase reporter system can be used in this case, in which Renilla luciferase (Rluc) is coexpressed with the Fluc reporter gene. Rluc expression (controlled by a constitutive promoter) serves as an internal control for the baseline response, which might be altered by receptor expression levels independent of receptor signaling properties. Fluc activity is normalized to Rluc activity to correct for this variability. In this case, the Rluc substrate coelenterazine-h should be added together with the Fluc substrate luciferin and Rluc and Fluc activity can be distinguished by measuring luminescence at 480 and 610 nm, respectively. Alternatively, both signals can be measured consecutively after quenching the first signal or by dividing the samples and individually incubating them with each substrate. Dual luciferase reporter systems are commercially available.

3.2.4 Ca^{2+} Mobilization

Many chemokine–CKR interactions result in a rapid, yet transient intracellular release of Ca^{2+} from ER stores. It has been postulated that this fast response directs chemotaxis, a key function of the chemokine system (Neptune & Bourne, 1997; White, Iqbal, & Greaves, 2013). Upon binding of the agonist, the $G\alpha_i$–$\beta\gamma$ heterotrimeric complex dissociates and is activated. The $\beta\gamma$-complex is thought to be primarily responsible for mobilization of intracellular calcium stores by direct stimulation of phospholipase C beta (PLC-β) and thus the inositol phosphates cascade. Alternatively, through the activation of phosphoinositide 3-kinase (PI3K), the $G\alpha_i$ subunit and also the $\beta\gamma$-complex can indirectly stimulate PLC-β and consequently promote Ca^{2+} release (Curnock, Logan, & Ward, 2002; White et al., 2013). Calcium signaling upon agonist binding can be detected in real time by application of cell permeable fluorescent dyes that recognize intracellular Ca^{2+} with high affinity, as exemplified by CXCL12-induced Ca^{2+} mobilization in CXCR4-expressing cells (Fig. 16). We prefer to use fluo-4 acetoxymethyl ester (Gee et al., 2000) as Ca^{2+} tracer because the dye has excellent characteristics in terms of rapid cell permeability, bright fluorescence upon calcium binding, and limited leakage upon inhibition of organic anion transporters by probenecid.

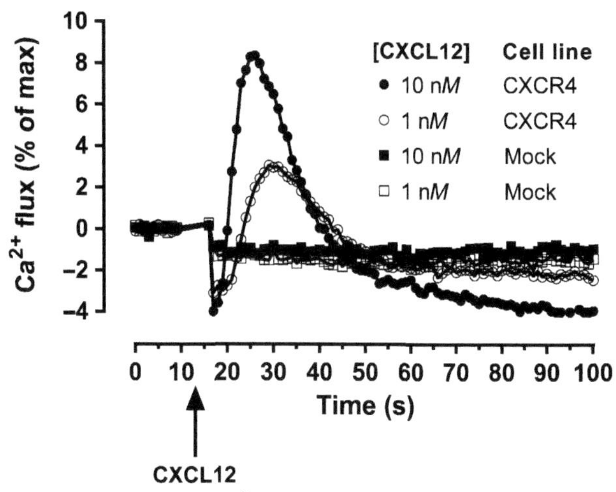

Figure 16 Chemokine-induced Ca^{2+} mobilization. Stimulation with 1 n*M* CXCL12 (arrow) induces a rapid and transient increase in Ca^{2+} mobilization in CXCR4-expressing leukemic lymphoblast K562 cells, as determined by a fluorescent Ca^{2+} dye. Stimulation with 10 n*M* CXCL12 results in a faster and larger response. CXCL12 stimulation is unable to induce Ca^{2+} mobilization in Mock-transfected K562 cells, indicating a CXCR4-specific effect. Data are plotted as percentage of the maximum signal as determined upon cell lysis.

Materials

- *Biological target*, cells stably transfected with CKR (e.g., SVEC or K562 cells) or cells endogenously expressing the receptor of interest,
- *Assay buffer*, HBSS, supplemented with 20 m*M* HEPES and 0.5% BSA.
- *Fluo-4 NW* (Thermo Fisher Scientific), cell permeable fluorescent calcium ester dye.
- *Probenecid water soluble* (Thermo Fisher Scientific), inhibitor of organic anion transporters, prevents leakage of dye to medium.
- *5% (v/v) Triton X-100/H_2O solution*, used to lyse cells and determine maximal calcium signal which is used as internal control for cell number variation and nonuniform dye loading.
  - *Optional: 10 μM ATP*, alternative internal control for maximal GPCR-dependent calcium response via purinergic receptors.
  - *Optional: 50 mM EDTA*, metal-chelating agent scavenges free Ca^{2+}, results in decrease of fluorescence and serves as negative control.
- *FlexStation or FLIPR microplate reader with integrated pipetting system* (Molecular devices), applied for ligand pipetting and real-time fluorescence measurement of calcium response.

Methods

- 10,000–100,000 cells/well are seeded in black 96-well plates with clear F-bottom (Greiner). Adherent cells are seeded in culture medium in poly-L-lysine-coated plates 1 day before the calcium assay. Suspension cells can be seeded on the assay day in assay buffer.
  - Note: The optimal cell number is cell line- and receptor-dependent and should be empirically determined. For example, for efficient CXCR4 calcium signaling in K562 cells (nonadherent cell line), we generally seed 100,000 cells/well.
- Medium is aspirated from adherent cells after O/N incubation at 37 °C/5% CO_2.
- Cells are loaded with fluo-4 NW (according to supplier instructions) dye and 2.5 m*M* probenecid solution (final concentration) and incubated for 30 min at 37 °C.
  - Optional: In case of antagonistic assay setup, cells can be preincubated with antagonist in parallel.
- In the meantime, the plate reader system is primed with assay buffer, ligand solutions, and 5% Triton X-100.
- Assay and ligand plates are put in the plate reader and calcium flux assay protocol is performed.
  - Example fluo-4 NW protocol (CXCR4 signaling in K562 cells).
    - Excitation: 485 nm; emission: 520 nm.
    - 10 s baseline measurement.
    - Chemokine/buffer injection, record fluorescent signal for 30–60 s.
      - Optimal temporal calcium tracing should be empirically determined for each experimental setup.
    - 5% Triton X-100 is injected to lyse cells and obtain the maximal internal signal, fluorescent trace is recorded for 10 s.

During data analysis of calcium mobilization experiments, it is important to consider that the different wells are measured in a consecutive manner, which might result in an increase of background fluorescence in time due to leakage of the dye. Furthermore, minor variations in cell number and/or dye loading can result in significant raw data variation. Thus, proper data normalization is imperative for accurate and quantitative results. For each well, the data are corrected for the baseline fluorescence and expressed in ΔRFU. Next, the calcium trace data are plotted as percentage of the maximal signal (Ca^{2+} influx due to cell lysis). In conclusion, calcium mobilization assays give an elegant real-time insight in CKR signaling and the automated setup and assay robustness enable HTS applications. However, due to the rapid and transient nature of calcium responses, calcium mobilization is often measured under hemi-equilibrium binding conditions. Hence, these calcium assays largely preclude the use of methods that assume equilibrium, such as a Schild plot analysis.

3.2.5 β-Arrestin Recruitment

β-Arrestin recruitment can be quantified using a BRET-based assay, as described in detail in chapter "Analysis of Arrestin Recruitment to Chemokine Receptors by Bioluminscence Resonance Energy Transfer" by N. Heveker. Alternatively, in the enzyme fragment complementation (EFC)-based β-arrestin recruitment PathHunter™ assay from DiscoverX, the receptor and β-arrestin are each fused to a β-galactosidase fragment, which are inactive on their own. However, when β-arrestin is recruited to the receptor, these fragments reconstitute to form an active β-galactosidase enzyme that generates a chemiluminescent signal (Fig. 17A).

Materials

- *Biological target*, PathHunter CHO-K1 cells stably expressing CKR-prolink and β-arrestin-EA (cultured in DMEM/F12, supplemented with 10% FBS, 1%, 250 μg/ml hygromycin, and 400 μg/ml G418).
- *Assay medium*, Optimem (Invitrogen) supplemented with 1% FBS.
- *Stimulation buffer*, PBS + 0.2% BSA.

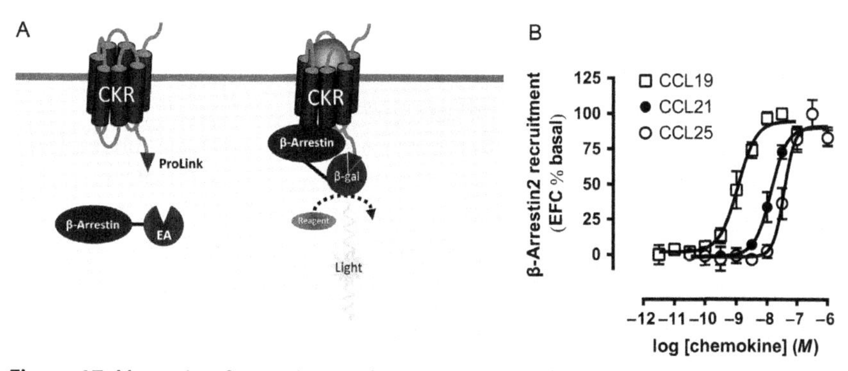

Figure 17 Measuring β-arrestin recruitment to activated CKRs using the enzyme fragment complementation (EFC)-based PathHunter assay. (A) PathHunter cell lines stably express a GPCR fused in frame with a 43-amino acid fragment of β-galactosidase called ProLink™ and β-arrestin fused to a N-terminal deletion mutant of β-galactosidase (EA: enzyme acceptor). β-Aarrestin recruitment to the activated CKRs results in the functional complementation of the two enzyme fragments, forming an active β-galactosidase enzyme that is capable of producing a chemiluminescent signal upon addition of the PathHunter™ detection reagent. (B) The PathHunter assay was used to detect β-arrestin2 recruitment to ACKR4 in response to increasing concentrations CCL19, CCL21, and CCL25. *Figure was adapted from Watts et al. (2013).*

- *Detection buffer*, Galacton Star, Emerald II, and Cell Assay (Lysis) buffer from the PathHunter detection kit.
- *Multilabel plate reader* (suitable to measure luminescence), such as Victor3 (PerkinElmer), PheraStar (BMG Labtechnologies), or Mithras (Berthold Technologies).

Methods

- Cells are seeded (14,000 cells/well) in white 384-well plates (20 μl/well).
- 24 h after seeding the cells, cell are incubated with agonist (4 μl/well) for 2 h at 37 °C.
  - Antagonists (4 μl/well) might be preincubated for 30–60 min (depending on their binding kinetics).
- Prepare detection reagent in the following ratio:
  - 760 μl Cell Assay (Lysis) buffer.
  - 40 μl Galacton Star.
  - 200 μl Emerald II.
- Add 12 μl/well detection reagent and incubate for 90 min at room temperature (protected from light).
- Luminescence is measured to quantify β-galactosidase activity in a multilabel plate reader.

This assay allows to measure receptor activation independent of G protein activation and is therefore suitable for ACKRs such as CXCR7. PathHunter™ cell lines for almost all human CKRs are available from DiscoverX and are mainly based on CHO-K1 cells and coexpressing β-arrestin2 fusion proteins. However, for some receptors, HEK293 and U2OS cells are available or cells that coexpress β-arrestin1 fusion proteins.

The PathHunter™ assay has been used to determine chemokine-induced β-arrestin recruitment (Fig. 17B) but also to functionally characterize antagonists for several CKRs, including CCR1 (Gilchrist et al., 2014), CCR4 (Santulli-Marotto et al., 2013), CXCR2 (de Kruijf et al., 2009; Maeda et al., 2015), and CXCR3 (Bernat, Admas, Brox, Heinemann, & Tschammer, 2014; Bernat et al., 2015).

A similar PathHunter™ internalization assay has been developed by DiscoverX, where the EA β-galactosidase fragments are fused to an early endosome marker.

3.2.6 Chemotaxis

Chemokines promote the directional migration of cells expressing CKRs, which can be measured in chemotaxis assays using a Boyden chamber. Cells are placed in the upper chamber, which is separated with a porous membrane from the lower chamber containing the chemoattractant (e.g., chemokines) and cells migrate toward the lower chamber.

Materials

- *Biological target,* we often use murine pre-B L1.2 cells (cultured in RPMI 1640 medium, containing 25 m*M* HEPES and GlutaMAX-1 and supplemented with 10% heat-inactivated FBS, 100 U/ml penicillin, 100 μg/ml streptomycin, 1% nonessential amino acids, 1 m*M* sodium pyruvate, and 50 μ*M* 2-mercaptoethanol), which are transfected with 10 μg CKR DNA per 1×10^7 cells using electroporation. Other migratory-competent cells can be used as well.
- *ChemoTx 96-well chemotaxis plate,* with 5 μ*M* pore size for L1.2 cells (NeuroProbe, Inc., Gaithersburg, MD). The pore diameter of the filter should be chosen depending on the cell type and should be large enough to allow active transmigration.

Methods

- The wells of the lower compartment of the chemotaxis 96-well plate are pretreated for 30 min with culture medium supplemented with 1% BSA (30 μl/well).
- Next, the medium with BSA is aspirated and a total volume of 30 μl/well chemokine or compound diluted in culture medium with 0.1% BSA is subsequently added to this lower compartment.
- The chemotaxis filter plate is placed on top of the 96-well bottom plate (the filter should be in direct contact with the chemoattractants in the lower compartment).
- 2.5×10^5 L1.2 cells/well in a total volume of 20 μl are pipetted on top of this upper compartment of the plate. The plate is incubated for 4–5 h at 37 °C and 5% CO_2 in a humidified chamber to avoid evaporation.
- 25 μl of migrated cells from the lower compartment are transferred to a white 96-well plate (resuspend cells first thoroughly). In addition, 25 μl of 1.0, 0.5, 0.25, 0.125, 0.0625, 0.03125, 0.015625, and 0.0078125×10^5 cells are plated in this white 96-well plate to prepare a calibration curve.
- Cell numbers are quantified by adding 1 μg/ml calcein AM (25 μl/well) and fluorescence (excitation: 495 nm; emission: 515 nm) is measured using a multilabel plate reader.

Although this assay measures the main feature of CKRs (i.e., chemotaxis), it does not resemble the *in vivo* situation. For example, the chemokines do not form a gradient in the lower compartment. Furthermore, the chemoattractants diffuse to the upper compartment in time, which ceases cell

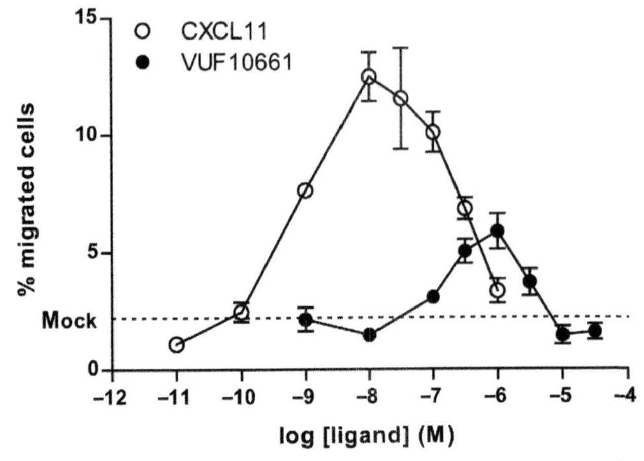

Figure 18 Chemokine-induced cell migration. CXCL11- and VUF10661-induced chemotaxis of CXCR3-expressing murine pre-B L1.2 cells, as measured in a Boyden chamber. A typical bell-shaped concentration–response curve is obtained. VUF10661 shows a lower maximal response and requires a higher concentration to reach the maximal effect as compared to CXCL11. *Figure was adapted from Scholten, Canals, Wijtmans, et al. (2012).*

migration (Toetsch, Olwell, Prina-Mello, & Volkov, 2009; Zhang et al., 2013). High concentrations of chemoattractants will limit chemotaxis, resulting in a typically bell-shaped concentration–response curve (Fig. 18). Inhibitors cause a dose-dependent right shift of this bell-shaped curve, sometimes along with a decreased maximum response.

Another popular assay to analyze cell migration is the scratch assay, in which a pin tool is used to introduce a scratch in a confluent monolayer of cells. Cells of the edges migrate to close the scratch and microscopy images before and during the migration are used to quantify the migration rate (Liang, Park, & Guan, 2007). The advantage of this assay is that it is simple and can record cell movement in real time. The scratch assay has been used to test inhibitors (Hulkower & Herber, 2011), but it is challenging to achieve reproducible and quantitative results (Kam, Guess, Estrada, Weidow, & Quaranta, 2008). More consistent results may be obtained using culture inserts (e.g., from Ibidi).

3.2.7 *Impedance*

Activation of CKRs leads to the modulation of the actin cytoskeleton and consequently changes cell morphology and cell adhesion. These changes can be measured by changes in electrical resistance (impedance) at the interface between cell and electrode using the xCELLigence system (ACEA Biosciences, Inc.). This technique allows to measure receptor activation as an integrated response rather than the activation of a specific signaling pathway (Rocheville & Jerman, 2009). However, specific inhibitors (e.g., PTX) or biased ligands might be used to dissect different signaling pathways (Stallaert, Dorn, van der Westhuizen, Audet, & Bouvier, 2012).

An advantage is that this technique does not require the labeling of cells with dyes, the tagging of proteins with fluorescent or luminescent probes, the labeling of the receptor with relatively large antibodies, or the recombinant expression of a biosensor (e.g., Epac-based BRET sensor) or reporter gene that may not capture the true physiology or pharmacology of the receptor. Hence, this technique allows for the noninvasive measurement of $G\alpha_i$-dependent signaling in the absence of FSK.

Materials

- *Biological target,* HEK293 cells stably transfected with CKR (stably transfected cell lines might produce more reproducible results as compared to transiently transfected cells) or cells that endogenously express the receptor of interest. When suspension cells are used, a coating is required that allows the cells to attach to the electrodes in the plate.
- *96-well E-plates,* with electrodes embedded in the bottom of the wells to measure impedance.
- *xCELLigence device station,* that holds the E-plate and uses a real-time cell electronic sensing (RT-CES) system for continuously monitoring cells.
- *Software,* that automatically converts the measured electrical resistance into cell index (CI) values by subtracting background electrical resistance (Yu et al., 2006).

Methods

- First, background electrical resistance is measured in the absence of cells by adding medium to the wells for the conversion of electrical resistance into CI values.
- Next, 5×10^4 HEK293 cells/well are plated in a 96-well E-plate in DMEM containing 10% FBS and penicillin and streptomycin.
- Cells are placed in the xCELLigence device station and allowed to attach and grow at 37 °C and 5% CO_2 while measuring CI in time. After cells have grown until a CI of 1.7–2.2 (~24 h after seeding the cells), the medium is replaced by 90 μl (when antagonist is added) or 95 μl (when only agonist is added) DMEM containing 0.1% FBS.
- After 3 h of adjustment to low-serum conditions, 5 μl/well ligand is added and impedance is continuously measured. Antagonists (5 μl/well)

might be preincubated prior to the addition of agonist. Antagonist incubation time depends on its binding kinetics.

Real-time changes in impedance can be measured to determine kinetics of chemokines or small molecules in the absence or presence of antagonists. Agonist stimulation results in an increased or decreased impedance that might return to baseline or stabilizes at a higher level. Peak values or the area under the curve can be used to generate dose–response curves from which ligand potencies and efficacies can be determined. Vehicle-treated cells should be taken along for baseline corrections, as opening the incubator might already cause changes in impedance.

This technique has been used to analyze CXCR3-mediated signaling in response to chemokines and a small molecule agonist (Watts, Scholten, Heitman, Vischer, & Leurs, 2012; Fig. 19). The selective CXCR3 antagonist NBI-74330 fully blocked CXCL10- and CXCL11-induced changes in impedance in CXCR3-expressing HEK293T cells but also revealed CXCR3-independent effects induced by CXCL9 and VUF10661. Interestingly, although this technique enables the measurement of β-arrestin

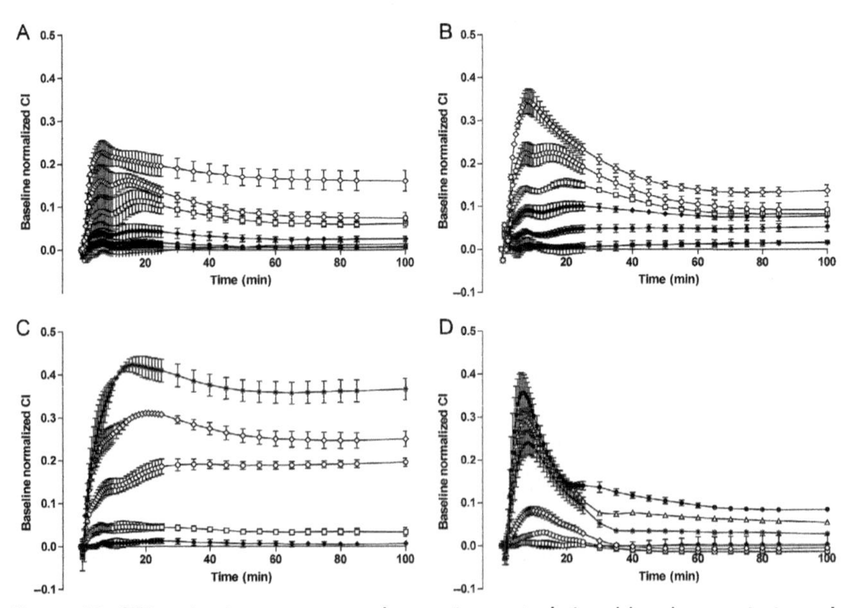

Figure 19 CKR activation as measured as an integrated signal by changes in impedance. Stimulation with increasing concentrations of CXCL9, CXCL10, CXCL11, and VUF10661 resulted in an increase in impedance in HEK293-CXCR3 cells. (+) 100 p*M*, (■) 1 n*M*, (•) 3.2 n*M*, (◆) 10 n*M*, (□) 32 n*M*, (○) 100 n*M*, (◊) 316 n*M*, (*) 1 μ*M*, (Δ) 3.2 μ*M*, and (•) 10 μ*M*. *Figure was adapted from Watts et al. (2012).*

recruitment to the niacin receptor GPR109A (Kammermann et al., 2011), changes in impedance only reflected G protein-dependent signaling of CXCR3 and not β-arrestin recruitment as was shown by the G_i-inhibitor pertussis toxin (PTX).

Besides ligand-induced signaling, other applications of this assay are the measurement of cell proliferation or cell viability (Kho et al., 2015). Chemotaxis can be measured in real time by combining this technique with the chemotaxis assay described in Section 3.2.6. Cells in the upper chamber of a CIM-plate migrate through the porous membrane into the chemoattractant-containing lower chamber and adhere to the microelectrode embedded in the bottom of the plate, leading to an increase in impedance (Iqbal et al., 2013).

An alternative label-free assay measures dynamic mass redistribution (DMR). GPCR activation results in the movement and rearrangement of signaling biomolecules, microfilament remodeling, and morphological changes, which can be measured by changes in the wavelength of refracted light (Fang, 2011). Similar to measuring impedance, this assay detects an integrated signal.

4. CONCLUSIONS AND FUTURE PERSPECTIVES

This chapter describes the different modes of action of CKR ligands and how to distinguish between them using binding and/or functional assays. However, experimental results can be ambiguous. For example, both orthosteric and allosteric modulators can elicit ligand displacement or insurmountable antagonism. In these cases, additional unique characteristics of orthosteric ligands and/or allosteric modulators should be identified to determine the ligand mode of action. Table 1 summarizes these characteristics. For example, a unique property of allosteric modulators is probe dependency. Yet, probe dependency might also complicate matters, as allosteric modulators might remain undetected when selecting the "wrong" probe. As most CKRs bind multiple chemokines, it is important not to limit the analysis to a single chemokine probe. Furthermore, since many small molecule ligands targeting CKRs are allosteric modulators, the topographically different ligand binding site should be characterized by using a traceable allosteric modulators and/or mutant receptors (Scholten et al., 2014, 2015).

Various assays are available to study CKR activation. However, many of these assays use tags or labels that might affect CKR responses. Efforts are

Table 1 Unique Characteristics or Orthosteric and Allosteric Ligands

Orthosteric Ligands	**Allosteric Modulators**
Only inhibit affinity/potency and/or efficacy of agonists	Inhibit or potentiate affinity/potency and/or efficacy
Cause an unsaturated parallel right shift of the agonist concentration–response curve	Cause a saturable change in affinity/potency and/or efficacy
Only decrease agonist efficacy when hemi-equilibrium conditions are obtained	Can have separate effects on affinity and efficacy
Inhibit affinity/potency and/or efficacy independently of the probe	The effect might be dependent on the probe
Cannot change the kinetics of a probe by definition as an orthosteric ligand does not allow a probe to bind due to competition for the binding site	Might change the kinetics of the probe (association and/or dissociation)

being made to reduce the size of these tags and labels. For example, NanoLuc (Nluc, Promega) is not only brighter but also smaller as compared to Fluc and Rluc (Fluc: 60.6 kDa, Rluc: 36.0 kDa, Nluc: 19.1 kDa) and could replace Fluc and Rluc in reporter gene and BRET assays, respectively. In a similar fashion, the small six-amino acid motif Cys-Cys-Pro-Gly-Cys-Cys used for fluorescent FlAsH labeling could be incorporated in the sequence of CKRs and replace, for example, large fused proteins such as yellow fluorescent protein (YFP) for imaging or BRET studies. The current trend of label-free assays such as measuring impedance or DMR enables to study GPCR responses in a relevant setting without artificial reporters. During the last decade, great progress has been made in novel fluorescence- and/or luminescence-based formats to detect CKR binding and may replace radioligands in the future to avoid the use of radioactivity. Fluorescent ligands may provide valuable knowledge on the mechanism of ligand binding, primarily due to real-time measurements and detection of receptor conformational changes. This is reflected by the number of fluorescent chemokines (e.g., CCL11, CCL19, CCL21, CXCL11, and CXCL12) and small molecules/peptides targeting chemokine receptors (e.g., AMD3100 and T-140 derivatives) that have been developed in recent years (Nomura et al., 2008; Sridharan, Zuber, Connelly, Mathew, & Dumont, 2014; Stoddart, Kilpatrick, Briddon, & Hill, 2015). The advancements in binding and functional assays are likely to contribute to more efficient and successful drug development campaigns and should result in more CKR-targeting drugs in the near future.

ACKNOWLEDGMENTS

The authors' work is supported by the Netherlands Organization for Scientific Research (NWO; Vici, ECHO grants), the Dutch Technology Foundation (STW), and European Union's Horizon2020 MSCA Programme under grant agreement 641833 (ONCORNET).

REFERENCES

Adage, T., Piccinini, A. M., Falsone, A., Trinker, M., Robinson, J., Gesslbauer, B., et al. (2012). Structure-based design of decoy chemokines as a way to explore the pharmacological potential of glycosaminoglycans. *British Journal of Pharmacology, 167*(6), 1195–1205. http://dx.doi.org/10.1111/j.1476-5381.2012.02089.x.

Allen, S. J., Crown, S. E., & Handel, T. M. (2007). Chemokine: Receptor structure, interactions, and antagonism. *Annual Review of Immunology, 25*, 787–820. http://dx.doi.org/10.1146/annurev.immunol.24.021605.090529.

Andrews, G., Jones, C., & Wreggett, K. A. (2008). An intracellular allosteric site for a specific class of antagonists of the CC chemokine G protein-coupled receptors CCR4 and CCR5. *Molecular Pharmacology, 73*(3), 855–867. http://dx.doi.org/10.1124/mol.107.039321.

Bachelerie, F., Ben-Baruch, A., Burkhardt, A. M., Combadiere, C., Farber, J. M., Graham, G. J., et al. (2014). International Union of Basic and Clinical Pharmacology. [corrected]. LXXXIX. Update on the extended family of chemokine receptors and introducing a new nomenclature for atypical chemokine receptors. *Pharmacological Reviews, 66*(1), 1–79. http://dx.doi.org/10.1124/pr.113.007724.

Beck, A., & Reichert, J. M. (2012). Marketing approval of mogamulizumab: A triumph for glyco-engineering. *mAbs, 4*(4), 419–425. http://dx.doi.org/10.4161/mabs.20996.

Bernat, V., Admas, T. H., Brox, R., Heinemann, F. W., & Tschammer, N. (2014). Boronic acids as probes for investigation of allosteric modulation of the chemokine receptor CXCR3. *ACS Chemical Biology, 9*(11), 2664–2677. http://dx.doi.org/10.1021/cb500678c.

Bernat, V., Brox, R., Heinrich, M. R., Auberson, Y. P., & Tschammer, N. (2015). Ligand-biased and probe-dependent modulation of chemokine receptor CXCR3 signaling by negative allosteric modulators. *ChemMedChem, 10*(3), 566–574. http://dx.doi.org/10.1002/cmdc.201402507.

Bernat, V., Heinrich, M. R., Baumeister, P., Buschauer, A., & Tschammer, N. (2012). Synthesis and application of the first radioligand targeting the allosteric binding pocket of chemokine receptor CXCR3. *ChemMedChem*, 7(8), 1481–1489. http://dx.doi.org/10.1002/cmdc.201200184.

Bradley, M. E., Dombrecht, B., Manini, J., Willis, J., Vlerick, D., De Taeye, S., et al. (2015). Potent and efficacious inhibition of CXCR2 signaling by biparatopic nanobodies combining two distinct modes of action. *Molecular Pharmacology, 87*(2), 251–262. http://dx.doi.org/10.1124/mol.114.094821.

Buntinx, M., Hermans, B., Goossens, J., Moechars, D., Gilissen, R. A., Doyon, J., et al. (2008). Pharmacological profile of JNJ-27141491 [(S)-3-[3,4-difluorophenyl)-propyl]-5-isoxazol-5-yl-2-thioxo-2,3-dihydro-1H-imida zole-4-carboxyl acid methyl ester], as a noncompetitive and orally active antagonist of the human chemokine receptor CCR2. *The Journal of Pharmacology and Experimental Therapeutics, 327*(1), 1–9. http://dx.doi.org/10.1124/jpet.108.140723.

Burg, J. S., Ingram, J. R., Venkatakrishnan, A. J., Jude, K. M., Dukkipati, A., Feinberg, E. N., et al. (2015). Structural biology. Structural basis for chemokine recognition and activation of a viral G protein-coupled receptor. *Science, 347*(6226), 1113–1117. http://dx.doi.org/10.1126/science.aaa5026.

Casarosa, P., Menge, W. M., Minisini, R., Otto, C., van Heteren, J., Jongejan, A., et al. (2003). Identification of the first nonpeptidergic inverse agonist for a constitutively active viral-encoded G protein-coupled receptor. *The Journal of Biological Chemistry, 278*(7), 5172–5178. http://dx.doi.org/10.1074/jbc.M210033200.

Cheng, Y., & Prusoff, W. H. (1973). Relationship between the inhibition constant (K1) and the concentration of inhibitor which causes 50% inhibition (I50) of an enzymatic reaction. *Biochemical Pharmacology, 22*(23), 3099–3108.

Clark-Lewis, I., Mattioli, I., Gong, J. H., & Loetscher, P. (2003). Structure-function relationship between the human chemokine receptor CXCR3 and its ligands. *The Journal of Biological Chemistry, 278*(1), 289–295. http://dx.doi.org/10.1074/jbc.M209470200.

Cox, M. A., Jenh, C. H., Gonsiorek, W., Fine, J., Narula, S. K., Zavodny, P. J., et al. (2001). Human interferon-inducible 10-kDa protein and human interferon-inducible T cell alpha chemoattractant are allotopic ligands for human CXCR3: Differential binding to receptor states. *Molecular Pharmacology, 59*(4), 707–715.

Curnock, A. P., Logan, M. K., & Ward, S. G. (2002). Chemokine signalling: Pivoting around multiple phosphoinositide 3-kinases. *Immunology, 105*(2), 125–136.

Daubeuf, F., Hachet-Haas, M., Gizzi, P., Gasparik, V., Bonnet, D., Utard, V., et al. (2013). An antedrug of the CXCL12 neutraligand blocks experimental allergic asthma without systemic effect in mice. *The Journal of Biological Chemistry, 288*(17), 11865–11876. http://dx.doi.org/10.1074/jbc.M112.449348.

De Clercq, E. (2009). The AMD3100 story: The path to the discovery of a stem cell mobilizer (Mozobil). *Biochemical Pharmacology*, 77(11), 1655–1664. http://dx.doi.org/10.1016/j.bcp.2008.12.014.

de Kruijf, P., van Heteren, J., Lim, H. D., Conti, P. G., van der Lee, M. M., Bosch, L., et al. (2009). Nonpeptidergic allosteric antagonists differentially bind to the CXCR2 chemokine receptor. *The Journal of Pharmacology and Experimental Therapeutics, 329*(2), 783–790. http://dx.doi.org/10.1124/jpet.108.148387.

de Munnik, S. M., Kooistra, A. J., van Offenbeek, J., Nijmeijer, S., de Graaf, C., Smit, M. J., et al. (2015). The viral G protein-coupled receptor ORF74 Hijacks beta-arrestins for endocytic trafficking in response to human chemokines. *PLoS One, 10*(4), e0124486. http://dx.doi.org/10.1371/journal.pone.0124486.

de Munnik, S. M., Smit, M. J., Leurs, R., & Vischer, H. F. (2015). Modulation of cellular signaling by herpesvirus-encoded G protein-coupled receptors. *Frontiers in Pharmacology, 6*, 40. http://dx.doi.org/10.3389/fphar.2015.00040.

Deng, H., Liu, R., Ellmeier, W., Choe, S., Unutmaz, D., Burkhart, M., et al. (1996). Identification of a major co-receptor for primary isolates of HIV-1. *Nature, 381*(6584), 661–666. http://dx.doi.org/10.1038/381661a0.

Dimberg, A. (2010). Chemokines in angiogenesis. *Current Topics in Microbiology and Immunology, 341*, 59–80. http://dx.doi.org/10.1007/82_2010_21.

Fang, Y. (2011). Label-free receptor assays. *Drug Discovery Today: Technologies*, 7(1), e5–e11. http://dx.doi.org/10.1016/j.ddtec.2010.05.001.

Fatkenheuer, G., Pozniak, A. L., Johnson, M. A., Plettenberg, A., Staszewski, S., Hoepelman, A. I., et al. (2005). Efficacy of short-term monotherapy with maraviroc, a new CCR5 antagonist, in patients infected with HIV-1. *Nature Medicine, 11*(11), 1170–1172. http://dx.doi.org/10.1038/nm1319.

Garcia-Perez, J., Rueda, P., Staropoli, I., Kellenberger, E., Alcami, J., Arenzana-Seisdedos, F., et al. (2011). New insights into the mechanisms whereby low molecular weight CCR5 ligands inhibit HIV-1 infection. *The Journal of Biological Chemistry, 286*(7), 4978–4990. http://dx.doi.org/10.1074/jbc.M110.168955.

Gee, K. R., Brown, K. A., Chen, W. N., Bishop-Stewart, J., Gray, D., & Johnson, I. (2000). Chemical and physiological characterization of fluo-4 Ca(2+)-indicator dyes. *Cell Calcium, 27*(2), 97–106. http://dx.doi.org/10.1054/ceca.1999.0095.

Gilchrist, A., Gauntner, T. D., Fazzini, A., Alley, K. M., Pyen, D. S., Ahn, J., et al. (2014). Identifying bias in CCR1 antagonists using radiolabelled binding, receptor internalization, beta-arrestin translocation and chemotaxis assays. *British Journal of Pharmacology*, *171*(22), 5127–5138. http://dx.doi.org/10.1111/bph.12835.

Gong, J. H., Uguccioni, M., Dewald, B., Baggiolini, M., & Clark-Lewis, I. (1996). RANTES and MCP-3 antagonists bind multiple chemokine receptors. *The Journal of Biological Chemistry*, *271*(18), 10521–10527.

Griffith, J. W., Sokol, C. L., & Luster, A. D. (2014). Chemokines and chemokine receptors: Positioning cells for host defense and immunity. *Annual Review of Immunology*, *32*, 659–702. http://dx.doi.org/10.1146/annurev-immunol-032713-120145.

Gruijthuijsen, Y. K., Casarosa, P., Kaptein, S. J., Broers, J. L., Leurs, R., Bruggeman, C. A., et al. (2002). The rat cytomegalovirus R33-encoded G protein-coupled receptor signals in a constitutive fashion. *Journal of Virology*, *76*(3), 1328–1338.

Gurevich, V. V., Pals-Rylaarsdam, R., Benovic, J. L., Hosey, M. M., & Onorato, J. J. (1997). Agonist-receptor-arrestin, an alternative ternary complex with high agonist affinity. *The Journal of Biological Chemistry*, *272*(46), 28849–28852.

Handel, T. M., Johnson, Z., Crown, S. E., Lau, E. K., & Proudfoot, A. E. (2005). Regulation of protein function by glycosaminoglycans—as exemplified by chemokines. *Annual Review of Biochemistry*, *74*, 385–410. http://dx.doi.org/10.1146/annurev.biochem.72.121801.161747.

Harrison, C., & Traynor, J. R. (2003). The [35S]GTPgammaS binding assay: Approaches and applications in pharmacology. *Life Sciences*, *74*(4), 489–508.

Hill, S. J., Williams, C., & May, L. T. (2010). Insights into GPCR pharmacology from the measurement of changes in intracellular cyclic AMP; advantages and pitfalls of differing methodologies. *British Journal of Pharmacology*, *161*(6), 1266–1275.

Hoffmann, C., Castro, M., Rinken, A., Leurs, R., Hill, S. J., & Vischer, H. F. (2015). Ligand residence time at G-protein-coupled receptors—Why we should take our time to study it. *Molecular Pharmacology*, *88*(3), 552–560. http://dx.doi.org/10.1124/mol.115.099671.

Hulkower, K. I., & Herber, R. L. (2011). Cell migration and invasion assays as tools for drug discovery. *Pharmaceutics*, *3*(1), 107–124. http://dx.doi.org/10.3390/pharmaceutics3010107.

Hulshof, J. W., Casarosa, P., Menge, W. M., Kuusisto, L. M., van der Goot, H., Smit, M. J., et al. (2005). Synthesis and structure-activity relationship of the first nonpeptidergic inverse agonists for the human cytomegalovirus encoded chemokine receptor US28. *Journal of Medicinal Chemistry*, *48*(20), 6461–6471. http://dx.doi.org/10.1021/jm050418d.

Iqbal, A. J., Regan-Komito, D., Christou, I., White, G. E., McNeill, E., Kenyon, A., et al. (2013). A real time chemotaxis assay unveils unique migratory profiles amongst different primary murine macrophages. *PLoS One*, *8*(3), e58744. http://dx.doi.org/10.1371/journal.pone.0058744.

Ishida, T., Joh, T., Uike, N., Yamamoto, K., Utsunomiya, A., Yoshida, S., et al. (2012). Defucosylated anti-CCR4 monoclonal antibody (KW-0761) for relapsed adult T-cell leukemia-lymphoma: A multicenter phase II study. *Journal of Clinical Oncology: Official Journal of the American Society of Clinical Oncology*, *30*(8), 837–842. http://dx.doi.org/10.1200/JCO.2011.37.3472.

Jahnichen, S., Blanchetot, C., Maussang, D., Gonzalez-Pajuelo, M., Chow, K. Y., Bosch, L., et al. (2010). CXCR4 nanobodies (VHH-based single variable domains) potently inhibit chemotaxis and HIV-1 replication and mobilize stem cells. *Proceedings of the National Academy of Sciences of the United States of America*, *107*(47), 20565–20570. http://dx.doi.org/10.1073/pnas.1012865107.

Johnson, Z., Kosco-Vilbois, M. H., Herren, S., Cirillo, R., Muzio, V., Zaratin, P., et al. (2004). Interference with heparin binding and oligomerization creates a novel anti-inflammatory strategy targeting the chemokine system. *Journal of Immunology*, *173*(9), 5776–5785.

Jorgensen, R., Martini, L., Schwartz, T. W., & Elling, C. E. (2005). Characterization of glucagon-like peptide-1 receptor beta-arrestin 2 interaction: A high-affinity receptor phenotype. *Molecular Endocrinology*, *19*(3), 812–823. http://dx.doi.org/10.1210/me.2004-0312.

Kam, Y., Guess, C., Estrada, L., Weidow, B., & Quaranta, V. (2008). A novel circular invasion assay mimics *in vivo* invasive behavior of cancer cell lines and distinguishes single-cell motility *in vitro*. *BMC Cancer*, *8*, 198. http://dx.doi.org/10.1186/1471-2407-8-198.

Kammermann, M., Denelavas, A., Imbach, A., Grether, U., Dehmlow, H., Apfel, C. M., et al. (2011). Impedance measurement: A new method to detect ligand-biased receptor signaling. *Biochemical and Biophysical Research Communications*, *412*(3), 419–424. http://dx.doi.org/10.1016/j.bbrc.2011.07.087.

Kenakin, T. (2009a). *A pharmacology primer: Theory, applications, and methods* (3rd ed.). Burlington, MA: Elsevier Academic Press.

Kenakin, T. (2009b). Quantifying biological activity in chemical terms: A pharmacology primer to describe drug effect. *ACS Chemical Biology*, *4*(4), 249–260. http://dx.doi.org/10.1021/cb800299s.

Kenakin, T. (2010). Being mindful of seven-transmembrane receptor 'guests' when assessing agonist selectivity. *British Journal of Pharmacology*, *160*(5), 1045–1047. http://dx.doi.org/10.1111/j.1476-5381.2010.00764.x.

Kenakin, T. (2013). New concepts in pharmacological efficacy at 7TM receptors: IUPHAR review 2. *British Journal of Pharmacology*, *168*(3), 554–575. http://dx.doi.org/10.1111/j.1476-5381.2012.02223.x.

Kenakin, T., & Miller, L. J. (2010). Seven transmembrane receptors as shapeshifting proteins: The impact of allosteric modulation and functional selectivity on new drug discovery. *Pharmacological Reviews*, *62*(2), 265–304. http://dx.doi.org/10.1124/pr.108.000992.

Kho, D., MacDonald, C., Johnson, R., Unsworth, C. P., O'Carroll, S. J., du Mez, E., et al. (2015). Application of xCELLigence RTCA biosensor technology for revealing the profile and window of drug responsiveness in real time. *Biosensors*, *5*(2), 199–222. http://dx.doi.org/10.3390/bios5020199.

Klarenbeek, A., Maussang, D., Blanchetot, C., Saunders, M., van der Woning, S., Smit, M., et al. (2012). Targeting chemokines and chemokine receptors with antibodies. *Drug Discovery Today: Technologies*, *9*(4), e237–e244. http://dx.doi.org/10.1016/j.ddtec.2012.05.003.

Koelink, P. J., Overbeek, S. A., Braber, S., de Kruijf, P., Folkerts, G., Smit, M. J., et al. (2012). Targeting chemokine receptors in chronic inflammatory diseases: An extensive review. *Pharmacology & Therapeutics*, *133*(1), 1–18. http://dx.doi.org/10.1016/j.pharmthera.2011.06.008.

Koenen, R. R., & Weber, C. (2010). Therapeutic targeting of chemokine interactions in atherosclerosis. *Nature Reviews. Drug Discovery*, *9*(2), 141–153. http://dx.doi.org/10.1038/nrd3048.

Kralj, A., Kurt, E., Tschammer, N., & Heinrich, M. R. (2014). Synthesis and biological evaluation of biphenyl amides that modulate the US28 receptor. *ChemMedChem*, *9*(1), 151–168. http://dx.doi.org/10.1002/cmdc.201300369.

Kubota, T., Niwa, R., Satoh, M., Akinaga, S., Shitara, K., & Hanai, N. (2009). Engineered therapeutic antibodies with improved effector functions. *Cancer Science*, *100*(9), 1566–1572. http://dx.doi.org/10.1111/j.1349-7006.2009.01222.x.

Kuritzkes, D., Kar, S., & Kirkpatrick, P. (2008). Fresh from the pipeline—Maraviroc. *Nature Reviews. Drug Discovery*, 7(1), 15–16. http://dx.doi.org/10.1038/Nrd2490.

Liang, C. C., Park, A. Y., & Guan, J. L. (2007). *In vitro* scratch assay: A convenient and inexpensive method for analysis of cell migration *in vitro*. *Nature Protocols*, *2*(2), 329–333. http://dx.doi.org/10.1038/nprot.2007.30.

Liu, C., Sandford, G., Fei, G., & Nicholas, J. (2004). Galpha protein selectivity determinant specified by a viral chemokine receptor-conserved region in the C tail of the human herpesvirus 8 g protein-coupled receptor. *Journal of Virology*, *78*(5), 2460–2471.

Maeda, Y., Kuroki, R., Haase, W., Michel, H., & Reilander, H. (2004). Comparative analysis of high-affinity ligand binding and G protein coupling of the human CXCR1 chemokine receptor and of a CXCR1-Galpha fusion protein after heterologous production in baculovirus-infected insect cells. *European Journal of Biochemistry*, *271*(9), 1677–1689. http://dx.doi.org/10.1111/j.1432-1033.2004.04064.x.

Maeda, K., Nakata, H., Koh, Y., Miyakawa, T., Ogata, H., Takaoka, Y., et al. (2004). Spirodiketopiperazine-based CCR5 inhibitor which preserves CC-chemokine/CCR5 interactions and exerts potent activity against R5 human immunodeficiency virus type 1 *in vitro*. *Journal of Virology*, *78*(16), 8654–8662. http://dx.doi.org/10.1128/JVI.78.16.8654-8662.2004.

Maeda, D. Y., Peck, A. M., Schuler, A. D., Quinn, M. T., Kirpotina, L. N., Wicomb, W. N., et al. (2015). Boronic acid-containing CXCR1/2 antagonists: Optimization of metabolic stability, *in vivo* evaluation, and a proposed receptor binding model. *Bioorganic & Medicinal Chemistry Letters*, *25*(11), 2280–2284. http://dx.doi.org/10.1016/j.bmcl.2015.04.041.

Martini, L., Hastrup, H., Holst, B., Fraile-Ramos, A., Marsh, M., & Schwartz, T. W. (2002). NK1 receptor fused to beta-arrestin displays a single-component, high-affinity molecular phenotype. *Molecular Pharmacology*, *62*(1), 30–37.

Martins-Green, M., Petreaca, M., & Wang, L. (2013). Chemokines and their receptors are key players in the orchestra that regulates wound healing. *Advances in Wound Care*, *2*(7), 327–347. http://dx.doi.org/10.1089/wound.2012.0380.

Maussang, D., Mujic-Delic, A., Descamps, F. J., Stortelers, C., Stigter-van Walsum, M., Vischer, H. F., et al. (2013). Llama-derived single variable domains (Nanobodies) directed against CXCR7 reduce head and neck cancer cell growth *in vivo*. *The Journal of Biological Chemistry*, *288*, 29562–29572. http://dx.doi.org/10.1074/jbc.M113.498436.

Montaner, S., Kufareva, I., Abagyan, R., & Gutkind, J. S. (2013). Molecular mechanisms deployed by virally encoded G protein-coupled receptors in human diseases. *Annual Review of Pharmacology and Toxicology*, *53*, 331–354. http://dx.doi.org/10.1146/annurev-pharmtox-010510-100608.

Mujic-Delic, A., de Wit, R. H., Verkaar, F., & Smit, M. J. (2014). GPCR-targeting nanobodies: Attractive research tools, diagnostics, and therapeutics. *Trends in Pharmacological Sciences*, *35*(5), 247–255. http://dx.doi.org/10.1016/j.tips.2014.03.003.

Muyldermans, S. (2013). Nanobodies: Natural single-domain antibodies. *Annual Review of Biochemistry*, *82*, 775–797. http://dx.doi.org/10.1146/annurev-biochem-063011-092449.

Nakata, H., Maeda, K., Miyakawa, T., Shibayama, S., Matsuo, M., Takaoka, Y., et al. (2005). Potent anti-R5 human immunodeficiency virus type 1 effects of a CCR5 antagonist, AK602/ONO4128/GW873140, in a novel human peripheral blood mononuclear cell nonobese diabetic-SCID, interleukin-2 receptor gamma-chain-knocked-out AIDS mouse model. *Journal of Virology*, *79*(4), 2087–2096. http://dx.doi.org/10.1128/JVI.79.4.2087-2096.2005.

Neptune, E. R., & Bourne, H. R. (1997). Receptors induce chemotaxis by releasing the betagamma subunit of Gi, not by activating Gq or Gs. *Proceedings of the National Academy of Sciences of the United States of America*, *94*(26), 14489–14494.

Nijmeijer, S., Leurs, R., Smit, M. J., & Vischer, H. F. (2010). The Epstein-Barr virus-encoded G protein-coupled receptor BILF1 hetero-oligomerizes with human CXCR4, scavenges Galphai proteins, and constitutively impairs CXCR4 functioning. *The Journal of Biological Chemistry*, *285*(38), 29632–29641. http://dx.doi.org/10.1074/jbc.M110.115618.

Nomura, W., Tanabe, Y., Tsutsumi, H., Tanaka, T., Ohba, K., Yamamoto, N., et al. (2008). Fluorophore labeling enables imaging and evaluation of specific CXCR4-ligand interaction at the cell membrane for fluorescence-based screening. *Bioconjugate Chemistry*, *19*(9), 1917–1920. http://dx.doi.org/10.1021/bc800216p.

Oldham, W. M., & Hamm, H. E. (2008). Heterotrimeric G protein activation by G-protein-coupled receptors. *Nature Reviews. Molecular Cell Biology*, *9*(1), 60–71. http://dx.doi.org/10.1038/nrm2299.

Patel, J., Channon, K. M., & McNeill, E. (2013). The downstream regulation of chemokine receptor signalling: Implications for atherosclerosis. *Mediators of Inflammation*, *2013*, 459520. http://dx.doi.org/10.1155/2013/459520.

Planaguma, A., Domenech, T., Pont, M., Calama, E., Garcia-Gonzalez, V., Lopez, R., et al. (2015). Combined anti CXC receptors 1 and 2 therapy is a promising anti-inflammatory treatment for respiratory diseases by reducing neutrophil migration and activation. *Pulmonary Pharmacology & Therapeutics*, *34*, 37–45. http://dx.doi.org/10.1016/j.pupt.2015.08.002.

Proudfoot, A. E., Power, C. A., & Schwarz, M. K. (2010). Anti-chemokine small molecule drugs: A promising future? *Expert Opinion on Investigational Drugs*, *19*(3), 345–355. http://dx.doi.org/10.1517/13543780903535867.

Qin, L., Kufareva, I., Holden, L. G., Wang, C., Zheng, Y., Zhao, C., et al. (2015). Structural biology. Crystal structure of the chemokine receptor CXCR4 in complex with a viral chemokine. *Science*, *347*(6226), 1117–1122. http://dx.doi.org/10.1126/science.1261064.

Rajagopal, S., Kim, J., Ahn, S., Craig, S., Lam, C. M., Gerard, N. P., et al. (2010). Beta-arrestin- but not G protein-mediated signaling by the "decoy" receptor CXCR7. *Proceedings of the National Academy of Sciences of the United States of America*, *107*(2), 628–632. http://dx.doi.org/10.1073/pnas.0912852107.

Raman, D., Sobolik-Delmaire, T., & Richmond, A. (2011). Chemokines in health and disease. *Experimental Cell Research*, *317*(5), 575–589. http://dx.doi.org/10.1016/j.yexcr.2011.01.005.

Richter, R., Casarosa, P., Standker, L., Munch, J., Springael, J. Y., Nijmeijer, S., et al. (2009). Significance of N-terminal proteolysis of CCL14a to activity on the chemokine receptors CCR1 and CCR5 and the human cytomegalovirus-encoded chemokine receptor US28. *Journal of Immunology*, *183*(2), 1229–1237. http://dx.doi.org/10.4049/jimmunol.0802145.

Richter, R., Jochheim-Richter, A., Ciuculescu, F., Kollar, K., Seifried, E., Forssmann, U., et al. (2014). Identification and characterization of circulating variants of CXCL12 from human plasma: Effects on chemotaxis and mobilization of hematopoietic stem and progenitor cells. *Stem Cells and Development*, *23*(16), 1959–1974. http://dx.doi.org/10.1089/scd.2013.0524.

Rocheville, M., & Jerman, J. C. (2009). 7TM pharmacology measured by label-free: A holistic approach to cell signalling. *Current Opinion in Pharmacology*, *9*(5), 643–649. http://dx.doi.org/10.1016/j.coph.2009.06.015.

Sachpatzidis, A., Benton, B. K., Manfredi, J. P., Wang, H., Hamilton, A., Dohlman, H. G., et al. (2003). Identification of allosteric peptide agonists of CXCR4. *The Journal of Biological Chemistry*, *278*(2), 896–907. http://dx.doi.org/10.1074/jbc.M204667200.

Salahpour, A., Espinoza, S., Masri, B., Lam, V., Barak, L. S., & Gainetdinov, R. R. (2012). BRET biosensors to study GPCR biology, pharmacology, and signal transduction. *Frontiers in Endocrinology*, *3*, 105. http://dx.doi.org/10.3389/fendo.2012.00105.

Sanni, S. J., Hansen, J. T., Bonde, M. M., Speerschneider, T., Christensen, G. L., Munk, S., et al. (2010). beta-Arrestin 1 and 2 stabilize the angiotensin II type I receptor in distinct high-affinity conformations. *British Journal of Pharmacology*, *161*(1), 150–161. http://dx.doi.org/10.1111/j.1476-5381.2010.00875.x.

Santulli-Marotto, S., Fisher, J., Petley, T., Boakye, K., Panavas, T., Luongo, J., et al. (2013). Surrogate antibodies that specifically bind and neutralize CCL17 but not CCL22. *Monoclonal Antibodies in Immunodiagnosis and Immunotherapy*, *32*(3), 162–171. http://dx.doi.org/10.1089/mab.2012.0112.

Schall, T. J., & Proudfoot, A. E. (2011). Overcoming hurdles in developing successful drugs targeting chemokine receptors. *Nature Reviews. Immunology*, *11*(5), 355–363. http://dx.doi.org/10.1038/nri2972.

Scholten, D. J., Canals, M., Maussang, D., Roumen, L., Smit, M. J., Wijtmans, M., et al. (2012a). Pharmacological modulation of chemokine receptor function. *British Journal of Pharmacology*, *165*(6), 1617–1643. http://dx.doi.org/10.1111/j.1476-5381.2011.01551.x.

Scholten, D. J., Canals, M., Wijtmans, M., de Munnik, S., Nguyen, P., Verzijl, D., et al. (2012b). Pharmacological characterization of a small-molecule agonist for the chemokine receptor CXCR3. *British Journal of Pharmacology*, *166*(3), 898–911. http://dx.doi.org/10.1111/j.1476-5381.2011.01648.x.

Scholten, D. J., Roumen, L., Wijtmans, M., Verkade-Vreeker, M. C., Custers, H., Lai, M., et al. (2014). Identification of overlapping but differential binding sites for the high-affinity CXCR3 antagonists NBI-74330 and VUF11211. *Molecular Pharmacology*, *85*(1), 116–126. http://dx.doi.org/10.1124/mol.113.088633.

Scholten, D. J., Wijtmans, M., van Senten, J. R., Custers, H., Stunnenberg, A., de Esch, I. J., et al. (2015). Pharmacological characterization of [3H]VUF11211, a novel radiolabeled small-molecule inverse agonist for the chemokine receptor CXCR3. *Molecular Pharmacology*, *87*(4), 639–648. http://dx.doi.org/10.1124/mol.114.095265.

Slack, R. J., Russell, L. J., Barton, N. P., Weston, C., Nalesso, G., Thompson, S. A., et al. (2013). Antagonism of human CC-chemokine receptor 4 can be achieved through three distinct binding sites on the receptor. *Pharmacology Research & Perspectives*, *1*(2), e00019. http://dx.doi.org/10.1002/prp2.19.

Sprenger, J. U., & Nikolaev, V. O. (2013). Biophysical techniques for detection of cAMP and cGMP in living cells. *International Journal of Molecular Sciences*, *14*(4), 8025–8046. http://dx.doi.org/10.3390/ijms14048025.

Springael, J. Y., Le Minh, P. N., Urizar, E., Costagliola, S., Vassart, G., & Parmentier, M. (2006). Allosteric modulation of binding properties between units of chemokine receptor homo- and hetero-oligomers. *Molecular Pharmacology*, *69*(5), 1652–1661. http://dx.doi.org/10.1124/mol.105.019414.

Sridharan, R., Zuber, J., Connelly, S. M., Mathew, E., & Dumont, M. E. (2014). Fluorescent approaches for understanding interactions of ligands with G protein coupled receptors. *Biochimica et Biophysica Acta*, *1838*(1 Pt. A), 15–33. http://dx.doi.org/10.1016/j.bbamem.2013.09.005.

Stallaert, W., Dorn, J. F., van der Westhuizen, E., Audet, M., & Bouvier, M. (2012). Impedance responses reveal beta(2)-adrenergic receptor signaling pluridimensionality and allow classification of ligands with distinct signaling profiles. *PLoS One*, 7(1), e29420. http://dx.doi.org/10.1371/journal.pone.0029420.

Stoddart, L. A., Kilpatrick, L. E., Briddon, S. J., & Hill, S. J. (2015). Probing the pharmacology of G protein-coupled receptors with fluorescent ligands. *Neuropharmacology*, *98*, 48–57. http://dx.doi.org/10.1016/j.neuropharm.2015.04.033.

Strange, P. G. (2010). Use of the GTPgammaS ([35S]GTPgammaS and Eu-GTPgammaS) binding assay for analysis of ligand potency and efficacy at G protein-coupled receptors. *British Journal of Pharmacology*, *161*(6), 1238–1249. http://dx.doi.org/10.1111/j.1476-5381.2010.00963.x.

Suffee, N., Hlawaty, H., Meddahi-Pelle, A., Maillard, L., Louedec, L., Haddad, O., et al. (2012). RANTES/CCL5-induced pro-angiogenic effects depend on CCR1, CCR5 and glycosaminoglycans. *Angiogenesis*, *15*(4), 727–744. http://dx.doi.org/10.1007/s10456-012-9285-x.

Swinney, D. C., Beavis, P., Chuang, K. T., Zheng, Y., Lee, I., Gee, P., et al. (2014). A study of the molecular mechanism of binding kinetics and long residence times of human CCR5 receptor small molecule allosteric ligands. *British Journal of Pharmacology*, *171*(14), 3364–3375. http://dx.doi.org/10.1111/bph.12683.

Tan, Q., Zhu, Y., Li, J., Chen, Z., Han, G. W., Kufareva, I., et al. (2013). Structure of the CCR5 chemokine receptor-HIV entry inhibitor maraviroc complex. *Science*, *341*(6152), 1387–1390. http://dx.doi.org/10.1126/science.1241475.

Thiele, S., & Rosenkilde, M. M. (2014). Interaction of chemokines with their receptors—From initial chemokine binding to receptor activating steps. *Current Medicinal Chemistry*, *21*(31), 3594–3614.

Thiele, S., Steen, A., Jensen, P. C., Mokrosinski, J., Frimurer, T. M., & Rosenkilde, M. M. (2011). Allosteric and orthosteric sites in CC chemokine receptor (CCR5), a chimeric receptor approach. *The Journal of Biological Chemistry*, *286*(43), 37543–37554. http://dx.doi.org/10.1074/jbc.M111.243808.

Toetsch, S., Olwell, P., Prina-Mello, A., & Volkov, Y. (2009). The evolution of chemotaxis assays from static models to physiologically relevant platforms. *Integrative Biology*, *1*(2), 170–181. http://dx.doi.org/10.1039/b814567a.

Ulvmar, M. H., Hub, E., & Rot, A. (2011). Atypical chemokine receptors. *Experimental Cell Research*, *317*(5), 556–568. http://dx.doi.org/10.1016/j.yexcr.2011.01.012.

Vela, M., Aris, M., Llorente, M., Garcia-Sanz, J. A., & Kremer, L. (2015). Chemokine receptor-specific antibodies in cancer immunotherapy: Achievements and challenges. *Frontiers in Immunology*, *6*, 12. http://dx.doi.org/10.3389/fimmu.2015.00012.

Verzijl, D., Pardo, L., van Dijk, M., Gruijthuijsen, Y. K., Jongejan, A., Timmerman, H., et al. (2006). Helix 8 of the viral chemokine receptor ORF74 directs chemokine binding. *The Journal of Biological Chemistry*, *281*(46), 35327–35335. http://dx.doi.org/10.1074/jbc.M606877200.

Viney, J. M., Andrew, D. P., Phillips, R. M., Meiser, A., Patel, P., Lennartz-Walker, M., et al. (2014). Distinct conformations of the chemokine receptor CCR4 with implications for its targeting in allergy. *Journal of Immunology*, *192*(7), 3419–3427. http://dx.doi.org/10.4049/jimmunol.1300232.

Vischer, H. F., Hulshof, J. W., Hulscher, S., Fratantoni, S. A., Verheij, M. H., Victorina, J., et al. (2010). Identification of novel allosteric nonpeptidergic inhibitors of the human cytomegalovirus-encoded chemokine receptor US28. *Bioorganic & Medicinal Chemistry*, *18*(2), 675–688. http://dx.doi.org/10.1016/j.bmc.2009.11.060.

Vischer, H. F., Siderius, M., Leurs, R., & Smit, M. J. (2014). Herpesvirus-encoded GPCRs: Neglected players in inflammatory and proliferative diseases? *Nature Reviews. Drug Discovery*, *13*(2), 123–139. http://dx.doi.org/10.1038/nrd4189.

Vischer, H. F., Watts, A. O., Nijmeijer, S., & Leurs, R. (2011). G protein-coupled receptors: Walking hand-in-hand, talking hand-in-hand? *British Journal of Pharmacology*, *163*(2), 246–260. http://dx.doi.org/10.1111/j.1476-5381.2011.01229.x.

Wang, J., & Knaut, H. (2014). Chemokine signaling in development and disease. *Development*, *141*(22), 4199–4205. http://dx.doi.org/10.1242/dev.101071.

Wang, J., Shiozawa, Y., Wang, J., Wang, Y., Jung, Y., Pienta, K. J., et al. (2008). The role of CXCR7/RDC1 as a chemokine receptor for CXCL12/SDF-1 in prostate cancer. *The Journal of Biological Chemistry*, *283*(7), 4283–4294. http://dx.doi.org/10.1074/jbc.M707465200.

Watson, C., Jenkinson, S., Kazmierski, W., & Kenakin, T. (2005). The CCR5 receptor-based mechanism of action of 873140, a potent allosteric noncompetitive HIV entry inhibitor. *Molecular Pharmacology, 67*(4), 1268–1282. http://dx.doi.org/10.1124/mol.104.008565.

Watts, A. O., Scholten, D. J., Heitman, L. H., Vischer, H. F., & Leurs, R. (2012). Label-free impedance responses of endogenous and synthetic chemokine receptor CXCR3 agonists correlate with Gi-protein pathway activation. *Biochemical and Biophysical Research Communications, 419*(2), 412–418. http://dx.doi.org/10.1016/j.bbrc.2012.02.036.

Watts, A. O., Verkaar, F., van der Lee, M. M., Timmerman, C. A., Kuijer, M., van Offenbeek, J., et al. (2013). beta-Arrestin recruitment and G protein signaling by the atypical human chemokine decoy receptor CCX-CKR. *The Journal of Biological Chemistry, 288*(10), 7169–7181. http://dx.doi.org/10.1074/jbc.M112.406108.

White, G. E., Iqbal, A. J., & Greaves, D. R. (2013). CC chemokine receptors and chronic inflammation—Therapeutic opportunities and pharmacological challenges. *Pharmacological Reviews, 65*(1), 47–89. http://dx.doi.org/10.1124/pr.111.005074.

Wijtmans, M., Scholten, D. J., de Esch, I. J. P., Smit, M. J., & Leurs, R. (2012). Therapeutic targeting of chemokine receptors by small molecules. *Drug Discovery Today: Technologies, 9*(4), e229–e236.

Willoughby, D., & Cooper, D. M. (2008). Live-cell imaging of cAMP dynamics. *Nature Methods, 5*(1), 29–36. http://dx.doi.org/10.1038/nmeth1135.

Wu, B., Chien, E. Y., Mol, C. D., Fenalti, G., Liu, W., Katritch, V., et al. (2010). Structures of the CXCR4 chemokine GPCR with small-molecule and cyclic peptide antagonists. *Science, 330*(6007), 1066–1071. http://dx.doi.org/10.1126/science.1194396.

Wyllie, D. J., & Chen, P. E. (2007). Taking the time to study competitive antagonism. *British Journal of Pharmacology, 150*(5), 541–551. http://dx.doi.org/10.1038/sj.bjp.0706997.

Yu, N., Atienza, J. M., Bernard, J., Blanc, S., Zhu, J., Wang, X., et al. (2006). Real-time monitoring of morphological changes in living cells by electronic cell sensor arrays: An approach to study G protein-coupled receptors. *Analytical Chemistry, 78*(1), 35–43. http://dx.doi.org/10.1021/ac051695v.

Zhang, J. H., Chung, T. D., & Oldenburg, K. R. (1999). A simple statistical parameter for use in evaluation and validation of high throughput screening assays. *Journal of Biomolecular Screening, 4*(2), 67–73.

Zhang, C., Jang, S., Amadi, O. C., Shimizu, K., Lee, R. T., & Mitchell, R. N. (2013). A sensitive chemotaxis assay using a novel microfluidic device. *BioMed Research International, 2013*, 373569. http://dx.doi.org/10.1155/2013/373569.

Zweemer, A. J., Nederpelt, I., Vrieling, H., Hafith, S., Doornbos, M. L., de Vries, H., et al. (2013). Multiple binding sites for small-molecule antagonists at the CC chemokine receptor 2. *Molecular Pharmacology, 84*(4), 551–561. http://dx.doi.org/10.1124/mol.113.086850.

CHAPTER TWENTY-TWO

Preparation and Characterization of Glycosaminoglycan Chemokine Coreceptors

Nikola Kitic*, Martha Gschwandtner*, Rupert Derler†, Tanja Gerlza†, Andreas J. Kungl*,†,1

*Department of Pharmaceutical Chemistry, Institute of Pharmaceutical Sciences, Karl-Franzens-University Graz, Graz, Austria
†Antagonis Biotherapeutics G.m.b.H., Graz, Austria
[1]Corresponding author: e-mail address: andreas.kungl@uni-graz.at

Contents

1. Introduction 518
2. Preparation and Characterization of GAGs 520
   2.1 Preparation of HS from Tissues 520
   2.2 Preparation of HS from Mammalian Cells 523
   2.3 Dot Blot Analysis to Estimate GAG Quantity/Concentration 524
   2.4 Depolymerization and Preparative Size-Exclusion Chromatography 524
   2.5 Strong Anion-Exchange–HPLC and MS Analyses 527
3. Methods for Studying Chemokine–GAG Interactions 529
   3.1 Gel Mobility Assay 530
   3.2 Isothermal Fluorescence Titration 531
   3.3 ELISA-Like Competition Assay 534
4. Concluding Remarks 536
Acknowledgment 537
References 537

Abstract

Interactions between chemokines and glycosaminoglycans (GAGs) are crucial for the physiological and pathophysiological activities of chemokines. GAGs are therefore commonly designated as chemokine coreceptors which are deeply involved in the chemokine-signaling network. Studying the interaction of chemokines with GAGs is therefore a major prerequisite to fully understand the biological function of chemokines. GAGs are, however, a very complex class of biomacromolecules which cannot be produced by conventional recombinant methods and which, if purchased from commercial suppliers, are often not subjected to rigorous quality control and therefore frequently differ in batch characteristics. This naturally impacts chemokine–GAG interaction studies. In order to standardize the quality of our GAG ligands, we have therefore established protocols for the preparation and characterization of GAGs from various

Methods in Enzymology, Volume 570
ISSN 0076-6879
http://dx.doi.org/10.1016/bs.mie.2015.09.018

cells and tissues, for which we give practical examples relating to the major GAG classes heparin, heparan sulfate, and chondroitin sulfate. We will also outline robust and sensitive protocols for chemokine–GAG interaction studies. By this means, a better and more common understanding of the involvement of GAGs in chemokine-signaling networks can be envisaged.

1. INTRODUCTION

Glycosaminoglycans (GAGs) are linear, in most cases sulfated, polysaccharides with molecular weights of around 10–100 kDa. This group of biomolecules has six members, namely: heparan sulfate (HS), heparin (HP), chondroitin sulfate (CS), dermatan sulfate (DS), keratan sulfate (KS), and hyaluronic acid (HA), the latter of which is the only unsulfated GAG (Gandhi & Mancera, 2008). Because HS/HP/CS/DS have been reported to be the main GAG ligands of chemokines, the techniques described in this chapter will be focused on the preparation and characterization of these GAGs. The polysaccharide chain of HS and HP is made of repeating disaccharide subunits consisting of an uronic acid linked to D-glucosamine (see Fig. 1). The uronic acid can occur be either α-L-iduronic acid or its epimer β-D-glucuronic acid. For the diversity of HS/HP, the sulfation and acetylation pattern is very important. The uronic acid can be 2-*O*-sulfated, whereas the D-glucosamine can be *N*-sulfated or *N*-acetylated. Both of these subunits

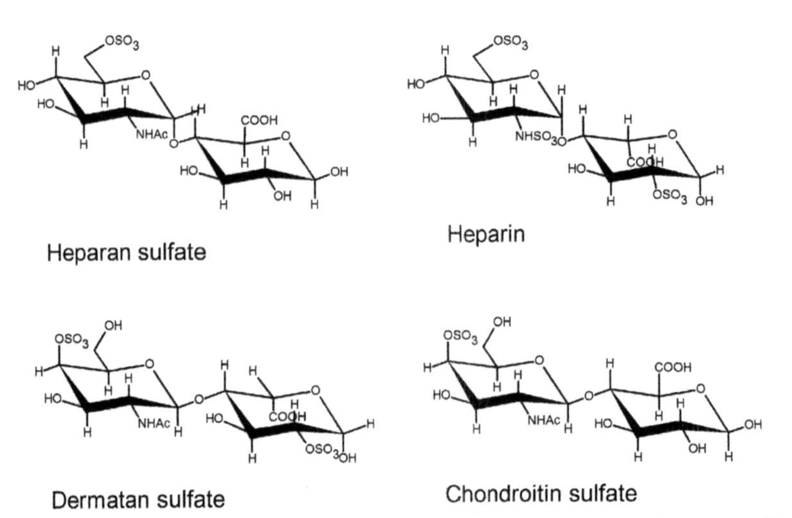

Figure 1 Structures of the disaccharide-building blocks of heparin (HP), heparan sulfate (HS), chondroitin sulfate (CS), and dermatan sulfate (DS).

can be further 6-*O*-sulfated, and the *N*-sulfated D-glucosamine can also be 3-*O*-sulfated. In general, HP exhibits a much higher sulfation degree than HS. HS is not only possessing a lower degree of sulfation in comparison to HP, the polysaccharide chain of HS is comprised of highly sulfated domains, which are separated from each other by unsulfated or lower sulfated domains (Rabenstein, 2002). CS and DS do not have the same building blocks as HS and HP (see Fig. 1). In their case, the repeating disaccharide subunits are made of a uronic acid linked to *N*-acetylgalactosamine. CS contains glucuronic acid and DS iduronic acid as part of their disaccharide subunits. Similar to HS/HP, CS/DS can also be sulfated at various positions. The *N*-acetylgalactosamine can carry a sulfate group on 4-*O*- and/or 6-*O*-position, and the uronic acid can be sulfated on position 2-*O* and less commonly on the 3-*O*-position (Silbert & Sugumaran, 2002).

Heparin is predominantly produced by mast cells as part of a proteoglycan (PG), where the GAG chains are covalently attached (*O*-linked) to a serglycin core protein. After synthesis, the heparin chains are cleaved, and the so obtained disperse mixture of polysaccharides is stored in the cytoplasmic granules of the cells. HS—as well as CS and DS—is also found as *O*-glycosidic part of PGs, but they can be attached to a variety of different core proteins (e.g., syndecans and glypicans). Almost all mammalian cells biosynthesize such PGs, which can be found as part of cell membranes or as components of the extracellular matrix (Rabenstein, 2002; Silbert & Sugumaran, 2002).

GAGs are involved in a variety of different chemokine-related processes *in vivo* (Gandhi & Mancera, 2008), and it is believed that almost all chemokines interact with GAGs due to the fact that most chemokines are basic and GAGs are highly acidic molecules. Questions that have been raised in this context include: What is the physiological impact of these interactions and how much do GAG-induced conformational changes play a biological role? Do chemokines recognize very specific GAG patterns? Are soluble chemokine–GAG complexes detectable in serum and are they biologically active? Is there a signal transfer induced following the interaction of chemokines and membrane-bound GAGs? Are chemokine–GAG interactions an interesting target for the development of novel therapeutic approaches? Some of these questions have already been answered, but many still remain open. For example, it is known today that at least some chemokines can distinguish between different GAG species which is expressed in different affinities for different GAG ligands (Rek, Krenn, & Kungl, 2009). Moreover, it was also shown that the interaction with GAGs induces the oligomerization of chemokines (Hoogewerf et al., 1997; Kuschert et al., 1999). In inflammatory processes, chemokines are responsible for the activation of immune cells and their guidance to the site of inflammation. The interaction with GAGs is a major prerequisite for this function. Through the interaction with cell-surface GAGs, chemokines are retained and presented on the surface of the endothelium which is a requirement for the formation of a chemotactic gradient. The GAG-induced oligomerization seems to be, at least for some chemokines, also a prerequisite for their chemotactic activity (Hoogewerf et al., 1997; Proudfoot et al., 2003). In the literature, there is also indication that the interaction with GAGs is important for chemokine secretion (Lebel-Haziv et al., 2014) and that by formation of a chemokine–GAG complex, chemokines can be protected from enzymatic cleavage (Ellyard et al., 2007). Chemokine–GAG interactions could therefore turn out to be interesting target sites for the development of novel anti-inflammatory drugs. So far, a limited number of different approaches were proposed to interfere in chemokine–GAG interactions, e.g., by GAG mimetics, by scaffold proteins, and by chemokine-derived peptides (Adage et al., 2012; Friand et al., 2009; Vanheule et al., 2015). Here, we present methods to prepare GAGs and to characterize their interactions with chemokines.

2. PREPARATION AND CHARACTERIZATION OF GAGs

2.1 Preparation of HS from Tissues

The preparation of GAGs from tissue samples or whole organs is a challenging task and yields are often low. The protocol described here is routinely applied in our labs and gives satisfying and reproducible results. It describes the preparation of HS from tissues but can be easily adapted for the preparation of other GAG species. We focused mainly on HS due to its importance in chemokine-mediated inflammatory processes (Parish, 2006). Preparation of HS is very important because most commercially available HS is derived from a single organism and a single organ, namely, porcine intestinal mucosa. The availability of HS from different tissues and cells is therefore a major prerequisite if tissue- or species-specific HS-related biological phenomena are investigated.

2.1.1 Required Materials

Azure A (Sigma, Cat. No. A6270)
Barium Acetate (Sigma, Cat. No. 32305)

Benzonase (Sigma, Cat. No. E1014)
BSA (Sigma, Cat. No. A7039)
Centricons 4 mL (Merck, Cat. No. UFC 800324)
Centricons 15 mL (Merck, Cat. No. UFC 900324)
Chloroform (Sigma, Cat. No. 372978)
Chondroitinase ABC (Sigma, Cat. No. C3667)
Cysteine (Sigma, Cat. No. W326305)
DEAA column material (GE Healthcare, Cat. No. 17-0709-10)
Ethylenediaminetetraacetic acid (EDTA) (Sigma, Cat. No. E5134)
Glycerol (Sigma, Cat. No. G5150)
Methanol (Sigma, Cat. No. 494437)
$MgCl_2$ (Sigma, Cat. No. M2393)
$NaBH_4$ (Merck, Cat. No. 1.06371.0100)
NaCl (Sigma, Cat. No. S9625)
NaOH (Sigma, Cat. No. S5881)
Papain (Sigma, Cat. No. P3375)
Sodium phosphate dibasic (Sigma, Cat. No. S5136)
Sodium phosphate monobasic (Sigma, Cat. No. S9638)
Tris–acetate (Trizma, Sigma, Cat. No. T1503)

(a) *Tissue homogenization*: prepare the tissue by washing it briefly with dH_2O to remove remaining blood, cut it into small pieces, and homogenize it with an appropriate blender or homogenizer. If homogenization is insufficient, mince the tissue further with mortar and pistil. When using tissues with a high lipid content (e.g., liver), remove fat by repeated washing with different chloroform/methanol mixtures (2:1, 1:1, 1:2, v/v) and subsequently dry it under vacuum (Warda et al., 2006).

(b) *Protein depletion*: add 4 mL of buffer (50 m*M* sodium phosphate at pH 6.5, 2 m*M* EDTA, and 2 m*M* cysteine) and 15 U papain per gram of wet weight tissue. Incubate for approximately 23 h at 65 °C (with shaking). Centrifuge for 1 h at 20,000 × *g* in order to remove residual particles, discard the pellet. Determine the exact volume of the supernatant and add NaOH and $NaBH_4$ to a final concentration of 0.5 and 0.1 *M*, respectively. Incubate overnight at 4 °C.

(c) *Anion-exchange chromatography*: adjust the solution by adding 6 *M* HCl to a pH of 6.5. Load the sample onto a DEAE column with an ÄKTAprime system (GE Healthcare) or similar. After loading is completed, wash the column with 0.1 *M* Tris–acetate buffer (pH 7) until the baseline is stable (monitor absorption at 280 nm). Wash with 10 column volumes (CV) of 0.1 *M* Tris–acetate buffer (pH 7) containing 300 m*M* NaCl and elute GAGs with 6 CV of the same buffer containing 2 *M* NaCl.

(d) *Removal of nucleic acids*: perform buffer exchange to 50 m*M* Tris–HCl buffer (pH 8) containing 2 m*M* $MgCl_2$ by using centricons 15 mL (with 3 kDa cutoff). After buffer exchange, the NaCl concentration should not be higher than 5 m*M*. Add 8 U benzonase per gram of wet weight tissue and incubate for 24 h at 37 °C (with slow shaking). Bring the solution to 60 m*M* sodium acetate, 0.02% BSA by the addition of the appropriate amount of a 40-fold stock (background buffer: 50 m*M* Tris–HCl buffer, pH 8).

(e) *Removal of CS*: add 7.5 mU of chondroitinase ABC per gram of wet weight tissue and digest for at least 16 h at 37 °C (with shaking). Adjust the pH to 7 by addition of acetic acid and centrifuge for 30 min at 16,000 × *g*, discard the pellet. Load the sample onto a DEAE column with an ÄKTAprime system (or similar) and proceed as described above. Using centricons 4 mL (3 kDa cutoff), perform a buffer exchange to HPLC grade H_2O. After the buffer exchange, the NaCl concentration should not be higher than 2 m*M*. Freeze-dry the sample.

(f) *Purity and identity check*: to check if the preparation contains other GAGs than HS perform an agarose gel electrophoresis (see Fig. 2). This

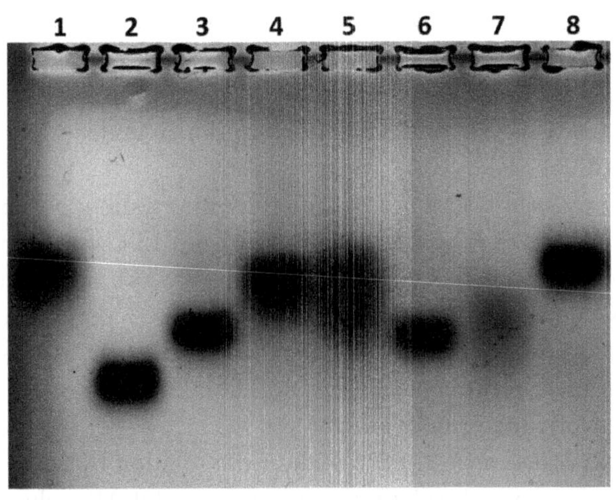

Figure 2 Example agarose gel of different GAGs. Lane 1: HS (Celsus Inc.), lane 2: CS (Carbosynth Inc.), lane 3: DS (Celsus Inc.), lane 4: HS (Iduron Ltd.), lane 6: HP (Iduron Ltd.), lane 7: enoxaparin natrium (Sigma), and lane 8: HS prepared in-house from porcine lung tissue.

method is adapted from a previously published protocol (van de Lest et al., 1994). Dissolve agarose in electrophoresis buffer (50 m*M* barium acetate, pH 5) to a final concentration of 1% and pour it into a gel-casting tray. Mix the samples from the preparation as well as the standard GAGs 1:1 with loading buffer (electrophoresis buffer containing 20% glycerol) and apply them to the gel. Run the gel for 1.5 h at 100 mV. Perform an azure A (0.05% Azure A in 5% acetic acid) stain.

2.2 Preparation of HS from Mammalian Cells

Alternatively to tissues, HS can be prepared from mammalian cell cultures. For this, at least 2.5 g cell pellet needs to be collected for GAG isolation which can be stored at −20 °C until usage.

2.2.1 Required Materials

$CaCH_3COO$ (Sigma, Cat. No. C1000)
$NaCH_3COO$ (Sigma, Cat. No. S8750)
Pronase (Roche, Cat. No. 10165921001)

(a) *Protein depletion*: add 2.7 mL pronase buffer (50 m*M* $CaCH_3COO$, 50 m*M* Tris Base, pH 7.5) per g cell pellet. Vortex the cell suspension and sonicate them to lyse the cell membrane. Add pronase to your lysed cells to a final concentration of 1 mg/mL. Incubate for 20 h at 50 °C (with shaking).

(b) *Removal of nucleic acids*: inactivate pronase by heating to 100 °C for 5 min and centrifuge at 13,000 × *g* for 10 min. Add benzonase buffer to the sample to adjust to a final concentration of 2 m*M* $MgCl_2$ and 0.01% BSA in 50 m*M* Tris Base (pH 7.5–8.0). Add 100 IU benzonase per gram cell pellet and incubate at 37 °C for 6 h.

(c) *Removal of CS*: add chondroitinase buffer to your sample to reach a final concentration of 60 m*M* $NaCH_3COO$, 0.02% BSA in 50 m*M* Tris Base (pH 7.5–8.0). Add 20 mIU chondroitinase ABC lyase per gram cell pellet and incubate at 37 °C for 16 h (with shaking).

(d) *Anion-exchange chromatography*: prepare your buffers in HPLC grade water. Connect your DEAE column to an ÄKTAprime system (GE Healthcare) or similar. Equilibrate your column with 4 CV 20 m*M* $NaCH_3COO$ (pH 6.0). Load your sample onto the column. Wash the column with 10 CV of 20 m*M* $NaCH_3COO$ (pH 6.0) containing 300 m*M* NaCl. Elute the GAGs with 6 CV of 20 m*M* $NaCH_3COO$, (pH 6.0) containing 2 *M* NaCl.

(e) *Buffer exchange and sample concentrating*: perform buffer exchange to your desired end buffer using Amicon Ultra centrifugal filters with regenerated cellulose and concentrate your samples to a finale volume between 100 and 200 μL. Purity and identity of the samples are checked as described above for tissue-derived GAGs.

2.3 Dot Blot Analysis to Estimate GAG Quantity/Concentration

2.3.1 Required Materials

Nylon transfer membrane (GE Healthcare, Cat. No. RPN 20313)

It is not straightforward to measure concentrations and amounts of GAG preparations, which are not enzymatically treated and are yet unfractionated. One of the easiest and quickest methods is the so-called dot blot analysis. For this purpose, 2 μL of standard concentrations (heparan sulfate) ranging from 2.5 to 25 μg are spotted onto positively charged nylon transfer membrane. Further spot three times 2 μL and five times 2 μL of your samples onto the membrane so that you have a spot with a total of 6 μL and a spot with 10 μL. Let the spots dry in between the spottings and after the last spot for at least 10–15 min. Perform an azure A (0.05% azure A in 5% acetic acid) stain, destain with water, and perform densitometry analysis of your samples. Calculate the concentration by comparing to the standard curve obtained with the HS standard.

2.4 Depolymerization and Preparative Size-Exclusion Chromatography

We and others have shown that chemokine–GAG interactions depend, in addition to overall charge and charge patterns, on the oligosaccharide length under investigation (Goger et al., 2002). To generate size-defined glycan oligosaccharides, preparative size-exclusion chromatography (SEC) is applied. Prior to chromatographic separation, full-length GAGs need to be digested in order to obtain artificial size-defined GAGs. Depolymerization of GAG oligosaccharides by bacterial lyases (heparinases) creates 4,5-unsaturated uronic acid residues at the nonreducing ends of the sugars that show a maximum UV-light absorption at a wavelength of 232 nm, which allows for detection and quantification of these GAGs (Jandik, Kruep, Cartier, & Linhardt, 1995). These size-defined GAGs can be further separated in charge defined fractions using strong anion exchange.

2.4.1 Required Materials

Acrylamide/bis-acrylamide solution (Carl Roth, Cat. No. 3029.1)

Ammonium Bicarbonate (Sigma, Cat. No. A6141)

Ammonium persulfate (Sigma, Cat. No. A3678)
Biogel P10 fine (Bio-Rad, Cat. No. 150-4144)
Boric acid (Sigma, Cat. No. 31146)
Cellulose nitrate filter (Sartorius, Cat. No. 11302-47-ACN)
Filtration unit (Wheaton, Cat. No. 635525)
Glycine (Sigma, Cat. No. G7126)
Heparinase I (Iduron, Cat. No. HEP-ENZ1)
Heparinase II (Iduron, Cat. No. HEP-ENZ2)
Heparinase III (Iduron, Cat. No. HEP-ENZ3)
Sucrose (Fluka, Cat. No. 84097)
Tetramethylethylenediamine (TEMED) (Sigma, Cat. No. T9281)

(a) *Depolymerization of heparin*: (heparin is used here as a typical example for the digestion protocol since this GAG is usually easily obtained at larger quantities; the same protocol can be applied, however, to HS as well as to CS, in the latter case, different enzymes must be applied) prepare heparinase buffer, i.e., 100 m*M* sodium acetate, 2 m*M* calcium acetate diluted in HPLC grade H_2O and pH adjusted to 7.5 with acetic acid. Weigh out 10 mg heparin and dissolve in 200 μL heparinase buffer. Add 20 mIU heparinase I and incubate for 4 h at 37 °C with slow agitation, then add 20 mIU heparinase III and incubate for further 4 h under the same conditions. Lastly, add 20 mIU heparinase II and incubate for 16 h at 37 °C. Then, heat sample to 99 °C for 5 min to deactivate the enzymes and store at −20 °C until usage (Powell, Ahmed, Yates, & Turnbull, 2010).

(b) *Preparative SEC*: glass columns with a length of 220 cm and an i.d. of 1 cm can be used. These columns are filled with biogel P10 fine. Prepare 0.1 *M* ammonium bicarbonate and stir until the ammonium bicarbonate is completely dissolved. Filtrate the buffer using a filtration unit with a cellulose nitrate filter connected to a water-jet pump or equivalent filtration equipment. Degas the solution for 30 min under stirring using a water-jet pump. Heat 500 mL degassed 0.1 *M* ammonium bicarbonate buffer on a heating plate to 100 °C. Weigh out 33 g biogel P10 and add it to the heated 0.1 *M* ammonium bicarbonate buffer under stirring with a spatula. Let the biogel P10 hydrate and cool down for 1 h without stirring. After hydration is complete, decant half of the supernatant and transfer the solution to a 1-L filter flask and attach it to a water-jet pump for 10 of minutes degassing. Add two bed volumes (375 mL) degassed 0.1 *M* ammonium bicarbonate buffer and swirl gently. Allow the gel to settle for approximately 20 min. Then remove the supernatant by suction or decanting and repeat all steps after degassing three times. Rinse the column with ammonium bicarbonate buffer and close the column exit. Apply a funnel on top of the column and fill it with 2 cm of degassed 0.1 *M* ammonium bicarbonate buffer. Carefully fill the gel into the column without introducing air bubbles until the column is filled to the top. Open the column outlet and let the column settle for 30 min. Remove the rest of the buffer on top of the column and fill up the column with gel. Overnight, close the column exit and put parafilm on top of the glass column. On the next day, open the outlet again and remove the rest of the buffer on top. Repeat the refilling on top until the column is completely packed, thus the upper end of the gel does not sink anymore. Put a stamp on the upper end of the column and connect it to an HPLC system. Apply a flow rate of 30 μL/min (0.1 *M* ammonium bicarbonate as mobile phase) and wait until the baseline is stable. Inject the LMWH or digested HS with a flow rate of 30 μL/min and collect fractions starting from minute 1200, every 40 min one. A single run takes approximately 6000 min. After collecting, concentrate your samples and pool the SEC fractions, which contain oligosaccharides of defined chain length (see Fig. 3).

(c) *High-resolution acrylamide gel electrophoresis*: to check the purity of dp fractions, acrylamide gel electrophoresis is performed. For this purpose, cast gels by mixing 9 mL 30% acrylamide/bis-acrylamide solution (37.5:1),

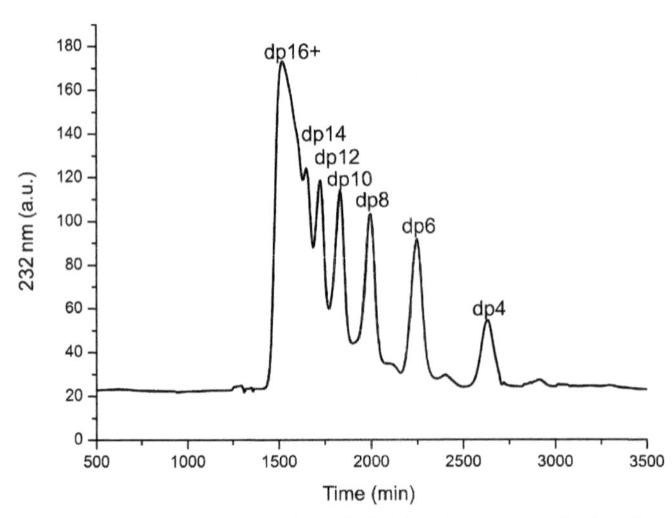

Figure 3 Size-exclusion chromatography of HP following enzymatic depolymerization. Indicated are the peaks of tetrasaccharides (dp4), hexasaccharides (dp6), octasaccharides (dp8), etc. (for a detailed experimental description, see text).

with 1 mL 10 × lower chamber buffer (1 *M* Trizma, 1 *M* boric acid and 0.1 *M* EDTA filled up to 1 L with dH_2O and pH adjusted to 8.3), 33 μL 10% ammonium persulfate (APS) and 6.6 μL TEMED. Pour gels vertically between glass plates separated by 1 mm spacers. Let gels polymerize and prepare the upper chamber buffer (0.2 *M* Trizma and 1.24 *M* glycine filled up to 1 L with dH_2O and pH adjusted to 8.3) and the sample buffer (50% sucrose in 0.1 *M* boric acid, 0.1 *M* Trizma, 0.01 *M* EDTA filled up to 100 mL with dH_2O and pH adjusted to 6.3). Mix samples 1:2 with sample buffer. Put the gel into Miniprotean Tetracell (Bio-Rad) or equivalent and fill the upper gel buffer into the upper chamber. Dilute the lower chamber buffer 1:10 and fill it in the lower chamber. Let the gels run for 2 h at 160 V. Prepare a tray with 100 mL Azure A solution. Stain the gel in the solution for 15 min, then destain with dH_2O (Gunay & Linhardt, 2003).

2.5 Strong Anion-Exchange–HPLC and MS Analyses

Strong anion-exchange chromatography (SAX) is used to further fractionate size-defined GAG oligosaccharides like the ones obtained by SEC (see above). Since chemokine–GAG interactions are strongly dependent on the overall charge of GAG ligands as well as on the charge/sulfate distribution on such ligands, the availability of size- and charge-fractionated GAG oligosaccharides is very important for chemokine–GAG interaction studies. SAX chromatography is a well established and widely used method for di- and oligosaccharide analysis of GAGs that has been in use for a long time (Linhardt, Rice, Kim, Engelken, & Weiler, 1988; Sharath et al., 1985; Turnbull, Fernig, Ke, Wilkinson, & Gallagher, 1992). Separation of different saccharides is achieved by elution of the negatively charged molecules from the positively charged column surface by applying a sodium chloride gradient at a pH of 3.5. This pH allows for protonation of the uronic acid carboxyl residues. Therefore, separation depends mainly on the number and position of sulfate groups within the molecules. Quantification and finally identification rely on retention times, peak areas, and MS analyses.

2.5.1 Required Materials

Ammonium hydroxide (Fluka, Cat. No. 44273)

Methanol (Merck, Cat. No. 1.06007.2500)

ProPac PA1 column (Thermo Scientific, Cat. No. 039658)

(a) *SAX chromatography*: from the different SAX columns available, a ProPac PA1 column was used here which provided good resolution and reproducible separation of GAGs. Freshly prepare and degas solvent A and solvent B. Solvent A: dH_2O (HPLC grade); solvent B: 2 *M* sodium chloride in dH_2O (HPLC grade), both adjusted to pH 3.5 with hydrochloric acid. Equilibration of the system is an important factor for reproducibility and should be carried out for a sufficient period of time prior to each experiment. Equilibrate with solvent A at the required flow rate, column temperature, and UV detection wavelength. Whenever the system runs on solvent A, impurities can accumulate on the column even when high-purity chemicals are used. Therefore, it is recommended to perform a blank run before starting an experiment in order to clean the column. A linear gradient from 0% to 70% solvent B over a time range of 42 min at a flow of 1 mL/min allows satisfyingly separation of GAG oligosaccharides. Sample load depends upon its complexity which increases with increasing depolymerization grade. For further subfractionation of a SEC-derived dp4 fraction according to charge, a 0.8-μg/μL solution (total 500 μL) was injected into the SAX system (see Fig. 4). By this means, three peaks according to the main charge fractions of dp4 were identified based on clearly separated retention times.

(b) *Mass spectrometry of GAG oligosaccharides*: MS can be used to confirm SAX–HPLC findings or to elucidate the compositional analysis a

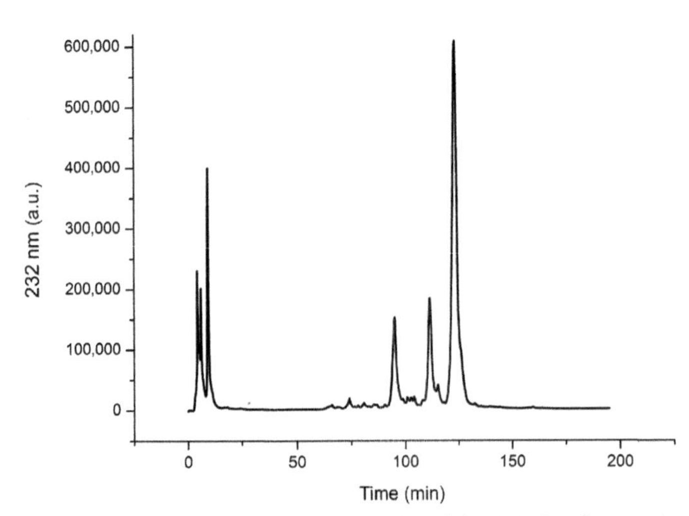

Figure 4 Strong anion-exchange chromatography of the SEC dp4 fraction (see Fig. 3). About 400 μg dp4 in 500 μL solvent A were injected (for a detailed experimental description, see text).

GAG sample. Due to the polyanionic nature of GAGs, the oligosaccharides will occur in different charge states in the spectrum, e.g., M^- through M^{3-} in the case of most HS/HP disaccharides. Therefore, the spectra of oligosaccharide mixtures might become rather complex with an increasing size and number of components. Separation and fractionation by SAX–HPLC (see above) might help reducing this complexity. Here, we describe an ESI-MS setup for a LCQ Deca XP+ system (Thermo Finnigan) that has been applied to analyze the HP tetrasaccharide (dp4) fraction. For other instruments, the parameters might have to be changed accordingly.

For all MS experiments, chemicals of highest possible purity should be used. Desalting of oligosaccharide samples to the best extent is an ultimate prerequisite for reliable MS results. Dilute the sample in a 60:40 mixture of dH_2O/methanol added with 0.1% (v/v) ammonium hydroxide to a concentration of approximately 0.1–0.5 nmol GAG per μL. Approximately 10 μL of this solution are enough to record a 30-min spectrum. Prior to MS analysis, we recommend tuning the instrument in negative ion mode. Calibration in negative ion mode is optional but not necessary. Apply the sample to a nanospray emitter needle. Set the instrument to negative ion mode and apply a spray voltage of 1.0–2.0 kV with a capillary temperature of 150–250 °C. In our experience, the signal-to-noise ratio of the spray improves when moving the emitter needle as far away as possible from the capillary. Depending on the question of the experiment, MS^2 analysis might be needed. For interpreting the data of MS^2 spectra, it might be useful to use bioinformatics tools such as GlycoWorkbench (Ceroni et al., 2008). Figure 5 shows a typical MS spectrum of a HP dp4 fraction. The m/z peak 357 corresponds to a nonacetylated 5-*O*-sulfated, and the m/z peak 383.7 to a nonacetylated 6-*O*-sulfated tetrasaccharide, respectively. Other, differently and multiple charged tetrasaccharides have not been found at significant concentrations in this HP preparation.

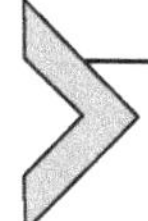

3. METHODS FOR STUDYING CHEMOKINE–GAG INTERACTIONS

In order to evaluate the quality and the chemokine-related functionality of GAG preparations (on top of elucidating their compositional analyses), chemokine interaction studies are applied here. Since chemokine–GAG interactions will strongly depend upon the availability of the chemokine-specific GAG sequence/epitope in the GAG sample under investigation,

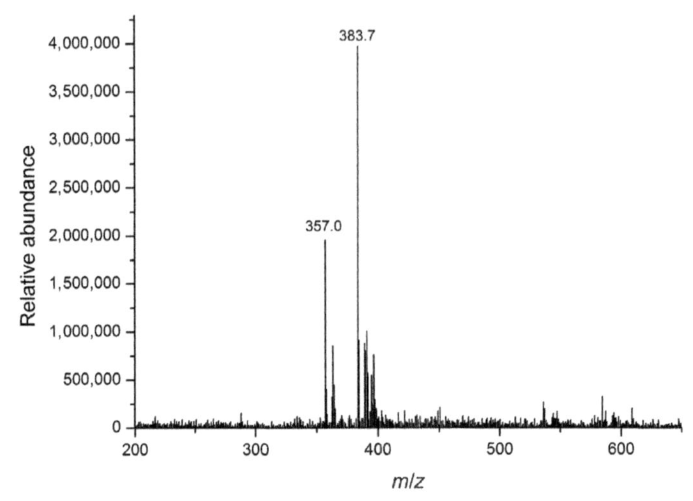

Figure 5 Ion-spray MS analysis of dp4 (for a detailed experimental description, see text).

measured affinities will reflect the proper ligand nature of a certain GAG preparation. In addition, competition experiments give an insight into the degeneracy of the availability of chemokine–GAG epitopes in a GAG preparation.

3.1 Gel Mobility Assay

A simple but fast way to check for chemokine–GAG interactions is the electrophoretic mobility shift assay. Due to the basic nature of most chemokines, they are not able to migrate to the anode in conventional native polyacrylamide gel electrophoresis. In contrast, GAGs are commonly highly negatively charged due to the large number of sulfate groups. If chemokines are thus bound to GAGs, the resulting complex becomes negatively charged and is able to migrate toward the anode, where it becomes detectable for protein as well as for glycan staining.

3.1.1 Required Materials

Acetic acid (Sigma, Cat. No. 27221)
$AgNO_3$ (Sigma, Cat. No. 85228)
Na_2CO_3 (Sigma, Cat. No. S7795)
$Na_2S_2O_3$ (Fluka, Cat. No. 72049)

Prepare the gel buffer (30 m*M* Tris–HCl, 3 m*M* EDTA, pH 7.3 with HCl) and the sample buffer (100 m*M* Tris–HCl, 1 m*M* EDTA, pH 7.4). Cast two gels by mixing 4.8 mL Rotiphorese 30, with 7.2 mL gel buffer, 45 μL 10%

APS, and 10 μL TEMED. Pour gels vertically between glass plates separated by 1 mm spacers. Let the gels polymerize and prepare a 10 × running buffer containing 400 m*M* Tris–HCl and 10 m*M* EDTA with the pH adjusted to 8 with HCl. Mix 1–10 μg of your chemokine with 5 μg of GAG and let them incubate for 15 min. Afterward, dilute your samples 1:2 with sample buffer. Also prepare two reference samples, one with GAG only and one with chemokine only. Put the gel into Miniprotean Tetracell (Bio-Rad) or equivalent, dilute the 10 × running buffer 1:10, and fill it into the Tetracell chamber. Let the gels run for 45 min at 100 V. To stain your chemokines, perform a silver stain according to EMBL Protocol. Transfer the gel to a staining dish and rinse with deionized water before fixing in 50% methanol, 5% acetic acid for 20 min at RT while shaking. Wash gel in 50% methanol for 10 min. After 10 min, discard to methanol waste and wash gel in H_2O three times for 10 min. Oxidize gel in 0.02% $Na_2S_2O_3$ for 1 min. Discard solution and wash gel twice in 100 mL H_2O for 1 min. Incubate gel in cold 0.1% $AgNO_3$ for 20 min at 4 °C. Discard solution and wash gel twice in 100 mL H_2O for 1 min. Develop gel in 0.04% formaldehyde, 2% Na_2CO_3 on the illuminator. Terminate staining in 5% acetic acid and use. Image the gel using the ChemiDoc XRS+ Imager or equivalent equipment. Prepare a tray with 100 mL Azure A solution (100 mg Azure A in 5% acetic acid). Stain the gel in the solution for 15 min, then destain with deionized water. Image the gel again and overlay the silver stain and the Azure A stain.

In Fig. 6, a typical gel mobility analysis of CCL2 binding to HS is presented. It clearly shows that upon increasing chemokine concentration, the HS band becomes weaker according to its immobilization in the chemokine–GAG complex, whereas the uncomplexed chemokine does not migrate at all toward the anode and is thus not detected in this setup.

3.2 Isothermal Fluorescence Titration

Isothermal fluorescence titration (IFT) is an in-solution method for determining the binding affinity (K_d values) between a protein and its ligand. It is based on measuring the dose-dependent change (quenching or dequenching) in fluorescence emission of an intrinsic or extrinsic fluorophore upon ligand addition. Due to the high sensitivity of the method, only very low amounts of chemokine (<1 μ*M*) are needed. The majority of chemokines contain a tryptophan residue which has the highest fluorescence quantum yield of all naturally occurring amino acids, i.e., much higher compared to the other aromatic amino acids tyrosine and phenylalanine

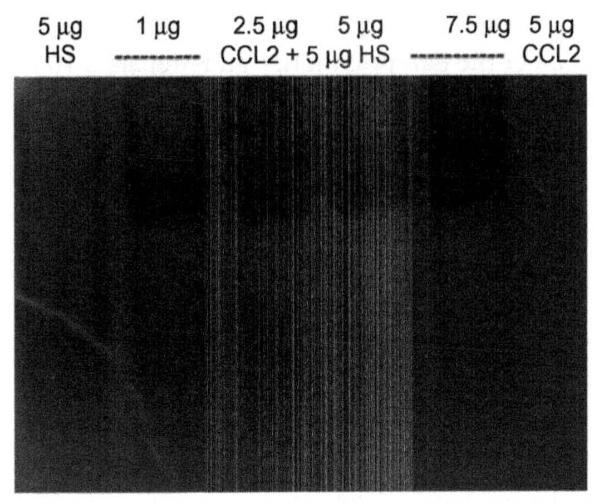

Figure 6 Gel mobility analysis of CCL2 binding to HP. Blue (gray in the print version) bands depict the migrated sugar stained with Azure A and brown (dark gray in the print version) bands correspond to CCL2 stained with silver nitrate. Band intensity increases with higher amounts of chemokines. CCL2 without GAG is not able to migrate into the gel due to its basic nature. Due to the sulfate groups of the GAGs, the CCL2/GAG complex is negatively charged and shifts toward the anode.

(Lakowicz, 2006). In addition, the fluorescence of a tryptophan residue is very sensitive to changes in the environment, and therefore, conformational changes due to ligand binding can lead to detectable changes of the emission intensity (Rek, Geretti, Goger, & Kungl, 2002). If the chemokine of interest does not contain a tryptophan residue, it has to be fluorescently engineered by coupling to a tag or genetically modified by replacing an amino acid with tryptophan. Another prerequisite for performing IFT measurements is the high purity of the interaction partners, since only minimal amounts of impurities can result in significant background signals, especially if the impurities contain fluorescent compounds like proteins with more than one tryptophan residue (Gerlza et al., 2014).

(a) *Fluorescence measurements*: switch on the spectrofluorometer (e.g., FP-6500, Jasco) at least 30 min before use (to avoid strong lamp shift) and set the external water bath to 20 °C. Provide a 700 n*M* solution of the chemokine to be analyzed in a total volume of 500 µL PBS (137 m*M* NaCl, 8 m*M* Na_2HPO_4, 2 m*M* NaH_2PO_4, pH 7.35) in the quartz cuvette (108F-QS, light path 10 × 4 mm, Hellma Analytics) and let the solution equilibrates until the fluorescence signal is stable, which is commonly achieved after 30 min. To record the spectrum, adjust the slits of the instrument to reach an intensity in the middle of the measurement range. Excite at 280 nm and record protein fluorescence emission spectra over the range of 300–400 nm at a scan speed of 500 nm/min. Use a data pitch of 0.5 nm, response of 1 s, sensitivity high and tick the auto-shutter control. Prepare a GAG solution (depending on the estimated affinity between 50 and 500 μ*M*) in dH_2O. Concentrated stock solutions ensure a minimal dilution of the protein solution due to ligand addition. After equilibration, record the protein spectrum and add 0.5 μL GAG solution. Mix gently and equilibrate the solution for 1 min in the instrument. Record the spectrum and add the next 0.5 μL of the ligand. Proceed until maximum quenching, corresponding to saturation of all binding sites, is reached. For determination of the background signal, proceed in the same way as described for determining the chemokine–GAG interaction, with the only exception of using PBS only instead of the chemokine solution. Clean the cuvette by thoroughly rinsing it with dH_2O and store it in concentrated nitric acid. Perform three measurements, subtract the background spectra from the respective chemokine spectra, and integrate the resulting curves. Calculate the normalized mean changes in fluorescence intensity according to the following equation:

$$-\frac{\Delta F}{F_0} = -\frac{A_i - A_0}{A_0},$$

where A_0 is the area recorded in the absence of ligand and A_i is the spectrum obtained after ligand addition.

(b) *Data analysis*: plot the normalized mean changes against the concentration of added GAG and analyze the resulting binding isotherms by nonlinear regression using, e.g., Origin (MicroCal Inc.) using the following equation describing a bimolecular association reaction, in which F_i is the initial and F_{max} is the maximum fluorescence value reached by the normalized mean changes in fluorescence intensity (see above), n is the assumed stoichiometry of the protein–ligand interaction, [chemokine], and [GAG] are the total concentrations of the interaction partners, and K_d is the dissociation constant.

$$F = F_i + F_{max}\left[\frac{K_d + [\text{chemokine}] + [\text{GAG}] - \sqrt{\left(K_d + [\text{chemokine}] + [\text{GAG}]\right)^2 - 4[\text{chemokine}][\text{GAG}]}}{2[\text{chemokine}]}\right]$$

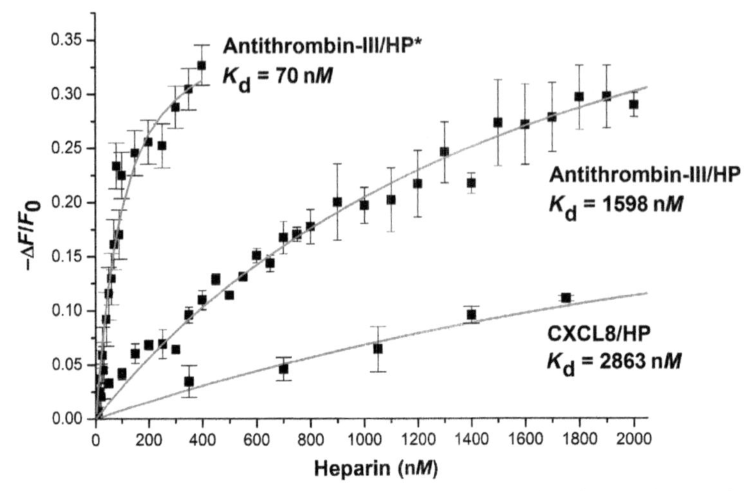

Figure 7 Binding isotherms of CXCL8 and antithrombin-III binding to low-molecular-weight HP. HP* refers to the antithrombin-III-specific heparin pentasaccharide/fondaparinux (Hricovíni et al., 2001) which has been used to record this binding isotherm.

This equation is derived from a general solution for a bimolecular association (Nomanbhoy & Cerione, 1996). Typically, the fitting parameters $F_i=0$, $c=700$, and $n=1$ are set as fixed, F_f and K_d are the variable parameters.

In Fig. 7, a typical binding isotherms for CXCL8 and HP are presented. As a comparison—and to show the potential of the method to discriminate between specific/high affinity and nonspecific interactions—the binding isotherms of antithrombin-III for its interaction with generic HP (low affinity) and with its very specific pentasaccharide (high affinity) is shown (Hricovíni et al., 2001).

3.3 ELISA-Like Competition Assay

With the ELISA-like competition assay, it is possible to determine and compare the competitive potential of different chemokines for the same GAG ligand. Instead of a capturing antibody, as it is used in a classical ELISA, GAGs are coated on specially prepared GAG-binding plates (Iduron) that adsorb GAGs without further chemical modifications while retaining their natural protein-binding characteristics. Biotinylated chemokines are incubated on these plates and can be further competed off with other nonlabeled chemokines, proteins, or mutants. Therefore, biotinylation of the chemokines is an essential first step. After competition, the nondisplaced biotinylated chemokine is detected with streptavidin horseradish peroxidase

(HRP). Streptavidin binds biotin noncovalently with high affinity. Similar to a typical ELISA, a substrate like tetramethylbenzidine (TMB) is added, and a color change is recorded (Goldsby, Kindt, Osborne, & Kuby, 2003). An intense color development indicates low, a weak signal means high competition. The intensity is then plotted against the concentration of the competitor, and IC50 values are calculated using an appropriate fit function. Among the affinity techniques described, the ELISA-Like Competition (ELICO) assay is the most complex one, but mimics the "real situation" in a human body best, because GAGs build a matrix with chemokines bound to it, that are then displaced by other wild-type or decoy chemokines.

3.3.1 Required Materials

EDC (Thermo Scientific, Cat. No. 22980)
EZ link pentylamine biotin (Thermo Scientific, Cat. No. 21345)
MES (Sigma, Cat. No. N2933)
Streptavidin HRP (Pierce, Cat. No. 21130)
Sulfuric acid (Sigma, Cat. No. 320501)
TMB (Sigma, Cat. No. T5525)
ZEBA desalting columns 7K MWCO (Pierce, Cat. No. 89889)

(a) *chemokine biotinylation*: by using EDC, a water-soluble carbodiimide cross-linker, the terminal primary amine of the reagent can be coupled to carboxyl groups of the chemokine through amide bond formation. It is important that the GAG-binding sites are not affected by biotinylation, otherwise the chemokine–GAG interaction is altered. For some chemokines, biotinylation of carboxyl groups is not possible because of their isoelectric point. The pentylamine reaction depends on acidic conditions around pH 5.5, but chemokines with a pI value between 5 and 6 precipitate under these conditions. Therefore, another biotinylation reaction is required. One possibility would be to biotinylated the amine groups, with consideration that it is important to protect chemokine–GAG-binding sites by incubating these chemokines with GAG fragments, prior to biotinylation (after biotinylation, the protecting bound GAG fragments have to be removed to reconstitute the chemokine-binding sites for application in this assay).

Exchange the buffer of your sample to 0.1 *M* MES using centricons with a cutoff of 3 kDa to provide optimum reaction conditions for biotinylation. Incubate the chemokine with 20 molar excess of EZ link pentylamine biotin and 10 molar excess of EDC for 2 h at RT, low agitation. For desalting, use ZEBA desalting columns (7K MWCO) according to manufacturer's protocol. Determine the biotinylation grade using a biotin quantification kit (Pierce) and the protein concentration by photometric measurements at 280 nm.

(b) *displacement experiments*: after chemokine biotinylation has been performed, dilute 2.5 μg GAG with 250 n*M* biotinylated chemokine, e.g., CXCL8 in PBS and coat it on specially prepared GAG-binding plates overnight at RT. To get rid of unbound biotinylated CXCL8, use an automatic plate washer (Tecan) or wash by hand. Afterward, incubate with different competitor concentrations starting from 200 μ*M*. Depending of the expected displacement potency, this starting concentration can be reduced to 50 μ*M* or increased if the competitor shows low displacement potency. Prepare eight different concentrations by diluting the start concentration 1:4 in PBS. As each concentration is measured in triplicates (100 μL/well) on one plate, prepare at least 500 μL of the starting concentration solution. Pipette 350 μL of each concentration into a separate 96-well plate, and by using a multichannel pipette, transfer them to your GAG-biotinylated chemokine/plate. Do not pipette air bubbles by sucking up the sample from the first plate. Incubate for 2 h at RT. To detect the remaining biotinylated CXCL8, wash your plate twice and then incubate it with high sensitivity streptavidin HRP diluted in 0.2% dry milk. After 1 h incubation at RT and removal of unbound streptavidin by washing, analyze the plate by adding the substrate TMB, which results in a blue color change. Wait until an appropriate color reaction has taken place (between 10 and 20 min) and stop the reaction with 2 *M* sulfuric acid. Read the absorbance at 450 nm in a Beckman Coulter DTX 800 Multimode Detector or equivalent equipment, with correction at 620 nm. Subtract the reference values (OD_{620}) from the sample values (OD_{450}) and calculate the mean and standard deviation of the replicates. Perform data analysis using specialized statistical software such as Origin (MicroCal) and calculating the IC50 value using a dose response function.

4. CONCLUDING REMARKS

In the past few decades, a lot of work has been done in the chemokine field, but until today, we still do not fully understand the chemokine network. The participation of GAGs in the chemokine network is still not sufficiently explored which depends also on the availability of tissue- and cell-derived GAGs. Here, we have described methods to prepare and to characterize GAG ligands for chemokines. This should allow for a more comprehensive picture of the involvement of GAGs in chemokine-related biological processes as well as to the development of novel drugs for the treatment of chemokine-related diseases (Proudfoot, Bonvin, & Power, 2015).

ACKNOWLEDGMENT

M.G. is recipient of a DOC Fellowship of the Austrian Academy of Sciences at the Institute of Pharmaceutical Chemistry, University of Graz.

REFERENCES

Adage, T., Piccinini, A. M., Falsone, A., Trinker, M., Robinson, J., Gesslbauer, B., et al. (2012). Structure-based design of decoy chemokines as a way to explore the pharmacological potential of glycosaminoglycans. *British Journal of Pharmacology, 167*, 1195–1205.

Ceroni, A., Maass, K., Geyer, H., Geyer, R., Dell, A., & Haslam, S. M. (2008). GlycoWorkbench: A tool for the computer-assisted annotation of mass spectra of glycans. *Journal of Proteome Research*, 7(4), 1650–1659.

Ellyard, J. I., Simson, L., Bezos, A., Johnston, K., Freeman, C., & Parish, C. R. (2007). Eotaxin selectively binds heparin. An interaction that protects eotaxin from proteolysis and potentiates chemotactic activity in vivo. *Journal of Biological Chemistry, 282*(20), 15238–15247.

Friand, V., Haddad, O., Papy-Garcia, D., Hlawaty, H., Vassy, R., Hamma-Kourbali, Y., et al. (2009). Glycosaminoglycan mimetics inhibit SDF-1/CXCL12-mediated migration and invasion of human hepatoma cells. *Glycobiology, 19*(12), 1511–1524.

Gandhi, N. S., & Mancera, R. L. (2008). The structure of glycosaminoglycans and their interactions with proteins. *Chemical Biology & Drug Design*, 72(6), 455–482.

Gerlza, T., Hecher, B., Jeremic, D., Fuchs, T., Gschwandtner, M., Falsone, A., et al. (2014). A combinatorial approach to biophysically characterise chemokine-glycan binding affinities for drug development. *Molecules, 19*(7), 10634.

Goger, B., Halden, Y., Rek, A., Mösl, R., Pye, D., Gallagher, J., et al. (2002). Different affinities of glycosaminoglycan oligosaccharides for monomeric and dimeric interleukin-8: A model for chemokine regulation at inflammatory sites. *Biochemistry, 41*, 1640–1646.

Goldsby, R., Kindt, T., Osborne, B., & Kuby, J. (2003). Enzyme-linked immunosorbent assay. *Immunology* (5th ed.). New York: Freeman.

Gunay, N. S., & Linhardt, R. J. (2003). Capillary electrophoretic separation of heparin oligosaccharides under conditions amenable to mass spectrometric detection. *Journal of Chromatography A, 1014*(1–2), 225–233.

Hoogewerf, A. J., Kuschert, G. S., Proudfoot, A. E., Borlat, F., Clark-Lewis, I., Power, C. A., et al. (1997). Glycosaminoglycans mediate cell surface oligomerization of chemokines. *Biochemistry, 36*(44), 13570–13578.

Hricovíni, M., Guerrini, M., Bisio, A., Torri, G., Petitou, M., & Casu, B. (2001). Conformation of heparin pentasaccharide bound to antithrombin III. *The Biochemical Journal, 359*, 265–272.

Jandik, K. A., Kruep, D., Cartier, M., & Linhardt, R. J. (1995). Accelerated stability studies of heparin. *Journal of Pharmaceutical Sciences, 85*(1), 45–51.

Kuschert, G. S., Coulin, F., Power, C. A., Proudfoot, A. E., Hubbard, R. E., Hoogewerf, A. J., et al. (1999). Glycosaminoglycans interact selectively with chemokines and modulate receptor binding and cellular responses. *Biochemistry, 38*(39), 12959–12968.

Lakowicz, J. R. (2006). *Principles of fluorescence spectroscopy* (3rd ed.). New York: Springer Science & Business Media.

Lebel-Haziv, Y., Meshel, T., Soria, G., Yeheskel, A., Mamon, E., & Ben-Baruch, A. (2014). Breast cancer: Coordinated regulation of CCL2 secretion by intracellular glycosaminoglycans and chemokine motifs. *Neoplasia, 16*(9), 723–740.

Linhardt, R. J., Rice, K. G., Kim, Y. S., Engelken, J. D., & Weiler, J. M. (1988). Homogeneous, structurally defined heparin-oligosaccharides with low anticoagulant activity inhibit the generation of the amplification pathway C3 convertase *in vitro*. *Journal of Biological Chemistry, 263*(26), 13090–13096.

Nomanbhoy, T. K., & Cerione, R. A. (1996). Characterization of the interaction between RhoGDI and Cdc42Hs using fluorescence spectroscopy. *Journal of Biological Chemistry, 271*(17), 10004–10009.

Parish, C. R. (2006). The importance of heparan sulphate at different stages during the entry of leukocytes into sites of inflammation. *Nature Reviews Immunology, 6*, 633–643.

Powell, A. K., Ahmed, Y. A., Yates, E. A., & Turnbull, J. E. (2010). Generating heparan sulfate saccharide libraries for glycomics applications. *Nature Protocols, 5*(5), 821–833.

Proudfoot, A. E., Bonvin, P., & Power, C. A. (2015). Targeting chemokines: Pathogens can, why can't we? *Cytokine, 74*(2), 259–267.

Proudfoot, A. E., Handel, T. M., Johnson, Z., Lau, E. K., LiWang, P., Clark-Lewis, I., et al. (2003). Glycosaminoglycan binding and oligomerization are essential for the in vivo activity of certain chemokines. *Proceedings of the National Academy of Sciences of the United States of America, 100*(4), 1885–1890.

Rabenstein, D. L. (2002). Heparin and heparan sulfate: Structure and function. *Natural Product Reports, 19*(3), 312–331.

Rek, A., Geretti, E., Goger, B., & Kungl, A. J. (2002). The biophysics of chemokine/glycosaminoglycan interactions. *Recent Research Developments in Biophysics and Biochemistry, 2*, 319–340.

Rek, A., Krenn, E., & Kungl, A. J. (2009). Therapeutically targeting protein-glycan interactions. *British Journal of Pharmacology, 157*, 686–694.

Sharath, M. D., Merchant, Z. M., Kim, Y. S., Rice, K. G., Linhardt, R. J., & Weiler, J. M. (1985). Small heparin fragments regulate the amplification pathway of complement. *Immunopharmacology, 9*(2), 73–80.

Silbert, J. E., & Sugumaran, G. (2002). Biosynthesis of chondroitin/dermatan sulfate. *IUBMB Life, 54*(4), 177–186.

Turnbull, J. E., Fernig, D. G., Ke, Y., Wilkinson, M. C., & Gallagher, J. T. (1992). Identification of the basic fibroblast growth factor binding sequence in fibroblast heparan sulfate. *Journal of Biological Chemistry, 267*(15), 10337–10341.

van de Lest, C. H., Versteeg, E. M., Veerkamp, J. H., & van Kuppevelt, T. H. (1994). Quantification and characterization of glycosaminoglycans at the nanogram level by a combined azure A-silver staining in agarose gels. *Analytical Biochemistry, 221*(2), 356–361.

Vanheule, V., Janssens, R., Boff, D., Kitic, N., Berghmans, N., Ronsse, I., et al. (2015). The positively charged COOH-terminal glycosaminoglycan binding CXCL9(74-103) peptide inhibits CXCL8-induced neutrophil extravasation and monosodium urate crystal-induced gout in mice. *Journal of Biological Chemistry, 290*, 21292–21304. jbc.M115.649855.

Warda, M., Toida, T., Zhang, F., Sun, P., Munoz, E., Xie, J., et al. (2006). Isolation and characterization of heparan sulfate from various murine tissues. *Glycoconjugate Journal, 23*(7–8), 555–563.

CHAPTER TWENTY-THREE

Production of Recombinant Chemokines and Validation of Refolding

Christopher T. Veldkamp[*,†,1], Chad A. Koplinski[*], Davin R. Jensen[*], Francis C. Peterson[*], Kaitlin M. Smits[‡,§,¶], Brittney L. Smith[‡,§,¶], Scott K. Johnson[||], Christina Lettieri[#], Wallace G. Buchholz[||], Joyce C. Solheim[‡,§,¶], Brian F. Volkman[*]

[*]Department of Biochemistry, Medical College of Wisconsin, Milwaukee, Wisconsin, USA
[†]Department of Chemistry, University of Wisconsin–Whitewater, Whitewater, Wisconsin, USA
[‡]Department of Biochemistry and Molecular Biology, University of Nebraska Medical Center, Omaha, Nebraska, USA
[§]Department of Pathology and Microbiology, University of Nebraska Medical Center, Omaha, Nebraska, USA
[¶]The Eppley Institute and the Fred and Pamela Buffett Cancer Center, University of Nebraska Medical Center, Omaha, Nebraska, USA
[||]Biological Process Development Facility, College of Engineering, University of Nebraska—Lincoln, Lincoln, Nebraska, USA
[#]Department of Pediatrics, Children's Hospital and Medical Center, University of Nebraska Medical Center, Omaha, Nebraska, USA
[1]Corresponding author: e-mail address: veldkamc@uww.edu

Contents

1. Introduction 540
2. Methods 544
   2.1 Chemokine Expression, Refolding, and Purification 544
   2.2 Mass Spectrometry 550
   2.3 Protein NMR Spectroscopy 552
   2.4 Verifying Biological Activity 555
3. Caveats and Limitations 558
4. Perspectives 558
   4.1 Applications 558
   4.2 Adaptation to a U.S. Food and Drug Administration Current Good Manufacturing Practices (cGMP) System 559
   4.3 Concluding Remarks 562
Acknowledgments 562
References 562

Methods in Enzymology, Volume 570
ISSN 0076-6879
http://dx.doi.org/10.1016/bs.mie.2015.09.031

Abstract

The diverse roles of chemokines in normal immune function and many human diseases have motivated numerous investigations into the structure and function of this family of proteins. Recombinant chemokines are often used to study how chemokines coordinate the trafficking of immune cells in various biological contexts. A reliable source of biologically active protein is vital for any *in vitro* or *in vivo* functional analysis. In this chapter, we describe a general method for the production of recombinant chemokines and robust techniques for efficient refolding that ensure consistently high biological activity. Considerations for initiating development of protocols consistent with Current Good Manufacturing Practices (cGMPs) to produce biologically active chemokines suitable for use in clinical trials are also discussed.

1. INTRODUCTION

Chemokine is shorthand for chemoattractant cytokine, a family of proteins that directs the trafficking of immune cells during normal immune function and participates in many disease states (Baggiolini, 2001). For example, the chemokines CCL19 and CCL21 recruit antigen-presenting dendritic cells and naïve T-cells that express the chemokine receptor CCR7 to the lymph nodes, thereby priming the immune responses (Forster, Davalos-Misslitz, & Rot, 2008). However, CCR7 and its ligands, CCL19 and CCL21, also play roles in the dissemination, migration, and metastasis of cancer cells that express CCR7 (Legler, Uetz-von Allmen, & Hauser, 2014). Other chemokines play similar roles in normal immune function and human pathologies including cardiovascular disease, arthritis, asthma, cancer, HIV-AIDS, and many others. Consequently, the ~50 members of the chemokine family and their G protein-coupled receptors are intensely studied as targets for drug development (Kufareva, Salanga, & Handel, 2015; O'Hayre, Salanga, Handel, & Hamel, 2010).

A supply of bioactive chemokine protein is essential to the study of chemokine function and to be fully active the protein must be properly folded. Figure 1 illustrates the differences in natural eukaryotic and recombinant bacterial production of chemokines, each of which is optimized to yield a natively folded, bioactive protein.

The chemokine fold (Fig. 1C) consists of a flexible N-terminus, an N-loop, occasionally a 3_{10} helix, an antiparallel three-stranded β-sheet, and a C-terminal α-helix. Within the antiparallel three-stranded β-sheet, the β1-strand is connected to β2-strand by the 30s loop and β2-strand is

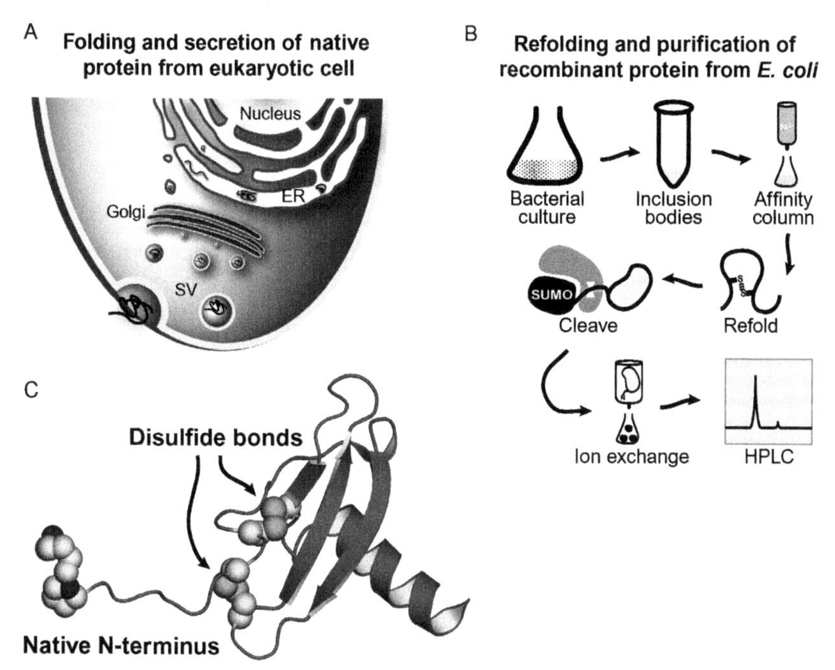

Figure 1 Recombinant chemokine production reproduces the natively folded bioactive protein secreted from eukaryotic cells. (A) Folding, transport, and secretion of chemokines in eukaryotic cells. (B) Schematic diagram of expression, purification, and refolding of recombinant chemokines from *E. coli.* (C) The 3D structure of human CXCL12 illustrates the conserved chemokine fold with structurally important disulfide bonds and functionally important native N-terminus.

linked to β3-strand by the 40s loop. Chemokines generally contain four cysteines that form two conserved disulfide bonds that stabilize a chemokine's tertiary structure. Two of these cysteines are located between the N-terminus and the N-loop while the others are located in the 30s loop and the β3-strand. The first cysteine in a chemokine sequence pairs with the third cysteine in the chemokine's 30s loop and the second cysteine in a chemokine sequence pairs with the fourth cysteine in the β3-strand. A few chemokines contain one fewer or one additional disulfide bond. The metamorphic chemokine XCL1 has only two cysteines that form one disulfide bond, which permits unfolding and interconversion between two unrelated folded structures, the canonical chemokine domain, or an all β-sheet structure (Tyler, Murray, Peterson, & Volkman, 2011). Other chemokines, like CCL21 or CCL28, contain six cysteines corresponding to a novel third disulfide in addition to the two conserved disulfides (Love et al., 2012; Thomas et al., 2015).

Production of functional chemokines can be inherently challenging due to the chemokine fold. For example, chemokines, whether produced by chemical synthesis or heterologous expression in *E. coli*, are initially unfolded, unoxidized, and nonfunctional (Clark-Lewis, 2000; Edgerton, Gerlach, Boesen, & Allet, 2000; Lu et al., 2009; Proudfoot & Borlat, 2000; Veldkamp et al., 2007). In order to become functional, synthetic, or recombinant chemokines must be refolded, a process that requires cysteine oxidation to achieve the correct pattern of conserved disulfide bonds (Clark-Lewis, 2000; Edgerton et al., 2000; Lu et al., 2009; Proudfoot and Borlat, 2000; Veldkamp et al., 2007). Approaches for refolding chemokines have included dialysis, infinite dilution, and on-column techniques (Clark-Lewis, 2000; Edgerton et al., 2000; Lu et al., 2009; Proudfoot and Borlat, 2000; Veldkamp et al., 2007). Another difficulty associated with producing functional recombinant chemokines is that *in vivo* chemokines are secreted proteins and the mature N-terminus that results from removal of the signal sequence of the chemokine is essential for activity. Hence, care must be taken so that the removal of any fusion protein or purification tags used in the production of recombinant chemokines yields a native, mature N-terminus (Lu et al., 2009; Proudfoot and Borlat, 2000; Veldkamp et al., 2007).

For example, we have previously produced both XCL1 and CXCL12 using a system in which an N-terminal hexahistidine tag used to purify the chemokine was removed using a TEV protease. In the case of XCL1, this protocol produced XCL1 without its N-terminal valine residue, a change that rendered the folded protein completely inactive. In the instance of CXCL12, this system left a gly-ser dipeptide preceding the mature, native N-terminal Lys residue producing a protein that was much less active. This is not unexpected, as the sequence within the N-terminus and particularly the N-terminal lysine residue of CXCL12 was shown by Crump et al. to be essential for chemokine activity (Crump et al., 1997). In fact, proteases that act on CXCL12 *in vivo*, like CD26/dipeptidyl peptidase, matrix metalloprotease 2, or exopeptidases, cleave the N-terminus of mature, native CXCL12 to yield inactive proteins (De La Luz Sierra et al., 2004; Segers et al., 2007; Sierra et al., 2003). CXCL12 variants that are resistant to proteolysis while retaining agonist properties have been developed as potentially useful tools (Segers et al., 2007). Some chemokines, like CXCL5, become more active agonist upon proteolysis of their mature N-terminus (Mortier et al., 2010; Nufer, Corbett, & Walz, 1999). Thus, there may be reasons for producing recombinant chemokines with mature, native N-termini or

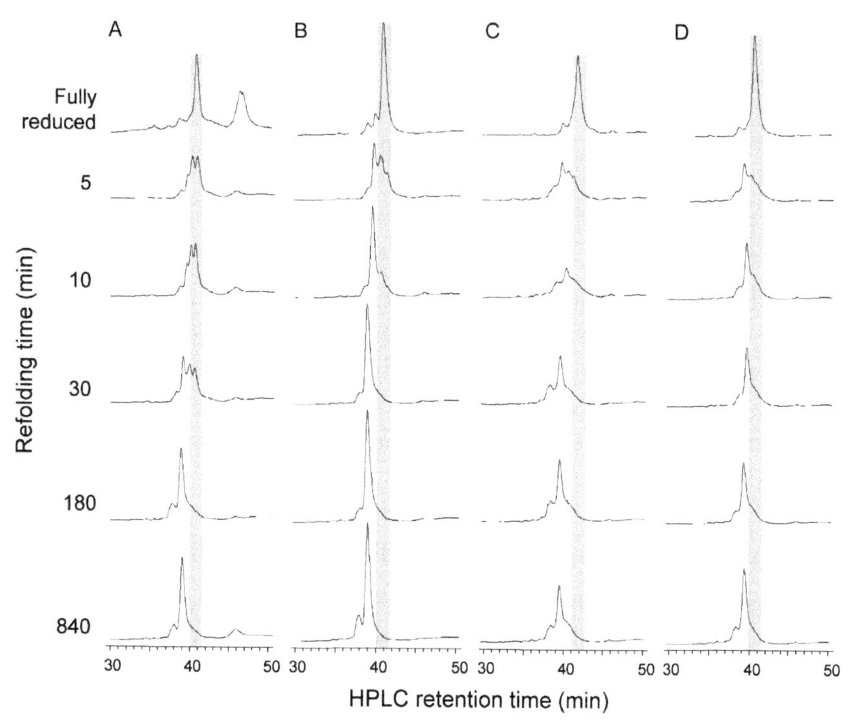

Figure 2 Kinetics of CXCL12 refolding in different redox buffers. Refolding of fully reduced CXCL12 (gray shading) was initiated by infinite dilution of the IMAC elution fractions into the refolding buffer and quenched by acidification. Disulfide formation was monitored by the shift to an earlier reverse phase HPLC retention time. Rates of refolding varied significantly for CXCL12 in (A) the absence of a redox agent or the presence of (B) cysteine/cystine, (C) β-mercaptoethanol/2-hydroxyethyl disulfide or (D) reduced/oxidized glutathione (GSH/GSSG).

varying N-terminal lengths and sequences. We and others have adopted the protein SUMO (SMT3) as a fusion protein for chemokine production in order to take advantage of the SUMO-specific protease ULP1 (Fig. 2B). As ULP1 recognizes the SMT3 fold and cleaves the amide linking the SMT3 C-terminus to the chemokine in the SMT3-chemokine fusion (Lu et al., 2009), this system allows for the production of chemokines with native or altered N-termini.

There are many approaches to producing functional recombinant chemokines from *E. coli* in addition to the ones referenced or presented here. The goal of this work is to describe a robust protocol we employ for many different chemokines, which draws from a variety of previously described approaches. We illustrate the impact of disulfide shuffling cocktails on the efficiency and yield of chemokine refolding and oxidation. Importantly, protocols used to assess the quality of the chemokines produced are also described. Finally, our recent observations from planning the development of cGMP protocols for chemokines are presented.

2. METHODS

This is a general protocol for purification of chemokines with expression sequences that include an N-terminal His_6 tag and a SMT3 cleavage site. Removal of the His_6-SMT3 from the chemokine with ubiquitin like protease-1 (ULP1 or SUMO-protease-1) results in a native, mature chemokine N-terminus that is suitable for all experimental applications (Lu et al., 2009).

2.1 Chemokine Expression, Refolding, and Purification

Although many purification tags, fusion proteins and cleavage options exist, we and others have found expression of a His_6-SMT3-chemokine produces insoluble protein of sufficient yields that refolds, in most cases, robustly and can be purified in a straightforward manner (Lu et al., 2009). While an abundance of refolding techniques have been used with chemokines, the infinite dilution refolding approach presented here works well for most chemokines and we present data on how solution conditions were optimized for refolding speed.

2.1.1 Required Materials

Synthetic gene coding for a SMT3-chemokine fusion protein cloned into the *Bsa*I and *Hind*III/*Pst*I of either pQE30 or pET28a.

Competent BL21 [pREP4] *E. coli* (for pQE30 expression plasmids) or BL21 (DE3) *E. coli* (for pET28a expression plasmids)

Buffer A: 50 m*M* sodium phosphate, 300 m*M* sodium chloride, 10 m*M* imidazole, 1 m*M* phenylmethylsulfonyl fluoride, 0.1% 2-mercaptoethanol, pH 8.0

Buffer AD: 6 *M* guanidine HCl, 50 m*M* sodium phosphate, 300 m*M* sodium chloride, 10 m*M* imidazole, 0.1% 2-mercaptoethanol, pH 8.0

Buffer BD: 6 *M* guanidine HCl, 100 m*M* sodium acetate, 300 m*M* sodium chloride, 10 m*M* imidazole, pH 4.5

Refolding buffers

Option 1: 100 m*M* Tris, 10 m*M* cysteine, 0.5 m*M* cystine, pH 8.0

Option 2: 100 m*M* Tris, 3 m*M* cysteine, 0.5 m*M* cystine, pH 8.0
Option 3: 100 m*M* Tris, 10 m*M* cysteine, 0.5 m*M* cystine, 10% glycerol, pH 8.0
Option 4: 100 m*M* Tris, 10 m*M* cysteine, 0.5 m*M* cystine, 150 m*M* NaCl, pH 8.0
Cation exchange binding: 100 m*M* Tris, 25 m*M* sodium chloride, pH 8.0
Cation exchange wash: 100 m*M* Tris, 50 m*M* sodium chloride, pH 8.0
Cation exchange elution: 100 m*M* Tris, 2 *M* sodium chloride, pH 8.0
RP-HPLC buffer A: Aqueous 0.1% trifluoroacetic acid
RP-HPLC buffer B: Aqueous 0.1% trifluoroacetic acid, 70% acetonitrile
RP-HPLC column: 218TP510
Lysogeny broth or [U-^{15}N] or [U-^{15}N/^{13}C] M9 minimal media (with 150 μg/mL ampicillin and 50 μg/ml kanamycin for pQE30 expression plasmids or 50 μg/mL kanamycin for pET28a expression plasmids)
2–4 mL of nickel affinity resin
S HyperD F cation exchange resin/column Pall Life Sciences
15 mL disposable column (Qiagen or Thermofisher)
1 M isopropyl-β-D-1-thiogalactopyranoside (IPTG)
French pressure cell or sonicator

2.1.2 Transformation, Expression, and Cell Harvesting

1. Transform the expression plasmid following the instructions in the QIA expressionist for pQE30 plasmids into BL21 [pREP4] *E. coli* or the pET system manual for pET28a plasmids into BL21 (DE3) *E. coli* (Novagen, 2005).
2. If desired, small-scale expression testing can be done at this point to optimize growth conditions (Waltner, Peterson, Lytle, & Volkman, 2005), but the conditions outlined below provide a good starting point.
3. Grow a 1 L culture at 37 °C using either Lysogeny broth or [U-^{15}N] or [U-^{15}N/^{13}C] M9 minimal media with appropriate antibiotics (150 μg/mL ampicillin/50 μg/ml kanamycin for pQE30 expression plasmids or 50 μg/mL kanamycin for pET28a expression plasmids). At an optical density at 600 nm of 0.6 induce expression with 1 mL of 1 *M* IPTG.
4. After 5 h collect the cells by centrifugation at 5000 × *g* for 10 min and store the cell pellet at −20 to −80 °C until further processing.

2.1.3 Nickel Affinity Chromatography

The His_6-SMT3-chemokine can be isolated from *E. coli* proteins and cell debris through nickel affinity chromatography.

1. Completely resuspend the cells by pipetting and repipetting the same 10 mL of buffer A until no cell pellet remains. Vortexing or a rocking incubation is also effective.
2. Lyse the resuspended cells with three passes through a French pressure cell (desired PSI 10,000) or with sonication pulsed at ~25% power for 10 s on and 10 s off for 10–15 min.
3. Collect the His_6-SMT3-chemokine inclusion body pellet by centrifugation at 15,000 × *g* for 30 min. The His_6-SMT3-chemokine, in most cases so far, is predominately insoluble and in the pellet, but both the supernatant and pellet should be saved if the location of His_6-SMT3-chemokine is unknown.
4. Dissolve the inclusion body pellet in 10 mL of buffer AD and clarify using centrifugation at 15,000 × *g* for 30 min. A 10 mL syringe with a 16-gauge needle can be used to help break up the inclusion body pellet along with incubating and shaking at 37 °C. Membrane fragments should be translucent.
5. Bind the supernatant containing the His_6-SUMO-chemokine to 2–4 mL of nickel affinity resin in batch mode using a 15 mL disposable column with rocking for 30 min at room temperature.
6. Place the column on a ring stand. Collect the flow through and wash the resin with four 10 mL portions of buffer AD.
7. Elute the His_6-SUMO-chemokine from the nickel affinity resin with 10 mL of buffer BD into 15 mL conical vials. Shake the vial, if bubbles form that do not quickly dissipate the fraction contains protein. Three 10 mL portions of buffer BD is usually enough to completely elute the column.

2.1.4 Refolding

Infinite dilution with a disulfide-reshuffling cocktail allows protein refolding. Selection of the disulfide reshuffling cocktail will affect refolding efficiency and total time because proper refolding is only achieved when cysteines are oxidized and in the correct disulfide pairs. Figure 2 shows the effect of various disulfide-reshuffling cocktails on CXCL12 refolding.

In our experience with a number of chemokines, cysteine/cystine provides optimal refolding kinetics, higher yields, and lower reagent cost than the alternatives.

1. Pool the eluted fractions that contain the His_6-SUMO-chemokine and determine the volume.
2. Refold by infinite dilution into a volume of refolding buffer that is 12 times the volume of the pooled elutions. Add the pooled elutions dropwise (45 drops per minute) with stirring to refolding buffer option 1, 2, 3, or 4. In most cases, refolding buffer option 1 is optimal.
3. Figure 3 shows that the cysteine/cystine redox pair can accelerate and improve the refolding of complex disulfide pairings.
4. Incubate with stirring overnight at 4 °C.

2.1.5 ULP1 Digestion

ULP1 digestion separates the His_6-SMT3 from the chemokine.

1. Double the volume of the refolding mixture with 100 m*M* Tris, pH 8.0. The concentration of the Guanidine HCl must be below 250 m*M* for the ULP1 digestion to be effective.
2. Add ULP1 and incubate with stirring at 4 °C.
3. Complete digestion should be observed through comparison of undigested and digested samples using SDS-PAGE.
4. Clarify the digestion through centrifugation at 5000 × *g* for 10 min or filtration (10,000 Da cut off).

2.1.6 Cation Exchange Chromatography

Isolation of the chemokine from the His_6-SMT3 protein is accomplished by cation exchange chromatography. This chromatography technique is especially useful due to the fact that most chemokines are positively charged. Under basic buffer conditions, the more positively charged chemokine will bind to the negatively charged cation exchange resin, while the negatively charged and slightly acidic His_6-SMT3 (theoretical pI 5.7) will not.

1. Equilibrate the column with cation exchange binding buffer.
2. Bind the clarified digestion to the column with a peristaltic pump at a flow rate of 2.5 mL/min.
3. Wash the column with five times the column volume of cation exchange wash buffer.
4. Elute the column with four times the column volume of cation exchange elution buffer.
5. Filter the elution using a 0.22 μm syringe filter.

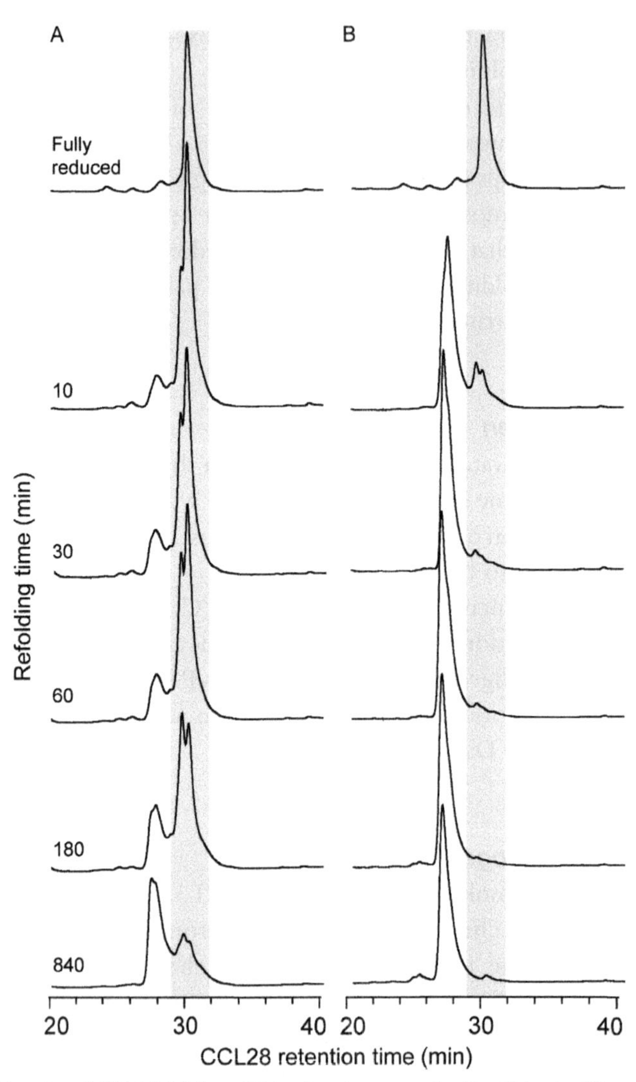

Figure 3 Kinetics of CCL28 folding. (A) In the absence of a disulfide shuffling redox pair, folding of CCL28 is slow and inefficient. (B) Refolding with cysteine/cystine accelerates the formation of both conserved disulfides as well as an additional disulfide linking Cys 30 to Cys 80.

2.1.7 Reverse Phase HPLC

Reverse phase HPLC (RP-HPLC) is the final cleanup step and leaves the protein in a volatile solvent suitable for lyophilization. Chemokines are functional when refolded and refolding involves cysteine oxidation and correct pairing in disulfide bonds. First, RP-HPLC is used to check for cysteine oxidation and disulfide bond formation. Then RP-HPLC is used to cleanup the sample by separating the major refolded species that has the correct disulfide pairings from small amounts of misfolded chemokine containing either incorrectly paired disulfide bonds or chemokine that is partially or completely reduced.

1. Disulfide formation check.
   a. Two 5 μL samples from the cation exchange elution are diluted by adding 745 μL of water. To one sample, 7.5 μL of 1 *M* dithiothreitol is added (final concentration ~10 m*M*) and 7.5 μL of water is added to the other. Both are incubated at 37 °C for 2 min and then acidified with a drop of 6 N HCl.
   b. These samples are run separately using a linear 30–60% RP-HPLC buffer B gradient (see Table 1). A difference in retention time for the major peaks in the samples with and without dithiothreitol indicates disulfide bond formation.
2. The refolded chemokine is isolated using RP-HPLC using a linear 30–60% buffer B gradient. See Table 1 for the gradient scheme and time program. A fraction collector that is programmable aids in collecting the refolded chemokine. In general, the major peak will have the correct disulfide pairings and will be functionally active in activity assays. The smaller peaks are misfolded chemokine or chemokine that is only partially or completely reduced.

2.1.8 Lyophilization

The elutions from the RP-HPLC are pooled and frozen at −80 °C and dried on a lyophilizer until dry (about 24 h) and stored at −20 °C.

Table 1 Reverse Phase HPLC Gradient Scheme

Time (min)	% RP-HPLC Buffer A	% RP-HPLC Buffer B
0	90	10
5	70	30
35	40	60
40	0	100
45	0	100
50	90	10
65	90	10

RP-HPLC buffer A: Aqueous 0.1% trifluoroacetic acid.
RP-HPLC buffer B: Aqueous 0.1% trifluoroacetic acid, 70% acetonitrile.

2.2 Mass Spectrometry

Mass spectrometry, in addition to RP-HPLC, serves as a powerful technique for assessing disulfide bond formation. Additionally, tandem mass spectrometry (MS-MS) can in some instances be used to define the pairing of cysteines, like the non-conserved disulfide bond in CCL28 that links C30 to C80 (Thomas et al., 2015), or confirm the locations of posttranslational modifications, like pyroglutamate formation in CCL2 as shown in Figs. 4 and 5, respectively.

2.2.1 *Required Materials*

Q Exactive™ Hybrid Quadrupole-Orbitrap mass spectrometer from ThermoFisher Scientific or comparable mass spectrometer
LC/MS grade acetonitrile, 0.1% formic acid
LC/MS grade H_20, 0.1% formic acid
Macro spin C4 tips, Nest group
Xtract and Xcalibur™ in QualBrowser from ThermoFisher Scientific or comparable software
ProSightLite (Fellers et al., 2015)

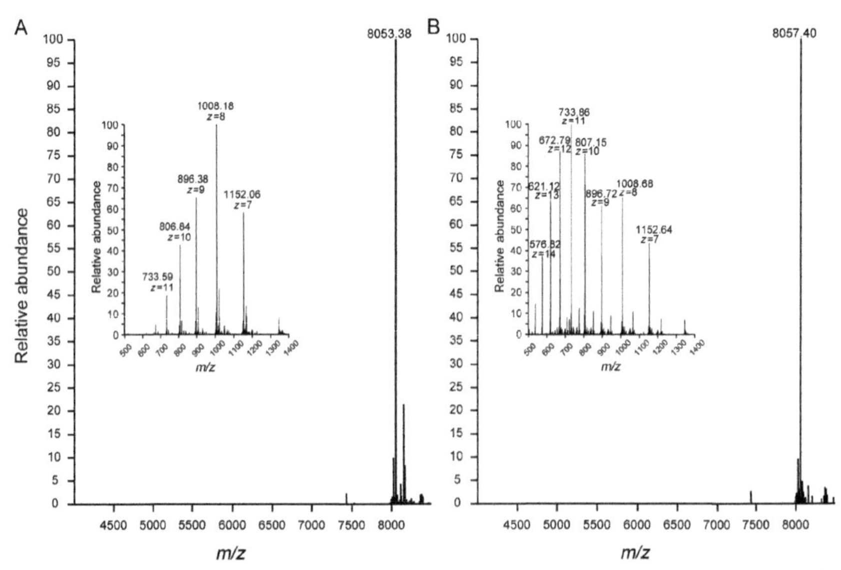

Figure 4 Using mass spectrometry to confirm identity and disulfide bond formation for a CXCL12 mutant. (A) The intact mass spectrum of folded and oxidized CXCL12 S-1 S4V with the charge state envelope inset. (B) The denatured and reduced protein, compared to the oxidized, and refolded protein the mass increased in size by 4 Da with all four cysteines being reduced and the charge state envelope has shifted to a lower *m/z*.

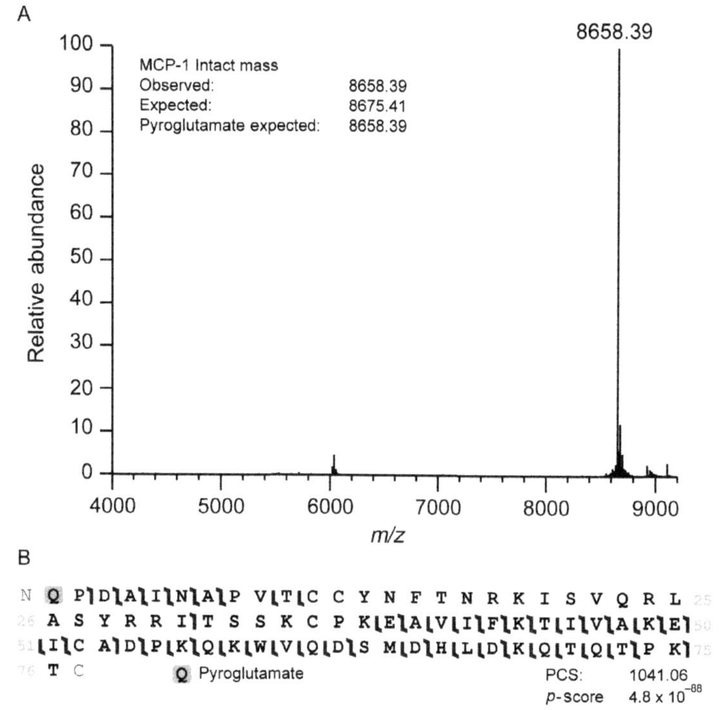

Figure 5 Top-down mass spectrometry analysis of CCL2. (A) Intact mass of purified recombinant CCL2. Pyroglutamate at the amino terminus explains the mass discrepancy. (B) Fragmentation of CCL2 shows amino terminal pyroglutamate ions with a *p*-value of 10×10^{-88}.

2.2.2 General Method for Detecting Disulfide Oxidation and Identity

For most chemokines, which contain four cysteines that form two conserved disulfide bonds, a measured mass to charge ratio will be four mass units less than the mass predicted based on protein sequence. Hence, measurement of a recombinant chemokine's mass to charge ratio using a sensitive and accurate mass spectrometer can confirm not only protein identity but also disulfide oxidation as seen in Fig. 4.

1. Dissolve lyophilized chemokine to a concentration of 1 μ*M* in aqueous 50% acetonitrile and 0.1% formic acid.
2. Analyze by direct infusion using a Q Exactive™ Hybrid Quadrupole-Orbitrap mass spectrometer (ThermoFisher Scientific). All analyses are performed with resolving power set to 140,000, four microscans for 30 s.
3. Deconvolute and deisotope with Xtract and Xcalibur™ in QualBrowser (ThermoFisher Scientific).
4. Compare measured mass to charge ratio to that predicted based on sequence.

2.2.3 Top-Down Approaches for the Analysis of Chemokines With Additional Disulfides or Posttranslational Modifications

In cases where the predicted mass calculated from the amino acid sequence does not match the observed mass, top-down sequencing is used to determine sites of posttranslational modifications as seen in Fig. 5.

1. Chemokine is dissolved or diluted into the denaturing/reducing buffer to reduce the mature chemokine's disulfide bonds. This is required for efficient fragmentation of the entire molecule.
2. After 10 min protein is desalted by applying to the Macro spin C4 tip, washing five times with 200 μL H_2O, 0.1% formic acid and eluting in 100 μL acetonitrile, 0.1% formic acid.
3. Protein concentration is determined and diluted to 1 μ*M* in 50% H_2O/acetonitrile, 0.1% formic acid.
4. Intact folded and oxidized protein was analyzed by direct infusion using a Q Exactive™ Hybrid Quadrupole-Orbitrap mass spectrometer (ThermoFisher Scientific) as before.
5. Top-down sequencing of the denatured and reduced protein was analyzed by direct infusion using a Exactive™ Hybrid Quadrupole-Orbitrap mass spectrometer with a resolving power of 70,000.
6. For CCL2/MCP1, fragmentation of the 1293 m/z peak was accomplished in the HCD cell at a power of 20 eV.
7. Top-down sequencing used ProSightLite to determine the pyroglutamate on the N-terminus.

2.3 Protein NMR Spectroscopy

NMR spectroscopy is a robust technique for evaluating the folding state of chemokines and proteins in general. Chemokines are properly folded when the cysteines present are oxidized and form properly paired disulfide bonds, a feature that is essential to maintain the chemokine fold. Properly folded chemokines generate an NMR spectrum with well dispersed chemical shift values (Fig. 6, top row), while unfolded chemokines display a collapsed NMR spectrum (Fig. 6, bottom row). A two-dimensional ^{1}H–^{15}N heteronuclear single quantum coherence (HSQC) or ^{1}H–^{15}N heteronuclear multiple–quantum coherence (HMQC) spectrum provides a "finger print" of the chemokine that contains a resonance for every backbone and side chain amide group. Comparison with a validated "finger print" or a prediction based on primary sequence allows for a rapid assessment of a chemokine's folding state. However, ^{1}H–^{15}N HSQC or HMQC spectroscopy requires high-field NMR instrumentation that may not be readily available and specialized

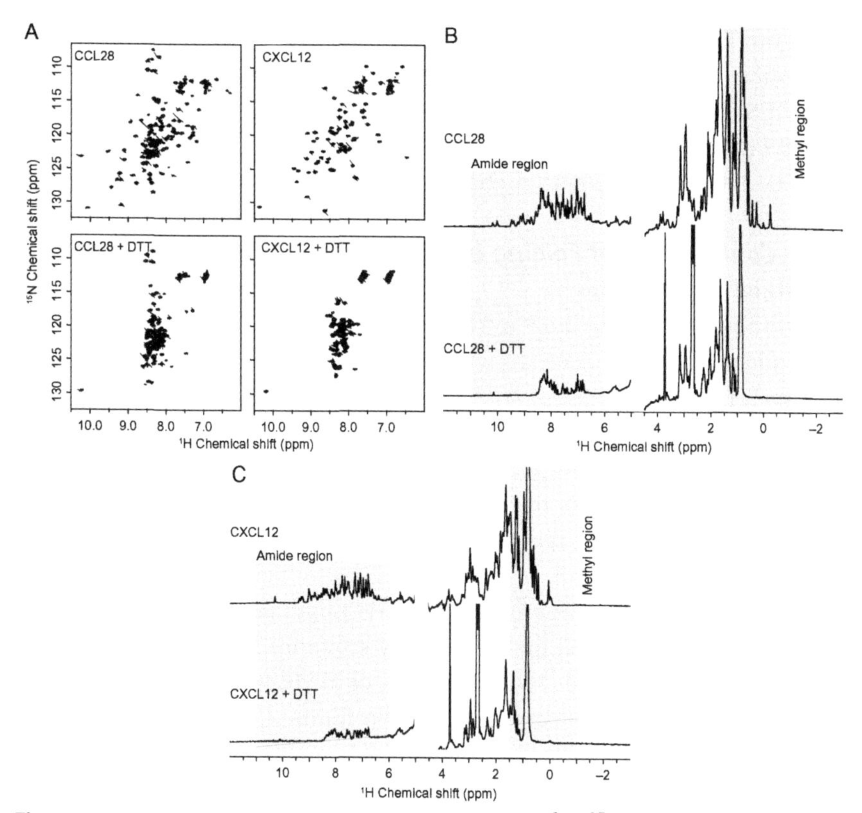

Figure 6 Using NMR to confirm chemokine folding. (A) ^{1}H–^{15}N HSQC spectra of CCL28 and CXCL12 with cysteines fully oxidized and properly paired (top row) or fully reduced using DTT (bottom row). One-dimensional ^{1}H spectra of CCL28 (B) and CXCL12 (C) with correct disulfide parings (top) or fully reduced using DTT (bottom).

growth media to enrich recombinant chemokines with the ^{15}N isotopic label. Alternatively, one-dimensional proton NMR spectroscopy can be utilized to evaluate the folding state of a chemokine without the requirement for specialized growth conditions or high-field NMR instrumentation.

2.3.1 Required Materials

Bruker high-field (≥500 MHz) NMR spectrometer running TopSpin or a comparable spectrometer equipped with a ^{1}H/^{13}C/^{15}N probe suitable for proteins

1D (p3919gp) and 2D HSQC (hsqcf3gpph19) or 2D HMQC (sfhmqcf3gpph) pulse sequences from the Bruker library or comparable pulse sequences

3 mm NMR tubes

Buffer: 25 m*M* deuterated 2-(N_morpholino)ethanesulfonic acid (MES), pH 6.8, containing 10% deuterium oxide

1 *M* dithiothreitol (DTT)

100–300 μg of lyophilized chemokine

2.3.2 Confirmation of Folding by Protein NMR

Oxidized chemokines

1. Dissolve chemokine in 25 m*M* MES, pH 6.8, containing 10% deuterium oxide to a final concentration of 100–200 μ*M*. A minimum sample volume of 200 μL is required.
2. Evaluate the folding state by collection of 2D or 1D NMR spectra using a Bruker high-field NMR spectrometer. All data will be collected at 25 °C with an appropriate number of scans.
3. Process the resulting spectrum with TopSpin (Bruker) or comparable software package.
4. Compare the spectrum with a validated "finger print" spectrum or with results predicted based on the primary sequence.
   a. Two-dimensional ^{1}H–^{15}N HSQC or HMQC spectra will contain a set of resonances equivalent to the number of backbone and side chain amide groups present in the chemokine. The folding state is evaluated based on the number of resonances present, and the uniformity of the shape and intensity of the resonances. Chemokines that are well behaved should display greater than 90% of the expected resonances (Fig. 6A, top).
   b. One-dimensional ^{1}H spectra are also useful for evaluating the folding state of chemokines but suffer from spectral overlap because of reduced dimensionality. To evaluate the folding we look at the peaks dispersion in the amide region and the methyl region of the one-dimensional ^{1}H spectrum (Fig. 6B and C, top).

Reduced chemokines

1. Reduce the chemokine disulfide bonds by adding DTT to the sample to a final concentration of 10 m*M* and incubating for 30 min at room temperature.
2. Evaluate the folding state of the reduced chemokine by collecting 2D or 1D NMR spectra as in step 2.
3. Compare the oxidized and reduced chemokine spectra to confirm that disulfide bond reduction caused the NMR signals to collapse (Fig. 6, bottom).

2.4 Verifying Biological Activity

There are numerous approaches for verifying biological activity of chemokines, an essential step prior to using recombinant chemokines for structural or functional research studies. Classic *in vitro* assays for biological activity can include calcium flux assays or chemotaxis assays (Kar et al., 2011; Love et al., 2012; Miller & Krangel, 1992; Neote, DiGregorio, Mak, Horuk, & Schall, 1993; Thomas et al., 2015). Calcium flux assays are those that detect the increase in the cytosol of the secondary messenger calcium that is released as a result of chemokine receptor activation (Love et al., 2012; Miller & Krangel, 1992; Neote et al., 1993). Chemotaxis assays monitor chemokine recruitment of cells expressing chemokine receptor across a barrier that is impermeable to cells without the chemokine functioning as an attractant (Kar et al., 2011; Miller & Krangel, 1992; Thomas et al., 2015). An example chemotaxis assay protocol for CCL21 is presented here.

2.4.1 Required Materials for a General Chemotaxis Assay

T2 cells (ATCC; 174 X CEM.T2 cells; Cat. No. CRL-1992) (Salter, Howell, & Cresswell, 1985)

L-glutamine (200 m*M*) (Invitrogen; Cat. No. 25030–081)

4-(2-hydroxyethyl)-1-piperazineethanesulfonic acid (HEPES) buffer (1 *M*) (Invitrogen; Cat. No. 15630–080)

Sodium pyruvate (100 m*M*) (Invitrogen; Cat. No. 11360)

Penicillin, streptomycin (10,000 U/ml penicillin and 10,000 μg/ml streptomycin) (Invitrogen; Cat. No. 15140–122)

Minimum essential medium (MEM) non-essential amino acids (100 ×) (Invitrogen; Cat. No. 11140–050)

CO_2 tissue culture incubator (set at 37 °C, 5% CO_2)

CCL21 produced as described above, or for use as controls purchased from a commercial source

Ultrapure water (such as can be produced by a Barnstead Nanopure system or several other equivalent water purification systems)

Microscope with digital imaging capability (Nikon Instruments Inc.; Eclipse 90i)

Iscove's Modified Dulbecco's Medium (IMDM) (Life Technologies; Cat. No. 12440)

Fetal bovine serum (FBS) (Atlantic Biologicals; Cat. No. S11150)

Sterile (autoclaved) 1 × phosphate-buffered saline (PBS), pH 7.4 (prepared by dilution of a 10 × solution purchased from Life Technologies; Cat. No. 70011–044)

Thincert tissue culture inserts for 24-well plates, 3-μm pore size (Greiner Bio One; Cat. No. 662630)
24-well flat-bottom plates (Corning; Cat. No. 3524)
Hema 3 STAT Pack Hematology Staining [Fixative, Solution I, Solution II] (Thermo Fisher Scientific; Cat. No. 123–869)
VectaMount™ Mounting Medium (Vector Laboratories; Cat. No. H-5000)
Microscope slides (3″ × 1″ × 1 mm) (Fisher Scientific; Cat. No. 12-544-3)
Glass coverslips (Fisher Scientific; Cat. No. 12-545-88)
Cotton-tipped applicators (Allegiance; Cat. No. C15053-006)
Scalpel (Personna Medical; Cat. No. 73–0111)
Tweezers

2.4.2 General Chemotaxis Assay for CCL21

1. Culture T2 cells to 70% confluency in IMDM with 15% fetal bovine serum, 2 m*M* L-glutamine, 10 m*M* HEPES buffer, 1 m*M* sodium pyruvate, 100 units per mL penicillin, 100 μg per mL streptomycin, and non-essential amino acids (used at 1 × final concentration).
2. Resuspend CCL21 in ultrapure water (at 600 ng/ml concentration for controls).
3. Collect and count the T2 cells and resuspend them in serum-free IMDM.
4. Fill 5 wells of a 24-well plate with 500 μl of the following (as the bottom chamber contents): 1 well with serum-free IMDM (negative control), 1 well with IMDM containing 15% vol/vol FBS and all the other media supplements mentioned in #1 above (positive control), and 3 wells with 600 ng/ml CCL21 suspended in serum- free IMDM (experimental triplicates).
5. Using tweezers, place tissue culture inserts into filled wells.
6. Seed 5×10^5 T2 cells (suspended in 180 μl of serum-free IMDM) on top of the tissue culture inserts (top chambers).
7. Incubate in a 37 °C, 5% CO_2incubator for 24 h.
8. Remove the plate from the incubator and wash the inserts with 1 × PBS (handling the inserts by using tweezers).
9. Fill 3 wells in a 24-well plate with 700 μl of each solution from the Hema 3 STAT pack (i.e., 1 well of Fixative, 1 well of Solution I, and 1 well of Solution II).

10. Label slides with the name of the cell line and the identity of the bottom chamber contents for each well.
11. Quickly dip insert in Fixative 30 times using tweezers.
12. Quickly dip insert in Solution I 30 times using tweezers.
13. Quickly dip insert in Solution II 30 times using tweezers.
14. Use a cotton-tipped applicator to very gently wipe the non-invaded cells off the top of the insert.
15. Put one drop of mounting media onto a labeled slide.
16. Carefully cut the insert out using a scalpel.
17. Place insert face down onto the drop of mounting media.
18. Cover insert with a glass coverslip.
19. Take three pictures of different fields of each insert using a camera-enabled microscope at a magnification of 100 ×.
20. Count cells and calculate the means, ranges, and standard deviations.

Example results from a CCL21 chemotaxis assay (performed as described above) are shown in Fig. 7. In this example, CCL21 prepared at the University of Nebraska Biological Process Development Facility was compared to commercially available CCL21 (Fig. 7).

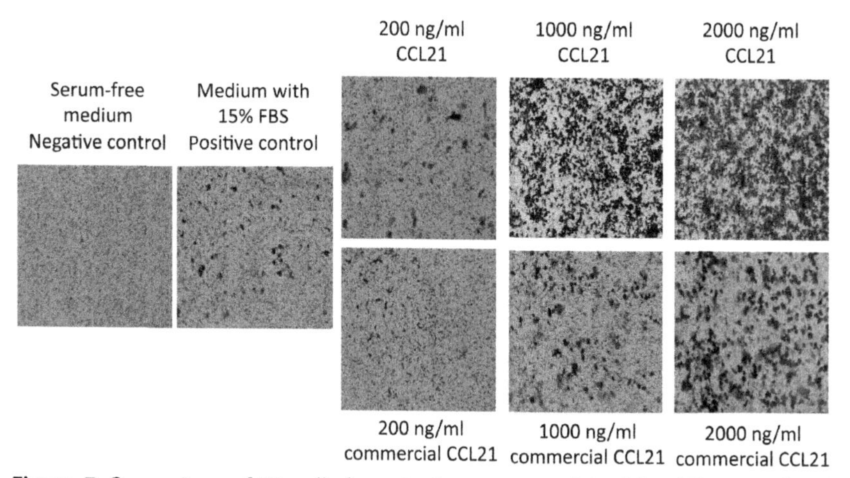

Figure 7 Comparison of T2 cell chemotactic response elicited by CCL21 produced at the University of Nebraska—Lincoln Biological Process Development Facility (top right panels) or by commercial CCL21 (bottom right panels) using the chemotaxis procedure outlined in Section 2.4. Fetal bovine serum (FBS), which contains a mixture of chemotactic components, was used as a positive control. The CCL21 wells contained no FBS.

3. CAVEATS AND LIMITATIONS

Here a general protocol for producing functional recombinant chemokines is presented that draws from many previously published approaches. The chemokine family has over 50 members and no one protocol can be expected to fit each chemokine, but this approach has worked for numerous chemokines. For example, XCL1, XCL2, CCL2, CCL11, CCL17, CCL20, CCL28, CXCL5, CXCL11, and CXCL12 have been prepared using the above protocol. CCL19 and CCL21 are examples where small modifications to the above protocol have resulted in functional chemokines. CCL19 requires the inclusion of 1 m*M* phenylmethylsulfonyl fluoride (PMSF) and 1 m*M* ethylenediaminetetraacetic acid in the refolding buffer as protease inhibitors and cysteine and cystine during the ULP1 digestion to ensure PMSF does not inhibit ULP1. CCL21 contains an extra disulfide bond in its extended C-terminus. If cysteine and cystine are used in the refolding buffer for CCL21 one must be on the look out for evidence of cystinylation as a side reaction that competes with formation of the third disulfide bond. Alternatively, for CCL21 cysteine and cystine can be left out of the refolding buffer. Other chemokines will likely present similar challenges requiring adaptation of the protocol to the specific situation. Nevertheless, we have found the approach described here to be robust and widely applicable.

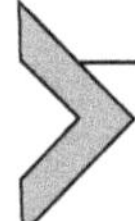

4. PERSPECTIVES

4.1 Applications

A thorough review of recombinant chemokines is beyond the scope of this chapter. However, many studies have indicated that chemokines themselves or chemokine variants may have therapeutic potential for a multitude of human diseases spanning from cardiovascular health to oncology. For example, CXCL12 and its variants have been proposed, based on animal models of human disease, to be a potentially rejuvenating therapy for myocardial infarction (Abbott et al., 2004; Askari et al., 2003; Koch et al., 2006; Saxena et al., 2008; Segers et al., 2007; Veldkamp et al., 2009) or as antimetastatic agents for colon or melanoma cancers (Drury et al., 2011; Takekoshi, Ziarek, Volkman, & Hwang, 2012). Additionally, CCL21 is an effective surgical neoadjuvant for treatment of mammary tumors in mouse models and has shown antitumor activity in other malignancies. (Ashour et al., 2007; Dubinett, Lee, Sharma, & Mule, 2010; Kar et al.,

2011; Kirk et al., 2001; Nomura & Hasegawa, 2000; Sharma et al., 2013; Turnquist et al., 2007). Incorporating therapeutic chemokines and other cytokines as adjuncts in immunotherapy protocols for cancer treatment has been efficacious in alleviating tumor immune evasion and increasing survival (Ardolino, Hsu, & Raulet, 2015; Wennerberg, Kremer, Childs, & Lundqvist, 2015).

The recurrent success of chemokines or chemokine variants in several animal models of human disease supports the potential of chemokines or chemokine-derived molecules for clinical use in medicine. Continued access to high quality recombinant chemokines that are produced for medicinal use is paramount to the advancement of their therapeutic uses. Additionally, clinical administration of chemokines would require that they be manufactured and formulated in facilities and according to protocols that satisfy the Current Good Manufacturing Practices set by the U.S. Food and Drug Administration.

4.2 Adaptation to a U.S. Food and Drug Administration Current Good Manufacturing Practices (cGMP) System

To adapt the protocols described here for use in a cGMP setting, many additional steps must be taken, which are briefly summarized here. Minimum specifications for all raw materials to be used in the process must be defined and met by the materials to be used. The reagents and raw materials should all be of United States Pharmacopoeia (USP)-grade or meet multicompendial criteria. (For testing of raw materials, multicompendial means that the quality control test results on the material have met the acceptance criteria set by multiple regulatory groups, such as USP, Japan Pharma, and equivalent agencies in the European Union and Canada.) Vendors normally will provide a Certificate of Analysis or Master Drug File number. If any products such as microbiological media are of animal origin, they must be certified as Bovine Spongiform Encephalopathy free.

Not only must the raw materials be cGMP compliant, but the equipment, laboratory space, and personnel all must comply with cGMP requirement to ensure the safety of the material being produced. All processing, upstream (microbial growth) and downstream (purification), must be conducted in cGMP-compliant space which requires environmental monitoring, periodic decontamination, and calibration and validation of the equipment. Personnel must periodically receive training for cGMP compliance and they must be individually qualified for each piece of equipment they use.

The developed process should be robust and shown to consistently produce acceptable quality material. Due to variability of expression, even within the same bacterial strain, an initial screen for high expressing isolates should be conducted. After introduction in the host strain, 50–100 colonies should be screened to identify the highest producer of the protein. Before beginning a cGMP production process, master cell banks must be established. At minimum, the master cell bank should consist of ~150–200 glycerol stock vials. An even better policy is to establish a working cell bank containing a similar number of vials (i.e., ~150–200), in addition to the master cell bank. Thorough characterization of the clone that will be used to express a biologic is required by FDA regulations. Therefore, both the master cell bank and the working cell bank should be tested for purity, identity, viability, plasmid retention, and sequence fidelity. Sequence fidelity should be confirmed by DNA sequencing. Using current DNA sequencing techniques, the complete sequence of the vector and insert can be determined with a high degree of accuracy. Restriction endonuclease mapping of the molecule can serve to quickly confirm the correct plasmid is harbored by the production strain. Though not a required test, plasmid copy number may also be determined.

After preliminary small-scale analysis of chemokine production has been conducted, further process development and scale-up to optimize conditions in a batch fermentor will be required to move to the cGMP level; conditions in a large fermentor are different from those in shake flasks or small fermentation vessels (e.g., mixing rate, dissolved oxygen, thermal control, pH control, etc.). In bench-top chemokine production dialysis tubing may be used, but not when moving to a larger scale (e.g., a 60-L fermentor). Cross-flow filtration can serve the same function and is recommended to achieve buffer exchange and concentration. Although this method exposes the protein to shear conditions as it is being circulated across a flat membrane sheet under pressure, no deleterious effects have been noted with small proteins or peptides.

When common resins and aqueous solvents were used to develop a purification process at the research bench, the purification methods can usually be scaled-up with minimal problems. However, if organic solvents have been used, scaling up to a cGMP process may be more complicated. For example, if acetonitrile is used for purification and there are no issues in the acute toxicity studies required by the FDA, then use of acetonitrile is not a problem and scale-up is straightforward. However, if toxicity is observed, it will not be clear if the observed toxicity is due to CCL21 or to acetonitrile residue. Thus, the risk/benefit of including acetonitrile in the purification process should be carefully weighed.

In order to establish that a process results in consistent product, analytical tests should be established and qualified by using them to test a reference standard. These tests are developed as in-process test methods to track the quantity and quality of product during optimization of the purification process. In-process analyses often involve using ion exchange HPLC, reverse phase HPLC, SDS-PAGE, Western blot, and/or ELISA assays.

Under cGMP guidelines, biotherapeutics destined for clinical trials require meeting stringent release testing criteria. For most products, purity should exceed 98% and there is some flexibility in selecting the specific set of release tests that characterize the identity, purity, function (usually), and important physical attributes of the product. For CCL21, release testing should include assays to confirm identity and function such as reducing and non-reducing SDS-PAGE, Western blot, N-terminal sequencing, potency, capillary isoelectric focusing (pI), amino acid analysis, intact mass spectroscopy, and peptide mapping. Release testing for safety must include an endotoxin assay (e.g., limulus amebocyte lysate, LAL), host cell protein, residual host DNA, and bioburden or sterility. Release assays to characterize physical attributes include color and appearance, pH, and conductivity. Many of the assays used in-process can also be used for release testing if they are qualified or compendial (i.e., follow a USP protocol). A subset of release tests are often also used to analyze stability of the purified product. Typically, stability studies are performed at the expected storage temperature and over the expected storage time of the product, e.g., −80 °C, 2 years. In some cases, forced degradation studies are conducted for shorter times at elevated temperatures. The analytical methods selected for stability studies must be able to detect low-level degradation of the product.

In adapting a chemokine production protocol to cGMP, it is also important to take into account the requirements for submitting an Investigational New Drug application to the Food and Drug Administration. Complete detailed records must be kept describing every step of the production and purification processes, and final formulation. This includes all the information regarding (1) raw materials and reagents (Certificate of Analysis or Master Drug File number, inventory and storage records, all product label information), (2) manufacturing (all related protocols must be written as Master Batch Records (MBRs) that are approved by Quality Assurance (QA) and followed precisely; deviations from the MBRs must be investigated by QA and the potential impact on the product explained), (3) quality control (all data on development and qualifications of analytical assays or verification of compendial methods must be included) and lastly, (4) equipment maintenance (records of equipment maintenance, qualification, and calibration).

4.3 Concluding Remarks

In conclusion, the supply of pure, natively folded chemokine proteins is vital for a variety of basic, translation and clinical research programs. Our structure–function studies and drug development efforts have motivated the development of the robust protocol for recombinant chemokine expression, refolding, purification, and validation presented here. Finally, many additional factors must be considered when manufacturing a product under cGMP guidelines that may not be obvious to personnel in research settings.

ACKNOWLEDGMENTS

Many approaches for preparing functional chemokines have been presented in the literature either as reviews or in individual research articles; we sincerely apologize to the numerous individuals whose work could not be discussed and/or cited due to space limitations. This work was supported by NIH grant 1R15CA159202-01 to C.T.V., NIH grants R01AI058072 and R01GM09738 to B.F.V., a grant to J.C.S. and W.B. from the Nebraska Research Initiative, and other funding to J.C.S. (NIH R03CA173223, NIH SPORE P50CA127297, the State of Nebraska through the Pediatric Cancer Research Group, and the Fred and Pamela Buffett Cancer Center Support Grant P30CA036727). We wish to thank Professor Rebekah Gundry for assistance and training in the mass spectrometry methods presented here.

REFERENCES

Abbott, J. D., Huang, Y., Liu, D., Hickey, R., Krause, D. S., & Giordano, F. J. (2004). Stromal cell-derived factor-1alpha plays a critical role in stem cell recruitment to the heart after myocardial infarction but is not sufficient to induce homing in the absence of injury. *Circulation, 110*(21), 3300–3305.

Ardolino, M., Hsu, J., & Raulet, D. H. (2015). Cytokine treatment in cancer immunotherapy. *Oncotarget, 6*(23), 19346–19347.

Ashour, A. E., Lin, X., Wang, X., Turnquist, H. R., Burns, N. M., Tuli, A., et al. (2007). CCL21 is an effective surgical neoadjuvant for treatment of mammary tumors. *Cancer Biology & Therapy, 6*(8), 1206–1210.

Askari, A. T., Unzek, S., Popovic, Z. B., Goldman, C. K., Forudi, F., Kiedrowski, M., et al. (2003). Effect of stromal-cell-derived factor 1 on stem-cell homing and tissue regeneration in ischaemic cardiomyopathy. *Lancet, 362*(9385), 697–703.

Baggiolini, M. (2001). Chemokines in pathology and medicine. *Journal of Internal Medicine, 250*(2), 91–104.

Clark-Lewis, I. (2000). Synthesis of chemokines. *Methods in Molecular Biology, 138*, 47–63. http://dx.doi.org/10.1385/1-59259-058-6:47.

Crump, M. P., Gong, J. H., Loetscher, P., Rajarathnam, K., Amara, A., Arenzana-Seisdedos, F., et al. (1997). Solution structure and basis for functional activity of stromal cell- derived factor-1; dissociation of CXCR4 activation from binding and inhibition of HIV-1. *EMBO Journal*, *16*(23), 6996–7007.

De La Luz Sierra, M., Yang, F., Narazaki, M., Salvucci, O., Davis, D., Yarchoan, R., et al. (2004). Differential processing of stromal-derived factor-1alpha and stromal-derived factor-1beta explains functional diversity. *Blood*, *103*(7), 2452–2459. http://dx.doi.org/10.1182/blood-2003-08-2857.

Drury, L. J., Ziarek, J. J., Gravel, S., Veldkamp, C. T., Takekoshi, T., Hwang, S. T., et al. (2011). Monomeric and dimeric CXCL12 inhibit metastasis through distinct CXCR4 interactions and signaling pathways. *Proceedings of the National Academy of Sciences of the United States of America*, *108*(43), 17655–17660. http://dx.doi.org/10.1073/pnas.1101133108.

Dubinett, S. M., Lee, J. M., Sharma, S., & Mule, J. J. (2010). Chemokines: Can effector cells be redirected to the site of the tumor? *Cancer Journal*, *16*(4), 325–335. http://dx.doi.org/10.1097/PPO.0b013e3181eb33bc.

Edgerton, M. D., Gerlach, L. O., Boesen, T. P., & Allet, B. (2000). Expression of chemokines in Escherichia coli. *Methods in Molecular Biology*, *138*, 33–40. http://dx.doi.org/10.1385/1-59259-058-6:33.

Fellers, R. T., Greer, J. B., Early, B. P., Yu, X., LeDuc, R. D., Kelleher, N. L., et al. (2015). ProSight lite: Graphical software to analyze top-down mass spectrometry data. *Proteomics*, *15*(7), 1235–1238. http://dx.doi.org/10.1002/pmic.201570050.

Forster, R., Davalos-Misslitz, A. C., & Rot, A. (2008). CCR7 and its ligands: Balancing immunity and tolerance. *Nature Reviews Immunology*, *8*(5), 362–371.

Kar, U. K., Srivastava, M. K., Andersson, A., Baratelli, F., Huang, M., Kickhoefer, V. A., et al. (2011). Novel CCL21-vault nanocapsule intratumoral delivery inhibits lung cancer growth. *PLoS One*, *6*(5), e18758. http://dx.doi.org/10.1371/journal.pone.0018758.

Kirk, C. J., Hartigan-O'Connor, D., Nickoloff, B. J., Chamberlain, J. S., Giedlin, M., Aukerman, L., et al. (2001). T cell-dependent antitumor immunity mediated by secondary lymphoid tissue chemokine: Augmentation of dendritic cell-based immunotherapy. *Cancer Research*, *61*(5), 2062–2070.

Koch, K. C., Schaefer, W. M., Liehn, E. A., Rammos, C., Mueller, D., Schroeder, J., et al. (2006). Effect of catheter-based transendocardial delivery of stromal cell-derived factor 1alpha on left ventricular function and perfusion in a porcine model of myocardial infarction. *Basic Research in Cardiology*, *101*(1), 69–77.

Kufareva, I., Salanga, C. L., & Handel, T. M. (2015). Chemokine and chemokine receptor structure and interactions: Implications for therapeutic strategies. *Immunology and Cell Biology*, *93*(4), 372–383. http://dx.doi.org/10.1038/icb.2015.15.

Legler, D. F., Uetz-von Allmen, E., & Hauser, M. A. (2014). CCR7: Roles in cancer cell dissemination, migration and metastasis formation. *The International Journal of Biochemistry & Cell Biology*, *54*, 78–82. http://dx.doi.org/10.1016/j.biocel.2014.07.002.

Love, M., Sandberg, J. L., Ziarek, J. J., Gerarden, K. P., Rode, R. R., Jensen, D. R., et al. (2012). Solution structure of CCL21 and identification of a putative CCR7 binding site. *Biochemistry*, *51*(3), 733–735. http://dx.doi.org/10.1021/bi201601k.

Lu, Q., Burns, M. C., McDevitt, P. J., Graham, T. L., Sukman, A. J., Fornwald, J. A., et al. (2009). Optimized procedures for producing biologically active chemokines. *Protein Expression and Purification*, *62*(2), 251–260. http://dx.doi.org/10.1016/j.pep.2009.01.017.

Miller, M. D., & Krangel, M. S. (1992). The human cytokine I-309 is a monocyte chemoattractant. *Proceedings of the National Academy of Sciences of the United States of America*, *89*(7), 2950–2954.

Mortier, A., Loos, T., Gouwy, M., Ronsse, I., Van Damme, J., & Proost, P. (2010). Posttranslational modification of the NH2-terminal region of CXCL5 by proteases or peptidylarginine deiminases (PAD) differently affects its biological activity. *The Journal of Biological Chemistry, 285*(39), 29750–29759. http://dx.doi.org/10.1074/jbc.M110.119388.

Neote, K., DiGregorio, D., Mak, J. Y., Horuk, R., & Schall, T. J. (1993). Molecular cloning, functional expression, and signaling characteristics of a C-C chemokine receptor. *Cell, 72*(3), 415–425.

Nomura, T., & Hasegawa, H. (2000). Chemokines and anti-cancer immunotherapy: Anti-tumor effect of EBI1-ligand chemokine (ELC) and secondary lymphoid tissue chemokine (SLC). *Anticancer Research, 20*(6A), 4073–4080.

Novagen (2005). *The pET System Manual* (11th ed.). Darmstadt, Germany: EMD Biosciences, Inc.

Nufer, O., Corbett, M., & Walz, A. (1999). Amino-terminal processing of chemokine ENA-78 regulates biological activity. *Biochemistry, 38*(2), 636–642. http://dx.doi.org/10.1021/bi981294s.

O'Hayre, M., Salanga, C. L., Handel, T. M., & Hamel, D. J. (2010). Emerging concepts and approaches for chemokine-receptor drug discovery. *Expert Opinion on Drug Discovery, 5*(11), 1109–1122. http://dx.doi.org/10.1517/17460441.2010.525633.

Proudfoot, A. E., & Borlat, F. (2000). Purification of recombinant chemokines from E. coli. *Methods in Molecular Biology, 138*, 75–87.

Salter, R. D., Howell, D. N., & Cresswell, P. (1985). Genes regulating HLA class I antigen expression in T-B lymphoblast hybrids. *Immunogenetics, 21*(3), 235–246.

Saxena, A., Fish, J. E., White, M. D., Yu, S., Smyth, J. W., Shaw, R. M., et al. (2008). Stromal cell-derived factor-1alpha is cardioprotective after myocardial infarction. *Circulation, 117*(17), 2224–2231. http://dx.doi.org/10.1161/CIRCULATIONAHA.107.694992 CIRCULATIONAHA.107.694992 [pii].

Segers, V. F., Tokunou, T., Higgins, L. J., MacGillivray, C., Gannon, J., & Lee, R. T. (2007). Local delivery of protease-resistant stromal cell derived factor-1 for stem cell recruitment after myocardial infarction. *Circulation, 116*(15), 1683–1692.

Sharma, S., Zhu, L., Srivastava, M. K., Harris-White, M., Huang, M., Lee, J. M., et al. (2013). CCL21 chemokine therapy for lung cancer. *International Trends in Immunity, 1*(1), 10–15.

Sierra, M. D., Yang, F., Narazaki, M., Salvucci, O., Davis, D., Yarchoan, R., et al. (2003). Differential processing of stromal-derived factor-1 {alpha} and {beta} explains functional diversity. *Blood, 103*(7), 2452–2459.

Takekoshi, T., Ziarek, J. J., Volkman, B. F., & Hwang, S. T. (2012). A locked, dimeric CXCL12 variant effectively inhibits pulmonary metastasis of CXCR4-expressing melanoma cells due to enhanced serum stability. *Molecular Cancer Therapeutics, 11*(11), 2516–2525. http://dx.doi.org/10.1158/1535-7163.MCT-12-0494.

Thomas, M. A., Buelow, B. J., Nevins, A. M., Jones, S. E., Peterson, F. C., Gundry, R. L., et al. (2015). Structure-function analysis of CCL28 in the development of post-viral asthma. *The Journal of Biological Chemistry, 290*(7), 4528–4536. http://dx.doi.org/10.1074/jbc.M114.627786.

Turnquist, H. R., Lin, X., Ashour, A. E., Hollingsworth, M. A., Singh, R. K., Talmadge, J. E., et al. (2007). CCL21 induces extensive intratumoral immune cell infiltration and specific anti-tumor cellular immunity. *International Journal of Oncology, 30*(3), 631–639.

Tyler, R. C., Murray, N. J., Peterson, F. C., & Volkman, B. F. (2011). Native-state interconversion of a metamorphic protein requires global unfolding. *Biochemistry, 50*(33), 7077–7079. http://dx.doi.org/10.1021/bi200750k.

Veldkamp, C. T., Peterson, F. C., Hayes, P. L., Mattmiller, J. E., Haugner, J. C., 3rd., de la Cruz, N., et al. (2007). On-column refolding of recombinant chemokines for NMR studies and biological assays. *Protein Expression and Purification, 52*(1), 202–209.

Veldkamp, C. T., Ziarek, J. J., Su, J., Basnet, H., Lennertz, R., Weiner, J. J., et al. (2009). Monomeric structure of the cardioprotective chemokine SDF-1/CXCL12. *Protein Science, 18*(7), 1359–1369.

Waltner, J. K., Peterson, F. C., Lytle, B. L., & Volkman, B. F. (2005). Structure of the B3 domain from Arabidopsis thaliana protein At1g16640. *Protein Science, 14*(9), 2478–2483.

Wennerberg, E., Kremer, V., Childs, R., & Lundqvist, A. (2015). CXCL10-induced migration of adoptively transferred human natural killer cells toward solid tumors causes regression of tumor growth in vivo. *Cancer Immunology, Immunotherapy, 64*(2), 225–235. http://dx.doi.org/10.1007/s00262-014-1629-5.

Quantitative Analysis of Dendritic Cell Haptotaxis

Jan Schwarz, Michael Sixt[1]
Institute of Science and Technology Austria (IST Austria), Klosterneuburg, Austria
[1]Corresponding author: e-mail address: michael.sixt@ist.ac.at

Contents

1. Introduction 567
2. Methods 570
   2.1 Lower Surface Preparation 571
   2.2 Upper Surface Preparation 575
   2.3 Haptotaxis Chamber Assembly 577
   2.4 Troubleshooting 579
3. Perspectives 580
Acknowledgments 581
References 581

Abstract

Chemokines are the main guidance cues directing leukocyte migration. Opposed to early assumptions, chemokines do not necessarily act as soluble cues but are often immobilized within tissues, e.g., dendritic cell migration toward lymphatic vessels is guided by a haptotactic gradient of the chemokine CCL21. Controlled assay systems to quantitatively study haptotaxis *in vitro* are still missing. In this chapter, we describe an *in vitro* haptotaxis assay optimized for the unique properties of dendritic cells. The chemokine CCL21 is immobilized in a bioactive state, using laser-assisted protein adsorption by photobleaching. The cells follow this immobilized CCL21 gradient in a haptotaxis chamber, which provides three dimensionally confined migration conditions.

1. INTRODUCTION

As the main antigen-presenting cells of the adaptive immune response, dendritic cells (DCs) play a crucial role in regulation and activation of adaptive immunity (Steinman & Banchereau, 2007). Under homeostatic conditions, DCs reside in peripheral tissues and continuously scan their

Methods in Enzymology, Volume 570
ISSN 0076-6879
http://dx.doi.org/10.1016/bs.mie.2015.11.004

environment for internal and external danger signals. After intercepting an inflammatory stimulus, DCs pause random migration and become highly phagocytic, thereby acquiring antigen (Reis e Sousa, 2006). Pathogen encounter then triggers a terminal differentiation program that is termed maturation. Maturation is accompanied by the production of cytokines, upregulation of costimulatory molecules, and the presentation of major histocompatibility–peptide complexes which are required to prime naïve T cells in the lymph node (Steinman & Banchereau, 2007). Additionally, the chemokine receptor CCR7 is highly increased, imparting responsiveness toward the CCR7 ligands CCL19 and CCL21 (Sallusto et al., 1998). In order to reach the lymphatic vessel and to eventually arrive in the T-cell area of the lymph node, the cells follow immobilized CCL21 gradients toward the vessels (Weber et al., 2013). This haptotactic migration is a rate-limiting step in the initiation of adaptive immune responses.

DCs feature a special type of amoeboid migration. In contrast to other leukocytes, migratory DCs exhibit even less adhesion to most substrates. This makes migration on two-dimensional (2D) surfaces inefficient, but allows fast locomotion in 3D confined environments (reviewed in Renkawitz & Sixt, 2010). The 3D meshwork conditions of the interstitium have been mimicked *in vitro*, for example, by collagen gel migration assays (Lämmermann et al., 2008). To examine DC migration in a qualitative and quantitative manner *in vitro*, it is advantageous to limit the dimensions of migration. Confinement from the top, for example with an agarose layer (Fig. 1A), allows the cells to migrate in a 3D-like configuration while pressed to a plane surface (Renkawitz et al., 2009). The surface can be covered with adhesive or nonadhesive surface coatings or functionalized specifically with chemokines to mimic haptotactic interstitial gradients.

The Piel group replaced the agarose layer with a second plane surface. Unlike agarose, that can be deformed by the cells and forms a soft cover, the stiff surface is positioned in a defined distance to the lower surface using

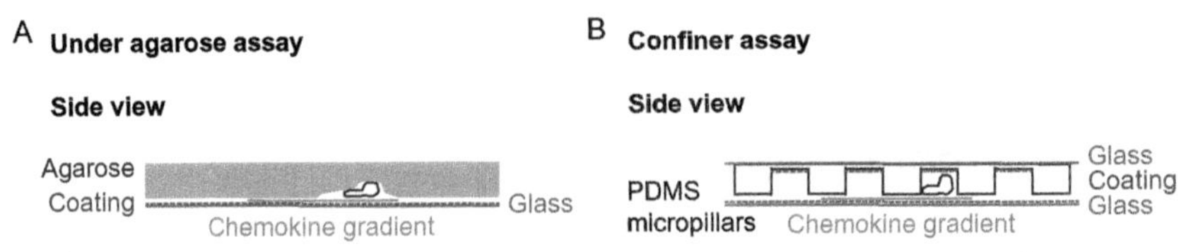

Figure 1 Three-dimensional migration assays. (A) Under agarose assay. Cells migrating under a sheet of agarose. (B) Confiner assay. Cells migrating between two glass slides. The spacing between glass slides is determined by polydimethylsiloxane (PDMS) micropillars.

polydimethylsiloxane (PDMS) micropillars as spacers (Fig. 1B; Le Berre et al., 2014). This setup provides a defined and controllable 3D migration chamber, which is ideally suited for the construction of a haptotaxis chamber, since both surfaces can be functionalized before chamber assembly. Furthermore, the cells can migrate without loosing contact to the functionalized substrate and will not migrate out of the focal imaging plane.

In vivo, interstitial CCL21 gradients are continuous, steep, and static gradients with lengths of about 100 μm (Weber et al., 2013). Many protein micropatterning techniques like mask-bound photolithography (Azioune, Storch, Bornens, Théry, & Piel, 2009; Blawas & Reichert, 1998) or microcontact printing (Whitesides, Ostuni, & Takayama, 2001) offer too low spatial resolution to mimic those homogenous, continuous gradients. Another disadvantage of some micropatterning techniques is the necessity of background blocking in the nonpatterned regions. This changes important substrate properties like adhesiveness between patterned and nonpatterned regions.

Laser-assisted protein adsorption by photobleaching (LAPAP) is one of the few techniques able to generate gradients in microscale resolution without the need for background blocking (Bélisle et al., 2008). The principle underlying LAPAP is the covalent immobilization of dye molecules, e.g., fluorescein, on surfaces by photoactivation (bleaching; Fig. 2A). With light intensity determining the amount of immobilized fluorescein, homogenous, continuous gradients can be generated using photomasks or movable lasers. Using a biotin–fluorescein heterodimer (B4F), any transparent surface can be biotinylated with arbitrary patterns and gradients (Fig. 2A). Streptavidin (SA) offers four binding sites for biotin and binds it with extremely high affinity. Therefore, the biotin pattern can be further functionalized by SA that, if fluorescently labeled, can be used to simultaneously visualize the gradient. Biotinylated chemokines (e.g., CCL21) are then attached to the SA-functionalized patterns (Fig. 2B).

CCL21 is immobilized to lymphatic endothelial cells and negatively charged extracellular matrix components by electrostatic interaction via its basic C-terminus (Fig. 2C). The N-terminal part of the chemokine is involved in receptor ligation and activation (Love et al., 2012). To avoid unspecific CCL21 binding we use a truncated version of CCL21. This version only contains the region responsible for receptor ligation (amino acids 24–98) and is lacking the basic C-terminus. In order to maximize chemokine binding to the patterned B4F/SA areas, the truncated CCL21 24–98 is monobiotinylated at the C-terminus.

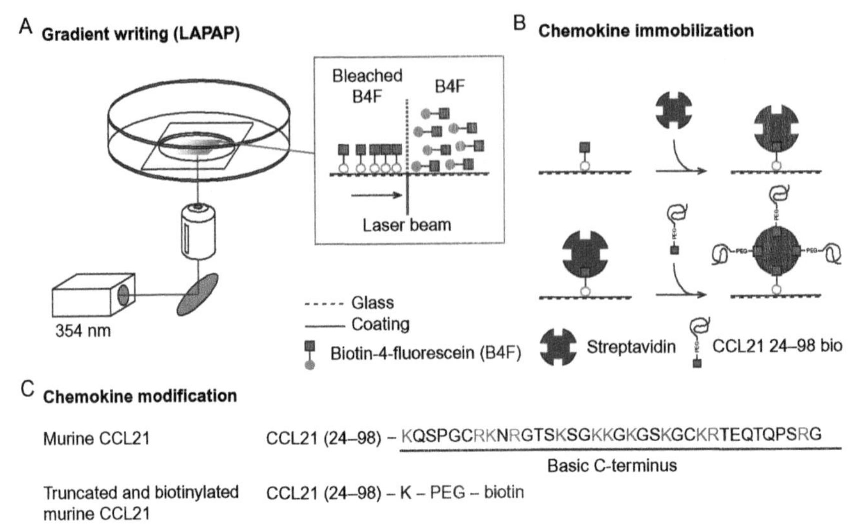

Figure 2 Chemokine micropatterning. (A) Gradient writing using laser-assisted protein adsorption by photobleaching (LAPAP). Biotin-4-fluorescein (B4F) is immobilized on a coated glass surface by a movable UV laser ($\lambda = 354$ nm). (B) Chemokine immobilization. Binding of streptavidin-Cy3 (SA-Cy3) functionalizes the B4F pattern. Free biotin-binding sites of the immobilized SA-Cy3 bind biotinylated CCL21 24–98 and present it in a bioactive state. (C) Chemokine modification. Basic CCL21 C-terminus is replaced by biotin attached to a polyethylene glycol (PEG) linker. (See the color plate.)

2. METHODS

In this chapter, we will explain in detail how both top and bottom parts of the chamber are manufactured, how migratory DCs are differentiated from bone marrow cultures, and how the haptotaxis chamber is assembled. Some parts are adapted from published sources. In those cases, the description will focus on the details we specifically changed. The respective methods are described in detail in the cited publications.

The haptotaxis chamber consists of two glass surfaces that are spaced by PDMS micropillars. The upper surface bears the PDMS micropillars to define the height of the chamber (Fig. 3A, upper surface). For the lower surface, a glass bottom dish is used which offers the chemokine pattern the cells are migrating on (Fig. 3A, lower surface). The lower surface and the PDMS micropillar bearing upper surface are pressed onto each other by a big, elastic

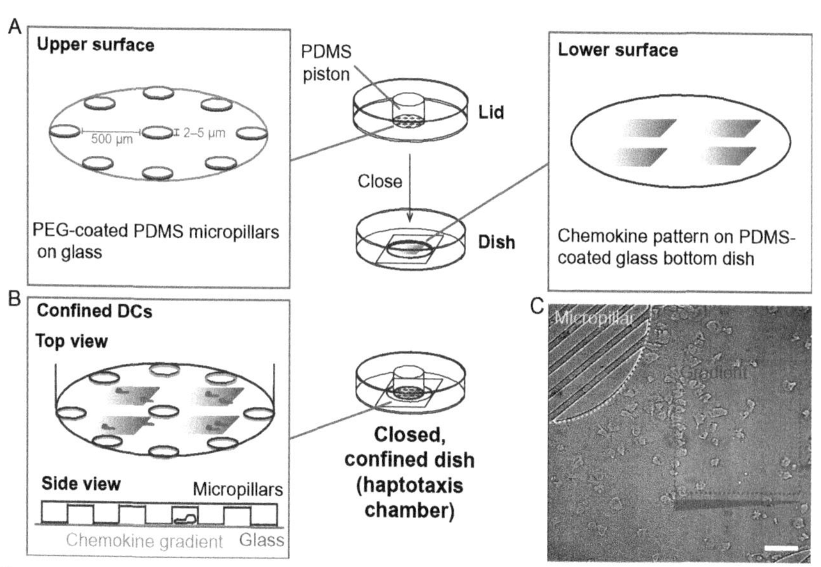

Figure 3 Haptotaxis chamber. (A) Upper and lower surface of the haptotaxis chamber. Lid with elastic PDMS piston and PEG-coated PDMS micropillars on a round glass slide and the glass bottom dish with CCL21 24–98 bio micropattern. Left magnified box: Round glass slide with PEG-coated PDMS micropillars. The distance of the pillars is 500 μm. The height of the micropillars is 2–5 μm. Right magnified box: CCL21 24–98 bio micropattern on the glass of a glass bottom dish. (B) Closed haptotaxis chamber. Top view and side view of the closed haptotaxis chamber. (C) Brightfield image of DCs migrating on a printed CCL21 24–98 biogradient in the haptotaxis chamber. Micropillar and gradient are indicated with dashed lines. Scale bar represents 50 μm. (See the color plate.)

PDMS piston which is glued into the lid of a glass bottom dish. Depending on the assay conditions, the lower surface can be polyethylene glycol (PEG) or BSA coated to avoid specific binding before LAPAP is used for functionalization with a haptotaxis-inducing chemokine pattern. To avoid cell adhesion to the upper surface, the PDMS micropillars are PEG coated. After fabrication of top and bottom parts the cells are added and the chamber is closed (Fig. 3B). The cells show haptotactic response within seconds and can be imaged directly after chamber assembly.

2.1 Lower Surface Preparation

2.1.1 Glass Slide Treatment and Dish Preparation

The lower surface of the haptotaxis chamber is a glass coverslip modified version of a 60 × 15 mm tissue culture dish. This setup allows convenient functionalization and washing of the cover slip, facilitates confinement, and enables imaging of the migrating cells. Commercially available glass bottom dishes can be used as well; however, they are less versatile than the custom-made version.

Required Materials

Iso-propanol (Sigma–Aldrich).
Ethanol (Sigma–Aldrich).
MΩ H_2O.
Spin coater (Laurell Technologies Corporation, WS 650-MZ-23NPP).
Planetary centrifugal mixer (Thinky, ARE250).
22 mm × 22 mm, #2 glass slides (Mentzel Gläser, Thermo Scientific).
PDMS Sylgard 184 Elastomere Kit (Dow Corning).
Transparent aquarium silicone sealant (Marina).
60 × 15 mm style nonpyrogenic polystyrene tissue culture dish (Falcon).
Plasma Cleaner (Harrick Plasma).
Oven (80 °C).
(Optional) PBS (pH 7.2 without $CaCl_2$, without $MgCl_2$, GIBCO).
(Optional) *PLL-PEG* (SuSos).
(Optional) BSA (3% m/v) in PBS (Sigma–Aldrich).
(Optional) Bovine fibronectin (Sigma–Aldrich).

1. Sonicate coverslips in *iso*-propanol and ethanol (each 20 min, sweeping sonication).
2. Rinse cover slips with MΩ water without letting them dry and blow-dry them with N_2 or wipe them carefully with tissues.
3. Mix silicone elastomer and curing reagent in a 10:1 ratio using a planetary centrifugal mixer.
4. Plasma clean dishes (2 min, high intensity) and immediately add 500 μL of the well mixed, bubble-free silicone elastomer/curing reagent mixture on the plasma-cleaned side of a glass coverslip.
5. Spin at 4000 rpm for 40 s with a prior acceleration of 200 rpm/s using the spin coater. The PDMS thickness is about 17 μm. If slides will be imaged using a TIRF microscope thinner PDMS layers should be used.
6. Bake slides for 6 h at 80 °C in an oven (step 6 in Section 2.1.1).
7. Drill a hole with a diameter of 17 mm in the middle of the bottom of a 60 × 15 mm Falcon Tissue culture dish (Fig. 4A). A diameter of 17 mm is ideal for gluing 22 × 22 mm coverslips onto the hole.
8. Glue the PDMS-coated glass slides on the bottom of the dishes using clear silicone aquarium sealant. The PDMS-covered side has to face the dish.
9. Dry dishes over night (min. 6 h) at rt or 1 h at 80 °C.

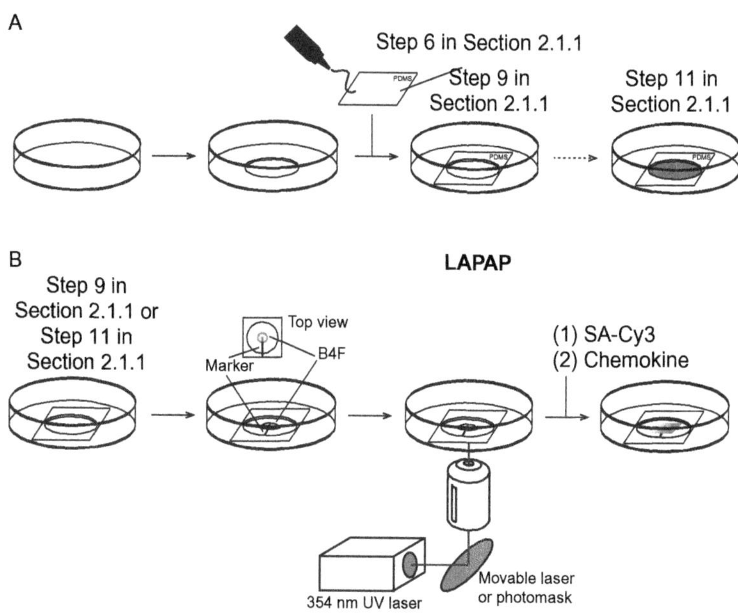

Figure 4 Production of the CCL21 24–98 bio patterned glass bottom dish. (A) Glass bottom dish production and modification. A hole is drilled into a 60 × 15 mm Falcon Tissue culture dish. A PDMS-coated coverslip (step 6 in Section 2.1.1) is glued onto the hole with the PDMS-coated surface facing into the dish (step 9 in Section 2.1.1). Surface coatings can be applied to the manufactured dish (step 11 in Section 2.1.1). (B) Chemokine micropatterning. A droplet of B4F is placed on the marked area of the glass bottom dish. Patterns are generated using LAPAP. SA-Cy3 and CCL24-98 bio are immobilized on the written B4F pattern.

10. (Optional for coated surfaces) plasma clean dishes (2 min, high intensity) and immediately incubated with PLL-PEG (0.5 mg/mL in HEPES), BSA, or fibronectin (100 μg/mL in PBS) for at least 1 h (drying of the solution is allowed).
11. (Optional for coated surfaces) wash with PBS (3 ×) and store under PBS at 4 °C (step 11 in Section 2.1.1).

2.1.2 Chemokine Printing

This part partially depends on the microscope setup used for writing. See original publication by Belisle et al. for detailed description of possible laser writing setups (Bélisle et al., 2008). We will focus on the staining and chemokine binding/handling procedures.

Required Materials

Microscope equipped with a movable laser or photomask and the corresponding software for gradient writing.
PBS (pH 7.2 without $CaCl_2$, without $MgCl_2$, GIBCO)
Biotin-4-Fluorescein (B4F, Sigma–Aldrich)
Parafilm
BSA (0.1% m/v) in PBS (Sigma–Aldrich)
CCL21 24–98 bio (custom synthesized from ALMAC; for more detailed information, see Section 2.4)
Streptavidin-Cy3 (SA-Cy3, Sigma–Aldrich)

1. Take prepared dish out of the oven and let them cool down to room temperature. If coated dishes are used, aspirate PBS.
2. Place 20 μL of B4F solution (150 μg/mL in PBS) in the middle of the PDMS-coated glass slide of the previously prepared dish (Fig. 4B). Close dish and seal it with parafilm to avoid drying of the B4F solution during laser writing.
3. Mark the middle of the B4F droplet with a marker on the glass side of the coverslip (Fig. 4B).
4. Focus on the PDMS surface next to the marked spot. Write gradients/ patterns (Fig. 4B).
5. Aspirate B4F and wash three times with PBS.
6. Add 20 μL of SA-Cy3 solution (10 μg/mL) and incubate at room temperature for 20 min in the dark.
7. Aspirate SA-Cy3 solution and wash three times with PBS. B4F/SA-Cy3 patterned dishes can be stored under PBS for up to 1 week at 4 °C in the dark. Seal the dishes with parafilm to avoid evaporation of the PBS.
8. Reconstitute the lyophilized CCL21 24–98 bio in 0.1% BSA in PBS to a final concentration of 25 μg/mL. Aliquots can be stored at −20 °C. Prior to use, the stock solution is diluted to 250 ng/mL in PBS.
9. Incubate SA-Cy3 patterned dishes with CCL21 24–98 bio (250 ng/mL) for 30 min at room temperature, subsequently washed three times with PBS and assemble haptotaxis chamber immediately.

This step needs to be timed well with the recovery period of the DCs. Only start incubation of the dishes with chemokine when the cells were recovered for at least 30 min.

2.2 Upper Surface Preparation

The haptotaxis chamber represents a modified version of the cell confiner established by the Piel group. Therefore, we will only briefly discuss how the individual components are manufactured. For a detailed description, please refer to Le Berre et al. (2014).

2.2.1 Micropillar Preparation

Required Materials

PDMS Sylgard 184 Elastomere Kit (Dow Corning)
Planetary centrifugal mixer (Thinky, ARE250)
Round cover glasses, #1, 12 mm diameter (Mentzel Gläser, Thermo Scientific)
Silicon waver with micropillars of desired height and spacing (Le Berre et al., 2014).
iso-propanol
Ethanol
Oven (80 °C)
Plasma Cleaner (Harrick Plasma)
Heating plate (95 °C)
PBS (pH 7.2 without $CaCl_2$, without $MgCl_2$, GIBCO)
PLL-PEG (SuSos, stock solution 2 mg/mL in HEPES; working solution 0.5 mg/mL)

1. Mix silicone elastomer and curing reagent in a 7:1 ratio using a planetary centrifugal mixer.
2. Carefully clean the silicon wafer with canned air. Then add about 4 mL of the bubble-free silicone elastomer/curing reagent mixture on the wafer.
3. Plasma clean round cover glasses at high intensity for 2 min. Then place them with the plasma-cleaned surface facing the silicone elastomer/curing reagent mixture on the wafer. Press them down to the silicon wafer and make sure to get rid of all the bubbles.
4. Bake PDMS on a heating plate for 15 min at 95 °C.
5. Remove PDMS-coated cover glasses carefully with *iso*-propanol using a razor blade.
6. Plasma clean PDMS micropillar containing surface of the round cover glass for 2 min at high intensity.
7. Add 50 μL PLL-PEG (100 μg/mL) and incubate at rt for at least 1 h (let it dry). Before usage, wash at least five times with PBS and store under PBS.

2.2.2 Lid with Soft PDMS Piston

Required Materials

PDMS Sylgard 184 Elastomere Kit (Dow Corning).
Planetary centrifugal mixer (Thinky, ARE250).
Aluminum mold for soft PDMS pistons.
Vacuum pump (Vacubrand, RZ6).
Vacuum desiccator.
Oven (80 °C).
60 × 15 mm style nonpyrogenic polystyrene tissue culture dish lid (Falcon).
Iso-propanol (Sigma–Aldrich).
Transparent aquarium silicone sealant (Marina).

To produce the soft PDMS piston (step 3 in Section 2.2.2) needed for confinement, the dimensions of the dish need to be considered. The pillar height (h_2 in Fig. 5A2) has to be 1 mm longer than the distance between glass surface and lid (h_1 in Fig. 5A1). This guarantees that the exerted force on the two surfaces is in an ideal range. The diameter of the pillar (*d* in Fig. 5A2) depends on the diameter of the hole and the used glass slide to

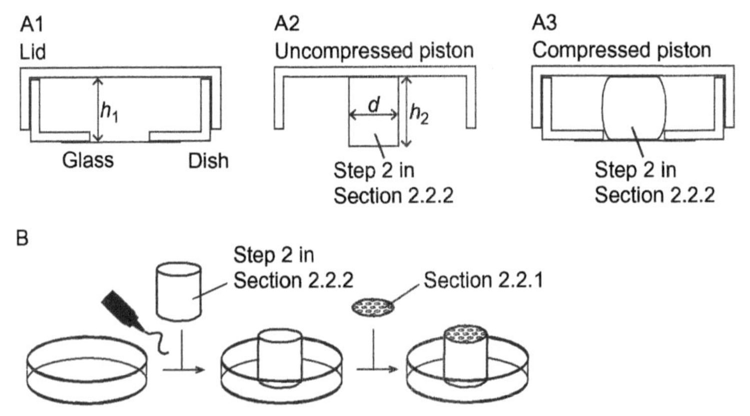

Figure 5 Lid production (A1) Side view of a closed glass bottom dish, where h_1 is the distance between glass slide and inner surface of the lid. (A2) Side view of lid with soft PDMS piston. Elastic PDMS piston (step 3 in Section 2.2.2) glued into the lid of the glass bottom dish, where h_2 is the height of the piston (step 3 in Section 2.2.2); *d* is the diameter of the piston (step 2 in Section 2.2.2); $h_2 = h_1 + 1$ mm. (A3) Closed glass bottom dish with compressed PDMS piston (step 3 in Section 2.2.2). Compression of the elastic piston (step 3 in Section 2.2.2) presses PDMS micropillar-coated round glass slide (Section 2.2.1) on the glass of the glass bottom dish. (B) Lid with soft PDMS piston preparation. Piston (step 2 in Section 2.2.2) is glued in the center of the inner part of the lid of the glass bottom dish. PDMS micropillar-coated round glass slide (Section 2.2.1) is placed on the elastic PDMS piston (step 3 in Section 2.2.2) with the micropillars facing away from the piston (step 3 in Section 2.2.2).

produce the micropillar-coated upper surface. For a 17 mm hole we use pillars of 10 or 12 mm diameter. To produce the pillars with the correct dimensions it is advantageous to use an aluminum mold (Le Berre et al., 2014).

1. Mix silicone elastomer and curing reagent in a 30:1 ratio using a planetary centrifugal mixer.
2. Pour mixture in the aluminum mold and degas in a vacuum desiccator until bubbles are completely gone.
3. Bake PDMS for 6 h at 80 °C, then remove soft PDMS pistons carefully using *iso*-propanol.
4. Glue soft PDMS piston (step 3 in Section 2.2.2) in the middle of the lid of the 60 × 15 mm nonpyrogenic polystyrene tissue culture dish using silicone aquarium sealant (Fig. 5B).
5. Place micropillar-coated, PEG-coated cover glass (Section 2.2.1) with the glass side facing the soft PDMS piston on the piston. The glass sticks to the PDMS without fixation.

2.3 Haptotaxis Chamber Assembly

2.3.1 Cell Preparation

DCs are generated from mouse bone marrow according to Lutz et al. (1999). In this work, we will focus on the isolation and enrichment of highly migratory DCs from bone marrow DC cultures.

Required Materials

LPS (Sigma–Aldrich).
R10 cell culture medium (RPMI 1640, supplemented with 10% fetal calf serum (FCS), 2 m*M* l-glutamin, 100 U/mL penicillin, 100 μg/mL streptomycin and 50 μ*M* 2-mercaptoethanol (all Invitrogen)).
GM-CSF supernatant from hybridoma cell culture.
Tissue culture dishes, 15 cm (VWR).
PBS (pH 7.2 without $CaCl_2$, without $MgCl_2$, GIBCO).
Table centrifuge.

1. Harvest the cell-containing supernatant earliest at day 8 of the bone marrow DC culture and concentrate them (5 min at 300 × *g*).
2. Resuspend approximately two million immature DCs in 25 mL 10% GM-CSF supernatant containing R10 medium. Transfer cell suspension in a 15 cm tissue culture dish.
3. For maturation, stimulate immature DCs by incubation with LPS (200 ng/mL) for at least 6 h (or overnight) at 37 °C and 5% CO_2.
4. After maturation, harvest cells containing supernatant without scratching or washing off nonmigratory, adhesive DCs and concentrate them (5 min at 300 × *g*).

5. Aspirate LPS-containing supernatant thoroughly and resuspend DCs in 1 mL R10. Incubate at 37 °C and 5% CO_2 for at least 1 h.
6. After chemokine incubation of the lower surface (Section 2.1.2), harvest the migratory DCs containing supernatant and centrifuge for 5 min at $300 \times g$. Resuspend in R10 to a concentration of about 4×10^4 cells/μL.

2.3.2 Chamber Assembly

Required Materials

Dish with CCL21 patterned surface (Section 2.1.2).
Lid with PEG-coated micropillars on soft PDMS pillar (Section 2.2.2).
Cell suspension (Section 2.3.1).
Fabric tape (Tesa, extra power perfect).

1. Aspirate PBS from the dish (Section 2.1.2) and add 4 μL of the migratory DC suspension (Section 2.3.1) on the marked area. Add 2 mL of R10 to the rim of the plastic dish. Avoid mixing with the cell suspension (Fig. 6B).
2. Aspirate PBS from the PDMS micropillars (Section 2.2.2) and add 4 μL of the migratory DC suspension (Section 2.3.1; Fig. 6B).
3. Lower the lid quickly, but in a parallel fashion onto the dish. Hold the soft pillar pressed down onto the patterned coverslip without creating shear stress. The lid should now touch the rim of the dish, closing it completely (Fig. 6C and D).

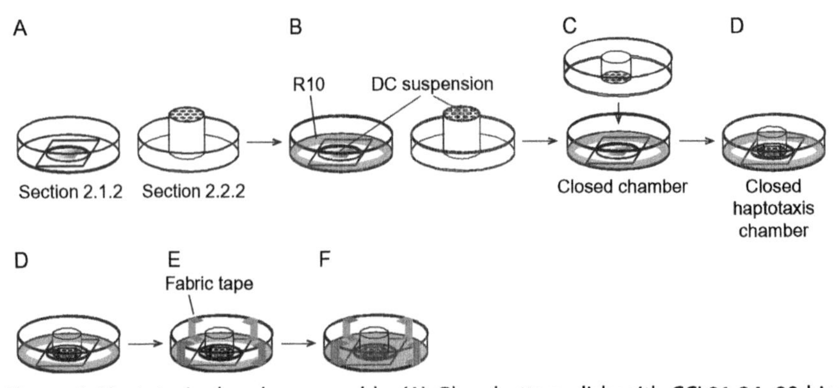

Figure 6 Haptotaxis chamber assembly. (A) Glass bottom dish with CCL21 24–98 bio patterned surface (Section 2.1.2) and lid with PEG-coated micropillars on soft PDMS piston (Section 2.2.2). (B) DC suspension is placed on the micropillars on soft PDMS piston (Section 2.2.2) and on the gradient patterned area of the glass bottom dish (Section 2.1.2). R10 cell culture medium is placed on the rim of the glass bottom dish (Section 2.2.2). (C) The glass bottom dish is closed to assemble the chamber. (D) Fabric tape is attached to keep the PDMS piston under pressure. (See the color plate.)

4. Fix lid on dish with stripes of fabric tape. Lid and dish should stay in touch and the soft PDMS pillar should visibly be compressed to connect the micropillars (upper surface) on the cover glass (lower surface). Always place tape at opposite sites to generate homogenous pressure on the micropillars (Fig. 6E).
5. Carefully shake the dish to wet the micropillars with the previously added R10 (Fig. 6F).
6. Image at 37 °C and 5% CO_2.

2.4 Troubleshooting

The preparation of the assay involves many single manufacturing steps, bearing the risk of small mistakes. In this section, we will focus on some sensitive parts of the protocol and how to avoid mistakes.

2.4.1 The Haptotaxis Chamber

Homogenous confinement is crucial to guarantee similar migration conditions under the whole confined surface. This is especially important if multiple gradients are imaged. The soft PDMS pistons exert pressure on the lower glass slide and the plastic lid of the dish. This pressure can, if too high, deform one or both of them leading to confinement only at the rim of the micropillar-coated slide. Deformation of the lower glass can be avoided by using glass thicker than 190 μm (coverslips thicker than #1.5). Replacement of the thin plastic lid by a thicker and more rigid plastic or glass lid can improve confinement homogeneity as well.

In some cases, medium or air bubbles can be stuck between both surfaces, preventing proper confinement. Gentle tapping with a pen or tweezers on the lower glass of the freshly confined chamber can remove the excessive air/medium.

PEG coating of the micropillars is important to avoid cell attachment to the upper, not chemokine-functionalized surface. In our protocol, we advise to use PLL-PEG, which can be toxic for the cells if present at high concentrations in cell suspensions. Therefore, careful washing of the PEG-coated micropillars is recommended.

2.4.2 Cell Treatment

Ideal quality of the cells is essential for optimal migration behavior in the haptotaxis chamber. To guarantee stable conditions and to minimize cell heterogeneity, it is advantageous to use cells that have been frozen at day 8 of the bone marrow culture.

After maturation, it is crucial to remove all the LPS-containing media and recover the cells for at least 1 h in fresh R10 and at high cell number. Recovery periods longer than 4 h reduce viability.

Depending on micropillar height, the volume between both surfaces is rather small and medium exchange with the surrounding medium reservoir is limited. If cell number is too high, toxic metabolites can harm the cells and influence their migration. Ideal cell numbers are dependent on the quality of the bone marrow culture and need to be titrated.

2.4.3 LAPAP/Chemokine Presentation

Upon CCL21 encounter, DCs start to spread their lamellipodium on the surface, adapting a "fried egg" like morphology. Stressed cells, however, contract and retract their protrusions. If cells contract instead of spread after confinement exchange all LAPAP reagents. B4F and SA-Cy3 decompose after a few months at 4 °C.

The success of the chemokine immobilization can be evaluated by antibody staining of the immobilized CCL21 (mouse CCL21 antibody; R&D, 10 μg/mL, 1 h at room temperature). Even without background block, the unspecific binding of truncated CCL21 to the PDMS surface is extremely low. However, to evaluate background levels, antibody staining of each batch of coverslips is recommended. Proper cleaning of the slides is necessary if glass slides without PDMS coating are used.

For immobilization, the chemokine needs to have a biotin tag. Considering the size of the chemokine (~12 kDa) versus SA (~50 kDa), it is beneficial to introduce a short PEG spacer between the chemokine and the SA-binding biotin tag. Furthermore, monobiotinylation can increase the chemokine concentration on the SA pattern.

3. PERSPECTIVES

LAPAP enables printing of chemokine patterns of arbitrary shape and intensity on a wide range of surface coatings. This opens a plethora of new possibilities to the field of haptotaxis, but also chemotaxis. With LAPAP not only defined gradients can be generated. The ability to create sharp borders allows the printing of even more complex patterns like staircase functions or sudden steps in gradients enabling to address fundamental mechanisms like cellular memory or polarization. The possibility to immobilize guidance cues to many surface coatings opens up new perspectives on the role of adhesion during haptotaxis.

Our haptotaxis assay was not only established to satisfy the requirements of DC haptotaxis. CCL19-biotin, CCL21-biotin, or CXCL12-biotin patterns allow to probe the migratory behavior of naïve or activated CD4+ T cells. Using other biotinylated ligands allows to employ other types of leukocytes, e.g., immobilization of biotinylated fMLP is suited to test haptotactic guidance of granulocytes, which is well explored in the chemotactic regime.

ACKNOWLEDGMENTS

This work was supported by the Boehringer Ingelheim Fonds, the European Research Council (ERC StG 281556), and a START Award of the Austrian Science Foundation (FWF). We thank Robert Hauschild, Anne Reversat, and Jack Merrin for valuable input and the Imaging Facility of IST Austria for excellent support.

REFERENCES

Azioune, A., Storch, M., Bornens, M., Théry, M., & Piel, M. (2009). Simple and rapid process for single cell micro-patterning. *Lab on a Chip*, *9*(11), 1640.

Bélisle, J. M., Correia, J. P., Wiseman, P. W., Kennedy, T. E., & Costantino, S. (2008). Patterning protein concentration using laser-assisted adsorption by photobleaching, LAPAP. *Lab on a Chip*, *8*(12), 2164.

Blawas, A. S., & Reichert, W. M. (1998). Protein patterning. *Biomaterials*, *19*(7–9), 595–609.

Lämmermann, T., Bader, B. L., Monkley, S. J., Worbs, T., Wedlich-Soldner, R., Hirsch, K., et al. (2008). Rapid leukocyte migration by integrin-independent flowing and squeezing. *Nature*, *453*(7191), 51–55.

Le Berre, M., Zlotek-Zlotkiewicz, E., Bonazzi, D., Lautenschlaeger, F., & Piel, M. (2014). Methods for two-dimensional cell confinement. *Methods in Cell Biology*, *121*, 213–229 (Volume 97, Elsevier Inc.).

Love, M., Sandberg, J. L., Ziarek, J. J., Gerarden, K. P., Rode, R. R., Jensen, D. R., et al. (2012). Solution structure of CCL21 and identification of a putative CCR7 binding site. *Biochemistry*, *51*(3), 733–735.

Lutz, M. B., Kukutsch, N., Ogilvie, A. L., Rössner, S., Koch, F., Romani, N., et al. (1999). An advanced culture method for generating large quantities of highly pure dendritic cells from mouse bone marrow. *Journal of Immunological Methods*, *223*(1), 77–92.

Reis e Sousa, C. (2006). Dendritic cells in a mature age. *Nature*, *6*, 476–483.

Renkawitz, J., Schumann, K., Weber, M., Lämmermann, T., Pflicke, H., Piel, M., et al. (2009). Adaptive force transmission in amoeboid cell migration. *Nature Cell Biology*, *11*(12), 1438–1443.

Renkawitz, J., & Sixt, M. (2010). Mechanisms of force generation and force transmission during interstitial leukocyte migration. *EMBO Reports*, *11*(10), 744–750.

Sallusto, F., Schaerli, P., Loetscher, P., Schaniel, C., Lenig, D., Mackay, C. R., et al. (1998). Rapid and coordinated switch in chemokine receptor expression during dendritic cell maturation. *European Journal of Immunology*, *28*(9), 2760–2769.

Steinman, R. M., & Banchereau, J. (2007). Taking dendritic cells into medicine. *Nature*, *449*(7161), 419–426.

Weber, M., Hauschild, R., Schwarz, J., Moussion, C., de Vries, I., Legler, D. F., et al. (2013). Interstitial dendritic cell guidance by haptotactic chemokine gradients. *Science*, *339*(6117), 328–332.

Whitesides, G. M., Ostuni, E., & Takayama, S. (2001). Soft lithography in biology and biochemistry. *Annual Review of Biomedical Engineering*, *3*, 335–373.

AUTHOR INDEX

Note: Page numbers followed by "*f*" indicate figures, and "*t*" indicate tables.

A

Abad, A., 14
Abadie, V., 55*t*, 65, 66*t*, 69
Abagyan, R., 390, 392, 397–400, 410–411, 459
Abbott, J.D., 558–559
Abe, K., 270–271
Abe, P., 91–92
Abhyankar, V.V., 37–38
Abola, E., 390–391
Abraham, G., 132–133
Abrol, R., 392
Adachi, R., 364, 381–382
Adage, T., 459–460, 519–520
Admas, T.H., 500
Adolphe, M., 341
Afonso, P.V., 209–210, 221–224
Aguila, B., 132, 141, 147–148
Ahmad, M., 276–277
Ahmed, Y.A., 525
Ahn, J., 500
Ahn, S., 121, 132–133, 147–149, 159–161, 164–165, 431, 434–436, 458–459
Aiyar, N., 393–394
Akinaga, S., 461–462
Alam, S.M., 235
Albert, M.L., 276–277
Albitar, M., 453–454
Alcami, A., 4, 6
Alcami, J., 463–464
Alessandrini, A.L., 270–271
Alexandrov, A.I., 251–253, 412–413
Ali, M., 215–216, 220
Ali, S., 310–330
Alizon, M., 393–394
Allavena, P., 442
Al-Lazikani, B., 3–4
Allen, P.M., 211
Allen, R.A., 422–423
Allen, S.J., 49–50, 91–92, 156–157, 163–164, 168, 234–235, 460
Allet, B., 91–92, 542
Alley, K.M., 500
Alouani, S., 49–50
Al-Rubeai, M., 243
Alves, I.D., 5
Amadi, O.C., 501–502
Amara, A., 360, 542–543
Amaral, F.A., 262–277
Amarandi, R.-M., 156–180
Ambrose, J.R., 161
Amiel, C., 69
Amraei, M., 132, 147–148
Anders, H.J., 4
Andersen, M.B., 170
Anderson, J., 50–52
Andersson, A., 555, 558–559
Andersson, A.C., 352–353
Andrew, D.P., 481
Andrews, G., 475
Andrews, J.D., 164–165
Andrich, J.M., 384–385
Angeli, V., 336
Angermann, B.R., 209–210, 221–224
Angers, S., 133, 146–147, 428–429
Anselmo, A., 89, 442–454
Antony, V.B., 270–271
Apfel, C.M., 504–505
Appay, V., 383–384
Apuzzo, T., 88–91, 93–95, 108*f*, 113–114
Aramini, A., 264–265
Archer, C.W., 264
Ardolino, M., 558–559
Arenzana-Seisdedos, F., 90–91, 112*f*, 360, 463–464, 542–543
Ariel, A., 88–89
Aris, M., 461–462
Arlt, M.J., 89–91, 93–94, 109–110
Armando, S., 148–149, 429, 431–433
Armelles, G., 14
Armstrong, R.A., 329
Artis, D.R., 394–395
Aschauer, H., 91–92
Ashour, A.E., 558–559

Askari, A.T., 558–559
Atienza, J.M., 503
Atkins, W.M., 255–257
Auberson, Y.P., 464–465, 500
Aubery, M., 341
Aucoin, M.G., 384–385
Audet, M., 133–134, 428–429, 431–434, 502–503
Audet, N., 132, 147–148
Auffray, C., 442–443
Aukerman, L., 558–559
Auvynet, C., 65, 66*t*, 69–70
Axup, J.Y., 392
Azioune, A., 569

B

Babcock, G.J., 5–6, 359, 380–381
Babu, M.M., 234–235
Bachelerie, F., 89–91, 156–157, 160–161, 164, 262, 264, 271, 276, 294, 390, 422–424, 427–428, 458
Bader, B.L., 568
Baggett, B.R., 383–384
Baggiolini, M., 90–91, 310, 390, 460, 540
Baillou, C., 69
Bakker, R.A., 160–161
Balabanian, K., 90–91, 111*f*
Baleux, F., 337–338
Balkwill, F.R., 48, 282, 311
Ball, A.M., 391–392
Banchereau, J., 567–568
Bannert, N., 235, 360, 380–381, 383–384
Bao, X., 337–338
Barak, L.S., 133, 141, 491–493
Baranski, T.J., 391–392
Baratelli, F., 555, 558–559
Barazza, A., 392
Barbas, C.F., 50–52
Barcelo, D., 14
Barcelos, L.S., 264–265
Bardi, G., 160–161
Barisas, B.G., 145–146
Barlic, J., 164–165
Barreira da Silva, R., 276–277
Barroso, R., 6–7
Barry, J.D., 89, 294
Bartek, J., 329
Barton, N.P., 487
Baskaran, H., 21
Basnet, H., 235, 360–361, 382–383, 397–400, 558–559
Bassoni, D.L., 132–133, 147, 157–160, 422–423, 436–437
Batlle, E., 282
Battmer, K., 10
Baumeister, P., 481
Bayburt, T.H., 255–257
Beall, C.J., 93–94
Beautrait, A., 132, 141, 147–148, 429, 431–433
Beavis, P., 474
Beck, A., 461–462
Beck-Sickinger, A.G., 90–91
Beebe, D.J., 37–40, 329–330
Behnke, C.A., 234–235
Bélisle, J.M., 569, 573–574
Bell, D.W., 42
Bellal, M., 147–148
Bellot, M., 132–134, 142–143, 146, 429
Ben-Baruch, A., 89–91, 156–157, 160–161, 164, 262, 264, 271, 276, 294, 360, 390, 422–423, 458, 519–520
Benned-Jensen, T., 165–166
Benoist, C., 209, 211, 226
Benovic, J.L., 89–91, 148–149, 282–284, 423–424, 465
Bensman, T., 264–265
Benton, B.K., 464–465
Benureau, Y., 422–424, 427–428
Berahovich, R., 121, 257
Berchiche, Y.A., 132–134, 139*f*, 140–143, 142*f*, 147–150, 422–423
Berger, E.A., 156
Berghmans, N., 519–520
Beringer, P.M., 264–265
Bernard, J., 503
Bernat, V., 464–465, 481, 500
Berthier, E., 38–40
Bertini, R., 393–394
Bertrand, L., 428–429
Besnard, J., 12–14
Betsholtz, C., 352–353
Beyermann, M., 392
Beyrau, M., 442–443
Bezos, A., 519–520
Bhandari, D., 282–284

Bianchi, G., 442
Bianchini, F.J., 271
Bidlack, J.M., 429
Bignon, A., 90–91, 112*f*
Bill, R.M., 390–391
Birnie, G.D., 25–27
Bishop, J.R., 339–341, 352–353
Bishop-Stewart, J., 496–498
Bisio, A., 534, 534*f*
Bjarnegard, M., 352–353
Black, J.W., 434–436
Blackburn, P.E., 140
Blair, E., 140
Blanc, S., 503
Blanchetot, C., 459–462, 485
Blanco, S., 89–91
Blaser, H., 89–91
Blawas, A.S., 569
Bleul, C.C., 4
Blomer, U., 10
Blow, N., 329–330
Boakye, K., 500
Bobanga, I.D., 48
Bockaert, J., 434
Bodkin, J.V., 442–443
Boerrigter, G., 160
Boesen, T.P., 91–92, 542
Boff, D., 262–277, 519–520
Boissonnas, A., 65, 66*t*, 69–70
Bokoch, G.M., 163–164
Boldajipour, B., 89–91
Bombosi, P., 89, 92–93
Bonazzi, D., 568–569, 575–577
Bonavita, O., 442–454
Bonde, M.M., 465
Bonecchi, R., 4, 89, 422–423, 442–454
Bongers, J., 276–277
Bonig, H., 282
Bonner, W.A., 330
Bonnet, D., 459–460
Bonneterre, J., 132–150
Bonsch, C., 66*t*, 68–69
Bonvin, C., 294–295
Bonvin, P., 73–85, 187–205, 536–537
Booth, P.J., 2–3
Borchers, M.T., 423–424
Bordignon, P.P., 442
Bordon, Y., 90–91
Borlat, F., 90–92, 189, 519–520, 542
Bornens, M., 569
Borroni, E.M., 89, 92–93, 157–159, 422–437, 442, 453–454
Borsatti, A., 442
Bosch, L., 461–462, 465, 485, 500
Bosinger, S.E., 164–165
Bosse, R., 425–427
Bouhelal, R., 424
Boulais, P.E., 139*f*, 140
Boularan, C., 132–150
Bourne, H.R., 170, 488, 496–498
Bourougaa, K., 147–148
Bouvier, M., 132–134, 143, 146–149, 428–429, 431–436, 502–503
Bowman, E.P., 329
Boyden, S., 20–21, 294–295, 311
Braber, S., 48, 459
Brackertz, D., 266–267
Bradley, M.E., 461–462
Brady, A.E., 3–4
Brand, M., 90–91
Brandenburg, G., 69
Braun, A., 89, 294–295, 299, 310–311
Braun, M., 93–94
Bräuner-Osborne, H., 160–161, 427–428
Braut-Boucher, F., 341
Bravo, R., 93–94
Brech, A., 282
Breckenridge, S., 93–94
Breit, A., 146–147
Breitfeld, D., 442
Brelot, A., 393–394
Brenet, F., 282
Brennecke, P., 89–91, 93–94, 109–110
Breton, B., 133–134, 143, 434–436
Briddon, S.J., 505–506
Bridger, G.J., 161, 170
Bridgford, J.L., 360, 382–383
Briggs, J.A., 11–12
Brigham-Burke, M., 5
Brocker, E.B., 294–296
Brodbelt, J.S., 374–375
Broers, J.L., 467
Brown, A., 383–384
Brown, A.F., 20–21
Brown, K.A., 496–498
Browning, D.D., 164

Brox, R., 464–465, 500
Broxmeyer, H.E., 93–94
Brubaker, S.A., 163–164
Bruggeman, C.A., 467
Brühl, H., 442–454
Bruinsma, M., 121–122
Brunner, T., 161
Brunowsky, W., 91–92
Buchanan, M.E., 4, 282, 311, 360
Buchholz, W.G., 540–562
Buchkovich, N.J., 282–283
Buck, E., 391–392
Buelow, B.J., 540–541, 550, 555
Buneker, C.K., 69–70
Buntinx, M., 479–481
Buracchi, C., 89, 92–93
Burdick, M.D., 90–91, 270–271
Burg, J.S., 69, 234–235, 390–391, 393–395, 460–461
Burkart, M.D., 369–370, 381
Burkhardt, A.M., 89–91, 156–157, 160–161, 164, 262, 264, 271, 276, 294, 310, 422–423, 458
Burkhart, M., 459
Burlion, A., 69
Burnett, J.C., 160
Burns, J., 121
Burns, J.M., 74, 257
Burns, M.C., 542–544
Burns, N.M., 558–559
Burrone, O.R., 276
Buschauer, A., 481
Busillo, J.M., 148–149
Busnelli, M., 422–437
Butler, S.J., 360, 382–384
Byers, M.A., 422–423

C

Cadene, M., 360–361, 371*f*, 372, 375–377, 380–382
Caffrey, M., 234–235
Cailhier, J.F., 270–271
Cain, R.J., 163–164
Calama, E., 474
Calle, A., 7, 14
Calloway, P.A., 422–423
Calvo, E., 276
Cameroni, E., 88–95, 108*f*, 113–114
Campanella, G.S., 132–133
Campanile, C., 89–91, 93–94, 109–110
Campbell, E.M., 393–395
Campbell, J.J., 132–133, 147, 157–160, 209, 226, 329, 422–423, 436–437
Camphausen, R.T., 359
Canals, M., 49–50, 156–157, 172, 360, 382–384, 393–394, 460, 479–481, 490*f*, 492*f*, 502*f*
Cancellieri, C., 422–424, 427–428
Cantarelli, E., 264–265
Canziani, G., 7
Capila, I., 337–338
Capuano, B., 160
Caput, D., 161
Card, A., 427–428
Carlow, D.A., 359
Carman, C.V., 423–424
Carmo, A.A., 270–271
Caron, M., 428–429
Caron, M.G., 133, 141, 431–433
Carter, V., 329–330
Cartier, M., 524
Casagrande, F., 234–235
Casarosa, P., 160–161, 164, 391–392, 459–460, 467
Cascio, G., 2–15
Cashen, A.F., 48–49
Castro, M., 474
Casu, B., 534, 534*f*
Catron, D., 4, 282, 311, 360
Catusse, J., 89
Cauwenberghs, S., 90–91
Caux, C., 336
Cayabyab, M., 359, 380–381
Cen, B., 160–161
Cerini, F., 48–70, 53*t*, 55*t*, 66*t*
Cerione, R.A., 534
Ceroni, A., 529
Cerovic, V., 89–91
Cervellera, M.N., 264–265
Cesta, M.C., 264–265
Ceyhan, G.O., 69–70
Chabot-Dore, A.J., 132, 147–148
Chae, P.S., 391–392
Chaffey, B.T., 313
Chait, B.T., 358, 360–361, 364, 371*f*, 372–378, 380–383

Chakera, A., 132–133
Chakravarty, L., 93–94
Chamberlain, J.S., 558–559
Champeil, P., 2–3
Chan, H.E., 453–454
Chang, R., 453–454
Chang, S., 121
Channon, K.M., 458–459, 488
Chapman, E., 369–370, 381
Charest, P.G., 147–148, 434
Charfi, I., 132, 147–148
Charles, R., 132, 141, 147–148
Charo, I.F., 390, 422
Chelsky, D., 133
Chemtob, S., 133–134, 143
Chen, G.Q., 366–368
Chen, M., 209–210
Chen, P.E., 485
Chen, Q., 392
Chen, S.-w.W., 395
Chen, W., 42
Chen, W.N., 496–498
Chen, Y., 382
Chen, Z., 234–235, 360, 460–461
Cheng, X., 282, 290
Cheng, Y., 477–478
Cheng, Z., 164–165
Cheng, Z.J., 160–161
Cheong, W.C., 42
Cherezov, V., 234–235, 390–391
Chevigne, A., 393–394
Chew, W.K., 222–224
Chiang, J.J., 384–385
Chichili, V., 123
Chien, E.Y., 234–235, 251–253, 384–385, 460–461
Chien, E.Y.T., 412–413
Childs, R., 558–559
Chini, B., 422–437
Chiswell, D.J., 52
Choe, H., 235, 360, 380–381, 383–384
Choe, S., 459
Choi, H.-J., 391–392
Choi, W.-T., 394–395
Chou, R.C., 209–210, 215–216, 226
Chow, K.Y., 90–91, 111*f*, 461–462, 485
Chow, M.T., 208
Chowdry, A.B., 65, 248, 248*f*
Choy, E.W., 164–165
Christensen, G.L., 465
Christopoulos, A., 157, 159–162, 434–436
Christou, I., 505
Chu, M., 234–235
Chuang, K.T., 474
Chuang, S., 361
Chun, E., 234–235
Chung, C.Y., 21–23, 22*f*, 25–27
Chung, S., 359–360, 380–381, 383–384
Chung, T.D., 482
Cihak, J., 447
Cimino, D.F., 164–165
Cirillo, R., 459–460
Cironi, P., 91–92, 100–102, 106, 112*f*
Citro, A., 264–265
Ciuculescu, F., 460
Clackson, T., 50–52
Claing, A., 132, 141, 147–148
Clarke, W.P., 132
Clark-Lewis, I., 50–52, 73–74, 90–91, 262–263, 310–311, 337–338, 460, 519–520, 542
Clayton, D.J., 360, 382–383
Coelho, F.M., 264–267, 269, 277
Coetzer, M., 68
Coggins, N., 121
Coggins, N.L., 90–91
Coin, I., 392
Colaço, N.C., 330
Colagioia, S., 264–265
Collins, S.J., 25–27
Colvin, R.A., 132–133
Combadiere, B., 55*t*, 65, 66*t*, 69
Combadiere, C., 69–70, 89–91, 93–94, 156–157, 160–161, 164, 262, 264, 271, 276, 294, 390, 422–423, 458
Comerford, I., 65, 140, 248, 248*f*
Condliffe, A., 208–209
Coniglio, S., 264–265
Conings, R., 393–394
Conklin, B.R., 170
Conn, P.M., 392
Connelly, S.M., 505–506
Conti, P.G., 465, 500
Cook, D.N., 89
Cooke, A., 157
Cooke, T.D., 266–267

Coombe, D.R., 6
Cooper, D.M., 491–493
Cooper, M.A., 2
Cooper, S., 93–94
Corbett, M., 542–543
Corbisier, J., 132–133, 140, 147, 159–160, 422–423, 429–436
Corich, M., 392
Cormier, E.G., 360
Correia, J.P., 569, 573–574
Corsi, M.M., 422, 442
Cortez-Retamozo, V., 219
Costa, B.R., 270–271
Costa, L., 24, 33–37, 34*f*, 38–39*f*
Costa, V.V., 264–267, 269, 277
Costagliola, S., 465
Costantino, S., 569, 573–574
Cotton, G., 6
Coulin, F., 310–311, 519–520
Coulon, V., 434
Cousin, P., 90–91
Cox, M.A., 465
Craig, S., 121, 132–133, 160–161, 235, 360, 380–381, 383–384, 458–459
Craven, C.J., 383–384
Cresswell, P., 555
Crick, F., 88
Crosby, W.L., 53–54
Crown, S.E., 49–50, 156–157, 234–235, 262–263, 458, 460
Crump, M.P., 360, 542–543
Cubedo, N., 89
Cukierman, E., 329
Cullere, X., 208
Cummings, R.D., 359
Cunningham, H.D., 422–423
Curnock, A.P., 496–498
Curnow, P., 2–3
Custers, H., 474*f*, 481, 487–488, 505
Czaplewski, L.G., 383–384
Czeloth, N., 299
Czodrowski, P., 392

D

da Silva Guerra, A.S.H., 330
Daaka, Y., 164–165
Dahinden, C.A., 90–91, 161
Dale, D.C., 264–265
Dalrymple, M.B., 143
Daly, J.W., 168
Dambly-Chaudiere, C., 89
D'Ambrosio, D., 442
D'Amico, G., 442
Danan, L.M., 368, 381–382
Das, B.B., 234–235
Dasen, B., 424
Daubeuf, F., 459–460
Davalos-Misslitz, A.C., 336, 540
Davis, D., 542–543
Dawson, P.E., 50–52, 90–91
De Clercq, E., 170, 460–461
de Esch, I.J., 474*f*, 481, 505
de Esch, I.J.P., 460–461
De Gaspari, E., 330
de Graaf, C., 148–149, 467, 492
de Kruijf, P., 48, 459, 465, 500
de la Cruz, N., 91–92, 235, 542
De la Cruz, N.B., 235, 360–361, 382–383, 397–400
De La Luz Sierra, M., 542–543
de la Torre, Y.M., 89
de Lima, M.do.C., 330
De Meester, I., 329
de Munnik, S.M., 148–149, 458–506, 490*f*, 492*f*, 502*f*
de Oliveira, L.C., 270–271
De Taeye, S., 461–462
de Vries, E.G., 282
de Vries, H., 481
de Vries, I., 310–311, 567–569
de Wit, R.H., 458–506
De, C.E., 89–91
de, K.P., 93–94
Debnath, B., 264–265
Decock, B., 393–394
Deems, R., 254–255
DeFea, K.A., 164–165, 171
Degorce, F., 427–428
Dehmlow, H., 504–505
Del Castillo, J., 330
Del Prete, A., 4
Delcayre, A., 243
Dell, A., 529
Della Rocca, G.J., 164–165
Delporte, C., 220
Demotz, S., 90–91

Denelavas, A., 504–505
Deng, H., 459
Deng, N., 37–38
Denis, C., 132–134, 142–143, 146, 429
Denisov, I.G., 255–257
Dennis, E.A., 254–255
Dennis, M., 133
Derler, R., 518–537
Deroo, S., 393–394
Dertinger, S.K., 21
Deruaz, M., 221
Descamps, F.J., 461–462
Deshane, J., 219
Despas, F., 133–134, 142–143, 146
Deterre, P., 69–70
Deuel, T.F., 73–74
Deupi, X., 234–235
Devi, S., 222–224
DeVico, A.L., 74
Devost, D., 429, 431–433
DeVries, M.E., 164–165
Dewald, B., 90–91, 93–94, 460
DeWire, S.M., 159–160, 164–165, 431, 434–436
D'Haese, J.G., 69–70
Dieu-Nosjean, M.C., 336
DiGregorio, D., 555
Dimaki, M., 294–295
Dimberg, A., 458
Dimitrova, S., 89, 92–93
Dioszegi, M., 6
DiSilvio, F.A., 359, 382–383
Dixon, J.B., 294–295
do Nascimento Malta, D.J., 330
Dogra, P., 93–94
Dohlman, H.G., 464–465
Doitsidou, M., 90–91
Domanska, U.M., 282
Dombrecht, B., 461–462
Domenech, T., 474
Doms, R.W., 7
Dona, E., 89, 294
Dong, C.-Z., 394–395
Dong, M., 391–392
Doni, A., 89
Donnadieu, A.C., 69–70
Donohue, J.F., 264–265
Doornbos, M.L., 481
Dores, M.R., 89–92
Dorfman, T., 384–385
Dorgham, K., 48–70, 53*t*, 55*t*, 66*t*
Dorn, J.F., 502–503
Dothager, R., 121–122
Dowthwaite, G.P., 264
Doyon, J., 479–481
Drake, M.T., 3–4, 160
Drury, L.J., 157–159, 382–383, 558–559
du Mez, E., 505
Dubinett, S.M., 558–559
Duchesnes, C.E., 394–395
Dufour, A., 208–209
Dukkipati, A., 69, 234–235, 390–391, 393–395, 460–461
Duma, L., 360, 382–383
Dumont, M.E., 505–506
Duncan, R., 380–381
Dunn, G.A., 20–21
Dunn, S., 187–205
Dunn, S.M., 188
Dupor, J., 89
Durr, U.H., 254–255
Dusetti, N., 276
Dutt, P., 163–164
Dwyer, M.P., 264–265
Dyer, D.P., 188
Dyrhaug, S.Y., 6–7
Dziejman, M., 91–92, 100–101

E

Early, B.P., 550
Eckart, A., 310–311
Edgerton, M.D., 91–92, 542
Edwards, J.C.W., 330
Edwards, S.W., 208–209
Egelhofer, M., 69
Eglen, R.M., 425–427
Ehlert, F.J., 434–436
Ehrlich, A., 121
Eidne, K.A., 132–133, 428–429
Elain, G., 442–443
Eldaly, H., 322
Elder, A., 170
Elgbratt, K., 339
Eller, K., 89, 294–295, 299, 310–311
Elling, C.E., 465
Ellmeier, W., 459

Ellwart, J., 442
Ellyard, J.I., 519–520
el-Sawy, T., 4
Elvin, P., 25–27
Emilie, D., 69–70
Emr, S.D., 282–283
Endo, N., 93–94
Engel, A., 255–257
Engel-Andreasen, J., 170
Engelken, J.D., 527
Eriksson, E.M., 157, 172
Escher, E., 132, 141, 147–148
Escola, J.M., 53*t*, 54, 55*t*, 58, 62–63, 65, 66*t*, 68
Esko, J.D., 262–263, 337–339
Espinoza, S., 491–493
Esteve, E., 147–148
Estrada, L., 502
Etzrodt, M., 219
Evans, G., 390–391

F

Fagni, L., 434
Fahmy, N.M., 4
Fairchild, R.L., 4
Fallahi-Sichani, M., 6–7
Falsone, A., 459–460, 519–520, 531–534
Fang, Y., 505
Farber, J.M., 89–91, 156–157, 160–161, 164, 262, 264, 271, 276, 294, 390, 422–423, 458
Farfel, Z., 170
Farmer, M., 73–74
Farzan, M., 235, 359–360, 380–381, 383–384
Fatkenheuer, G., 460–461
Faust, N., 222–224
Fazzini, A., 500
Fei, G., 491
Feinberg, E.N., 69, 234–235, 390–391, 393–395, 460–461
Feldman, B.J., 143
Fellers, R.T., 550
Fenalti, G., 234–235, 384–385, 460–461
Fenn, T.D., 143
Ferguson, S.S., 133, 164–165, 431–433
Fernandez, E.J., 49–50
Fernig, D.G., 527
Ferrara, P., 161
Field, M.E., 164–165
Fiette, L., 276–277
Fievez, V., 393–394
Filizola, M., 397–400, 410–411
Finana, F., 132, 429
Fine, J., 465
Finn, A., 4, 390
Fischer, N., 73–85, 189
Fischer, T., 284
Fish, J.E., 558–559
Fisher, J., 500
Flory, C.M., 93–94
Flyckt, R., 68
Fogg, D., 442–443
Folkerts, G., 48, 459
Follo, M., 89
Fong, A.M., 164–165, 235
Font, J., 341
Ford, L.B., 89–91
Forghani, R., 215–216, 220
Fornwald, J.A., 542–544
Forssmann, U., 93–94, 460
Forster, R., 294–307, 336, 540
Forudi, F., 558–559
Fotiadis, D., 255–257
Fox, B.A., 234–235
Fox, C.R.J., 330
Fraile-Ramos, A., 465
Fraser, A.R., 6
Fratantoni, S.A., 459
Fredman, G., 88–89
Fredriksson, R., 3–4, 234–235
Fredriksson, S., 352–353
Freedman, N.J., 431
Freeman, C., 519–520
Friand, V., 519–520
Fridlender, Z.G., 442–443
Friedl, P., 113–114, 294–296
Friess, H., 69–70
Frimpong, K., 9
Frimurer, T.M., 170, 479–481
Fritchley, S.J., 313
Fuchs, T., 531–534
Fujii, N., 133–134, 139*f*, 140, 142–143
Fujikawa, Y., 364, 381–382
Fukada, S.Y., 266–267
Fukuda, M., 337–338

Fuller, S.D., 11–12
Fulton, A.L., 329–330
Fung, J.J., 391–392
Furstenberg, A., 66*t*, 68–69
Fusetani, N., 329
Fuster, M.M., 262–263, 336–353

G

Gabel-Eissens, S.J., 359, 382–383
Gabriel, D., 424
Gaertner, H.F., 48–70, 53*t*, 55*t*, 66*t*
Gainetdinov, R.R., 491–493
Galandrin, S., 132–134, 142–143, 146
Gales, C., 132–150, 159–160, 422–423, 429–436
Gallagher, J.T., 524, 527
Gallagher, R.E., 25–27
Galliardt, H., 145–146
Galliera, E., 422, 442
Gallo, R.C., 25–27, 74
Gambhir, S., 121
Gambhir, S.S., 143
Gammon, S., 121–122
Gan, A.T., 42
Gandhi, N.S., 518–520
Ganju, R.K., 163–164
Gannon, J., 542–543, 558–559
Ganser, A., 10
Gao, F., 392
Gao, Y., 37–38
Garcia, C.C., 264–265
Garcia, M.L., 380–381
Garcia-Gonzalez, V., 474
Garcia-Perez, J., 463–464
Garcia-Sanz, J.A., 461–462
Garcia-Zepeda, E.A., 337–338
Gardner, M.R., 384–385
Garfa, M., 442–443
Garneau, P., 133–134, 428–429, 431–433
Garoff, H., 7
Garrett, D.S., 383–384
Gaspar, E.B., 330
Gasparik, V., 459–460
Gati, C., 234–235
Gaudin, F., 69–70
Gauntner, T.D., 500
Ge, N., 4, 282, 311, 360
Ge, Y., 337–338
Gee, K.R., 496–498
Gee, P., 474
Geiser, T., 161
Geissmann, F., 442–443
George, N., 91–92, 106
George, S., 384–385
Gerard, C., 132–133, 147, 157–160, 390, 422–423, 436–437
Gerard, C.J., 209, 226
Gerard, N., 121
Gerard, N.P., 132–133, 147, 157–161, 359, 380–381, 422–423, 436–437, 458–459
Gerarden, K.P., 382–383, 540–541, 555, 569
Geretti, E., 531–534
Gerisch, G., 5
Gerlach, L.-O., 91–92, 161, 542
Gerlza, T., 518–537
Gesslbauer, B., 459–460, 519–520
Gesty-Palmer, D., 132, 141, 147–148, 164–165, 434
Getschman, A.E., 382–383
Geyer, H., 529
Geyer, R., 529
Ghadiri, A., 53*t*, 55*t*, 65, 66*t*, 69–70
Ghirlando, R., 6–7
Ghosh, E., 390–391
Ghysen, A., 89
Gichinga, C., 68
Giedlin, M., 558–559
Gierschik, P., 93–94
Gilchrist, A., 500
Gildenberg, M., 254–255
Gilissen, R.A., 479–481
Gill, S.E., 6
Gilliland, C.T., 172, 433–434
Gillitzer, M.L., 359, 382–383
Giordano, F.J., 558–559
Girvin, M.E., 254–255
Gizzi, P., 459–460
Godbey, S.W., 270–271
Goddard, W.A., 392
Goger, B., 524, 531–534
Golan, D.E., 91–92, 100–102, 106, 112*f*
Goldman, C.K., 558–559
Goldsby, R., 534–535
Gomariz, R. P., 2–15
Gomes, J.H., 269

Gong, J.H., 90–91, 160–161, 360, 460, 542–543
Gonsiorek, W., 465
Gonzalez-Pajuelo, M., 461–462, 485
Goossens, J., 479–481
Gorochov, G., 48–70, 55*t*, 66*t*
Gossens, K., 359
Gouaux, E., 366–368
Gouwy, M., 90–91, 393–394, 542–543
Gowen-MacDonald, W., 159–160, 434–436
Graf, T., 222–224
Graham, G.J., 89–91, 140, 156–157, 160–161, 164, 262, 264, 271, 276, 294, 390, 422–423, 442, 453–454, 458
Graham, T.L., 542–544
Granot, Z., 442–443
Gravel, S., 132–133, 139*f*, 140–141, 142*f*, 147–150, 157–159, 382–383, 422–423, 558–559
Grawrisch, K., 3
Gray, D., 496–498
Greaves, D.R., 132–133, 496–498
Green, L.C., 68
Green, M.R., 322–323
Greer, J.B., 550
Gregory, K.J., 159–160
Grespan, R., 266–267
Grether, U., 504–505
Grez, M., 10
Griffith, J.W., 4, 48, 208, 276, 390, 442, 458
Griffith, M.T., 234–235
Griffiths, A.D., 52–54
Grinkova, Y.V., 255–257
Grisshammer, R., 6–7, 235
Grosdidier, A., 337–338
Gross, S.P., 21
Grover, A.K., 188
Gruijthuijsen, Y.K., 464–465, 467
Grunbeck, A., 392
Grzesiek, S., 360, 382–383
Gschwandtner, M., 518–537
Guabiraba, R., 264–265
Guan, J.L., 502
Gueneau, F., 74, 189
Gueret, V., 243
Guerrini, M., 534, 534*f*
Guess, C., 502
Gullberg, M., 352–353
Gunay, N.S., 526
Gundry, R.L., 540–541, 550, 555
Guo, F., 4
Guo, J., 208–209
Guo, K., 330
Gupta, M., 121–122
Gupta, S.K., 393–394
Gurevich, V.V., 465
Gustafsdottir, S.M., 352–353
Gustavsson, M., 234–257, 390–416
Gutkind, J.S., 162, 390, 459
Guyon, E., 65, 66*t*, 69–70
Gvozdenovic, A., 89–91, 93–94, 109–110
Gvozdenovic-Jeremic, J., 6–7

H

Ha, H., 264–265
Haase, W., 465
Hachet-Haas, M., 459–460
Haddad, O., 459–460, 519–520
Haddox, J.L., 20–21
Hadingham, T.C., 6
Haege, S., 88–89
Hafith, S., 481
Hagemann, I.S., 391–392
Hagihara, K., 310–311
Hajek, P., 132–133
Hakimi, A.A., 282
Halden, Y., 524
Hall, N.E., 159–160
Hall, P., 360, 382–384
Hamdan, F.F., 133–134, 428–429, 431–433
Hamel, D.J., 6, 74, 91–92, 156, 235, 540
Hamilton, A., 464–465
Hamm, H.E., 162, 489–491
Hamma-Kourbali, Y., 519–520
Hammer, D.A., 161, 359, 422–423
Han, G.W., 234–235, 360, 460–461
Hanai, N., 461–462
Handel, T.M., 6, 49–50, 73–74, 91–92, 156–157, 162–164, 168, 172, 234–257, 262–263, 310–311, 336–353, 360, 382, 390–416, 433–434, 458, 460, 519–520, 540
Hanes, M.S., 65, 248, 248*f*
Hansell, C.A., 89–91
Hansen, J.T., 132, 429, 465

Hansen, M., 178–179
Hanson, M.A., 234–235, 251–253, 412–413
Hansson, M., 339
Harada, S., 93–94
Haraldsen, G., 89
Harding, P.J., 6
Hardwick, J.A., 270–271
Haribabu, B., 161, 235
Harriague, J., 90–91, 111*f*
Harrison, C., 491
Harris-White, M., 558–559
Hartigan-O'Connor, D., 558–559
Hartley, O., 48–70, 55*t*, 66*t*
Hartmann, T.N., 89
Harvey, J.R., 322
Hasegawa, H., 558–559
Hashimoto, K., 310–311
Haslam, S.M., 529
Hastrup, H., 465
Hatakeyama, S., 337–338
Hatse, S., 89–91, 161, 170
Hattermann, K., 132–133
Hattori, M., 3–4
Haugner, J.C., 91–92, 235, 360–361, 382–383, 397–400, 542
Hauschild, R., 310–311, 567–569
Hauser, M.A., 540
Haussinger, D., 360, 382–383
Haybaeck, J., 347
Hayes, P.L., 91–92, 235, 542
He, Z., 37–38
Hebert, C.A., 394–395
Hecher, B., 531–534
Heck, S., 222–224
Heguy, A., 282
Heijnen, C.J., 172
Heimann, A., 55*t*, 62–63, 65, 66*t*, 67–69
Heimer, E.P., 276–277
Heinemann, F.W., 500
Heinrich, J.N., 93–94
Heinrich, M.R., 459, 464–465, 481, 500
Heinrikson, R.L., 73–74
Heitman, L.H., 156–159, 162, 422–423, 434–437, 504–505, 504*f*
Hellewell, P.G., 266–267
Henderson, P.J.F., 390–391
Henne, W.M., 282–283
Henson, G.W., 161
Heo, J., 93–94
Herber, R.L., 502
Hermand, P., 53*t*, 55*t*, 65, 66*t*, 69–70
Hermans, B., 479–481
Herren, S., 459–460
Herzenberg, L.A., 330
Heveker, N., 132–150, 142*f*, 393–394, 422–423
Hewit, K.D., 6
Hickey, R., 558–559
Higgins, L.D., 383–384
Higgins, L.J., 542–543, 558–559
Hilairet, S., 133
Hildebrandt, A., 14
Hill, S.J., 474, 493, 505–506
Hiller, S., 255–257
Hinds, M.G., 360, 382–383
Hine, D.W., 330
Hintzen, G., 299
Hirahara, K., 329
Hirai, T., 234–235, 393–394
Hirakawa, J., 337–338
Hiramatu, K., 161
Hirsch, K., 568
Hjortø, G.M., 156–180
Hlawaty, H., 459–460, 519–520
Ho, H., 264–265
Hockemeyer, K., 20–42, 34*f*, 38–39*f*
Hoenderdos, K., 208–209
Hoepelman, A.I., 460–461
Hoffhines, A.J., 368, 375, 381–382
Hoffman, T.L., 7
Hoffmann, C., 143, 474
Hofmeister, W., 24, 33–37, 34*f*, 38–39*f*
Hofsteenge, J., 358–359
Hohmann, J.G., 3–4
Holden, L.G., 69, 234–235, 382, 384–385, 390–416, 399*f*, 411*f*, 415*f*, 460–461
Holgado, B.L., 6–7
Holleman, J., 282–283
Hollingsworth, M.A., 558–559
Holmes, W.E., 264
Holst, B., 465
Holst, P.J., 160–161
Holt, J.A., 141
Homburger, V., 434
Homey, B., 4, 282, 310–311, 360
Homola, J., 2

Honing, S., 4–5
Honma, S., 270–271
Hoogenboom, H.R., 52
Hoogewerf, A.J., 49–50, 310–311, 519–520
Hopkins, A.L., 3–4, 12–14
Hori, T., 234–235
Horuk, R., 50–52, 65–67, 235, 390, 555
Hosey, M.M., 465
Hosohata, K., 392
Hott, J.W., 270–271
Hou, X., 361
Hou, X.Y., 69–70
Houlberg, K., 270–271
Howell, D.N., 555
Hoxie, J.A., 4
Hricovíni, M., 534, 534*f*
Hruby, V.J., 5
Hrvatin, S., 91–92, 100–102, 106, 112*f*
Hsu, C.H., 42
Hsu, J., 558–559
Hu, M., 42
Hu, W., 160–161
Huang, A.Y., 48
Huang, C., 123*f*
Huang, C.C., 255–257
Huang, M., 555, 558–559
Huang, M.C., 42
Huang, Y., 558–559
Hub, E., 89, 294–295, 299, 310–311, 458–459
Hubbard, R.E., 310–311, 519–520
Huber, T., 392
Hudson, P., 52
Hulett, H.R., 330
Hulkower, K.I., 502
Huls, G., 282
Hulscher, S., 459
Hulshof, J.W., 459
Humpert, M.L., 90–91, 112*f*
Hunter, C.A., 161, 422–423
Hurd, E.R., 266–267
Huskens, D., 170
Husmann, K., 89–91, 93–94, 109–110
Huszagh, A., 132–133, 140, 147, 159–160, 422–423, 429–436
Huttenlocher, A., 37–40
Huttenrauch, F., 4–5
Huttner, W.B., 358–359
Hwa, J., 2–3
Hwang, S.T., 157–159, 382–383, 558–559

I

Iga, M., 53*t*, 55*t*, 65, 66*t*, 69–70
Igarashi, T., 93–94
IJzerman, A.P., 156–159, 162, 422–423, 434–437
Ikeda, Y., 147
Illy, C., 425–427
Imai, T., 90–91, 93–94, 235
Imai, Y., 337–338
Imbach, A., 504–505
Imberty, A., 337–338
Inagaki, S., 6–7
Infantino, S., 90–91, 109, 111*f*
Ingber, D.E., 24–25
Ingersoll, M.A., 276–277
Ingram, J.R., 69, 234–235, 390–391, 393–395, 460–461
Iqbal, A.J., 496–498, 505
Irene del Molino del Barrio, 310–330
Irimia, D., 42
Irving, P.E., 161
Isaacs, J.D., 319–320
Ishida, T., 461–462
Ishigatsubo, Y., 93–94
Ishii, K., 93–94
Ishii, S., 270–271
Islam, S.A., 442
Ito, T., 270–271
Iwai, Y., 209, 212
Iwakura, Y., 209, 226, 228, 276
Iwata, S., 390–391
Izumi, M., 369–370, 381

J

Jackson, D.G., 336
Jacobs, J.P., 209, 226
Jacobs, S., 284
Jacquelin, S., 69–70
Jaggi, M., 339
Jahnichen, S., 461–462, 485
Jaiman, D., 390–391
Jain, S.K., 339
James, D., 234–235
Jamieson, T., 89
Jandik, K.A., 524

Janetopoulos, C., 24, 33–37, 34*f*, 38–39*f*
Jang, S., 501–502
Janssens, R., 519–520
Jarvius, J., 352–353
Jarvius, M., 352–353
Jasin, H.E., 266–267
Jastrzebska, B., 255–257
Jen, C.H., 360, 375, 380–382
Jenh, C.H., 465
Jenkinson, S., 460–461, 464–465, 487
Jensen, D.R., 382–383, 540–562, 569
Jensen, P.C., 165–166, 170, 479–481
Jeon, N.L., 21
Jeremic, D., 531–534
Jerman, J.C., 502–503
Ji, H., 208–209, 211
Jia, L., 7
Jia, W., 368, 381–382
Jian, Z., 208–209
Jiang, X., 24–25
Jilani, I., 453–454
Jin, H., 69, 383–384
Jochheim-Richter, A., 460
Joh, T., 461–462
Johansson, E.L., 339
John, A.E., 132–133
John, B., 161, 422–423
John, T.J., 20–21
Johns, S.C., 336–353
Johnson, I., 496–498
Johnson, K.S., 52
Johnson, L.A., 336
Johnson, M.A., 460–461
Johnson, R., 505
Johnson, S.K., 540–562
Johnson, Z., 73–74, 162, 262–263, 310–311, 337–338, 458–460, 519–520
Johnsson, K., 91–92, 106
Johnsson, N., 91–92, 106
Johnston, K., 519–520
Join-Lambert, O., 442–443
Joly, E., 133, 428–429
Jones, C., 475
Jones, M.L., 93–94
Jones, S.E., 540–541, 550, 555
Jongejan, A., 459, 464–465
Jorgensen, R., 465
Jude, K.M., 69, 234–235, 390–391, 393–395, 460–461
Julius, D., 170
Jung, S., 442–443
Jung, Y., 458–459
Junger, S., 132–133, 157–159, 163–164, 422–423

K

Kagiampakis, I., 69
Kajikawa, O., 6
Kalams, S.A., 337–338
Kalatskaya, I., 132–133
Kam, Y., 502
Kammermann, M., 504–505
Kamp, V.M., 442–443
Kanaoka, Y., 209–210
Kanniess, F., 264–265
Kanouchi, H., 310–311
Kantarci, A., 88–89
Kaptein, S.J., 467
Kar, S., 48–49, 460–461
Kar, U.K., 555, 558–559
Karasavvas, N., 93–94
Kardash, E., 89–91
Karlshøj, S., 156–180
Karpova, D., 282
Kataoka, N., 310–311
Kataura, M., 442
Katritch, V., 234–235, 384–385, 392, 460–461
Kattenhorn, L.M., 384–385
Kavelaars, A., 172
Kawaguchi, Y., 364, 381–382
Kawamoto, S., 93–94
Kawamura, T., 89–92, 172, 234–235, 391–394, 397–400, 399*f*, 433–434
Kay, B.K., 50–52
Kayal, S., 442–443
Kazmierski, W., 460–461, 464–465, 487
Ke, Y., 527
Keen, J.H., 282–284
Keim, P.S., 73–74
Kelay, P., 89, 294–295, 299, 310–311
Kelleher, N.L., 550
Kellenberger, E., 463–464
Kelly, L.M., 222–224
Kelvin, A.A., 164–165

Kenakin, T., 434–436, 460–461, 463–466, 482–483, 485–487
Kenakin, T.P., 157, 159–162
Kennedy, T.E., 569, 573–574
Kent, S.B., 90–91
Kenyon, A., 505
Kerr, I.B., 25–27
Kessler, W., 102
Kett, W.C., 6
Khan, I.M., 264
Khmelinskii, A., 89, 294
Kho, D., 505
Khorana, H.G., 2–3
Kiafard, Z., 299
Kickhoefer, V.A., 555, 558–559
Kiedrowski, M., 558–559
Kiermaier, E., 359, 382–383
Kiilerich-Pedersen, K., 294–295
Kilpatrick, L.E., 505–506
Kim, C.H., 392
Kim, D., 337–339, 350–353
Kim, J., 121, 132–133, 156, 160–161, 458–459
Kim, K.S., 90–91
Kim, M., 37–38
Kim, N.D., 208–230, 276
Kim, T., 37–38
Kim, Y.S., 527
Kindt, T., 534–535
Kinstrie, R., 90–91
Kipari, T., 270–271
Kiprilov, E., 359, 380–381
Kirby, J.A., 310–330
Kirk, C.J., 558–559
Kirkpatrick, P., 48–49, 460–461
Kirpotina, L.N., 500
Kitamura, K., 329
Kitaura, M., 93–94
Kitic, N., 518–537
Klarenbeek, A., 459–462
Klco, J.M., 391–392
Kledal, T.N., 156, 160–161, 164
Klein, A., 270–271
Klingler, D., 157, 172
Knapik, G.P., 427–428
Knaut, H., 89, 458–459
Knochel, W., 93–94
Knowles, I.W., 20–21
Kobayashi, H., 329
Kobilka, B., 132
Kobilka, B.K., 234–235
Kobilka, T.S., 234–235
Kocan, M., 143
Koch, A.E., 156
Koch, F., 577–578
Koch, K.C., 558–559
Koelink, P.J., 48, 459
Koenen, R.R., 156, 459–460
Kofuku, Y., 234–235, 393–394
Koh, Y., 464–465
Kohout, T.A., 132–133, 157–159, 163–164, 422–423
Kolakowski, L.F., 65–67
Kolattukudy, P.E., 93–94
Kolbeck, R., 170
Kolchinsky, P., 359, 380–381
Kollar, K., 460
Kondur, H.R., 384–385
Kong, X., 383–384
Kooistra, A.J., 148–149, 467, 492
Koplinski, C.A., 540–562
Kosco-Vilbois, M.H., 74, 189, 459–460
Kostenis, E., 163–164, 170
Kownatzki, E., 91–92
Kozasa, T., 164
Kraft, R., 132–133
Kralj, A., 459
Krangel, M.S., 555
Krause, D.S., 558–559
Krauss, G., 166
Krausslich, H.G., 11–12
Kremer, L., 461–462
Kremer, V., 558–559
Kremmer, E., 442
Krenn, E., 519–520
Kroegel, C., 270–271
Kruep, D., 524
Kruizinga, R.C., 282
Kuang, W.J., 264
Kubota, T., 461–462
Kuby, J., 534–535
Kuenzi, G., 53*t*, 54, 55*t*, 58, 62–63, 65, 66*t*, 68
Kufareva, I., 65, 69, 162, 234–235, 248, 248*f*, 360, 382–385, 390–416, 399*f*, 411*f*, 415*f*, 459–461, 540

Kufferath, I., 347
Kuhmann, S.E., 48
Kuhn, D.E., 93–94
Kuijer, M., 90–91, 499*f*
Kukutsch, N., 577–578
Kumagai, H., 147
Kumar, A.H., 69–70
Kumar, S., 394–395
Kumar, V., 123
Kumari, P., 390–391
Kumasaka, T., 234–235
Kunert, R., 69
Kungl, A.J., 518–537
Kunji, E.R.S., 390–391
Kunkel, S.L., 270–271
Kunze, A., 89, 294
Kuritzkes, D., 48–49, 460–461
Kurogi, K., 364, 381–382
Kuroki, R., 465
Kurt, E., 459
Kuschert, G.S., 519–520
Kuschert, G.S.V., 310–311
Kuszewski, J., 383–384
Kuta, E.G., 394–395
Kuusisto, L.M., 459
Kwasny, D., 294–295

L

Lagane, B., 90–91, 111*f*, 422–424, 427–428
Lagerstrom, M.C., 3–4, 234–235
Laguri, C., 90–91
Lai, M., 487–488, 505
Laird, M.E., 276–277
Lakowicz, J.R., 531–534
Lam, B.K., 209–210
Lam, C., 91–92, 121
Lam, C.M., 132–133, 147–149, 159–161, 434–436, 458–459
Lam, P.C.-H., 391–392
Lam, V., 491–493
Lamanna, W.C., 337–339
Lambeir, A.-M., 329
Lambert, N.A., 162, 166
Lambros, T., 276–277
Lämmermann, T., 209–210, 221–224, 295–296, 299, 568
Lammers, J.W., 442–443
Landay, A., 68
Landegren, U., 347
Lane, J.R., 160
Lane, W.S., 360, 381–382
Lang, R., 442
Laporte, S.A., 141
Laranjeira, L.P.M., 330
Lark, M.W., 160
Larsen, N.B., 178–179
Larsen, O., 156–161, 422–423
Lau, E.K., 73–74, 262–263, 310–311, 337–338, 458, 519–520
Laus, R., 243
Lautenschlaeger, F., 568–569, 575–577
Lawrence, R., 338–341, 350–353
Lazarides, A.A., 255–257
Le Berre, M., 568–569, 575–577
Le Brocq, M.L., 6
Le Gouill, C., 133–134, 143, 429, 431–433
le Maire, M., 2–3
Le Minh, P.N., 465
Leary, J.A., 360, 368, 375, 380–382
Lebecque, S., 336
Lebel-Haziv, Y., 519–520
Lebon, G., 234–235
Lechuga, L.M., 14
Lederman, M.M., 68
Leduc, M., 133–134, 143
LeDuc, R.D., 550
Lee, B.-K., 392
Lee, D., 161, 422–423
Lee, I., 474
Lee, J., 264
Lee, J.J., 423–424
Lee, J.M., 558–559
Lee, R.T., 501–502, 542–543, 558–559
Leff, P., 434–436
Lefkowitz, R.J., 3–4, 132, 141, 147–148, 157–159, 164–165, 234–235, 382–383, 423–424, 431, 434, 436–437
Legault, M., 428–429
Legler, D.F., 310–311, 540, 567–569
Leick, M., 89
Lemos, H.P., 266–267
Lenaerts, J.-P., 393–394
Lenig, D., 567–568
Lennard, T.W.J., 322
Lennartz-Walker, M., 481

Lennertz, R., 558–559
Leonard, W.J., 383–384
Lerchenberger, M., 310–311
Lesuffleur, A., 3
Lettieri, C., 540–562
Leuchowius, K.J., 352–353
Leurs, R., 156, 160–161, 164, 235, 390–391, 422–423, 458–506, 504*f*
Levacher, B., 69
Levoye, A., 90–91, 112*f*
Lewellis, S.W., 89
Lewin, S., 121
Lewin, S.A., 90–91
Lewis, G.K., 74
Leyte, A., 358–359
Li Jeon, N., 21
Li, C.M., 42
Li, D., 37–38
Li, F., 3–4, 422–423
Li, H.D., 93–94
Li, J., 234–235, 360, 460–461
Li, P., 69
Li, R., 382
Li, W., 359–360, 380–381, 383–384
Li, X., 91–92
Li, Y.M., 282, 290
Li, Z., 102, 243
Liang, C.C., 502
Liberati, D., 264–265
Licata, F., 69–70
Liehn, E.A., 90–91, 558–559
Liekens, S., 89–91
Lievens, D., 90–91
Lifson, J.D., 68
Lim, H.D., 465, 500
Lim, L.S., 42
Lima, S.H., 93–94
Limberg, B.J., 132–133
Limbird, L.E., 3–4
Lin, A.J., 91–92, 100–102, 106, 112*f*
Lin, C.H., 369–370, 381
Lin, F., 21, 161, 422–423
Lin, F.T., 4–5
Lin, S.W., 360
Lin, W.H., 37–38
Lin, X., 558–559
Linderman, J.J., 6–7
Lindley, I., 91–92
Lindquist, N.C., 3
Ling, B., 68
Ling, L., 243
Linhardt, R.J., 337–338, 524, 526–527
Lipfert, J., 132–133
Lipp, M., 442
Lippner, D.R., 359, 382–383
Lira, S.A., 4
Littman, D.R., 442–443
Liu, C., 491
Liu, C.C., 384–385
Liu, D., 394–395, 558–559
Liu, F., 91–92
Liu, J., 4, 234–235
Liu, Q., 3
Liu, R., 459
Liu, W., 234–235, 384–385, 390–391, 460–461
Liu, X., 243, 380–381
Liu, Y., 23–27, 23*f*, 30*f*, 32–33*f*, 40, 382
LiWang, P., 73–74, 262–263, 310–311, 337–338, 519–520
Liwang, P.J., 69
LiWang, P.J., 383–384
Llorente, J., 157
Llorente, M., 461–462
Locati, M., 4, 89, 164, 294, 422–437, 442–454
Lodi, P.J., 383–384
Lodowski, D.T., 234–235
Loening, A.M., 143
Loetscher, M., 160–161
Loetscher, P., 93–94, 160–161, 177–178, 360, 460, 542–543, 567–568
Logan, M.K., 496–498
Loghmani, F., 270–271
Lohse, M.J., 113–114, 143
Lokuta, M.A., 37–38
Lolis, E., 49–50
Looi, X.L., 42
Loos, T., 393–394, 542–543
Lopes, F., 269, 277
Lopez, L., 160
Lopez, R., 474
Lortat-Jacob, H., 337–338
Losy, J., 156
Louedec, L., 459–460
Love, M., 382–383, 540–541, 555, 569

Lowe, J.B., 310–311
Lowell, C.A., 208
Lowman, H.B., 50–52
Lu, L., 93–94
Lu, Q., 542–544
Lu, Z., 91–92, 100–101
Lucas, A., 93–94
Lucas, P., 2–15
Luckow, B., 447
Ludden, P.J., 368, 381–382
Ludeman, J.P., 358–361, 382–384
Luini, W., 442
Lukacs, N.W., 266–267
Lukas, C., 329
Lukas, J., 329
Luker, G., 120–122
Luker, G.D., 89, 120–128
Luker, K.E., 89–91, 120–128
Lundgren, A., 339
Lundin, L.G., 3–4, 234–235
Lundqvist, A., 558–559
Luongo, J., 500
Lusso, P., 360, 382–383
Luster, A.D., 4, 48–49, 88–89, 91–92, 100–101, 132–133, 156, 208–230, 262–263, 276, 390, 442, 458
Lustig, K.D., 170
Lusti-Narasimhan, M., 49–50
Luther, R., 219
Lüttichau, H.R., 160–161, 164
Luttrell, L.M., 164–165
Lutz, M.B., 577–578
Lyddiatt, A., 243
Lytle, B.L., 545

M

Ma, L., 164–165
Ma, S.H., 69–70
Maass, K., 529
MacDonald, C., 505
Macdonald-Bravo, H., 93–94
MacGillivray, C., 542–543, 558–559
Mack, M., 442–454
Mackay, C.R., 4, 567–568
Mackay, F., 88–89, 91–92
Mackay, I.R., 266–267
Madaan, A., 339
Madani, N., 394–395
Maeda, D.Y., 500
Maeda, H., 329
Maeda, K., 464–465
Maeda, N., 4
Maeda, Y., 465
Magistrelli, G., 74, 189
Magnussen, H., 264–265
Mahabaleshwar, H., 89–93, 113
Mahajan, S., 93–94
Maheswaran, S., 42
Mahmood, U., 208–209, 211
Maia, M.B.S., 330
Maillard, L., 459–460
Mak, J.Y., 65–67, 555
Makhoul, J.A., 391–392
Malagoli, B.G., 269
Malik, G., 319–320
Malik, R., 282–284
Malins, L.R., 380–381
Malmgaard-Clausen, M., 170
Malouf, C., 139*f*, 140
Mamdouh, Z., 310–311
Mamon, E., 519–520
Mancardi, S., 276
Mancera, R.L., 518–520
Manfredi, J.P., 464–465
Manice, L.A., 132–133
Manini, J., 461–462
Mantovani, A., 4, 89, 156–157, 164, 294, 393–394, 422–423, 442
Marangoni, F., 221
Marchese, A., 172, 282–290
Marelli-Berg, F., 310–311
Marinissen, M.J., 162
Markley, J.L., 378
Marodon, G., 69
Marsh, M., 465
Martellucci, S.A., 161
Martin, K., 69–70
Martin, R., 89–91
Martinez, A.C., 6–7
Martinez-Martin, N., 89–91
Martinez-Munoz, L., 2–15
Martini, L., 160–161, 465
Martinius, H., 69
Martins-Green, M., 458
Masi, T.J., 93–94
Masri, B., 491–493

Massague, J., 282
Massara, M., 442–454
Mathew, E., 505–506
Mathis, D., 209, 211, 226
Matsuo, M., 464–465
Matthies, H.J., 133–134, 142–143, 146
Matthiesen, M.E., 37–38
Mattioli, I., 160–161, 460
Mattmiller, J.E., 91–92, 235, 542
Maudsley, S., 164–165
Mauriz, E., 14
Maussang, D., 49–50, 156, 393–394, 459–462, 479–481, 485
Mavin, E.R., 330
May, L.T., 493
Mayadas, T.N., 208
Maynard, J.A., 3
Mazzon, C., 453–454
McArdle, C.A., 157
McArdle, C.S., 25–27
McCafferty, J.D., 50–52
McColl, S.R., 65, 248, 248*f*, 339
McCoy, J., 91–92, 100–101
McCulloch, C.V., 6, 140
McDevitt, P.J., 542–544
McDonald, B., 209–210
McDonnell, J.M., 6
McEver, R.P., 359
McGovern, K.W., 164–165
McInnes, I.B., 90–91
McKeating, J., 383–384
McLean, K.A., 160–161
McLean, M.A., 255–257
McLean, P., 89
McLoughlin, S.M., 91–92
McNeill, E., 458–459, 488, 505
McPherson, J., 157
Meddahi-Pelle, A., 459–460
Meiser, A., 481
Melikian, A., 121, 257
Melikyan, G., 69
Mellado, M., 2–15
Mellor, P., 322
Melotti, A., 48–70, 53*t*, 55*t*, 66*t*
Menezes, G.B., 209–210
Menge, W.M., 459
Mercalli, A., 264–265
Merchant, Z.M., 527
Mercier, J.F., 146–147
Mertens, K., 358–359
Merzaban, J.S., 359
Meshel, T., 519–520
Met, O., 178–179
Meucci, O., 148–149
Meyer, J., 163–164
Miao, Z., 121, 257
Michel, H., 390–391, 465
Michnick, S., 121
Middleton, J., 89, 92–93
Mierke, D.F., 391–392
Mihalko, L., 121
Mihalko, L.A., 89–91
Mikulski, Z., 337–339, 350–353
Milasta, S., 140
Mileni, M., 251–253, 412–413
Millard, C.J., 360, 382–383
Millecamps, M., 132, 147–148
Miller, D.L., 391–392
Miller, L.J., 160, 392, 434–436, 466
Miller, M., 310–311
Miller, M.D., 555
Miller, W.E., 164–165
Milligan, G., 140, 163–164
Milliken, D., 311
Milling, S.W., 89–91
Minina, S., 89–91
Minisini, R., 459
Mirolo, M., 89
Mirzabekov, T., 359, 380–381
Misra, R.S., 423–424
Mitchell, G.F., 266–267
Mitchell, R.N., 501–502
Mittal, S., 21
Miura, N.N., 215–216
Miyabe, C., 208–230
Miyabe, Y., 208–230
Miyakawa, T., 464–465
Miyamoto, D.T., 42
Mnie-Filali, O., 132, 147–148
Moechars, D., 479–481
Moelants, E.A.V., 159
Moepps, B., 88–114, 108*f*, 111*f*
Mohammed, K.A., 270–271
Mohaupt, M., 299
Mok, S.C., 4
Mokrosinski, J., 479–481

Mol, C.D., 234–235, 384–385, 460–461
Moller, J.V., 2–3
Monach, P.A., 209, 211
Monaghan, P., 391–392
Monkley, S.J., 568
Montaner, S., 390, 459
Montes, M., 393–394
Montoya, A., 14
Montpas, N., 132–150
Moore, K.L., 358–361, 368, 374–375, 380–382
Moots, R.J., 208–209
Moraes, I., 390–391
Moresco, J., 294–295
Moriconi, A., 264–265
Morizumi, T., 255–257
Morrow, V., 140
Mortier, A., 157–159, 263, 276–277, 393–394, 542–543
Mortrud, M.T., 3–4
Moseman, E.A., 337–338
Moser, B., 90–91, 177–178
Mosier, D.E., 68
Mösl, R., 524
Motoki, M., 93–94
Motoshima, H., 234–235
Moulon, C., 90–91
Moussion, C., 310–311, 567–569
Mu, L., 37–38
Mubarak, K.K., 270–271
Mudde, L., 89, 92–93
Mueller, D., 558–559
Mueller, W., 88–89, 91–92
Muff, R., 89–91, 93–94, 109–110
Muir, T.W., 90–91
Mujic-Delic, A., 461–462
Mule, J.J., 558–559
Muljadi, R.C., 208–209
Muller, A., 4, 282, 360
Müller, A., 311
Muller, S.M., 145–146
Muller, W.A., 310–311, 317–318
Mulligan, M.S., 93–94
Munch, J., 460
Mundell, S.J., 157, 282
Mungalpara, J., 170
Muniz-Medina, V., 434–436
Munk, S., 465
Munoz, E., 521
Munoz, L.M., 6–7
Munteanu, M., 66*t*, 68–69
Murdoch, C., 4, 390
Murooka, T.T., 221
Murphy, J.E., 157, 172
Murphy, K., 329
Murphy, P.M., 4, 89, 156, 161, 163–164, 264, 294, 394–395
Murray, N.J., 540–541
Muslmani, M., 74, 189
Muyldermans, S., 461–462
Muzio, V., 459–460
Myszka, D.G., 2, 5–7

N

Naemi, F.M.A., 329–330
Nagai, K., 329
Naganathan, S., 392
Nagaraj, V.J., 3
Nagel, F., 284
Nagengast, W.B., 282
Nagrath, S., 42
Naik, S., 121–122
Nakajima, T., 93–94
Nakamura, E., 310–311
Nakao, Y., 329
Nakata, H., 464–465
Nalesso, G., 487
Namkung, Y., 429, 431–433
Nandagopal, S., 161, 422–423
Napimoga, M.H., 266–267
Narazaki, M., 542–543
Narula, S.K., 465
Nascimento, D.C., 271
Nashaat, Z.N., 393–395
Nasreddine, S., 69–70
Nasreen, N., 270–271
Naumann, U., 89–93, 113
Naus, S., 359
Navarro, G., 6–7
Navenot, J.-M., 161
Navis, M., 160–161, 164
Navratilova, I., 5–7, 12–14
Nebuloni, M., 89, 92–93
Nedellec, R., 68
Nederpelt, I., 481
Negrete-Virgen, J.A., 243

Nelson, R.D., 20–21
Neote, K., 65–67, 555
Neptune, E.R., 488, 496–498
Neutze, R., 390–391
Nevins, A.M., 540–541, 550, 555
New, D.C., 422–423
Newstead, S., 390–391
Newton, A., 219
Nguyen, C.M., 21
Nguyen, D.H., 235
Nguyen, P., 490*f*, 492*f*, 502*f*
Nibbs, R., 89, 294
Nibbs, R.J., 89–91, 294, 442
Nice, E., 123*f*
Nicholas, J., 491
Nicholas, S.L., 132–133, 157–159, 163–164, 422–423
Nichols, D.E., 132
Nicholson, C., 89
Nickoloff, B.J., 558–559
Nickols, H.H., 3–4
Niehrs, C., 358–359
Nielsen, M.C., 170
Nigrovic, P.A., 209
Nijmeijer, S., 148–149, 235, 390–391, 460, 465, 467, 492
Nikiforovich, G.V., 391–392
Nikolaev, V.O., 491–493
Ning, X., 208–209
Nitzki, A., 4–5
Niwa, R., 461–462
Nixon, C., 89
Nobles, K.N., 147–149
Nogueira, C.R., 270–271
Nomanbhoy, T.K., 534
Nomura, T., 558–559
Nomura, W., 505–506
Noppen, S., 90–91, 393–394
Norskov-Lauritsen, L., 427–428
Northup, J.K., 6–7
Nothnagel, H.J., 234–235
Nourshargh, S., 442–443
Novagen, 545
Novick, S., 434–436
Nowak, M., 90–91
Noyon, C., 220
Nuber, S., 143
Nufer, O., 542–543
Nygaard, R., 170
Nzeusseu Toukap, A., 220

O

O'Hayre, M., 310–311
Oakley, R.H., 141
O'Boyle, G., 330
O'Carroll, S.J., 505
O'Cearbhaill, E.D., 37–38
Odde, S., 264–265
Odemis, V., 132–133
Odobasic, D., 208–209
Offord, R.E., 49–52, 53*t*, 54, 55*t*, 58, 62–63, 65, 66*t*, 68
Ogata, H., 464–465
Ogilvie, A.L., 577–578
O'Hara, M., 140
O'Hayre, M., 156, 163–164, 168, 540
Ohba, K., 505–506
Ohl, L., 299, 442
Ohmura, K., 208–209, 211
Ohyama, T., 329
Oishi, S., 139*f*, 140
Okamoto, T., 310–311
Okubo, T., 93–94
Oldenburg, K.R., 482
Oldfield, S.F., 264
Oldham, W.M., 162, 489–491
Oligny-Longpre, G., 132
Oliveira, R.D., 271
Olsen, M.H., 178–179
Olson, W.C., 360
Olsson, C., 352–353
Olwell, P., 501–502
Omagari, A., 161
Onorato, J.J., 465
Opella, S.J., 254–255
Oppermann, M., 4–5
Orry, A., 392
Orsini, M.J., 282
Ortiz, A.M., 14
Ortiz-Lopez, A., 209, 226
Osborne, B., 534–535
O'Shannessy, D.J., 5
Oskarsson, T., 282
Ostuni, E., 24–25, 569
O'Sullivan, K.M., 208–209
Otto, C., 459

Ouyang, Q., 37–38
Ouyang, Y.B., 360–361, 381–382
Overbeek, S.A., 48, 459
Overington, J.P., 3–4
Overney, J., 294–295
Owen, P., 50–52
Ozawa, T., 3–4

P

Padmanabhan, S., 161
Paing, M.M., 172
Pakianathan, D.R., 394–395
Palczewski, K., 234–235, 255–257
Pals-Rylaarsdam, R., 465
Pan, M., 121–122
Pan, Y., 282, 290
Panavas, T., 500
Pancera, M., 6
Pande, J., 188
Panitz, N., 90–91
Pankov, R., 329
Papy-Garcia, D., 519–520
Pardo, L., 464–465
Pardon, E., 391–392
Parent, J.L., 282
Parent, S., 428–429
Paris, H., 132, 429
Parish, C.R., 310–311, 519–520
Park, A.Y., 502
Park, C.K., 5
Park, P.S., 255–257
Park, S.H., 234–235, 254–255
Parkhill, A.L., 429
Parmentier, M., 12, 132–133, 140, 147, 159–160, 422–423, 429–436, 465
Partida-Sanchez, S., 423–424
Paslin, D.A., 93–94
Pasqualini, F., 89, 92–93
Patel, D.D., 164–165, 235
Patel, J., 458–459, 488
Patel, P., 481
Patterson, G.H., 145–146
Paulmurugan, R., 121
Paulsson, J., 352–353
Payne, R.J., 371–372, 380, 382–385
Pease, J.E., 394–395
Peck, A.M., 500
Peck, K., 5
Pei, G., 164–165
Pelayo, J.-C., 157, 172
Pellegrini, S., 264–265
Pellegrino, A., 160–161
Pellequer, J.-L., 395
Pelletier, J., 133–134, 428–429, 431–433
Pelletier, M.E., 132–133, 140–141, 142*f*, 147–150, 422–423
Pennings, E.J.M., 369
Percherancier, Y., 133–134, 142–143, 429
Perez-Bercoff, D., 55*t*, 62–63, 65, 66*t*, 67–69
Perroy, J., 434
Perry, S.J., 132–133, 157–159, 163–164, 422–423
Persuh, M., 360
Pertel, T., 113–114
Peterson, F.C., 91–92, 235, 358, 360–361, 364, 372–378, 380–383, 397–400, 540–562
Petit, N., 69
Petitou, M., 534, 534*f*
Petley, T., 500
Petreaca, M., 458
Petrosiute, A., 48
Petryniak, B., 337–338
Pfeifer, M.J., 424
Pfister, R.R., 20–21
Pfleger, K.D., 143, 428–429
Pflicke, H., 568
Pham, V., 392
Phillips, R.M., 481
Piatak, M., 68
Piccinini, A.M., 459–460, 519–520
Pichon, J., 341
Pick, H., 91–92, 106
Piel, M., 568–569, 575–577
Pienta, K.J., 458–459
Pierce, K.L., 164–165, 234–235
Pietras, K., 352–353
Pillarisetti, K., 393–394
Pillay, J., 442–443
Pinho, V., 264–267, 269–271, 277
Pinon, D.I., 391–392
Pinto, L.G., 271
Piston, D.W., 145–146
Pitcher, J.A., 431
Pitt, E.A., 93–94

Pittman, K., 209–210
Piwnica-Worms, D., 121–122
Piwnica-Worms, H., 121–122
Pla, R., 88–89
Plachy, J., 447
Planaguma, A., 474
Plettenberg, A., 460–461
Poget, S.F., 254–255
Poh, Y.K., 37–38
Pont, M., 474
Pontier, S., 429
Poole, D.P., 157
Popovic, Z.B., 558–559
Postea, O., 90–91
Poupel, L., 53*t*, 55*t*, 65, 66*t*, 69–70
Powell, A.K., 525
Power, C., 187–205
Power, C.A., 49–50, 156, 188, 310–311, 460–461, 519–520, 536–537
Pozniak, A.L., 460–461
Premont, R.T., 164–165, 234–235
Prestegard, J.H., 337–338
Prevot, S., 69–70
Prina-Mello, A., 501–502
Princen, K., 89–91, 170
Projahn, D., 90–91
Proost, P., 157–159, 263, 276–277, 329, 393–394, 542–543
Prossnitz, E.R., 164–165
Proudfoot, A.E.I., 4, 6, 49–52, 73–85, 90–92, 156–157, 162, 187–205, 262–263, 310–311, 337–338, 436–437, 447, 458–461, 519–520, 536–537, 542
Pruenster, M., 89–93, 113
Prusoff, W.H., 477–478
Pulli, B., 215–216, 220
Pye, D., 524
Pyen, D.S., 500

Q

Qian, W., 42
Qin, L., 69, 89–92, 234–235, 382, 384–385, 390–416, 399*f*, 411*f*, 415*f*, 460–461
Qin, S., 163–164
Quach, T., 93–94
Quaranta, V., 502
Queiroz-Junior, C.M., 269
Quiding-Jarbrink, M., 339
Quie, P.G., 20–21
Quinn, M.T., 500
Quirin, C., 89, 294

R

Rabenstein, D.L., 337–338, 518–519
Radresa, O., 429, 431–433
Raiborg, C., 282–284
Rajagopal, K., 164–165
Rajagopal, S., 121, 132–133, 147–149, 157–161, 422–423, 434–437, 458–459
Rajarathnam, K., 90–91, 360, 542–543
Ramamoorthy, A., 254–255
Raman, D., 48, 459
Rammos, C., 558–559
Ranchalis, J.E., 3–4
Randolph, G.J., 336, 442–443
Ransohoff, R.M., 390, 422
Raschle, T., 255–257
Rasmussen, S.G., 234–235
Rasmussen, S.G.F., 391–392
Rat, P., 341
Raulet, D.H., 558–559
Ravn, U., 74, 189
Ray, P., 89, 121
Raz, E., 89–93, 113
Reddi, K., 339–341, 352–353
Redman, S.N., 264
Reeves, P.J., 2–3
Regan-Komito, D., 505
Reichert, J.M., 461–462
Reichert, W.M., 569
Reichman-Fried, M., 89–91
Reilander, H., 465
Reilly, D., 50–52
Reinhart, G., 132–133, 157–159, 163–164, 422–423
Reis e Sousa, C., 567–568
Reisine, T., 425–427
Rek, A., 519–520, 524, 531–534
Remick, D., 330
Remy, I., 121
Ren, J., 69–70
Renkawitz, J., 568
Rennard, S.I., 264–265
Rettig, M.P., 48–49
Revankar, C.M., 164–165
Reykjalin, E., 93–94

Reynolds, E., 89
Rhee, S.W., 21
Ribeiro, S., 235
Ribka, E.P., 68
Ricart, B.G., 161, 422–423
Rice, A.J., 255–257
Rice, G.C., 264
Rice, K.G., 527
Rich, R.L., 2
Richardson, R.M., 164–165
Richmond, A., 20–42, 22–23*f*, 30*f*, 32–33*f*, 41*f*, 48, 459
Richter, R., 460
Ridderstrale, K., 352–353
Ridley, A.J., 163–164
Ries, J., 90–91
Rinas, U., 102
Rinken, A., 474
Ritchie, T.K., 255–257
Rivero, G., 157
Robertson, H., 319–320
Robia, S.L., 282
Robinson, J., 359, 459–460, 519–520
Robinson, M.R., 374–375
Roby, P., 425–427
Roche, M., 380–381
Rocheville, M., 502–503
Rode, R.R., 382–383, 540–541, 555, 569
Rodero, M., 53*t*, 55*t*, 65, 66*t*, 69–70
Rodgers, S.D., 359
Rodrigues, D.H., 264–267, 269, 277
Rodriguez, M., 3
Rodriguez-Frade, J.M., 2–15
Rodriguez-Panadero, F., 270–271
Roederer, M., 445–446
Roes, I., 37–38
Roffe, E., 264–265
Rogers, L., 93–94
Rogers, M., 20–42
Rogowski, M., 360, 382–383
Roland El Ghazal, 336–353
Rollins, B.J., 390
Romagnani, P., 4
Romani, N., 577–578
Rominger, D.H., 159–160, 434–436
Ronsse, I., 519–520, 542–543
Rosenbaum, D.M., 234–235
Rosenbaum, J.S., 132–133
Rosenblatt, M., 391–392
Rosenkilde, M.M., 156–180, 422–423, 460, 479–481
Rossi, D., 4
Rossitto-Borlat, I., 48–70, 66*t*
Rössner, S., 577–578
Rot, A., 73–74, 89, 140, 163–164, 294, 310, 336–338, 458–459, 540
Roth, C.B., 234–235
Roudabush, F.L., 164–165
Rouleau, N., 425–427
Roumen, L., 49–50, 156, 393–394, 460, 479–481, 487–488, 505
Rousseau, F., 188
Rozlosnik, N., 294–295
Rucker, J., 7
Rueda, P., 463–464
Rummel, P.C., 170
Ruscetti, F.W., 25–27
Russell, L.J., 487
Russo, R.C., 89, 92–93, 264–265
Ryseck, R.P., 93–94

S

Saadi, W., 21
Sachpatzidis, A., 464–465
Sachs, D., 264–267, 269, 277
Sackmann, E.K., 329–330
Sadik, C.D., 209, 226, 228, 276
Sadir, R., 337–338
Sai, J., 20–42, 22–23*f*, 30*f*, 32–33*f*, 41*f*
Saini, V., 393–395
Saito, H., 337–338
Sakai, Y.I., 330
Sakmar, T.P., 358–385, 371*f*, 392
Salahpour, A., 133, 146–147, 491–493
Salanga, C.L., 65, 156, 162–164, 168, 172, 248, 248*f*, 310–311, 336–353, 360, 382, 393–394, 433–434, 540
Sallusto, F., 442, 567–568
Salogni, L., 90–91
Salon, J.A., 234–235
Salter, R.D., 555
Salvucci, O., 542–543
Sambrook, J., 322–323
Samrakindi, M., 121–122
Sanchez, A., 14
Sanchez-Alcaniz, J.A., 88–89

Sanchez-Weatherby, J., 390–391
Sandberg, J.L., 382–383, 540–541, 555, 569
Sanders, K.L., 270–271
Sandford, G., 491
Sanfiz, A., 358–385, 371*f*
Sanni, S.J., 465
Santini, F., 282–284
Santo, N.V., 235, 360, 380–381, 383–384
Santulli-Marotto, S., 500
Sarrazin, S., 337–339
Sarris, A.H., 93–94
Sato, M., 147
Satoh, M., 461–462
Sauliere, A., 132, 143, 429
Saunders, M., 459–462
Savino, B., 89, 92–93, 157–159, 422–423, 442
Sawatzky, D.A., 270–271
Saxena, A., 558–559
Scammells, P.J., 160
Schaak, S., 429
Schaefer, W.M., 558–559
Schaerli, P., 567–568
Schall, T.J., 65–67, 156–157, 436–437, 459–460, 555
Schallmeiner, E., 352–353
Schaniel, C., 567–568
Schenkel, A.R., 310–311
Scherr, M., 10
Schertler, G.F., 234–235
Schiering, C., 90–91
Schioth, H.B., 3–4, 234–235
Schmidt, B., 121
Schmidt, B.T., 90–91
Schmit, J.-C., 393–394
Schneider, J., 145–146
Schnitzler, C.E., 359–360, 380–381, 383–384
Schols, D., 89–91, 161, 329
Scholten, D.J., 49–50, 156, 393–394, 422–423, 460–461, 474*f*, 479–481, 487–488, 490*f*, 492*f*, 502*f*, 504–505, 504*f*
Schour, L., 93–94
Schroeder, J., 558–559
Schuler, A.D., 500
Schult-Dietrich, P., 69
Schultz, P.G., 384–385, 392
Schulz, S., 88–89, 284
Schumann, K., 568
Schutyser, E., 393–394
Schutz, D., 91–92
Schwartz, T.W., 156, 160–161, 164, 170, 465
Schwarz, J., 310–311, 567–581
Schwarz, M.K., 156, 460–461
Scott, M.G., 147–148
Scotton, C.J., 311
Seamon, K.B., 168
Seddon, A.M., 2–3
Sedgwick, A.D., 330
Seeber, R.M., 132–133, 143
Segers, V.F., 542–543, 558–559
Seibert, C., 235, 358–385, 371*f*, 397–400
Seidel, T., 145–146
Seifert, J.M., 91–92
Seifert, R., 433–434
Seifried, E., 460
Seki, E., 337–338
Senard, J.M., 143
Sengupta, R., 148–149
Seo, W., 359
Sequist, L.V., 42
Seung, E., 209–210, 215–216, 226
Sexton, P.M., 159–160, 392
Shannon, L., 422–423
Sharath, M.D., 527
Sharma, S., 558–559
Sharp, J.S., 337–338
Shaw, J., 188
Shaw, R.M., 558–559
Shcherbo, D., 121
Shen, X., 383–384
Shenoy, S.K., 3–4, 132, 141, 147–148, 164–165, 431, 434
Shepard, L.W., 164
Shi, G., 423–424
Shibayama, S., 464–465
Shieh, A.C., 294–295
Shimizu, K., 501–502
Shimizu, T., 270–271
Shiozawa, Y., 458–459
Shitara, K., 461–462
Shoham, M., 361
Shonberg, J., 160
Shu, W., 282

Shukla, A.K., 132, 141, 147–149, 390–391, 434
Si, G., 37–38
Siani, M.A., 329
Siderius, M., 156, 459
Sidhu, S.S., 50
Sieber, O.F., 20–21
Sielaff, I., 6, 74
Sierra, M.D., 542–543
Silbert, J.E., 518–519
Simmons, R.L., 20–21
Simonis, C., 447
Simpson, C.V., 140
Simpson, L.S., 360, 382–383
Simson, L., 519–520
Singer, M.S., 359
Singh, A.T., 339
Singh, R.K., 558–559
Sionov, R.V., 442–443
Siu, F.Y., 392
Sivaraman, J., 123
Sixt, M., 295–296, 299, 567–581
Skach, A., 147
Skelton, N.J., 394–395
Skerlj, R.T., 161
Slack, R.J., 487
Slagsvold, T., 282
Sligar, S.G., 255–257
Slight, I., 133–134, 142–143
Slocombe, P.M., 422–423
Smit, M.J., 48–50, 93–94, 148–149, 156, 160–161, 164, 235, 390–391, 393–394, 458–506
Smith, B.L., 540–562
Smith, E.W., 382
Smith, M., 121–122
Smith, S., 422–423
Smith, T.H., 89–92
Smits, K.M., 540–562
Smyth, J.W., 558–559
Sobolik-Delmaire, T., 48, 459
Soderberg, O., 352–353
Sodroski, J., 5–7
Soejima, M., 364, 381–382
Soergel, D.G., 160
Sogah, D., 235, 360, 380–381, 383–384
Sogawa, Y., 329
Soh, S., 427–428
Sohy, D., 12
Sokol, C.L., 4, 48, 208, 262–263, 276, 390, 442, 458
Sokolenko, S., 384–385
Solheim, J.C., 540–562
Somma, P., 442–454
Sommers, C.I., 20–21
Soria, G., 519–520
Soriano, S.F., 6–7
Soto, H., 4, 282, 311, 360
Soubias, O., 3
Souto, F.O., 271
Souza, A.L., 264–265
Sozzani, S., 4, 442
Spector, S.A., 9
Speerschneider, T., 465
Spieth, K., 93–94
Sprenger, J.U., 491–493
Springael, J.Y., 12, 132–133, 140, 147, 159–160, 422–423, 429–436, 460, 465
Springer, T.A., 4
Sridharan, R., 505–506
Srinivasan, R.S., 338–339, 350–353
Sriramarao, P., 262–263
Srivastava, M.K., 555, 558–559
Stallaert, W., 502–503
Stamp, G., 311
Standker, L., 460
Stanta, G., 276
Staren, D.M., 393–395
Stark, K., 310–311
Starks, D., 68
Staropoli, I., 463–464
Staszewski, S., 460–461
Steele, J.M., 89
Steen, A., 156–161, 170, 422–423, 479–481
Stein, T., 5
Steinman, R.M., 567–568
Stenlund, P., 5–6
Stenmark, H., 282–284
Stephens, B., 89–92, 382–383
Stephens, B.S., 234–235, 391–394, 397–400, 399*f*
Sterjovski, J., 380–381
Stevens, R.C., 234–235, 251–253, 390–391, 412–413
Stewart, P.D.S., 390–391

Steyaert, J., 234–235
Stigter-van Walsum, M., 461–462
Stoddart, L.A., 505–506
Stone, M.J., 358–361, 371–372, 380, 382–385
Stone, S.R., 358–359
St-Onge, G., 132–133, 140–141, 142*f*, 147–150, 422–423
Storch, M., 569
Stortelers, C., 461–462
Stott, S.L., 42
Straight, P.D., 91–92
Strange, P.G., 429, 491
Stremler, M., 21–23, 22*f*, 25–27
Strieter, R.M., 4, 270–271
Strong, A.E., 90–91
Struthers, R.S., 132–133, 157–159, 163–164, 422–423
Struyf, S., 90–91, 157–159, 161, 329
Stumm, R., 284
Stunnenberg, A., 474*f*, 481, 505
Su, J., 558–559
Suffee, N., 459–460
Sugumaran, G., 518–519
Sukman, A.J., 542–544
Summers, B., 121
Summers, B.C., 257
Sun, J., 37–38, 132–133
Sun, L., 69–70
Sun, P., 521
Sun, T., 392
Sun, Y., 160–161, 164–165
Sun, Y.P., 88–89
Suomalainen, M., 7
Surfus, J., 38–40
Sutherland, E.R., 264–265
Sutherland, J.N., 3
Suwa, Y., 93–94
Svane, I.M., 178–179
Svendsen, W.E., 294–295
Svennerholm, A.M., 339
Swanberg, S.L., 91–92, 100–101
Swartz, M.A., 294–295, 336
Sweet, R.G., 330
Swinney, D.C., 474
Szczuciński, A., 156
Szekanecz, Z., 156
Szewczyk, M.M., 188

T

Tacke, F., 442–443
Tafuri, W.L., 266–267
Tager, A.M., 209–210, 215–216, 226
Tai, H.H., 393–394
Tak, T., 442–443
Takakura, H., 3–4
Takaoka, Y., 464–465
Takayama, S., 24–25, 569
Takekoshi, T., 157–159, 382–383, 558–559
Talbot, J., 271
Taleski, D., 382–383
Talmadge, J.E., 558–559
Talvani, A., 266–267
Tamamura, H., 133–134, 142–143, 161
Tan, C.M., 3–4
Tan, J.H., 360, 382–384
Tan, M., 282, 290
Tan, Q., 234–235, 360, 460–461
Tanabe, Y., 505–506
Tanaka, H., 93–94
Tanaka, M., 3–4
Tanaka, T., 505–506
Tani, K., 93–94
Tanino, Y., 6
Tarbashevich, K., 90–91
Tate, C.G., 234–235
Taub, D.D., 235
Tavares, L.P., 270–271
Teixeira, M.M., 262–277
Temple, B.R.S., 172
Tenaillon, L., 424
Teramoto, T., 364, 381–382
Terasawa, H., 234–235, 393–394
Terekhov, A., 24, 33–37, 34*f*, 38–39*f*
Terrillon, S., 147–148, 434
Tesmer, J.J., 255–257
Thelen, M., 6–7, 88–114, 108*f*, 164, 294
Thelen, S., 88–91, 93–95, 108*f*, 112*f*, 113–114
Théry, M., 569
Thian, F.S., 234–235
Thiele, S., 156–161, 170, 422–423, 460, 479–481
Thierry, A.C., 90–91
Thiriot, A., 337–338
Thomas, B.E., 391–392

Thomas, M.A., 540–541, 550, 555
Thomas, R.A., 393–394
Thompson, A.A., 234–235
Thompson, D.A., 329, 360
Thompson, G.L., 157
Thompson, S.A., 487
Thomsen, A.R., 427–428
Thueringer, A., 347
Thuret, A., 147–148
Tian, H., 392
Tian, S., 394–395
Tian, Y., 234–235, 422–423
Tiffany, H.L., 93–94, 264
Tighe, M., 423–424
Timmer-Bosscha, H., 282
Timmerman, C.A., 90–91, 499*f*
Timmerman, H., 160–161, 164, 464–465
Tobin, A.B., 159–160
Toetsch, S., 501–502
Tohgo, A., 164–165
Toida, T., 521
Tokunou, T., 542–543, 558–559
Tominaga, S., 234–235, 393–394
Tomlinson, I.M., 53–54
Toner, M., 21
Toraskar, J., 156–159, 162, 422–423, 434–437
Torres, N.M., 157–159
Torri, G., 534, 534*f*
Trakimas, D., 121
Traynor, J.R., 491
Trejo, J., 172, 282–284, 433–434
Trinker, M., 459–460, 519–520
Trinquet, E., 427–428
Truty, J., 338–339, 350–353
Trzaskowski, B., 392
Tsadik, E., 93–94
Tsang, M.L., 383–384
Tschammer, N., 459, 464–465, 481, 500
Tsuboi, K., 337–338
Tsukrov, D.I., 42
Tsutsumi, H., 505–506
Tuladhar, A., 384–385
Tuli, A., 558–559
Turnbull, J.E., 525, 527
Turner, E.C., 69–70
Turnquist, H.R., 558–559
Twomey, B.M., 422–423
Tyler, R.C., 540–541
Tzouros, M., 90–91, 112*f*

U

Uchimura, K., 359
Ueda, A., 93–94
Ueda, T., 234–235, 393–394
Uetz-von Allmen, E., 540
Uguccioni, M., 90–91, 93–94, 460
Uhl, B., 310–311
Uike, N., 461–462
Ulich, T.R., 330
Ulkus, L., 42
Ulrich, E.L., 378
Ulvmar, M.H., 89, 294–295, 299, 310–311, 458–459
Unsworth, C.P., 505
Unutmaz, D., 459
Unzek, S., 558–559
Urban, J.D., 132
Urizar, E., 465
Utard, V., 459–460
Utsunomiya, A., 461–462
Uy, G.L., 48–49

V

Våbenø, J., 170
Vacchini, A., 422–437
Vago, J.P., 270–271
Valentin, G., 89, 294
Valle, A., 264–265
Van Damme, J., 159, 263, 276–277, 393–394, 542–543
van de Lest, C.H., 522
Van de Water, L., 21
van den Heuvel, J., 102
van der Goot, H., 459
van der Lee, M.M., 90–91, 465, 499*f*, 500
van der Westhuizen, E.T., 434–436, 502–503
van der Woning, S., 459–462
van Dijk, M., 464–465
Van Durm, J.J., 429
Van Dyke, T.E., 88–89
van Heteren, J., 459, 465, 500
van Hoffen, E., 442–443
Van Horn, R.D., 270–271
van Kempen, G.M.J., 369

van Kuppevelt, T.H., 522
van Offenbeek, J., 90–91, 148–149, 467, 492, 499*f*
van Schijndel, H.B., 358–359
van Senten, J.R., 474*f*, 481, 505
van Soest, R.W.M., 329
Vandercappellen, J., 90–91
Vanharanta, S., 282
Vanheule, V., 519–520
Varas, F., 222–224
Vasilieva, N., 359–360, 380–381, 383–384
Vasina, E.M., 90–91
Vassart, G., 465
Vassilatis, D.K., 3–4
Vassy, R., 519–520
Veazey, R.S., 68
Vecile, E., 276
Veelken, H., 89
Veerkamp, J.H., 522
Veerman, K.M., 359
Vega, B., 7, 14
Vela, M., 461–462
Veldkamp, C.T., 91–92, 157–159, 235, 358–385, 397–400, 540–562
Venkatakrishnan, A.J., 69, 234–235, 390–391, 393–395, 460–461
Venkiteswaran, G., 89
Verbeet, M.P., 358–359
Verbsky, J., 38–40
Verheij, M.H., 459
Verkaar, F., 90–91, 461–462, 499*f*
Verkade-Vreeker, M.C., 487–488, 505
Verma, R., 339
Vermeire, K., 89–91, 170
Vernier, M., 424
Versteeg, E.M., 522
Verzijl, D., 160–161, 164, 464–465, 490*f*, 492*f*, 502*f*
Viale, A., 282
Vicari, A., 336
Victorina, J., 459
Vidi, P., 120
Vieira, S.M., 266–267
Viejo-Borbolla, A., 6, 89–91
Vilgelm, A., 24, 33–37, 34*f*, 38–39*f*
Villalobos, V., 121–122
Villares, R., 2–15
Vines, C.M., 164–165
Viney, J.M., 481
Viola, A., 48–49, 156
Violin, J.D., 132, 141, 147–148, 157–160, 382–383, 434
Vischer, H.F., 156, 235, 390–391, 422–423, 458–506, 504*f*
Vishnivetskiy, S.A., 255–257
Visser, T., 442–443
Vlerick, D., 461–462
Vo, L., 50–52
Vogel, H., 91–92, 106
Vold, R.R., 254–255
Volkman, B.F., 358, 360–361, 364, 372–378, 380–383, 393–395, 540–562
Volkmer-Engert, R., 133–134, 142–143
Volkov, Y., 501–502
Volpe, S., 88–91, 93–95, 108*f*, 113–114
von Andrian, U.H., 163–164, 294, 310
von Schaewen, M., 384–385
von Tscharner, V., 161
von Zastrow, M., 132
Voyno-Yasenetskaya, T., 164
Vrieling, H., 481
Vroon, A., 172

W

Wacker, D., 234–235, 390–391
Wagner, A., 384–385
Wagner, G., 255–257
Wagner, L., 337–338
Wahlby, C., 352–353
Wain, J.H., 319–320
Walbaum, D., 270–271
Walden, H.R., 330
Waldhoer, M., 160–161, 164
Walker, B.D., 337–338
Walker, G.M., 21–23, 22*f*, 25–27, 40, 41*f*
Walsh, C.T., 91–92, 106
Waltman, B.A., 42
Waltner, J.K., 545
Walz, A., 177–178, 542–543
Walz, T., 255–257
Wang, C., 69, 234–235, 382, 384–385, 390–395, 411*f*, 415*f*, 416, 460–461
Wang, D., 234–235
Wang, H., 464–465
Wang, J., 89, 458–459
Wang, L., 262–263, 458

Wang, Q., 3–4
Wang, S., 3
Wang, S.J., 21
Wang, W., 3, 208–209
Wang, X., 337–338, 503, 558–559
Wang, Y., 4, 121, 222–224, 257, 458–459
Ward, M.J., 270–271
Ward, P.A., 93–94
Ward, S., 310–311
Ward, S.G., 496–498
Warda, M., 521
Warren, J.S., 93–94
Warrington, A.E., 3
Waterhouse, P., 53–54
Watson, C., 434–436, 460–461, 464–465, 487
Watson, S., 270–271
Watts, A., 6
Watts, A.O., 90–91, 422–423, 465, 499*f*, 504–505, 504*f*
Watts, V., 120
Weatherbee, J.A., 383–384
Webb, D., 37–38
Weber, C., 156, 459–460
Weber, M., 90–91, 140, 310–311, 567–569
Wedderburn, J., 360, 382–384
Wedlich-Soldner, R., 568
Wehrman, T.S., 132–133, 147, 157–160, 422–423, 436–437
Wei, H., 164–165
Wei, Y., 282, 290
Weibrecht, I., 352–353
Weidow, B., 502
Weigelin, B., 113–114
Weigl, S.A., 317–318
Weiler, J.M., 527
Weiner, J.J., 558–559
Weinstein, H., 132
Welker, R., 11–12
Wells, J.A., 391–392
Wells, T.N., 49–52
Wen, Y., 38–40
Wennerberg, E., 558–559
Wenzel-Seifert, K., 433–434
Werth, K., 89, 294–307, 310–311
Westler, W.M., 378
Weston, C., 487
Whalen, E.J., 132, 141, 147–148, 157–160, 164–165, 382–383, 434, 436–437
Whiles, J.A., 254–255
Whistler, J.L., 160–161
White, G.E., 496–498, 505
White, J.F., 6–7
White, M.D., 558–559
Whitesides, G.M., 21, 24–25, 569
Whorton, M.R., 255–257
Wicomb, W.N., 500
Widlanski, T.S., 360, 382–383
Wijtmans, M., 49–50, 156, 393–394, 460–461, 474*f*, 479–481, 487–488, 490*f*, 492*f*, 502*f*, 505
Wikswo, J.P., 20–42, 22–23*f*, 30*f*, 32–33*f*, 41*f*
Wilk, T., 11–12
Wilkins, P.P., 359
Wilkinson, M.C., 527
Willet, J.D.P., 330
Williams, C., 493
Williams, M.J., 359
Williams, R., 264
Williams, S.C., 53–54
Williams, T.J., 394–395
Willis, J., 461–462
Willoughby, D., 491–493
Willoughby, D.A., 330
Wilson, J.L., 311
Winkler, J.S., 90–91
Winter, G., 52
Winter, J., 50–52
Wipke, B.T., 211
Wirthmueller, U., 93–94
Wiseman, P.W., 569, 573–574
Wisler, J.W., 160
Wittelsberger, A., 391–392
Wolf, M., 177–178
Wong, C.H., 369–370, 381
Wood, W.I., 264
Worbs, T., 568
Woznica, I., 391–392
Woznica, K., 6
Wreggett, K.A., 475
Wright, H.L., 208–209
Wright, P.L., 359–360, 380–381, 383–384
Wu, A.M., 143
Wu, B., 234–235, 384–385, 460–461

Wu, D., 161, 422–423
Wu, J., 3, 123*f*
Wu, L., 4
Wu, T., 37–38
Wu, Y.L., 160–161
Wunderlin, M., 93–94
Wuyts, A., 329
Wyatt, R., 359, 380–381
Wyatt, R.T., 6
Wyllie, D.J., 485

X

Xiang, Z., 392
Xiao, K., 132, 147–149
Xie, B., 427–428
Xie, J., 521
Xie, N., 123*f*
Xie, P., 164
Xie, X., 424
Xu, C., 37–38
Xu, D., 339–341, 352–353
Xu, F., 390–391
Xu, L., 164–165
Xu, X., 391–392
Xu, Y., 91–92, 100–102, 106, 112*f*
Xue, F., 4

Y

Yamada, K.M., 329
Yamagami, S., 93–94
Yamaguchi, Y., 337–338
Yamamoto, K., 461–462
Yamamoto, N., 505–506
Yanagisawa, M., 147
Yang, F., 542–543
Yang, H.B., 42
Yang, M., 164
Yang, O.O., 91–92, 100–101, 337–338
Yang, S.T., 38–40
Yang, Y., 3, 163–164
Yansura, D., 50–52
Yarchoan, R., 542–543
Yates, E.A., 525
Yatim, N., 276–277
Ye, S., 392
Yeheskel, A., 519–520
Yi, E.S., 330
Yin, J., 91–92, 106
Yin, S., 330
Yin, X., 89–92, 336–353
Yoshida, S., 461–462
Yoshie, O., 282, 310, 422–423
Yoshiura, C., 234–235, 393–394
Youngman, K.R., 329
Yu, M., 42
Yu, N., 503
Yu, S., 558–559
Yu, S.R., 90–91
Yu, T.Y., 255–257
Yu, X., 550
Yu, Y., 264–265, 375
Yu, Y.R., 164–165
Yu, Z., 368, 381–382
Yung, L.Y., 422–423

Z

Zaratin, P., 459–460
Zatloukal, B., 347
Zatloukal, K., 347
Zavodny, P.J., 465
Zeng, H., 3–4
Zerwes, H.G., 89–93, 113
Zhang, C., 501–502
Zhang, F., 521
Zhang, F.K., 69–70
Zhang, J., 133
Zhang, J.H., 482
Zhang, P., 91–92
Zhang, R., 424
Zhang, W., 4
Zhang, W.-B., 161
Zhang, X., 38–40, 390–391
Zhang, Y., 42
Zhao, C., 69, 234–235, 382, 384–385, 391–395, 397–400, 399*f*, 411*f*, 415*f*, 416, 460–461
Zhao, J., 160–161
Zhao, P., 157, 172
Zhao, X., 123*f*
Zhen, J.H., 69–70
Zheng, X.T., 42
Zheng, Y., 69, 234–257, 382, 384–385, 390–416, 411*f*, 415*f*, 460–461, 474
Zhou, B.P., 282, 290
Zhou, H., 393–394
Zhou, Z., 91–92, 100–102, 106, 112*f*

Zhu, G., 170
Zhu, J., 503
Zhu, J.Z., 360–361, 382–383
Zhu, L., 558–559
Zhu, X., 37–38
Zhu, Y., 234–235, 360, 460–461
Ziarek, J.J., 157–159, 382–383, 393–395, 540–541, 555, 558–559, 569
Zicha, D., 20–21
Zidar, D.A., 66*t*, 68–69, 157–159, 382–383, 422–423
Ziff, M., 266–267
Zigmond, S.H., 20–21
Ziltener, H.J., 359
Zimmerman, B., 132, 141, 147–148
Zlotek-Zlotkiewicz, E., 568–569, 575–577
Zlotnik, A., 4, 282, 310, 422–423
Zmijewski, J.W., 219
Zolnerciks, J.K., 255–257
Zsak, M., 89, 92–93
Zuber, J., 505–506
Zuchtriegel, G., 310–311
Zweemer, A.J., 422–423, 434–437, 481
Zweemer, A.J.M., 156–159, 162
Zwirner, J., 299

SUBJECT INDEX

Note: Page numbers followed by "*f*" indicate figures, and "*t*" indicate tables.

A

ACKRs. *See* Atypical chemokine receptors (ACKRs)
Acrylamide gel electrophoresis, 526
Adenylyl cyclase (AC), 163–164, 488
Adhesion and transmigration
  procedures, 345–347
  required materials, 344
  results, 347, 348*f*
Agonist-induced responses, 483–484
Air pouch, 319–320, 319*f*, 321*f*, 330
Alanine mutants, ELISAs
  materials, 194
  phage ELISA, 194
  secreted proteins and/or of proteins present in periplasm, 196
Allosteric modulators, 485–487, 486*f*, 506*t*
AlphaScreen technology, Gαi protein activation, 425–427, 426*f*
Amaxa transfection, 324–325
Amplified luminescent proximity homogeneous assay (Alpha) Screen, 424
Anion-exchange chromatography, 521, 523
Antigen-induced arthritis (AIA), 266–269
  protocol, 268–269, 269*f*
  required material, 268
Arrestin dissociation rate, 134
Arrestin fusions, 133–134
β-Arrestin recruitment assay
  materials, 172
  protocol, 172
β-Arrestins, 121–122, 123*f*, 159, 458–459, 488
β-Arrestins signaling
  measurement
    constitutive recruitment, 433–434
    ligand-dependent recruitment, 431–433, 432*f*
  required materials, 431
Arthritis histological scoring
  decalcification, 213
  dehydration, 214
  deparaffinization, 213
  evaluation, 214–215, 215*f*
  hematoxylin and eosin staining, 214
Articular lavage, 269–270, 270*f*
Artificial cysteine, 397
Association binding, 471
Atypical chemokine receptors (ACKRs), 55*t*, 65–67, 66*t*, 89, 121–122, 123*f*, 156–157, 235–236, 237*f*, 295–296, 299
Autoimmune diseases, 156
Avidin/biotin, 5–6

B

Backbone -backbone contacts, 396–397
Bacmid purification, 400–401
Bacteria, recombinant tagged chemokines, 103–106
Baculovirus expression system, 400
Baculovirus-infected insect cells, 93–100
Baculovirus production
  equipment, 237–238
  protocol, 239–243, 241*f*
  reagents, 238
Baculovirus stocks, 401
Biased agonism, 132, 382–383
Biased receptor, 132–133
Biased signaling, 434–436, 435*f*, 482–483
Bias platform, 157
Binding assay optimization
  equilibrium saturation, 482
  hydrophobic nature, 482
  radioligand concentration, 482
  radioligand depletion, 481
  reversible two-state model, 481
  signal-to-noise ratio, 482
  Z'-factor, 482
Binding theory
  equilibrium, 463
  fractional occupancy, 463
  Langmuir adsorption isotherm, 463

Binding theory (*Continued*)
ligands, 462–463
pharmacological parameter, 463
receptor occupancy, 463
reversible two-state model, 462–463
Biolayer interferometry (BLI)
assay, 79
chemokine binding, inhibition of
data analysis, 83
experiment, 82
heparan sulfate, 81–85, 82*f*
HS, biotinylation of, 77–80, 78*f*
monoclonal antibodies binding, 77–80
Biological target, 467
Bioluminescence
live-cell imaging, 126–127
materials, 125–126
Bioluminescence resonance energy transfer (BRET)
arrestin-dependent signaling pathways, 132
arrestin recruitment assays
optimal luciferase quantity, 136–138, 137*f*
photon emission, 136
polyethylenimine method, 135
G proteins conformational changes, 428–430
interpretation and limitations
arrestin recruitment *vs.*chemokine receptors, 147
beta arrestin, 132
$BRET_{50}$, 146–147
$BRET_{max}$, 146
different ligands *vs.* chemokine receptor, 147–148, 148*f*
donor-encoding vector, 136–138
receptor mutants, 148–149
ligand-dependent β-arrestin recruitment, 431–433
methods, 133–145
$BRET^2$ uses:noise ratio, 143–145
constitutive arrestin recruitment, 140
donor, optimal quantity of, 136–138
dose-response experiments, 140
general protocol, 134–136, 137*f*
materials, 133–134
probe position, switching, 142–143, 144*f*
saturation, 138–140, 139*f*
time-course experiments, 141, 142*f*
troubleshooting, 141–145
redundant, 132–133
Biotinylated heparan sulfate, 77–81, 78*f*
Biphasic transition, 412–413, 414*f*
BLI. *See* Biolayer interferometry (BLI)
Bolton–Hunter techniques, 467
Bone marrow chimeric (BMC) mice, 210
Bone marrow neutrophil
adoptive transfer, 226–228
fluorescently labeling, 226, 228
isolation of, 226–228
Bovine serum albumin (BSA), 366
Boyden chamber, 500–501, 502*f*
BRET. *See* Bioluminescence resonance energy transfer (BRET)
Buffer exchange and sample concentrating, 524
Bystander BRET, 138, 139*f*

C

Calcium flux assays, 555
Calcium phosphate transfection, 165–166, 170
Calcium trace data, 498–499
cAMP-responsive element (CRE), 489*f*, 495*f*, 488. *See also* Cyclic adenosine monophosphate (cAMP)
Candidate residue pairs
artificial cysteine, 397
chemokine residues, 396–397
cherry-picking, 396–397
conformational sampling, 396–397
efficient cross-linking, 397–400, 399*f*
$E_{SS, scatter plot of}$, 397–400, 399*f*
in silico, 396–400
mutant pair characterization, 397–400
polypeptide chain, 398*f*
protein misfolding, 397
residue geometry, 397
Cardiac puncture, blood collection, 444
Cation exchange chromatography, 547
Caveats, 380–381
CCL21, 578*f*, 580
CCR2, 141–143, 147–148
CCR2-dependent monocytes/macrophages
pleural cavity, 270–271, 272*f*

protocol, 271, 272*f*
required materials, 271
CCR5 pharmacology, 68–69
CD26, 276–277
Cell-based assays, 126
Cell biology, 68–69
Cell-cell interaction
endothelial cell loading, 35
extracellular matrix, 36
fibroblast loading, 36
spheroid formation, 33, 35
tumor microenvironment, 31–37
tumor spheroid loading, 36–37
Cell libraries selection, phage display technology
cell lines, 62–63
considerations, 60–62, 61–62*f*
high-affinity cell surface binding, 64–65
internalizing ligands, 63–64
required materials, 63
Cell migration
ImageJ, 303–305, 304*f*
Imaris (Bitplane), 305–307, 306*f*
Cell polarity *vs.* cell turning, 40–41
Cell proliferation, 505
Cell recovery, chemokine biology
joint cavity, 269–270
pleural cavity, 272
Cell staining
and counting
cytospin, 321
flow cytometry, 320–321
materials, 296
Chemiluminescent signal, 499–500, 499*f*
Chemoattractant cytokine, 540
Chemokine(s), 132–134, 140, 310, 312–314, 319–320, 360, 382–384, 458
analogs, 66*t*
health, disease, 48–49
HIV prevention, 68
receptors, 50, 51*f*
structure and activity, 49–50, 49*f*
biotinylated HS, 79
biotinylation, 535
CCL19, 540
CCL21, 540
coated cell surfaces, 80–81
cysteines, 540–541
disulfide bond, 540–541
endothelial surfaces, 73–74
Escherichia coli, 74
expression, 544–549
cation exchange chromatography, 547
lyophilization, 549
materials, 544–545
nickel affinity chromatography, 546
purification, 544–549
refolding, 546–547, 548*f*
reverse phase HPLC, 548–549, 549*t*
transformation and cell harvesting, 545
ULP1 digestion, 547
fold, 540–541, 541*f*, 552–553, 553*f*
GAGs, 74
gradients, 295–296, 310–311
heparan sulfate, 77–80, 78*f*
heparin-sepharose affinity chromatography, 73–74
inhibition, 81–84, 82*f*
interfaces
conformational mechanisms, 394–395
conserved cysteine motif, 393–394, 394*f*
cysteines, 392–393
distinct epitopes, 393–394
double-cysteine motif, 392–393
flexible N-terminus, 392–393, 393*f*
structural determinants, 394–395
three-strand β-sheet, 392–393
ligands, 120, 140
oxidation, 543–544
purification, 544–549
receptor signaling
agonist, 484–485
allosteric effect, 484–485
allosteric modulators, 485–487, 486*f*
effect of, 484–485, 484*f*
full receptor occupancy, 484–485, 484*f*
mode of action, 485
mutagenesis studies, 487–488
NAM, 484–485
noncompetitive orthosteric antagonism, 485, 486*f*, 487–488
PAM, 484–485
rightward shift, 485
saturable effect, 487–488
Schild equation, 485

Chemokine(s) (*Continued*)
Schild plot, 485–487, 486*f*
receptor therapeutics
AMD3100/plerixafor, 460–461, 461*f*
CKRs, 459–461
cognate receptors, 459–460
glyco-engineered CCR4-targeting antibody, 461–462
GPCR expression, 461–462
ionic interactions, 460
ligand design, 460–461
maraviroc, 460–461, 461*f*
orthosteric ligand binding site, 460
pharmacological intervention, 459–460
structure-based virtual screening, 460–461
two-step binding model, 460
refolding, 543–544
system, 156–159, 162
Chemokine binders, phage display
pharmacophore of
alanine mutants, ELISAs, 193–197
Evasin cDNAs, 190–193
material, 189
phage ELISA and soluble ELISA, 189, 190*f*
selectivity of
phage display libraries, construction of, 198–203
selections, 203–204, 204*t*
Chemokine-binding proteins (CKBP), 74–77, 188–189
Chemokine biology
antigen-induced arthritis, 266–269
CCR2-dependent monocytes/macrophages, 270–271
cell recovery
joint cavity, 269–270
pleural cavity, 272
cell type identification
immunofluorescence, confocal microscopy, 274–275
morphology (optical microscopy), 273–274
CXCR2-dependent neutrophil recruitment, 264–265
leukocytes, 262–264, 276–277
limitations, 275–277
tibiofemoral joint, injection, 265–266
Chemokine CCL2. *See* Monocyte chemotactic protein-1 (MCP-1)
Chemokine/chemokine receptor complexes
baculovirus production
equipment, 237–238
protocol, 239–243, 241*f*
reagents, 238
design of constructs
chemokine constructs, 237, 237*f*
receptor constructs, 236–237, 237*f*
expression
equipment, 243–244
protocol, 244–245, 246*f*
reagents and solutions, 244
purification
buffers, 246–247
equipment, 245–246
protocol, 247–250, 248*f*
reagents, 246
receptor/chemokine complex purity, characterization of
buffers, 250
equipment, 250
reagents and solutions, 250
SDS-PAGE/Western Blot, 250–251
size exclusion chromatography (SEC), 251, 252*f*
thermal unfolding, 251–253
reconstitution
bicelles, 254–255
buffers, 253–254
equipment, 253
nanodiscs, 255–257, 256*f*
reagents and solutions, 253
Chemokine-mediated dendritic-endothelial cell interactions
heparan sulfate
DC adhesion and transmigration, 343–347
DC migration toward LEC, 341–343
perspectives, 350–353
visualizing, 347–350
Chemokine-mediated migration
chemoattractant cytokines, 310
chemotaxis assays, 310–311
GAGs, 310–311

in vitro chemotaxis
adherent cells, diffusion gradient of, 312–314
counting migrated cells, 315–316
suspension cells, diffusion gradient of, 314
in vivo chemotaxis
cytospin, cell staining and counting, 321
flow cytometry, cell staining and counting, 320–321
leukocyte recruitment, murine air pouch, 319–321
transendothelial, 317–318
leukocytes, 310–311
mammalian transfectants, chemokine receptors
cloning, 322–323
stable transfection cells selection, 325–329
vector transfection, 323–325
migrated cells counting
beads, 315–316
hemocytometer, 315
microscopy, 315, 315*f*
tissue development, 310
Chemokine receptors, 120, 123, 140, 142–143, 147, 311, 322, 359, 382–384
applications, 12–14
blood collection, mice
cardiac puncture, 444
retro-orbital blood collection, 444–445
cell surface markers, staining of, 445–449, 446*t*, 448*t*
flow cytometry analysis
blood cells preparation, mice, 445
compensation procedures, 450–451, 450*t*
data acquisition, 451
data presentation and analysis, 451–453
gating strategy, 451, 452–453*f*
LSR Fortessa cytometer, 449, 449*t*
quality and reproducibility controls, 449–450
GPCR family, 3–4
LVPs, 8*f*, 9–10
materials, 7–8
retroviral particles, 8–9, 8*f*
sensor chip, immobilization, 4–6, 5*f*
VPs, 5*f*, 6–12
biosensor surfaces, 12, 13*f*
levels on, 11
receptor number quantitation, 11–12
titration, 10
Chemokine receptor trafficking, confocal immunofluorescence microscopy
cell culture and transfection
cell counting and plating, 286
materials, 284–285
passaging and maintaining HeLa cells, 285
siRNA, transfection with, 286
cell preparation
blocking and antibody incubation, 288
fixation and permeabilization, 288
mounting, 289
washing, 288–289
cover slips and stimulation
coverslip preparation, 286–287
passaging cells onto coverslips, 287
stimulating cells, 287–288
microscope image acquisition, 289–290
Chemokine recognition site 1 (CRS1), 393–394, 399*f*, 411*f*
Chemotactic force, shearing force, 40, 41*f*
Chemotaxis
chamber, 22*f*
materials, 24
procedures, 25
classic transwell assay, 177–178
Ibidi® μ-slide chemotaxis 3D, 178–180, 179*f*
Cheng-Prusoff equation, 477–478
Cherry-picking, 396–397
3-[(3-Cholamidopropyl)-dimethylammonio]-1-propane-sulfonate (CHAPS), 2–3
Chondroitin sulfate (CS), 74, 518*f*, 522–523
Chronic obstructive pulmonary disease (COPD), 459–461
Circulating tumor cells (CTCs), 42
Classic transwell assay
chemoattractant, 177–178
chemotactic migration, 177–178
Clear sharp peak, 412–413, 413*f*

Click beetle luciferase (CLuc)
 chemokine signaling, 122
 green and red spectral variants, 121
 protein interactions, 121
 real-time analysis, 123*f*
Cloning
 bacmid purification, 400–401
 codon optimization, 400
 Escherichia coli cells, 400–401
 FLAG tag, 400
 metal affinity purification, 400
 pFastBac™ vector, 400
 recombinant bacmids, 400–401
 Sf9, 400
CLuc. *See* Click beetle luciferase (CLuc)
CM5 sensor chip *vs.* N-terminal cysteine thiol, 4–5
Codon optimization, 400
Coelenterazine 400A, 143
Coelenterazine-h, 493–494
Coexpression
 multiplicity of infection (MOI), 244–245
 receptor/chemokine coexpression, 245
 in Sf9 cells, 235–236
Cognate receptors, 459–460
Cold ligands, 473*f*, 467, 471. *See also* unlabeled ligand
Collision-induced dissociation (CID), 374–375
Colony-picking, 325–327
Competitive displacement curve, 477–478, 479*f*
Complementation reporter constructs
 cDNA, 122
 CLuc, 123–125
 plasmids, 122
 reagents and equipment, PCR, 122
 vectors, 122
Complement-dependent cytotoxicity, 461–462
Complex monodispersity, 412
Concentration-response curves, 482–483, 483*f*
Confocal immunofluorescence microscopy, 282–284
Conventional molecular cloning, 322–323
Corning® Matrigel® Matrix, 296
Coupled signaling pathways, 482–483
Covalently trapped complexes, 401–402
CRE. *See* cAMP-responsive element (CRE)
CRS1. *See* Chemokine recognition site 1 (CRS1)
Crystallography, 390–391, 394–395
Current good manufacturing practices (cGMP) system, 559–562
CXC chemokine, 264
CXCL12, 235–236, 242, 244–245, 246*f*
CXCR4, 121–122, 123*f*, 140, 142–143, 148–149
CXCR7, 132–133, 139*f*, 140
CXCR2-dependent neutrophil recruitment, 264–265
CX3CR1, fractalkine receptor, 69–70
Cyclic adenosine monophosphate (cAMP), 488, 489*f*, 491–494
 AlphaScreen technology, 425–427, 426*f*
 G proteins signaling, 423–430
Cysteine. *See also* disulfide trapping
 mutagenesis, 396–397
 natural cross-linking agent
  cellular compartments, 395–396
  chemokine receptors, 395–396
  dissociation energy, 395
  disulfide bonds, 395
  thiol side chain, 395
  TM domain, 395–396
 oxidation, 548–549
 trapping, 391

D

Data analysis, 533
Dendritic cell haptotaxis
 chamber assembly
  cell preparation, 577–578
  LAPAP/chemokine presentation, 580
  materials, 578–579
  troubleshooting, 579–580
 lower surface preparation
  chemokine printing, 573–574
  glass slides treatment and dish preparation, 571–573
 perspectives, 580–581
 upper surface preparation
  lid with soft PDMS piston, 576–577, 576*f*
  micropillar preparation, 575

Dendritic cells (DCs)
adhesion and transmigration
procedures, 345–347
required materials, 344
results, 347, 348*f*
migration
procedures, 342–343
required materials, 341–342
results, 343, 344*f*
trafficking, 336, 337*f*, 339–341
Dermatan sulfate (DS), 74, 518–519, 518*f*
Diacylglycerol (DAG), 488, 489*f*
Differential scanning fluorimetry (DSF), 403
Diffusion gradient assay, 311–312, 311*f*
Diffusion gradient chemotaxis, 311–312
Displacement binding, 472
allosteric interaction, 479–481, 480*f*
allosteric radioligands, 481
cognate chemokines, 481
competitive binding, 479–481, 479*f*
competitive displacement curve, 477–478, 479*f*
fractional receptor occupancy, 477–478
metal ion chelators, 479–481
negative sigmoidal curve, 477–478
nonlinear regression, 477–478
PAMs, 479–481
radioligand concentration, 477, 478*f*
unlabeled orthosteric ligands, 477
Displacement experiments, 536
Dissociation binding, 471
Dissociation constant, 83–84
Dissociation energy, 395
Dissociation kinetics, 404–407
Disulfide cross-link, 403–404
Disulfide shuffling, 396, 398*f*, 411–412
Disulfide trapping
candidate residue pairs, selection of, 396–400
characterization of
covalent complexes, flow cytometry, 404–408
experimental approaches, 405*t*
SDS-PAGE, 408–412
SEC and CPM-DSF, 412–416, 415*f*
western blotting, 408–412
chemokine complex, 396–400
cysteine residues, 391
generation of
baculovirus stocks, 401
cloning, 400–401
membrane preparation, 402
receptor complexes, purification of, 403
Sf9 coexpression, 401–402
irreversible covalent binary protein complexes, 391
mutated proteins, 396
natural cross-linking agent, cysteine, 395–396
receptor, architecture of, 392–395
Dithiothreitol (DTT), 370
DNA sequencing, 125
Donor plasmid, 136–138
Dose calibrator, 468
Dose-response curve, 323–324
Dot blot technique, 524
Double-cysteine motif, 392–393
Draining lymph nodes (DLN), 336, 337*f*
Dual-color luciferase complementation
β-arrestin 2, 121
bioluminescence
live-cell imaging, 126–127
materials required, 125–126
chemokine receptors signaling for, 120
click beetle luciferase, 121
complementation reporter constructs
CLuc, 123–125, 123–124*f*
materials, 122
data analysis, 128
luciferase enzymes, 120
N- and C-terminal fragments, 121
protein fragment complementation assays, 120
protein interactions, 121
Dual luciferase reporter system, 496
Duffy antigen receptor for chemokines (DARC). *See* Atypical chemokine receptors (ACKRs)
Dynamic mass redistribution (DMR), 505

E

Elastase, 208–209
Electroporation
amaxa transfection, 324–325
flow cytometry, 325
Electrospray ionization (ESI), 373–374

ELISA
alanine mutants
materials, 194
phage ELISA, 194
secreted proteins and/or of proteins present in periplasm, 196
heparin-bound chemokine by
anti-chemokine monoclonal antibody, 75–77, 75*f*
in vitro experiments, 75–77
low-molecular-weight, 75–77
materials, 75
methods, 76
phage ELISA, 194
ELISA-like competition assay, 534–536
Endogenous agonists, 460, 463
Endogenous signaling networks, 459
Endoproteinases Asp-N, 376
Endosomal sorting complex required for transport (ESCRT) pathway, 282–283
Endosome, 282–284
Endothelial cells
chemokines, binding to
anti-chemokine antibodies, 80–81
binding assay, 81
cell culture, 80
GAGs, 80–81
HUVEC, 80–81
human umbilical vein endothelial cells (HUVEC), 80–81
in vivo transendothelial chemotaxis, 317
loading, tumor microenvironment, 35
lymphatic endothelial cells (LECs), 348*f*
Energy transfer efficiency, 145–146, 145*f*, 148*f*
Enhanced chemiluminescence (ECL), 175
Enzymatic labeling, tagged chemokines, 100–108
Enzyme fragment complementation (EFC), 499–500
Epac-based sensors, 491–493
ERK phosphorylation
HRP-coupled antibody., 175
MAPKs., 175
SDS-PAGE., 175
western blotting., 175
Escherichia coli, 74
chemokine, 74
cloning, 400–401
expression
inclusion bodies, solubilization, 364–365
materials, 364
refolding, 366–367
size-exclusion chromatography, 366
transformation and expression, 364–365
recombinant chemokines, 542
TG1
library construction, 58–60
required materials, 58
Evasin cDNAs, cloning of
cloning, 191
materials, 190
site-directed mutagenesis, alanine scanning by, 192
Evasin proteins, 188–189
Extracellular loop (ECL), 392–393

F

Fibroblast loading, 36
Firefly luciferase (Fluc), 494
FLAG-tagged receptors, 173
Flow cytometry, 122, 126, 133–134, 327, 328*f*
blood cells preparation, mice, 445
compensation procedures, 450–451, 450*t*
covalent complexes, detection of
cell distribution, 407–408
cross-linked ACKR3, 407*f*
cross-linked CXCR4, 406*f*
CRS1 disulfide cross-linked complexes, 411*f*
C-terminal tag, 407–408
disulfide-trapped complexes, 404
fluorescent antibodies, 404
prolonged incubation, 404–407
receptor-bound chemokine, 404
single-receptor mutant, 407–408
data acquisition, 451
data presentation and analysis, 451–453
gating strategy, 451, 452–453*f*
LSR Fortessa cytometer, 449, 449*t*
quality and reproducibility controls, 449–450

Fluc activity, 496
Fluorescence-activated cell sorting (FACS), 316
Fluorescence measurements, 532
Fluorescence minus one (FMO), 445–446
Fluorescent chemokines, monitoring scavenging with, 93–108
Fluorescent protein-tagged chemokines
  cell culture, 95
  insect culture, 94
  purification, 94–95
  recombinant baculoviruses, 95
  recombinant chemokines, 97–98
  recombinant fluorescent chemokine-encoding baculoviruses, 95–97, 98*f*
  recombinant fluorescent chemokine fusion proteins, 99–100
Fluorometric microvolume assay technology, 80–81
Fluorophore-*vs*.-protein (FTP), 453–454
Forskolin (FSK), 491, 492*f*
Förster radius (R0), 145–146
4D live-cell imaging
  cell migration
    quantitative analysis, ImageJ, 303–305, 304*f*
    semiautomated cell tracking, Imaris, 305–307
  cells preparation, 296, 299–300
  chambers
    filling, 296, 300–302, 301*f*
    preparation, 296–299, 298*f*
  materials and equipment, 296–297
  time-lapse imaging, 302
FSK. *See* Forskolin (FSK)
Functional assays
  β-arrestin recruitment
    BRET-based assay, 499–500
    chemiluminescent signal, 499–500, 499*f*
    GRKs, 488, 489*f*
    materials, 499
    methods, 500
    PathHunter cell lines, 499, 499*f*
    G protein activation, 500
  Ca^{2+} mobilization, 496–499
    assay robustness, 498–499
    βg-complex, 496–498
    calcium trace data, 498–499
    CXCL12 stimulation, 496–498, 497*f*
    data normalization, 498–499
    fluo-4 acetoxymethyl ester, 496–498, 497*f*
    hemi-equilibrium binding conditions, 498–499
    materials, 497
    methods, 498
    PLC-β, 496–498
    transient intracellular, 496–498
  cAMP production, 491–494
    BRET, 491, 492*f*
    coelenterazine-h, 493–494
    Epac-based sensors, 491–493
    fluorescence-based kits, 494
    materials, 492
    methods, 493
    Rluc8, 491, 492*f*
    signal-to-noise, 494
    single-chain Epac sensor, 491–493
  chemotaxis, 500–502
    bell-shaped concentration-response curve, 501–502, 502*f*
    Boyden chamber, 500–501, 502*f*
    chemoattractants, 501–502
    materials, 501
    methods, 501
    scratch assay, 502
  GTPgS-binding assay
    advantage, 491
    CXCL11, 489–491, 490*f*
    disadvantage, 491
    downstream effector proteins, 489–491
    $G\alpha_{i/o}$-coupled receptors, 491
    Gα subunit, 489–491
    isolated membranes, 489–491
    materials, 490
    method, 490
    VUF10661, 489–491, 490*f*
  impedance
    actin cytoskeleton, 502–503
    agonist stimulation, 504
    cell proliferation, 505
    CKR activation, 504–505, 504*f*
    CXCR3-mediated signaling, 504–505
    label-free assay, 505
    luminescent probes, 503–504

Functional assays (*Continued*)
materials, 503
methods, 503
real-time changes, 504
receptor activation, 502–503
xCELLigence system, 502–503
reporter gene assay
CKR-mediated activation, 494
CRE-luciferase reporter gene assay, 494–496, 495*f*
dual luciferase reporter system, 496
Fluc activity, 496
luciferase transcription, 494
materials, 494
methods, 495
plasmid encoding, 494
Rluc activity, 496
translational machinery, 496

G

Gamma counter, 468
Gα protein activation
G proteins conformational changes, 428–430
detection, 430
required materials, 430
second messengers, measurement
AlphaScreen technology, 425–427
HTRF technology, 427–428
required materials, 424–425
Gaussia luciferase (Gluc), 494
Gel electrophoresis, 416
Glu-C, 376
Glucagon-like peptide-1 (GLP-1), 465
Glycosaminoglycan-bound chemokines
anti-chemokine antibodies, 74
BLI, inhibition of, 81–84
chemokine-binding proteins, 74
classes of, 74
endothelial cells
binding assay, 81
cell culture, 80
materials, 80
haptotaxis, 73–74
heparin-sepharose affinity chromatography, 73–74
recognition of
BLI, HS-bound chemokine by, 77–80
ELISA, heparin-bound chemokine by, 75–77
Glycosaminoglycan (GAGs) chemokine coreceptors
dot blot analysis, 524
HS preparation
mammalian cells, 523–524
tissues, 520–523
interactions
ELISA-like competition assay, 534–536
gel mobility assay, 530–531, 532*f*
isothermal fluorescence titration, 531–534
size-exclusion chromatography, 524–527, 526*f*
strong anion-exchange chromatography, 527–529, 530*f*
Glycosaminoglycans (GAGs), 162, 310–311, 458
retroviral particles, 8–9
sensor chip, immobilization, 5*f*, 6
G protein binding, nonselective assay
GTPgS binding, 166
membrane preparations, 166
protocol, 167–168
G protein-coupled receptor (GPCR), 156–157, 540, 458–459, 488
chemokine receptors, 3–4
LVPs, 9–10
VPs, 6–12
G protein signaling
activation
conformational changes, 428–430
second messengers, measurement, 424–428
cAMP assay, 168–169
materials, 169
protocol, 169
SPA-IP3 assay
materials, 170
protocol, 170–171
Gradient switch
materials, 28
procedures, 28
Granulocyte/macrophage colony-stimulating factor (GM-CSF), 339, 340*f*
GraphPad Prism, 77, 84

H

Haptotaxis, 73–74
Haptotaxis chamber assembly
cell preparation, 577–578
LAPAP/chemokine presentation, 580
materials, 578–579
troubleshooting, 579–580
Hemocytometer, 314–315
Heparan sulfate (HS), 74, 337–338, 518*f*
dendritic cells
adhesion and transmigration, 343–347
migration toward LEC, 341–343
required materials
mammalian cells, 523–524
tissues, 520–523
visualizing, 347–350
Heparan sulfate-bound chemokine, 77–80, 78*f*
Heparin-bound chemokine, 75–77, 75*f*
Heparin (HP) depolymerization, 518*f*, 525
Heparin-sepharose affinity chromatography, 73–74
Herpesviridae, 390
High-affinity radioligand, 467
His_6-SMT3-chemokine, 544, 546
HIV prevention, 68
HL-60 cells, 25–27
cell tracking and data analysis, 30–31, 32*f*
chemokine gradient, 30*f*
polarization, 31
chemotaxis chamber
materials, 26
procedures, 26
CXCR2, 25–27, 41*f*
gradient switch
materials, 28
procedures, 23*f*, 29
Homogenous time-resolved fluorescence (HTRF), 424
HS. *See* Heparan sulfate (HS)
HTRF technology, Gαq protein activation, 427–428
Human cytomegalovirus (HCMV), 459
Human umbilical vein endothelial cells (HUVEC), 80–81
Hyaluronic acid (HA), 74
Hydrodynamic volume, 413–414

I

Ibidi® μ-slide chemotaxis 3D
biased effect, CCL19 and CCL21, 178–179
collagen polymerization, 178–179
IBMX. *See* Isobutylmethylxanthine (IBMX)
Immobilized-metal ion affinity chromatography (IMAC), 101
Immune cell trafficking *in vivo*, 209–210
Immunization, 268, 269*f*
Immunofluorescence, confocal microscopy
material, 274
protocol, 274–275
Inflammatory arthritis
histological scoring
decalcification, 213
dehydration, 214
deparaffinization, 213
evaluation, 214–215, 215*f*
hematoxylin and eosin staining, 214
K/BxN serum transfer model
isolation of serum, 211–212
serum transfer, 211–212
mixed BMC mice experiments, 228–229
MP-IVM, 221–224, 223*f*
neutrophil adoptive transfer, 224–228, 225*f*
neutrophil migration into joint
immunohistochemistry (IHC), 216–217, 216*f*
synovial fluid or synovial tissue, FCM, 217–220, 219*f*
paw thickness, arthritis clinical scoring and measurement of, 212, 213*f*
Inositol-1-phosphate (IP1), 427–428
Inositol 1,4,5-trisphosphate (IP_3), 163–164, 488
Intact cell binding, 466
Internalization assay
confocal microscopy-based, 174–175
ELISA-based, 173
Intra-articular injection, 265, 267*f*
Intrakine, HIV protect and vaccine, 69
Intramolecular disulfide bonds, 392–393, 394*f*, 395–396
In vitro chemotaxis
Boyden chamber, 311
diffusion gradient assay

In vitro chemotaxis (*Continued*)
adherent cells, 312–314
suspension cells, 314
In vivo chemotaxis, 318–321
In vivo models. *See* Chemokine biology
In vivo transendothelial chemotaxis
EA.hy-926 monolayer cells, 317, 318*f*
endothelial cells, 317
proinflammatory changes, 317
Ionic interactions, 460
^{125}I-radiolabeling, 468, 470
Isobutylmethylxanthine (IBMX), 491–493
Isothermal fluorescence titration (IFT), 74, 531–534, 534*f*

J

Joint cavity. *See also* Cell recovery, Chemokine biology
protocol, 269–270, 270*f*
required materials, 269
Jurkat cells, 314

K

K/BxN serum transfer model
isolation of serum, 211–212
serum transfer, 211–212
Keratin sulfate (KS), 74
Kinetic binding
affinity, 474
allosteric modulators, 475
association rate constants, 474–475
dissociation rate constants, 474–475
drug target-residence time, 474
equilibrium, 473
radioligand, 473–475, 473–474*f*
tracer ligand dissociation, 475

L

Label-free assay, 505
Langmuir adsorption isotherm, 463
Laser-assisted protein adsorption by photobleaching (LAPAP), 569, 570*f*, 573*f*, 580
Lentiviral particles (LVPs), 8*f*, 9–10
chemokine receptors
applications, 12–14
in SPR, 6–12
VP biosensor surfaces, 12
Leukocyte trafficking, 88
Library design and construction, phage display technology
chemokine walking, 54–57, 55*t*
construction, 58–60
diversity, 53, 53*t*
diversity feasible, 53–54
partial diversity, 54, 55*t*
phage display system, 52, 52*f*
required materials, 58
Ligand bias, 382–383
bias plot cellular responses, 159–160, 159*f*
chemokine monomers *vs.* dimers, 157–159
chemokine system, 157–159
posttranslational modifications, 159
G protein signaling, 159
Ligand binding, 531–534
Ligand-dependent β-arrestin recruitment, 431–433, 432*f*
Ligand-dependent induction, 123, 126
Ligand efficacy, 482–483, 483*f*
Linear regression analysis, 82*f*, 84
Lipid reagents, transfection, 324
LPS-stimulated macrophages, 315, 315*f*
Luciferase complementation assays, 121
D-Luciferin, 127–128
Lymphatic endothelial cells (LECs), 348*f*
dendritic cells
adhesion and transmigration, 343–347
migration, 341–343
monolayer, 345–346
Lyophilization, 549
Lysosomes, 282–283, 287

M

MALDI-TOF. *See* Matrix-assisted laser desorption/ ionization-time-of-flight (MALDI-TOF)
Mammalian cells expression
anti-protein C immunoaffinity purification, 362–363
cell culture and transfection, 362
materials, 361–362
Mammalian transfectants
chemokine receptor cloning, 322–323
materials, 322
stable transfection cells selection, 325–329
vector transfection, 323–325

Mass spectrometry
disulfide oxidation and identity, detecting, 550*f*, 551
ESI, 377–378
MALDI-TOF, 376–377
materials, 375–376, 550
proteolytic cleavage, 376
top-down approaches, 551*f*, 552
Matrix-assisted laser desorption/ ionization-time-of-flight (MALDI-TOF), 373–374, 376–377
Mediated intracellular signaling
G protein-dependent signaling
AC, 163–164
Chemokine receptor signaling pathways, 163–164, 163*f*
signal transducers, 162
G protein-independent, 164–165
β-arrestins, 164–165
canonical signaling pathways, 164–165
MAPK pathway, 164–165
signaling via 7TM receptors, 162
Melting transitions, 416
Membrane isolation, 466, 469
Metal affinity purification, 400, 402
Metal ion chelators, 479–481
Methylated bovine serum albumin (mBSA), 268–269
Microfluidic devices
cell-cell interaction in tumor microenvironment, 31–37
cell polarity *vs.* cell turning, 40–41
chemotactic force, 40
chemotaxis chamber, 24–25
circulating tumor cells, 42
designes, 33–37, 38–39*f*
gradient switch, 23*f*, 27–31
HL-60 cells
chemotaxis chamber, 25–27
gradient switch, 28–31
limitations, 37–40
3D microbioreactors, 42
shearing force, 40
Migrated cells counting
beads, 315–316
hemocytometer, 315
microscopy, 315, 315*f*
Migrating cells, monitoring scavenging in, 113–114
Mitogen-activated protein kinase (MAPK), 163–164
Modified pET-17B
bacterial culture, 102
purification, 102
Molecular docking, 397–400
Molecular pharmacology
chemokine receptors
binding, pharmacological quantification, 462–482
signaling, pharmacological quantification, 482–505
therapeutics, 459–462
Monoclonal antibody, 75–80, 75*f*, 78*f*
Monocyte chemotactic protein-1 (MCP-1), 271
Monocytes, 442–443, 445–446, 446*t*, 448*t*, 452*f*
Monophasic melting transition, 414–415
Mono Q anion-exchange chromatography, 368–370
Morphology (optical microscopy)
protocol, 273–274
required materials, 273
Multiphoton intravitalmicroscopy (MP-IVM), 209–210, 221–224, 223*f*
Multiple promising crosslinks, 416
Multiplicity of infection (MOI), 244–245, 401, 402*f*
Murine air pouch, 319–321, 319*f*
Murine leukemia virus (MLV)
retroviral particles, 8–9
VPs, biosensor surfaces, 12
Mutagenesis, 190–193
Mutant pair characterization, 397–400
Myeloperoxidase (MPO), 208–209, 220–221

N

NAM. *See* Negative allosteric modulator (NAM)
Nanobody/VHH platform, 461–462
Negative allosteric modulator (NAM), 463–464, 484–485, 484*f*
Neutrophil adoptive transfer, 224–228, 225*f*
Neutrophil extracellular traps (NETs), 208–209
Neutrophil migration into joint
immunohistochemistry, 216–217, 216*f*

Neutrophil migration into joint (*Continued*)
synovial fluid or synovial tissue, FCM, 217–220, 219*f*
N-hydroxysuccinimide/(1-ethyl-3-(3-dimethylamino-propyl)-carbodiimide (NHS/EDC), 5
Nickel affinity chromatography, 546
Non-Hodgkin's lymphoma, 460–461
Nonspecific covalent adducts, 395–396
N-terminal chemokine receptor peptides, *in vitro* sulfation of
enzymatic sulfation, 370
materials, 370
Nuclear factor of activated T cells (NFAT), 488
Nucleic acids removal, 522–523

O

Odyssey IR imaging system, 408–409
Off-target effects, 488
Oligosaccharides, mass spectrometry, 528
Optimal orientations, 126
Orthosteric ligand, 463–465, 506*t*
Orthosteric ligand affinity, 463–464
Orthosteric *vs.* allosteric ligand binding
allosteric modulators, 463–464
allosterism, 463–464
dissociation rate, 463–464
G protein coupling, 465
high-affinity chemokine binding, 465
maraviroc, 463–464
orthosteric ligand, 463–465
receptor expression levels, 464–465
receptor internalization, 464–465
steric hindrance, 463–464

P

PAM. *See* Positive allosteric modulator (PAM)
PathHunter®, 171
PathHunter assay, 499–500, 499*f*
Paw thickness, arthritis clinical scoring and measurement of, 212, 213*f*
Peptide sulfation, 361, 363
Peripheral blood mononuclear cell (PBMC), 314, 315*f*
Pertussis toxin (PTX)
AlphaScreen technology, 265
ligand-dependent β-arrestin recruitment, 269–270
Phage display
atypical chemokine receptor DARC (ACKR1), 65–67
chemokine analogs, 66*t*
health, disease, 48–49
receptors, 50, 51*f*
structure activity, 49–50, 49*f*
chemokine binders, pharmacophore of
alanine mutants, ELISAs, 193–197
Evasin cDNAs, 190–193
material, 189
phage ELISA and soluble ELISA, 189, 190*f*
chemokine binders, selectivity of
phage display libraries, construction of, 198–203
selections, 203–204, 204*t*
libraries, construction
insert, digestion of, 199
materials, 198
phage particles, 201
random mutagenesis, 198
vector, digestion of, 199
methods, 50–65, 51*f*
libraries on cells, 60–65
library design, construction, 52–60
perspectives
CCR5 pharmacology, 68–69
and cell biology, 68–69
chemokine, HIV prevention strategy, 68
CX3CR1, fractalkine receptor, 69–70
intrakine, HIV protect and vaccine, 69
Phage ELISA, 194
Pharmacological parameter, 463
Pharmacological quantification
chemokine receptor binding
association binding, 471
binding assay optimization, 481–482
binding theory, 462–463
displacement, 472, 477–481
dissociation, 471
harvesting and measurement, 472
kinetic, 473–475
membrane isolation, 469
orthosteric *vs.* allosteric ligand, 463–465

^{125}I-radiolabeling, 470
saturation, 471, 475–477
chemokine receptor signaling
functional assays, 488–505
functional studies, ligands in, 484–488
Phosphate-buffered saline (PBS), 466
Phosphatidylinositol-4,5-bisphosphate (PIP_2), 488
3'-Phosphoadenosine-5'-phosphate (PAP), 368
3'-Phosphoadenosine-5'-phosphosulfate (PAPS)
analysis, ion-pair RP-HPLC, 369
purification
materials, 368
mono Q anion-exchange chromatography, 369–370
stock solutions preparation, 368
Phosphodiesterase (PDE), 491
Phospholipase C (PLC), 488
Phospholipase C beta (PLC-β), 496–498
Photoaffinity labeling, 392
Plasmid constructs, 124
Pleural cavity. *See also* Cell recovery, chemokine biology
morphology (optical microscopy), 273–274
protocol, 272, 272*f*
required materials, 272
pMS101C vector, 190–193
Polymerase chain reaction (PCR), 364
Polymorphonuclear neutrophils (PMN), 442–443, 445–446, 446*t*, 448*t*
Porcine intestinal mucosa, 75–77
Positive allosteric modulator (PAM), 484–485, 484*f*
Posttranslational modifications, 358–359, 382–383
Potency, 482–483, 483*f*
Preparative SEC, 525, 526*f*
Probe switching, 142–143, 144*f*
Probing biased signaling, chemokine receptors
β-arrestin recruitment assay, 171–172
bias types, 157–162, 158*f*
chemotaxis
classic transwell assay, 177–178
Ibidi® μ-slide chemotaxis 3D, 178–180, 179*f*
ELISA or confocal microscopy, 172–175
ERK phosphorylation, 175–177
general methods, 165–166
G protein binding, nonselective assay, 166–168
G protein signaling, selective assays, 168–171
leukocyte migration, 156
mediated intracellular signaling
G Protein-dependent, 162–164
G protein-independent, 164–165
Protein depletion, 521, 523
Protein estimation assay, 470
Protein fragment complementation assays, 120
Protein kinase A (PKA), 488
Protein kinase C (PKC), 488
Protein NMR spectroscopy
chemokines folding state, 552–553
finger print, 552–553
materials, 378, 553–554
^{15}N-^{1}H HSQC Spectroscopy, 378–379, 379*f*
oxidized chemokines, 553*f*, 554
reduced chemokines, 554
Protein-protein complexes, 390–391
Proteolytic degradation, 410
P-selectin glycoprotein ligand-1 (PSGL-1), 359
PSGL-1. *See* p-selectin glycoprotein ligand-1 (PSGL-1)

R

Radiolabeled chemokines, monitoring scavenging with, 92–93
Radioligands
depletion, 478–479, 481
ligand-receptor interactions, 465
nonspecific binding, 465–466
Radionuclide, 470
Receptor:arrestin complex, 147–148, 148*f*
Receptor bias
chemokine system, 160–161
scavenger chemokine receptors, 160–161
Receptor internalization, 458–459, 464–465
Receptor occupancy, 463
Recombinant bacmids, 400–401

Recombinant chemokines, 468
biological activity verification, 555–557
caveats and limitations, 558
CCR7, 540
chemoattractant cytokine, 540
chemokine fold, 540–541
CXCL12 refolding, kinetics of, 543*f*
cysteine oxidation, 542
disulfide shuffling cocktails, 543–544, 546
E. coli, 542
expression, 544–549
G protein-coupled receptors, 540
mass spectrometry, 550–552, 550–551*f*
natively folded bioactive protein, 540, 541*f*
N-terminal lysine residue, 542–543
perspectives
applications, 558–559
cGMP system, 559–562
protein NMR spectroscopy, 552–554, 553*f*
purification, 544–549
refolding, 544–549
robust protocol, 543–544, 562
Recombinant fluorescent chemokine fusion proteins, 99–100
Recombinant proteins, 392
Recombinant tagged chemokines, 103–106
Redundant, 156–157
Refolding, chemokine expression, 546–547, 548*f*
Region-of-interest analysis software, 128
Relative density, 411
Renilla luciferase (Rluc), 496
Residue proximities, 391–392
Resonance units (RUs) detection, 4–5
Retro-orbital blood collection, blood collection, 444–445
Retroviral particles, 8–9, 8*f*
Reversed-phase HPLC
analytical, 372
materials, 372
semipreparative, 373
Reverse phase HPLC, 548–549, 549*t*
Rluc activity, 496
RLuc-arrestin/receptor-YFP, 133–134
Rotor-Gene Q-pure detection software, 412–413

S

Saturable effect, 487–488
Saturation binding, 471
diverging receptor expression, 475–477
hypothetical data, 475–477, 476*f*
radioligand concentrations, 475, 476*f*
specific binding curve, 475–477
Scatchard plot, 475–477
Scavenging activity
fluorescent chemokines, 89–108
fluorescent protein-tagged chemokines, 93–100
microscopy and flow cytometry, 108–114
radiolabeled chemokines, 92–93
tagged chemokines, 100–108
Schild equation, 485
Scintillation fluid, 468
Scratch assay, 502
SDS-PAGE
cross-linked complex, analysis of, 409*f*, 410
cross-linking efficiency, 409*f*, 410
molecular weight shift, 408
polyacrylamide gel, 408
relative band intensity, 408
Self-assembled monolayer (SAM), 14
Sensor chip, immobilization, 4–6, 5*f*
7-transmembrane. *See* G protein-coupled receptor (GPCR)
7-transmembrane domain receptors (7TMRs), 132, 390–393
chemokine system, 156–157
endogenous bias, 157
G proteins conformational changes, 428–430
ligand-dependent β-arrestin recruitment, 431–433
Shearing force, chemotactic force, 40, 41*f*
Signal-to-noise ratio, 482
Signal transduction pathways, 120
Single-cell dilution, 327–329
Site-directed mutagenesis, 190–193
Size-exclusion chromatography (SEC), 251, 252*f*, 403, 524–527, 526*f*
Sodium dodecyl sulfate (SDS-PAGE), 175
Split luciferase, 120

Spodoptera frugiperda (Sf9) coexpression, 237, 244
band densitometry, 402*f*
biomass, 401
chemokine molecules, 401–402
disulfide bond formation, 401–402
disulfide-trapped complexes, 401
metal affinity purification, 402*f*
MOI, 401, 402*f*
Spontaneous dissociation, 391
Stable transfectants, 311, 330
Stable transfection cells selection
colony-picking, 325–327
single-cell dilution, 327–329
Stem cell mobilization, 460–461
Stimulus bias, 157
Strong anion-exchange (SAX)
chromatography, 527, 528*f*
required materials, 527–529
Sulfopeptides, 371–373, 378–380
Sulfotyrosine, 358–360, 380, 382
Sulfotyrosine peptide, 371–375
Surface plasmon resonance (SPR), 2–3, 74
applications, 12–14
determination of levels on VP, 11
GPCR family, 3–4
LVPs, 8*f*, 9–10
materials, 7–8
retroviral particles, 8–9, 8*f*
sensor chip, immobilization, 4–6, 5*f*
VPs, 5*f*, 6–12
biosensor surfaces, 12, 13*f*
receptor number, 11–12
titration, 10
Synovium, joint cavity, 264
Syringe, knee joint, 265, 266*f*

T

Tagged chemokine, 107–108
Tagged chemokines-encoding DNAs, 103
Tandem mass spectrometry (MS-MS), 374–375, 550
T-cell leukemia, 461–462
T cells, 442
Thermal denaturation, 412–413
Thiol side chain, 395
Three-dimensional migration assays, 568, 568*f*
3D Microbioreactors, 42
Tibiofemoral joint, injection
intra-articular injection, 265, 267*f*
protocol
injection into the joint, 265–266, 267*f*
syringe preparation, 265, 266*f*
required materials, 265
Time-course experiments, 135
arrestin association kinetics, 141, 142*f*
automated injector, 141
functional selectivity, 141
Time-lapse imaging, 302, 306*f*
Timelapse microscopy recordings, 178–179
Tissue bias
cell dependent signal sorting, 161
chemokine system, 161–162
Tracer ligand dissociation, 475
Transendothelial chemotaxis assay, 311*f*
Transendothelial migration (TEM), 208
Transfectant colonies, 325–326, 326*f*
Transient transfection, 126
Tumor microenvironment
cell-cell interaction, 31–37
endothelial cell loading, 35
extracellular matrix, 36
fibroblast loading, 36
spheroid formation, 33, 35
tumor spheroid loading, 36–37
Tumor spheroid loading, 36–37
Two-photon polymerization, 178–179
Two-step binding model, 460
Tyrosine sulfation, 358–359, 368
Tyrosylprotein sulfotransferase (TPST) enzymes
caveats and limitations, 380–381
characterization, protein NMR, 378–379, 379*f*
cosubstrate PAPS, 358, 360
expression
E. coli, 363–367
mammalian cells, 361–363
hemophilia A, 358–359
mammalian cells, purification, 361–363
mass spectrometry, 373–378
N-terminal chemokine receptor peptides, 370, 371*f*
PAPS
analysis, 368–370

Tyrosylprotein sulfotransferase (TPST)
enzymes (*Continued*)
purification, 368–370
perspectives, 381–385
protein-protein interactions, 358–359
P-selectin glycoprotein ligand-1
(PSGL-1), 359
refolding, 363–367
reversed-phase HPLC, 371–373, 374*f*
sulfotyrosine residues, 359, 375–376

U

Ubiquitin, 282–283
Ultra-thin layer sample preparation method,
376–377
Unbound chemokine, 84
United States Pharmacopoeia (USP), 559
Unlabeled competitor ligand, 407–408
Unlabeled ligand, 477–481
Untreated macrophages, 315, 315*f*
UV spectroscopy, 368, 370

V

Vehicle-treated cells, 504
Vesicular stomatitis virus (VSV), 9
Viral particles (VPs), 5*f*, 6–12
biosensor surfaces, 12, 13*f*
receptor number quantitation, 11–12
titration, 10
Viral proteins, 10
Visualizing chemokine
procedures, 349–350
required materials, 347–349
results, 350, 351*f*

W

Western blotting, 126–127
disulfide-trapped constructs, 409
Flag-tagged receptor, 408–409
HA-tagged chemokine bands, 408–409
Whole tissue, monitoring scavenging in, 113

Y

YbbR13-tagged chemokine, 106–107
Yellow fluorescent protein (YFP), 505–506
emission channel, 136

Z

Zeocin sensitivity, 324
Z'-factor, 482

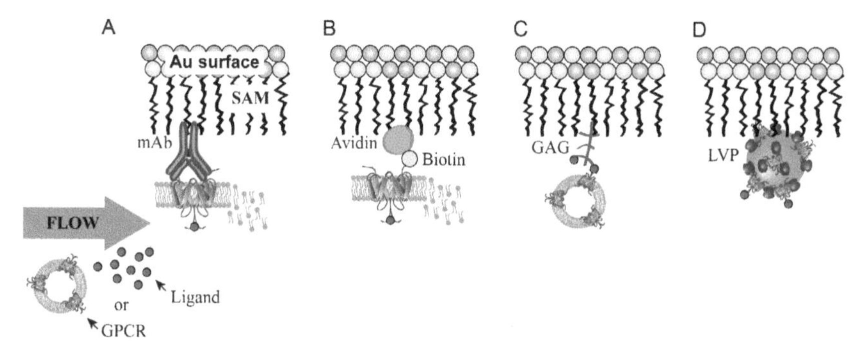

José Miguel Rodríguez-Frade *et al.*, Figure 1 Strategies for GPCR attachment to dextran matrix. The interactive surface of a sensor chip consists of a self-assembled dextran monolayer (SAM) that bears functional groups to allow interaction of different compounds. For GPCR capture on the sensor chip surface, several strategies are used including direct binding to the dextran surface of monoclonal antibodies (A), avidin (B), GAG (C), or LVP (D).

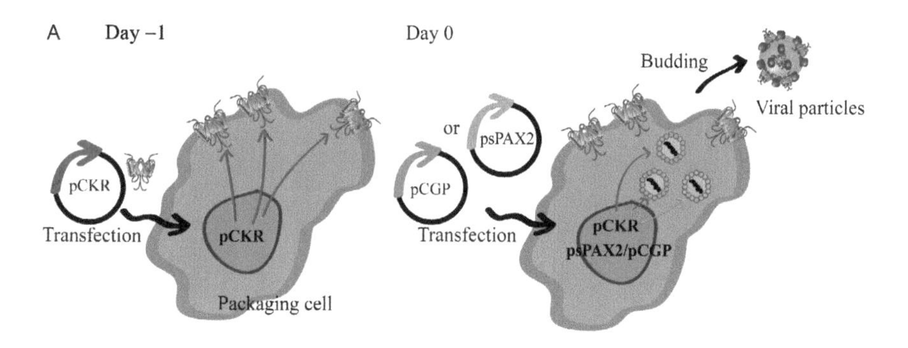

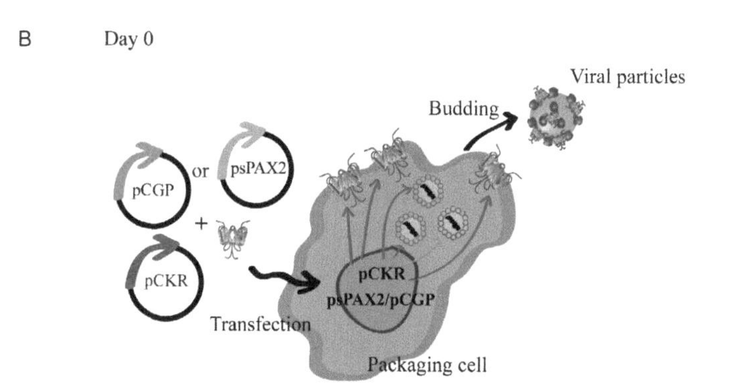

José Miguel Rodríguez-Frade *et al.*, Figure 2 Generation of retroviral and lentiviral particles. Packaging cells are transfected with a chemokine receptor-containing plasmid (pCKR) in advance (A) or simultaneously (B) with the vectors needed to generate retroviral (pCGP) or lentiviral particles (psPAX2). During budding, viral particles incorporate the cell membrane as a coating that retains the structure and functional properties of the receptor.

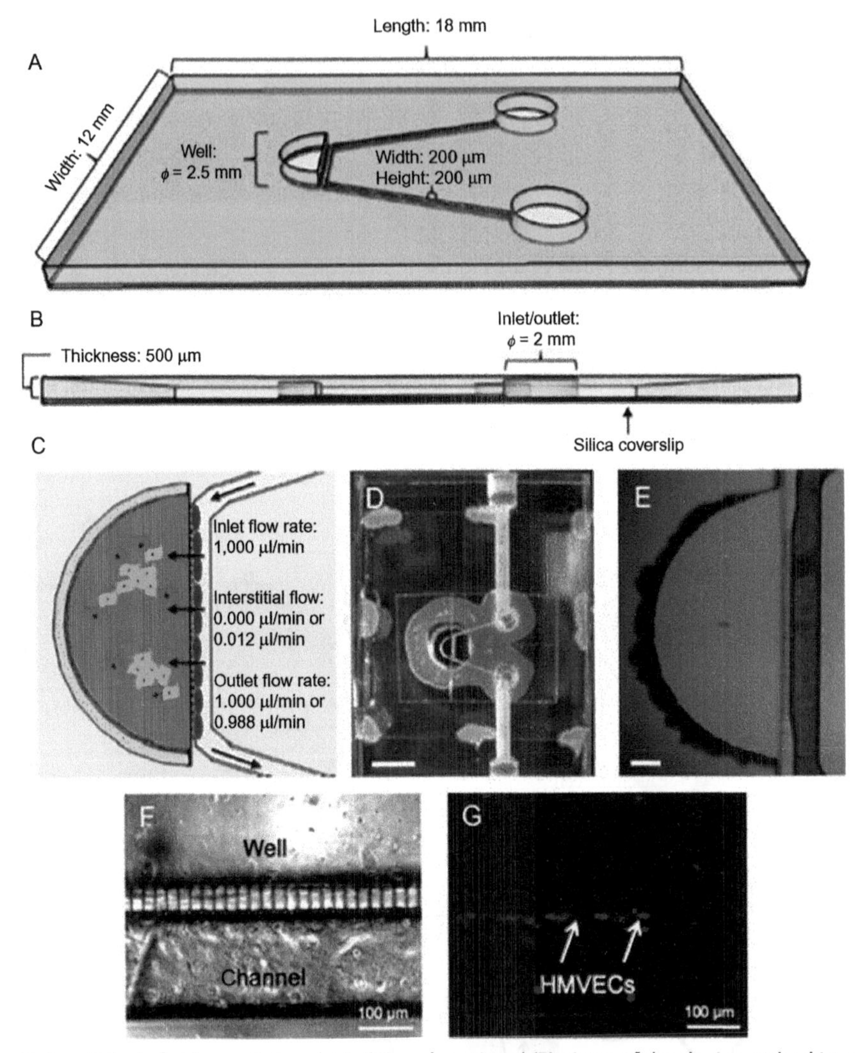

Jiqing Sai *et al.*, Figure 6 Isometric (A) and sectional (B) views of the device etched in a silica chip and sealed to a coverslip. A zoomed in orthogonal view of the semicircular well and channel (C) displays spatial orientation of cancer cell spheroids (green), fibroblasts (red), and microvascular endothelial cells (blue) during an experiment along with rates of inlet flow, outlet flow, and interstitial flow. Images of the fabricated device (D and E) give a size reference with scale bars 5 mm and 200 μm, respectively. (D) Assembled device glued to the acrylic manifold. Inlet and outlet port supply is sealed by red o-rings on the top and bottom of the diagram. (E) A view of the cell well, interstitial flow channel, and bridge between the two containing 6 μm flow channels every 30 μm. (F) A simulated vascular wall through endothelial cell monolayer is created by HMVECs labeled with Cell Tracker Blue cultured on channel side of porous membrane spatially mimic blood vessel wall. 20 × magnification of channel in bright field shows some visibility of cells in monolayer. (G) 20 × magnification of DAPI channel visually confirms HMVECs are beginning to form a monolayer. *Originally published in Hockemeyer et al. (2014).*

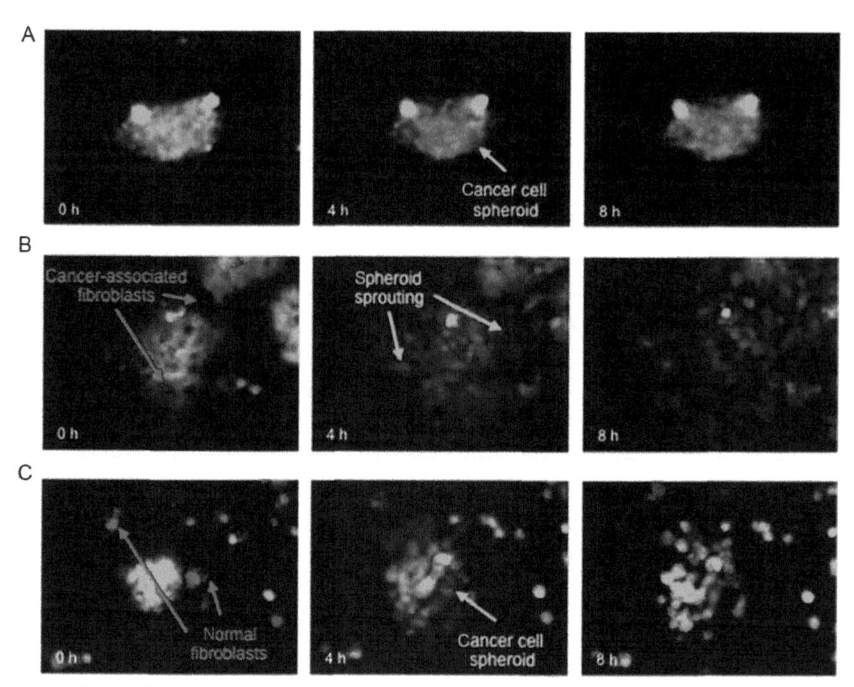

Jiqing Sai *et al.*, Figure 7 Spheroid sprouting with cultured spheroids alone, in the presence of cancer-associated fibroblasts, or in the presence of normal tissue-associated fibroblasts. Shortly after polymerization of the reconstituted basement membrane, the device was imaged over 12 h. Cancer cells are GFP-expressing and fibroblasts were labeled with Cell Tracker Red. (A) Cancer cell spheroids alone did not exhibit sprouting. (B) Cancer cells sprouted into the surrounding matrix when cultured with CAFs. (C) When cultured with NAFs, cancer cells moved around the spheroid or within small clusters, but very little migration occurred outside of local movement. *Originally published in Hockemeyer et al. (2014).*

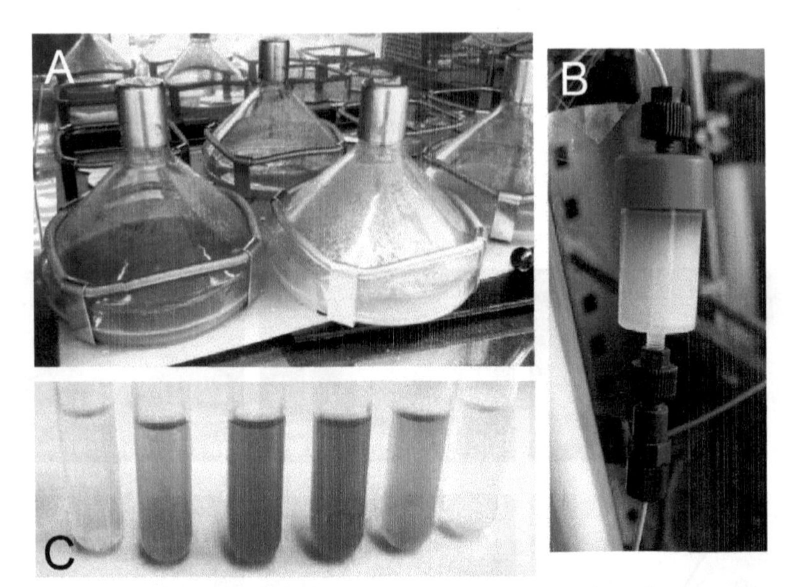

Barbara Moepps and Marcus Thelen, Figure 1 Production and purification of fluorescent chemokines from baculovirus-infected insect cells. (A) Insect cells are grown in Fernbach culture flasks. (B) The culture medium of baculovirus-infected insect cells is applied to affinity chromatography using a HiTrap™ Heparin HP column. (C) Fractions containing the fluorescent chemokine are collected. As shown, all production and purification steps can be visually monitored.

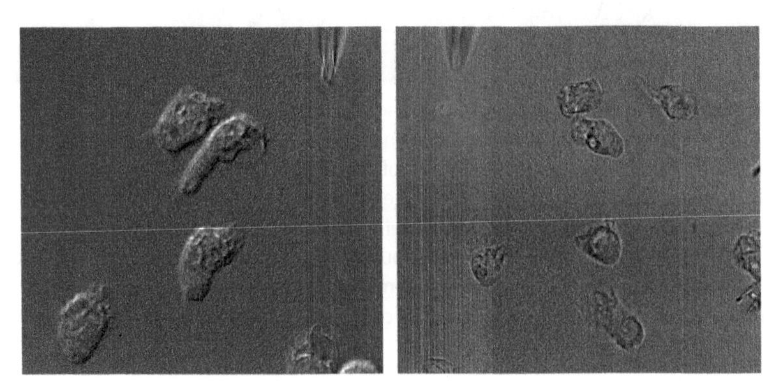

Barbara Moepps and Marcus Thelen, Figure 2 CCL2-mCherry uptake by migrating monocytes. Human monocytes were isolated from peripheral blood (Volpe et al., 2012). Left: 100 n*M* CCL2-mCherry was loaded in the pipette; right: a mixture of 1 μ*M* CCL2 and 1 μ*M* CCL20-venus. Note that monocytes take up only CCL2-mCherry and not CCL20-venus (excluding pinocytosis as mechanism of uptake). CCL2-mCherry is stored in endosomes located in the posterior part of the cells. A selected frame from a time-lapse video is shown.

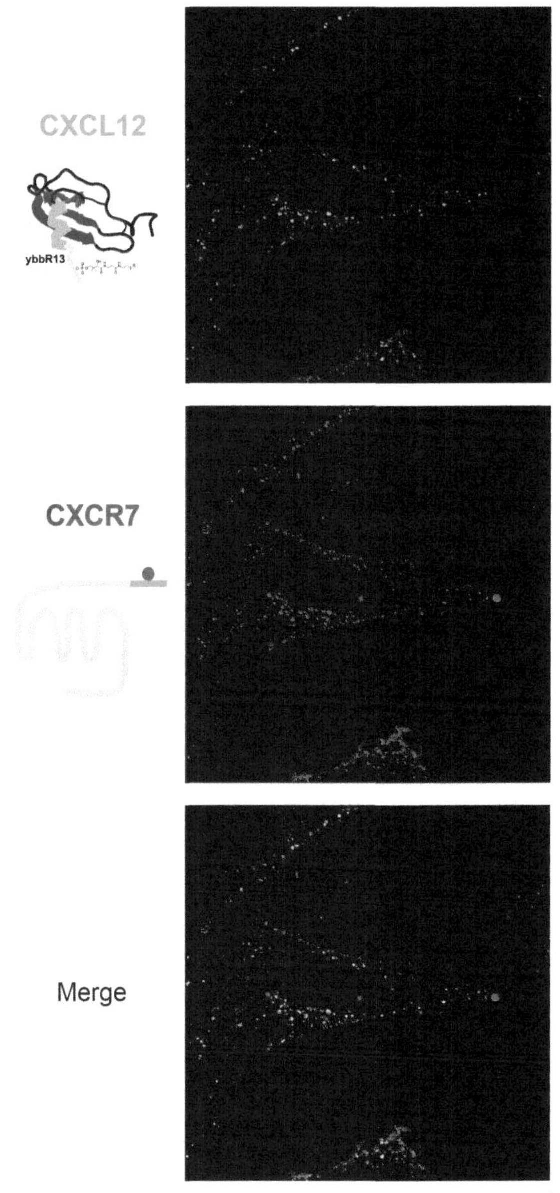

Barbara Moepps and Marcus Thelen, Figure 5 Scavenging of CXCL12_YbbR13_Atto700 by ACKR3. HeLa cells were transiently transfected with human ACKR3 tagged at the N-terminus with the peptidyl carrier protein sequence S6 (Zhou et al., 2007) and grown on glass bottom cover slips as described elsewhere (Humpert et al., 2012). After 48 h, cells were washed and enzymatically labeled at 17 °C with phosphopantetheinyl transferase (Sfp) (Humpert et al., 2012) and CoAAtto565 (Humpert et al., 2012). After labeling, cells were exposed to 50 n*M* CXCL12_YbbR13_Atto700 for 30 min at 37 °C. Cells were fixed with 3.7% PFA and images taken with a confocal laser scanning microscope.

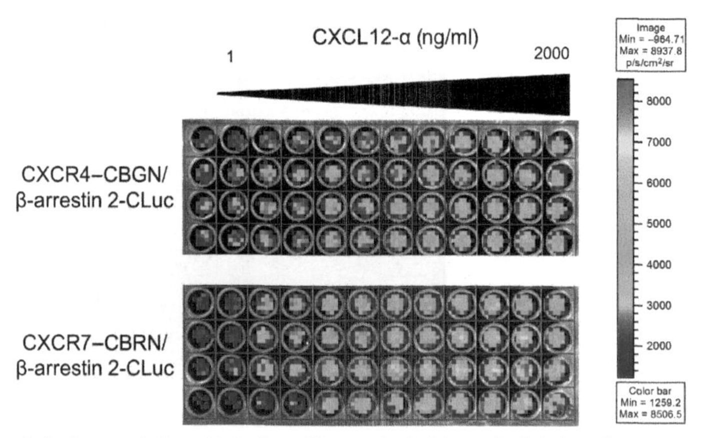

Kathryn E. Luker and Gary D. Luker, Figure 2 Cell-based click beetle green and red complementation assay for CXCL12-dependent activation of CXCR4–CBGN and CXCR7 (ACKR3)–CBRN association β-arrestin 2-CLuc. MDA-MB-231 breast cancer cells coexpressing CXCR4–CBGN and β-arrestin 2-CLuc or CXCR7 (ACKR3)–CBRN and β-arrestin 2-CLuc were seeded at 1.5×10^4 cells per well in black wall 96-well plates. We incubated cells with vehicle only (far left column) or increasing concentrations of CXCL12-α (1–2000 ng/ml). The figure shows a representative bioluminescence image of green (top) and red (bottom) bioluminescence obtained 12 and 14 min, respectively, after adding CXCL12. The grid overlay is used to quantify photons per well by region-of-interest analysis. Bioluminescence is depicted as a pseudocolor display with red and blue defining high and low values for photon flux.

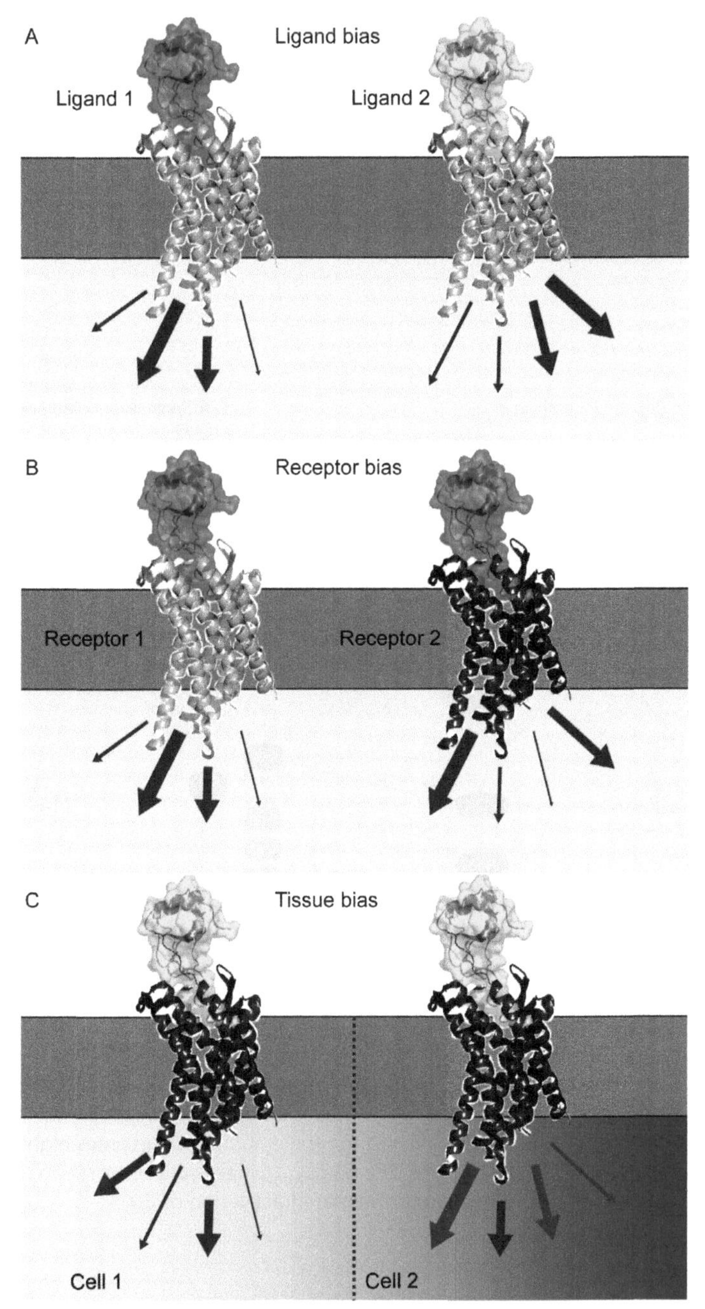

Roxana-Maria Amarandi *et al.*, Figure 1 Bias in the chemokine system and other 7TMRs. Biased signaling describes the ability of a receptor to induce different signaling pathways or cellular events. Thereby, different ligands may activate different pathways via the same receptor (ligand bias, A), or the same ligand induces different outcomes at different receptors (receptor bias, B). Also the cell or tissue that "hosts" the ligand:receptor interaction can modulate the induced signaling pathway (tissue bias, C).

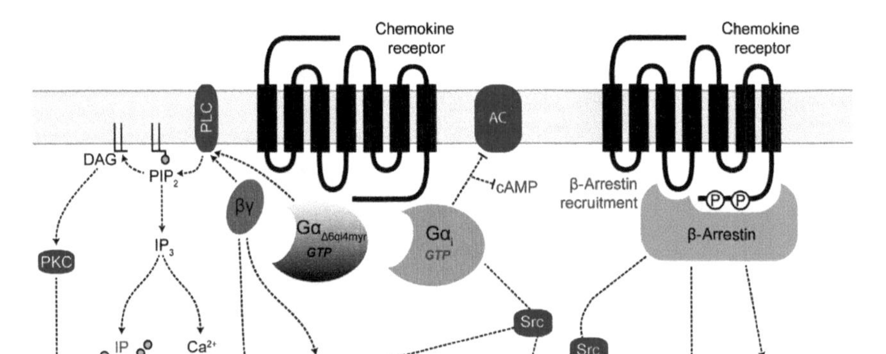

Roxana-Maria Amarandi ***et al.*****, Figure 3** Chemokine receptor signaling pathways. Rough overview of the signaling pathways mediated by chemokine receptors that lead to migration or contribute to cell survival and proliferation. For outcomes highlighted in pink, assays are described in Section 4. The $G\alpha_{\Delta 6qi4myr}$ is an artificial G protein used in an IP_3 turnover assay to measure receptor-mediated activation of $G\alpha_i$.

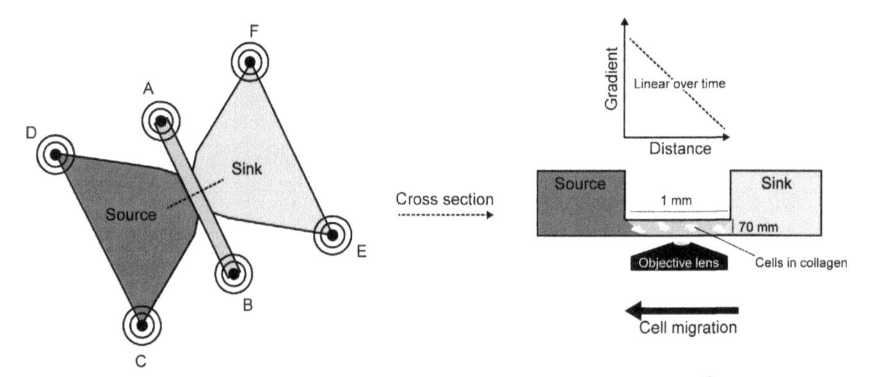

Roxana-Maria Amarandi *et al.*, Figure 4 Schematic drawing of the Ibidi® μ-Slide migration chamber. On the left, the chamber is shown from top with inlets A–F. Note that the channel containing cells and collagen is in the center and between inlets A and B. On the right, the cross section of the chamber is shown. Cells will migrate along the short edge of the chamber, which is approximately 1 mm wide.

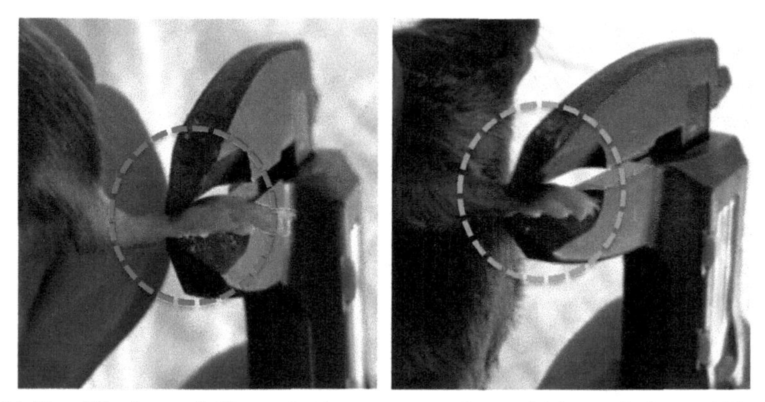

Yoshishige Miyabe *et al.*, Figure 1 Measurement of paw thickness. Left panel: The red circle shows the measurement of paw thickness of the forefoot. Right panel: The green circle shows measurement of paw thickness from the instep to plantar region in the hind limb.

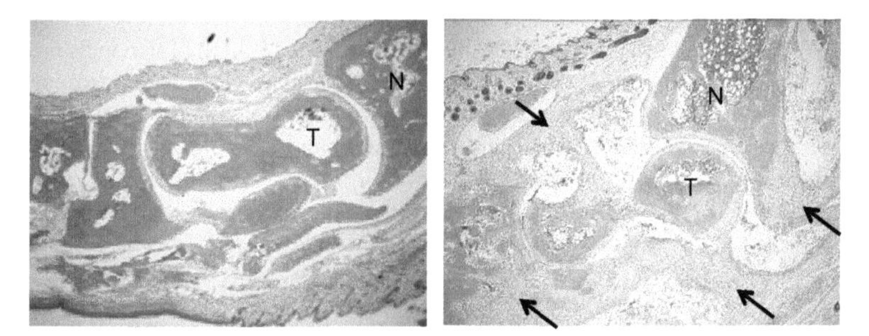

Yoshishige Miyabe *et al.*, Figure 2 Evaluation of the histological score of joints. H&E stained sections are evaluated histologically and inflammation is scored using the following criteria: 0 = no inflammation, 1 = focal inflammation, and 2 = severe and diffuse inflammatory infiltration. Left panel shows an example of control ankle. Right panel shows an example of diffuse inflammatory infiltration throughout the ankle. Arrow indicates the inflammatory cell infiltration into the joints. T, Taulus; N, Naviculare.

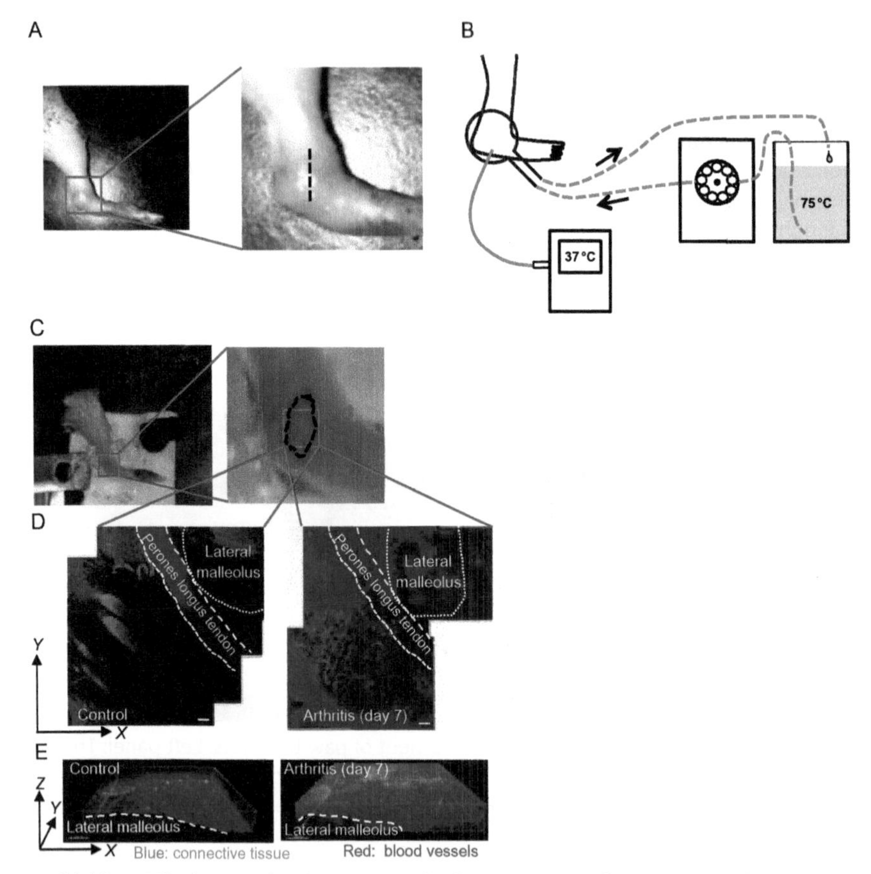

Yoshishige Miyabe *et al.*, Figure 5 Multiphoton intravital microscopy for imaging joints. (A) Depilated area (left panel). Dashed line suggests position of skin incision (right panel). (B) Schematic of the observed joint and temperature control system. (C) Positioning of experimental animal after surgical preparation (left panel). Dash lined suggests the observed area after surgical preparation (right panel). (D and E) Pictures of the joints in control (left panel) and arthritic mice (right panel). Blue color suggests connective tissue. Red color represents a blood vessel stained with Dextran. Angiogenesis and hyperplasia of synovial tissue is seen in the arthritic joint, compared with control.

A

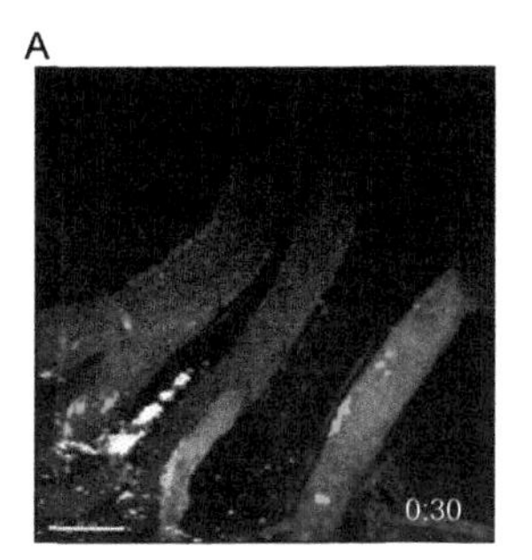

B

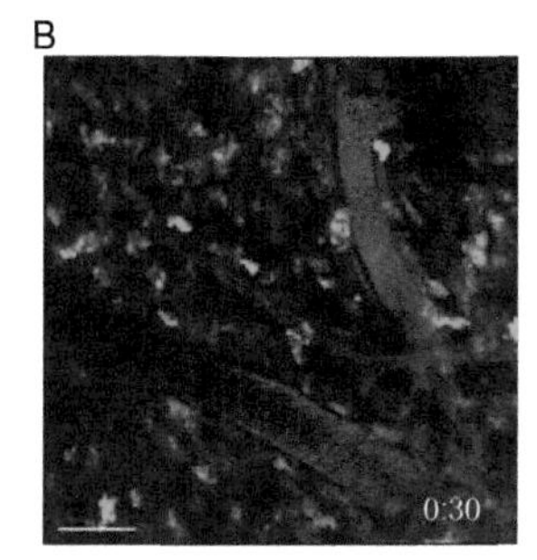

WT Neutrophils Connective tissue Qdots: blood vessels Time in hours: min after starting image

C

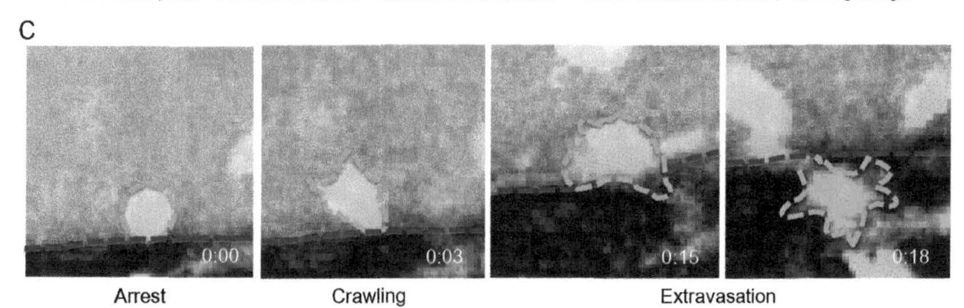

WT Neutrophils Connective tissue Qdots : blood vessels Time in hours: min after arrest

Yoshishige Miyabe *et al.*, Figure 6 Imaging the joints of live mice. Imaging of the ankle joint of control (A) and arthritic LysM-GFP mice on day 7 after arthritogenic serum injection (B). Green color suggests neutrophils or macrophages. Blue color suggests connective tissue. White color represents blood vessel stained with Q tracker 655 (Qdots). Many extravasated neutrophils are seen in the arthritic joint (B), but not the control joint (A). Imaging depth is typically 100–150 μm below the skin surface. For multiphoton excitation and second harmonic generation, a MaiTai Ti:sapphire laser (Newport/Spectra-Physics) was tuned to between 830 and 920 nm for optimized excitation of the fluorescent probes used. For four-dimensional recordings of cell migration, stacks of 11 optical sections (512×512 pixels) with 4 μm *z*-spacing were acquired every 15 s to provide imaging volumes of 40 μm in depth. Emitted light and second harmonic signals were detected through 455/50 nm, 525/50 nm, 590/50 nm, and 665/65 band-pass filters with non-descanned detectors. (C) Arthritic mice (day 7 after arthritogenic serum injection) were imaged to analyze neutrophils migration cascade in the inflamed joint of live mice. Arrest is characterized by a round shape cell that remains in the same position for at least 30 s. A crawling cell is described as an amoeboid shaped cell and crawls inside of a blood vessel. Transendothelial migration describes a cell that has left the blood vessel lumen and has entered into the tissue. GFP represents WT neutrophils, Qdots blood vessels, and blue suggests connective tissue. Green broken line defines the outline of a neutrophil, and blue broken line suggests the boarder of blood vessels.

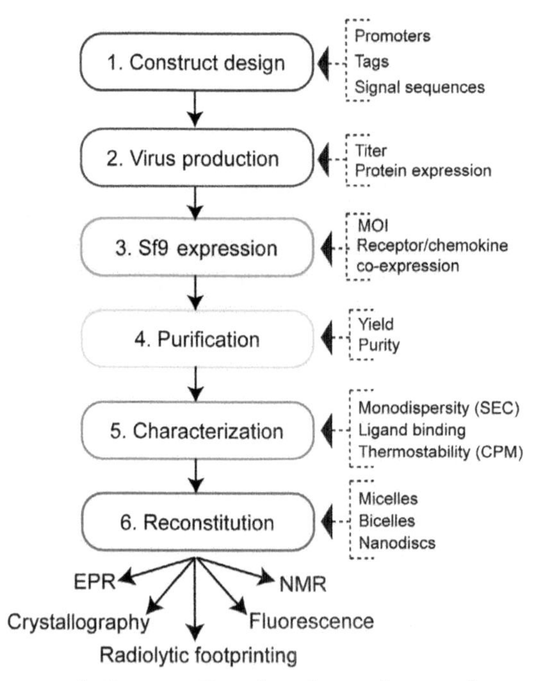

Martin Gustavsson *et al.*, Figure 1 Flow chart for production of complexes.

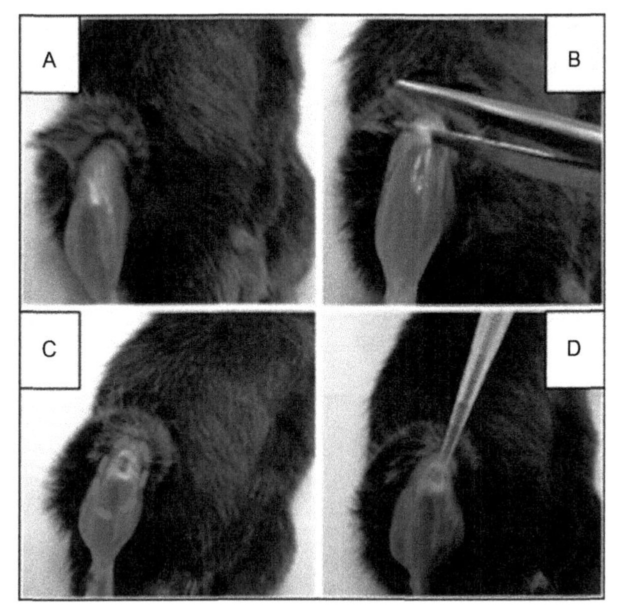

F.A. Amaral *et al.*, Figure 4 Procedure for cell recovery from the tibiofemoral joint. Remove the skin under the knee (A). Cut the patellar tendon carefully with a delicate scissor (B). Open the articular cavity using tweezers (C). Dashed line can be removed for enzymatic assays (D). Wash the joint cavity with diluent (5 μL, 2×) to recover accumulated cells.

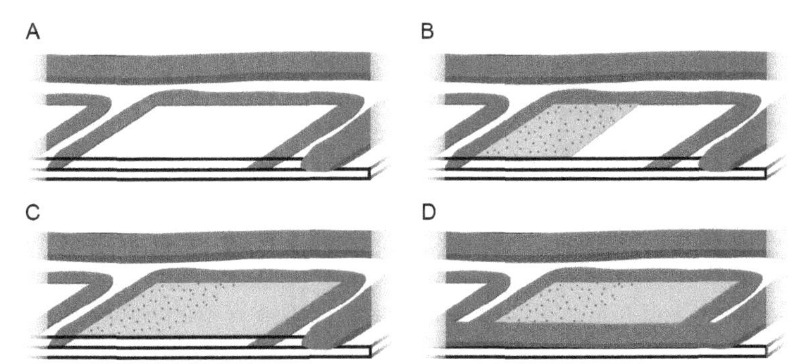

Kathrin Werth and Reinhold Förster, Figure 2 Filling of chamber: (A and B) first half of the chamber is filled with Matrigel matrix containing cells and chemokine, (C) after polymerization of the gel, the second half of the chamber is filled with Matrigel containing chemokine only, and (D) the chamber is sealed with paraffin.

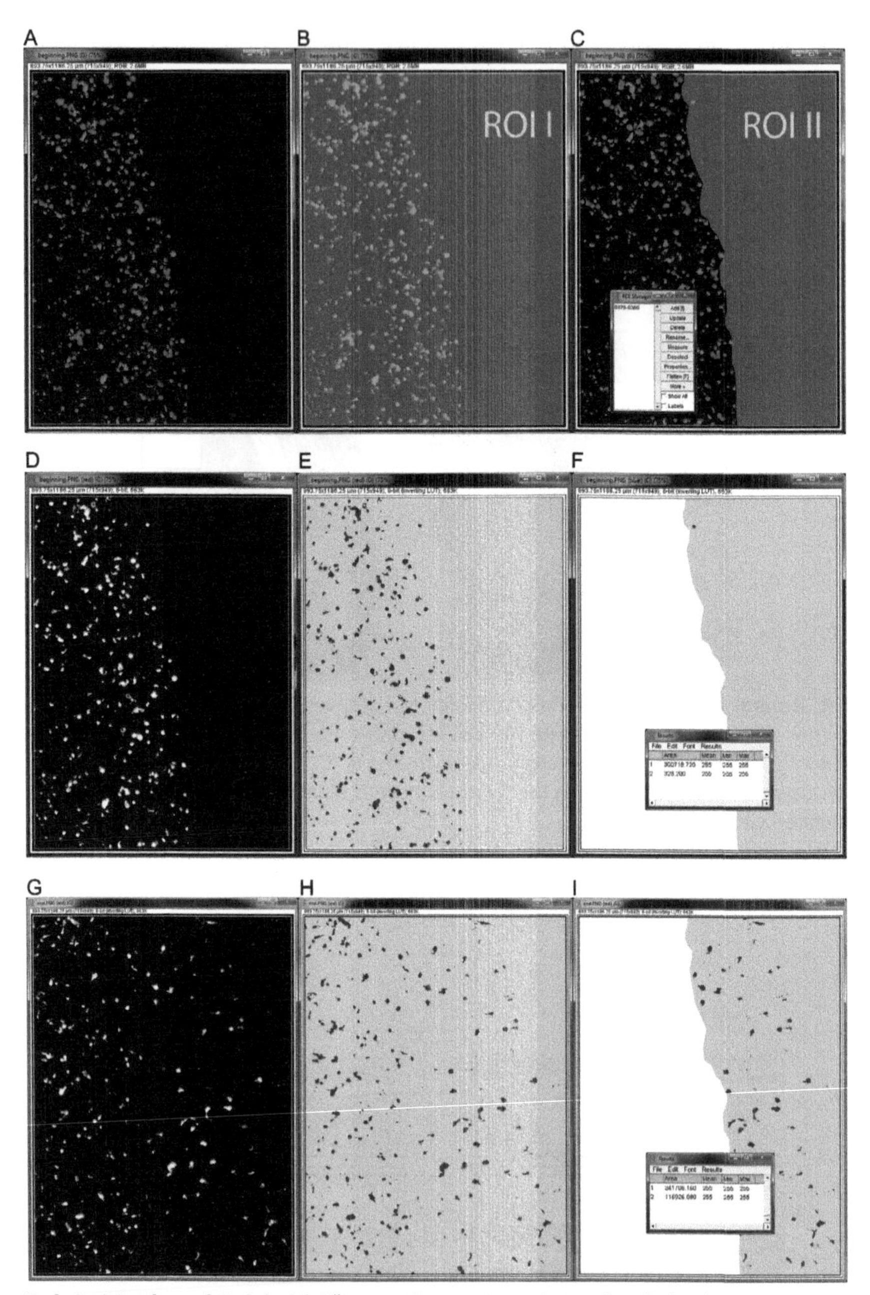

Kathrin Werth and Reinhold Förster, Figure 3 Analysis of cell displacement using ImageJ: (A) first frame of the movie; (B and C) definition of ROI I and ROI II, with the latter being added to the ROI manager; (D) channel depicting the reporter cells is separated from the others using the split channels command; (E) this channel is now inverted and binarized, and all black pixels can be automatically counted in ROI I; (F) after clearing signals outside of ROI II, black pixels can be counted in ROI II; and (G–I) the same operation has to be applied to the last frame of the movie.

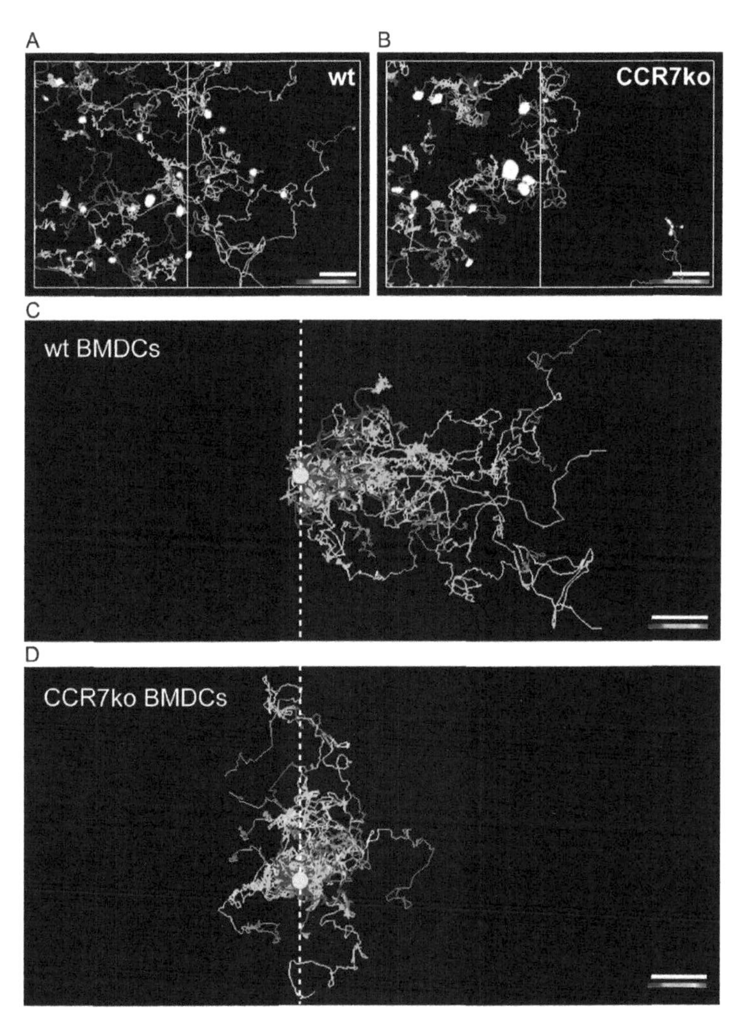

Kathrin Werth and Reinhold Förster, Figure 4 Depiction of cell tracks in Imaris using a time-based color code: (A and B) in original movie, after processing as described in Section 8 step 2; (C and D) after defining a common starting point (white dot). Dotted line indicates *y*-axis at $x=0$. Images show highly directional migration for wild type in contrast to *Ccr7*-deficient bone marrow-derived dendritic cells. Scale bar 100 μm.

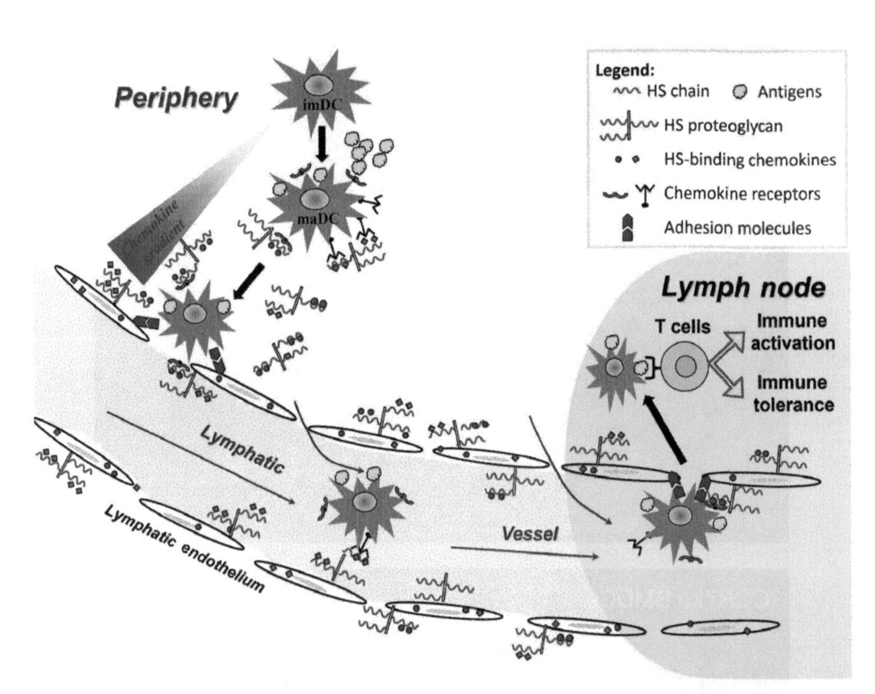

Xin Yin *et al.*, Figure 1 Schematic representation of DC traffic from the periphery to the draining LN (DLN) and the potential mechanisms for HS in this process. Immature DCs (imDC) pick up antigen in the periphery, undergo maturation to become mature DCs (maDC), and are associated with the upregulation of certain chemokine receptors, such as CCR7. Following HS-mediated chemokine gradients, lymphatic flow, and adhesion events that may also be mediated by lymphatic endothelial HS, antigen-loaded DCs travel from the periphery to the DLN, where they present antigen to T cells and modulate the balance between immune activation and immune tolerance. Importantly, HS on soluble proteoglycans may also mediate chemokine–receptor interactions *in trans*.

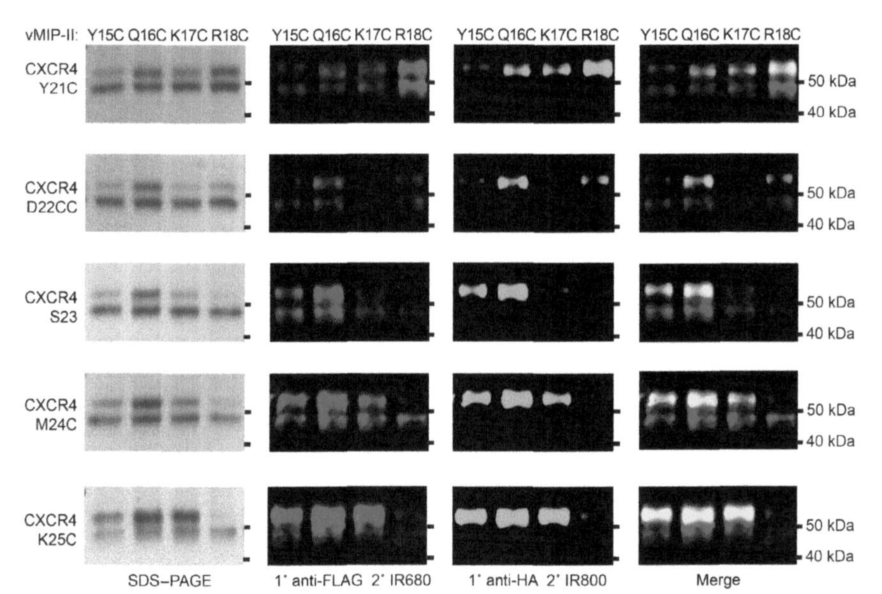

Irina Kufareva *et al.*, Figure 9 SDS-PAGE and Western blotting analysis of CRS1 disulfide crosslinked complexes of CXCR4 and vMIP-II. Left column: Coomassie stained SDS-PAGE gels of five different CXCR4 Cys mutants with four different vMIP-II Cys mutants. Second column: detection of the same SDS-PAGE gel by Western with an anti-FLAG antibody against a tag on the receptor. Third column: detection with an anti-HA antibody against a tag on the chemokine. Fourth column: merge of the anti-Flag and anti-HA Westerns. *Figure adapted from Qin et al. (2015).*

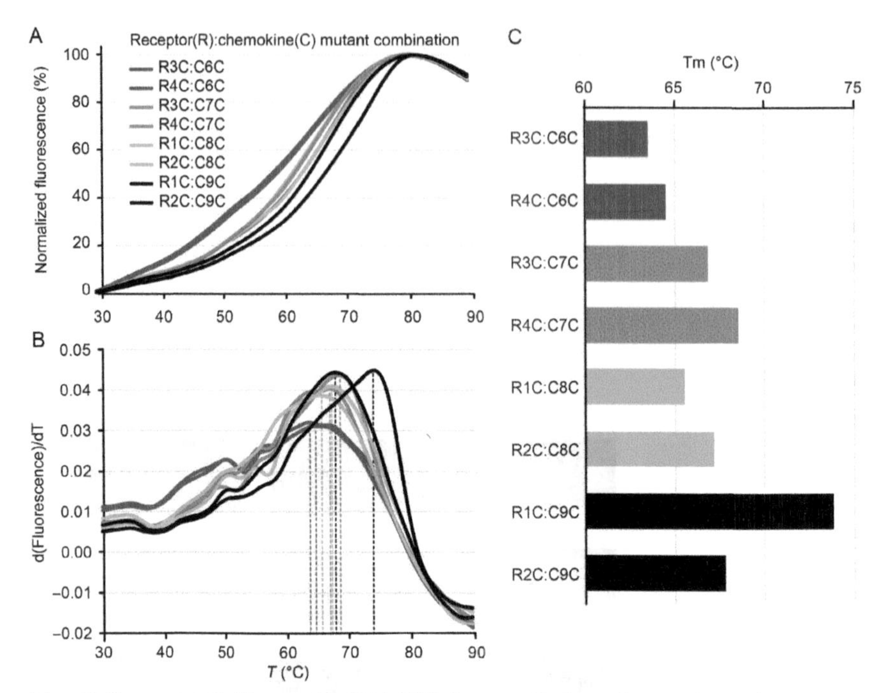

Irina Kufareva *et al.*, Figure 10 CPM-DSF characterization of several cysteine mutant pairs for a CC chemokine and its receptor. (A) Melting curves plotting the increasing fluorescence of the CPM dye as a function of temperature. (B) First derivative curves of (A) demonstrate the positions of the peaks used to calculate T_m. (C) Melting temperatures are calculated from the derivative curves. One mutant combination stands out as having a significantly higher T_m.

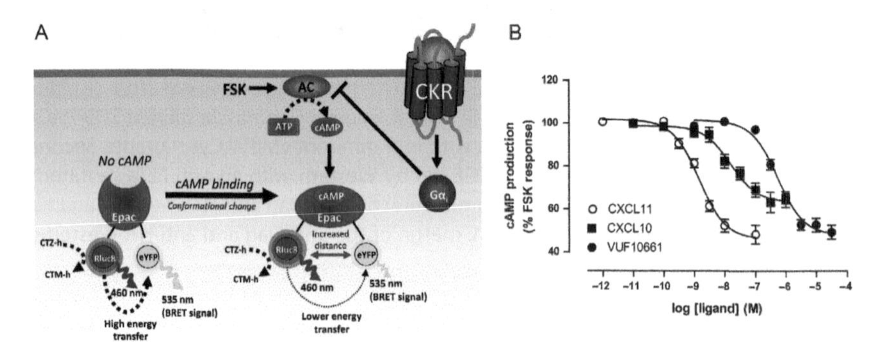

Raymond H. de Wit *et al.*, Figure 14 BRET-based Epac sensor to measure CKR-mediated inhibition of cAMP production. (A) In the absence of a CKR agonist, energy is transferred from Rluc8 to eYFP upon oxidation of the Rluc8 substrate coelenterazine-h (CTZ-h) to coelenteramide-h (CTM-h). Forskolin (FSK)-induced activation of adenylyl cyclase (AC) results in the production of cAMP, which binds to Epac and subsequently induces a conformational change, resulting in a decreased BRET. CKR activation inhibits the production of cAMP and consequently increases BRET between Rluc8 and eYFP. (B) The chemokines CXCL10 and CXCL11 and the small molecule VUF10661 inhibit FSK-induced cAMP accumulation in a dose-dependent manner in CXCR3-expressing HEK293 cells. *Figure was adapted from Scholten, Canals, Wijtmans, et al. (2012).*

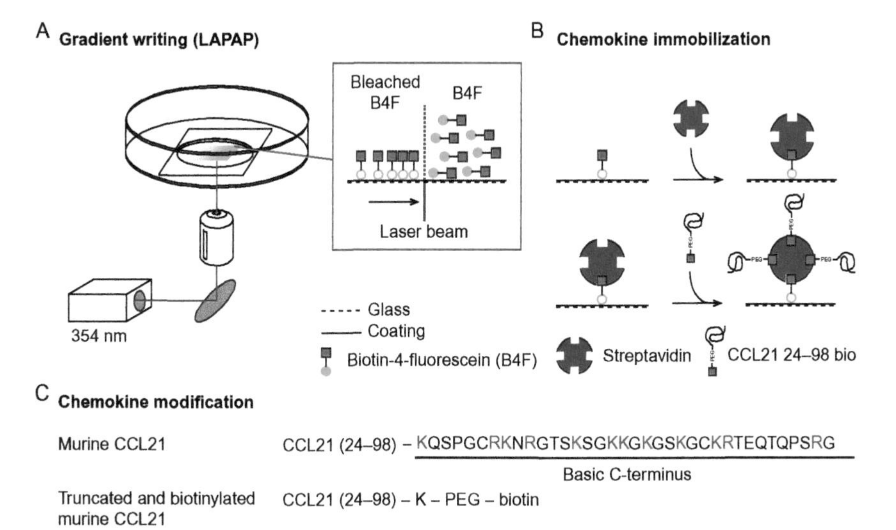

Jan Schwarz and Michael Sixt, Figure 2 Chemokine micropatterning. (A) Gradient writing using laser-assisted protein adsorption by photobleaching (LAPAP). Biotin-4-fluorescein (B4F) is immobilized on a coated glass surface by a movable UV laser ($\lambda = 354$ nm). (B) Chemokine immobilization. Binding of streptavidin-Cy3 (SA-Cy3) functionalizes the B4F pattern. Free biotin-binding sites of the immobilized SA-Cy3 bind biotinylated CCL21 24–98 and present it in a bioactive state. (C) Chemokine modification. Basic CCL21 C-terminus is replaced by biotin attached to a polyethylene glycol (PEG) linker.

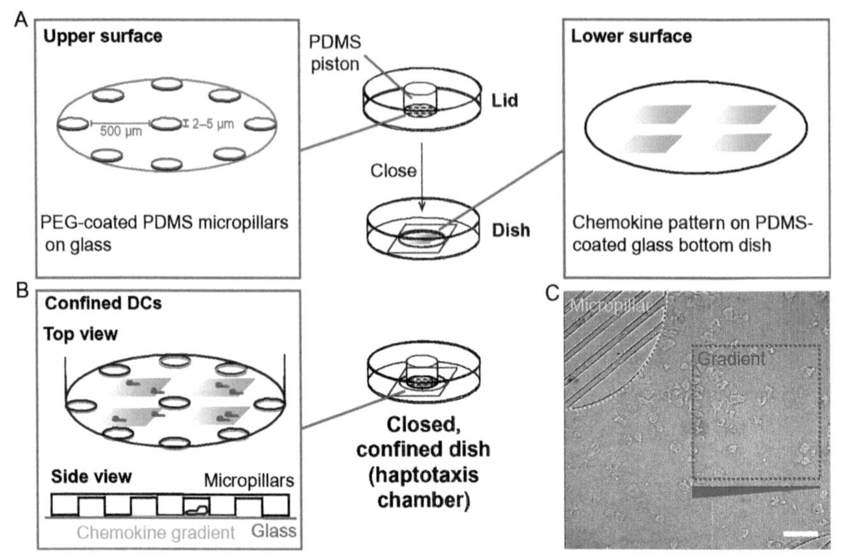

Jan Schwarz and Michael Sixt, Figure 3 Haptotaxis chamber. (A) Upper and lower surface of the haptotaxis chamber. Lid with elastic PDMS piston and PEG-coated PDMS micropillars on a round glass slide and the glass bottom dish with CCL21 24–98 bio micropattern. Left magnified box: Round glass slide with PEG-coated PDMS micropillars. The distance of the pillars is 500 μm. The height of the micropillars is 2–5 μm. Right magnified box: CCL21 24–98 bio micropattern on the glass of a glass bottom dish. (B) Closed haptotaxis chamber. Top view and side view of the closed haptotaxis chamber. (C) Brightfield image of DCs migrating on a printed CCL21 24–98 biogradient in the haptotaxis chamber. Micropillar and gradient are indicated with dashed lines. Scale bar represents 50 μm.

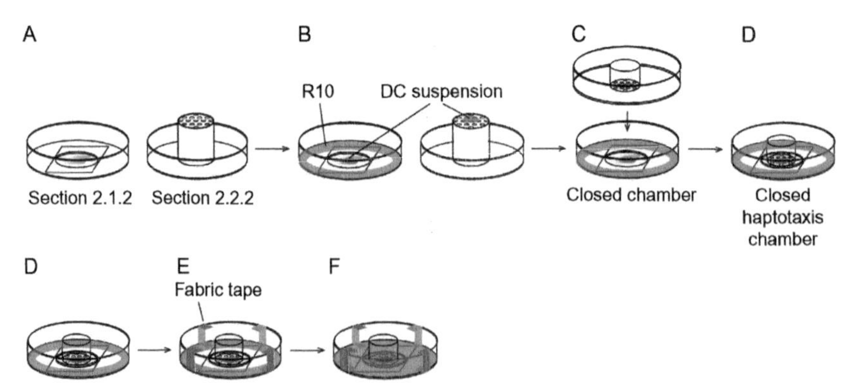

Jan Schwarz and Michael Sixt, Figure 6 Haptotaxis chamber assembly. (A) Glass bottom dish with CCL21 24–98 bio patterned surface (Section 2.1.2) and lid with PEG-coated micropillars on soft PDMS piston (Section 2.2.2). (B) DC suspension is placed on the micropillars on soft PDMS piston (Section 2.2.2) and on the gradient patterned area of the glass bottom dish (Section 2.1.2). R10 cell culture medium is placed on the rim of the glass bottom dish (Section 2.2.2). (C) The glass bottom dish is closed to assemble the chamber. (D) Fabric tape is attached to keep the PDMS piston under pressure.

Edwards Brothers Malloy
Ann Arbor MI. USA
March 8, 2016